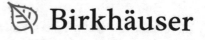

Anton Baranov • Sergei Kisliakov • Nikolai Nikolski
Editors

50 Years with Hardy Spaces

A Tribute to Victor Havin

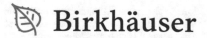 Birkhäuser

Editors
Anton Baranov
Department of Mathematics and Mechanics
Saint Petersburg State University
Saint Petersburg, Russia

Sergei Kisliakov
Russian Academy
Steklov Institute of Mathematics
Saint Petersburg, Russia

Nikolai Nikolski
Université de Bordeaux
Department of Mathematics & Informatics
Talence, France

ISSN 0255-0156   ISSN 2296-4878   (electronic)
Operator Theory: Advances and Applications
ISBN 978-3-030-09638-0   ISBN 978-3-319-59078-3   (eBook)
https://doi.org/10.1007/978-3-319-59078-3

Mathematics Subject Classification (2010): 01-06, 26, 28, 30, 37, 41, 42, 46, 47

Printed on acid-free paper

This book is published under the imprint Birkhäuser, www.birkhauser-science.com by the registered company Springer International Publishing AG part of Springer Nature.
The registered company address is: Gewerbestrasse 11, 6330 Cham, Switzerland

Contents

Operator Theory:
Advances and Applications, Vol. 261, vii–vii
© Springer International Publishing AG, part of Springer Nature 2018

Preface

This volume is dedicated to the memory of Victor Petrovich Havin (March 7, 1933 – September 21, 2015), whose personality illuminated and inspired the Saint Petersburg mathematical community (and many others) during many decades.

This volume contains Victor Havin's biographical sketch with analysis of his influence on mathematics, the text of his address on the occasion of receiving the Doctor Honoris Causa degree from Linköping University (reproduced here with kind permission of that university), a complete list of Havin's publications, and 20 articles dedicated to his memory. It should be mentioned that besides Konstantin Dyakonov's pure reminiscence note, the contributions by James Brennan and Albrecht Böttcher also contain some warm personal comments on their contacts with V. P.

The editors are deeply indebted to many friends, colleagues, and Havin's former students for assistance of various kind, and especially to Evgenia Malinnikova and Aleksei Poltoratskii for accepting the task to compile mathematical commentaries about Havin's influence on certain fields of analysis (these commentaries are incorporated into V.P.'s biomathography below).

Our sincere thanks are extended to James Brennan, Burglind Jöricke, Vladimir Maz'ya, Andrei Lodkin, and Anatolii Podkorytov for supplying us with a number of Havin's photographs and giving permission for their publication in this volume (V. Maz'ya also helped us in getting permission from Linköping University to reproduce Havin's Honoris Causa address). The rest of the photographs included in the volume come from Havin's family archives (many thanks to Havin's widow Valentina Afanasievna) and Nikolai Nikolski's personal archive.

We are also indebted to N. Shirokov for various help with the preparation of the volume, to A. Podkorytov for helping us to complete Havin's reference list, and to P. Mozolyako for re-typing for this edition a long quotation from a Havin article. Our special thanks are extended to A. Khrabrov who did all TeXnical work for the volume.

Finally, we sincerely thank the Birkhäuser editorial staff for including the volume in their "OT" series, and for friendly patience during the book's preparation.

The editors

Part I
Havin's Biomathography

Operator Theory:
Advances and Applications, Vol. 261, 3–55
© Springer International Publishing AG, part of Springer Nature 2018

Victor Petrovich Havin,
A Life Devoted to Mathematics

Victor Havin (in Russian – Виктор Петрович Хавин) was one of a few charismatic leaders of the Saint Petersburg analysis community during the past 50 years. Being an outstanding mathematician, he has founded and modelled the face of the modern Saint Petersburg analysis school, building a true hotbed of talents, which still continues to bring forth new generations of bright scholars. Here we try to trace his way, describe his profound impact on our community, and simply sketch a few features of his unforgettable personality.

Below, we widely use the texts of Havin's two interviews from "Reminiscences Collections" of the Mathematics Department of St. Petersburg (Leningrad) University [133], [134], as well as Havin's obituary in [Ob2016]. Here and in what follows, references such as [133] are linked with Havin's List of Publications (placed immediately after this article), whereas references such as [Ob2016] are linked with this article's own bibliography.

This article is compiled by A.D. Baranov, S.V. Kislyakov, and N.K. Nikolski.

Contents

Havin in front of the main entrance
to St. Petersburg University, around 2008.

Life

Havin was born on the 7th of March, 1933, in Saint Petersburg (named Petrograd from 1914 to 1924 for "patriotic reasons", and next changed to Leningrad by the Bolsheviks from 1924 to 1991). His family originated from Saint Petersburg (Havin's grandfather was a pharmacist having permission to reside at the empire capital as a "useful Jew").

His father Pyotr Yakovlevich Havin was a philologist, slavist, and one of the founders of the Journalism Division (later a department) at Leningrad University; being a Communist Party member, he served during WWII at the Red Army propaganda department, and was decorated with several medals [HPYa]. The rest of the family succeeded in avoiding the apocalypse of the "Leningrad siege", having been evacuated to Tashkent (now, Uzbekistan).

Pyotr Yakovlevich Havin, (1901–1967).

Havin's mother Dina Yakovlevna Havina was a musician, a violinist of the Leningrad Philharmonic Orchestra and the Mikhailovsky (Maly) Opera Theater Ensemble.

After returning to St. Petersburg, Havin completed his education at one of the city's best high schools, the former "First St. Petersburg Gymnasium". In 1950, graduating with a gold medal, he entered the Department of Mathematics and Mechanics[2] of Leningrad University. For young Havin, it was not so obvious a decision: At that time, speaking fluently at least three foreign languages (German, French, and English), he was much inclined (according to his own words, see [133]) to continue at the Languages Department of the University. But suddenly, in March–April 1950, three months before the entrance examinations, his father strictly prohibited him even to dream of linguistics and ordered him to make a choice between physics and mathematics as future studies. This was a protective reaction of a university professor to the publication of Josef Stalin's "scientific" opus "*Marxism and Problems of Linguistics*" (March 1950).

Indeed, several generations of Russians knew that every state attack on intelligentsia started with publications of ideological "philosophic" articles by the Communist party authorities, and followed with dismissals,

"Session of the Social Sciences Division of the USSR Academy of Sciences devoted to the One Year Anniversary of the publication of I.V. Stalin's genius work "Marxism and the problems of linguistics." Symposium of the documents", Moscow, 1951.

---

[2]Sometimes, we refer to this department as "Mat-Mekh" (this is the standard Russian abbreviation transliterated). The Moscow counterpart is "Mekh-Mat."

purges, exiles, and personal disappearances. To list only a few such attacks, recall Stalin's propaganda mouthpiece "Правда" ("Truth") campaign "Muddle Instead of Music" against D. Shostakovich's "Lady Macbeth of Mtsensk", or T. Lysenko's genetics dispersal, or A. Zhdanov's campaign against the writers A. Akhmatova and M. Zoshchenko, and next, the "cosmopolitanism" antisemitic campaign, and even – preparing the future anti-cybernetics purge – decreting (in statistics) the law of large numbers and random deviations as "false theories." By Havin's own evidence, after a short hesitation, he chose mathematics (about which decision he wrote jokingly [133]: *"Finally, I was grateful for this to both my father and Stalin."*)

Photo 1954, Havin's archive.

Havin's university years, 1950–1955, fell on the beginning of extensive science developments in the USSR: The famous Soviet projects for atomic bombs (first test explosion in 1949) and ballistic missiles (first successful test in 1957) were powerful hidden sources for massive investments in science and education, giving excellent results for a period of about 30 years (1950–1980). Affected by the total Soviet propaganda, especially aimed at the young generations, Havin entered the university as a convinced pupil of the Youth Communist League ("Komsomol") and spent (as did many, many others – practically all students) a lot of time in various campaigns common for the league (including regular voluntary journeys for physical jobs at Soviet collective farms). But times change: Stalin's death (1953) was followed by a political instability period and finished with "Khrushchev's destalinisation", which relieved all aspects of Soviet life.

For Victor Havin, it was mathematics that was already above all: The first talk to a student seminar in 1951, the first graduate course attended in 1952, regular participation and first talks to the Division of Analysis research seminar (under G. Fichtenholtz, L. Kantorovich, future Nobel Prize winner in economics, and V. Smirnov) in 1953. In fact, the last one was a series of talks on duality theory, novel and influential at that time, followed by a survey on "analytic functionals" by P. Lévy; later on, Havin was invited to repeat the talks at the famous Gelfand seminar in Moscow.

Vladimir Smirnov, a Full member of the Russian Academy of Sciences, being impressed by the bright talents of the young student, strongly recommended him for graduate studies ("aspirantura"). At that time, this recommendation was not a standard natural deal as it is today, but a civic act against local activists, in view of the Soviet state antisemitism politics (it suffices to recall a grievous "doctors-

Gleb Akilov (1924–1986), Associate Professor (Docent) of the Math. Department, who greatly influenced young Havin, both mathematically and morally (Akilov was a convinced dissident toward the Stalin / Soviet regime, even during the most hardened times). One of the principal architects of the analysis presentation reform mentioned below.

Professor Grigorii Fichtenholtz (1888–1959) gave the principal analysis courses at the Department until the end of the 1950s, very warmly remembered by V. Havin, [133].

killers plot" campaign (1952–1953) – a kind of an anti-semitic national hysteria, which was still very fresh in memory).

Anyway, Havin was accepted for graduate studies and prepared his PhD thesis during 1955–1958 under L. Kantorovich (and, informally, V. Smirnov).

It is curious to notice that, after all, Havin happened to be a kind of "autodidact" in mathematics: Despite a high-level mathematical surrounding (to the personalities mentioned above we can add also D. Faddeev, A. Markov (Junior), V. Zalgaller, I. Natanson, ...), nobody directed the young PhD student in his choice of future research themes. For instance (according to Havin's own reminiscence), Kantorovich said to him: "You are free to choose your research directions," hinting at his own involvement in mathematical methods in economics at that time. Perhaps, this explains why in its earlier period, Havin's research was quite dispersed and looked rather like "a search for a genre" (see below).

Havin defended his thesis in November 1959, having already 6 papers published in refereed journals and being the author of several remarkable results (such

Vladimir Smirnov (1887–1974), Member of the Academy of Sciences, the principal XX c. representative of the St. Petersburg Analysis school (ascending to V. Steklov, A. Lyapunov, and ultimately to P. Chebyshev), viewed by many as the "conscience of scientific Petersburg" for his high moral standards.

Leonid Kantorovich (1912–1986), a pioneer of the applications of extremal problems of functional analysis (linear programming) in economics. Nobel Prize in economics (1975), an author (with G. Akilov) of a popular textbook in functional analysis.

as a solution to V. Golubev's problem from the 1920s). Having at hand a state assignment for a teaching position at one of the Petersburg technology universities, Havin was refused with an unmotivated *"We don't need you"*, which was repeated in many other institutions [133]. Prof. Dmitry Faddeev (the father of Ludwig Faddeev) saved this desperate situation by insisting with his authority in front of the university administration in order to open a position for Havin. Thus started Havin's long carrier at St. Peterburg University: 1959–1962 as an assistant professor, 1962–1970 as an associate ("docent"), 1971–2015 as a full professor (Head of the Analysis Division in 1997–2004).

The first of the photos below is an official photograph from Havin's promotion case to the Associate Professor position (1962), and the second one is taken in the fields of a state farm (probably in 1963), where professors regularly accompanied their students for (forced) manual work. Havin remembered, [133], that even to be considered for his promotion to the Full Professor position, he was obliged to serve such a "field mission", despite a conflicting invitation for the same period to give a plenary talk at a large conference in complex analysis held in Kharkov (now, in

V.P. Havin, 1962, on the occasion of promotion to an Associate Professor position.

V.P. Havin and R.A. Lyakh, an official Math. Department representative, at the agricultural work campaign, fall 1963.

Ukraïne). (He was released from the farm job for a few days to give a talk and then returned into the fields, where he received the news that he was promoted indeed. . .).

The entire life of Havin was linked with the Department of Mathematics and Mechanics of St. Petersburg (Leningrad) University. Very rapidly his scientific activities and values overcame the university scale and became a key point for the mathematical life of the entire city with its tens of science and technology universities and research institutions. Below, we briefly mention the principal forms of these activities:

- university teaching, both on the undergraduate and graduate levels,
- individual research work,
- research seminars and mentoring of numerous doctoral and post-doctoral students,
- other forms of spreading mathematical culture (editing, refereeing, etc.).

Havin's research and related seminar and mentoring work are described below, in separate sections. Here we briefly trace the variety of his other activities.

In teaching, in the 1960s, Havin (with Gleb Akilov) modernized the course of analysis at Leningrad University, in particular, having included the Lebesgue integral and integration of differential forms in $\mathbb{R}^n$ in the basic part. At that time, this was a revolutionary step, made even earlier than it was done at the Sorbonne, or Moscow University, or other first-grade universities in the world. Since then, he delivered this course 17 times and published textbooks on the basis of his lectures, [1, 2, 3, 5]. He also delivered plenty of special lecture courses in Leningrad-Saint Petersburg and, as an invited professor, in many cities of the former Soviet Union and abroad (for instance, already in a different epoch, he taught at McGill University in Montreal during several semesters in 1995–2002). The profound and always modern content of Havin's lectures, as well as their extremely vivid and transparent presentation, always gathered full lecture rooms and seduced most brilliant students.

Havin's famous *research seminar* started in 1962 with four students (S. Preobrazhensky, V. Solovyov, S. Vinogradov, and N. Nikolski), but grew very rapidly in a true factory of young talents. Havin ranked this form of mathematical life very highly and invested in it a lot of mastery, time, and efforts; we describe this below in a separate section.

Surely, besides seminars, *other kinds of contacts* are also very important for a normal development of mathematical research – conferences, meetings, correspondence, visits. Unfortunately, in the former Soviet Union, where Havin passed most of his life, all these forms of contacts were quite restricted. The mobility and mutual visits, even between so large and close centers as Moscow and St. Petersburg (Leningrad), were limited, and represented exceptions rather than rules. Conferences were rare – for instance, the so-called "All Union Mathematical congresses" (analogs of the Joint Meetings of the AMS) were organized only thrice over the entire USSR history (1930, 1934, 1961), and more or less regular conferences in harmonic and complex analysis started in 1964 (one per year,

Harold S. Shapiro,
KTH (Stockholm).

with the only exceptional international meeting in Yerevan in 1965). Trips of Russian mathematicians abroad were absolute exceptions until the 1980s, and visits of foreign mathematicians (at least to a "provincial" city of Leningrad...) were very rare as well. Havin himself remembered in [133] one of the first such visits by Harold S. Shapiro (University of Michigan and KTH, Stockholm) in the middle of the 1960s: it was extremely important, for Havin and for the community, helping to break an information blockade and hear some new ideas. In particular, Shapiro gave a mini-course on extremal problems in Hardy spaces. In fact, it was done behind the administration back – by Soviet rules, such a course required special permission from the KGB, which was not obtained in Shapiro's case...

As for personal trips abroad, Havin had his first permissions to leave the country as late as in 1974 and 1976, and for "socialist countries" only (Cuba, Czechoslovakia). He waited until the late 1980s (during Gorbachev's "perestroika") for making and receiving regular exchange visits to/from abroad.

Yet another way to serve the mathematical community – besides the individual research, running seminars, and guiding PhD students – is to participate in *editing activities*. Havin made an enormous impact on this side of math life, as one of the key actors for two notable Russian editing projects – the Springer *"Encyclopaedia of Mathematical Sciences"* and *"Linear and Complex Analysis Problem Books"*, 1–3. The former one was a large ambitious project started at the Moscow Institute for Scientific and Technical Information (VINITI) under the general management of Prof. Revaz Gamkrelidze, and finished with Springer-Verlag, which published about 100 full volumes during the decade 1990–2000. V. Havin proposed a conception, made the planning and then realized two subseries of the *Encyclopaedia* (in collaboration with the third author of the present article): *"Commutative Harmonic Analysis"* (CHA), and *"Complex Analysis"*. Havin himself wrote the principal survey article for the first volume of CHA – an impressive panorama of two centuries of developments in harmonic analysis. The series has had an obvious success and soon became the desk books for specialists and consumers of harmonic analysis.

The second of the projects mentioned above, the *"Linear and Complex Analysis Problem Books"*, was a "byproduct" of Havin's research seminar (see below), and the three consecutive editions of the book were attempts to present the state of the art in the fields covered by the seminar in a problem solving form. The idea was to ask a large group of experts (in fact, hundreds) joined in an invisible collective working on a common range of problems about questions that eluded their understanding and looked most attractive, influential, and important. Havin, with collaborators, repeated the query three times – in 1978, 1984, and 1994 – and received 99 answers in 1978 (published as a separate issue of the Steklov Institute journal [104] under the title *"99 Unsolved Problems in Linear and Complex Analysis"*), 199 answers in 1984 (published as vol. 1043 in Springer Lecture Notes series [108]) and 341 answers in 1994, published in a two volume set [115], [116]. The problem statements were endowed with short surrounding surveys, quick independent expert commentaries, a joint subject index, etc. The reader can easily imagine how huge the work behind was... (not only inventing the conception and then classifying hundreds of mini-articles received, but simply realizing an enormous correspondence with the authors, for the first two editions – in pre-Internet times...; in particular, handwritten answers were received from Gabor Szegö, Mary Cartwright, Rolf Nevanlinna). At least two of the authors of this text, who actively participated in Havin's team preparing these collections, can remember a number of funny stories related to the job. Here are two of them. The first of these problem books (1978) was edited in the (former) Soviet Union where all printed materials had to be censored; finishing the preparations, the third of the authors (being one of the editors of [104]) brought the typed manuscript (in English) to the censor office and waited for a stamp in a hall before a high closed door; a clerk appeared – "a print-out cannot be adopted since there is no Russian translation" – it was a catastrophe since the publication was supposed to be shown and publicized during the ICM 1978 in Helsinki, and we were on the deadline date to be on time...; returning to the Steklov Institute and gathering Havin's seminar, we asked everybody to immediately sacrifice at place 2–3 hours of personal time and to realize a handwritten Russian translation of some 3–5 pages of the text – 30 to 40 active mathematicians postponed their affairs planned and started to write... (installing themselves on the desks, on chairs, on staircases, ...), two typist girls typed the manuscript until midnight, and the next morning the more than 200 pages long translation was submitted to the censorship; then stamped, printed, and presented to the Analysis Section of the ICM. The second story is more funny: Several years later, at the end of the 1980s when visits from behind the "iron curtain" became more regular, Donald Sarason started his talk at the Steklov Institute saying "*I am talking today on Problem* 100"... – and explained, looking at the puzzled faces, that this was in the appreciation of the "*Problem Book* 1", which had stopped at the 99th problem.

During a break between lectures, McGill University
(V. Havin, P. Koosis, S. Smirnov).

Boris Yeltsin on a tank in Moscow streets announcing
the failure of the August putsch (August 1991).

Yet another of Havin's big writing projects, which dominated his life during a decade, 1980–1990, was his fundamental monograph (with B. Jöricke) "*The Uncertainty Principle in Harmonic Analysis*" (Springer-Verlag, 1994, see [4]). Since the monograph is rather a huge research memoir, we speak of it below, in the "Research" section of this biography.

Havin's academic life was diversified from the end of the 1980s, with Gorbachev's "thaw" and the accompanying opening of the country. A flow of famous mathematicians rushed into the seminar (see in the "Seminar" chapter below), and Havin himself could then partly realize several honorable invitations from Western universities, a stock of which (before unrealizable) decorated his desk from long ago. He visited, lectured and worked in Europe (Check Republic, France, Germany, Norway), Canada, and USA. For instance, he taught at McGill University in Montreal on a regular basis during several semesters in 1995–2002.

However, in general, the 1991 break-down of the Soviet Union was a long awaited (but unexpected) overthrow of the whole Russian life, which dramatically

V. Havin receiving the Doctor Honoris Causa Diploma
from Linköping University (1993).

changed all living conditions: Taking over a kind of freedom (in particular, freedom
to leave the country), the citizens happened to have, in exchange, a complete life
disorder, galloping inflation, criminality growth, halts of salary payments, ...

In particular, very shortly, seminar rooms and student classes became de-
serted... Having decided to still stay in Russia, Havin survived several sad sep-
arations similar to that on the photo below: In May 1991, waiting for the last
joint seminar at Steklov Institute of Mathematics... (with the third author of the

Waiting for the last joint seminar at
Steklov Institute of Mathematics, May 1991.

present article). In return, in that period, Havin's scientific and teaching activity (finally) got well-deserved (though, to our mind, quite late and not full-scale) recognition. He got several national and international awards, among which the degree of *Doctor Honoris Causa* from Linköping University (Sweden), 1993. His remarkable address [71] to the Academic Council of the University is reproduced in this book.

V. Havin was also elected the *Spencer Lecturer* of Kansas State University (USA, 1996) and the *Onsager Professor* at Trondheim University (Norway, 2000). In 2004, his subtle results (with his student J. Mashreghi) about admissible majorants for model subspaces were awarded the *Robinson Prize* of the Canadian Mathematical Society.

Havin at his desk.

Finally (... it is so difficult to become "a prophet in his own homeland"...), Havin got the *Chebyshev Prize* of the St. Petersburg Government (2011) and the *Honorary Professorship* at the St. Petersburg University. He became also an *Honored Scientist of Russian Federation* in 2003 and was awarded the *Order of Friendship* in 2011.

Despite his standard and smooth-looking professor-like career, Havin's personal itinerary was a big character trial: Not speaking on the social overthrows mentioned above, he survived an early dramatic loss of his first wife (taken by a fast-running cancer at 40), leaving him with two young boys, one of whom perished later, and all that in very modest living conditions (until the beginning of the 1980s).

But, as Havin wrote himself in a memorial article on his early departed and most beloved pupil Stanislav Vinogradov, "before all of that were mathematics".

Mathematics

I. Early years 1958–1967: Looking for a track[1]

The initial period of V. Havin's research from 1958 until 1967, conventionally, can be classified as a *"soft analysis"* period, meaning that during that time, he was mostly interested in analysis problems that could be solved with *functional analysis techniques*. Certainly, this track was predetermined by his first student steps in mathematics: Being influenced by the Akilov–Kantorovich functional analysis school, and next introduced to the first research mathematical text with F. Pelligrino's survey [Pel1951] on analytic functionals (in topological locally convex spaces), V. Havin naturally was looking for applications of his new knowledge of *duality theory* to analysis problems.

Techniques coming from duality theory consist mostly in (integral) representations of linear functionals over linear topological spaces, and (Hahn–Banach type) extensions, as well as in various boundedness estimates with respect to different topologies (norms). Havin's results of that period give excellent illustrations of these techniques. The starting point of all applications of duality is to view a function, say g, not as a mapping defined pointwise but by its action on other functions, usually considering a functional Φ_g defined by complex integration (or, Cauchy duality)

$$\Phi_g(f) =: \langle f, g \rangle =: \frac{1}{2\pi i} \int_\gamma f(z)g(z)\, dz,$$

γ being a convenient rectifiable curve. Specific properties of g come from properties of Φ_g, as boundedness with respect to particular (semi)norms, extension properties, etc. Havin's papers [6]–[22] fit faithfully this strategy. Let us describe certain specific results.

Separation of singularities, [7], [87], [97]

Given two open sets $\Omega_j \subset \widehat{\mathbb{C}} = \mathbb{C} \cup \{\infty\}$ ($j = 1, 2$), the problem is to decide

> *whether for every holomorphic function $f \in \mathrm{Hol}(\Omega)$, $\Omega =: \Omega_1 \cap \Omega_2$, there exist $f_j \in \mathrm{Hol}(\Omega_j)$ such that $f = f_1\big|_\Omega + f_2\big|_\Omega$.*

The problem was first solved by H. Poincaré [Poi1892] (in the framework of discussions with E. Borel and M. Fréchet (and others) on the concept of holomorphic continuations) for the particular case of $\Omega_1 = \widehat{\mathbb{C}} \setminus [0, 1]$, $\Omega_2 = \widehat{\mathbb{C}} \setminus \{(-\infty, 0] \cup [1, \infty]\}$, and next appeared as quite a partial case of the additive P. Cousin problem (1895). Probably, it was solved completely in H. Cartan's theory of Cousin problems (1930–1950, with a difficult exact localization inside numerous publications of the "Cartan Seminars," Paris); see [Hör1966] for a modern presentation of this solution (Chapter 1 for the case of one complex variable; the solution makes use of Whitney–Schwartz C^∞ decompositions of unity and elementary facts about $\bar{\partial}$-equations).

---

[1] Compiled by Nikolai Nikolski, nikolski@math.u-bordeaux.fr

Independently of these advancements, N. Aronszajn published a 150 pages long Acta. Math. paper [Aro1935] devoted to a (positive) solution of the problem for the most general case; the proof involved Runge's theorem on rational approximation followed by a very delicate separation of poles responsible for the singularities of f on $\mathbb{C} \setminus \Omega_j$, $j = 1, 2$. In future publications (see [97] for a survey), the papers [Aro1935] and [Poi1892] were overlooked until being rediscovered at the very end of the 20th century (the construction of [Poi1892] was however reproduced in G. Valiron's book [Val1954], where [Aro1935] was also mentioned but not disclosed...).

Henri Cartan at the Yerevan conference
(Armenia) in 1965.

Havin's note [7] contains a record short (1/2 printed page) and clear proof of the most general Aronszajn statement by using the following classical duality result by S. Da-Silva and G. Köthe: for every closed set $F \subset \widehat{\mathbb{C}}$, $(\mathrm{Hol}(F))^* = \mathrm{Hol}(\widehat{\mathbb{C}} \setminus F)$ with respect to the Cauchy duality mentioned above (here we suppose that if $\infty \in \omega$, then $\mathrm{Hol}(\omega)$ means the space of all holomorphic functions f on (an open neighborhood of) ω with $f(\infty) = 0$).

Havin's proof. Take $f \in \mathrm{Hol}(\Omega)$ and observe that $\Phi_f | \mathrm{Hol}(F_j)$, where $F_j = \widehat{\mathbb{C}} \setminus \Omega_j$, $j = 1, 2$, are well defined and continuous; identifying $x \in \mathrm{Hol}(F_1) \cap \mathrm{Hol}(F_2)$ with the diagonal element $\{x, x\} \in D$ in $\mathrm{Hol}(F_1) \times \mathrm{Hol}(F_2)$, one can extend $\Phi_f | D$ to a continuous linear functional Φ on $\mathrm{Hol}(F_1) \times \mathrm{Hol}(F_2)$; since the dual to the last space is identified naturally with $(\mathrm{Hol}(F_1))^* + (\mathrm{Hol}(F_2))^*$, we get $f_j \in \mathrm{Hol}(\Omega_j)$ such that $\Phi\{x, y\} = \Phi_{f_1}(x) + \Phi_{f_2}(y)$, and finally taking the Cauchy kernel $x = y = k_w =: \frac{1}{z-w}$ ($w \in \Omega$) we obtain $f(w) = \Phi_f(k_w) = \Phi\{k_w, k_w\} = \Phi_{f_1}(k_w) + \Phi_{f_2}(k_w) = f_1(w) + f_2(w)$, which gives the required decomposition of f. $\square$

After (and in parallel to) Havin's paper [7] (in fact, the Russian title of [7] says "isolation" rather than "separation" of singularities), various subclasses of $\mathrm{Hol}(\Omega)$, $\Omega = \Omega_1 \cap \Omega_2$ were examined for separation of singularities with specific restrictions on the functions behavior: Let us associate with any (or with certain) open sets Ω a holomorphic space $X(\Omega)$; then the question is whether one has the separation property

$$X(\Omega) = X(\Omega_1)\big|_\Omega + X(\Omega_2)\big|_\Omega,$$

and whether separation can be done by a linear procedure. That is, denoting $R(f_1, f_2) = f_1\big|_\Omega + f_2\big|_\Omega$ for $f_j \in X(\Omega_j)$, the question is whether the map R: $X(\Omega_1) \times X(\Omega_2) \to X(\Omega)$ is "onto," and if "yes," whether there exists a linear right inverse S such that $RS = \mathrm{id}_{X(\Omega)}$. (By the way, in Poincaré's partial case of Ω_j such a right inverse does not exist, as was shown by B. Mityagin and G. Henkin [MH1971] by using their ε-entropy estimates). Several spaces were examined, such as Bergman $X = L_a^p$, Hardy $X = H^p$ etc. (see [97] for a short survey), with a particular interest in the case of $X(\Omega) = H^\infty(\Omega)$, the space of bounded holomorphic functions on Ω. V. Havin (with coauthors) returned to this problem several times [87, 92, 97], discovering that a positive answer depends on the bounded solvability of some $\bar{\partial}$-equation coming from a Whitney–Schwartz decomposition for Ω_j and showing a series of configurations where the singularities can be separated boundedly (and linearly), and where this cannot be done (depending on the transversality of the boundaries $\partial\Omega_j$, $j = 1, 2$).

V.V. Golubev problem and Havin's regular sets

In 1961, V. Havin published the following theorem proved already in his thesis and which happened to answer a question raised as long before as in V.V. Golubev's thesis (1918) (the latter fact had been noticed by V.I. Smirnov when he had considered a draft of Havin's thesis).

V.V. Golubev (1884–1954), having started with analytic functions, became an applied mathematician in military uniform, a key figure in the organization of mathematical war work during WWII, the Head of the Division of Mathematics in N.E. Zhukovskii Airforce Academy, and the Dean of the Moscow University Mekh-Mat Department.

V.V. Golubev (1884–1954) was a former student of D.F. Egorov at Moscow University and a specialist in automorphic functions. The problem in question can be found on p. 111 in the book [Gol1961] of his selected works. The question was to find an analog of the Taylor series representation of a function $f \in \mathrm{Hol}(\widehat{\mathbb{C}} \setminus \{0\})$, $f(z) = f(\infty) + \sum_{k \geq 1} \frac{a_k}{z^k}$, for functions with singularities on a given simple rectifiable arc K.

In order to state the result, we need the following definition of regularity for a set (we will say, *Havin's regularity*): a compact set $K \subset \mathbb{C}$ is said to be *Havin regular* if for every $R > 0$ there exists an open neighborhood $V \supset K$ such that every function f analytic on a neighborhood of K and having the property that

$$\limsup_{k \to \infty} \frac{|f^{(k)}(w)|^{1/k}}{k!} =: R(w) \leq R \quad \text{for every} \quad w \in K,$$

(i.e., the Taylor series of f at w converges at least in the disc $|z - w| < 1/R(w)$) coincides with a function $g \in \mathrm{Hol}(V)$ on an open neighborhood $W \subset V$ of K: $f|_W = g|_W$.

Havin's Theorem ([9], [16], see also [59]). *Let K be a (Havin) regular compact set and μ a positive Borel measure on K satisfying $\mathrm{Clos}(K \setminus E) = K$ for every Borel $E \subset K$ with $\mu(E) = 0$. Then every $f \in \mathrm{Hol}(\widehat{\mathbb{C}} \setminus K)$ is representable by the following formula*

$$f(z) = f(\infty) + \sum_{n \geq 1} \int_K \frac{y_n(\zeta)\, d\mu(\zeta)}{(\zeta - z)^n}, \qquad z \in \widehat{\mathbb{C}} \setminus K,$$

where $y_n \in L^2(K, \mu)$ and $\lim_{n \to \infty} \|y_n\|_{L^2(\mu)}^{1/n} = 0$ ("a Golubev series," following Havin).

Notice that there is an obvious relationship between Golubev series and separation of singularities: once the Havin–Golubev series formula is established, we easily represent f in the form $f_1|_\Omega + f_2|_\Omega$, where $\Omega = \widehat{\mathbb{C}} \setminus K$ and f_1, f_2 come from the decomposition $\int_K = \int_{K_1} + \int_{K \setminus K_1}$ for integrals, $K = K_1 \cup K_2$; then, obviously $f_j \in \mathrm{Hol}(\widehat{\mathbb{C}} \setminus K_j)$, $j = 1, 2$.

Havin's regularity of a locally connected compact set (in particular, an answer to Golubev's question) can be checked easily, but for merely connected compacta it requires some effort and was verified in [Var1981]. Havin's regularity of K is essential for Golubev series representations of functions from $\mathrm{Hol}(\widehat{\mathbb{C}} \setminus K)$ (as well as a kind of connectedness – for regularity); V. Havin gave the following example [59]: Every sequence tending to 0 is non-regular – indeed, if

$$K = \{0\} \cup \{z_j : j = 1, 2, \dots\}, \qquad \lim_j z_j = 0$$

and (for the sake of notational simplicity) $|z_{j+1}| < |z_j|$ ($\forall j$), then the functions $f_j(z) = 1$ for $|z| > |z_j|$ and $f_j(z) = 0$ for $|z| < |z_j|$, are in $\mathrm{Hol}(K)$ and $R(w) = 0$ for every $w \in K$ (and every function f_j), but there is NO neighborhood of K where *all* f_j are analytic (by Montel's compactness principle). Golubev series representations are also impossible for $f \in \mathrm{Hol}(\widehat{\mathbb{C}} \setminus K)$, for example, for a function f having simple poles with residues 1 at every z_j. Several other comments and references were given in [59] (to papers by V. Trutnev, A. Baernstein II, W.R. Zame, A. Vitushkin, and others).

On the other side of the uncertainty principle

One can say that Havin's paper [11] was his first attempt at the "uncertainty principle in harmonic analysis", a major theme of all his mathematics, but ... by approaching from the reverse side of the coin. A form of the uncertainty principle (see also below for more details) says that, given two transformations A and B, and three spaces X, Y, Z, we need to find relations among them guaranteeing

$$(h \in X, Ah \in Y, B\hat{h} \in Z) \Longrightarrow h = 0,$$

where $\hat{h}$ stands for the Fourier transform of h. Often, A and B are restriction transformations, $Ah = h|E$, $B\hat{h} = \hat{h}|F$, and Y and Z are simply $\{0\}$.

Havin's impact on the uncertainty principles will be described later on (in the reviews by A. Poltoratski and A. Baranov) and now, we observe that an opposite side to such principles consists of any kind of "independence" of Ah and $B\hat{h}$, up to giving them arbitrary values $Ah = f$, $B\hat{h} = g$ prescribed in advance. V. Havin proved in [11] (in particular) that

given a sequence $d_k \geq 0$, $\sum_k d_k^2 = \infty$, and an exponent $0 < p < 2$ (say $p = 1$), there exists a function $h \in C(|z| \leq 1) \cap \text{Hol}(\widehat{\mathbb{C}} \setminus [1, \infty))$ such that $\sum_{k \geq 0} d_k^{2-p} |\hat{h}(k)|^p = \infty$.

Stanislav Vinogradov (1941–1997).

The result was next improved by S. Vinogradov [Vin1970] showing that one can find such a function h even in $C(|z| \leq 1) \cap \text{Hol}(\widehat{\mathbb{C}} \setminus \{1\})$. Many other, even more spectacular results were obtained later by Havin's two former students S. Hruschev and S. Vinogradov [HrV1984]; see also [86] for a survey and numerous references.

It should be mentioned that the existence of merely a continuous function $h \in C(|z| = 1)$ making the series $\sum_{k \geq 0} d_k^{2-p} |\hat{h}(k)|^p$ divergent follows, of course, from a (subsequent) domination result of [dLKK1977], and the same fact for the disc algebra $h \in C(|z| \leq 1) \cap \mathrm{Hol}(|z| < 1)$ – from [Kis1981].

Cauchy integrals on rectifiable curves

From the beginning of the 1960s, V. Havin had a vast program of studying the properties of Cauchy integrals of (complex) measures,

$$C\mu(z) = \frac{1}{2\pi i} \int \frac{d\mu(\zeta)}{\zeta - z}, \qquad z \in \mathbb{C} \setminus \mathrm{supp}(\mu).$$

This program included the treatment of them as a tool for representing holomorphic functions $f = C\mu$ or as the mapping $\mu \mapsto C\mu$ on specific classes of measures (particularly, for $d\mu = h\,ds$, $h \in L^p(\gamma, ds)$ for the arc length measure ds on a rectifiable curve γ), as well as the analysis of the properties of $C\mu$ on its own, especially the existence of nontangential boundary limits as $z \to \mathrm{supp}(\mu)$ and/or, equivalently, the existence of singular principal values p.v.$(C\mu)(z)$ of $C\mu$ as $z \in \mathrm{supp}(\mu)$. The papers [6, 10, 12–14, 17–19, 23], in one way or another, are related to this theme. The author of these commentaries remembers that in the very beginning of Havin's research seminar (see the "Seminar" section below), one could often see Havin walking in Mat-Mekh corridors with a thick book of N.N. Luzin's newly reprinted PhD thesis [Luz1915] at hand – V.P. believed at that time (and propagated his belief around him) in one of the principal ideas of Luzin that purely real variable methods are most adapted in order to treat the convergence and boundedness of singular integrals – an idea which was novel in Luzin's time but did not become quite widespread even in the 1960s (the most known and simple proofs of the classical Hilbert and M. Riesz theorems on the $L^p(\mathbb{T})$ boundedness of

$$h \longmapsto \frac{1}{2\pi i} \mathrm{p.v.} \int_{\mathbb{T}} \frac{h(\zeta)\,d\zeta}{\zeta - z}$$

for $1 < p < \infty$, were done with complex analysis techniques). Havin's dream for this period was to treat similar (open) problems for $L^p(\gamma, ds)$ for an arbitrary rectifiable γ. Havin made an important reduction showing in [19] that the existence almost everywhere on γ of the integrals

$$C_\gamma h(z) = \frac{1}{2\pi i} \mathrm{p.v.} \int_\gamma \frac{h(\zeta)\,ds(\zeta)}{\zeta - z}, \qquad z \in \gamma,$$

for every continuous h implies a similar fact for

$$\frac{1}{2\pi i} \mathrm{p.v.} \int_\gamma \frac{d\mu(\zeta)}{\zeta - z}$$

with an arbitrary complex measure μ. As is well known, a complete solution of the problem waited until 1977–1983 when it was solved due to efforts of A. Calderón and his followers (R. Coifman, A. McIntosh, T. Meyer, T. Murai, G. David, . . .).

This was a dramatic race for the final result: first, Calderón proved (1977) that C_γ is bounded on $L^2(\gamma)$ for every Lipschitz curve γ with sufficiently small Lipschitz constant, then five years later Coifman, McIntosh and Meyer showed in 1981 (with a very difficult proof) that the same is true for an arbitrary Lipschitz γ, and finally David derived from the last mentioned result that C_γ is bounded on $L^p(\gamma)$, $1 < p < \infty$, if and only if γ is an Ahlfors curve (that is $|\{\zeta \in \gamma : |\zeta - z| < r\}| \leq Cr$ for every $z \in \gamma$, $r > 0$), and moreover, if C_γ is bounded in $L^p(\gamma)$ for some p, $1 < p < \infty$, then it is bounded for all p.

As for Havin's initial question on the mere existence of singular integrals $\frac{1}{2\pi i}$ p.v.$\int_\gamma \frac{d\mu(\zeta)}{\zeta - z}$ (or, equivalently, the existence of nontangential boundary limits as $z \to \gamma$), from these results it follows that the mentioned principal values (and nontangential limits) exist almost everywhere on γ for every rectifiable curve. Moreover, it is interesting to note that, as it was shown in E.M. Dyn′kin's survey [Dyn1987], p. 216, already Havin's paper [19], *in fact*, contains a direct reduction of the general case of a rectifiable curve γ to Calderón's case of Lipschitz γ with arbitrarily small Lipschitz constant (so that, avoiding the forthcoming complicated arguments by R. Coifman, A. McIntosh, T. Meyer, and G. David).

A separate series of Havin's papers [6, 10, 17, 18, 33] (with his students S.A. Vinogradov and M.G. Goluzina) deals with properties of holomorphic functions on an open set D representable by the Cauchy integrals $C\mu$ of complex measures. In particular, this class of functions (denote it by $CI(D)$) was compared with the usual Hardy spaces $H^p(D)$, $0 < p < 1$, the conformal invariance of $CI(D)$ was proved, and the multipliers of $CI(D)$ (i.e., holomorphic functions φ such that "$f \in CI(D) \Rightarrow \varphi f \in CI(D)$") were described. Many interesting properties of multipliers were discovered; later on, this class of functions was identified as the class of symbols of bounded Hankel operators acting on the Hardy algebra $H^\infty(D)$, $D = \{|z| < 1\}$, see [HrV1984].

Looking back, one can say that this initial period of Havin's mathematics (commented above) represented a search for his own track in analysis. It was crowned with an influential survey [21] on vector spaces of holomorphic functions, saying "Goodbye" to Havin's "soft analysis" period and marking the passage to the "hard analysis" (a discussion on the concepts "soft/hard" can be found in T. Tao [Tao2007]; it can be shortened by saying that "hard analysis has more inequalities than identities").

The next important period for Havin's mathematical evolution was his long collaboration with V.G. Maz′ya on approximation problems and potential theory.

II. Approximation in the mean and potential theory: 1967–1974

We briefly describe this important theme quoting Havin's own presentation from [85]. Yet another survey of the same theme is presented in J. Brennan's paper in this volume. And now we reproduce an excerpt from Havin's paper [85] in the collection of papers dedicated to V.G. Maz′ya's anniversary.

"On the 60th anniversary of my friend Vladimir Maz'ya I am happy to remember a fortunate time long ago when we worked on some problems in function theory (1965–1972). Our themes were

1. L^p-approximation by analytic and harmonic functions,
2. Uniqueness properties of analytic functions,
3. The Cauchy problem for the Laplace equation,
4. Non-linear potential theory.

A characteristic feature of everything we did was the heavy use of potential-theoretic methods and ideas. Even if a problem didn't contain anything potential-theoretic in its statement (as was the case with the first and to some extent with the second theme), then potential theory would emerge by itself in the solution. Working on (1) and (2) we were compelled to invent a "non-linear potential theory" (among the initial works on the subject are [31, 39], but this is another story, not to be discussed here; see [AH1996]). As to theme (3), the problem was potential theoretic from the outset, but it is related to and suggested by traditional themes of pure function theory (quasianalyticity, moment problem, and weighted polynomial approximation in the spirit of S.N. Bernstein) and so corresponds completely to the title of this article.

V.G. Maz'ya and V.P. Havin.

Approximation theory

I became a student of the Department of Mathematics and Mechanics of Leningrad State University in 1950. One of the major events in analysis of that time was Mergelyan's proof of his theorem about uniform polynomial approximation on plane compacta. The statement (not the proof!) of the result is simple, so I was able to understand it and be impressed by it rather early. My further interest in approximation properties of analytic functions was sparked by my father's 1953 New Year's present to me, a one year "Uspekhi" subscription. I had already started taking my first course of complex analysis. One of the rare articles I was educated enough to understand was [Mer1953]. It was about weighted L^2-approximation by polynomials. Approximation in the mean (with its subtle phenomena making it so different from uniform approximation) puzzled and attracted me for several years after.

At that time approximation in the complex plane meant "polynomial" or "rational" approximation and was generally perceived as the development and outgrowth of the Weierstrass theorem on approximation by polynomials on an interval. But the $\bar{\partial}$-ideology was in the air, and the time was ripe to understand (at last) that analytic functions are solutions of the Cauchy–Riemann system (*generalized* solutions at that). By the Runge theorem rational uniform approximation on a compact set $K \subset \mathbb{C}$ is equivalent to uniform approximation by functions analytic n e a r K, that is by solutions of a concrete system of differential equations.

An event that prepared me (without my realization) for this point of view and future cooperation with Vladimir Maz'ya was my candidate (=Ph.D. preliminary) exam in 1956. L.V. Kantorovich, my adviser, made me study a pile of books including Sobolev's book [Sob1963] published by our university in 1950. This great book was extremely hard to read as the density of misprints and small mistakes was exceedingly high. The subject had nothing to do with the theme of my Ph.D. thesis, my permanent concern at the time. So I considered studying this book as a great nuisance, but had to obey my adviser, and so I spent one semester struggling with the text. At the beginning of the sixties the two themes (polynomial approximation in the mean and Sobolev spaces) somehow crossed in my head and lead me to the following simple observation:

Let G be a plane domain, $1 < p < +\infty$; a function $f \in L^p(G)$ is analytic iff $\bar{\partial} f = 0$ (distributionally), that is $\int_G f \bar{\partial} \varphi = 0$ for any $\varphi \in C_0^\infty$ with supp $\varphi \subset G$ ($\varphi \in \mathcal{D}(G)$ for short). This fact makes it possible to apply duality and reduce the L^p-approximation by analytic functions on subsets of G to some uniqueness questions addressed to the closure in $L^q(G)$, $q = p/(p-1)$, of functions $\bar{\partial}\varphi$, $\varphi \in \mathcal{D}(G)$. But (using the Calderón–Zygmund estimates of the simplest singular integrals) this closure can be shown to coincide with $\bar{\partial} \mathring{W}_q^1(G)$, $\mathring{W}_q^1(G)$ being the Sobolev space of functions φ with grad $\varphi \in L^1(G)$ vanishing (in a due sense) on the boundary of G. This scheme looked especially promising for $p = 2$, since the spaces $W_2^1(G)$ are within reach of classical potential theory whose terms and methods were ready to provide simple proofs of Vitushkin-like approximation theorems by rational functions in L^2. The case $1 < p < 2$ is even simpler, since W_q^1 consists of continuous functions, but for $p \in (2, +\infty)$ an adequate "potential theory" was needed. From this point of view the classical L^∞-setting (uniform rational approximation) turns out to be the hardest, an extreme case crowning the easier L^p-scale of problems accessible to "purely" real means: the capacities used in L^p-approximation theorems are much more "real" and explicit than the analytic capacity involved in the uniform approximation theory (very "complex" in both senses of this word and resisting reduction to more palpable characteristics).

I was lucky to know V. Maz'ya. He was in his twenties when we met, but a true expert and master of potential theory and Sobolev spaces. He helped me a lot having taught me the "fine" W_p^l-theory ("fonctions précisées" de Deny–Lions [DL1953]). Very soon I was able to give the above vague ideas a definite form and describe the sets $E \subset \mathbb{C}$ such that L^2-closure of functions analytic near E coincides with the set of all $L^2(E)$-functions analytic in the interior of E [25]. The description was given in terms of the Cartan fine topology. This result is now a very particular case of subsequent results due to Bagby and Hedberg (see [AH1996]).

Then we started working together on problems of polynomial approximation in L^p-spaces. To describe these problems let us first look at a compact set $K \subset \mathbb{C}$ dividing the plane, so that $\mathbb{C} \setminus K$ has a nonempty bounded component g. Then any sequence of polynomials convergent u n i f o r m l y on K converges uniformly on $K \cup g$ as well. This fact is an obstacle for the uniform polynomial approximation on K (e.g., $(z - a)^{-1}$ cannot be uniformly approximated by polynomials on K if $a \in g$), and according to classical results it is the only obstacle if the approximated function is continuous on K and analytic in its interior. Suppose now our set K is "thick" enough; then the same can be said on sequences of polynomials converging in $L^p(K)$ for finite values of p (e.g., if K is a non-degenerate circular annulus). But this time not only topological, but some quantitative characteristics of K (of its "thickness") come into play. This phenomenon discovered by M.V. Keldyš can be illustrated by the so-called "crescent domains" K (see Figure 1); we do not suppose K to be compact anymore, this restriction being not necessary for the L^p-approximation: $K = G \setminus (g \cup \gamma)$, where G, g are Jordan domains with the boundaries Γ, γ such that $g \subset G$, $\Gamma \cap \gamma = \{p\}$. Keldyš showed that for the circular crescent ($\Gamma = \{|z| = 1\}$, $\gamma = \{|z - a| = 1 - a\}$, $0 < a < 1$) the $L^2(K)$-convergence of a sequence of polynomials implies their convergence in g, and $\mathcal{P}$, the set of all polynomials, is not dense in $L_a^2(K)$, the set of all $L^2(K)$-functions analytic in K; thus the circular crescent is "thick". On the other hand $\mathcal{P}$ is dense in $L_a^2(K)$ if the arc lengths $l_1(r)$, $l_2(r)$ (see Figure 1) decrease very rapidly as $r \to 0$.

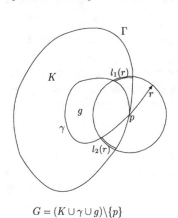

$G = (K \cup \gamma \cup g) \setminus \{p\}$

FIGURE 1

M.M. Dzhrbashyan and A.A. Shaginyan have proved that $\mathrm{Clos}_{L^2(K)}\,\mathcal{P} = L_a^2(K)$ iff

$$\int \log l(r)\,dr = -\infty, \qquad l(r) = l_1(r) + l_2(r),$$

provided γ and Γ satisfy some (rather strong) qualitative regularity conditions (see [Mer1953] for historical information and references).

In [26, 41] the approach based on Sobolev spaces and sketched above was used. The density problem of $\mathcal{P}$ in L_a^p, $1 < p < +\infty$, was reduced to a real variable property of functions in the Sobolev class $\mathring{W}_q^1(G)$. This approach resulted in a new insight into the problem. We were able to replace G in Figure 1 by an *arbitrary* bounded domain and g by its *arbitrary* Jordan subdomain with γ satisfying some smoothness conditions; all restrictions imposed on $\gamma \cap \Gamma$ (where Γ is the boundary of G) were dropped, this set is not necessarily a singleton anymore. We obtained a lot of information on the closure of $L_a^p(G)$ in $L^p(K)$. In the case when $\mathrm{Clos}_{L^p(G)}\,\mathcal{P} = L_a^p(G)$ this information yields sharp conditions ensuring the equality $\mathrm{Clos}_{L^p(K)}\,\mathcal{P} = L_a^p(K)$. Here I only want to state the above-mentioned real variable property of $\mathring{W}_q^1(G)$ (with respect to K), the most essential point of [26] (see also §3 of [41]):

> Let μ be the harmonic measure of g on γ (w.r. to a point of g); any precised $\varphi \in \mathring{W}_q^1(G)$ satisfies
>
> $$\int_\gamma \log|\varphi|\,d\mu = -\infty.$$

(For $p \in (1,2)$ "precised" means just "continuous".) This property of the triple (G, g, p) expresses certain "thinness" of K in a rather implicit form: elements of $\mathring{W}_q^1(G)$ are "continuous" (this is almost literally true if $q \geq 2$, and can be made precise for $q \leq 2$ using q-capacities, see [AH1996], [39], [Maz1985]); but $\varphi \in \mathring{W}_q^1$ is zero on Γ and, thanks to "continuity" remains small on parts of γ that are close to $\gamma \cap \Gamma$ so that the geometric mean of φ along γ w.r. to the probability measure μ is zero. It is just this degree of thinness of K that ensures the desired approximation properties of $L_a^p(G)$ in $L^p(K)$. This very indirect characterization of the thinness of K can be made much more explicit in concrete situations using techniques of the Sobolev spaces and p-capacities (see [26, 41]). In [45] our approach was generalized to the L^p-approximation by solutions of some elliptic systems in $\mathbb{R}^n$. In [28, 29] it was used to obtain a quite explicit description of $\mathrm{Clos}_{L^p(K)}\,\mathcal{P}$ for a crescent K with regular γ and Γ (see Figure 1); this description is especially telling for the Keldyš circular crescent. An important progress in L^p-approximation by polynomials was made in [Br1973]–[Br1994]. And in [24, 37] an approach similar to that of [26, 41] was applied to the L^p-approximation by harmonic functions in $\mathbb{R}^n$. But this time the main difficulty was caused by the lack of uniqueness theorems for harmonic vector fields in $\mathbb{R}^n$, $n \geq 3$, and we had to use our uniqueness results from [43]."

Many other aspects of the tandem collaboration Maz'ya–Havin, as well as the corresponding results, very influential for the subsequent evolution of the subjects, are described in the rest of the survey [85] and in J. Brennan's paper in this volume (see the photo below, Brennan and Havin talking). We can also notice that

V. Havin worked on multi-dimensional analogs of the approximation properties of analytic functions; see Part V of the present review below.

The next two themes of Havin's mathematical heritage (Parts III and IV below) may seem to be devoted to quite particular questions, but in fact, in their time, they were challenging important questions of 20th century analysis and Havin's solutions of them were very emotionally appreciated by the community and gave rise to important activities in the field (see details in Parts III and IV).

V.P. Havin and J. Brennan, 2008.

III. Smoothness up to the boundary of an analytic function compared to the smoothness of its modulus[2]

Let F be a function analytic in the unit disk $\mathbb{D}$ and continuous up to the boundary. Suppose that $|F|$ satisfies the Hölder condition of order α in the closed disk. What can be said about the Hölder continuity of F itself? In general, the answer is: Nothing. It turns out, however, that the $(\alpha/2)$-Hölder condition is always guaranteed for F under certain assumptions on its zeros, and this is sharp. We shall refer to this feature as the *smoothness drop with coefficient* $\frac{1}{2}$ in what follows.

The simplest assumption ensuring this smoothness drop is the *total* absence of zeros, including the asymptotic zeros on the boundary. A precise form of this requirement is that F should be outer; i.e., its canonical Nevanlinna–Smirnov factorization should not involve a nontrivial inner factor. In this case and for $0 < \alpha < 1$, the statement about $\left(\frac{1}{2}\right)$-smoothness drop is attributed to Carleson and Jacobs, who, however, never published an account of that work.

The first publications on the subject in question were the short announcement [32] by Havin and Shamoyan and the detailed exposition [34] by Havin alone (1970–1971). Havin and Shamoyan were not aware of the work mentioned in the

---

[2]Compiled by Sergey Kislyakov, skis@pdmi.ras.ru

preceding paragraph and obtained their results absolutely independently of their predecessors. In fact, they refined the statement of Carleson and Jacobs considerably: In [32], for instance, it was only required that F were zero free; i.e., a singular inner factor in the canonical factorization was allowed. In [34], moreover, even internal zeros were admitted, provided they did not accumulate to the unit circle tangentially (it is fairly easy to realize that this condition on zeros cannot be further relaxed). In all these cases, smoothness drop with coefficient $\frac{1}{2}$ occurs.

Next, also in [34], Havin proved that if the continuity modulus of $|F|$ is dominated (on the boundary) by a certain function ω, then (say, again on the boundary) F has continuity modulus dominated by $\omega(\delta^{1/2})$. Surely, this was done under standard regularity assumptions on ω; among other things, these assumptions tell us that, still, we deal here with smoothness between 0 and 1.

For power-type Hölder conditions of arbitrary order $\alpha \in (0, \infty)$, the same $\left(\frac{1}{2}\right)$-smoothness drop was established only some 15 years after Havin and Shamoyan by Shirokov, see [Sh1988].

It is interesting that the theme was revived in recent years. To discuss these events, a remark is in order. As can be seen already from [34], the case of an outer function F in the disk remains principal in all the refinements mentioned above (adjoining an inner part under some natural conditions is easy). So, in what follows, we assume that F is outer (though some information beyond this case is available); i.e., $F = |F| \exp(i\mathcal{H}(\log|F|))$ on the boundary, where $\mathcal{H}$ is the harmonic conjugation operator.

Now, in the paper [VKM2013] published in 2013 it was shown that the smoothness drop discussed above has *local* nature. Specifically, if $|F|$ is α-Hölder at one point only on the unit circle, then F is $(\alpha/2)$-Hölder at the same point. Unfortunately, this was done only for $\alpha \in (0, 2]$. Also, there is a mild case of cheating in the above statement, because now the Hölder condition on F can be ensured only in some integral sense. However, the statement does imply the claims of Havin–Shamoyan–Carleson–Jacobs, see [VKM2013] for the details.

Surely, I discussed this result with Havin during the preparation of the paper [VKM2013] and somewhat later. He was vividly interested, but my general impression was that, around 2013, he viewed his work about smoothness drop as a mere episode. In particular, despite his interest, he never exclaimed: «Во как!» (an expression whose meaning is difficult to convey in English in an equally short way; some blend of "That's how it is!" and "Who could imagine this is so!"; those who knew Havin often heard this «Во как!» when he was excited by some mathematical fact). Now we can only guess about the reasons for this attitude, but maybe Havin explained them himself as early as in 1971 by giving that particular title to his paper [34]. Indeed, at least for smoothness smaller than 1, the general recipe in this subject is fairly banal and looks like this: "Do with $\mathcal{H}$ what is usually done with singular integrals, and you'll win inevitably." By the way, the procedures in [VKM2013] also follow this recipe.

However, there is a subtlety to be mentioned. It turns out that the obstruction for an *uncontrollable* smoothness drop is fairly delicate, namely, this is the integrability of $\log|F|$ on the circle. In fact, Havin presented in [34] many specific and interesting quantitative relations that lie behind the smoothness drop, and he investigated quite thoroughly *the involvement of the absolute continuity modulus of the measure with density* $|\log|F||$ in these relations. We do not reproduce anything of that, but we quote quite a transparent observation in the same vein made by Shirokov in 2013, see [Sh2013]: if $\log|F| \in L^p(\mathbb{T})$ with $p \in [1,\infty)$, then smoothness drop with coefficient $\frac{p}{p+1}$ occurs.

The last feature is present also in the "local" setting; see [VKM2013]. See also [Med2015] for a refinement where the condition imposed on $\log|F|$ is its membership in a fixed rearrangement invariant function space on the circle in place of L^p (that paper is also about the local setting).

Finally, quite recently (see [Sh2016]) Shirokov showed that the $\left(\frac{1}{2}\right)$-smoothness drop occurs for zero free functions analytic in the ball of $\mathbb{C}^n$ and continuous up to its boundary. This was done by reduction to the case of the disk.

IV. The quotient L^1/H^1 is weakly sequentially complete[3]

V.P. Havin established the fact mentioned in the title of this section in the short note [40], published in 1973. The same was proved by Mooney independently and more or less simultaneously, but Mooney's paper [Moo1972] appeared somewhat earlier.

Though, informally, Havin perceived the result as a statement about convergence rather than a fact about Banach spaces, it immediately became quite popular among Banach space analysts. Whereas only several years before the main bulk of specific examples in Banach space theory had been confined to various spaces of measurable functions (such as L^p, Orlicz spaces, etc.; surely, this also had included various sequence spaces), in the period in question some tools arose that enabled the researchers to look more thoroughly at other classical examples, first and foremost at certain spaces of analytic or smooth functions. For instance, it was already well known at the time that the disk-algebra $C_A = \{f \in C(\mathbb{T}) : \hat{f}(n) = 0 \text{ for } n < 0\}$ (this is the space of restrictions to $\mathbb{T}$ of functions analytic in the unit disk and continuous in its closure) is not linearly homeomorphic to the space $C(K)$ for any compact set K, but it was conjectured that, nevertheless, C_A should share many properties of $C(K)$-spaces. The weak sequential completeness of the dual was the first candidate on the list. Since the dual of C_A splits into the direct sum of L^1/H^1 and the space of singular measures on $\mathbb{T}$, Havin's theorem resolved this conjecture in the positive.

Usually, an abstract question, if addressed to a space of (say) holomorphic functions, invokes a blend of abstract considerations and very specific methods of classical analysis. Precisely this happened in Havin's paper [40]. The classical result

---

[3]Compiled by Sergey Kislyakov, skis@pdmi.ras.ru

that lay at the basis of the proof was a quantitative refinement of the well-known construction of a function in the disk-algebra that has a peak on a given subset e of the circle with zero Lebesgue measure. Later it was called Havin's lemma. In that lemma, the set e in question is of positive (rather that zero) measure and the analytic function in question does not have a peak on it, but it is somewhat close to doing so. Moreover, Havin was even able to construct a sort of a partition of unity modeled on the pattern $\chi_e + \chi_{\mathbb{T} \setminus e} = 1$, but with *analytic* summands. These summands are close in a sense to the characteristic functions in the above formula, and the deviations are controlled in terms of the measure of e only and tend to zero as $|e| \to 0$. The proof involved standard estimates of the harmonic conjugation operator on Lebesgue spaces.

Havin's result led to some further developments. This happened in two stages. At the first stage, Havin's lemma *itself* was employed by various people to carry some more properties of $C(K)$ spaces over to the disk-algebra. See Sections 6–8 in the monograph [Pe1976] for the exposition and all credits. However, at the second stage it turned out that the matter is much softer: Almost no subtle classical analysis is really needed. See Bourgain's papers [Bou1983, Bou1984]. As can be seen from the titles, Bourgain was able to prove the weak sequential completeness of the dual and other related properties for some spaces of analytic or smooth functions of *several* variables. He succeeded in doing so precisely because his method did not require anything anywhere near as subtle as, say, the weak type $(1, 1)$ inequality for harmonic conjugation. An even softer approach (still based eventually on Bourgain's ideas) to this range of questions was suggested later by Saccone, see [Sa1997].

V. Approximation properties of harmonic vector fields[4]

Following quite a widespread opinion (shared by V. Havin), the harmonic vector fields are *the* true several variables analog of holomorphic functions in one variable. A vector field $v = (v_1, \ldots, v_n)$ defined on an open subset O of $\mathbb{R}^n$ is said to be harmonic if its Jacobi matrix J_v is symmetric and has zero trace, i.e.,

$$\frac{\partial v_j}{\partial x_k} = \frac{\partial v_k}{\partial x_j}, \quad k \neq j, \ 1 \leq k, j \leq n \quad \text{and} \quad \frac{\partial v_1}{\partial x_1} + \cdots + \frac{\partial v_n}{\partial x_n} = 0. \tag{1}$$

Equivalently, v is harmonic if it is locally the gradient of a harmonic function. Harmonic vector fields are a well-known object in harmonic analysis, and they play an essential role in the real approach to Hardy spaces [FeSt1972].

In dimension two the above system is the usual Cauchy–Riemann system for the pair of functions $(v_1, -v_2)$ and a vector field $v = (v_1, v_2)$ is harmonic if and only if $f = v_1 - iv_2$ is complex analytic. For $n = 3$ system (1) can be written as

$$\operatorname{curl} v = 0, \quad \operatorname{div} v = 0. \tag{2}$$

---

[4]Compiled by Eugenia Malinnikova, eugenia.malinnikova@math.ntnu.no

This system appears in mathematical descriptions of various physical phenomena, for example in hydrodynamics, electrostatics, and gravitational theory. It is natural to look for higher-dimensional versions of various results of complex analysis.

V.P. Havin frequently stressed the importance of the trinity "uniqueness – normal families – approximation" in complex analysis and its applications to harmonic analysis. In the early 1990s he was studying those questions in the setting of harmonic fields (and more generally, harmonic differential forms). It seemed that approximation was the easiest of the three concepts above to address, see [80].

Runge and Hartogs–Rosenthal theorems for harmonic vector fields

Many classical approximation theorems originally formulated for analytic and harmonic functions have been generalized to solutions of elliptic systems and elliptic equations. In particular, in the early 1990s N. Tarkhanov obtained a series of results for systems with a surjective symbol and P.V. Paramonov investigated C^m-approximation by harmonic functions on compact subsets of $\mathbb{R}^n$. Both works can be regarded as generalizations of theorems on rational approximation. The idea is to approximate an arbitrary solution uniformly on some compact set by "rational" solutions, which satisfy the system or equation everywhere in $\mathbb{R}^n$ except for a discrete singular set as it is done in classical complex analysis. These results are closely related to approximation of harmonic fields but cannot be applied directly. Harmonic gradients form a proper subclass of harmonic fields and it turns out that system (1) is not a system with a surjective symbol.

The classical Runge theorem on rational approximation for analytic functions can be stated as follows: For a compact set $K \subset \mathbb{C}$ and an open set O that intersects all components of the complement $\mathbb{C} \backslash K$, any function f analytic in a neighborhood of K can be uniformly approximated on K by rational functions with poles in O. One way to prove it is to use the Cauchy integral formula, approximate the integral by a combination of Cauchy kernels and move the poles into O.

The first question of rational approximation of harmonic vector fields was to find elementary building blocks that should play the role of the Cauchy kernel in higher dimension. The answer was given by V.P. Havin and S.K. Smirnov [84] and V.P. Havin and A. Presa [68].

First, as in dimension two, the gradient of the fundamental solution of the Laplace operator is a natural example of a harmonic field with a point singularity. For any $a \in \mathbb{R}^3$ the field

$$\mathrm{Coul}_a(x) = \nabla \frac{1}{|x - a|}$$

is harmonic in $\mathbb{R}^3 \backslash \{a\}$. It is called the Coulomb field of a. The crucial difference between dimension two and higher is that there are harmonic fields in higher dimensions with singularities on curves that cannot be dissected.

Let γ be a closed loop in $\mathbb{R}^3$, $\gamma : [0, l] \mapsto \mathbb{R}^3$; we assume it is Lipschitz, so that $|\gamma'| = 1$ a.e., and $\gamma(0) = \gamma(l)$. Then the Biot–Savart field of γ is defined by

$$\mathrm{BS}_\gamma(x) = \mathrm{curl} \int_0^l \frac{\gamma'(s)}{|\gamma(s) - x|} \, ds.$$

This field is harmonic in $\mathbb{R}^3 \setminus \Gamma$, where $\Gamma = \gamma[0, l]$. Further, BS_γ cannot be written as the sum of two harmonic fields v_1 and v_2, where one is harmonic in $\mathbb{R}^3 \setminus \gamma[0, c]$ and the other in $\mathbb{R}^3 \setminus \gamma[c, l]$.

It turns out that any harmonic field can be decomposed into an integral of Coulomb and Biot–Savart fields and the following approximation theorem holds.

Theorem. *Let E be a compact subset of $\mathbb{R}^n$, U an open set containing E, and K a smooth compact manifold with boundary such that $E \subset K \subset U$. Let $\{\gamma_j\}_{j=1}^J$ be closed loops in $\mathbb{R}^n \subset K$ that form a basis for 1-homologies of $\mathbb{R}^n \setminus K$. Finally, let O be an open subset of $\mathbb{R}^n \setminus K$ that intersects any component of $\mathbb{R}^n \setminus K$. Then any vector field v harmonic in U can be approximated uniformly on E by linear combinations of BS_{γ_j} and Biot–Savart and Coulomb fields with singularities in O.*

This theorem was proved by V.P. Havin and A. Presa [68, 79] by using duality arguments and the Hahn–Banach theorem; later a constructive proof was given in [81]. Both proofs involve the Alexander–Pontryagin duality.

Another classical theorem of rational approximation in complex analysis, the Hartogs–Rosenthal theorem, states that if K is a compact subset of $\mathbb{C}$ of zero measure then any continuous function on K can be uniformly approximated by rational functions. It is clear that in higher dimension there are compact sets of zero measure that contain cycles and on which some vector fields cannot be approximated by harmonic ones. A set K is called invisible if in some coordinate system the areas of the projections of K onto all coordinate two-dimensional planes are zero. It was proved by Havin and Presa that on an invisible set any continuous vector field can be approximated by harmonic ones. The first proof given in [79] was by duality arguments. The second constructive proof was given in [81]. It should be mentioned that these results were proved in a more general settings of harmonic differential forms in $\mathbb{R}^n$. The construction in Runge's theorem was further simplified by S. Dager and A. Presa in [DaP1996], and the results were generalized to pairs of harmonic forms in [HP2005] and to harmonic differential forms on smooth Riemannian manifolds in [Mal1999].

Bishop's locality principle

In classical analytic approximation, the following locality principle plays a very important role. Let K be a compact subset of $\mathbb{C}$. If for each point $z \in K$ there exists some ball B centered at z and such that any function continuous on $B \cap K$ can be approximated by rational functions on $B \cap K$, then any function continuous on K can be approximated by analytic functions (see for example [Gam1969]). Vitushkin's characterization of sets that admit rational approximation, given in terms of analytic capacity, relies heavily on this principle. It was proved by Havin and Smirnov that for approximation by harmonic vector fields the locality principle fails [84]. They gave the following counterexample. There exists a continuous function F on $\{(x,y) : x^2 + y^2 \le 1\}$ vanishing on the boundary C of the circle and such that the graph $K = \{(x,y,F(x,y)) : x^2 + y^2 \le 1\}$ contains no rectifiable curves except for the arcs of C. Then fields with non-zero circulation along the boundary circle cannot be approximated by harmonic ones. At the same time locally K contains no rectifiable arcs except for those in C and any continuous vector field on $K \cap B$ can be approximated by harmonic fields. The key point of the proof is the following statement [84]. If μ is a compactly supported vector-valued measure with zero divergence (in the sense of distributions), then its support contains a non-degenerate rectifiable arc. Such measures are called solenoids and a much stronger result on decomposition of solenoids into convex combination of elementary ones was obtained by Smirnov in [Smi1993], which was his master thesis.

New results on approximation by harmonic vector fields were recently obtained by a former student of V.P. Havin, M. Dubashinskii, see [Du2011, Du2013, Du2016]. The related topic of C^m-approximation also develops further; see the recent paper [Fed2012] of K. Fedorovskiy, and references therein.

VI. Uncertainty principle in harmonic analysis[5]

The uncertainty principle (UP) in harmonic analysis is a classical subject that stems from the ideas of Norbert Wiener in the 1920s. The original postulate of the principle can be formulated as follows.

A function (measure, distribution) f and its Fourier transform $\hat{f}$ cannot be simultaneously small.

As a classical example one should mention the Heisenberg uncertainty principle, which, in its simplest form, may be expressed as the inequality

$$\|xf(x)\|_{L^2(\mathbb{R})}\|t\hat{f}(t)\|_{L^2(\mathbb{R})} \ge \|f\|_{L^2(\mathbb{R})}^2/4\pi.$$

However, the "smallness" in this broad statement can be given a number of completely different mathematical meanings (smallness of support, fast decay at infinity, etc.), many of them leading to deep and important problems. At present, this area of study has grown far beyond its original borders and now includes studies of bases and frames in Banach spaces, harmonic analysis on groups, and other branches; see for instance [FS1997]. The Fourier transform in the general statement of the UP can be replaced with other transforms; see our discussion of [52, 89] below. Instead of the smallness of f and $\hat{f}$ one can study other "dual" properties of the given function, like asymptotics and sizes of zero sets in classes of entire functions.

As in several other fields of analysis, the contribution of V.P. Havin in the area of the UP is broad, diverse and significant. It includes not only original results [52, 89, 90, 91, 94] but also promoting and teaching the subject of the UP to many

---

[5]Compiled by Alexei Poltoratski, alexeip@math.tamu.edu

of his colleagues and friends in the St. Petersburg analysis group. Due to his efforts, a lot of classical problems of the UP were studied and solved by the members of the said group and a number of outstanding talks on the UP were delivered at the Havin–Nikolski seminar.

Surveying all of the achievements of Havin's colleagues and students in the area of the UP obtained under his direct influence would require a separate volume. Being unable to undertake such a task here, let us concentrate on V.P. Havin's own work on the subject.

A well-known theorem by M. Riesz shows the following "non-Fourier" variation of the theme of the UP. Let μ be a finite measure on $\mathbb{R}^n$ such that

$$\int_{\mathbb{R}^n} \frac{d\mu(t)}{1 + |t|^{n-a}} < \infty$$

for any $0 < a < n$. The (Riesz) potential of μ is defined as

$$U_a^\mu(x) = \int_{\mathbb{R}^n} \frac{d\mu(t)}{|t - x|^{n-a}}.$$

Riesz' result says that if μ and $U^m u_a$ both vanish on the same non-empty open set then μ is zero identically. In his paper [52] Havin made a significant improvement of this theorem in the one-dimensional case. The main result of [52] says that if the measure is absolutely continuous with Hölder density of order $1 - a + \varepsilon$, then the open set in Riesz' statement can be replaced with an arbitrary Borel set of positive measure. The problem of extending this result to $\mathbb{R}^n$ and improving the Hölder condition to continuous densities stood open for many years and was studied by V.P. Havin together with several of his outstanding students. The problem was finally solved in the negative by Havin and D. Beliaev (Havin's "mathematical grandson") in [89]. The sophisticated construction presented in [89] used some of the techniques of J. Bourgain and T. Wolff.

It should be mentioned that it was Havin who introduced the surprising (for that time) idea that the UP holds for many convolution operators and posed the problem of description of kernels with this property. This idea was developed into the comprehensive theories of "epsilon-local" and "antilocal" operators with numerous natural examples (see Part II, Chapter 5 of [4] for details).

One of the deepest parts of the UP is the famous Beurling–Malliavin (BM) theory, which produced the celebrated theorem on completeness for families of exponential functions $\{e^{i\lambda_n z}\}$ in $L^2(0,1)$. The original

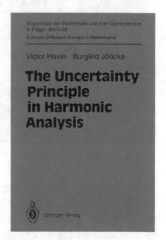

proofs, which appeared in the 1960s [BM1962, BM1962], continue to fascinate harmonic analysts to this day. A significant part of V.P. Havin's work in the UP concerned the study and promotion of the BM-theory to his students and colleagues. Over the years he has made a number of essential improvements and extensions for various parts of the theory, which opened the possibilities for further studies, still pursued by many mathematicians.

A principal ingredient of the BM-theory is the so-called Beurling–Malliavin multiplier theorem, which says that for any entire function f of Cartwright class there exists a Paley–Wiener function m of arbitrarily small exponential type (the multiplier) such that fm is bounded on the real line. Equivalently, the statement can be reformulated as follows: For any Lipschitz function w, $0 < w < 1$, whose logarithm is Poisson-summable, there exists a Paley–Wiener function m of arbitrarily small exponential type such that $|m| < w$ on $\mathbb{R}$. In this way the theorem becomes a statement on "*admissible majorants*" for the Paley–Wiener spaces. In a series of papers by Havin and J. Mashreghi [90, 91] this statement was extended to much broader classes of spaces of analytic functions, the model spaces K_θ, with θ satisfying some natural restrictions. Some of these results were later improved by Havin and A. Baranov [95]. The methods developed in those papers led to a completely new "real" proof of the Beurling–Malliavin multiplier theorem contained in the paper of Havin with Mashreghi and F. Nazarov [94].

The work of V.P. Havin in the area of the UP culminated with the seminal book "*The Uncertainty Principle in Harmonic Analysis*" [4] written jointly with his student Burglind Jöricke (a shorter survey, presenting a digest of the book is contained in [88]). The book covers classical results of the UP and gives many of them new updated proofs, many of which were improved by the authors themselves, including a full treatment of the BM-theory. Some of the theorems are given up to three different proofs. The results are presented in two groups, those that can be proved with real methods and those that require complex analysis (with a non-empty intersection between the groups).

Burglind Jöricke

The book with B. Jöricke, in fact, was a large *research* project. It was written during the 1980s (finished about 1990) in a tight collaboration with a number of people in Havin's seminar (mostly, Havin's pupils), especially with (of course) Burglind Jöricke, Fedor Nazarov, Sergei Hruschev, Alexander Borichev, Alexander Volberg. Their (collaborators) own results entered the book as sections and chapters, and they passed long hours of mathematical work at Havin's three piece apartment on Golodai Island. The thirst for perfection was insatiable, and the resulting text is destined to be a reference source for experts for many years to come. Together with the book by V.P. Havin's friend and colleague at McGill University, P. Koosis, "*The Logarithmic Integral*" [Ko1988], the book by Havin and Jöricke [4] is among the most significant texts in the field of the UP.

VII. Admissible majorants and the Beurling–Malliavin Multiplier Theorem[6]

Among the topics of particular interest for V.P. Havin, a prominent place was occupied by the Beurling–Malliavin theory, considered by him to be one of the greatest achievements of 20th-century harmonic analysis. It deals with uniqueness/nonuniqueness properties for functions with bounded spectra. This includes, in particular, the following

Problem. *Let $a > 0$ and let $f \in L^2(\mathbb{R})$, $\operatorname{supp} f \subset [-a, a]$. For which functions $w \geq 0$, does the estimate $|\widehat{f}| \leq w$ imply that $f = 0$ a.e.?*

If there exists a nonzero function f with $|\widehat{f}| \leq w$, the function w is said to be an *admissible majorant* (for the given a); this notion and the term were introduced by Havin. A necessary condition for w to be an admissible majorant is the convergence of the logarithmic integral (i.e., the summability of $\log w$ with respect to the Poisson measure on the line). This is a criterion for being an admissible majorant in the Hardy space H^2 (i.e., for functions with semi-bounded spectrum).

For a wide class of majorants, the solution is given by the famous Beurling–Malliavin multiplier theorem (or First Beurling–Malliavin theorem, BM1 in what follows) [BM1962]: *if the logarithmic integral converges and the function $\Omega = -\log w$ is Lipschitz on $\mathbb{R}$, then w is an admissible majorant for any $a > 0$.*

The term "multiplier theorem" refers to the fact that the function $1/w$ (which should be thought of as large) can be multiplied by an entire function with an arbitrary small spectrum so that the product is bounded. This theorem is a crucial step in the proof of another famous result – second Beurling–Malliavin theorem [BM1967], which computes the radius of completeness for a system of exponentials.

The multiplier theorem is a deep result. Several different proofs are known (Beurling–Malliavin, Koosis, de Branges), but none of them is easy.

All these proofs used essentially complex analytic methods, especially the theory of entire functions, because the Fourier transform maps functions with bounded spectra to the Paley–Wiener space of entire functions of exponential type. In a series of papers [90, 91, 94] Havin, jointly with his former students J. Mashreghi and F. Nazarov, found a *real* proof of BM1. A vivid account of the trend of getting rid of complex methods in essentially real results ("taking off the complex mask") can be found in [94].

Another advantage of Havin's approach to BM1 is that it applies to more general spaces of analytic functions (model subspaces $K_\Theta = H^2 \ominus \Theta H^2$ of the Hardy space associated with various inner functions Θ, or de Branges spaces) generalizing the Paley–Wiener space.

---

[6]Compiled by Anton Baranov, anton.d.baranov@gmail.com

Havin's approach to the multiplier theorem

The first step in Havin's approach to BM1 was to write a certain "parametrization of admissible majorants" [90]. Namely, let $w \geq 0$ with Poisson-summable $\Omega = -\log w$. Then w is an admissible majorant if and only if there exists a bounded function $m \geq 0$ with $mw \in L^2(\mathbb{R})$ and $\log m$ Poisson-summable such that

$$at + \widetilde{\Omega}(t) = \widetilde{\log m}(t) + \pi k$$

a.e. on $\mathbb{R}$, where k is a measurable function with integer values, and $\widetilde{\Omega}$ denotes the (regularized) Hilbert transform of Ω. A similar result holds for spaces K_Θ with at replaced by $\frac{1}{2} \arg \Theta(t)$. This representation is based on an observation due to K. Dyakonov [Dya1990]: $h = |f|$ for $f \in K_\Theta$ if and only if $h^2 \Theta \in H^1$.

Though essentially a reformulation of the initial problem, this representation turns out to be very useful. Using it, Havin and Mashreghi found admissibility conditions in terms of $\widetilde{\Omega}$:

If $\widetilde{\Omega}$ is Lipschitz, then w is an admissible majorant for any $a > \|(\widetilde{\Omega})'\|_\infty$.

This result is a special case of a much more general sufficient admissibility condition applicable to general de Branges spaces (i.e., spaces K_Θ with meromorphic Θ).

But $\Omega \in Lip$ does not imply that $\widetilde{\Omega} \in Lip$ and so this result itself did not yield BM1. Using an ingenious combinatorial construction, Nazarov proved the following "regularization" theorem:

If $\Omega \in Lip$, then for any $\varepsilon > 0$ there exists Poisson-summable Lipschitz $\Omega_1 \geq \Omega$ such that $\|(\widetilde{\Omega}_1)'\|_\infty < \varepsilon$.

Combined with the result of Havin and Mashreghi cited above, this gives a real variable and, probably, the shortest proof of the multiplier theorem. Indeed, now $w_1 = \exp(-\Omega_1)$ is admissible and so is w.

Nazarov's regularization theorem with necessary background from [90, 91] appeared in the paper "Beurling–Malliavin Multiplier Theorem: the Seventh Proof" [94]. This title emphasized the fact that several (about six) different proofs of BM1 were known previously and at the same time made an allusion to M.A. Bulgakov's famous novel *"The Master and Margarita."* As an epigraph Victor Petrovich, who knew and admired classical Russian literature, used the following quotation (recall that Bulgakov/Woland speaks on the proofs of the devil's existence. . .):

«. . . *На это существует седьмое доказательство и уж самое надежное! И вам оно сейчас будет предъявлено.*»
("*. . . Yet the seventh proof of that exists, reliable beyond doubt! And it will be shown you in a while."*)

Further developments

The papers by Havin and Mashreghi cited above contain also a number of results about admissible majorants in de Branges spaces of entire functions. In particular, it was shown that the theory bifurcates into two essentially different situations: *Fast* (*linear or super-linear*) growth of $\arg \Theta$ and *slow* (*sub-linear*) growth of $\arg \Theta$ on $\mathbb{R}$.

In [95] some refinements of the results of [90, 91] were obtained while in [96] some model cases of slow growth of the argument were studied (e.g., the case of inner functions Θ of the form $z_n = n^\alpha + i$ or $z_n = \pm n^\alpha + i$, $n \in \mathbb{N}$), and some links with quasianalyticity problems were found. Yu. Belov [Be2008] considered the case of the fast growth of the argument ($|\arg \Theta(t)| \asymp |t|^\alpha$, $\alpha > 1$). In an interesting paper [Be2009] Belov found sufficient conditions for w to be a two-sided majorant, that is, for the existence of f in a given model space for which $|f| \asymp w$.

Seminar and Pupils

One more result of Victor Havin's mathematical activity deserves a detailed description – we mean the joint University/Steklov Institute analysis seminar. The seminar worked on a weekly basis under the guidance of Havin since 1963, and it is a "must-visit" for all analysts in Saint Petersburg. We can even say that, with years of evolution, it became a cultural event for the whole city's mathematical community counting many hundreds of members. All young mathematicians of "Havin's school" (see below), many of whom received world renown, started their work within that seminar. Let us recall some steps of its development.

As has already been mentioned (see the "Life" section above), Havin's seminar and research group started about 1963 with four people, and the first semester (or more) was devoted to detailed presentation of two books – N.N. Luzin's thesis on trigonometric series and singular integrals (1915, second commented edition, 1951), and the Russian translation of K. Hoffman's *"Banach spaces of analytic functions"* (Prentice-Hall, Inc., 1962) having appeared shortly before. Possessing no regular time and place for an improvised seminar (which was not included either in Department timetables or in Havin's teaching assignment), this tiny group met every Monday at the huge Auditorium 66 (designed for 300–400 people...) at the old Mat-Mekh Department building in the 10th Line of Vassilievsky Island. Very soon, somebody more important drove us out, and the seminar started to meet in the Astronomy Wing of the Department: Mounting to the second floor, you follow a long snake-curved corridor passing beside a toilet designed in XIX century style with an advertisement *"Do not drink from the urinals"*..., in order to arrive at a little room whose ceiling is supported by wooden scaffolding to prevent falling on heads and whose tiny blackboard was strongly hostile towards supporting any traces of chalk... This was where Harold Shapiro started his lecture series (mentioned above) on extremal problems in Hardy spaces addressing about five-to-ten highly excited young people.

Hardy spaces became the principal framework for Havin's seminar for long years, forming a kind of unavoidable entrance tickets to the community – knowledge of the "bible books" by I. Privalov, K. Hoffman and (later on) J. Garnett, P. Koosis was obligatory for novices. By chance, but mostly by an unmistakable mathematical taste, Havin chose in Hardy spaces one of the principal and most productive trends in analysis of the XXth century. Indeed, Hardy spaces are not only particular classes of holomorphic functions with limited integral growth to the boundary (as G. Hardy defined them in 1915), but a really ubiquitous mathematical object that turns out to be the principal tool for tens of branches of analysis and its applications, such as (weighted) polynomial approximation, orthogonal polynomials, Toeplitz matrices and their applications to Ising model, Wiener–Hopf equations and Birkhoff–Wiener–Hopf factorization, signal processing, stationary processes, BMO function classes and related "maximal functions" ideology, analytic probability theory, etc. Finally, how can we explain this efficiency and irreplaceability of Hardy spaces? Perhaps, by some enigmatically well-hidden structural reasons of "mathematics of one variable". . . , but surely because so many different structures cross-fertilized each other when meeting at this subject. The holomorphy as orthogonality (which provides leading examples for the study of functions with restricted Fourier spectrum, and for various uncertainty principles), causality for signals and past/future structures for stationary processes, a nontrivial multiplicative structure (= the canonical Smirnov–Riesz–Nevanlinna factorization) related to the invariant subspaces, partial order, models for Hilbert space operators (Szőkefalvi-Nagy–Foias and de Branges models),. . . and the list of course can be continued (even though, probably, a mere selection of only one of the above reasons suffices to develop a consistent mathematical field). In short, this tremendously genuine Havin's subject choice gave a free field for the development for hundreds of his followers – direct disciples, next students of his students, and so on.

Havin gave a great significance to maintenance of his seminar and spent a lot of time, energy and talent, without measuring, in order to keep it at top level. The seminar very soon was transformed into a workshop for forming high professionals in mathematical analysis, and for this, it functioned in a special way. On the eve of every new semester, Havin took a meeting with his closest collaborators (usually, 1–2 persons) in order to discuss and choose the principal theme (just one) for the seminar for the coming four months (the third of the authors of the present text always took part in such meetings). The theme could be, for example, "Weighted estimates of singular integrals," or "Polynomial/rational mean approximation", or "Invariant subspaces of linear operators," or "Sz.-Nagy–Foias function model", etc. After detailing subjects for 10–12 future meetings (!), the proposals/invitations were sent to possible speakers. The seminar gathered for 14–15 meetings per semester, for two hours talks, with a break in the middle. Often (very often) the talk was preceded by a home "repetition" with Havin himself, or with an expert in the seminar of his choice. And so on, and so on. . . for over 30 years. . .

Numerical and qualitative seminar growth was very fast, starting with 4 participants around 1963, then passing to 5–6 participants in 1965, to about 10 in 1970, in order to reach at least 25 to 1975, and over 40–50 every Monday since 1980. . . From 1976, the seminar gathered at room 311 of the Steklov Mathematical Institute building on 27 Fontanka River: Around a round oak table of about 3 meters diameter (descending to A.A. Markov, senior) were sitting Havin himself, followed by his pupils S. Vinogradov, N. Nikolski, A. Aleksandrov, S. Hruschev, and then E. Dyn′kin, S. Kislyakov, N. Shirokov, A. Volberg,. . . and then N. Makarov, P. Kargaev, V. Vasyunin, and later on F. Nazarov, S. Smirnov, B. Jöricke, E. Malinnikova . . . , but we have no intention to give here a complete list. Better, we reproduce an excerpt from a paper by Lars Hedberg (University of Linköping) who was a frequent speaker at the St. Petersburg analysis seminar, [Hed2002]:

"When the history of mathematics in Russia during the second half of the twentieth century gets written, the St. Petersburg Analysis Seminar deserves a prominent place. It was started by Havin, then a young "docent", in the early 1960s with a few students, including Nikolski and Vinogradov. After a few years. . . , the seminar grew so that in the mid-eighties it had 40–50 active participants, a

Lars Inge Hedberg (1935–2005), Margarita Hedberg, and Victor Havin, at a Garden party of Doctor Honoris Causa ceremony at Linköping University (Sweden, 1993).

surprisingly large number of which are now mathematicians of world renown. To quote the contribution of J.-P. Kahane in . . . (*Hegberg meant the paper* [Kah2000] *in the same Vinogradov memorial volume – Ed.*): "*It is sad to see the bright school to which* [*Vinogradov*] *belonged dispersed all over the world. On the other hand, never was the Leningrad school of analysis better known than now, precisely because its members are everywhere.*"

Jean-Pierre Kahane (who played an important role in establishing links between Havin's seminar and the European math community), Victor Havin, and Nikolai Nikolski relaxing after Kahane's talk to the seminar (1989). Havin takes one of the first issues of a newly created St. Petersburg Mathematical Journal.

How can one explain this success? An outside observer can guess at a combination of factors: The . . . St. Petersburg mathematical and intellectual tradition, . . . [but also] the personality of Victor Havin. . . ".

As has already been mentioned, the flow of Western seminar visitors started growing after a shy raising of the "iron curtain" at the beginning of the 1980s. It suffices to recall the famous Louis de Branges three-months stay in Leningrad in 1984, which resulted in correction of his proof of the Bieberbach conjecture, see [BDDM1986] about that story; a funny part of the story, nowhere mentioned, is on a little panic in our large (but completely "Soviet" in its conscience) seminar when de Branges started his first talk saying "I propose you to sign a joint protest petition to Presidents of the US and the USSR against an absolutely insufficient financement of Functional Analysis research in both countries..." (at that time, the idea of any joint petition with US citizens was treated criminal in Soviet Union..., provided it was not KGB inspired...). The more the political climat changed in the country with Gorbachev's "perestroika", the more Western speakers passed in front of the mentioned Markov's round table... An incomplete list includes (together with already mentioned) H. Shapiro, A. Shields, Y. Domar, J. Ecalle, W. Hayman, A. Böttcher, G. Dales, P. Casazza, J. Zemanek, Th. Bagby, P. Duren, P. Koosis, P. Gauthier, J. Conway, J. M. Anderson, L. Hedberg, M. Essen, L. de Branges, Ch. Davis, L. Ehrenpreis, J.-P. Kahane, G. Pisier, D. Sarason, J. Korevaar, R. Rochberg, B. Fuglede, E. Amar, C. Berenstein, ...

P. Duren (center), V. Havin, and N. Nikolski in front of the entrance to Steklov Institute (St. Petersburg Division), 27 Fontanka River.

B. Fuglede (University of Copenhagen, now emeritus) and V. Havin after a seminar talk, 1988.

Most of Havin's 31 Ph.D. students (following the famous Mathematics Genealogy Project – see below Havin's Math genealogy tree) passed under his guidance through the seminar described above. Speaking on the international seminar success, we can mention, first, 3 of Havin's (former) pupils winning the Salem

V. Havin and S. Smirnov, Geneva 2006.

Prize (A. Aleksandrov, F. Nazarov, S. Smirnov), and then his 150 scientific descendants (following the MGP) which, in its turn, includes already 9 Salem Prize winners: The three above, and then A. Volberg, S. Treil, N. Makarov, S. Petermichl, Zhen Dapeng, D. Chelkak, a beautiful list crowned with Stanislav Smirnov's Fields Medal of 2010 (by the way, Stas Smirnov being Havin's master diploma student, is at the same time his mathematical great-grandson...).

Below, we list some of the chief mathematical results obtained in the framework of Havin's seminar, but before we give some more pictures, not always serious...

Sometimes around the seminar:

V. Tkachenko (the speaker of the day; FTINT, Kharkov), V. Havin, E. Dyn′kin, S. Kisliakov, V. Peller (late 1970s).

E. Dyn′kin, S. Hruschev, and V. Havin during a seminar break (early 1980s).

... and sometimes after (the seminar)...

"... and next, imagine, I split the integral in three..." (Yu. Brudnyi (mirror), N. Nikolski, I. Ostrovskii, and V. Havin at rest...; 1980s).

V. Perov's (Russian genre painter) original "The Hunters at rest" (1871).

We finish this section with something more serious – an abridged list of some outstanding results obtained and first presented at Havin's seminar:

- 1980: Hankel Operators of Schatten–von Neumann classes and their applications to function classes, approximations, stochastic processes, and so on (V. Peller; then V. Peller – S. Hruschev, S. Treil, V. Vasyunin, ...).
- 1982: Existence of Inner Functions in the ball of $\mathbb{C}^n$ (A. Aleksandrov, Salem Prize 1982).

- 1985: LogLog Distribution Law for Bloch Functions and Derivatives of Conformal Mappings (with subsequent breakthroughs in Löwner evolutions, DLA asymptotics and dynamic systems) – N. Makarov, Salem Prize 1986.
- 1988, 2002–2014: Matrix and Operator Corona Problem (solution of the Sz.-Nagy problem, $1/\delta^2 \cdot \log\log(1/\delta)$ lower estimate, a curvature solvability criterion, etc.) – S. Treil, Salem Prize 1993.
- 1987–2016: Clark's Measures and other analytic techniques for Model and de Branges Spaces – A. Aleksandrov, V. Havin – J. Mashreghi, N. Makarov – A. Poltoratskii (with applications to spectral string theory, completeness problems, etc.), A. Baranov, D. Yakubovich, Yu. Belov, . . .
- 1995–2015: The Bellman Function Method in Harmonic Analysis: for the study of the weighted and matrix weighted Hilbert transform by F. Nazarov (Salem Prise 1999), S. Treil, and A. Volberg (Salem Prise 1988) in 1995, next in a paper by S. Treil and F. Nazarov in 1996; then the study of two weight boundedness around 1999 by A. Volberg and S. Petermichl (Salem Prize 2006); then the work of V. Vasyunin and his students, . . .
- 1997–2004: Calderón–Zygmund Operators on nonhomogeneous spaces – F. Nazarov – S. Treil – A. Volberg's series: "Tb" theorem without doubling condition, solution to Vitushkin's "rectifiability" conjecture (1955), applications to two weight estimates for CZOs, next to Birkhoff–Wiener–Hopf factorization in full generality, to Tolsa's solution to the Painlevé conjecture (100 years old), etc. . . .
- 2001–2010: S. Smirnov's proof of the conformal invariance of scaling limits for Ising planar models, percolation, and other critical lattice models in statistical mechanics related to the Schramm–Löwner Evolution – Salem Prize (2001), – Fields Medal (2010).

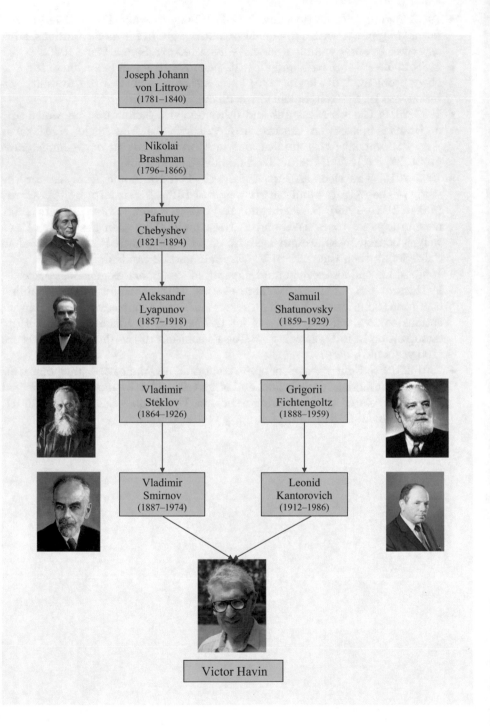

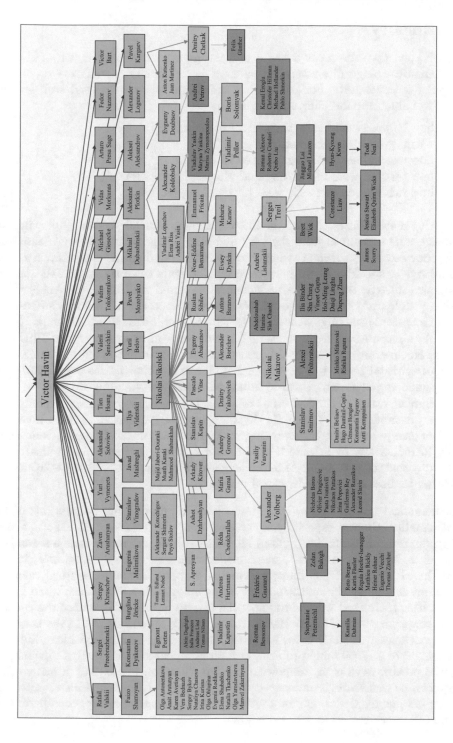

Personality

This is, perhaps, the most delicate and intimate part of our version of Havin's mathematical biography where we collect a few fragmentary witnesses on his unforgettable personality. For us, Havin's individual traits were (and still are) extremely influential and charming, especially

how he professed his mathematical professorship,
where were his mathematical tastes,
how musical sentiments saturated Havin's life,
how tasteful were his writing and philosophy preferences,
how rich and charismatic were his speaking habits...

Teaching mathematics. All of Havin's colleagues, and all students following his courses, were convinced from the very first minutes listening V.P.'s presentations that they saw a God-given teacher – so fascinating and enticing was Havin's emotional style of speaking about mathematics. The years when Havin gave the general analysis course at the Math Department, all talented students chose Analysis Division ("Kafedra") for graduate courses. See also Konstantin Dyakonov's (a former Havin's student) moving witness on V.P.'s teaching, below in this volume.

At a more advanced level, with his numerous undergraduate and PhD students, Havin continued to inspire one's imagination by involving his pupils in a kind of spiritual team, but never pressed on them, either by his own knowledge, or by galloping or forcing them to follow his own understanding rhythm.

Similar warm and intellectual relations linked V. P. with his university colleagues, members of the "Analysis Division" (counting about 20 people). Investing a lot of energy and time in improving courses in analysis at all levels (from freshman introductions to advanced post-doctoral courses), Havin maintained close everyday contact with many of his colleagues, as in the photo, during a "Kafedra" meeting to discuss some pedagogical questions.

Mathematical tastes. Meanwhile, Havin knew doubts in the foundations of his mathematical choices, especially at the beginning of his career in the 1960s, a kind of an "uncertainty complex". Often he talked about pure analysis as a secondary discipline, "crawling analysis", and considered himself to be a "dilettante" in any domain where he did not make a principal impact. These self-critical opinions were not shared by the mathematical community, and, for instance, when in the mid 1980s, I. Gelfand wanted to know the (freshly appeared) proof of the Bieberbach conjecture he invited V. Havin (rather than a narrow expert in the field...). V.P. was however strictly convinced that hard analysis is irreplaceable in the final settlement of any valuable problem, and often repeated that "analysis techniques should be always as ready as powder kept dry". In this sense, and in many other respects, he was strongly influenced by G. Hardy's "*A Mathematician's Apology*" which became his desk book for a visible period of time (as it happened later with some other conceptual books, see below).

During an Analysis Division meeting, 1978
– Boris Makarov, Lydia Florinskaia, Michail Solomyak,
Alexei Potepun, Victor Havin.

As has already been mentioned (and as is clear from the above description of Havin's results), V.P.'s appreciation of different branches of analysis varied from "soft" to "hard" passing by

- topological (and normed) holomorphic vector spaces,
- sheaves of holomorphic functions,
- interpolation by holomorphic functions (there was a long period when V. P. ranked interpolation, in the sense of describing the traces $X|E$ of holomorphic spaces X on sparse subsets E of the domain, as the principal theme of the Seminar; there was a kind of a "cult" of Carleson type free interpolation – and the very term "free interpolation" meaning "free of any traces of holomorphy" was invented in Havin's seminar (maybe, even by Havin himself); the virtuosity of interpolation techniques in papers written by S. Vinogradov, S. Hruschev, E. Dyn'kin, and many others, transformed Havin's seminar for many years in a world capital of free interpolation. . . ; in particular, Havin– Vinogradov research surveys [44], [46] determined the problematic of the field for many years ahead),
- "real methods" in complex analysis (maximal functions, etc.),
- potential theoretic methods (especially, in a long-lasting collaboration with V. Maz'ya),
- uncertainty principle in harmonic analysis,
- holomorphic vector field theory,
- de Branges spaces of entire functions and Beurling–Malliavin theory.

Havin's classification of human activities:

1. Music 2. Mathematics 3. Writing 4. Philosophy

It seemed that plastic arts – painting, sculpture, architecture – did not take much place in V.P.'s heart and head...

Music dominated in Havin's emotional life: D. Shostakovich's modern classics as his fourth, and next thirteenth symphonies were among Havin's beloved pieces (... he shared, however D. Faddeev–V. Smirnov's opinion that *"... it's Mahler, everything is just Mahler..."*). From the classic repertory, Mozart and Bach were at the first line, and in romantic music he preferred Tchaikovsky. In Havin's own word, he was "simply a music-holic having no music education", distancing himself from extreme expert opinions (like *"... odd Beethoven symphonies are boring..."*, following his quotation from Vladimir Smirnov in [134]). Havin's home collection of music (mostly on vinyl discs) was large and various, and his marches to Leningrad Philarmonia Grand Hall were very frequent, merely usual life events...

In Writing, the Russian classic Leo Tolstoy, especially his *"War and Peace"* (which V.P. studied as a Bible), *"The Kingdom of God is within you"*, *"Letters and Diaries"*, were (for latest Havin) above all, with a breakaway gap from the next... Earlier in time – Pushkin, Tyutchev (especially, his socially resonant poems and texts), Bulgakov, and Grossman always were placed not far from his desk... Vassily Grossman's *"Life and Fate"* was appreciated over Pasternak's *"Doctor Zhivago"* (V.P. mentioned several times that he could identify himself with Viktor Shtrum, the principal hero of Grossman's novel), and in poetry I. Brodsky was regarded as the greatest Russian poet in the XXth century. However, from his "Komsomol youth", Havin kept warm feelings towards "Soviet classics": To V. Mayakovsky (from whom V.P. could quote tens of pages by heart), and then – to I. Ilf and E. Petrov with their satiric *"The Twelve Chairs"* and *"The Little Golden Calf"*. From "Khrushchev Thaw" time, many things came to Havin's shelves, as A. Voznesensky (early poems) and D. Granin; A. Solzhenitsyn's classics (especially, *"The Gulag Archipelago"*), of course, bowled over Havin's vision of the Soviet regime but later on he was repelled by Solzhenitsyn's antisemitisme.

What was quite rare in "Soviet intelligentsia", Havin preferred to read Western writers in originals, not in translations, especially in German and French. He liked to share his impressions on reading German classics (Göthe, Heine,...) and recite them, and then T. Mann (who was of interest for an evolution from antisemitic to anti-nazism...), H. Ibsen (with *"An Enemy of the People"* (in German) – a warning (expressed as early as in the XIX century) against blind "crowd opinions"), L. Feuchtwanger (with his *"Jud Süss"* and an unknown in Soviet Union description (... but not the phenomenon itself!) of "court Jews" in Germany of the 18th century, later on used by Göbbels propaganda), and then many J.-P. Sartre novels (in French), etc.

Havin often quoted famous literary maxims and formulae in everyday speaking when they responded to his feelings of the moment...:

"... *Но пораженья от победы*
Ты сам не должен отличать."

from Boris Pasternak, or Aleksander Pushkin's

"*Не дорого ценю я громкие права...*
Иная, лучшая потребна мне свобода:
Зависеть от царя, зависеть от народа –
Не всё ли нам равно? Бог с ними. ..."

("Из Пиндемонти"), or – from Mikhail Bulgakov's "*The Master and Margarita*":

"*Never beg for anything, especially from those who is stronger than you*"

(Woland to Margarita), or, anew from A. Pushkin,

"*Плюнь, барин, поцелуй злодею ручку*"

(Grinev and Savel'ich scene before Pugatchev). The last three citations very well mirror some conciliatory and conformist traits of Havin's character (... let's say, typical for most Soviet survivors). However, in certain circumstances, V.P. showed a sharp insistence; for instance, if he needed permission to recruit a new Jewish student (which, with his own Jewishness, occurred very seldom, say – beyond his youth years – never) or organize a Ph.D. defense overcoming the "2/5 rule", Havin took his suit and a necktie and went to the Dean (or a Party Committee member) office, which action was called a "*Walk to Canossa*" (or "Humiliation of Canossa") as Henry IV (of "Holy Roman Empire") went to Pope Gregory VII, in 1077... (By the way, for explanation, it should be mentioned that to that period (of Havin's professional activity) a widespread antisemitism in the Soviet Union was so habitual that the very words "Jew, Jewish" became almost obscene... – as one does not talk of a rope in the house of a hanged man...).

In Philosophy, from Russian thinkers, Havin enjoyed L. Tolstoy's nonviolence (which seemed to be the only answer to the universal and omnipotent evil of the everyday Soviet reality) but also the ruthless P. Tchaadaev's Russian life analysis (especially, as presented in his famous "1*st Philosophical Letter*"). In 20th century philosophy, he was somewhat enthusiastic about existentialism of J.-P. Sartre and Hannah Arendt's profound analysis of the totalitarianism phenomenon.

By a play of circumstances, Havin was seriously carried away by A.A. Lyubishchev's behaviorist philosophy of time-management and "work as a refuge" (as it was presented in D. Granin's documentary novel "*This strange life*" (c. 1974)). During a dramatic period of his life (see above), V.P. literally followed Lyubishchev's maxim – "*Systematic work is a stretched rope holding on which you can cross the desert of life*".

He made every endeavour to rank his time schedule into "First Rank Time," "Second Rank Time," "Third Rank Time,"... (following one of Lyubischev's principal recipes...) – for research in mathematics, for reviews and letters, for reading, etc. During long years, Havin carried on a kind of a diary – an everyday (with no omissions!) nightly report on what was done and what should be improved, as well as on the analysis of his life observations (up to joining to the text long clippings from his favourite intellectual acquisitions of the day, including mathematical thinkings).

Havin's only philosophical writing of his own (published until now) is [71], a short essay on perceptional and philosophical values of a mathematical vision of life. The idea is that mathematics is not only to state, demonstrate or establish something, but in fact, also to teach to doubt, by showing some conceptual impossibilities for certain mental (or even material world) designs. In Havin's personal life, doubts, perhaps, prevailed over affirmations...

Last Words

Unfortunately, as it often happens, a formal list of merits, achievements, and outstanding qualities does not convey the depth, beauty, and significance of the personality we have lost. The life of V.P. Havin has come on tragic periods of Russian history – wild outbursts of Stalinism, state policy of anti-semitism, a permanent fear of repressions, etc., but Viktor Petrovich maintained the "innocent" enthusiasm towards life, and the belief in the value of talent and human good. Havin's life, work, and personality are still awaiting thoughtful analysis.

Havin invented many beautiful ideas and proved many sharpest theorems, publishing about 100 research papers. But why, however, is his personality so important and unforgettable for so many people in the world? There are a lot of mathematicians who published more... even 200, ...500... papers (notice: at least 300 000 new theorems are published yearly...). We are convinced that V.P. also could increase his publication list many times, but instead he invested his talent and energy in us, in the community. Prophesying the main streams of analysis in the second half of 20th century since 1960, in particular "Hardy classes obsession/philosophy" (which celebrated 100 years in 2015, 50 years in Havin's Seminar...), and being an incomparably emotional leader, Havin succeeded to set in motion a successful mathematical life for tens and hundreds of young people of his seminar, which is – if you wish – his Best Theorem. Invariably, he attracted students and his colleagues by his amicability, openness, and intellectual generosity. He was a remarkably kind and decent person, and a very rare pure unmercenary who never even understood what money is. Being a natural, artless stoic (in the face of the life's hardness, including unprecedented insults against Jewish self-identity during the Soviet era), Havin passed a hard and merciless historic period without any moral fault.

The radiant memory of Viktor Petrovich Havin will always remain with us.

Havin, listening a talk
at St. Petersburg Analysis Conference, 2008.

References

[AH1996]   D.R. Adams, L.-I. Hedberg, *Function Spaces and Potential Theory*, Springer, 1996.

[Aro1935]   N. Aronszajn, *Sur les décompositions des fonctions analytiques uniformes et sur leur applications*, Acta Math. **65** (1935), 1–156.

[BDDM1986]   *The Bieberbach Conjecture*, Proc. Symposium on the occasion of the proof (A. Baerstein, D. Drasin, P. Duren, eds.), Amer. Math. Soc. Surveys and Monographs, Vol. 21, Providence, R.I., 1986.

[BeH2001]   D.B. Beliaev, V.P. Havin, *On the uncertainty principle for M. Riesz potentials*, Ark. Mat. **39** (2001), no. 2, 223–243.

[Be2008]   Yu.S. Belov, *Admissibility criteria for model subspaces with fast growth of the argument of the generating inner function*, Zap. Nauchn. Sem. S.-Peterburg. Otdel. Mat. Inst. Steklov. (POMI) **345** (2007), 55–84; English transl.: J. Math. Sci. **148** (2008), no. 6, 813–829.

[Be2009]   Yu.S. Belov, *Model functions with nearly prescribed modulus*, Algebra i Analiz **20** (2008), no. 2, 3–18; English transl.: St. Petersburg Math. J. **20** (2009), no. 2, 163–174.

[BM1962]   A. Beurling, P. Malliavin, *On Fourier transforms of measures with compact support*, Acta Math. **107** (1962), 291–309.

[BM1967]   A. Beurling, P. Malliavin, *On the closure of characters and the zeros of entire functions*, Acta Math. **118** (1967), 79–93.

[Bou1983]   J. Bourgain, *On weak completeness of the dual of spaces of analytic and smooth functions*, Bull. Soc. Math. Belg., Sér. B, **35** (1983), 111–118.

[Bou1984]   J. Bourgain, *The Dunford–Pettis properties for the ball-algebra, the polydisc-algebras, and the Sobolev spaces*, Studia Math. **72** (1984), 245–253.

[Br1973]   J.E. Brennan, *Invariant subspaces and weighted polynomial approximation*, Ark. Mat. **11** (1973), 167–189.

[Br1985]   J.E. Brennan, *Weighted polynomial approximation, quasianalyticity, analytic continuation*, J. Reine Angew. Math. **357** (1985), 23–50.

[Br1994]   J.E. Brennan, *Weighted polynomial approximation and quasianalyticity on general sets*, Algebra i Analiz **6** (1994), 69–89. English transl.: St. Petersburg Math. J. **6** (1995) no. 4, 763–779.

[DaP1996]   S.R. Dager, A.S. Presa, *On an analogue of the Runge theorem for harmonic differential forms*, Zap. Nauchn. Sem. S.-Peterburg. Otdel. Mat. Inst. Steklov. (POMI) **232** (1996), 109–117; English transl.: J. Math. Sci. (New York) **92** (1998). no. 1, 3613–3618.

[Dau1992]   I. Daubechies, *Ten Lectures on Wavelets*, CBMS-NSF Regional Conference Series in Applied Mathematics, SIAM, 1992

[dLKK1977]   K. de Leeuw, Y. Katznelson, J.-P. Kahane, *Sur les coefficients de Fourier des fonctions continues*, C.R. Acad. Sci. Paris Sér. A-B, **285** (1977), 16, A1001–A1003.

[DL1953]   J. Deny, J.-L. Lions, *Les espaces du type de Beppo Levi*, Ann. Inst. Fourier **5** (1953/1954), 305–370.

[Du2011]   M.B. Dubashinskii, *On uniform approximation by harmonic and almost harmonic vector fields*, Zap. Nauchn. Sem. S.-Peterburg. Otdel. Mat. Inst. Steklov. (POMI) **389** (2011), 58–84; English transl.: J. Math. Sci. (New York) **182** (2012), no. 5, 617–629.

[Du2013]   M.B. Dubashinskii, *On a method for the approximation of vector fields by gradients*, Algebra i Analiz **25** (2013), no. 1, 3–36; English transl.: St. Petersburg Math. J. **25** (2014), no. 1, 1–22.

[Du2016]   M. Dubashinskii, *Periods of L^2-forms in an infinite-connected planar domain*, C. R. Math. Acad. Sci. Paris **354** (2016), no. 11, 1060–1064.

[Dya1990]   K.M. Dyakonov, *Moduli and arguments of analytic functions from subspaces in H^p that are invariant under the backward shift operator*, Siberian Math. J. **31** (1990), no. 6, 926–939.

[Dyn1987]   E.M. Dyn'kin, *Methods of the theory of singular integrals: Hilbert transform and Calderón–Zygmund theory*, in "Itogi Nauki i Techniki, Contemp. Problems Math.", vol. 15 (eds. V. Khavin and N. Nikolski), 197–292, VINITI, Moscow; English transl.: *Encyclopaedia of Math. Sci.*, vol. 15, 167–260, Springer-Verlag, Berlin, 1991.

[Fed2012]   K.Yu. Fedorovskiy, *On the C^m-approximability of functions by polynomial solutions of elliptic equations on compact plane sets*, Algebra i Analiz **24** (2012), no. 4, 201–219. English transl.: St. Petersburg Math. J. **24** (2013), no. 4, 677–689.

[FeSt1972]   Ch. Fefferman, E.M. Stein, *H^p spaces of several variables*, Acta Math. **7** (1972), 137–193.

[FS1997]   G.B. Folland, A. Sitaram, *The uncertainty principle: A mathematical survey*, J. Fourier Anal. Appl. **3** (1997), no. 3, 207–238.

[Gam1969]   T.M. Gamelin, *Uniform Algebras*. Prentice Hall, Englewood Cliffs, 1969.

[Gol1961]   V.V. Golubev, *Single-valued Analytic Functions. Automorphic Functions*, Moscow, "Fizmatgiz Editors", 1961.

[Hed2002]   L.-I. Hedberg, A review of *"Complex Analysis, Operator Theory, and Related Topics: the S.A. Vinogradov Memorial Volume"* (V.P. Havin, N.K. Nikolski, eds.), Algebra i Analiz **14** (2002), no. 4, 229–237; English transl.: St. Petersburg Math. J. **14** (2003), no. 4.

[HPYa]   P.Ya. Havin, *Autobiography*, www.jf.spbu.ru/museum/271/294.html

[HP2005]   A. Herrera Torres, A. Presa, *Approximation properties of holomorphic pairs of differential forms*, Complex Var. Theory Appl. **50** (2005), no. 2, 89–101.

[Hör1966]   L. Hörmander, *An Introduction to Complex Analysis in Several Variables*, North-Holland, Amsterdam, 1966.

[HrV1984]   S.V. Hruschev, S.A. Vinogradov, *Free interpolation in the space of uniformly convergent Taylor series*, Lect. Notes Math. **864**, Springer, 1984, 171–213.

[Kah2000]   J.-P. Kahane, *Multiplicative chaos and multimeasures*, in "Complex Analysis, Operator Theory, and Related Topics: the S.A. Vinogradov Memorial Volume" (V.P. Havin, N.K. Nikolski, eds.), Operator Theory: Advances and Applications, vol. 113, Birkhäuser, Basel, 2000, 115–126.

[Kis1981]   S.V. Kislyakov, *Fourier coefficients of boundary values of functions that are analytic in the disc and bidisc*, Trudy Math. Inst. Steklov **155** (1981), 77–94.

[Ko1988]   P. Koosis, *The Logarithmic Integral, Vol. I & II*, Cambridge Univ. Press, Cambridge, 1988.

[Luz1915]   N.N. Luzin, *L'intégrale et séries trigonométriques*, PhD Thesis, Moscow University, 1915 (reproduced in Matem. Sbornik **30** (1916), no. 1, 1–242); 2nd edition: 1951, Izd. Akad. Nauk SSSR (Russian).

[Mal1999]   E.V. Malinnikova, *Uniform approximation by harmonic differential forms on compact subsets of a Riemannian manifold*, Algebra i Analiz **11** (1999), no. 4, 115–138; English transl.: St. Petersburg Math. J. **11** (2000), no. 4, 625–641.

[Maz1985]   V.G. Maz'ya, *Sobolev Spaces*, Springer, 1985.

[Med2015]   A.N. Medvedev, *Drop of the smoothness of an outer function compared to the smoothness of its modulus, under restrictions on the size of boundary values*, Zap. Nauchn. Sem. St. Petersburg Otdel. Mat. Inst. Steklov (POMI) **443** (2015), 101–115; English transl.: J. Math. Sci. (New York) **215** (2016), 608–616.

[Mer1953]   S.N. Mergelyan, *On the completeness of systems of analytic functions*, Uspekhi Mat. Nauk **8** (1953), 3–63; English transl.: AMS Transl. **19** (1962), 109–166.

[MH1971]   B.S. Mitjagin, G.M. Henkin, *Linear problems of complex analysis*, Uspekhi Mat. Nauk **26** (1971), no. 4 (160), 93–152; English transl.: Russian Math. Surveys **26** (1971), no. 4, 99–164.

[Moo1972]   M.C. Mooney, *A theorem on bounded analytic functions*, Pacific J. Math. **43** (1972), 457–463.

[Ob2016]   *Obituary: To the memory of Viktor Petrovich Havin*, Vestnik of St. Petersburg State University. Series 1. Mathematics. Mechanics. Astronomy, vol. **3(61)** (2016), no. 1, 173–177 (Russian).

[Par1990]   P.V. Paramonov, *Harmonic approximations in the C^1-norm*, Mat. Sb. **181** (1990), no. 10, 1341–1365; English transl.: Mathematics of the USSR-Sbornik **71** (1992), no. 1, 183–207.

[Par1993]   P.V. Paramonov, *C^m-approximations by harmonic polynomials on compact sets in $\mathbb{R}^n$*, Mat. Sb. **184** (1993), no. 2, 105–128; English transl.: Sbornik Math. **78** (1994), no. 1, 231–251.

[Pel1951]   F. Pelligrino, *La théorie des fonctionnelles analytiques et ses applications*, pp. 357–477, an *Appendix* to the book of P. Lévy, *Problèmes concrets d'analyse fonctionnelle*, 1951, Paris.

[Pe1976]   A. Pełczyński, *Banach Spaces of Analytic Functions and Absolutely Summing Operators*, Regional conference series in mathematics, **30**. AMS, Providence, Rhode Island, 1976.

[Poi1892]   H. Poincaré, *Sur les fonctions à espaces lacunaires*, Amer. J. Math. **14** (1892), 201–221.

[Sa1997]   S. Saccone, *The Pełczyński property for tight subspaces*, J. Funct. Anal. **148** (1997), no. 1, 86–116.

[Sh1988]   N.A. Shirokov, *Analytic functions smooth up to the boundary*, Lecture Notes in Math., vol. **1312**, Springer, Berlin, 1988.

[Sh2013]   N.A. Shirokov, *Sufficient conditions for the Hölder smoothness of a function*, Algebra i Analiz **25**, no. 3 (2013), 200–206; English transl.: St. Petersburg Math. J. **25** (2014), no. 3, 507–511.

[Sh2016]   N.A. Shirokov, *Smoothness of a function holomorphic in the ball and of its modulus on the sphere*, Zap. Nauchn. Sem. St. Petersburg Otdel. Mat. Inst. Steklov (POMI) **447** (2016), 123–128.

[Smi1993]   S.K. Smirnov, *Decomposition of solenoidal vector charges into elementary solenoids, and the structure of normal one-dimensional flows*, Algebra i Analiz **5** (1993), no. 4, 206–238; English transl.: St. Petersburg Math. J. **5** (1994), no. 4, 841–867.

[Sob1963]   S.L. Sobolev, *Applications of Functional Analysis in Mathematical Physics*, Providence, AMS, 1963.

[Tao2007]   T. Tao, *Soft analysis, hard analysis, and the finite convergence principle*, https://terrytao.wordpress.com/2007/05/23.

[Tar1993]   N.N. Tarkhanov, *Approximation on compacta by solutions of systems with a surjective symbol*, Uspekhi Mat. Nauk **48** (1993), no. 5, 107–146; English transl.: Russian Math. Surveys **48** (1993), no. 5, 103–145.

[Val1954]   G. Valiron, *Fonctions analytiques*, Presses Universitaires de France, Paris, 1954.

[Var1981]   A.L. Varfolomeev, *Analytic extension from a continuum to its neighborhood*, Zapiski Nauchn. Seminarov LOMI **113** (1981), 27–40; English transl.: J. Sov. Math. **22** (1983), no. 6, 1709–1718.

[VKM2013]   A.V. Vasin, S.V. Kislyakov, A.N. Medvedev, *Local smoothness of an analytic function compared to the smoothness of its modulus*, Algebra i Analiz **25**, no. 3 (2013), 52–85; English transl.: St. Petersburg Math. J. **25** (2014), no. 3, 397–420.

[Vin1970]   S.A. Vinogradov, *Interpolation theorems of Banach–Rudin–Carleson and norms of embedding operators for certain classes of analytic functions*, Zapiski Nauchn. Seminarov LOMI **19** (1970), 6–54 (Russian).

Operator Theory:
Advances and Applications, Vol. 261, 57–66
© Springer International Publishing AG, part of Springer Nature 2018

List of Publications of Victor Havin

I. Books, research papers, and surveys

Books

[1] *An Elementary Introduction to Theory of the Integral* (with G.P. Akilov, B.M. Makarov), Leningrad University Publishers, Leningrad, 1969, 349 pp. (in Russian).

[2] *Theory of Measure and Integral, II. The Integral* (with L.V. Florinskaia), Leningrad University Publishers, Leningrad, 1975, 210 pp. (in Russian).

[3] *Fundamentals of Mathematical Analysis. Differential and Integral Calculus of Functions of One Real Variable.* Leningrad University Publishers, Leningrad, 1989, 445 pp. (in Russian).

[4] *The Uncertainty Principle in Harmonic Analysis* (with B. Jöricke), Ergebnisse der Mathematik und ihrer Grenzgebiete (3), **28**, Springer-Verlag, Berlin, 1994, 543 pp.

[5] *Fundamentals of Mathematical Analysis. Differential and Integral Calculus of Functions of One Real Variable.* Second edition. Lan Publishers, Saint Petersburg, 1998, 448 pp. (in Russian).

Research papers and surveys

[6] *On analytic functions representable by an integral of Cauchy–Stieltjes type*, Vestnik Leningrad. Univ. Ser. Mat. Meh. Astr. **13** (1958), no. 1, 66–79.

[7] *The separation of the singularities of analytic functions*, Dokl. Akad. Nauk SSSR **121** (1958), 239–242.

[8] *Analytic continuation of power series and Faber polynomials*, Dokl. Akad. Nauk SSSR (N.S.) **118** (1958), 879–881.

[9] *On a problem of V.V. Golubev*, Dokl. Akad. Nauk SSSR **126** (1959) 511–513.

[10] *The rate of growth of functions of the H_p class and the multiplication of integrals of the Cauchy–Stieltjes type*, Dokl. Akad. Nauk SSSR **127** (1959), 757–759.

[11] *On the norms of some operations in the space of polynomials*, Vestnik Leningrad. Univ. **14** (1959), no. 19, 47–59.

[12] *On the space of bounded regular functions*, Dokl. Akad. Nauk SSSR **131** (1960), 40–43; translated in Soviet Math. Dokl. **1** (1960), 202–204.

[13] *Some estimates of analytic capacity* (with S.Ja. Havinson), Dokl. Akad. Nauk SSSR **138** (1961), 789–792.

[14] *On the space of bounded regular functions*, Sibirsk. Mat. Z. **2** (1961), 622–638.

[15] *Moment problems and analytical continuation of power series*, in: "Investigations in modern problems of functions of complex variable", Fizmatgiz Publishers, Moscow, 1961, pp. 113–121 (in Russian).

[16] *An analog of Laurent series*, in: "Investigations in modern problems of functions of complex variable", Fizmatgiz Publishers, Moscow, 1961, pp. 121–131 (in Russian).

[17] *Relations between certain classes of functions regular in the unit circle*, Vestnik Leningrad. Univ. **17** (1962), no. 1, 102–110.

[18] *Analytic representation of linear functionals in spaces of harmonic and analytic functions which are continuous in a closed region*, Dokl. Akad. Nauk SSSR **151** (1963), 505–508.

[19] *Boundary properties of integrals of Cauchy type and of conjugate harmonic functions in regions with rectifiable boundary*, Mat. Sb. (N.S.) **68 (110)** (1965), 499–517.

[20] *Removability of singularities of analytic functions and transfer of masses* (with R.E. Valskii), Sibirsk. Mat. Z. **7** (1966), 55–60.

[21] *Spaces of analytic functions*, Math. Analysis, Akad. Nauk SSSR Inst. Naucn. Informacii, Moscow, 1966, 76–164.

[22] *Duality of spaces of analytic functions and certain of its applications.*, Contemporary Problems in Theory Anal. Functions (Proc. Internat. Conf., Erevan, 1965), "Nauka," Moscow, 1966, 311–314.

[23] *Continuity in L_p of an integral operator with the Cauchy kernel*, Vestnik Leningrad. Univ. **22** (1967), no. 7, 103–108.

[24] *Approximation in the mean by harmonic functions* (with V.G. Maz'ya), Zap. Naucn. Sem. Leningrad. Otdel. Mat. Inst. Steklov. (LOMI) **5** (1967), 196–200.

[25] *Approximation by analytic functions in the mean*, Dokl. Akad. Nauk SSSR **178** (1968), 1025–1028.

[26] *Approximation in the mean by analytic functions* (with V.G. Maz'ya), Vestnik Leningrad. Univ. **23** (1968), no. 13, 62–74.

[27] *The Cauchy problem for Laplace's equation* (with V.G. Maz'ya), Vestnik Leningrad. Univ. **23** (1968), no. 7, 146–147.

[28] *Polynomial approximation in the mean in certain non-Carathéodory regions. I*, Izv. Vyssh. Uchebn. Zaved. Matematika **9 (76)** (1968), 86–93.

[29] *Polynomial approximation in the mean in certain non-Carathéodory regions. II*, Izv. Vyssh. Uchebn. Zaved. Matematika **10 (77)** (1968), 87–94.

[30] *On the uniqueness theorem of L. Carleson for analytic functions with finite Dirichlet integral* (with V.G. Maz'ya), Problems of Math. Anal. **2**: Linear Operators and Operator Equations, 153–156, Izdat. Leningrad. Univ., Leningrad, 1969.

[31] *A nonlinear analogue of the Newtonian potential, and metric properties of (p, l)-capacity* (with V.G. Maz'ya), Dokl. Akad. Nauk SSSR **194** (1970), 770–773.

[32] *Analytic functions with a Lipschitzian modulus of the boundary values* (with F.A. Shamoyan), Zap. Naucn. Sem. Leningrad. Otdel. Mat. Inst. Steklov. (LOMI) **19** (1970), 237–239.

[33] *Multipliers and divisors of Cauchy–Stieltjes type integrals* (with S.A. Vinogradov and M.G. Goluzina), Zap. Naucn. Sem. Leningrad. Otdel. Mat. Inst. Steklov. (LOMI) **19** (1970), 55–78.

[34] *A generalization of the Privalov–Zygmund theorem on the modulus of continuity of the conjugate function*, Izv. Akad. Nauk Armjan. SSR Ser. Mat. **6** (1971), no. 2-3, 252–258; ibid. **6** (1971), no. 4, 265–287.

[35] *Impossibility of the Carleman approximation of functions continuous on the unit circle by the boundary values of functions analytic and uniformly continuous in the unit disk*, Zap. Naucn. Sem. Leningrad. Otdel. Mat. Inst. Steklov. (LOMI) **22** (1971), 161–170.

[36] *The factorization of analytic functions that are smooth up to the boundary*, Zap. Naucn. Sem. Leningrad. Otdel. Mat. Inst. Steklov. (LOMI) **22** (1971), 202–205.

[37] *Approximation in the mean by harmonic functions* (with V.G. Maz'ya), Investigations on linear operators and the theory of functions, III. Zap. Naucn. Sem. Leningrad. Otdel. Mat. Inst. Steklov. (LOMI) **30** (1972), 91–105.

[38] *On the theory of nonlinear potentials and (p, l)-capacity* (with V.G. Maz'ya), Vestnik Leningrad. Univ. Mat. Meh. Astronom. **13** (1972), no. 3, 46–51.

[39] *A nonlinear potential theory* (with V.G. Maz'ya), Uspekhi Mat. Nauk **27** (1972), no. 6, 67–138.

[40] *Weak completeness of the space L^1/H_0^1*, Vestnik Leningrad. Univ. Mat. Meh. Astronom. **13** (1973), no. 3, 77–81.

[41] *Application of the (p, l)-capacity to certain problems of the theory of exceptional sets* (with V.G. Maz'ya), Mat. Sb. (N.S.) **90 (132)** (1973), 558–591.

[42] *The spaces H^∞ and L^1/H_0^1*, Investigations on linear operators and the theory of functions, IV. Zap. Naucn. Sem. Leningrad. Otdel. Mat. Inst. Steklov. (LOMI) **39** (1974), 120–148.

[43] *The solutions of the Cauchy problem for the Laplace equation (uniqueness, normality, approximation)* (with V.G. Maz'ya), Trudy Moskov. Mat. Obsch. 30 (1974), 61–114.

[44] *Free interpolation in H^∞ and in certain other classes of functions. I* (with S.A. Vinogradov), Investigations on linear operators and the theory of functions, V. Zap. Naucn. Sem. Leningrad. Otdel. Mat. Inst. Steklov. (LOMI) **47** (1974), 15–54.

[45] *Approximation in L^p by the solutions of certain systems of linear differential equations*, Collection of articles dedicated to the memory of Academician V.I. Smirnov. Vestnik Leningrad. Univ. Mat. Meh. Astronom. **1** (1975), no. 1, 150–158.

[46] *Free interpolation in H^∞ and in certain other classes of functions. II* (with S.A. Vinogradov), Investigations on linear operators and theory of functions, VI. Zap. Naucn. Sem. Leningrad. Otdel. Mat. Inst. Steklov. (LOMI) **56** (1976), 12–58.

[47] *On the Carleson formula for the Dirichlet integral of an analytic function* (with A.B. Aleksandrov and A.E. Dzrbasjan), Vestnik Leningrad. Univ. Mat. Mekh. Astronom. (1979), no. 4, 8–14.

[48] *The uncertainty principle for operators that commute with a shift. I* (with B. Jöricke), Investigations on linear operators and the theory of functions, IX. Zap. Nauchn. Sem. Leningrad. Otdel. Mat. Inst. Steklov. (LOMI) **92** (1979), 134–170.

[49] *Letter to the editor* (with S.A. Vinogradov), IX. Zap. Nauchn. Sem. Leningrad. Otdel. Mat. Inst. Steklov. (LOMI) **92** (1979), 317.

[50] *On a class of uniqueness theorems for convolutions* (with B. Jöricke), Complex analysis and spectral theory (Leningrad, 1979/1980), Lecture Notes in Math., **864**, Springer, Berlin-New York, 1981, 143–170,

[51] *The uncertainty principle for operators that commute with a shift. II* (with B. Jöricke), Investigations on linear operators and the theory of functions, XI. Zap. Nauchn. Sem. Leningrad. Otdel. Mat. Inst. Steklov. (LOMI) **113** (1981), 97–134.

[52] *The uncertainty principle for one-dimensional Riesz potentials*, Dokl. Akad. Nauk SSSR **264** (1982), no. 3, 559–563.

[53] *Free interpolation and the Dirichlet problem* (with A.O. Derviz), Vestnik Leningrad. Univ. Mat. Mekh. Astronom. 1983, 84–86.

[54] *Analogues of the Carleman–Goluzin–Krylov interpolation formula* (with I.V. Videnskii and E.M. Gavurina), Operator theory and function theory, no. 1, 21–32, Leningrad. Univ., Leningrad, 1983.

[55] *Free interpolation and the Dirichlet problem* (with A.O. Derviz), Investigations on linear operators and the theory of functions, XIII. Zap. Nauchn. Sem. Leningrad. Otdel. Mat. Inst. Steklov. (LOMI) **135** (1984), 51–65.

[56] *Equivalent norms in L^p-spaces of analytic functions and free interpolation of harmonic functions* (with B. Jöricke), Vestnik Leningrad. Univ. Mat. Mekh. Astronom. 1984, no. 4, 103–105.

[57] *Jones' interpolation formula*, Appendix I, in P. Koosis, *Introduction to H_p spaces*, Mir, Moscow, 1984, 329–337.

[58] *Weak completeness of the space $L^1/H^1(0)$*, Appendix II, in P. Koosis, *Introduction to H_p spaces*, Mir, Moscow, 1984, 338–346.

[59] *Golubev series and the analyticity on a continuum*, Linear and Complex Analysis Problem Book (V.P. Havin, S.V. Hruscev and N.K. Nikolski, eds.), Lecture Notes in Math. **1043**, Springer-Verlag, Berlin, 1984, 670–673.

[60] *Traces of harmonic functions and comparison of L^p-norms of analytic functions* (with B. Jöricke), Math. Nachr. **123** (1985), 225–254.

[61] *The uncertainty principle for operators that commute with a shift. III* (with B. Jöricke), Geometric questions in the theory of functions and sets, Kalinin. Gos. Univ., Kalinin, 1985, 62–80.

[62] *A remark on Taylor series of harmonic functions*, Multidimensional complex analysis, Akad. Nauk SSSR Sibirsk. Otdel., Inst. Fiz., Krasnoyarsk, 1985, 192–197.

[63] *The uncertainty principle in harmonic analysis* (with B. Jöricke), Proc. GDR Mathematicians' Congress (Rostock, 1986), 72–74, Wilhelm-Pieck-Univ., Rostock, 1986.

[64] *The length of a harmonic vector* (with M. Giesecke), Vestnik Leningrad. Univ. Mat. Mekh. Astronom. 1987, no. 3, 33–38.

[65] *Methods and structure of commutative harmonic analysis*, Commutative harmonic analysis, I. Current problems in mathematics. Fundamental directions, **15**, 6–133, Vsesoyuz. Inst. Nauchn. i Tekhn. Inform., Moscow, 1987. Translation in Encyclopaedia Math. Sci., **15**, 1–111, Springer-Verlag, Berlin, 1991.

[66] *V.I. Smirnov's results in complex analysis and their subsequent developments* (with N.K. Nikolski), in: "V.I. Smirnov, Selected works. Complex Analysis. Mathematical Diffraction Theory" (V.M. Babich, N.K. Nikolski, V.P. Havin, eds.), Leningrad University Publishers, 1988, pp. 111–145 (in Russian).

[67] *The Poisson kernel is the only approximate identity that is asymptotically multiplicative on H^∞* (with H. Wolff), Zap. Nauchn. Sem. Leningrad. Otdel. Mat. Inst. Steklov. (LOMI) **170** (1989), 82–89; translation in J. Soviet Math. **63** (1993), no. 2, 159–163.

[68] *Approximation properties of harmonic vector fields and differential forms* (with A. Presa Sague), Methods of approximation theory in complex analysis and mathematical physics (Leningrad, 1991), 149–156, Lecture Notes in Math., **1550**, Springer, Berlin.

[69] *The uncertainty principle in harmonic analysis* (with B. Jöricke). Commutative harmonic analysis, III, Current problems in mathematics. Fundamental directions, 181–260, Vsesoyuz. Inst. Nauchn. i Tekhn. Inform., Moscow, 1991. Translation in Encyclopaedia Math. Sci., **72**, 177–259, Springer-Verlag, Berlin, 1995.

[70] *Weighted approximation by trigonometric sums and the Carleman–Goluzin–Krylov formula* (with V.A. Bart), Zap. Nauchn. Sem. St. Peterburg. Otdel. Mat. Inst. Steklov. (POMI) **206** (1993), 5–14; translation in J. Math. Sci. **80** (1996), no. 4, 1873–1879.

[71] *Mathematics as a source of certainty and uncertainty*, Address given at Linköping University on occasion of the awarding of a doctorate Honoris Causa, Report LiTH-MAT-R-93-03, Linköping University, 1993.

[72] *Szegő–Kolmogorov–Krein theorems on weighted trigonometric approximation, and Carleman-type formulas* (with V.A. Bart), Ukrain. Mat. Zh. **46** (1994), no. 1-2, 100–127; translation in Ukrainian Math. J. **46** (1994), no. 1-2, 101–132.

[73] *On the definition of $H^p(\mathbb{R}^n)$* (with A.B. Aleksandrov), in: "Linear and complex analysis. Problem book 3. Part I" (V.P. Havin and N.K. Nikolski, eds.), Lecture Notes in Mathematics, **1573**, Springer-Verlag, Berlin, 1994, p. 464.

[74] *Sets of uniqueness for analytic functions with finite Dirichlet integral* (with S.V. Hruschev), in: "Linear and complex analysis. Problem book 3. Part II" (V.P. Havin and N.K. Nikolski, eds.), Lecture Notes in Mathematics, **1574**, Springer-Verlag, Berlin, 1994, pp. 216–218.

[75] *Analytic functions stationary on a set, the uncertainty principle for convolutions, and algebras of Jordan operators* (with B. Jöricke and N. Makarov), in: "Linear and complex analysis. Problem book 3. Part II" (V.P. Havin and N.K. Nikolski, eds.), Lecture Notes in Mathematics, **1574**, Springer-Verlag, Berlin, 1994, pp. 219–222.

[76] *Approximation and capacities. Introduction* (with J. Brennan and A. Volberg), in: "Linear and complex analysis. Problem book 3. Part II" (V.P. Havin and N K. Nikolski, eds.), Lecture Notes in Mathematics, **1574**, Springer-Verlag, Berlin, 1994, pp. 74–75.

[77] *Uniqueness, moments, normality. Introduction* (with J. Brennan and A. Volberg), in: "Linear and complex analysis. Problem book 3. Part II" (V.P. Havin and N.K. Nikolski, eds.), Lecture Notes in Mathematics, **1574**, Springer-Verlag, Berlin, 1994, p. 208.

[78] *Uniqueness and free interpolation for logarithmic potentials and the Cauchy problem for the Laplace equation in R^2* (with A. Aleksandrov, J. Bourgain, M. Giesecke, and Yu. Vymenets), Geom. Funct. Anal. **5** (1995), no. 3, 529–571.

[79] *Uniform approximation by harmonic differential forms in Euclidean space* (with A. Presa Sage), Algebra i Analiz **7** (1995), no. 6, 104–152; translation in St. Petersburg Math. J. **7** (1996), no. 6, 943–977.

[80] *Approximation properties of harmonic vector fields and differential forms*, Potential theory–ICPT 94 (Kouty, 1994), 91–102, de Gruyter, Berlin, 1996.

[81] *Uniform approximation by harmonic differential forms. A constructive approach* (with E.V. Malinnikova), Algebra i Analiz **9** (1997), no. 6, 156–196; translation in St. Petersburg Math. J. **9** (1998), no. 6, 1149–1180.

[82] *Jones' interpolation formula*, Appendix I, in P. Koosis, *Introduction to H_p spaces*. Second edition. Cambridge Tracts in Mathematics, **115**. Cambridge University Press, Cambridge, 1998, 263–270.

[83] *Weak completeness of the space $L^1/H^1(0)$*, Appendix II, in P. Koosis, *Introduction to H_p spaces*. Second edition. Cambridge Tracts in Mathematics, **115**. Cambridge University Press, Cambridge, 1998, 271–277.

[84] *Approximation and extension problems for some classes of vector fields* (with S.K. Smirnov), Algebra i Analiz **10** (1998), no. 3, 133–162; translation in St. Petersburg Math. J. **10** (1999), no. 3, 507–528.

[85] *On some potential theoretic themes in function theory*, The Mazya anniversary collection, Vol. 1 (Rostock, 1998), 99–110, Oper. Theory Adv. Appl., **109**, Birkhäuser, Basel, 1999.

[86] *Stanislav Aleksandrovich Vinogradov, his life and mathematics* (with N.K. Nikolski), Complex analysis, operators, and related topics, 1–18, Oper. Theory Adv. Appl., **113**, Birkhäuser, Basel, 2000.

[87] *Bounded separation of singularities of analytic functions* (with A.H. Nersessian), Entire functions in modern analysis (Tel-Aviv, 1997), 149–171, Israel Math. Conf. Proc., 15, Bar-Ilan Univ., Ramat Gan, 2001.

[88] *On the uncertainty principle in harmonic analysis*, Twentieth century harmonic analysis – a celebration (Il Ciocco, 2000), 3–29, NATO Sci. Ser. II Math. Phys. Chem., **33**, Kluwer Acad. Publ., Dordrecht, 2001.

[89] *On the uncertainty principle for M. Riesz potentials* (with D.B. Beliaev), Ark. Mat. **39** (2001), no. 2, 223–243.

[90] *Admissible majorants for model subspaces of H^2. I. Slow winding of the generating inner function* (with J. Mashreghi), Canad. J. Math. **55** (2003), no. 6, 1231–1263.

[91] *Admissible majorants for model subspaces of H^2. II. Fast winding of the generating inner function* (with J. Mashreghi), Canad. J. Math. **55** (2003), no. 6, 1264–1301.

[92] *Separation of singularities of analytic functions with preservation of boundedness*, Algebra i Analiz **16** (2004), no. 1, 293–319; translation in St. Petersburg Math. J. **16** (2005), no. 1, 259–283.

[93] *On a theorem of I.I. Privalov on the Hilbert transform of Lipschitz functions* (with Yu.S. Belov), Mat. Fiz. Anal. Geom. **11** (2004), no. 4, 380–407.

[94] *The Beurling–Malliavin multiplier theorem: the seventh proof* (with J. Mashreghi and F.L. Nazarov), Algebra i Analiz **17** (2005), no. 5, 3–68; translation in St. Petersburg Math. J. **17** (2006), no. 5, 699–744.

[95] *Admissible majorants for model subspaces and arguments of inner functions* (with A.D. Baranov), Funktsional. Anal. i Prilozhen. **40** (2006), no. 4, 3–21; translation in Funct. Anal. Appl. **40** (2006), no. 4, 249–263.

[96] *Majorants of meromorphic functions with fixed poles* (with A.D. Baranov and A.A. Borichev), Indiana Univ. Math. J. **56** (2007), no. 4, 1595–1628.

[97] *Uniform estimates in the Poincaré–Aronszajn theorem on the separation of singularities of analytic functions* (with A.H. Nersessian and J. Ortega-Cerdà), J. Anal. Math. **101** (2007), 65–93.

[98] *The Beurling–Malliavin multiplier theorem and its analogs for the de Branges spaces* (with Yu. Belov), Operator Theory (D. Alpay, ed.), SpringerReference, Springer, Basel, 2015, 1–24.

[99] *Boundedness of variation of a positive harmonic function along the normals to the boundary* (with P.A. Mozolyako), Algebra i Analiz **28** (2016), no. 3, 67–110; translation in St. Petersburg Math. J. **28** (2017), no. 3.

II. Translations, editing, and biographical notes

Translations and editing

[100] W.K. Hayman, *Multivalent functions.* Translated from the English by V.P. Havin. Edited by I.E. Bazilevic. Izdat. Inostr. Lit., Moscow, 1960.

[101] J.A. Jenkins, *Univalent functions and conformal mapping.* Translated from the English by V.P. Havin. Izdat. Inostr. Lit., Moscow, 1962.

[102] W. Rudin, *Principles of mathematical analysis.* Translated from the English by V.P. Havin. Mir, Moscow, 1966; second edition: Mir, Moscow, 1976; third edition: Lan Publishers, Saint Petersburg, 2004.

[103] L. Carleson, *Selected problems in the theory of exceptional sets.* Translated from the English by V.P. Havin. Edited by V.G. Maz'ya. Mir, Moscow, 1971.

[104] *Investigations on linear operators and the theory of functions. 99 unsolved problems in linear and complex analysis.* Compiled and edited by N.K. Nikolskii, V.P. Havin and S.V. Hruschev. Zap. Nauchn. Sem. Leningrad. Otdel. Mat. Inst. Steklov. (LOMI) **81** (1978). Nauka, Leningrad, 1978.

[105] *Complex analysis and spectral theory.* Papers from the Seminar held in Leningrad, 1979/1980. Edited by V.P. Khavin and N.K. Nikolskii. Lecture Notes in Mathematics, **864**. Springer, Berlin, 1981.

[106] J. Garnett, *Bounded analytic functions.* Translated from the English by E.M. Dyn'kin. Translation edited and with a preface by V.P. Havin. Mir, Moscow, 1984.

[107] P. Koosis, *Introduction to H^p spaces. With an appendix on Wolff's proof of the corona theorem.* Translated from the English by V.V. Peller and A.G. Tumarkin. Translation edited and with a preface and appendices by V.P. Havin.

[108] *Linear and complex analysis problem book. 199 research problems.* Edited by V.P. Havin, S.V. Hruschev and N.K. Nikolskii. Lecture Notes in Mathematics, **1043**, Springer-Verlag, Berlin, 1984.

[109] V.I. Smirnov, *Selected works. Complex analysis. Mathematical theory of diffraction.* With a biography of Smirnov by O.A. Ladyzhenskaya. Compiled and with supplementary material by V.M. Babich, N.K. Nikolskii and V.P. Havin. Leningrad. Univ., Leningrad, 1988.

[110] *Commutative harmonic analysis. I. General survey. Classical aspects.* Current problems in mathematics. Fundamental directions, **15**. Vsesoyuz. Inst. Nauchn. i Tekhn. Inform., Moscow, 1987. Translation by D. Khavinson and S.V. Kislyakov. Translation edited by V.P. Havin and N.K. Nikolskii. Encyclopaedia of Mathematical Sciences, **15**, Springer-Verlag, Berlin, 1991.

[111] *Commutative harmonic analysis. II. Group methods in commutative harmonic analysis.* Current problems in mathematics. Fundamental directions, **25**, Vsesoyuz. Inst. Nauchn. i Tekhn. Inform., Moscow, 1988. Translated by D. Dynin and S. Dynin. Translation edited by V.P. Havin N.K. Nikolskii. Encyclopaedia of Mathematical Sciences, **25**. Springer-Verlag, Berlin, 1998.

[112] *Commutative harmonic analysis. IV. Harmonic analysis in R^n.* Current problems in mathematics. Fundamental directions. **42**. Vsesoyuz. Inst. Nauchn. i Tekhn. Inform., Moscow, 1989. Translation by J. Peetre. Translation edited by V.P. Havin and N.K. Nikolskii. Encyclopaedia of Mathematical Sciences, **42**, Springer-Verlag, Berlin, 1992.

[113] *Commutative harmonic analysis. III. Generalized functions. Applications.* Current problems in mathematics. Fundamental directions. Vsesoyuz. Inst. Nauchn. i Tekhn. Inform., Moscow, 1991. Translation by R. Cooke. Translation edited by V.P. Havin and N.K. Nikolskii. Encyclopaedia of Mathematical Sciences, **72**, Springer-Verlag, Berlin, 1995.

[114] *Complex analysis. I. Entire and meromorphic functions. Polyanalytic functions and their generalizations.* Akad. Nauk SSSR, Vsesoyuz. Inst. Nauchn. i Tekhn. Inform., Moscow, 1991. Translation by V.I. Rublinetskii and V.A. Tkachenko. Translation edited by A.A. Gonchar, V.P. Havin and N.K. Nikolski. Encyclopaedia of Mathematical Sciences, **85**, Springer-Verlag, Berlin, 1997.

[115] *Linear and complex analysis. Problem book 3. Part I.* Edited by V.P. Havin and N.K. Nikolski. Lecture Notes in Mathematics, **1573**, Springer-Verlag, Berlin, 1994.

[116] *Linear and complex analysis. Problem book 3. Part II.* Edited by V.P. Havin and N.K. Nikolski. Lecture Notes in Mathematics, **1574**, Springer-Verlag, Berlin, 1994.

[117] *Complex analysis, operators, and related topics. The S.A. Vinogradov memorial volume.* Edited by V.P. Havin and N.K. Nikolski. Operator Theory: Advances and Applications, **113**, Birkhäuser Verlag, Basel, 2000.

[118] S.A. Vinogradov, *Interpolation problems for analytic functions continuous in the closed disk and for functions whose sequence of coefficients is in l^p.* Translated from the 1968 Russian original by V.P. Havin. Oper. Theory Adv. Appl., **113**, Complex analysis, operators, and related topics, 23–29, Birkhäuser, Basel, 2000.

[119] S.A. Vinogradov, *Free interpolation in spaces of analytic functions.* Translated from the 1983 Russian original by V.P. Havin. Oper. Theory Adv. Appl., **113**, Complex analysis, operators, and related topics, 31–42, Birkhäuser, Basel, 2000.

Biographical notes

[120] *Boris Zaharovich Vulih* (*on the occasion of his sixtieth birthday*) (with D.A. Vladimirov, M.K. Gavurin, G.I. Natanson, I.V. Romanovskii), Vestnik Leningrad. Univ. Mat. Meh. Astronom. **7** (1973), no. 2, 158–159.

[121] *Grigorii Mikhailovich Fikhtengolts* (*on the 100th anniversary of his birth*) (with S.A. Vinogradov, D.A. Vladimirov, M.K. Gavurin, S.M. Ermakov, B.M. Makarov, G.I. Natanson, M.Z. Solomyak, and D.K. Faddeev), Vestnik Leningrad. Univ. Mat. Mekh. Astronom. 1988, no. 3, 3–6.

[122] *Gleb Pavlovich Akilov* (with A.D. Aleksandrov, A.M. Vershik, V.V. Ivanov, A.G. Kusraev, S.S. Kutateladze, B.M. Makarov, and Yu.G. Reshetnyak), Uspekhi Mat. Nauk **43** (1988), no. 1, 181–182; translation in Russian Math. Surveys **43** (1988), no. 1, 221–223.

[123] *Recollections of G.P. Akilov* (*Leningrad, the 1950s and 1960s*) (with A.I. Veksler, D.A. Vladimirov and B.M. Makarov), Optimizatsiya **48 (65)** (1990), 35–40.

[124] *Viktor Solomonovich Videnskii* (*on the occasion of his eightieth birthday*) (with V.V. Zhuk, V.N. Malozemov and G.I. Natanson), Uspekhi Mat. Nauk **57** (2002), no. 5 (347), 182–186; translation in Russian Math. Surveys **57** (2002), no. 5, 1033–1038.

[125] *Semen Yakovlevich Khavinson* (*obituary*) (with A.G. Vitushkin, A.A. Gonchar, M.V. Samokhin, V.M. Tikhomirov, P.L. Ul'yanov, V.Ya. Eiderman), Uspekhi Mat. Nauk **59** (2004), no. 4, 186–192; translation in Russian Math. Surveys **59** (2004), no. 4, 777–785.

[126] *Garald Isidorovich Natanson* (with O.L. Vinogradov, V.V. Zhuk and V.L. Fainshmidt), Proceedings of the St. Petersburg Mathematical Society, Vol. X, 237–252, Amer. Math. Soc. Transl. Ser. 2, **214**, Amer. Math. Soc., Providence, RI, 2005.

[127] *Semyon Yakovlevich Khavinson* (*May* 17, 1927–*January* 30, 2003), Selected topics in complex analysis, 1–22, Oper. Theory Adv. Appl., **158**, Birkhäuser, Basel, 2005.

[128] *Vladimir Gilelevich Maz'ya* (*On the occasion of his 70th anniversary*) (with M.V. Anolik, Yu.D. Burago, Yu.K. Dem'yanovich, S.V. Kislyakov, G.A. Leonov, N.F. Morozov, S.V. Poborchii, N.N. Ural'tseva, N.A. Shirokov), Vestnik St. Petersburg University. Mathematics **41** (2008), no. 4, 287–289.

[129] *Vladimir Gilelevich Maz'ya* (*to his 70th birthday*) (with M.S. Agranovich, Yu.D. Burago, B.R. Vainberg, M.I. Vishik, S.G. Gindikin, V.A. Kondrat'ev, V.P. Maslov, S.V. Poborchii, Yu.G. Reshetnyak, M.A. Shubin), Uspekhi Mat. Nauk **63** (2008), no. 1, 183–189; translation in Russian Math. Surveys **63** (2008), no. 1, 189–196.

[130] *Viktor Fedorovich Osipov* (with Z.L. Abzhandadze, V.V. Zhuk, G.A. Leonov, B.M. Makarov, G.M. Hitrov, N.A. Shirokov), Vestnik St. Petersburg University, Series 1. Mathematics. Mechanics. Astronomy, 2011, no. 1, 173–174.

[131] *Odinets Vladimir Petrovich* (*on his 65th birthday*) (with A.M. Vershik, O.Ya. Viro, V.N. Isakov, G.A. Leonov, M.J. Pratussevitch, N.A. Shirokov), Vladikavkaz. Mat. Zh. **12** (2010), no. 4, 79–81.

[132] *Mikhail Shlemovich Birman* (*obituary*) (with V.M. Babich, V.S. Buslaev, A.M. Vershik, S.G. Gindikin, S.V. Kislyakov, A.A. Laptev, V.A. Marchenko, N.K. Nikol'skii, L.A. Pastur, B.A. Plamenevskii, M.Z. Solomyak, T.A. Suslina, N.N. Ural'tseva,

L.D. Faddeev, D.R. Yafaev), Uspekhi Mat. Nauk **65** (2010), no. 3, 185–190; translation in Russian Math. Surveys **65** (2010), no. 3, 569–575.

[133] *Interview*. In "Matmeh LGU–SPbGU. 1960s and not only: from the beginning to recent days. A collection of reminiscences." Copy-R Group, Saint Petersburg, 2012, 113–136.

[134] *Reminiscences on Vladimir Ivanovich Smirnov*. In "Matmeh LGU–SPbGU. 1960s and not only: from the beginning to recent days. A collection of reminiscences." Copy-R Group, Saint Petersburg, 2012, 137–145.

Operator Theory:
Advances and Applications, Vol. 261, 67–73
© Springer International Publishing AG, part of Springer Nature 2018

Mathematics as a Source of Certainty and Uncertainty

V.P. Havin

Ladies and gentlemen,

It is difficult for me to express adequately my appreciation of the honour which has been conferred upon me by inviting me to address this audience. So I immediately start with the topic of this lecture, but before I have to stress that its genre is unusual for me. Unlike in lectures any mathematician is used to deliver, I'm not going to try to increase your store of knowledge. I'll rather express some feelings and a certain mood. For some technical reasons the sources I could use in preparing this lecture were scarce. This explains the excessive use of my personal experience and memory for which I apologize in advance.

"Certainty" seems to be the most appropriate word to express what the non-mathematical public thinks of mathematics. This opinion could be supported by quotations of famous mathematicians, though there is no lack of scepticism in what mathematicians think and say about their subject. But the non-mathematical public has no doubt that mathematics is a realm of certainty. This feeling is well expressed by the following saying of the French painter Georges Braque: "Art upsets, science reassures" (if we replace the word "science" by "mathematics").

This popular (and completely mistaken) opinion determined my destiny. It is thanks to it that I became a mathematician. The decision was actually taken not by me, but by my father, a philologist whom I fully obeyed at the time, though my mathematical achievements had been nothing more than good high school marks. According to my natural inclinations I'd rather become a philologist too. But I graduated from high school in 1950, the year when Stalin also got interested in philology and wrote a booklet "Marxism and problems of linguistics." This event, memorable for everybody of my generation, had been preceded by several ideological campaigns when politicians taught (sometimes with the use of violence) writers, historians, musicians, and literary critics what is good or bad. Impressed

Address given at Linköping University June 4, 1993, on the occasion of the awarding of an honorary doctorate to prof. Havin. It was published as Linköping University Report LiTH-MAT-R-93-03. Reproduced with a kind permission of the Linköping University Mathematical Department.

by this practice, my father forbade me even to think about the humanities and *ordered* me to become a mathematician. Now, 43 years later I'm grateful to him for this risky decision. But he was completely mistaken in what concerns his main point, namely, search for certainty. Of course, politicians hardly can prescribe to mathematicians which theorems to prove. Nevertheless, mathematics is the worst place to look for certainty and definiteness. It is, to the contrary, a realm of uncertainty. This uncertainty, its malignant and beneficial sides, are the subject of this lecture.

I'll describe how one acquires and then loses the comforting feeling of certainty usually ascribed to mathematics. Then I turn to a kind of uncertainty which I consider as beneficial, and which is supported by mathematical thought in an essential way.

My first impressions from mathematics fully agreed with the opinion of Braque. This science reassured indeed. I was fascinated by the preciseness and expressive power of the language. Mathematicians are much more delicate, cautious, and I'd say, nervous word users than anybody else. Unlike others, they are really bothered with the *meaning* of words they are pronouncing or writing down. I was deeply impressed by the capacity of mathematical language to describe – in an unequivocal and very expressive way – and fix very numerous and heterogeneous ideas, ranging from analysis to probability, from classical to quantum mechanics, from economics to linguistics.

The convincing power of mathematical proofs seemed overwhelming, irresistible to me, exceeding by far any proof in any domain where verbal proofs play an essential role, be it physics, history, or law.

There is one more component of this special certainty insidiously conquering any young mathematician as he gradually opens up his profession. It is something spontaneously felt by any mathematician in spite of the fact that some would deny it and notwithstanding well-founded criticism of logicians and philosophers. It is a deeply rooted, visceral belief in (or rather a sensation of) the existence of a special mathematical reality "hard as rock" according to Hardy who praised this transcendental world in his "Mathematician's Apology" claiming that the mathematical reality is more real than the physical one. It is a belief making an active mathematician insensible to remarkable results of logic which warn him against this "naive superstition." He is just unable to question the existence of things whose properties it is his task to conceive and whose very real resistance depriving him of sleep and rest he permanently tries to overcome. He reacts to theorems of Gödel and Cohen, to critical attitudes of intuitivists and constructivists with a mixture of respect and vague feelings of guilt, forgetting all this in the everyday communication with the stubbornly existing mathematical reality. As to me, the loss of certainty came not from logic. Its origin was of lower, much more earthly and practical level. My first doubts can be squeezed to quite silly, childish questions: "What for?" "What are the aims of mathematics?" "What is good and what is bad in it?" "What are the criteria of value?" Now I know that these questions just cannot be answered. But then, in the fifties, I have been really upset, after it became clear to me that nobody can provide me with a satisfactory and comforting answer. Eventually I had to accept this situation and to live on. Now I can say that mathematics, being a beautiful, miraculous science, is, at the same time, subject to fashion and cult of power. Its value criteria (at least those applied in practice) are very often determined by market forces and whimsical, arbitrary, and irrational opinions and tastes. I could illustrate this sad assertion by several funny stories and almost every mathematician could add his. Let me only briefly mention a curious fate of the Cantor set which I choose as a symbol of a domain inhabited by species usually called "bad sets" or "bad functions." For the generation of my teachers they symbolized progress. I was brought up in deep respect of these objects and related ideas and techniques. But soon after I've graduated from the university I knew that these favourite things of my teachers had become obsolete, a mark of backwardness. It became fashionable to say that "bad functions do not exist." The term "Theory of functions of a real variable" acquired an abusive nuance. At that time I often heard from my colleagues that the attention paid in the twenties and thirties to bad sets and functions in Russia and Poland was a kind of decadence and degeneration distracting mathematics from its true destination which is to solve problems of physical origin. But what do we see now? The Cantor set is fashionable again! Masses of people are really obsessed by it (and similar objects) claiming that physics (physics!) just would perish without them. Luxurious volumes of pictures are being printed and successfully sold, the Cantor set and its relatives got a new name, they are "fractals" now; they are not "bad", but "beautiful" (everybody knows the title "The beauty of fractals"). Of course, this boom is related to really deep discoveries in the theory of dynamical systems, new understanding of chaos. Grimaces of vanity and fashion, market tendencies in mathematics coexist with significant development of thought, only masking and distorting it. The above description does not contradict the well-known metaphor comparing mathematics with an orchestra whose participants don't know each other being separated by distances and interdisciplinary barriers, but the orchestra is nevertheless perfectly concerted, producing divine music, as if it were led by an invisible Conductor. This *is* true, but this image can be perceived only from afar, and nobody has ever seen the score. The whereabouts of the Conductor and his plans are obscure, and nobody, no group, no organization can claim his role.

But in real life a mathematician is often in a situation where he has to judge, to accept or reject. Those who are obliged to accept or reject papers for publication or applications for a job deserve to be pitied. The total lack of formal, algorithmic criteria of selection makes their situation highly unpleasant. If your department got 500 applications, then usually it is not hard to reject 450 according to reasonable and sound considerations. But what if you have only 2 positions, and the rest of applicants consists of good, serious specialists but does not contain, say, a Gauss and a Hilbert? Then you make a clever face and say that a class of spaces, the favourite theme of candidate X, is not worth considering, or that the theorem of Y is good but not a breakthrough, and a theorem of Z *is* a breakthrough, but the number of complex variables is one, and this is old-fashioned.

I'm not criticizing. I have no proposals. I'm describing. In such situations there is no way to escape subjective conclusions conforming with personal tastes. The only thing which could be avoided is to pretend that you possess the objective truth and are motivated by superscientific considerations.

I remember a lecture delivered by P. Aleksandrov, the famous topologist, in Leningrad, somewhere at the end of the sixties. Its title was "Criteria of value in mathematics." He analyzed, one after another, three criteria: applicability, fashion, degree of difficulty – only to reject them all. He proposed instead something very indefinite like "a feeling of a new horizon". This is by far not an algorithmic solution. But I prefer it to the terrible practice when the works of a person are judged according to the journal which published them. This is very algorithmic indeed and frees you from reading mathematics which requires concentration, energy and time.

So far about the criteria of value in mathematics and understanding its own aims and necessity. It is uncertainty in its purest and very unpleasant form.

But, to conclude with an optimistic note, let us turn to a beneficial kind of uncertainty inherent in mathematics and contrasting its malignant forms.

Mathematics is often and deservedly lauded for its applications to other sciences. It is impossible to deny these merits of mathematics whose very existence always was determined (and still is) by a subtle interplay of exterior and interior incentives. But I want to emphasize not the applications, but a capability to create sound and reasonable *doubts* and *uncertainty*, things which are in a short supply, but very necessary nowadays. In this connection we could remember again great achievements of logicians, but I'll dwell on much more elementary, almost high school matters.

Any mathematician, unlike (unfortunately) other people, knows (not only knows, but has it in his flesh and bones) that not everything which has got a name exists in reality. Mathematicians are *professionally* obsessed by existence problems. And not only by the existence of an *object* (a solution, a function or a set with prescribed properties), but by existence of a solving procedure when the algorithm in question has to satisfy certain requirements.

Normal people, not trained in this school of professional doubt, confronted with any problem, rarely suspect it can be unsolvable. They just start solving it. This is normal. And this is awful. Consider the following series of isomorphic proposals.

Let us trisect an angle using compass and ruler only, let us construct a perpetuum mobile, or "socialism"; let us do away with inflation and unemployment. Mathematics contains a powerful sobering potential suggesting how cautiously you must react to these appeals. Creating and propagating reasonable uncertainty and doubt, mathematics is capable to calm down, to cool away many dangerous and contagious enthusiasms.

In the sixties it became fashionable to jeer and sneer about compass and ruler problems in high school ("Why compass and ruler, why not something else?"). The jokers seem to be the same people who produced awful high school geometry books with axioms of a vector space preceding triangle and circle. There are several strong arguments in defense of the compass-ruler problems, but I emphasize only one: it so happened that just these ancient problems served as the material for discoveries whose contribution to culture is tremendous and whose results need to be inculcated into the mass psyche, to become a commonplace: *Not every problem can be solved.* This is the main reason to include these problems into high school teaching. They can be explained to every schoolboy and schoolgirl producing a salutary pedagogical influence. Being acquainted with the procedure of bisection of an angle, it is natural to start thinking about trisection. Why not? These problems are so similar! The non-solvability of the second is highly not obvious. Nevertheless it *is* unsolvable and this can be rigorously *proved*. Mathematics abounds in results of this kind when something seems to be within one's reach but eventually turns out to be impossible. But denying the possibility to find or do something, mathematics yields some consolations in the form of *approximate* solutions, *optimization* algorithms suggesting the ideology of *compromise*. A person brought up in this spirit hardly can join a crowd crying like mad "liberté, egalité, fraternité" only to start mass killings afterwards. An easy reasoning will lead this person to the conclusion that the terms of this triad are not compatible with each other, and it is better to look for something approximate, but feasible.

For ages the general human aspiration was to catch and freeze everything as *notions*, creating all-embracing and all-explaining systems of thought. Isn't it clear now that this is only possible with relatively trivial things? The real complexity of world can be only *approximately* described, and this description cannot manage with *notions*, it needs *images*. Mathematics is a source of a lot of images, not less expressive than images of poetry. Penetrating your heart they are capable to influence your world perception.

Let me use two quotations, one due to a famous sociologist, and another to a humorist. "Many of the greatest things man has achieved are not the result of consciously directed thought, and still less the product of a deliberately coordinated effort of many individuals, but of a process in which the individual plays a part which he can never fully understand" (von Hayek). The second quotation is much shorter: "No snowflake in an avalanche feels responsible" (Jerzy Lec). But in my feeling, vanity and *futility* of individual efforts hardly can be expressed with a greater force than by the following "uncertainty theorem": the value of the Lebesgue integral does not depend on values of the integrand on any prescribed set of zero length.

The theorem is a flagrant expression of senselessness of such notions as "cause," "guilt," or "responsibility" applied to results of sufficiently massive, integral character. Meanwhile every Russian traveller is being asked daily: "What do you think about Gorbachev or Yeltsin?" as if these men (or anybody else) can be considered as causing or governing immense, cosmic changes going on in Russia. Returning to the "damned questions"[1] of mathematics ("what for?" "what

---

[1]Literal translation of a Russian expression denoting the most fundamental problems concerning man's essence and existence.

is good or bad?") we can again use the above theorem as an instructive image. Results of work of the mathematical community at any moment before the 2nd World War could be expressed as a SUM of individual efforts. By the end of the sixties the set of terms of this sum became practically infinite (though, probably, still countable). But now this sum definitely has become an integral. I think this is a *Lebesgue* integral of personal efforts, though the presence of an infinite set of point masses may be arguable. I'd rather admit a singular *continuous* component with no distinguishable separate points. But let us agree, at least, that this image has a right to exist.

It can cool the incandescence of passions and weaken prohibitive trends. Nobody can or must feel or claim responsibility for mathematics as a whole. Its sense, its message are as inconceivable as life. It is an integral, and preoccupations inspired by fashion are *vanitas vanitatum*.

Powerful images carrying mighty expressive charge are connected with analytic functions and their antipodes, so-called "bad" functions. Creators of analysis were spontaneously convinced that all functions are analytic (even before this term had been coined). This spirit has been weakened as a result of the "string dispute" at the end of the XVIII century. But in a milder form this frame of mind generally persisted even in the XIXth century.

Nature needs analytic functions only. Such was the healthy, elemental view of people believing in God, predeterminacy and predictability of Being. A property drastically opposing an analytic function to a "bad" one is *uniqueness*.

Past and future of a process described by an analytic function are completely determined by its course during a second, or a one millionth part of a second. This mathematical image is apt to create mixed emotions. An analytic function is a symbol of highest perfection as is a favourite melody or line of poetry. Starting first notes or words it is impossible to continue differently from the classical sample. But at the same time the analyticity is a severe verdict, an inflexible prediction, impossible to contest. If an analytic curve $y = f(t)$ coincides with the parabola $y = t^2$ on $(0,1)$, then these two curves are doomed to coincide forever, no choice is possible. There is something very significant in the reluctance of old classics to accept the possibility to represent "an arbitrary" curve as a sum of trigonometrical series, such representation being "a formula". But all functions defined by formulas have to be analytic and cannot change their course in an arbitrary way.

Oscillations of fashion around "bad" and "good" functions mentioned above reproduce, in a sense, the old "string dispute". In spite of its vagueness, abundance of terms not duly defined, absurd claims and personal biases, this dispute includes something really important. Human beings can be divided into two categories. The first one believes (or feels) that the world is described by analytic functions. For the second everything is expressed by Lebesgue measurable functions. At any moment their course is absolutely unpredictable and, hence, can (in principle) be changed in any desired way. So, this second attitude implies *certainty*, those people feel they are masters the of world. Of course, no argument is thinkable here. We are dealing not with clear statements to be proved or disproved. We are dealing with different psychological approaches to reality, with conflicting world orientations. I dare to express my deeply personal, non-verifiable, non-arguable confidence in the analyticity of the world. Chaotic behaviour results from the interaction of innumerable analytic processes. This irrational feeling is warranted by some rigorously proved mathematical facts. However wild a function of time might look, it is, eventually, the sum of a series of polynomials or a difference of two analytic functions.

I used these elementary examples to show how some images so familiar to any mathematician can suggest the noble habit of doubt and strengthen the feeling of beneficial uncertainty.

This eulogy of uncertainty and doubt I'm finishing to deliver is not something unusual nowadays. The mood I tried to express is gaining more and more room, undermining certainties and self-confidence of conceited leaders, making it harder for politicians to subdue masses by cheap incantations devoid of any real content. After I've already sent the title of this lecture to professor Hedberg, in a Montreal airport I bought "Le Monde" of the 21st of April with an article of Edgar Morin, French Socialist, "La pensée socialiste en ruine," and was surprised to read the following lines (a newspaper is the last place where I could dream of finding something useful for this lecture): "In the opinion of Marx science is a *source of certainty*. But today we know that sciences yield *local certainties*, but theories are scientific insofar as they can be refuted, that is, are *not certain*. And, in what concerns fundamental questions, the scientific cognition runs into bottomless *uncertainties*. For Marx, the scientific *certainty* eliminated philosophical interrogation. Today we see that scientific progress only animates fundamental philosophical problems."

The attitudes I expressed become more and more banal which is illustrated by their frequent appearance even in the mass media. The more banal, the more commonplace they become, the more is our hope for the eventual improvement of the world, more human relations between human beings. And I hope that the experience accumulated in mathematics, joint with the experience of everyday practice, history, philosophy, positive sciences, religion and art will contribute to making these attitudes of beneficial uncertainty a commonplace indeed.

V.P. Havin

Operator Theory:
Advances and Applications, Vol. 261, 75–80
© Springer International Publishing AG, part of Springer Nature 2018

Remembering Victor Petrovich Havin

Konstantin M. Dyakonov

> Когда погребают эпоху,
> Надгробный псалом не звучит.
>
> Анна Ахматова

Victor Petrovich Havin will probably be remembered for many things. Of course, he was an eminent mathematician – quite a towering figure in his field – and the founding father of a renowned analysis school. He was also a wonderful lecturer (certainly the best I've ever listened to), a truly charismatic teacher, and a man of great personal charm. This last quality is, in fact, crucial in understanding why he was able to attract and influence so many; yet it seems to be the least describable part, too. Indeed, that mysterious (though unmistakably recognizable) substance known as personal charm – or glamour, or charisma, whatever – defies description. Why do some people have it, while others have none? God knows, but he won't tell.

I find it equally hard to explain why V.P.'s lectures were so inspiring, but I have a clear memory of being immediately fascinated by his personality and his subject, mathematical analysis, when I first heard him speak. (That was 35 years ago, so he was 48, younger than I am now. The audience consisted of over 100 freshers aged 17 plus epsilon, crammed in a large lecture theater at *Mat-Mekh*, in Old Peterhof.) Somehow, he taught *as one who has authority*, and you had the feeling that if this highly intelligent, eloquent and overall extraordinary person is so excited about the matter at hand, then that's probably the sort of stuff you should be doing – what's good enough for Havin is good enough for me.

The mathematical content of V.P.'s lectures would often be quite technical: a host of ε's and δ's at the initial stage, then lots of *hard analysis* proofs involving lengthy calculations, combinatorial arguments, sophisticated estimates of various unwieldy integrals and whatnot at a more advanced level. (Quite frankly, at times his heavy notation and terminology seemed somewhat too dense to me, and I found myself thinking that certain terms – such as ε-*допуск*, say – could be safely dispensed with.) However, even those technical matters were presented in a truly artistic manner, everything was extremely well organized, and you never felt bored.

Supported in part by grant MTM2014-51834-P from El Ministerio de Economía y Competitividad (Spain) and grant 2014-SGR-289 from AGAUR (Generalitat de Catalunya).

On the contrary, it was always clear that something interesting and important was
going on.

Besides, whenever possible, V.P. would begin with an informal and/or heuris-
tic "hand-waving argument" to unveil the simple truth (*сермяжная правда*, he
used to say) underlying this or that result; the formal proof that ensued would then
become much more digestible. For example, to convince us that $f \circ g$ is continuous
provided both f and g are, V.P. would first write down something like

$$a \approx b \implies g(a) \approx g(b) \implies f(g(a)) \approx f(g(b))$$

by way of explanation, and then convert this "quasi-proof" into a rigorous ε-δ
argument. Likewise, brief informal comments were offered to clarify the implicit
function theorem (whose full proof was quite long and exhausting, if I recall well),
the change of variable theorem (involving the Jacobian) for integrals in $\mathbb{R}^n$, and
other real analysis facts. Remarkably, though, when it came to complex analysis,
V.P. confessed that he could see no simple explanation for Cauchy's theorem (the
one that says

$$\oint_{\partial\Omega} f(z)dz = 0$$

for f holomorphic on $\overline{\Omega}$), no *homespun truth* behind it that would make the result
essentially obvious. Rather, he viewed it as an amazing fact with an air of magic
about it, a miracle that continued to surprise him ever since he had come across it
a few decades before. Now he was genuinely elated at rediscovering this and other
gems of the complex realm (for that's where his heart belonged), and he knew how
to convey elation to his disciples.

At the same time, it wasn't Havin's idea to conquer the audience by making
the presentation entertaining. After all, he was there to provide us with food
for thought, and the student was supposed to toil in the sweat of his/her brow
(this last point was repeatedly emphasized). And not only that, the student was
assumed to be obsessed with mathematics above all other things! Needless to say,
this was not always the case in reality, and occasionally the students' ignorance
and/or apathy seemed to grind V.P.'s gears. He would then address some bitter
words of disappointment to those guilty (or, by extension, to all those he had in
front).

As any lecturer, V.P. had some home-made jokes in his toolbox, but not
many. For instance, he would warn us – with a straight face – that "the triangle
inequality" (*неравенство треугольника*) should not be misinterpreted as "Triangle's inequality," i.e., a result due to the hypothetical mathematician named
Triangle (*Треугольник*), rather an unlikely interpretation indeed. Many an important formula on the blackboard was boxed, supplied with a skull and crossbones
symbol (☠), and much woe was prophesied to those poor buggers among us who
would even begin to hesitate, albeit for a tiny fraction of a second, in reproducing
it at the exam if required. Any such student would be mercilessly flunked, V.P.
told us in a menacing tone. The formulas you had to forget in order to deserve such

capital punishment included the basic Taylor–Maclaurin expansions of elementary functions, the Euler identities

$$\cos z = \frac{e^{iz} + e^{-iz}}{2}, \qquad \sin z = \frac{e^{iz} - e^{-iz}}{2i}$$

and so forth.

What really mattered, though, was V.P.'s unique emotional style, his individualistic – sometimes strikingly original – view on many mathematical phenomena, and the highly expressive metaphorical language he used to describe them. For instance, the zeros of analytic functions were said to be *contagious* (if there are too many, they propagate like plague until they absorb everything), and the Bürmann–Lagrange inversion formula was claimed to make the *golden dream of mankind* come true (namely, that every equation be solvable). Conversely, V.P. would occasionally resort to mathematical language to explain certain phenomena outside mathematics. One such digression, related again to the identity principle for analytic functions, concerned the nature of great works of art (e.g., in poetry or music), and V.P.'s point was that *analyticity* is a key feature of any true masterpiece. Indeed, once you have a germ (say, the first line of a poem or a few opening bars of a musical piece), it is bound to extend on its own, and in the only possible way; the continuation is unique, and you have no further control over it. Thus, according to V.P., if it begins with

> Буря мглою небо кроет,

then the next line, and in fact the rest of the poem, is predetermined: it just *has to be* what it is. Consequently, the poet doesn't have to sit there thinking what to write next, for he is but a tool employed by that divine force, the analytic extension, to do the job.

Havin's own language was poetic – in his lectures, writings, and daily life – and pertinent quotations from (or allusions to) various authors would often serve him to reinforce it. For instance, when running out of time and forced to skip certain details, he said he was "setting his heel on the throat of his own song," an obvious reference to Mayakovsky's poem "At the Top of My Voice." On another occasion, and unexpectedly enough, V.P. borrowed Mayakovsky's lines

> Но поэзия –
>
> пресволочнейшая штуковина:
>
> существует –
>
> и ни в зуб ногой

to describe the evasive nature of mathematics (the only change suggested was to replace *поэзия* by *математика*); actually, this was incorporated into a toast that V.P. made at a banquet.

Elsewhere – namely, in one of his appendices to Paul Koosis' book "Introduction to H_p Spaces" – V.P. invoked Tyutchev's famous line

> Мысль изреченная есть ложь

("A thought, once uttered, is a lie") to support his feeling that the essence of a theorem can seldom, if ever, be grasped from its statement alone. This remarkable claim seems to be typical of Havin's mathematical philosophy, and I take the liberty of reproducing a bit of his text in Koosis' translation: *"Confined within the narrow limits of its formulation, a theorem does not tell* all *of the truth about itself and is perhaps only fit for inclusion in a handbook. Its true meaning is inseparable from the proof (or proofs, if there are several of them)."*

One quotation that comes to mind in connection with Havin himself is Weierstrass' often cited maxim that "a mathematician who is not somewhat of a poet, will never be a perfect mathematician." Well, V.P. undoubtedly met the criterion.

It was my good fortune to be a Ph.D. student of Havin in my post-university years. Strictly speaking, I was rather a *соискатель*, a fairly vague status implying that I had no formal link to the university, and hence to V.P., while working on my thesis. Nonetheless, he took care of me with great kindness, and his fatherly attitude went far beyond what a supervisor is normally expected to offer. Once in a while I would ring him up, then come to his place, we would sit down at his enormous desk with piles of books, manuscripts, reprints, preprints, letters, photos and whatnot on it, and talk about mathematics. This could go on for hours. I would tell him about any progress I was (or wasn't) able to make in the past weeks, and his vivid reaction – sometimes approving, even enthusiastic, sometimes perhaps tinged with skepticism – meant a lot to me. V.P.'s excellent taste and keen intuition were as impressive as his great fund of knowledge, and talking to him would broaden your horizons. Moreover, he had that special flair, a kind of perfect pitch in mathematics, which allowed him to tell you, with a reliable degree of certainty, whether your stuff was attractive and likely to have interesting connections with other things. Some of the insights and associations that occurred to him along the way were really illuminating.

As a "mathematical son" of V.P., you could always count on his advice, reassurance (in case you got stuck) and – most importantly – encouragement. Yes, encouragement is what you need badly when you are a beginner with little or no self-confidence, with no achievements yet to boast of; and V.P. did his best to encourage you generously whenever you happened to make a perceptible step forward. Sometimes you were lucky enough to do more – crack the problem completely, or perhaps chance upon an unexpected twist that led to a sexy result – and V.P. would be really pleased. He would speak highly of your result, then probably do some advertising by telling other people about it and/or getting you to speak at the seminar... To see him excited or surprised by what you had done was very gratifying indeed, and his praise made you feel euphoric.

I did not see much of V.P. after I had moved to Spain (which happened around 2000), but I kept paying him a visit once a year, usually towards the end of August, while on vacation in St. Petersburg. V.P. and Valentina Afanas'evna, his wife, would kindly offer me a cup of coffee or a glass of wine; then V.P. and I would spend a few hours talking in his room (following the tradition, and perhaps abusing his friendly attention, I went on testing my new theorems on him); alternatively, weather permitting, we might go for a walk. As always, V.P. irradiated that peculiar warmth which made his company so enjoyable; it was invariably a pleasure to be by his side and listen to what he had to say. When asked about his health – which left very much to be desired – V.P. would respond briefly and switch to some other topic. He would speak of the current state of things at *Mat-Mekh* (being often displeased with the changes that modern times had brought about), of what it was like in former days (his reminiscences and anecdotes involved great names of the past, such as V.I. Smirnov, V.A. Rokhlin, A.D. Alexandrov, S.N. Mergelyan), or he might touch upon his own recent work, usually done in collaboration with this or that Ph.D. student. During my last visit, in August 2015, he explained to me – in a fairly detailed way – his joint paper with Pavel Mozolyako on the normal variation of a positive harmonic function.

On a less serious note, I recall a quasi- (or pseudo?) mathematical puzzle, which V.P. once offered me by way of entertainment: compute the product

$$\sin\alpha \cdot \sin\beta \cdot \sin\gamma \cdot \cdots \cdot \sin\omega.$$

(Spoiler: the answer is 0, since one of the factors is $\sin\pi$.)

Further topics of our conversations – or his monologues – included politics, literature, and most notably, music. In fact, V.P. had a lifelong passion for music and spoke about it knowledgeably. He admired Mahler's symphonies and Shostakovich's quartets, knew a large number of operas thoroughly, if not by heart. I also remember him saying that Beethoven's even-numbered symphonies deserve more attention than the (overwhelmingly more famous) odd-numbered ones; V.P. traced this opinion back to V.I. Smirnov, his own mathematical father, and seemed to concur with it. I, for one, still don't feel fully convinced. . .

By the way, V.P. had a fine bass voice and sang beautifully. His repertoire seemed to consist chiefly of *блатные песни* – the songs attributed to, and describing the miserable life of, the criminal dregs of society. (I have always suspected that this kind of folk culture is largely produced by intellectuals, maybe even distinguished professors.) Although I had no chance to witness any live performance by V.P., I did have an occasion to listen to him on tape. The record was provided by the late Evsey Matveevich (Seva) Dyn'kin when I was visiting him in Haifa, back in 1998, and he played it on his tape recorder for me. I had a strange feeling when V.P.'s voice entered the room: the genre and subject matter were somewhat unusual (I would have rather expected V.P. to besing harmonic vector fields or the like), but after all, his account of a bank holdup was no less emotional:

> *Взяли в сберкассе мы сумму немалую –*
> *Двадцать пять тысяч рублей . . .*

He was a lovable and unforgettable human being. We shall not look upon his like again.

Konstantin M. Dyakonov
Departament de Matemàtiques i Informàtica
Universitat de Barcelona, IMUB and BGSMath
Gran Via, 585
E-08007 Barcelona, Spain

  and

Institució Catalana de Recerca i Estudis Avançats (ICREA)
Pg. Lluís Companys, 23
E-08010 Barcelona, Spain
e-mail: `konstantin.dyakonov@icrea.cat`

Part II
Contributed Papers

Operator Theory:
Advances and Applications, Vol. 261, 83–95

Interpolation by the Derivatives of Operator Lipschitz Functions

A.B. Aleksandrov

Dedicated to the memory of V.P. Havin

Abstract. Let Λ be a discrete subset of the real line $\mathbb{R}$. We prove that for every bounded function φ on Λ there exists an operator Lipschitz function f on $\mathbb{R}$ such that $f'(t) = \varphi(t)$ for all $t \in \Lambda$. The same is true for the set of operator Lipschitz functions f on $\mathbb{R}$ such that f' coincides with the non-tangential boundary values of a bounded holomorphic function on the upper half-plane. In other words, for every bounded function φ on Λ there exists a commutator Lipschitz function f on the closed upper half-plane such that $f'(t) = \varphi(t)$ for all $t \in \Lambda$. The same is also true for some non-discrete countable sets Λ. Furthermore, we consider the case where Λ is a subset of the closed upper half-plane, $\Lambda \not\subset \mathbb{R}$. Similar questions for commutator Lipschitz functions on a closed subset $\mathfrak{F}$ of $\mathbb{C}$ are also considered.

Mathematics Subject Classification (2010). Primary 47A56; Secondary 30H05.

Keywords. Operator Lipschitz functions, commutator Lipschitz functions, interpolation.

1. Introduction

In this section we present definitions and some results concerning operator Lipschitz functions. All these results can be found in more detailed form in the recent survey [3].

Let $\mathfrak{F}$ be a closed subset of the complex plane $\mathbb{C}$. A continuous function f defined on $\mathfrak{F}$ is said to be *operator Lipschitz* if there exists a constant c such that

$$\|f(M) - f(N)\| \leq c\|M - N\| \tag{1.1}$$

This work was completed with the support of RFBR grant 14-01-00198.

for every normal operators M and N with spectra in $\mathfrak{F}$. We denote by $\mathrm{OL}(\mathfrak{F})$ the set of all operator Lipschitz functions on $\mathfrak{F}$. Let $\|f\|_{\mathrm{OL}(\mathfrak{F})}$ denote the minimal constant $c \geq 0$ satisfying (1.1).

A continuous function f defined on $\mathfrak{F}$ is said to be *commutator Lipschitz* if there exists a constant c such that

$$\|f(N)R - Rf(N)\| \leq c\|NR - RN\| \tag{1.2}$$

for every bounded operator R and every normal operator N with spectrum in $\mathfrak{F}$. We denote by $\mathrm{CL}(\mathfrak{F})$ the set of all commutator Lipschitz functions on $\mathfrak{F}$. Let $\|f\|_{\mathrm{CL}(\mathfrak{F})}$ denote the minimal constant $c \geq 0$ satisfying (1.2).

It is not difficult to verify that

$$\|f(M)R - Rf(N)\| \leq \|f\|_{\mathrm{CL}(\mathfrak{F})}\|MR - RN\|$$

for every bounded operator R and every normal operators M and N such that $\sigma(M), \sigma(N) \subset \mathfrak{F}$. Hence, $\mathrm{CL}(\mathfrak{F}) \subset \mathrm{OL}(\mathfrak{F})$ and $\|f\|_{\mathrm{OL}(\mathfrak{F})} \leq \|f\|_{\mathrm{CL}(\mathfrak{F})}$ for all $f \in \mathrm{CL}(\mathfrak{F})$.

The equality $\mathrm{CL}(\mathfrak{F}) = \mathrm{OL}(\mathfrak{F})$ holds if and only if $\overline{z} \in \mathrm{CL}(\mathfrak{F})$, see [8].

It is known that every function $f \in \mathrm{CL}(\mathfrak{F})$ has the derivative $f'(z)$ as a function of complex variable at each non-isolated point $z \in \mathfrak{F}$, see [6]. Moreover, if $\mathfrak{F}$ is not bounded, then f has the derivative at infinity defined as follows: $f'(\infty) \stackrel{\mathrm{def}}{=} \lim_{z \to \infty} z^{-1}f(z)$, see [6].

Thus, if $\mathrm{CL}(\mathfrak{F}) = \mathrm{OL}(\mathfrak{F})$, then the function $\overline{z}$ is differentiable in the complex sense at each non-isolated point of $\mathfrak{F}$. E. Kissin and V.S. Shulman proved in [8] that $\mathrm{CL}(\mathfrak{F}) = \mathrm{OL}(\mathfrak{F})$ if $\mathfrak{F}$ is a compact curve of class C^2.

It follows from results of H. Kamowitz [7] that the equality $\mathrm{CL}(\mathfrak{F}) = \mathrm{OL}(\mathfrak{F})$ is fulfilled together with the equality $\|f\|_{\mathrm{OL}(\mathfrak{F})} = \|f\|_{\mathrm{CL}(\mathfrak{F})}$ for all $f \in \mathrm{CL}(\mathfrak{F})$ if and only if $\mathfrak{F}$ is a subset of a straight line or of a circle.

Thus $\mathrm{CL}(\mathbb{R}) = \mathrm{OL}(\mathbb{R})$ with the equality $\|f\|_{\mathrm{OL}(\mathbb{R})} = \|f\|_{\mathrm{CL}(\mathbb{R})}$ for all $f \in \mathrm{CL}(\mathbb{R})$, and $\mathrm{CL}(\mathbb{T}) = \mathrm{OL}(\mathbb{T})$ with the equality $\|f\|_{\mathrm{OL}(\mathbb{T})} = \|f\|_{\mathrm{CL}(\mathbb{T})}$ for all $f \in \mathrm{CL}(\mathbb{T})$, where $\mathbb{T}$ denote the unit circle.

Let us note that the result of B.E. Johnson and J.P. Williams [6] (mentioned above) concerning the differentiability of commutator Lipschitz functions implies the following statement.

Let u be the Poisson integral of f', where $f \in \mathrm{OL}(\mathbb{R})$. Then $f'(t)$ is a non-tangential boundary value of u at each point t of the extended real line $\widehat{\mathbb{R}} \stackrel{\mathrm{def}}{=} \mathbb{R} \cup \{\infty\}$. The same is true in the case of the unit circle $\mathbb{T}$, see [1] for details.

We are going to use the symbols $\mathrm{OL}(\mathfrak{F})$ and $\mathrm{CL}(\mathfrak{F})$ for arbitrary subsets of $\mathbb{C}$ putting $\mathrm{OL}(\mathfrak{F}) \stackrel{\mathrm{def}}{=} \mathrm{OL}(\mathrm{clos}\,\mathfrak{F})$ and $\mathrm{CL}(\mathfrak{F}) \stackrel{\mathrm{def}}{=} \mathrm{CL}(\mathrm{clos}\,\mathfrak{F})$.

Let $f \in \mathrm{CL}(\mathbb{D})$, where $\mathbb{D}$ denotes the open unit disk in $\mathbb{C}$. Then $f|\mathbb{T} \in \mathrm{CL}(\mathbb{T}) = \mathrm{OL}(\mathbb{T})$, and $\|f\|_{\mathrm{OL}(\mathbb{T})} = \|f\|_{\mathrm{CL}(\mathbb{T})} \leq \|f\|_{\mathrm{CL}(\mathbb{D})}$. E. Kissin and V.S. Shulman [9] proved that a function $f \in \mathrm{OL}(\mathbb{T})$ can be represented in the form $f = F|\mathbb{T}$, where $F \in \mathrm{CL}(\mathbb{D})$ if and only if f belongs to the disk algebra. This result was generalized by the author [2]. Moreover, it follows from results of [2] that $\|f\|_{\mathrm{OL(T)}} = \|F\|_{\mathrm{CL(D)}}$.

Similar results hold in the case of the upper half-plane $\mathbb{C}_+ \overset{\text{def}}{=} \{z \in \mathbb{C} : \operatorname{Im} z > 0\}$, see [2]. Clearly, $f|\mathbb{R} \in \mathrm{OL}(\mathbb{R})$ and $\|f\|_{\mathrm{OL}(\mathbb{R})} = \|f\|_{\mathrm{CL}(\mathbb{R})} \le \|f\|_{\mathrm{CL}(\mathbb{C}_+)}$ for every $f \in \mathrm{CL}(\mathbb{C}_+)$. Moreover, a function $f \in \mathrm{OL}(\mathbb{R})$ can be represented in the form $f = F|\mathbb{R}$ for $F \in \mathrm{CL}(\mathbb{C}_+)$ if and only if f' is the boundary values of a bounded holomorphic function in $\mathbb{C}_+$. Furthermore, in this case $\|f\|_{\mathrm{OL}(\mathbb{R})} = \|F\|_{\mathrm{CL}(\mathbb{C}_+)}$, see [2].

Let $\mathfrak{F}'$ denote the set of all limit points of a closed set $\mathfrak{F}$, $\mathfrak{F} \subset \mathbb{C}$. Denote by $\mathrm{CL}'(\mathfrak{F})$ the set of all function h on $\mathfrak{F}'$ of the form $h(z) = f'(z)$, where $f \in \mathrm{CL}(\mathfrak{F})$. Put $\|h\|_{\mathrm{CL}'(\mathfrak{F})} \overset{\text{def}}{=} \inf\{\|f\|_{\mathrm{CL}(\mathfrak{F})} : f' = h \text{ everywhere on } \mathfrak{F}'\}$. Put $\mathrm{CL}'(\mathfrak{F}) = \{0\}$ if $\mathfrak{F}' = \varnothing$.

Denote by $\mathscr{M}(\mathfrak{F})$ the set of complex Radon measures on $\mathbb{C} \setminus \mathfrak{F}$ such that

$$\|\mu\|_{\mathscr{M}(\mathfrak{F})} \overset{\text{def}}{=} \sup_{z \in \mathfrak{F}} \int_{\mathbb{C}\setminus\mathfrak{F}} \frac{d|\mu|(\zeta)}{|\zeta - z|^2} < +\infty.$$

In general, the Cauchy integral of $\mu \in \mathscr{M}(\mathfrak{F})$,

$$\widehat{\mu}(z) \overset{\text{def}}{=} \int_{\mathbb{C}\setminus\mathfrak{F}} \frac{d\mu(\zeta)}{\zeta - z},$$

is not defined for $z \in \mathfrak{F}$. But we can consider the formal derivative of the Cauchy integral of μ putting

$$\widehat{\mu}'(z) \overset{\text{def}}{=} \int_{\mathbb{C}\setminus\mathfrak{F}} \frac{d\mu(\zeta)}{(\zeta - z)^2}$$

for $z \in \mathfrak{F}$. Clearly, $\widehat{\mu}'(z)$ is well defined for all $\mu \in \mathscr{M}(\mathfrak{F})$ and all $z \in \mathfrak{F}$. Moreover, we can define a modified Cauchy integral. For that we fix a point $z_0 \in \mathfrak{F}$, and put

$$\widehat{\mu}_{z_0}(z) \overset{\text{def}}{=} \int_{\mathbb{C}\setminus\mathfrak{F}} \left(\frac{1}{\zeta - z} - \frac{1}{\zeta - z_0} \right) d\mu(\zeta)$$

for $\mu \in \mathscr{M}(\mathfrak{F})$ and $z \in \mathfrak{F}$. It follows from the Cauchy–Bunyakovsky inequality that $\widehat{\mu}_{z_0}(z)$ is well defined for $z \in \mathfrak{F}$. It is easy to see that if z is a non-isolated point of $\mathfrak{F}$, then $\widehat{\mu}'(z)$ coincides with the derivative of $\widehat{\mu}_{z_0}$ at z. Thus $\widehat{\mu}'(z) = (\widehat{\mu}_{z_0})'(z)$ at each non-isolated point z of $\mathfrak{F}$.

Denote by $\widehat{\mathscr{M}}(\mathfrak{F})$ the set of functions representable in the form $f = c + \widehat{\mu}_{z_0}$, where $\mu \in \mathscr{M}(\mathfrak{F})$ and c is a constant. Put

$$\|f\|_{\widehat{\mathscr{M}}(\mathfrak{F})} \overset{\text{def}}{=} \inf\{\|\mu\|_{\mathscr{M}(\mathfrak{F})} : \mu \in \mathscr{M}(\mathfrak{F}),\ f - \widehat{\mu}_{z_0} = \text{const on } \mathfrak{F}\}.$$

It is easily seen that the definitions of the space $\widehat{\mathscr{M}}(\mathfrak{F})$ and the seminorm $\|f\|_{\widehat{\mathscr{M}}(\mathfrak{F})}$ do not depend on a fixed point $z_0 \in \mathfrak{F}$.

It was proved in [1] that

$$\widehat{\mathscr{M}}(\mathfrak{F}) \subset \mathrm{CL}(\mathfrak{F}) \quad \text{and} \quad \|f\|_{\mathrm{CL}(\mathfrak{F})} \le \|f\|_{\widehat{\mathscr{M}}(\mathfrak{F})} \tag{1.3}$$

for all $f \in \widehat{\mathscr{M}}(\mathfrak{F})$.

Denote by $\widehat{\mathscr{M}'}(\mathfrak{F})$ the set of functions f on $\mathfrak{F}$ of the form $f = \widehat{\mu}'$, where $\mu \in \mathscr{M}(\mathfrak{F})$. Put

$$\|f\|_{\widehat{\mathscr{M}'}(\mathfrak{F})} \overset{\text{def}}{=} \inf\{\|\mu\|_{\mathscr{M}(\mathfrak{F})} : \mu \in \mathscr{M}(\mathfrak{F}), \ f = \widehat{\mu}'\}.$$

It follows from (1.3) that for every $\mu \in \widehat{\mathscr{M}}(\mathfrak{F})$ the restriction of $\widehat{\mu}'$ to $\mathfrak{F}'$ belongs to $\mathrm{CL}'(\mathfrak{F})$ and $\|\widehat{\mu}'\|_{\mathrm{CL}'(\mathfrak{F})} \leq \|f\|_{\widehat{\mathscr{M}'}(\mathfrak{F})}$.

2. Interpolation by functions in $\widehat{\mathscr{M}'}(\mathfrak{F})$ on discrete sets

Let $\partial\mathfrak{F}$ denote the boundary of a closed subset $\mathfrak{F}$, $\mathfrak{F} \subset \mathbb{C}$. We say that a point $z_0 \in \partial\mathfrak{F}$ is an $\mathscr{M}$-*regular* boundary point of $\mathfrak{F}$ if for every $\varepsilon > 0$ and every neighborhood U of z_0 there exists a positive measure $\mu \in \mathscr{M}(\mathfrak{F})$ such that

$$\int_{\mathbb{C}\backslash\mathfrak{F}} \frac{d\mu(\zeta)}{|\zeta - z_0|^2} > 1 - \varepsilon, \quad \text{and} \quad \int_{\mathbb{C}\backslash\mathfrak{F}} \frac{d\mu(\zeta)}{|\zeta - z|^2} < \max(\varepsilon, \mathbb{1}_U(z))$$

for all $z \in \mathfrak{F}$, where $\mathbb{1}_U$ denotes the characteristic function of the set U.

We first state a sufficient condition for a point z_0 to be an $\mathscr{M}$-regular boundary point of $\mathfrak{F}$.

Lemma 2.1. *Let $z_0 \in \mathfrak{F}$, where $\mathfrak{F}$ is a closed subset of $\mathbb{C}$. Suppose that*

$$\limsup_{z \to z_0} \frac{\mathrm{dist}(z, \mathfrak{F})}{|z - z_0|} > 0.$$

Then z_0 is an $\mathscr{M}$-regular boundary point of $\mathfrak{F}$.

Proof. Let us fix a positive ε, $\varepsilon < 1$, and a neighborhood U of z_0. Denote by $\mathscr{M}_\varepsilon^U(\mathfrak{F})$ the set of all $\mu \in \mathscr{M}(\mathfrak{F})$ such that $\mu \geq 0$, the support of μ is finite, and

$$\int_{\mathbb{C}\backslash\mathfrak{F}} \frac{d\mu(\zeta)}{|\zeta - z|^2} < \max(\varepsilon, \mathbb{1}_U(z)).$$

Take $\sigma > 0$ such that $\limsup_{z \to z_0} \dfrac{\mathrm{dist}(z, \mathfrak{F})}{|z - z_0|} > \sigma$. Then we can construct a sequence $\{z_n\}_{n=1}^\infty$ in $\mathbb{C} \backslash \mathfrak{F}$ such that $\lim_{n \to +\infty} z_n = z_0$ and $\dfrac{\mathrm{dist}(z_n, \mathfrak{F})}{|z_n - z_0|} > \sigma$ for all $n \geq 1$. It is easy to see that if $\mu \in \mathscr{M}_\varepsilon^U(\mathfrak{F})$ and $t = \int_{\mathbb{C}\backslash\mathfrak{F}} |\zeta - z_0|^{-2}\, d\mu(\zeta)$, then $\mu_n = \mu + \frac{1-t}{2}(\mathrm{dist}(z_n, \mathfrak{F}))^2 \delta_{z_n} \in \mathscr{M}_\varepsilon^U(\mathfrak{F})$ for sufficiently large n. Note that

$$\int_{\mathbb{C}\backslash\mathfrak{F}} \frac{d\mu_n(\zeta)}{|\zeta - z_0|^2} = t + \frac{1-t}{2} \cdot \frac{(\mathrm{dist}(z_n, \mathfrak{F}))^2}{|z_n - z_0|^2} > t + \frac{1-t}{2}\sigma^2 \tag{2.1}$$

Put

$$t_0 \overset{\text{def}}{=} \sup\left\{\int_{\mathbb{C}\backslash\mathfrak{F}} \frac{d\mu(\zeta)}{|\zeta - z_0|^2} : \mu \in \mathscr{M}_\varepsilon^U(\mathfrak{F})\right\}.$$

It follows from (2.1) that $t_0 = 1$. $\qquad\square$

Corollary 2.2. *Let $\mathfrak{F}$ be a closed subset of $\mathbb{C}$. Suppose there exists a triangle Δ with a vertex at $z_0 \in \mathfrak{F}$ such that the interior of Δ does not contain any point in $\mathfrak{F}$. Then z_0 is an $\mathcal{M}$-regular boundary point of $\mathfrak{F}$.*

Denote by $\ell^\infty(\Lambda)$ the space of all bounded functions on a set Λ. Put $\|\varphi\|_{\ell^\infty(\Lambda)}$ $\overset{\text{def}}{=} \sup_\Lambda |\varphi|$, where $\varphi \in \ell^\infty(\Lambda)$.

A subset Λ of $\mathfrak{F}$ is said to be *interpolation* for $\widehat{\mathcal{M}'}(\mathfrak{F})$ if for every $\varphi \in \ell^\infty(\Lambda)$ there exists $\mu \in \mathcal{M}(\mathfrak{F})$ such that $\widehat{\mu}'(\zeta) = \varphi(\zeta)$ for all $\zeta \in \Lambda$.

Let $a \geq 1$. We say that a subset Λ of $\mathfrak{F}$ is *a-interpolation* for $\widehat{\mathcal{M}'}(\mathfrak{F})$ if for every $\varphi \in \ell^\infty(\Lambda)$ and every $\varepsilon > 0$ there exists $\mu \in \mathcal{M}(\mathfrak{F})$ such that $\widehat{\mu}'(\zeta) = \varphi(\zeta)$ for all $\zeta \in \Lambda$ and $\|\mu\|_{\mathcal{M}(\mathfrak{F})} \leq (a + \varepsilon)\|\varphi\|_{\ell^\infty(\Lambda)}$.

In other words, Λ is a-interpolation for $\widehat{\mathcal{M}'}(\mathfrak{F})$ if for every $\varphi \in \ell^\infty(\Lambda)$ with $\|\varphi\|_{\ell^\infty(\Lambda)} < 1$ there exists $\mu \in \mathcal{M}(\mathfrak{F})$ such that $\widehat{\mu}'(\zeta) = \varphi(\zeta)$ for all $\zeta \in \Lambda$ and $\|\mu\|_{\mathcal{M}(\mathfrak{F})} \leq a$.

It it easy to see that Λ is interpolation for $\widehat{\mathcal{M}'}(\mathfrak{F})$ if and only if there exists $a \geq 1$ such that Λ is a-interpolation for $\widehat{\mathcal{M}'}(\mathfrak{F})$.

Theorem 2.3. *Let Λ be a discrete subset of $\partial\mathfrak{F}$, where $\mathfrak{F}$ is a closed subset of $\mathbb{C}$. Suppose that each point in Λ is an $\mathcal{M}$-regular boundary point of $\mathfrak{F}$. Then Λ is a 1-interpolation set for $\widehat{\mathcal{M}'}(\mathfrak{F})$.*

We need the following lemma.

Lemma 2.4. *Let z_0 be an $\mathcal{M}$-regular boundary point of a closed set $\mathfrak{F}$, $\mathfrak{F} \subset \mathbb{C}$. Then for every $\varepsilon > 0$ and every neighborhood U of z_0 there exists a Radon measure $\mu \in \mathcal{M}(\mathfrak{F})$ such that $\widehat{\mu}'(z_0) = 1$ and*

$$\int_{\mathbb{C}\backslash\mathfrak{F}} \frac{d|\mu|(\zeta)}{|\zeta - z|^2} < \varepsilon + \mathbb{1}_U(z)$$

for all $z \in \mathfrak{F}$.

Proof. It follows immediately from the definition of $\mathcal{M}$-regular boundary points that there exists a positive measure $\nu \in \mathcal{M}(\mathfrak{F})$ such that

$$\int_{\mathbb{C}\backslash\mathfrak{F}} \frac{d\nu(\zeta)}{|\zeta - z_0|^2} = 1, \quad \text{and} \quad \int_{\mathbb{C}\backslash\mathfrak{F}} \frac{d\nu(\zeta)}{|\zeta - z|^2} < \varepsilon + \mathbb{1}_U(z)$$

for all $z \in \mathfrak{F}$. Now the required Radon measure μ can be defined as follows:
$$d\mu(\zeta) = \frac{(\zeta - z_0)^2 \, d\nu(\zeta)}{|\zeta - z_0|^2}. \qquad \square$$

We need also the following well-known statement.

Remark 2.5. *Let $a \geq 1$ and let Λ be a subset of a closed set $\mathfrak{F}$, $\mathfrak{F} \subset \mathbb{C}$. Then Λ is a-interpolation for $\widehat{\mathcal{M}'}(\mathfrak{F})$ if and only if for every $\varphi \in \ell^\infty(\Lambda)$ and every $\varepsilon > 0$ there exists $\mu \in \mathcal{M}(\mathfrak{F})$ such that $|\widehat{\mu}'(\zeta) - \varphi(\zeta)| < \varepsilon$ for all $\zeta \in \Lambda$ and $\|\mu\|_{\mathcal{M}(\mathfrak{F})} \leq (a + \varepsilon)\|\varphi\|_{\ell^\infty(\Lambda)}$.*

Proof of Theorem 2.3. It suffices to prove that for every $\varepsilon > 0$ and every function $\varphi \in \ell^\infty(\Lambda)$ with $\|\varphi\|_{\ell^\infty(\Lambda)} < 1$ there exists a Radon measure $\mu \in \mathscr{M}(\mathfrak{F})$ such that $\|\mu\|_{\mathscr{M}(\mathfrak{F})} \leq 1$ and $\|\widehat{\mu}' - \varphi\|_{\ell^\infty(\Lambda)} < \varepsilon$.

We may assume that $\varepsilon < 1 - \|\varphi\|_{\ell^\infty(\Lambda)}$. Let $\{z_n\}_{n=1}^N$ $(1 \leq N \leq +\infty)$ be an enumeration of Λ. The set Λ is discrete. Hence, for every integer n, $1 \leq n \leq N$, we can take a neighborhood U_n of z_n in such a way that the U_n be pairwise disjoint.

By Lemma 2.4 for each integer n, $1 \leq n \leq N$, there exists a Radon measure $\mu_n \in \mathscr{M}(\mathfrak{F})$ such that $\widehat{\mu}_n'(z_n) = 1$ and $\int_{\mathbb{C}\backslash\mathfrak{F}} |\zeta - z|^{-2} \, d|\mu_n|(\zeta) < 2^{-n}\varepsilon + \mathbb{1}_{U_n}(z)$ for all $z \in \mathfrak{F}$.

Put $\mu \overset{\text{def}}{=} \sum_{n=1}^N \varphi(z_n)\mu_n$. We have

$$\int_{\mathbb{C}\backslash\mathfrak{F}} \frac{d|\mu|(\zeta)}{|\zeta - z|^2} \leq \sum_{n=1}^N (2^{-n}\varepsilon + \mathbb{1}_{U_n}(z))|\varphi(z_n)| \leq (1+\varepsilon)\|\varphi\|_{\ell^\infty(\Lambda)} < 1 - \varepsilon^2$$

for all $z \in \mathfrak{F}$ and

$$|\widehat{\mu}'(z_n) - \varphi(z_n)| = \left| \sum_{k=1}^N \varphi(z_k)\widehat{\mu}_k'(z_n) - \varphi(z_n) \right| \leq \|\varphi\|_{\ell^\infty(\Lambda)} \sum_{k\neq n} |\widehat{\mu}_k'(z_n)| \leq \varepsilon\|\varphi\|_{\ell^\infty(\Lambda)}.$$

Hence, $\|\widehat{\mu}' - \varphi\|_{\ell^\infty(\Lambda)} \leq \varepsilon\|\varphi\|_{\ell^\infty(\Lambda)} < \varepsilon$. $\square$

Remark 2.6. Let Λ be a subset of a closed set $\mathfrak{F}$, $\mathfrak{F} \subset \mathbb{C}$. Suppose that Λ is 1-interpolation for $\widehat{\mathscr{M}'}(\mathfrak{F})$. Then for each connected component U of the interior of $\mathfrak{F}$ the intersection $U \cap \Lambda$ consists at most of one point.

Proof. Suppose ζ and ξ belong to $U \cap \Lambda$. Let V be a bounded simply connected domain such that $\{\zeta, \xi\} \subset V \subset U$. By the Riemann mapping theorem there exists a biholomorphic mapping f from V into the open disk $\mathbb{D}$ such that $f(\zeta) = 0$. It remains to note that the Schwarz lemma implies that if Λ is a-interpolation for $\widehat{\mathscr{M}'}(\mathfrak{F})$, then $a \geq |f(\xi)|^{-1} > 1$. $\square$

3. Completely interpolation sets for $\widehat{\mathscr{M}'}(\mathfrak{F})$

We say that a subset Λ of $\mathfrak{F}$ is *completely a-interpolation* for $\widehat{\mathscr{M}'}(\mathfrak{F})$ if for every $\varphi \in \ell^\infty(\Lambda)$, every $\varepsilon > 0$ and every closed subset $\mathfrak{F}_0$ of $\mathfrak{F}$, $\mathfrak{F}_0 \cap \Lambda = \varnothing$, there exists $\mu \in \mathscr{M}(\mathfrak{F})$ such that $\widehat{\mu}'(\zeta) = \varphi(\zeta)$ for all $\zeta \in \Lambda$, $|\widehat{\mu}'(\zeta)| < \varepsilon$ for all $\zeta \in \mathfrak{F}_0$ and $\|\mu\|_{\mathscr{M}(\mathfrak{F})} \leq (a+\varepsilon)\|\varphi\|_{\ell^\infty(\Lambda)}$.

A subset Λ of $\mathfrak{F}$ is said to be *completely interpolation* for $\widehat{\mathscr{M}'}(\mathfrak{F})$ if it is completely a-interpolation for $\widehat{\mathscr{M}'}(\mathfrak{F})$ for some $a \geq 1$.

Remark 3.1. Let Λ be a subset of a closed set $\mathfrak{F}$, $\mathfrak{F} \subset \mathbb{C}$. Suppose that Λ is completely interpolation for $\widehat{\mathscr{M}'}(\mathfrak{F})$. Then $\Lambda \subset \partial\mathfrak{F}$.

Proof. Suppose $\Lambda \not\subset \partial\mathfrak{F}$. Then there exists $\zeta_0 \in \Lambda$ such that ζ_0 is an interior point of $\mathfrak{F}$. Clearly, ζ_0 is an isolated point of Λ. Hence there exists a closed disk K centered at ζ_0 such that K is contained in the interior of $\mathfrak{F}$ and $K \cap \Lambda = \{\zeta_0\}$. Put $\mathfrak{F}_0 \stackrel{\text{def}}{=} \partial K$. It remains to note that if $|\widehat{\mu}'| \leq \varepsilon$ on $\mathfrak{F}_0$, then $|\widehat{\mu}'(\zeta_0)| \leq \varepsilon$. $\qquad\square$

Now it is clear that a finite subset Λ of $\mathfrak{F}$ is completely interpolation for $\widehat{\mathscr{M}'}(\mathfrak{F})$ if and only if $\Lambda \subset \partial\mathfrak{F}$. On the other hand, each finite subset of $\mathfrak{F}$ is interpolation for $\widehat{\mathscr{M}'}(\mathfrak{F})$ provided $\mathfrak{F} \neq \mathbb{C}$.

Theorem 3.2. *Let Λ be a discrete subset of $\partial\mathfrak{F}$, where $\mathfrak{F}$ is a closed subset of $\mathbb{C}$. Suppose that each point in Λ is an $\mathscr{M}$-regular boundary point of $\mathfrak{F}$. Then Λ is a completely 1-interpolation set for $\widehat{\mathscr{M}'}(\mathfrak{F})$.*

To prove this theorem it suffices to note that we can require in the proof of Theorem 2.3 that $U_n \cap \mathfrak{F}_0 = \varnothing$ for all n.

The following theorem yields in a sense a complete description of 1-interpolation subsets Λ of a sufficiently regular closed set $\mathfrak{F}$.

Theorem 3.3. *Let $\mathfrak{F}$ be a closed subset of $\mathbb{C}$. Suppose that all boundary points in $\mathfrak{F}$ are $\mathscr{M}$-regular. Then for every subset Λ of $\partial\mathfrak{F}$ the following statements are equivalent:*

(i) *Λ is 1-interpolation for $\widehat{\mathscr{M}'}(\mathfrak{F})$;*
(ii) *Λ is completely 1-interpolation for $\widehat{\mathscr{M}'}(\mathfrak{F})$;*
(iii) *Λ is discrete.*

To prove this theorem we need auxiliary assertions. Let $\mu \in \mathscr{M}(\mathfrak{F})$, where $\mathfrak{F}$ is a closed subset of $\mathbb{C}$. Put

$$
(\mathfrak{D}\widehat{\mu})(z, w) \stackrel{\text{def}}{=} \begin{cases} \dfrac{\widehat{\mu}(z) - \widehat{\mu}(w)}{z - w} & \text{if } z, w \in \mathfrak{F}, z \neq w, \\[2mm] \widehat{\mu}'(z) & \text{if } z = w \in \mathfrak{F}. \end{cases}
$$

Clearly,

$$
(\mathfrak{D}\widehat{\mu})(z, w) = \int_{\mathbb{C}\setminus\mathfrak{F}} \frac{d\mu(\zeta)}{(\zeta - z)(\zeta - w)}.
$$

Let A be a matrix. Denote by $\|A\|_{\mathfrak{M}}$ the norm of this matrix as a Schur multiplier.

Lemma 3.4. *Let $\{\zeta_j\}$ be a sequence in a closed subset $\mathfrak{F}$ of $\mathbb{C}$. Then*
$$
\|\{(\mathfrak{D}\widehat{\mu})(\zeta_j, \zeta_k)\}\|_{\mathfrak{M}} \leq \|\mu\|_{\mathscr{M}(\mathfrak{F})}.
$$

Proof. Let $d\mu = h\,d|\mu|$ for some $h \in L^\infty(|\mu|)$. It is well known that if $A = \{(u_j, v_k)\}$ for some families $\{u_j\}$ and $\{v_k\}$ in a Hilbert space $\mathcal{H}$, then $\|A\|_{\mathfrak{M}} \leq \sup_j \|u_j\|_{\mathcal{H}} \sup_k \|v_k\|_{\mathcal{H}}$ (see, for example, [3]). We can take $L^2(|\mu|)$ as $\mathcal{H}$, and put $u_j(\zeta) \stackrel{\text{def}}{=} h(\zeta)(\zeta - \zeta_j)^{-1}$ and $v_k(\zeta) \stackrel{\text{def}}{=} (\overline{\zeta} - \overline{\zeta_j})^{-1}$. Then

$$
\|\{(\mathfrak{D}\widehat{\mu})(\zeta_j, \zeta_k)\}\|_{\mathfrak{M}} \leq \sup_j \|u_j\|_{L^2(|\mu|)} \sup_k \|v_k\|_{L^2(|\mu|)} \leq \|\mu\|_{\mathscr{M}(\mathfrak{F})}. \qquad\square
$$

Lemma 3.5. *Let $\mu \in \mathcal{M}(\mathfrak{F})$. Suppose that there exists a sequence $\{\zeta_n\}_{n=0}^{\infty}$ in $\mathfrak{F}$ such that $\lim_{n \to +\infty} \zeta_n = \zeta_0$, $\widehat{\mu}'(\zeta_0) = 1$ and $\widehat{\mu}'(\zeta_n) = -1$ for all $n \geq 1$. Then $\|f\|_{\mathcal{M}(\mathfrak{F})} \geq \sqrt{2}$.*

Proof. By Lemma 3.4 we have

$$\left\| \begin{pmatrix} \mathfrak{D}\widehat{\mu}(\zeta_0, \zeta_0) & \mathfrak{D}\widehat{\mu}(\zeta_0, \zeta_n) \\ \mathfrak{D}\widehat{\mu}(\zeta_n, \zeta_0) & \mathfrak{D}\widehat{\mu}(\zeta_n, \zeta_n) \end{pmatrix} \right\|_{\mathfrak{M}} \leq \|\mu\|_{\mathcal{M}(\mathfrak{F})}$$

for all $n \geq 0$. Letting $n \to +\infty$, we get

$$\|\mu\|_{\mathcal{M}(\mathfrak{F})} \geq \left\| \begin{pmatrix} 1 & 1 \\ 1 & -1 \end{pmatrix} \right\|_{\mathfrak{M}} = \sqrt{2}. \qquad \square$$

Corollary 3.6. *Let $\mathfrak{F}$ be a closed subset of $\mathbb{C}$. Suppose that Λ is an a-interpolation set for $\widehat{\mathcal{M}}'(\mathfrak{F})$, where $a < \sqrt{2}$. Then Λ is discrete.*

Proof of Theorem 3.3. The implication (iii) $\implies$ (ii) follows from Theorem 3.2. The implication (ii) $\implies$ (i) is evident. The implication (i) $\implies$ (iii) follows from Corollary 3.6. $\qquad \square$

4. Interpolation by functions in $\widehat{\mathcal{M}}'(\mathfrak{F})$ on non-discrete subsets of $\mathfrak{F}$

The following lemma allows us to construct examples of non-discrete interpolation sets for the space $\widehat{\mathcal{M}}'(\mathfrak{F})$.

Lemma 4.1. *Let $\mathfrak{F}$ be a closed subset of $\mathbb{C}$. Suppose that a subset Λ of $\mathfrak{F}$ is represented in the form $\Lambda = S \cup T$ in such a way that $S \cap \operatorname{clos} T = \varnothing$, S is completely a-interpolation for $\widehat{\mathcal{M}}'(\mathfrak{F})$ and T is completely b-interpolation for $\widehat{\mathcal{M}}'(\mathfrak{F})$, where $a, b \geq 1$. Then Λ is completely $(ab + a + b)$-interpolation for $\widehat{\mathcal{M}}'(\mathfrak{F})$.*

Proof. Let $\varphi \in \ell^{\infty}(\Lambda)$ with $\|\varphi\|_{\ell^{\infty}(\Lambda)} \leq 1$ and let $\mathfrak{F}_0$ be a closed subset of $\mathfrak{F}$ such that $\mathfrak{F}_0 \cap \Lambda = \varnothing$. There exists $\mu \in \mathcal{M}(\mathfrak{F})$ such that $\widehat{\mu}'(\zeta) = \varphi(\zeta)$ for all $\zeta \in T$, $|\widehat{\mu}'(\zeta)| < \varepsilon$ for all $\zeta \in \mathfrak{F}_0$ and $\|\mu\|_{\mathcal{M}(\mathfrak{F})} \leq (b + \varepsilon)\|\varphi\|_{\ell^{\infty}(T)} \leq b + \varepsilon$. There exists $\nu \in \mathcal{M}(\mathfrak{F})$ such that $\widehat{\nu}'(\zeta) = \varphi(\zeta) - \widehat{\mu}'(\zeta)$ for all $\zeta \in S$, $|\widehat{\nu}'(\zeta)| < \varepsilon$ for all $\zeta \in \mathfrak{F}_0 \cup \operatorname{clos} T$ and $\|\nu\|_{\mathcal{M}(\mathfrak{F})} \leq (a + \varepsilon)\|\varphi - \widehat{\mu}'\|_{\ell^{\infty}(S)} \leq (a + \varepsilon)(b + 1 + \varepsilon)$. Then $|\widehat{\mu}'(\zeta) + \widehat{\nu}'(\zeta) - \varphi(\zeta)| < 2\varepsilon$ for all $\zeta \in \Lambda$, $|\widehat{\mu}'(\zeta) + \widehat{\nu}'(\zeta)| < 2\varepsilon$ for all $\zeta \in \mathfrak{F}_0$ and $\|\widehat{\mu}' + \widehat{\nu}'\|_{\mathcal{M}(\mathfrak{F})} \leq (a + \varepsilon)(b + 1 + \varepsilon) + b + \varepsilon$. $\qquad \square$

Let Λ be a subset of $\mathbb{C}$. Put $\Lambda_{[0]} \overset{\text{def}}{=} \Lambda$ and $\Lambda_{[n+1]} \overset{\text{def}}{=} \Lambda_{[n]} \cap \Lambda'_{[n]}$, where $\Lambda'_{[n]}$ denotes the set of all limit points of $\Lambda_{[n]}$.

Theorem 4.2. *Let Λ be a subset of $\partial\mathfrak{F}$, where $\mathfrak{F}$ is a closed subset of $\mathbb{C}$. Suppose that each point in Λ is an $\mathcal{M}$-regular boundary point of $\mathfrak{F}$ and $\Lambda_{[n]} = \varnothing$ for some positive integer n. Then Λ is completely a_n-interpolation for $\widehat{\mathcal{M}}'(\mathfrak{F})$ with $a_n = 2^n - 1$.*

Proof. We use induction on n. The case of $n = 1$ follows from Theorem 3.2. Suppose the theorem has been proved for $n - 1$. Applying this for the set $\Lambda_{[1]}$

we deduce that $\Lambda_{[1]}$ is completely a_{n-1}-interpolation. Note that the set $\Lambda \setminus \Lambda_{[1]}$ is completely a_1-interpolation by Theorem 3.2. It remains to apply Lemma 4.1 and to note that $a_n = a_1 a_{n-1} + a_1 + a_{n-1}$. □

5. Interpolation by functions in $\mathrm{CL}'(\mathfrak{F})$

Let $\mathfrak{F}$ be a closed subset of $\mathbb{C}$ and let $a \geq 1$. We say that a subset Λ of $\mathfrak{F}'$ is *a-interpolation* for $\mathrm{CL}'(\mathfrak{F})$ if for every $\varphi \in \ell^\infty(\Lambda)$ and every $\varepsilon > 0$ there exists $f \in \mathrm{CL}(\mathfrak{F})$ such that $f'(\zeta) = \varphi(\zeta)$ for all $\zeta \in \Lambda$ and $\|f\|_{\mathrm{CL}(\mathfrak{F})} \leq (a + \varepsilon)\|\varphi\|_{\ell^\infty(\Lambda)}$.

A subset Λ of $\mathfrak{F}'$ is said to be *interpolation* for $\mathrm{CL}'(\mathfrak{F})$ if for every $\varphi \in \ell^\infty(\Lambda)$ there exists $f \in \mathrm{CL}(\mathfrak{F})$ such that $f'(\zeta) = \varphi(\zeta)$ for all $\zeta \in \Lambda$.

It is easy to see that a subset Λ of $\mathfrak{F}'$ is interpolation for $\mathrm{CL}'(\mathfrak{F})$ if and only if Λ is a-interpolation for $\mathrm{CL}'(\mathfrak{F})$ for some $a \geq 1$.

We say that a subset Λ of $\mathfrak{F}'$ is *completely a-interpolation* for $\mathrm{CL}'(\mathfrak{F})$ if for every $\varphi \in \ell^\infty(\Lambda)$, every $\varepsilon > 0$ and every closed subset $\mathfrak{F}_0$ of $\mathfrak{F}'$, $\mathfrak{F}_0 \cap \Lambda = \varnothing$, there exists $f \in \mathrm{CL}(\mathfrak{F})$ such that $f'(\zeta) = \varphi(\zeta)$ for all $\zeta \in \Lambda$, $|f'(\zeta)| < \varepsilon$ for all $\zeta \in \mathfrak{F}_0$ and $\|f\|_{\mathrm{CL}(\mathfrak{F})} \leq (a + \varepsilon)\|\varphi\|_{\ell^\infty(\Lambda)}$.

A subset Λ of $\mathfrak{F}'$ is said to be *completely interpolation* for $\mathrm{CL}'(\mathfrak{F})$ if it is a-interpolation for some $a \geq 1$.

Theorem 3.2 and statement (1.3) yield the following result.

Theorem 5.1. *Let Λ be a discrete subset of $\partial\mathfrak{F}$, where $\mathfrak{F}$ is a closed subset of $\mathbb{C}$. Suppose that each point in Λ is a non-isolated $\mathscr{M}$-regular boundary point of $\mathfrak{F}$. Then Λ is completely 1-interpolation for $\mathrm{CL}'(\mathfrak{F})$.*

Let us consider the case where $\Lambda \not\subset \partial\mathfrak{F}$.

Theorem 5.2. *Let Λ be a subset of $\mathfrak{F}'$, where $\mathfrak{F}$ is a closed subset of $\mathbb{C}$. Suppose that Λ is 1-interpolation for $\mathrm{CL}'(\mathfrak{F})$. Then for each connected component U of the interior of $\mathfrak{F}$ the intersection $U \cap \Lambda$ consists of at most one point. Moreover, if $U \cap \Lambda \neq \varnothing$ for some connected component U of the interior of $\mathfrak{F}$, then $\Lambda \cap \partial U = \varnothing$.*

Proof. The first statement can be proved in the same way as Remark 2.6. Let us prove the second statement. Let $\zeta \in U \cap \Lambda$. It suffices to prove that the set $\{\zeta, \xi\}$ is not 1-interpolation for $\mathrm{CL}'(\mathfrak{F})$ for any $\xi \in \partial U$. Suppose that $\{\zeta, \xi\}$ is 1-interpolation for $\mathrm{CL}'(\mathfrak{F})$ for some $\zeta \in \partial U$. Then for any positive integer n there exists a function $f_n \in \mathrm{CL}(\mathfrak{F})$ such that $f_n(\zeta) = 0$, $f'_n(\zeta) = 1$, $f'_n(\xi) = 0$ and $\|f_n\|_{\mathrm{CL}(\mathfrak{F})} < 1 + \frac{1}{n}$. Clearly, $f'_n \to 1$ uniformly on compact subsets of U. Hence, $f_n(z) \to z - \zeta$ pointwise on U and even on $\mathrm{clos}\,U$ because the functions f_n are equicontinuous on $\mathrm{clos}\,U$. We have

$$\|\{(\mathfrak{D}f_n)(\zeta,\xi)\}\|_{\mathfrak{M}} \leq \|f_n\|_{\mathrm{CL}(\mathfrak{F})} < 1 + \frac{1}{n}$$

for every integer $n \geq 1$ (see, for example, [3]). Letting $n \to \infty$ we get $\sqrt{2} = \left\|\left(\begin{smallmatrix} 1 & 1 \\ 1 & -1 \end{smallmatrix}\right)\right\|_{\mathfrak{M}} \leq 1$, a contradiction. □

Remark 5.3. Let Λ be a subset of $\mathfrak{F}'$, where $\mathfrak{F}$ is a closed subset of $\mathbb{C}$. Suppose that Λ is completely interpolation for $\mathrm{CL}'(\mathfrak{F})$. Then $\Lambda \subset \partial\mathfrak{F}$.

This remark can be proved in the same way as Remark 3.1.

Now we state an analog of Theorem 3.3 for $\mathrm{CL}'(\mathfrak{F})$.

Theorem 5.4. *Let $\mathfrak{F}$ be a closed subset of $\mathbb{C}$. Suppose that all boundary points in $\mathfrak{F}$ are $\mathscr{M}$-regular. Then for every subset Λ of $\partial\mathfrak{F}'$ the following statements are equivalent:*
  (i) *Λ is 1-interpolation for $\mathrm{CL}'(\mathfrak{F})$;*
 (ii) *Λ is completely 1-interpolation for $\mathrm{CL}'(\mathfrak{F})$;*
(iii) *Λ is discrete.*

This theorem can be proved in the same way as Theorem 3.3 but instead of Lemma 3.4 we need the following well-known statement:

$$\|\{(\mathfrak{D}f)(\zeta_j, \zeta_k)\}\|_{\mathfrak{M}} \leq \|f\|_{\mathrm{CL}(\mathfrak{F}')} \leq \|f\|_{\mathrm{CL}(\mathfrak{F})}$$

for every sequence $\{\zeta_j\}$ in $\mathfrak{F}'$ (see, for example, [3]).

Theorem 5.5. *Let Λ be a subset of $\partial\mathfrak{F}'$, where $\mathfrak{F}$ is a closed subset of $\mathbb{C}$. Suppose that each point in Λ is an $\mathscr{M}$-regular boundary point of $\mathfrak{F}$ and $\Lambda_{[n]} = \varnothing$ for some positive integer n. Then Λ is completely a_n-interpolation for $\mathrm{CL}'(\mathfrak{F})$ with $a_n = 2^n - 1$.*

This theorem is reduced to Theorem 4.2 with the help of statement (1.3).

6. Interpolation by functions in $\mathrm{CL}'(\mathbb{C}_+)$

Consider first the case where $\Lambda \subset \mathbb{R}$. Clearly, each point in $\mathbb{R}$ is an $\mathscr{M}$-regular boundary point of the half-plane $\{z \in \mathbb{C} : \operatorname{Im} z \geq 0\}$. Hence, Theorem 5.4 and Theorem 5.5 yield the following results.

Theorem 6.1. *Let $\Lambda \subset \mathbb{R}$. Then the following statements are equivalent:*
  (i) *Λ is 1-interpolation for $\mathrm{CL}'(\mathbb{C}_+)$;*
 (ii) *Λ is completely 1-interpolation for $\mathrm{CL}'(\mathbb{C}_+)$;*
(iii) *Λ is discrete.*

Theorem 6.2. *Let $\Lambda \subset \mathbb{R}$. Suppose that $\Lambda_{[n]} = \varnothing$, where n is a positive integer. Then Λ is completely a_n-interpolation for $\mathrm{CL}'(\mathbb{C}_+)$ with $a_n = 2^n - 1$.*

Note that $\mathrm{CL}'(\mathbb{C}_+) \subset \mathrm{OL}'(\mathbb{R}) = \mathrm{CL}'(\mathbb{R})$. Thus, certain analogs of Theorem 6.1 and Theorem 6.2 hold for the space $\mathrm{OL}'(\mathbb{R})$ of derivatives of operator Lipschitz functions on $\mathbb{R}$.

Let us note that Theorem 5.2 and Theorem 6.1 imply the following result.

Theorem 6.3. *Let $\Lambda \subset \mathbb{R} \cup \mathbb{C}_+$. Suppose that Λ is 1-interpolation for $\mathrm{CL}'(\mathbb{C}_+)$. Then either Λ is a discrete subset of $\mathbb{R}$ or Λ is a singleton in $\mathbb{C}_+$.*

Now we consider the case where $\Lambda \subset \mathbb{C}_+$. Clearly, $\mathrm{CL}'(\mathbb{C}_+) \subset H^\infty$, where H^∞ denotes the space of all bonded holomorphic functions on the upper half-plane $\mathbb{C}_+$.

Hence, every interpolation set for $\mathrm{CL}'(\mathbb{C}_+)$ is interpolation for H^∞. The famous Carleson theorem describes the interpolation sets for H^∞, see, for example, [5].

Let $\alpha > 0$ and $t \in \mathbb{R}$. Put $\Gamma_\alpha(t) \stackrel{\text{def}}{=} \{z \in \mathbb{C} : |\operatorname{Re} z - t| < \alpha \operatorname{Im} z\}$. Such sets are called Stolz domaines in $\mathbb{C}_+$.

The following result shows that there exist Carleson interpolation sets which are not interpolation for $\mathrm{CL}'(\mathbb{C}_+)$.

Theorem 6.4. *Let $\Lambda \subset \mathbb{C}_+$. Suppose that Λ is interpolation for $\mathrm{CL}'(\mathbb{C}_+)$. Then the intersection Λ with each Stolz domaine is finite.*

Proof. Suppose that $\Lambda_0 \stackrel{\text{def}}{=} \Lambda \cap \Gamma_\alpha(t)$ is infinite for some $\alpha > 0$ and $t \in \mathbb{R}$. Then Λ_0 has a limit point z_0 in the extended complex plane $\widehat{\mathbb{C}} \stackrel{\text{def}}{=} \mathbb{C} \cup \{\infty\}$. Clearly, $z_0 = t$ or $z_0 = \infty$. It is known (see §1) that each function $g \in \mathrm{CL}'(\mathbb{C}_+)$ has a finite non-tangential boundary value at any point $x \in \widehat{\mathbb{R}}$. To get a contradiction it suffices to note that there exists $\varphi \in \ell^\infty(\Lambda_0)$ such that the limit $\lim\limits_{\zeta \to z_0} \varphi(\zeta)$ does not exist. $\square$

The author does not know whether the space $\mathrm{CL}'(\mathbb{C}_+)$ is an algebra. To avoid this problem we need a sufficiently large subset X of $\mathrm{CL}'(\mathbb{C}_+)$ such that X is an algebra. Such a space X appears in the sufficient condition obtained by J. Arazy, T. Barton and Y. Friedman [4].

Let X be the set of functions f holomorphic in $\mathbb{C}_+$ such that

$$\|f\|_X \stackrel{\text{def}}{=} \sup_{t \in \mathbb{R}} \frac{2}{\pi} \int \frac{(\operatorname{Im} z)|f'(z)|\,dx\,dy}{|z - t|^2} < +\infty.$$

Every function $f \in X$ has finite non-tangential boundary values everywhere on $\widehat{\mathbb{R}}$ and $\|f - f(\infty)\|_{\mathrm{CL}'(\mathbb{C}_+)} \leq \|f\|_X$, see [4], [1] and [3]. Note that,

$$X \subset \widehat{\mathscr{M}'}(\operatorname{clos} \mathbb{C}_+) + \mathbb{C} \subset \mathrm{CL}'(\mathbb{C}_+) \subset H^\infty, \tag{6.1}$$

see [1]. It is easy to see that X is an algebra.

We say that a subset Λ of $\mathbb{C}_+$ satisfies the uniform Frostman condition if

$$\sup_{t \in \mathbb{R}} \sum_{\zeta \in \Lambda} \frac{\operatorname{Im} \zeta}{|\zeta - t|} < +\infty.$$

It is well known that such Λ is interpolation for H^∞ if and only if Λ is *separated*, i. e. $\inf\limits_{\zeta,\xi \in \Lambda} \dfrac{|\zeta - \xi|}{|\zeta - \bar\xi|} > 0$.

Theorem 6.5. *Let Λ be a subset of $\mathbb{C}_+$ satisfying the uniform Frostman condition. Then the following statements are equivalent:*

(i) *Λ is interpolation for X;*
(ii) *Λ is interpolation for $\mathrm{CL}'(\mathbb{C}_+)$;*
(iii) *Λ is interpolation for H^∞;*
(iv) *Λ is separated.*

94   A.B. Aleksandrov

Proof. Note that (i) $\implies$ (ii) $\implies$ (iii) in view of (6.1). As it was noted above (iii) $\iff$ (iv). It remains to prove that (iv) $\implies$ (i). By the famous Carleson theorem Λ is interpolation for H^∞ if and only if $(\mathrm{Im}\,\zeta)|B'(\zeta)| \geq \delta > 0$ for some δ and all $\zeta \in \Lambda$, where B is the Blaschke product in $\mathbb{C}_+$ with simple zeros $\{\zeta\}_{\zeta\in\Lambda}$. Let $\varphi \in \ell^\infty(\Lambda)$ with $\|\varphi\|_{\ell^\infty(\Lambda)} = 1$. Put

$$f(z) \overset{\text{def}}{=} \sum_{\zeta\in\Lambda} \frac{\varphi(\zeta)B(z)}{B'(\zeta)(z-\zeta)}.$$

It is easy to see that $f \in H^\infty$ and

$$\|f\|_{H^\infty} \leq \delta^{-1} \sup_{t\in\mathbb{R}} \sum_{\zeta\in\Lambda} \frac{\mathrm{Im}\,\zeta}{|\zeta - t|} = c(\Lambda).$$

Clearly, $f(\zeta) = \varphi(\zeta)$ for all $\zeta \in \Lambda$. It remains to prove that $f \in X$. Put $H^\infty \cap BH^\infty_- \overset{\text{def}}{=} \{h \in L^\infty(\mathbb{R}) : h \in H^\infty \text{ and } B\overline{h} \in H^\infty\}$. Note that $f \in H^\infty \cap BH^\infty_-$. It follows from Theorem 7.4 in [1] that $H^\infty \cap BH^\infty_- \subset \mathrm{CL}'(\mathbb{C}_+)$ but it is clear from the proof of Theorem 7.4 in [1] that $H^\infty \cap BH^\infty_- \subset X$. $\square$

Let us consider now the case where $\Lambda \subset \mathbb{C}_+ \cup \mathbb{R}$. We can define completely interpolation sets for X in the same way as we have done in the case of spaces $\mathscr{M}'(\mathfrak{F})$.

Theorem 6.6. *Let Λ be a discrete subset of $\mathbb{R}$. Then Λ is completely interpolation for X.*

This theorem can be proved in the same way as Theorem 2.3 and Theorem 3.2 but instead Lemma 2.4 we should use the following lemma.

Lemma 6.7. *For every closed subset $\mathfrak{F}_0$ of $\mathrm{clos}\,\mathbb{C}_+$, for every $t_0 \in \mathbb{R} \setminus \mathfrak{F}_0$, for every $\varepsilon > 0$ and every neighborhood U of t_0 there exists $f \in X$ such that $f(\infty) = 0$, $f(t_0) = 1$, $|f| < \varepsilon$ everywhere on $\mathfrak{F}_0$ and*

$$\frac{2}{\pi} \int \frac{(\mathrm{Im}\,z)|f'(z)|\,dxdy}{|z-t|^2} < \varepsilon + C\mathbb{1}_U(t)$$

for all $t \in \mathbb{R}$, where C is an absolute constant.

Proof. It suffices to consider the case where $t_0 = 0$. Put $f_\sigma(z) \overset{\text{def}}{=} -\frac{\sigma^2}{(z+\sigma i)^2}$, where $\sigma > 0$. It is easy to see that we can take as f the function f_σ for sufficiently small σ. $\square$

Remark 6.8. We can view Lemma 6.7 as an analog of Lemma 2.4. Almost all results of §2–§4 concerning interpolation by functions in $\mathscr{M}'(\mathfrak{F})$ (in the case of $\mathfrak{F} = \mathrm{clos}\,\mathbb{C}_+$) are based mainly on Lemma 2.4. Using Lemma 6.7 instead Lemma 2.4 we can obtain corresponding results concerning interpolation by functions in $X \subset \mathscr{M}'(\mathrm{clos}\,\mathbb{C}_+) + \mathbb{C}$. But it should be noted that we have some constant C in Lemma 6.7. This does not allow us to affirm, for example, that Λ is completely 1-interpolation in Theorem 6.6 and Λ is completely a_n-interpolation with $a_n = 2^n - 1$ in Corollary 6.9 below.

Corollary 6.9. *Let* $\Lambda \subset \mathbb{R}$. *Suppose that* $\Lambda_{[n]} = \varnothing$, *where* n *is a positive integer. Then* Λ *is completely interpolation for* X.

Theorem 6.10. *Let* $\Lambda \subset \mathbb{C}_+ \cup \mathbb{R}$. *Suppose that* $(\Lambda \cap \mathbb{R})_{[n]} = \varnothing$ *for some positive integer* n, $\Lambda \cap \mathbb{C}_+$ *is interpolation for* H^∞, *and* $\Lambda \cap \mathbb{C}_+$ *satisfies the uniform Frostman condition. Then* Λ *is interpolation for* X *and consequently for* $\mathrm{CL}'(\mathbb{C}_+)$.

Proof. Let B be the Blaschke product with simple zeros $\{\zeta\}_{\zeta \in \Lambda \cap \mathbb{C}_+}$. By Theorem 6.5 there exists $f \in X$ such that $f(\zeta) = \varphi(\zeta)$ for all $\zeta \in \Lambda \cap \mathbb{C}_+$. By Corollary 6.9 there exists $g \in X$ such that $g(\zeta) = \overline{B(\zeta)}\varphi(\zeta) - \overline{B(\zeta)}f(\zeta)$ for all $\zeta \in \Lambda \cap \mathbb{R}$. Clearly, $(f + Bg)|\Lambda = \varphi$. It remains to note that $f + Bg \in X$ because X is an algebra and $B \in H^\infty \cap BH^\infty_- \subset X$, see the end of the proof of Theorem 6.5. $\square$

Remark 6.11. All results of this section have natural analogs for the unit disk $\mathbb{D}$.

References

[1] A.B. Aleksandrov, *Operator Lipschitz functions and model spaces*, Zap. Nauchn. Sem. POMI, **416**, 2013, 5–58 (Russian); English transl., J. Math. Sci. (N.Y.), **202**:4 (2014), 485–518.

[2] A.B. Aleksandrov, *Commutator Lipschitz functions and analytic continuation*, Zap. Nauchn. Sem. POMI, **434**, 2015, 5–18 (Russian); English transl., J. Math. Sci. (N.Y.), **215**:5 (2016), 543–551.

[3] A.B. Aleksandrov and V.V. Peller, *Operator Lipschitz functions*, Uspekhi Mat. Nauk, **71**:4(430) (2016), 3–106 (Russian); English transl.: Russ Math Surv, **71**:4 (2016), 605–702.

[4] J. Arazy, T. Barton, and Y. Friedman, *Operator differentiable functions*, Int. Equat. Oper. Theory, **13**, 462–487 (1990).

[5] J.B. Garnett, *Bounded analytic functions*. Academic Press, New York, 1981.

[6] B.E. Johnson and J.P. Williams, *The range of a normal derivation*, Pacific J. Math. **58** (1975), 105–122.

[7] H. Kamowitz, *On operators whose spectrum lies on a circle or a line*, Pacific J. Math. **20** (1967), 65–68.

[8] E. Kissin and V.S. Shulman, *Classes of operator-smooth functions. I Operator-lipschitz functions*, Proc. Edinburgh Math. Soc. **48** (2005), 151–173.

[9] E. Kissin and V.S. Shulman, *On fully operator Lipschitz functions*. J. Funct. Anal. **253**:2 (2007), 711–728.

A.B. Aleksandrov
dom 32, korp. 2, kv. 47
Novoizmailovskii pr.
Saint Petersburg 196191, Russia
e-mail: `alex@pdmi.ras.ru`

Operator Theory:
Advances and Applications, Vol. 261, 97–120
© Springer International Publishing AG, part of Springer Nature 2018

Discrete Multichannel Scattering with Step-like Potential

Isaac Alvarez-Romero and Yurii Lyubarskii

Dedicated to the bright memory of Victor Petrovich Havin

Abstract. We study direct and inverse scattering problem for systems of interacting particles that have web-like structure. Such systems consist of a finite number of semi-infinite chains attached to the central part formed by a finite number of particles. We assume that the semi-infinite channels are homogeneous at infinity, but the limit values of the coefficients may vary from one chain to another.

Mathematics Subject Classification (2010). Primary 34L25. Secondary 81U40, 47B36, 05C50.

Keywords. Inverse scattering problem, Jacobi matrices, quantum graphs.

1. Introduction

The aim of this article is to study the direct and inverse problems for small oscillations near the equilibrium position for a system of particles $\mathcal{A} = \{\alpha, \beta, \dots\}$ which interact with each other and perhaps with an external field. The interaction is described by the matrix

$$\mathcal{L} = (L(\alpha, \beta))_{\alpha, \beta \in \mathcal{A}}.$$

We say that the particles α and β interact with each other if $L(\alpha, \beta) \neq 0$ and we assume that each particle interacts with at most a finite number of its neighbors: $\#\{\beta, L(\alpha, \beta) \neq 0\} < \infty$ for each $\alpha \in \mathcal{A}$.

Our system has a "web-like" structure: it includes a finite set of "channels", i.e., semi-infinite chains of particles attached to a "central part" formed by a finite number of interacting particles.

This research has been supported by the Norwegian Research Council project DIMMA 213638.

Given a set of particles $\mathcal{X}$ we denote by $\mathcal{M}(\mathcal{X})$ and $l^2(\mathcal{X})$ the spaces of all functions on $\mathcal{X}$ and square summable functions on $\mathcal{X}$ respectively. The matrix $\mathcal{L}$ is related to the Hessian matrix of the potential energy near the equilibrium position, so we always assume that all $L(\alpha, \beta)$ are real and

$$\mathcal{L} > 0,$$

here $\mathcal{L}$ is viewed as an operator in $l^2(\mathcal{A})$. In what follows we do not distinguish between matrices and the corresponding linear operators.

After separation of variables one arrives at the spectral problem

$$\mathcal{L}\xi = \lambda\xi; \ \xi = \{\xi(\alpha)\} \in l^2(\mathcal{A}). \tag{1.1}$$

which can be regarded as a discrete version of spectral problems for quantum graphs, see [3, 7, 9, 18] as well as later articles [10, 11, 12].

The web-like structure of the system allows us to treat the spectral problem (1.1) as a scattering problem: the points of continuous spectrum correspond to frequencies of incoming and outcoming waves which are propagating along the channels; the points of discrete spectrum correspond to proper oscillations of the system. The spectral data includes transmission and reflection coefficients: we consider waves incoming along one of the channels and observe (at infinite ends of the channels) how such waves come through the system. The direct problem is to determine the spectral data in terms of the characteristics of the system. The inverse problem deals with recovering the characteristics of the system by the spectral data. This is a classical setting of the scattering problem, see, e.g., [17].

This work is a continuation of [13] that considered the case of the same wave propagation speed along all channels. Now we assume that each channel has its own speed. The problem then becomes more complicated: different points of continuous spectrum may have different multiplicities; we do not have a single scattering matrix for the whole spectrum; next, the generalized eigenfunctions may decay exponentially along some of the channels. These eigenfunctions need to be treated specially, it does not make sense to observe the phase of a wave which decays exponentially at infinity, only their amplitudes should be taken into account. Respectively, for such waves, only absolute values of the transmission coefficients can be included to spectral data.

As in [13] the problem can be reduced to a system of difference Schrödinger equations on a semi-infinite discrete string with the initial conditions related to the way the channels are attached to the central part. In our setting the potentials have different limits for different equations in this system. In the classical case of an infinite string this corresponds to a step-like potential, such problem for the continuous case was studied for example in [2, 4, 8], for the Jacobi operators the case of step-like quasi-periodic potential was considered in [6]. We modify the techniques of these articles, especially those in [4], in order to make it applicable to the graph setting.

The scattering data are determined as the set of scattering coefficients, eigenvalues, and also as normalization constants of the corresponding eigenfunctions.

These scattering data are associated with the data which can be measured in an experiment at infinite ends of the channels.

We solve the direct problem, i.e., description of scattering data from physical characteristic of the system, and (under additional assumptions) the inverse problem, i.e., recovering the characteristics of the channels from given spectral data, that is, we reconstruct the matrix $\mathcal{L}$ on the channels. For this purpose, we reduce the problem to a discrete analog of the Marchenko equation, which is known to have a unique solution and actually admits numerical implementation.

In this article we do not discuss the possibility of recovering the data related to the central part of the graph. This would have required additional sparsity conditions of the central part. We refer the reader to the recent book [16], which contains examples of such conditions.

The article is organized as follows: in the next section we describe the system, derive the *boundary condition* which allows us to treat the problem as a system of equations on a string. In Section 3 we consider the characteristics of the channels, introduce the Jost solutions and also describe the spectral data related to the continuous spectrum. In Section 4 we collect some known results, see, e.g., [14, 15, 19], as well as some new ones about solutions of finite difference equations. These results will be used in the sequel. In particular in Section 5 we construct the special solutions, i.e., solutions which correspond to a wave incoming along one of the channels. Using these solutions we study the structure of discrete spectrum, this is done in Section 6.

As it has already been mentioned, no scattering matrix can be defined for the whole spectrum, however the scattering coefficients possess some symmetry which plays the crucial role in our construction. This symmetry is described in Section 7. In Section 8 we return to discrete spectrum and connect each eigenvalue of the problem with a normalized matrix of eigenfunctions which gives the energies, completing the set of spectral data.

In Section 9 we collect all previous results and finally obtain equations of the inverse scattering problem. Section 10 contains some concluding remarks.

Similar problems related to scattering on quantum graphs have been studied intensively during the recent decades. We refer the reader to [3], [7], [9]–[12] and references therein. The scattering with step-like potential on the real line was considered in [2] and [4] for the continuous case and in [6], [5], [20] for the Jacobi operators.

2. Geometry of the system and the boundary condition

We consider the systems $\mathcal{A}$ which have a "web-like" structure. Namely $\mathcal{A} = \mathcal{A}_0 \cup \mathcal{A}_1$ where $\mathcal{A}_1$ is a central part, and $\mathcal{A}_0$ is a union of a finite number of semi-infinite channels. For such web-like systems the inverse spectral problem can be treated as an inverse scattering problem: sending a wave along one of the channels and observing how it passes through the system, we are trying to reconstruct the characteristics of the system, i.e., the values $L(\alpha, \beta)$.

Definition. A sequence of particles $\sigma = \{\alpha(p)\}_{p=0}^{\infty}$ is a *channel* if, for $p > 0$, the particle $\alpha(p)$ interacts with the particles $\alpha(p-1)$ and $\alpha(p+1)$ only (and, perhaps, with the external field), while $\alpha(0)$ interacts with $\alpha(1)$ and some other particles in $\mathcal{A}$ which do not belong to σ.

We will use the following notation:

- the set of all channels is $\mathcal{C}$, the channels will be denoted by σ, ν, γ etc.;
- the particles in $\sigma \in \mathcal{C}$ are $\sigma(0)$, $\sigma(1)$, $\sigma(2)$, ..., the point $\sigma(0)$ is called the attachment point of σ;
- $\Gamma := \{\sigma(0)\}_{\sigma \in \mathcal{C}}$, $\mathcal{A}_0 := \bigcup_{\sigma \in \mathcal{C}} \bigcup_{k=1}^{\infty} \sigma(k)$, $\mathcal{A}_1 := \mathcal{A} \setminus \mathcal{A}_0$.
- For $\sigma \in \mathcal{C}$, $k = 1, 2, \ldots$ we also denote

$$-\mathfrak{b}_\sigma(k-1) = L(\sigma(k-1), \sigma(k)), \qquad \mathfrak{a}_\sigma(k) = L(\sigma(k), \sigma(k)), \qquad (2.1)$$

so equation (1.1) on the channel σ takes the form:

$$-\mathfrak{b}_\sigma(k-1)\xi(\sigma(k-1)) + \mathfrak{a}_\sigma(k)\xi(\sigma(k)) - \mathfrak{b}_\sigma(k)\xi(\sigma(k+1)) = \lambda\xi(\sigma(k)). \qquad (2.2)$$

We assume that the number of particles in the central part as well as the number of channels are finite:

$$M := \#\mathcal{A}_1 < \infty, \qquad \#\mathcal{C} < \infty.$$

Also (for simplicity) we assume $\sigma(0) \neq \nu(0)$, $\sigma, \nu \in \mathcal{C}$, $\sigma \neq \nu$.

2.1. Boundary conditions

Let $\xi \in \mathcal{M}(\mathcal{A})$ be a solution to (1.1). Then it meets (2.2) and also, for each $\alpha \in \mathcal{A}_1$,

$$\lambda\xi(\alpha) - \sum_{\beta \in \mathcal{A}_1} L(\alpha, \beta)\xi(\beta) = \sum_{\beta \in \mathcal{A}_0} L(\alpha, \beta)\xi(\beta).$$

The only pairs $(\alpha, \beta) \in \mathcal{A}_1 \times \mathcal{A}_0$ for which $L(\alpha, \beta) \neq 0$ are of the form $(\sigma(0), \sigma(1))$, $\sigma \in \mathcal{C}$, so with account of (2.1) this relation can be written as

$$\lambda\xi(\alpha) - \sum_{\beta \in \mathcal{A}_1} L(\alpha, \beta)\xi(\beta) = \begin{cases} -\mathfrak{b}_\nu(0)\xi(\nu(1)), & \alpha = \nu(0) \in \Gamma, \\ 0, & \alpha \in \mathcal{A}_1 \setminus \Gamma. \end{cases} \qquad (2.3)$$

Consider the matrix

$$\mathcal{L}_1 = (L(\alpha, \beta))_{\alpha, \beta \in \mathcal{A}_1}.$$

Being a truncation of $\mathcal{L}$, the matrix $\mathcal{L}_1$ also is strictly positive. Let $0 < \lambda_1 \leq \lambda_2 \leq \cdots \leq \lambda_M$ and $p_1, \ldots, p_M$ be its eigenvalues and the corresponding normalized eigenvectors. We may choose p_j's to be real-valued. For $\lambda \notin \{\lambda_j\}_{j=1}^{M}$ the operator $\mathcal{L}_1 - \lambda I$ is invertible and

$$(\mathcal{L}_1 - \lambda I)^{-1} = R(\lambda) = (r(\alpha, \beta; \lambda))_{\alpha, \beta \in \mathcal{A}_1}, \qquad r(\alpha, \beta; \lambda) = \sum_{j=1}^{M} \frac{p_j(\alpha)p_j(\beta)}{\lambda_j - \lambda}.$$

Relation (2.3) can be then written as

$$\xi(\alpha) = \sum_{\nu \in \mathcal{C}} r(\alpha, \nu(0); \lambda) \mathfrak{b}_\nu(0) \xi(\nu(1)); \ \ \alpha \in \mathcal{A}_1. \tag{2.4}$$

In particular for $\alpha = \sigma(0)$, $\sigma \in \mathcal{C}$ we obtain

$$\xi(\sigma(0)) = \sum_{\nu \in \mathcal{C}} r(\sigma(0), \nu(0); \lambda) \mathfrak{b}_\nu(0) \xi(\nu(1)). \tag{2.5}$$

After introducing the vector notation

$$\vec{\xi}(k) := (\xi(\sigma(k)))_{\sigma \in \mathcal{C}}, \ k = 0, 1, \ldots, \ \ \mathcal{R}(\lambda) = (r(\sigma(0), \nu(0); \lambda))_{\sigma, \nu \in \mathcal{C}},$$

we can rewrite (2.5) as

$$\vec{\xi}(0) = \mathcal{R}(\lambda) \ \mathrm{diag}\{\mathfrak{b}_\nu(0)\}_{\nu \in \mathcal{C}} \vec{\xi}(1). \tag{2.6}$$

This relation links the values of a solution ξ on $\mathcal{A}_0$ and on Γ. It plays the role of the boundary condition for vector-valued scattering problem.

The following statement holds.

Theorem 2.1 (See Theorem 1 in [13]). *Let $\lambda \notin \{\lambda_j\}_{j=1}^M$. A function $\xi \in \mathcal{M}(\mathcal{A}_0 \cup \Gamma)$ admits extension to a function $\xi \in \mathcal{M}(\mathcal{A})$ satisfying (1.1) if and only if it satisfies (2.2) and also the boundary condition (2.5). The extension is unique and is defined by (2.4).*

3. Characteristics of the channels and spectral data

3.1. Jost solutions

We assume that the channels are asymptotically homogeneous at infinity. Namely, for each $\sigma \in \mathcal{C}$ there exist b_σ and a_σ such that

$$\mathfrak{a}_\sigma(k) \to a_\sigma, \ \ \mathfrak{b}_\sigma(k) \to b_\sigma \text{ as } k \to \infty.$$

Moreover

$$\sum_{k=1}^\infty k\{|\mathfrak{a}_\sigma(k) - a_\sigma| + |\mathfrak{b}_\sigma(k) - b_\sigma|\} < \infty. \tag{3.1}$$

This relation provides the existence of Jost solutions on each channel $\sigma \in \mathcal{C}$. It is well known, see, e.g., [16, 19], that for each $\sigma \in \mathcal{C}$ there is a family $\{e_\sigma(k, \theta)\}_{k=0}^\infty$ of functions holomorphic inside the open disk $\mathbb{D}$, continuous up to the boundary and, for each $\theta \in \mathbb{D}$ and $k \geq 1$,

$$-\mathfrak{b}_\sigma(k-1)e_\sigma(k-1, \theta) + \mathfrak{a}_\sigma(k)e_\sigma(k, \theta) - \mathfrak{b}_\sigma(k)e_\sigma(k+1, \theta) = \lambda_\sigma(\theta)e_\sigma(k, \theta), \tag{3.2}$$

where

$$\lambda_\sigma(\theta) = a_\sigma - b_\sigma\left(\theta + \theta^{-1}\right). \tag{3.3}$$

In addition

$$e_\sigma(k, \theta) = \theta^k(1 + o(1)) \text{ as } k \to \infty$$

uniformly with respect to $\theta \in \overline{\mathbb{D}}$.

The functions $e_\sigma(k, \theta)$ admit the representation

$$e_\sigma(k, \theta) = c_\sigma(k) \sum_{m \geq k} a_\sigma(k, m)\theta^m, \quad k = 0, 1, \ldots. \tag{3.4}$$

Here

$$c_\sigma(k) = \prod_{p=k}^{\infty} \frac{b_\sigma}{b_\sigma(p)}, \quad a_\sigma(k, k) = 1 \text{ and } \lim_{k \to \infty} \sum_{m \geq k+1} |a_\sigma(k, m)| = 0.$$

For $k \geq 1$ the coefficients $\mathfrak{a}_\sigma(k), \mathfrak{b}_\sigma(k)$ can be expressed in terms of the coefficients of the functions $e_\sigma(k, \theta)$

$$\frac{\mathfrak{a}_\sigma(k) - a_\sigma}{b_\sigma} = a_\sigma(k-1, k) - a_\sigma(k, k+1);$$

$$\frac{\mathfrak{b}_\sigma^2(k)}{b_\sigma^2} = \frac{\mathfrak{a}_\sigma(k) - a_\sigma}{b_\sigma} a_\sigma(k, k+1) + a_\sigma(k, k+2) - a_\sigma(k-1, k+1) + 1.$$

These relations can be obtained by substituting the representation (3.4) in (3.2) and then comparing the coefficients with the same powers of θ.

In this article we assume that the coefficients $\mathfrak{a}_\sigma(k), \mathfrak{b}_\sigma(k)$ approach their limit values faster than (3.1). Namely, we assume that, for some $\epsilon > 0$,

$$\sum_{k=1}^{\infty} (1 + \epsilon)^k \{|\mathfrak{a}_\sigma(k) - a_\sigma| + |\mathfrak{b}_\sigma(k) - b_\sigma|\} < \infty.$$

Under this condition the functions $e_\sigma(k, \theta)$ admit holomorphic continuation to some neighborhood of the unit disk $\mathbb{D}$, see, e.g., Chapter 10 in [19]. This allows us to avoid additional technicalities which are necessary in the general case.

3.2. Spectrum, scattering data corresponding to the continuous spectrum

As the parameter θ runs through the unit circle $\mathbb{T}$, the corresponding value $\lambda_\sigma(\theta)$ defined by (3.3) runs through the segment

$$I_\sigma = [a_\sigma - 2b_\sigma, a_\sigma + 2b_\sigma].$$

We assume for simplicity that $\bigcup_{\sigma \in C}(a_\sigma - 2b_\sigma, a_\sigma + 2b_\sigma)$ is a connected set and choose $a, b \in \mathbb{R}$ so that

$$I = \bigcup_{\sigma \in C} I_\sigma = [a - 2b, a + 2b].$$

We need the Zhukovskii mappings $W, W_\sigma : \mathbb{D} \to \mathbb{C}$:

$$W : \omega \mapsto a - b(\omega + \omega^{-1}), \quad W_\sigma : \theta \mapsto a_\sigma - b_\sigma(\theta + \theta^{-1}), \tag{3.5}$$

$$\theta_\sigma(\omega) = W_\sigma^{-1} \circ W(\omega). \tag{3.6}$$

Further we choose the set $J_\sigma \subset [-1, 1]$, so that θ_σ maps $\mathbb{D}$ onto $\mathbb{D}_\sigma := \mathbb{D} \setminus J_\sigma$ and let

$$T_\sigma^+ = \theta_\sigma^{-1}(\mathbb{T}), \quad T_\sigma^- = \theta_\sigma^{-1}(J_\sigma) = \mathbb{T} \setminus T_\sigma^+.$$

The set J_σ consists of one or two segments attached to the points ± 1.

The functions $e_\sigma(k, \theta_\sigma(\omega))$ are holomorphic in $\mathbb{D}$. Moreover, for $\omega \in T_\sigma^+$, the functions $e_\sigma(k, \theta_\sigma(\omega)^{-1})$ are well defined.

By the special solution which corresponds to $\sigma \in \mathcal{C}$ we mean the function $\psi^\sigma(\alpha, \omega)$, $\omega \in \mathbb{D}$, $\alpha \in \mathcal{A}$ with the following properties.

1. For each $\alpha \in \mathcal{A}_0$ the function $\psi^\sigma(\alpha, \cdot)$ is holomorphic in a neighborhood of $\overline{\mathbb{D}}$, except, perhaps a finite set $\mathcal{O} \subset \mathbb{D}$ where it has poles.

2. For each $\omega \in \overline{\mathbb{D}} \setminus \mathcal{O}$, the function $\psi^\sigma(\cdot, \omega) \in \mathcal{M}(\mathcal{A})$ meets the equation

$$\mathcal{L}\psi^\sigma(\cdot, \omega) = \lambda \psi^\sigma(\cdot, \omega), \quad \lambda = a + b(\omega + \omega^{-1}).$$

3. For $\gamma \neq \sigma$ there exist functions $s_{\sigma\gamma}(\omega)$ meromorphic in a neighborhood of $\overline{\mathbb{D}}$ and a function $s_{\sigma\sigma}(\omega)$ continuous on T_σ^+ such that the function ψ^σ has the following representation along the channels:

$$\psi^\sigma(\gamma(k), \omega) = \begin{cases} e_\sigma(k, \theta_\sigma(\omega)^{-1}) + s_{\sigma\sigma}(\omega)e_\sigma(k, \theta_\sigma(\omega)), & \gamma = \sigma, \\ s_{\gamma\sigma}(\omega)e_\gamma(k, \theta_\gamma(\omega)), & \gamma \neq \sigma; \end{cases} \quad \omega \in T_\sigma^+, \quad (3.7)$$

and for $\omega \in \mathbb{D}$

$$\psi^\sigma(\gamma(k), \omega) = \begin{cases} \theta_\sigma(\omega)^{-k}(1 + o(1)), \text{ as } k \to \infty, & \gamma = \sigma, \\ s_{\gamma\sigma}(\omega)e_\gamma(k, \theta_\gamma(\omega)), & \gamma \neq \sigma; \end{cases} \quad \omega \in \overline{\mathbb{D}} \setminus T_\sigma^+. \quad (3.8)$$

The *scattering coefficients* $s_{\sigma\gamma}(\omega)$ are the elements of the scattering matrix corresponding to the problem (1.1). In Section 5 we will prove the existence and uniqueness of such special solutions for each $\sigma \in \mathcal{C}$.

The solution ψ^σ corresponds to the wave incoming along the channel σ and distributing through the whole system. This wave is well defined if $\omega \in T_\sigma^+$ (respectively $\lambda \in I_\sigma$), for these values equation (2.2) along the channel σ admits two independent bounded solutions corresponding to in- and outcoming waves. For $\omega \in T_\sigma^+ \cap T_\gamma^+$ the wave incoming along σ generates an outcoming wave along the channel γ, this wave can be observed at infinity. For $\omega \in T_\sigma^+ \cap T_\gamma^-$ the wave incoming along σ generates an exponentially decaying wave γ, observation of the phase of such a wave at infinity is virtually impossible, one can measure the absolute values only. These arguments explain the definition of the spectral data.

Definition 3.1. By continuous spectral data corresponding to the channel $\sigma \in \mathcal{C}$ we mean the set of functions

$$\mathcal{S}_\sigma = \{s_{\gamma\sigma}(\omega) : \omega \in T_\sigma^+ \cap T_\gamma^+; \quad \gamma \in \mathcal{C}\} \cup \{|s_{\gamma\sigma}(\omega)| : \omega \in T_\sigma^+ \cap T_\gamma^-; \gamma \in \mathcal{C}\}.$$

By full continuous spectral data we mean

$$\mathcal{S} = \bigcup_{\sigma \in \mathcal{C}} \mathcal{S}_\sigma.$$

Later in Sections 6 and 8 we will discuss the discrete spectrum, which corresponds to the eigenfunctions of the operator $\mathcal{L} : l^2(\mathcal{A}) \to l^2(\mathcal{A})$.

4. Properties of solutions of finite difference equations

We need to establish some properties of solutions to (1.1) as well as solutions to difference equations along the channels. In this section we collect known (see, e.g., [14, 15, 19]) properties of solutions of the finite-difference equation as well as some new statements related to solutions of (1.1)

$$-\mathfrak{b}(k-1)x(k-1) + \mathfrak{a}(k)x(k) - \mathfrak{b}(k)x(k+1) = \lambda x(k), \quad k = 1, 2, \ldots \quad (4.1)$$

with real coefficients $\mathfrak{a}(k)$, $\mathfrak{b}(k)$. We also assume that $\mathfrak{b}(k) > 0$ in order that the Jost solutions be well defined, see, e.g., [19].

Given two functions $x = x(k)$, $y = y(k)$ on the integers we define their *Wronskian* $\{x, y\}$ as

$$\{x, y\}(k) = x(k)y(k+1) - x(k+1)y(k), \quad k = 0, 1, \ldots .$$

1. Let x, y be solutions to (4.1). Then, for all N,

$$\mathfrak{b}(N)\{x, y\}(N) - \mathfrak{b}(0)\{x, y\}(0) = 0 \quad (4.2)$$

and

$$\mathfrak{b}(N)\{x, \overline{x}\}(N) - \mathfrak{b}(0)\{x, \overline{x}\}(0) = (\lambda - \overline{\lambda}) \sum_{k=1}^{N} |x(k)|^2. \quad (4.3)$$

If in addition $\lambda = \lambda(\omega)$ and $x = x(k, \omega)$ are differentiable functions of a parameter ω, then

$$\mathfrak{b}(N)\{\dot{x}, \overline{x}\}(N) - \mathfrak{b}(0)\{\dot{x}, \overline{x}\}(0) = \dot{\lambda} \sum_{k=1}^{N} |x(k)|^2 + (\lambda - \overline{\lambda}) \sum_{k=1}^{N} \dot{x}(k)\overline{x(k)}, \quad (4.4)$$

here and in what follows the dot denotes the derivative with respect to ω.

Let

$$\sum_{k=1}^{\infty} k(|\mathfrak{b}(k) - b| + |\mathfrak{a}(k) - a|) < \infty,$$

and let $e(\theta) = e(k, \theta)$ be the corresponding Jost solutions. Then relation (4.3) yields

$$\mathfrak{b}(0)\{e(\theta), \overline{e}(\theta)\}(0) = \begin{cases} b(\overline{\theta} - \theta), & |\theta| = 1; \\ b(\overline{\theta} - \theta)(|\theta|^{-2} - 1) \sum_{k=1}^{\infty} |e(k, \theta)|^2, & |\theta| < 1. \end{cases} \quad (4.5)$$

2. We will consider Wronskians, which correspond to various channels $\sigma \in \mathcal{C}$. Given two functions $\xi, \eta \in \mathcal{M}(\mathcal{A})$ and $\sigma \in \mathcal{C}$ we denote

$$\{\xi, \eta\}_\sigma(k) := \{\xi(\sigma(k))\eta(\sigma(k+1)) - \xi(\sigma(k+1))\eta(\sigma(k))\}$$

Remark. If both ξ and η are solutions of (1.1), from (4.2) it follows that the quantity $\mathfrak{b}_\sigma(k)\{\xi, \eta\}_\sigma(k)$ depends on σ only.

Lemma 4.1. *Let* $\omega \in \mathbb{T} \cup (-1,1)$ *and* $\sigma \in \mathcal{C}$ *be such that* $\theta_\sigma(\omega) \in (-1,1)$. *Then*

$$-\mathfrak{b}_\sigma(0)\{\dot{e}_\sigma,(\theta_\sigma(\omega))e_\sigma(\theta_\sigma(\omega))\}_\sigma(0) = \dot{\lambda}\sum_{k=1}^{\infty} e_\sigma(k,\theta_\sigma(\omega))^2. \qquad (4.6)$$

This is an immediate consequence of (4.4).

Lemma 4.2. *Let* $\xi,\eta \in \mathcal{M}(\mathcal{A})$ *be solutions to* (1.1). *Then*

$$\sum_{\sigma \in \mathcal{C}} \mathfrak{b}_\sigma(0)\{\xi,\eta\}_\sigma(0) = 0 \qquad (4.7)$$

Proof. For $\alpha \in \mathcal{A}_1$ we have

$$\sum_{\beta \in \mathcal{A}_1} L(\alpha,\beta)\xi(\beta) + \sum_{\beta \in \mathcal{A}_0} L(\alpha,\beta)\xi(\beta) = \lambda\xi(\alpha),$$

$$\sum_{\beta \in \mathcal{A}_1} L(\alpha,\beta)\eta(\beta) + \sum_{\beta \in \mathcal{A}_0} L(\alpha,\beta)\eta(\beta) = \lambda\eta(\alpha).$$

We multiply these relations by $\eta(\alpha)$ and $\xi(\alpha)$ respectively. Summation with respect to $\alpha \in \mathcal{A}_1$ gives

$$\sum_{\alpha,\beta \in \mathcal{A}_1} L(\alpha,\beta)\xi(\beta)\eta(\alpha) + \sum_{\alpha \in \mathcal{A}_1, \beta \in \mathcal{A}_0} L(\alpha,\beta)\xi(\beta)\eta(\alpha) = \lambda \sum_{\alpha \in \mathcal{A}_1} \xi(\alpha)\eta(\alpha),$$

$$\sum_{\alpha,\beta \in \mathcal{A}_1} L(\alpha,\beta)\eta(\beta)\xi(\alpha) + \sum_{\alpha \in \mathcal{A}_1, \beta \in \mathcal{A}_0} L(\alpha,\beta)\xi(\alpha)\eta(\beta) = \lambda \sum_{\alpha \in \mathcal{A}_1} \xi(\alpha)\eta(\alpha).$$

The right-hand sides in these relations coincide as well as the first terms on the left-hand sides (since $L(\alpha,\beta) = L(\beta,\alpha)$). Therefore

$$\sum_{\alpha \in \mathcal{A}_1, \beta \in \mathcal{A}_0} L(\alpha,\beta)\left(\xi(\beta)\eta(\alpha) - \xi(\alpha)\eta(\beta)\right) = 0.$$

This proves the lemma since the only option for $L(\alpha,\beta) \neq 0$ for $\alpha \in \mathcal{A}_1$, $\beta \in \mathcal{A}_0$ is $\alpha = \sigma(0)$, $\beta = \sigma(1)$ for some $\sigma \in \mathcal{C}$ and in this case $L(\sigma(1),\sigma(0)) = -\mathfrak{b}_\sigma(0)$. $\quad\square$

3. We need a special statement in order to calculate the energy of eigenfunctions of the operator $\mathcal{L}$.

Lemma 4.3. *Let a function* $\xi(\alpha) = \xi(\alpha,\omega) \in \mathcal{M}(\mathcal{A})$ *be differentiable with respect to* ω *in a neighborhood of* $\hat{\omega} \in \overline{\mathbb{D}}$, *let* $\operatorname{Im} \lambda(\hat{\omega}) = 0$, *and let* $\xi(\alpha)$ *satisfy the equation*

$$\lambda(\omega)\xi(\alpha,\omega) = \sum_{\beta \in \mathcal{A}} L(\alpha,\beta)\xi(\beta,\omega). \qquad (4.8)$$

Let also a function $\eta(\alpha)$ *meet this equation for* $\omega = \hat{\omega}$. *Then*

$$\dot{\lambda}(\hat{\omega}) \sum_{\alpha \in \mathcal{A}_1} \xi(\alpha,\hat{\omega})\overline{\eta(\alpha)} = \sum_{\sigma \in \mathcal{C}} \mathfrak{b}_\sigma(0)\{\dot{\xi}(\hat{\omega}),\overline{\eta}\}_\sigma(0).$$

In particular

$$\dot{\lambda}(\hat{\omega}) \sum_{\alpha \in \mathcal{A}_1} |\xi(\alpha,\hat{\omega})|^2 = \sum_{\sigma \in \mathcal{C}} \mathfrak{b}_\sigma(0)\{\dot{\xi}(\hat{\omega}),\overline{\xi(\hat{\omega})}\}_\sigma(0).$$

Proof. The statement is the same as in Lemma 2.1 in [13], we repeat here the construction.

Differentiate (4.8) with respect to ω:

$$\dot{\lambda}(\omega)\xi(\alpha,\omega) + \lambda(\omega)\dot{\xi}(\alpha,\omega) = \sum_{\beta\in\mathcal{A}} L(\alpha,\beta)\dot{\xi}(\beta,\omega).$$

Furthermore,

$$\lambda(\hat{\omega})\overline{\eta(\alpha)} = \overline{\lambda(\hat{\omega})\eta(\alpha)} = \sum_{\beta\in\mathcal{A}} L(\alpha,\beta)\overline{\eta(\beta)}.$$

Combining these equations, we obtain

$$\dot{\lambda}(\hat{\omega})\xi(\alpha,\hat{\omega})\overline{\eta(\alpha)} = \sum_{\beta\in\mathcal{A}} L(\alpha,\beta)[\dot{\xi}(\beta,\hat{\omega})\overline{\eta(\alpha)} - \dot{\xi}(\alpha,\hat{\omega})\overline{\eta(\beta)}],$$

and

$$\dot{\lambda}(\hat{\omega})\sum_{\alpha\in\mathcal{A}_1}\xi(\alpha,\hat{\omega})\overline{\eta(\alpha)} = \sum_{\alpha\in\mathcal{A}_1}\sum_{\beta\in\mathcal{A}_1} L(\alpha,\beta)[\dot{\xi}(\beta,\hat{\omega})\overline{\eta(\alpha)} - \dot{\xi}(\alpha,\hat{\omega})\overline{\eta(\beta)}]$$
$$+ \sum_{\alpha\in\mathcal{A}_1}\sum_{\beta\in\mathcal{A}_0} L(\alpha,\beta)[\dot{\xi}(\beta,\hat{\omega})\overline{\eta(\alpha)} - \dot{\xi}(\alpha,\hat{\omega})\overline{\eta(\beta)}].$$

The first summand on the right-hand side of this relation vanishes because it is anti-symmetric in α and β. In the second summand the only non-zero coefficient $L(\alpha,\beta)$ appears in the case $\alpha = \sigma(0)$, $\beta = \sigma(1)$ for some $\sigma \in \mathcal{C}$ and $L(\sigma(0),\sigma(1)) = -\mathfrak{b}_\sigma(0)$. $\qquad\square$

5. Construction of the special solutions

Consider the diagonal matrices

$$B := \text{diag}\{b_\sigma\}_{\sigma\in\mathcal{C}}, \quad B(0) := \text{diag}\{\mathfrak{b}_\sigma(0)\}_{\sigma\in\mathcal{C}},$$

and also the matrix-functions in $\bar{\mathbb{D}}$

$$\mathcal{E}(k,\omega) := \text{diag}\{e_\sigma(k,\theta_\sigma(\omega))\}_{\sigma\in\mathcal{C}}, \quad P(k,\omega) := \text{diag}\{p_\sigma(k,\omega)\}_{\sigma\in\mathcal{C}},$$

here the functions $\{p_\sigma(k,\omega)\}_{k=0}^\infty$ satisfy the following equation on the channels:

$$-\mathfrak{b}_\sigma(k-1)p_\sigma(k-1,\omega) + \mathfrak{a}_\sigma(k)p_\sigma(k,\omega) - \mathfrak{b}_\sigma(k)p_\sigma(k+1,\omega) = \lambda(\omega)p_\sigma(k,\omega)$$

for $k = 1, 2, \ldots, \sigma \in \mathcal{C}$, the boundary conditions being

$$p_\sigma(0,\omega) = 1, \quad p_\sigma(1,\omega) = 0, \quad \sigma \in \mathcal{C}. \qquad (5.1)$$

The functions $\mathcal{E}(k,\omega)$ and $P(k,\omega)$ are holomorphic in a neighborhood of $\bar{\mathbb{D}}$, except, perhaps zero, where $P(k,\omega)$ may have poles.

Moreover, it is well known, see [16, 19], that the Jost solutions form a fundamental system of solutions of the finite-difference equation (4.1). If $|\theta| = 1$ and $\theta \neq \pm1$, it follows from (4.5) that $e(\theta)$, $\bar{e}(\theta) = e(\theta^{-1})$ are independent solutions of (4.1) and any other solution $\{x(k,\theta)\}_{k\geq0}$ can be expressed as $x(k,\theta) =$

$m(\theta)e(k,\theta)+n(\theta)e(k,\theta^{-1})$, with $m(\theta), n(\theta)$ independent of k. Thus, for $\omega \in T_\sigma^+$ the function $p_\sigma(k,\omega)$ may be expressed in terms of the corresponding Jost solutions:

$$p_\sigma(k,\omega) = \frac{\mathsf{b}_\sigma(0)}{\mathsf{b}_\sigma} \frac{e_\sigma(k,\theta_\sigma(\omega))e_\sigma(1,\theta_\sigma(\omega)^{-1}) - e_\sigma(k,\theta_\sigma(\omega)^{-1})e_\sigma(1,\theta_\sigma(\omega))}{\theta_\sigma(\omega)^{-1} - \theta_\sigma(\omega)}. \quad (5.2)$$

In order to construct the special solutions we need an auxiliary operator-function $T(\omega) : l^2(\mathcal{C}) \to l^2(\mathcal{C})$ defined as

$$T(\omega) = \mathcal{E}(0,\omega) - \mathcal{R}(\lambda(\omega))B(0)\mathcal{E}(1,\omega), \quad \omega \in \mathbb{D}. \quad (5.3)$$

This function is holomorphic in a neighborhood of $\overline{\mathbb{D}}$, except the set $\mathcal{O}$ of poles of the function $\mathcal{R}(\lambda(\omega))$.

Denote

$$\Delta_\sigma(\omega) = \begin{cases} \mathsf{b}_\sigma, & |\theta_\sigma(\omega)| = 1; \\ \mathsf{b}_\sigma(|\theta_\sigma(\omega)|^{-2} - 1)\sum_{k=1}^\infty |e_\sigma(k,\theta_\sigma(\omega))|^2, & |\theta_\sigma(\omega)| < 1, \end{cases}$$

and

$$\Delta(\omega) = \mathrm{diag}\{\Delta_\sigma(\omega)\}_{\sigma \in \mathcal{C}}, \quad \Phi(\omega) = \mathrm{diag}\{\bar{\theta}_\sigma(\omega) - \theta_\sigma(\omega)\}_{\sigma \in \mathcal{C}}. \quad (5.4)$$

The operator $\Delta(\omega)$ is positive uniformly with respect $\omega \in \mathbb{D}$, i.e., for some $C > 0$,

$$\langle \Delta(\omega)\mathbf{x}, \mathbf{x} \rangle \geq C\|\mathbf{x}\|^2, \quad \mathbf{x} \in l^2(\mathcal{C}), \quad \omega \in \mathbb{D}.$$

Next, relation (4.5) now reads

$$B(0)\{\mathcal{E}(0,\omega)\mathcal{E}(1,\omega)^* - \mathcal{E}(0,\omega)^*\mathcal{E}(1,\omega)\} = \Phi(\omega)\Delta(\omega), \quad (5.5)$$

Lemma 5.1 (See Lemma 3.1 in [13]). *The following inequality holds for all $\omega \in \mathbb{D} \setminus \mathcal{O}$, $\mathbf{x} = (x_\sigma)_{\sigma \in \mathcal{C}} \in l^2(\mathcal{C})$:*

$$|\langle \mathcal{E}(1,\omega)^* B(0)T(\omega)\mathbf{x}, \mathbf{x} \rangle| \geq |\mathrm{Im}\langle \mathcal{E}(1,\omega)^* B(0)T(\omega)\mathbf{x}, \mathbf{x} \rangle|$$

$$\geq C \sum_{\sigma \in \mathcal{C}} |\bar{\theta}_\sigma(\omega) - \theta_\sigma(\omega)||x_\sigma|^2. \quad (5.6)$$

Proof. We have

$$2\,\mathrm{Im}\langle \mathcal{E}(1,\omega)^* B(0)T(\omega)\mathbf{x}, \mathbf{x} \rangle = \langle [\mathcal{E}(1,\omega)^* B(0)T(\omega) - T(\omega)^* B(0)\mathcal{E}(1,\omega)]\mathbf{x}, \mathbf{x} \rangle$$

and, by (5.3),

$$\mathcal{E}(1,\omega)^* B(0)T(\omega) - T(\omega)^* B(0)\mathcal{E}(1,\omega)$$
$$= B(0)(\mathcal{E}(0,\omega)\mathcal{E}(1,\omega)^* - \mathcal{E}(0,\omega)^*\mathcal{E}(1,\omega))$$
$$+ (B(0)\mathcal{E}(1,\omega))^*(\mathcal{R}(\lambda(\omega))^* - \mathcal{R}(\lambda(\omega)))(B(0)\mathcal{E}(1,\omega))$$

The lemma now follows from (5.4) – (5.5) and from the fact that

$$\mathcal{R}(\lambda(\omega))^* - \mathcal{R}(\lambda(\omega)) = b(\bar{\omega} - \omega)(|\omega|^{-2} - 1)\underbrace{\left(\sum_{l=1}^M \frac{p_l(\sigma(0))p_l(\nu(0))}{|\lambda_l - \lambda(\omega)|^2}\right)_{\sigma,\nu \in \mathcal{C}}}_{\Delta_1(\omega)},$$

here $\Delta_1(\theta)$ is a non-negative operator. $\qquad \square$

Corollary 5.1. *The operators $T(\omega)$ are invertible for all non-real ω in the open disk $\mathbb{D} \setminus \{\mathcal{O} \cup \{\pm 1\}\}$.*

Proof. Indeed, fix an $\omega \in \mathbb{D} \setminus [-1, 1]$ and denote $\delta(\omega) = \inf_\sigma \{|\operatorname{Im} \theta_\sigma(\omega)|\} > 0$. Relation (5.6) now reads

$$\|T(\omega)\mathbf{x}\| > C\delta(\omega)\|\mathbf{x}\|,$$

which yields the invertibility of $T(\omega)$.   □

Since $T(\omega)$ is an analytic function, we can now claim that $T(\omega)$ is invertible in the vicinity of $\overline{\mathbb{D}}$, except perhaps at a finite set of points.

Consider the analytic matrix functions:

$$D(\omega) := \operatorname{diag}\{\theta_\sigma(\omega)^{-1} - \theta_\sigma(\omega)\}$$

and

$$U(k, \omega) = \left[-P(k, \omega) + \mathcal{E}(k, \omega)T(\omega)^{-1}\right] BB(0)^{-1}D(\omega)\mathcal{E}(1, \omega)^{-1}. \tag{5.7}$$

Lemma 5.2. *Consider the vector function on $\mathcal{A}_0 \cup \Gamma$*

$$\psi^\sigma(k, \omega) = (\psi^\sigma(\gamma(k), \omega))_{\gamma \in \mathcal{C}} := U(k, \omega)\mathbf{n}_\sigma, \tag{5.8}$$

where $\mathbf{n}_\sigma = (\delta_{\sigma\gamma})_{\gamma \in \mathcal{C}} \in l^1(\mathcal{C})$. This function satisfies the boundary condition (2.6) and, for $\omega \in T_\sigma^+$, admits a representation of the form (3.7) along the channels. Thus by (2.4) it may be extended to a special solution of (1.1).

Proof. The representation (3.7) for $\omega \in T_\sigma^+$ is simply a consequence of (5.7) and (5.2).

In order to prove that $\psi^\sigma(k, \omega)$ meets the boundary condition for $\omega \in T_\sigma^+$, we prove that this condition is met by the whole matrix-function $U(k, \omega)$:

$$U(0, \omega) = \mathcal{R}(\lambda(\omega))B(0)U(1, \omega), \quad \omega \in \mathbb{D} : \tag{5.9}$$

This implies that $\psi^\sigma(k, \omega)$ also meets the boundary condition. Relation (5.9) is immediate: after factoring out the inessential factor $BB(0)^{-1}D(\omega)\mathcal{E}(1, \omega)^{-1}$ in the definition (5.7) it becomes

$$-I + \mathcal{E}(0, \omega)T(\omega)^{-1} = \mathcal{R}(\lambda(\omega))B(0)\mathcal{E}(1, \omega)T(\omega)^{-1},$$

which is precisely the definition of $T(\omega)$.   □

So far the special solution is not defined at a point $\omega_0 \in T_\sigma^+$ in case ω_0 is a pole of $U(k, \omega)$. This case will be considered in the following sections: in Section 6 we prove that all poles of $U(k, \omega)$ are simple, later in Section 8 we show that, if $\omega_0 \in T_\sigma^+$, the function $U(k, \omega)\mathbf{n}_\sigma$ is continuos at ω_0 even if ω_0 is a pole of $U(k, \omega)$. In particular

$$\operatorname{Res}_{\omega_0} u_{\nu\sigma}(\omega) = 0, \quad \nu \in \mathcal{C}. \tag{5.10}$$

6. Singularities of $U(k, \omega)$

The matrix function $U(k, \omega)$ is analytic in the vicinity of $\bar{\mathbb{D}}$, in particular it has a finite number of poles which belong to $\bar{\mathbb{D}}$.

Lemma 6.1. *There is a finite set $\Omega \subset \bar{\mathbb{D}} \setminus \{0\}$ such that all poles of the matrix function $U(k, \omega) \in \bar{\mathbb{D}}$ belong to $\Omega \cup \{0\}$. In the origin $U(k, \omega)$ has pole of order k, all poles in Ω are simple.*

Proof. The proof follows the pattern of Lemma 4.1 in [13]. We rewrite (5.7) as

$$U(k, \omega) = - P(k, \omega)BB(0)^{-1}D(\omega)\mathcal{E}(1, \omega)^{-1}$$
$$+ \mathcal{E}(k, \omega)T(\omega)^{-1}BB(0)^{-1}D(\omega)\mathcal{E}(1, \omega)^{-1}. \tag{6.1}$$

Therefore all poles of $U(k, \omega)$ in $\bar{\mathbb{D}}$ belong to the finite set $\Omega \cup \{0\}$ where Ω includes all poles of $T(\omega)^{-1}$ and $\mathcal{E}(1, \omega)^{-1}$ in $\bar{\mathbb{D}}$.

That at the origin the functions $U(0, \omega)$ and $U(1, \omega)$ have poles of order 0 and 1 respectively follows from the initial conditions for P. For $k \geq 2$ the main contribution to singularity of $U(k, \omega)$ at zero comes from the first term on the right-hand side of (6.1) because $P(k, \omega)$ is a polynomial of degree $k - 2$ with respect to $\lambda = a - b(\omega + \omega^{-1})$.

It now suffices to study the singularities of the second term on the right-hand side in (6.1). Let Ω be the set of all such singularities in $\bar{\mathbb{D}}$. These singularities comes from the poles of $T(\omega)^{-1}$, that is, when $\det(T(\omega)) = 0$ and also from the zeros of $\det(\mathcal{E}(1, \omega)) = \prod_{\sigma \in \mathcal{C}} e_\sigma(1, \theta_\sigma(\omega))$.

Actually $U(k, \omega)$ cannot have poles outside $\mathbb{T} \cup (-1, 1)$. This is due to the fact that the poles of $T(\omega)^{-1}$ are located in $\mathbb{T} \cup (-1, 1)$ and $e_\sigma(1, \theta_\sigma(\omega)) = 0$ only if $\theta_\sigma(\omega) \in (-1, 1)$, see (4.5), and thus $\omega \in \mathbb{T} \cup (-1, 1)$.

Let $\hat{\omega} \in (-1, 0) \cup (0, 1)$. We are going to use (5.6) as ω approaches $\hat{\omega}$. We have $|\theta_\sigma(\omega) - \theta_\sigma(\bar{\omega})| \asymp |\omega - \bar{\omega}|$, so for any $\mathbf{y} \in l^2(\mathcal{C})$ relation (5.6) with

$$\mathbf{x} = T(\omega)^{-1}BB(0)^{-1}D(\omega)\mathcal{E}(1, \omega)^{-1}\mathbf{y}$$

gives

$$|\langle \mathcal{E}(1, \omega)^* \mathcal{E}(1, \omega)^{-1}BD(\omega)\mathbf{y}, T(\omega)^{-1}BB(0)^{-1}D(\omega)\mathcal{E}(1, \omega)^{-1}\mathbf{y}\rangle|$$
$$\geq C|\omega - \bar{\omega}| \|T(\omega)^{-1}BB(0)^{-1}D(\omega)\mathcal{E}(1, \omega)^{-1}\mathbf{y}\|^2.$$

Since $\mathcal{E}(1, \omega)^* \mathcal{E}(1, \omega)^{-1}$ is unitary and also $|\det D(\omega)|$ stays bounded from below near $\hat{\omega}$, the Schwarz inequality gives

$$\|T(\omega)^{-1}BB(0)^{-1}D(\omega)\mathcal{E}(1, \omega)^{-1}\mathbf{y}\| \leq \mathrm{Const} \frac{\|\mathbf{y}\|}{|\omega - \bar{\omega}|} \quad \text{as } \omega \to \hat{\omega},$$

this is possible only in case $\hat{\omega}$ is a simple pole of $T(\omega)^{-1}BB(0)^{-1}D(\omega)\mathcal{E}(1, \omega)^{-1}$.

In the case when $\hat{\omega} \in \mathbb{T}$, we argue in a similar way, it suffices to let ω approach $\hat{\omega}$ in a way that $|\theta_\sigma(\omega) - \hat{\theta}_\sigma(\omega)| \asymp |\omega - \hat{\omega}|$. $\square$

7. Relation for the scattering coefficients

The special solution $\psi^\sigma(\alpha, \omega)$ is now well defined for all $\omega \in \mathbb{T} \setminus \Omega$. Thus the non-diagonal scattering coefficients $s_{\sigma,\gamma}(\omega)$ $\sigma \neq \gamma$ are also well defined for all $\omega \in \mathbb{T} \setminus \Omega$. Together with $\psi^\sigma(\omega)$ they are analytic in the vicinity of $\overline{\mathbb{D}}$.

The scattering coefficients $s_{\sigma,\sigma}(\omega)$ are so far well defined for $\omega \in T_\sigma^+$ only. For $\omega \in T_\sigma^-$ the corresponding value $\theta_\sigma(\omega)$ belongs to the unit disk. One can construct (similarly to how this is done in Theorem 1.4.1 in [1] for the continuous case) a real-valued solution $e_\sigma^{(1)}(k, \theta_\sigma)$ of (3.2) such that

$$e_\sigma^{(1)}(k, \theta_\sigma) = \theta_\sigma^{-k}(1 + o(1)), \ k \to \infty. \tag{7.1}$$

This choice is not unique since by adding any multiple of e_σ we obtain a solution which still meets this relation. However we fix some choice of functions $e_\sigma^{(1)}(k, \theta_\sigma)$. The functions p_σ can be represented as

$$p_\sigma(k, \omega) = \frac{\mathfrak{b}_\sigma(0)}{\mathfrak{b}_\sigma} \frac{e_\sigma(k, \theta_\sigma(\omega)) e_\sigma^{(1)}(1, \theta_\sigma(\omega)) - e_\sigma^{(1)}(k, \theta_\sigma(\omega)) e_\sigma(1, \theta_\sigma(\omega))}{\theta_\sigma(\omega)^{-1} - \theta_\sigma(\omega)}, \tag{7.2}$$

which yields

$$\psi^\sigma(\sigma(k), \omega) = \theta_\sigma(\omega)^{-k}(1 + o(1)), \ k \to \infty$$

as it should be for a special solution.

Lemma 7.1. *For each* $\sigma, \gamma \in \mathcal{C}$, $\sigma \neq \gamma$ *we have*

$$b_\gamma(\theta_\gamma(\omega)^{-1} - \theta_\gamma(\omega)) s_{\gamma\sigma}(\omega) = b_\sigma(\theta_\sigma(\omega)^{-1} - \theta_\sigma(\omega)) s_{\sigma\gamma}(\omega), \ \omega \in \mathbb{D} \setminus \Omega. \tag{7.3}$$

In addition, the scattering coefficients $s_{\gamma\sigma}(\omega)$ *and* $s_{\sigma\gamma}(\omega)$ *are continuous up to* $\mathbb{T} \setminus \Omega$.

Proof. Relation (7.3) follows from (4.7) for $\xi = \psi^\sigma(\omega)$, $\eta = \psi^\gamma(\omega)$ if one takes into account that according (3.7) and (3.8) we have

$$\mathfrak{b}_\nu(0)\{\psi^\sigma, \psi^\gamma\}_\nu(0) = \begin{cases} -b_\sigma(\theta_\sigma(\omega)^{-1} - \theta_\sigma(\omega)) s_{\sigma\gamma}(\omega), & \nu = \sigma; \\ b_\gamma(\theta_\gamma(\omega)^{-1} - \theta_\gamma(\omega)) s_{\gamma\sigma}(\omega), & \nu = \gamma; \\ 0, & \nu \neq \sigma, \gamma. \end{cases}$$

The continuity of the scattering coefficients $s_{\gamma\sigma}(\omega)$ follows from the definition of $U(k, \omega)$ and the fact that it has finitely many poles contained in Ω. $\qquad \square$

Corollary 7.1. *Let* $\hat{\omega} \in T_\sigma^+ \cap T_\nu^+$ *be a pole of* $U(k, \omega)$. *Then*

$$\mathrm{Res}_{\hat\omega} u_{\sigma\nu}(k, \omega) = \mathrm{Res}_{\hat\omega} u_{\nu\sigma}(k, \omega) = 0.$$

Proof. Since all poles of $U(k, \omega)$ on $\mathbb{T}$ are simple, the limit

$$\vec{\xi}(k) = \{\xi(\gamma(k))\}_{\gamma \in \mathcal{C}} = \lim_{\omega \to \hat\omega} (\omega - \hat\omega)\{\psi^\sigma(\gamma(k), \omega)\}_{\gamma \in \mathcal{C}}$$

exists and can be extended to a solution $\{\xi(\alpha)\}_{\alpha \in \mathcal{A}}$ of (1.1) with $\lambda = \lambda(\hat\omega)$.

We also have

$$\xi(\gamma(k)) = \mathrm{Res}_{\hat\omega} u_{\sigma\gamma}(k, \omega) = e_\gamma(k, \theta_\gamma(\hat\omega)) \lim_{\omega \to \hat\omega} (\omega - \hat\omega) s_{\sigma\gamma}(\omega),$$

i.e., on each channel $\gamma \in \mathcal{C}$ the corresponding solution is represented by an outgoing wave. The function $\{\eta(\alpha)\}_{\alpha \in \mathcal{A}}$, $\eta(\alpha) = \overline{\xi(\alpha)}$ also solves this problem. It remains to apply Lemma 4.2 in order to see that such solutions cannot contain non-decaying waves. $\qquad\square$

Corollary 7.2. *Let $\omega \in T_\sigma^+ \cup T_\nu^+$, $\sigma, \nu \in \mathcal{C}$ and $\sigma \neq \nu$. Then*

$$s_{\nu\sigma}(\omega^{-1}) = \overline{s_{\nu\sigma}}(\omega)$$

In particular $|s_{\nu\sigma(\omega)}|^2 = s_{\nu\sigma}(\omega)s_{\nu\sigma}(\omega^{-1})$.

Proof. We have $u_{\nu\sigma}(k, \omega) = e_\nu(k, \theta_\nu(\omega))s_{\nu\sigma}(\omega)$ and by the construction of the matrix $U(k, \omega)$, we know that

$$u_{\nu\sigma}(k, \omega^{-1}) = \overline{u_{\nu\sigma}}(k, \omega), \quad \omega \in T_\sigma^+ \cup T_\nu^+$$

and the corollary follows. $\qquad\square$

8. Discrete spectrum of the operator $\mathcal{L}$

Lemma 8.1. *Let $\hat{\omega} \in \bar{\mathbb{D}} \setminus \{0\}$ be a pole of $U(k, \omega)$. Then $\lambda(\hat{\omega}) = a - b(\hat{\omega} + \hat{\omega}^{-1})$ is an eigenvalue of (1.1). If in addition $\hat{\omega} \in T_\sigma^+$, then all elements $u_{\sigma,\nu}(k, \omega)$, $\nu \in \mathcal{C}$ are regular at $\hat{\omega}$.*

Proof. Let $\hat{\omega} \in \bar{\mathbb{D}} \setminus \{0\}$ be a (simple) pole of $U(k, \omega)$. Denote

$$a_\nu(\omega) = \frac{\theta_\nu(\hat{\omega}) - \theta_\nu(\hat{\omega})^{-1}}{\theta_\nu(\omega) - \theta_\nu(\omega)^{-1}}(\omega - \hat{\omega}), \ \nu \in \mathcal{C}; \quad A(\omega) = \mathrm{diag}\{a_\nu(\omega)\}_{\nu \in \mathcal{C}}.$$

We then have

$$\mathrm{Res}_{\hat{\omega}}U(k, \omega) = \lim_{\omega \to \hat{\omega}} U(k, \omega)A(\omega).$$

For each $\nu \in \mathcal{C}$ and for each $\omega \neq \hat{\omega}$ the νth column of $U(k, \omega)A(\omega)$

$$\vec{\phi}^\nu(k, \omega) = (\phi^\nu(\sigma(k), \omega))_{\sigma \in \mathcal{C}} = U(k, \omega)A(\omega)\mathbf{n}_\nu$$

can be extended into $\mathcal{A}_1$ to a solution of (1.1) with $\lambda = \lambda(\omega)$ according (2.4):

$$\phi^\nu(\alpha, \omega) = \sum_{\gamma \in \mathcal{C}} r(\alpha, \gamma(0); \lambda)\mathbf{b}_\gamma(0)\phi^\nu(\gamma(1), \omega); \ \alpha \in \mathcal{A}_1.$$

Since $U(k, \omega)$ has a simple pole at $\hat{\omega}$, the following limit exists:

$$\vec{\phi}^\nu(k, \hat{\omega}) = \lim_{\omega \to \hat{\omega}} \vec{\phi}^\nu(k, \omega).$$

Claim. *The vector $\vec{\phi}^\nu(k, \hat{\omega})$ also can be extended to a solution of the problem (1.1).*

In case $\lambda(\hat{\omega})$ is not a pole of $\mathcal{R}(\lambda(\omega))$, the extension is straightforward. By (2.4), if $\lambda(\hat{\omega})$ is at the same time an eigenvalue of $\mathcal{L}_1$ one can apply the same reasonings as in Lemma 4.3 in [13]. We omit the details.

Let now

$$T(\omega)^{-1} = (\tau_{\sigma\nu}(\omega))_{\sigma, \nu \in \mathcal{C}},$$

and denote

$$h_{\sigma\nu}(\omega) = -\frac{b_\nu}{b_\nu(0)}\frac{\theta_\nu(\hat\omega) - \theta_\nu(\hat\omega)^{-1}}{e_\nu(1,\theta_\nu(\omega))}(\omega - \hat\omega)\tau_{\sigma\nu}(\omega). \tag{8.1}$$

It follows from (5.7) that for $\sigma \neq \nu$ we have

$$\text{Res}_{\hat\omega} u_{\sigma\nu}(k,\cdot) = \phi^\nu(\sigma(k),\hat\omega) = e_\sigma(k,\hat\omega)m_{\sigma\nu}(\hat\omega), \tag{8.2}$$

here we denote $m_{\sigma\nu}(\hat\omega) = h_{\sigma\nu}(\hat\omega)$.

If $e_\nu(1,\theta_\nu(\hat\omega)) \neq 0$, the representation (8.2) is valid for $\sigma = \nu$ as well. Assume that $e_\nu(1,\theta_\nu(\hat\omega)) = 0$. Then

$$p_\nu(k,\hat\omega)e_\nu(0,\theta_\nu(\hat\omega)) = e_\nu(k,\theta_\nu(\hat\omega)), \tag{8.3}$$

because the expressions on the two sides satisfy the same recurrence equation and the same initial conditions, and also it follows from Lemma 4.1 that $\dot e_\nu(1,\theta_\nu(\hat\omega)) \neq 0$ because $e_\nu(1,\theta_\sigma(\hat\omega))$ and $\dot e_\nu(1,\theta_\sigma(\hat\omega))$ cannot vanish simultaneously. Thus we again obtain (8.2), yet now

$$m_{\nu\nu}(\hat\omega) = \frac{b_\nu}{b_\nu(0)}\frac{\theta_\nu(\hat\omega) - \theta_\nu(\hat\omega)^{-1}}{\{e_\nu(\hat\omega),\dot e_\nu(\hat\omega)\}_\nu(0)} + h_{\nu\nu}(\hat\omega).$$

The representation (8.2) is now valid for all $\nu,\sigma \in \mathcal{C}$.

If $\hat\omega \in \mathbb{D}$ it follows from (8.2) that $\phi^\nu(\alpha,\hat\omega)$ is an eigenfunction of $\mathcal{L}$ with $\lambda(\hat\omega)$ as the eigenvalue. If $\hat\omega \in \mathbb{T}$, we may have $\theta_\sigma(\hat\omega) \in \mathbb{T}$, i.e., $\hat\omega \in T_\sigma^+$ for some $\sigma \in \mathcal{C}$. It follows from Corollary 7.1 that for such σ we have $m_{\sigma\nu}(\hat\omega) = 0$, in particular

$$\text{Res}_{\hat\omega} u_{\sigma\nu}(k,\cdot) = \phi^\nu(\sigma(k),\hat\omega) = 0, \ \hat\omega \in \mathbb{T}, \theta_\sigma(\hat\omega) \in \mathbb{T}.$$

and, again $\phi^\nu(\alpha,\hat\omega)$ is an eigenfunction of $\mathcal{L}$ with $\lambda(\hat\omega)$ as the eigenvalue. $\quad\square$

Consider the matrix

$$\mathfrak{m}(\hat\omega) = (m_{\sigma\nu}(\hat\omega))_{\sigma,\nu\in\mathcal{C}}.$$

Properties of $\mathfrak{m}(\hat\omega)$ are summarized in the statement below

Lemma 8.2. *Let $\hat\omega \in \Omega$. Then*

$$\text{Res}\,U(k,\hat\omega) = \mathcal{E}(k,\hat\omega)\mathfrak{m}(\hat\omega), \ k = 0,1,2,\ldots. \tag{8.4}$$

The diagonal elements $m(\nu,\nu;\hat\omega)$ satisfy

$$\|\phi^\nu(\hat\omega)\|^2 = -\frac{b_\nu(1 - \theta_\nu(\hat\omega)^{-2})}{b(1 - \bar{\hat\omega}^{-2})}\theta_\nu(\hat\omega)m_{\nu\nu}(\hat\omega), \tag{8.5}$$

where $\phi^\nu = \phi^\nu(\alpha,\hat\omega)$ is the eigenvector of $\mathcal{L}$ corresponding to the eigenvalue $\lambda(\hat\omega)$ and such that

$$\phi^\nu(\sigma(k),\hat\omega) = e_\sigma(k,\theta_\sigma(\hat\omega))m_{\sigma\nu}(\hat\omega) \ \sigma \in \mathcal{C}, \ k \geq 0. \tag{8.6}$$

Proof. Relations (8.4) and (8.6) were already established in Lemma 8.1. It remains to prove (8.5).

We apply Lemma 4.3 with $\xi(\alpha) = \phi^\nu(\alpha, \hat{\omega})$:

$$\dot{\lambda}(\hat{\omega}) \sum_{\alpha \in \mathcal{A}_1} |\phi^\nu(\alpha, \hat{\omega})|^2 = \sum_{\sigma \in \mathcal{C}} \mathfrak{b}_\sigma(0)\{\dot{\phi}^\nu(\hat{\omega}), \overline{\phi^\nu(\hat{\omega})}\}_\sigma(0), \tag{8.7}$$

and calculate the Wronskians on the right-hand side of this formula.

We have

$$\phi^\nu(\sigma(k), \omega) = p_\sigma(k, \omega)h_\nu(\omega)\delta_{\sigma\nu} + e_\sigma(k, \omega)h_{\sigma\nu}(\omega), \tag{8.8}$$

with

$$h_\nu(\omega) = \frac{b_\nu}{\mathfrak{b}_\nu(0)} \frac{\theta_\nu(\hat{\omega}) - \theta_\nu(\hat{\omega})^{-1}}{e_\nu(1, \theta_\nu(\omega))}(\omega - \hat{\omega}),$$

and $h_{\sigma,\nu}$ is already defined in (8.1).

Let $e_\nu(1, \theta_\nu(\hat{\omega})) \neq 0$. Then, for all $\sigma, \nu \in \mathcal{C}$,

$$h_\nu(\hat{\omega}) = 0, \quad \phi^\nu(\sigma(k), \hat{\omega}) = e_\sigma(k, \theta_\sigma(\hat{\omega}))h_{\sigma\nu}(\hat{\omega}), \tag{8.9}$$

and

$$m_{\sigma\nu}(\hat{\omega}) = h_{\sigma\nu}(\hat{\omega}), \quad \mathfrak{m}(\hat{\omega}) = (m_{\sigma\nu}(\hat{\omega}))_{\sigma,\nu \in \mathcal{C}} \tag{8.10}$$

We use (8.8), (8.9), (8.10), (8.4) and that $\dot{p}_k(\omega) = 0$ for $k = 0, 1$ as follows from (5.1):

$$\dot{\phi}^\nu(\sigma(k), \omega) = p_\sigma(k, \omega)\dot{h}_\nu(\omega)\delta_{\sigma\nu} + \dot{e}_\sigma(k, \omega)h_{\sigma\nu}(\omega) + e_\sigma(k, \omega)\dot{h}_{\sigma\nu}(\omega), \quad k = 0, 1. \tag{8.11}$$

Moreover,

$$\dot{h}_\nu(\hat{\omega}) = \frac{b_\nu}{\mathfrak{b}_\nu(0)} \frac{\theta_\nu(\hat{\omega}) - \theta_\nu(\hat{\omega})^{-1}}{e_\nu(1, \theta_\nu(\hat{\omega}))}.$$

Since $e_\nu(1, \theta_\nu(\hat{\omega})) \neq 0$, we also have $h_\nu(\hat{\omega}) = 0$ and, according to (4.6) and (8.2),

$$\mathfrak{b}_\sigma(0)\{\dot{\phi}^\nu(\hat{\omega}), \overline{\phi^\nu(\hat{\omega})}\}_\sigma(0)$$
$$= b_\nu(\theta_\nu(\hat{\omega}) - \theta_\nu(\hat{\omega})^{-1})\overline{m_{\sigma\nu}(\hat{\omega})}\delta_{\sigma\nu}$$
$$+ \mathfrak{b}_\sigma(0)\{\dot{e}_\sigma(\hat{\omega}), e_\sigma(\theta_\sigma(\hat{\omega}))\}_\sigma(0)|m_{\sigma\nu}(\hat{\omega})|^2 \tag{8.12}$$
$$= b_\nu(\theta_\nu(\hat{\omega}) - \theta_\nu(\hat{\omega})^{-1})\overline{m_{\sigma\nu}(\hat{\omega})}\delta_{\sigma\nu} - \dot{\lambda}(\hat{\omega}) \sum_{k=1}^{\infty} |\phi^\nu(\sigma(k), \theta_\sigma(\hat{\omega}))|^2.$$

We can now return to (8.7) in order to obtain the normalization condition (8.5) for the matrix $\mathfrak{m}$:

In the case $e_\nu(1, \hat{\omega}) = 0$ representation (8.8) is still valid, yet

$$h_\nu(\hat{\omega}) = \frac{b_\nu}{\mathfrak{b}_\nu(0)} \frac{\theta_\nu(\hat{\omega}) - \theta_\nu(\hat{\omega})^{-1}}{\{e_\nu, \dot{e}_\nu\}_\nu(0)} e_\nu(0, \hat{\omega}) \neq 0.$$

Taking (8.3) into account we again obtain (8.2), yet now

$$m_{\sigma\nu}(\hat{\omega}) = \frac{b_\nu}{\mathfrak{b}_\nu(0)} \frac{\theta_\nu(\hat{\omega}) - \theta_\nu(\hat{\omega})^{-1}}{\{e_\nu(\hat{\omega}), \dot{e}_\nu(\hat{\omega})\}_\nu(0)}\delta_{\sigma\nu} + h_{\sigma\nu}(\hat{\omega}).$$

Relation (8.11) is still valid and for $\nu \neq \sigma$ we arrive at relation (8.12).

For $\sigma = \nu$ we have

$$\mathfrak{b}_\nu(0)\{\dot{\phi}^\nu(\hat{\omega}), \overline{\dot{\phi}^\nu(\hat{\omega})}\}_\nu(0) = \mathfrak{b}_\nu(0)\{\dot{e}_\nu(\hat{\omega})h_2(\hat{\omega}), \overline{e_\nu(\hat{\omega})}\}_\nu(0)\overline{m}_{\nu\nu}(\hat{\omega})$$

$$= \mathfrak{b}_\nu(0)\{\dot{e}_\nu(\hat{\omega}), \overline{e_\nu(\hat{\omega})}\}_\nu(0)|m_{\nu\nu}(\hat{\omega})|^2 - \mathfrak{b}_\nu(\theta_\nu(\hat{\omega}) - \theta_\nu(\hat{\omega})^{-1})\overline{m}_{\nu\nu}(\hat{\omega}),$$

and we again arrive at (8.12).

Now one can complete the proof in the same way as if $e_\nu(1, \theta_\nu(\hat{\omega})) \neq 0$.   □

Remark. The matrix $\overline{\mathfrak{m}(\hat{\omega})}$ corresponds to the point of the discrete spectrum $\lambda(\hat{\omega})$. Its columns are normalized eigenfunctions. This normalization is defined by relation (8.5) and is therefore unique.

We will see in the next section that $m_{\nu\nu}(\hat{\omega})$, i.e., the energies of the normalized eigenfunctions, are the quantities which participate in the equations for the inverse scattering problem.

9. Equations of the inverse scattering problem

9.1. In order to obtain equations of the inverse scattering problem we introduce the matrix function

$$\Delta_l(\omega) := \mathrm{diag}\left\{\theta_\nu(\omega)^{l-1}\frac{d\theta_\nu(\omega)}{d\omega}\right\}_{\nu \in \mathcal{C}}, \quad l \in \mathbb{Z} \tag{9.1}$$

and consider the integral

$$\mathcal{J}(l, k) = (j_{\nu\sigma}(l, k))_{\nu,\sigma \in \mathcal{C}} = \frac{1}{2\pi\mathrm{i}}\int_{\mathbb{T}}\Delta_l(\omega)U(k, \omega)d\omega.$$

Since $\mathbb{T}$ may contain (simple) poles of $U(k, \omega)$ this integral as well as all integrals in this section are treated in the principal value sense. We will calculate this integral in two ways: through the residues, this would correspond to the contribution of the discrete spectrum, and through the scattering coefficients, this would correspond to the contribution of the continuous spectrum.

Comparing two different expressions for $\mathcal{J}(l, k)$ leads one to the equations of the inverse scattering problem.

Let as before $U(k, \omega) = (u_{\nu\sigma}(k, \omega))_{\nu,\sigma}$. Then

$$\begin{aligned}
j_{\nu\sigma}(l, k) &= \frac{1}{2\pi\mathrm{i}}\int_{\mathbb{T}}\theta_\nu(\omega)^{l-1}u_{\nu\sigma}(k, \omega)\frac{d\theta_\nu(\omega)}{d\omega}d\omega \\
&= \frac{1}{2\pi\mathrm{i}}\int_{\mathbb{T}}\theta_\nu(\omega)^{l-1}\psi_\nu^\sigma(k, \omega)\frac{d\theta_\nu(\omega)}{d\omega}d\omega,
\end{aligned} \tag{9.2}$$

here $\psi^\nu(\cdot, \omega)$ is the special solution, defined in (5.8).

Assume first that $\sigma \neq \nu$. Then, according to (3.7) and (3.8),

$$j_{\nu\sigma}(l, k) = \frac{1}{2\pi\mathrm{i}}\int_{\mathbb{T}}\theta_\nu(\omega)^{l-1}e_\nu(k, \theta_\nu(\omega))s_{\sigma\nu}(\omega)\frac{d\theta_\nu(\omega)}{d\omega}d\omega.$$

Consider the function

$$q_{\nu\sigma}(n) = \frac{1}{2\pi i}\int_{\mathbb{T}} s_{\sigma\nu}(\omega)\theta_{\nu}(\omega)^{n-1}\frac{d\theta_{\nu}(\omega)}{d\omega}\,d\omega. \tag{9.3}$$

Relation (3.4) now yields

$$j_{\nu\sigma}(l,k) = c_{\nu}(k)\sum_{m\geq k} q_{\nu\sigma}(m+l)a_{\nu}(k,m), \tag{9.4}$$

this is the desired expression.

Remark. The functions $q_{\nu\sigma}$ cannot be determined from the spectral data generally speaking. However relation (9.4) determines the **structure** of the equation of the inverse scattering problem. Later in Section 9.2 we will use this structure to get rid of the functions which cannot be observed from the spectral data.

Let now $\nu = \sigma$ and $T(\omega)^{-1} = (\tau_{\sigma\nu}(\omega))_{\sigma,\nu}$. Relations (5.7) and (9.1) yield

$$j_{\sigma\sigma}(k,l)$$
$$= \frac{1}{2\pi i}\int_{\mathbb{T}}\left\{\left[-p_{\sigma}(k,\omega) + e_{\sigma}(k,\theta_{\sigma})\tau_{\sigma\sigma}(\omega)\right]\frac{b_{\sigma}(\theta_{\sigma}^{-1}-\theta_{\sigma})}{b_{\sigma}(0)e_{\sigma}(1,\theta_{\sigma})}\theta_{\sigma}(\omega)^{l-1}\frac{d\theta_{\sigma}(\omega)}{d\omega}\right\}d\omega$$
$$= \frac{1}{2\pi i}\left(\int_{T_{\sigma}^{+}} + \int_{T_{\sigma}^{-}}\right)\{\cdot\} = j_{\sigma\sigma}^{+}(k,l) + j_{\sigma\sigma}^{-}(k,l). \tag{9.5}$$

As ω runs over T_{σ}^{+} the function $\theta_{\sigma}(\omega)$ runs over the whole $\mathbb{T}$. Let $\omega(\theta_{\sigma})$ be the inverse function. Relations (9.2) together with (3.7) yield

$$j_{\sigma\sigma}^{+}(k,l) = \frac{1}{2\pi i}\int_{\mathbb{T}}\left[e_{\sigma}(k,\theta_{\sigma}^{-1}) + s_{\sigma\sigma}(\omega(\theta_{\sigma}))e_{\sigma}(k,\theta_{\sigma})\right]\theta_{\sigma}^{l-1}\,d\theta_{\sigma}.$$

Next, it follows from corollary 7.1 that $s_{\sigma,\sigma}(\omega)$ is bounded on $\mathbb{T}_{\sigma}^{+}$.
    Consider the Fourier series of $s_{\sigma\sigma}(\omega(\theta_{\sigma}))$:

$$s_{\sigma\sigma}(\omega(\theta_{\sigma})) = \sum_{n=-\infty}^{\infty}\tilde{s}_{\sigma}(n)\theta^{-n}; \quad \tilde{s}_{\sigma}(n) = \frac{1}{2\pi i}\int_{\mathbb{T}}s_{\sigma\sigma}(\omega(\theta_{\sigma}))\theta_{\sigma}^{n-1}\,d\theta_{\sigma}.$$

Together with (3.4) this yields

$$e_{\sigma}(k,\theta_{\sigma}^{-1}) + s_{\sigma\sigma}(\omega(\theta_{\sigma}))e_{\sigma}(k,\theta_{\sigma})$$
$$= c_{\sigma}(k)\sum_{m}\left\{a_{\sigma}(k,m) + \sum_{n}a_{\sigma}(k,n)\tilde{s}_{\sigma}(m+n)\right\}\theta_{\sigma}^{-m},$$

and finally

$$j_{\sigma\sigma}^{+}(k,l) = c_{\sigma}(k)\left[a_{\sigma}(k,l) + \sum_{n=-\infty}^{\infty}a_{\sigma}(k,n)\tilde{s}_{\sigma}(l+n)\right]. \tag{9.6}$$

We now study $j_{\sigma\sigma}^-(k,l)$. Denote

$$a_1(k,\omega) = -p_\sigma(k,\omega)\frac{b_\sigma}{b_\sigma(0)}e_\sigma(1,\theta_\sigma)^{-1}(\theta_\sigma^{-1} - \theta_\sigma);$$

$$a_2(\omega) = \tau_{\sigma,\sigma}(\omega)\frac{b_\sigma}{b_\sigma(0)}e_\sigma(1,\theta_\sigma)^{-1}(\theta_\sigma^{-1} - \theta_\sigma).$$

We then have

$$\psi_\sigma^\sigma(k,\omega) = a_1(k,\omega) + e_\sigma(k,\theta_\sigma(\omega))a_2(\omega),$$

and

$$j_{\sigma\sigma}^-(k,l) = \frac{1}{2\pi i}\int_{T_\sigma^-} a_1(k,\omega)\theta_\sigma(\omega)^{l-1}\frac{d\theta_\sigma}{d\omega}\,d\omega$$

$$+ \frac{1}{2\pi i}\int_{T_\sigma^-} a_2(\omega)e_\sigma(k,\theta_\sigma(\omega))\theta_\sigma(\omega)^{l-1}\frac{d\theta_\sigma}{d\omega}\,d\omega.$$

As ω runs over T_σ^-, the function $\theta_\sigma(\omega)$ runs twice in the opposite directions over $J_\sigma = \theta_\sigma(T_\sigma^-) \subset \mathbb{R}$ and the values $\theta_\sigma(\omega)$ and $\theta_\sigma(\omega^{-1})$ coincide. Respectively $e_\sigma(k,\theta_\sigma(\omega)), p_\sigma(k,\omega) \in \mathbb{R}$ and $e_\sigma(k,\theta_\sigma(\omega)) = e_\sigma(k,\theta_\sigma(\omega^{-1})), p_\sigma(k,\omega) = p_\sigma(k,\omega^{-1})$, thus $a_1(k,\omega) = a_1(k,\omega^{-1})$.

Moreover, $a_2(\omega) = \overline{a_2(\omega^{-1})}$, since $\psi^\sigma(k,\omega) = \overline{\psi^\sigma(k,\omega^{-1})}$ for $\omega \in \mathbb{T}$.

Therefore

$$j_{\sigma\sigma}^-(k,l) = \frac{1}{2\pi i}\int_{\Gamma_\sigma} [a_2(\omega) - a_2(\omega^{-1})]e_\sigma(k,\theta_\sigma)\theta_\sigma^{l-1}\frac{d\theta_\sigma}{d\omega}\,d\omega, \qquad (9.7)$$

where $\Gamma_\sigma = T_\sigma^- \cap \mathbb{C}_+$.

The function $a_1(k,\omega)$ satisfies equation (2.2) on the channel σ and relations (7.1), (7.2) yield

$$a_1(k,\omega) = \theta_\sigma(\omega)^{-k}(1 + o(1)).$$

Therefore

$$b_\sigma(0)\{\psi^\sigma(\omega), \psi^\sigma(\omega^{-1})\}_\sigma(0) = b_\sigma(\theta_\sigma(\omega)^{-1} - \theta_\sigma(\omega))[a_2(\omega) - a_2(\omega^{-1})].$$

On the other hand by (4.7)

$$b_\sigma(0)\{\psi^\sigma(\omega), \psi^\sigma(\omega^{-1})\}_\sigma(0) = -\sum_{\nu \in \mathcal{C}\setminus\{\sigma\}} b_\nu(0)\{\psi^\sigma(\omega), \psi^\sigma(\omega^{-1})\}_\nu(0).$$

We have

$$\{\psi^\sigma(\omega), \psi^\sigma(\omega^{-1})\}_\nu(0) = 0, \ \omega \in T_\nu^-.$$

For $\omega \in T_\nu^+ \cap T_\sigma^-$ we obtain

$$b_\nu(0)\{\psi^\sigma(\omega), \psi^\sigma(\omega^{-1})\}_\nu(0)$$

$$= b_\nu(0)\{s_{\nu\sigma}(\omega)e_\nu(\theta_\nu(\omega)), s_{\nu\sigma}(\omega^{-1})e_\nu(\theta_\nu(\omega)^{-1})\}_\nu(0)$$

$$\times |s_{\nu\sigma}(\omega)|^2 b_\nu(0)\{e_\nu(\theta_\nu(\omega)), e_\nu(\theta_\nu(\omega)^{-1})\}_\nu(0)$$

$$= |s_{\nu\sigma}(\omega)|^2 b_\nu(\theta_\nu(\omega)^{-1} - \theta_\nu(\omega)) = -|s_{\sigma\nu}(\omega)|^2\frac{(b_\sigma(\theta_\sigma^{-1} - \theta_\sigma))^2}{b_\nu(\theta_\nu^{-1} - \theta_\nu)}.$$

Now, for $\theta \in J_\sigma$, we define $w_\sigma(\theta) = W^{-1} \circ W_\sigma(\theta)$, here the functions W, W_σ are defined by (3.5) and the branch is chosen so that $w_\sigma(\theta) \in \Gamma_\sigma$. Let

$$N_\sigma(\theta) = \{\nu; \; w_\sigma(\theta) \in T_\nu^+\}, \; \theta \in J_\sigma.$$

Denote

$$\Phi_\sigma(\theta) = \sum_{\nu \in N_\sigma(\theta)} |s_{\sigma\nu}(w_\sigma)|^2 \left(\frac{b_\sigma(\theta^{-1} - \theta)}{b_\nu(\theta_\nu(w_\sigma)^{-1} - \theta_\nu(w_\sigma))} \right), \quad w_\sigma(\theta) \in T_\sigma^-.$$

We finally obtain

$$a_2(\omega) - a_2(\omega^{-1}) = \Phi_\sigma(\theta_\sigma(\omega)), \tag{9.8}$$

this function is expressed via the spectral data.

Remark. All functions $|s_{\nu\sigma}(\omega)|$ participating in (9.8) are well defined and continuous. In addition, $s_{\nu\sigma}(\omega^{-1}) = \overline{s_{\nu\sigma}}(\omega)$ for $\omega \in T_\nu^+$.

Relation (9.7) now takes the form

$$j_{\sigma\sigma}^-(k, l) = \frac{1}{2\pi i} \int_{J_\sigma} e_\sigma(k, \theta) \Phi_\sigma(\theta) \theta^{l-1} d\theta,$$

and with account (3.4) we obtain

$$j_{\sigma\sigma}^-(k, l) = c_\sigma(k) \sum_n a_\sigma(k, n) q_{\sigma\sigma}(n + l), \tag{9.9}$$

where

$$q_{\sigma\sigma}(n) = \frac{1}{2\pi i} \int_{J_\sigma} \Phi_\sigma(\theta) \theta^{n-1} d\theta. \tag{9.10}$$

Combining (9.5), (9.6), and (9.9) we finally obtain

$$j_{\sigma\sigma}(k, l) = c_\sigma(k) \left[a_\sigma(k, l) + \sum_{n=-\infty}^{\infty} a_\sigma(k, n)\big(\tilde{s}_\sigma(l + n) + q_{\sigma\sigma}(n + l)\big) \right], \tag{9.11}$$

the functions $\tilde{s}_\sigma(\cdot)$ and $q_{\sigma\sigma}(\cdot)$ are defined through the scattering data.

9.2.

Let now

$$\Theta(\omega) = \text{diag}\{\theta_\sigma(\omega)\}_{\sigma \in \mathcal{C}}.$$

We use the representation (3.4) for $\mathcal{E}(k, \omega)$,

$$\mathcal{E}(k, \omega) = C(k) \sum_{m=-\infty}^{\infty} A(k, m)\Theta(\omega)^m.$$

The coefficients $C(k)$ and $A(k, m)$ are the diagonal matrices

$$C(k) = \text{diag}\{c_\sigma(k)\}_{\sigma \in \mathcal{C}}, \; A(k, m) = \text{diag}\{a_\sigma(k, m)\}_{\sigma \in \mathcal{C}},$$

and also $A(k, m) = 0$ for $m < k$.

It follows from (9.4) and (9.11) that

$$J(l,k) = \frac{1}{2\pi i} \int_{\mathbb{T}} \Delta_l(\omega) U(k,\omega) d\omega = C(k) \left\{ A(k,l) + \sum_{l=-\infty}^{\infty} A(k,m) Z(l+m) \right\},$$
$$(9.12)$$

here the integral is taken as a principal value and the matrix $Z(n) = (z_{\nu\sigma}(n))_{\nu,\sigma \in \mathcal{C}}$ is given by the relations

$$z_{\nu\sigma}(n) = q_{\nu\sigma}(n), \quad \nu \neq \sigma, \tag{9.13}$$
$$z_{\sigma\sigma}(n) = \tilde{s}_\sigma(n) + q_{\sigma\sigma}(n), \tag{9.14}$$

here $\tilde{s}_\sigma(n)$ are the Fourier coefficients of the reflection coefficient $s_{\sigma\sigma}$ with respect to θ_σ and the function $q_{\nu\sigma}$ is defined in (9.3) and (9.10).

Let now $\Omega \subset \bar{\mathbb{D}}$ be the set of all poles of $U(k,\omega)$. It follows from (5.7) that $\overline{U(k,\omega)} = U(k,\bar{\omega})$, thus for each $\hat{\omega} \in \Omega \cap \mathbb{T}$ the point $\bar{\hat{\omega}}$ also belongs to Ω. Moreover, for such $\hat{\omega}$ we have $\overline{\text{Res}_{\hat{\omega}} U(k,\omega)} = \text{Res}_{\bar{\hat{\omega}}} U(k,\omega)$. Denote $\tilde{\Omega} = \{\hat{\omega} \in \Omega : |\hat{\omega}| < 1\} \cup \{\hat{\omega} \in \mathbb{T} \cap \Omega : \text{Im} \hat{\omega} > 0\}$. We use (8.4) and (9.1):

$$J(k,l) = \sum_{\hat{\omega} \in \tilde{\Omega}} \Delta_l(\hat{\omega}) \mathcal{E}(k,\hat{\omega}) \mathfrak{m}(\hat{\omega}) = C(k) \sum_m A(k,m) M(m+l),$$

where

$$M(n) = \sum_{\hat{\omega} \in \tilde{\Omega}} \text{diag}\left\{ \frac{d\theta_\sigma}{d\omega}(\hat{\omega}) \right\}_{\sigma \in \mathcal{C}} \Theta(\hat{\omega})^{n-1} \text{Re}\, \mathfrak{m}(\hat{\omega}). \tag{9.15}$$

By comparing this with (9.12) and taking into account that $A(k,m) = 0$ for $m < k$ and $A(k,k) = I$, we obtain the system of equations

$$A(k,m) + \sum_{s=k}^{\infty} A(k,s) F(s+m) = 0, \quad m = k+1, k+2, \dots,$$

where

$$F(n) = Z(n) - M(n). \tag{9.16}$$

Since $A(k,k) = I$, this relation can be written as

$$F(k+m) + A(k,m) + \sum_{s=k+1}^{\infty} A(k,s) F(s+m) = 0, \quad k = 1, 2, \dots, \quad m > k. \tag{9.17}$$

The matrices $A(k,m)$ are diagonal. The diagonal elements of $F(n)$ can be expressed through the spectral data. The non-diagonal elements of the matrices $F(n)$ vanish, as follows from the lemma below.

Lemma 9.1. *The matrices $F(n)$ are diagonal for all $n \geq 1$. Specifically, $F(n) = \text{diag}\{f_\sigma(n)\}_{\sigma \in \mathcal{C}}$, and the diagonal entries $f_\sigma(n)$ are determined by $a_\sigma(k,m)$, $m > k \geq \left[\frac{n-1}{2}\right]$ only.*

This statement was proved in [13], (Lemma 5.1) and we omit the proof.

Theorem 9.1. *The following properties hold for the systems under consideration.*

1) *Equations* (9.17) *split into a system of independent scalar equations*

$$f_\nu(k+m) + a_\nu(k,m) + \sum_{s=k+1}^{\infty} a_\nu(k,s)f_\nu(s+m) = 0, \ m \geq k+1 \geq 1, \qquad (9.18)$$

here $f_\nu(n)$ *is defined in* (9.16), (9.15), (9.14).

2) *For each* $k \geq 0$ *equations* (9.18) *have unique solutions* $a_\sigma(k,m)$.

Proof. The first statement follows directly from relation (9.16) which defines $F(n)$ and Lemma 9.1. It follows from the same lemma that the functions $f_\nu(n)$, $n \geq 1$ are uniquely defined by the coefficients $\mathfrak{b}_\nu(k)$, $\mathfrak{a}_\nu(k)$, $k \geq 1$. Therefore (9.18) coincide with the standard equations for the inverse scattering problem for equation (2.2) with boundary condition $\xi(\nu(0)) = 0$. It is well known (see for example [15]) that the later have a unique solution. $\square$

10. Concluding remarks

In the case where the continuous spectrum $I = \bigcup_{\sigma \in \mathcal{C}}[a_\sigma - 2b_\sigma, a_\sigma + 2b_\sigma]$ splits into a number of disjoints intervals, one can repeat the procedure separately for each connected component of I.

If, say, $I^{(0)}$ is a connected component of I and σ is a channel corresponding to this component, then each wave incoming along σ generates decaying waves in all channels which correspond to other connected components of I. We omit the details.

So far we have discussed reconstruction of the part $\mathcal{A}_0$ of the whole system, that is, the channels. This information is, generally speaking, insufficient for reconstruction the whole matrix $\mathcal{L}$. However if the matrix $\mathcal{L}_1$ corresponding to the "central" part of the system is sufficiently sparse and also we know the matrix $B(0) = \mathrm{diag}\{\mathfrak{b}_\sigma(0)\}_{\sigma \in \mathcal{C}}$ which realizes connections between the channels and the central part of the system, the whole matrix $\mathcal{L}$ can be recovered from the scattering data. We refer the reader to Chapter 11 in [16], where statements of such type are obtained.

Acknowledgment

We thank V.A. Marchenko, who suggested the problem, and with whom we had numerous fruitful discussions. We also thank I. Egorova, who read a preliminary version of the article and made quite useful remarks.

References

[1] Z.S. Agranovich, V.A. Marchenko, *The inverse problem of scattering theory*, Gordon and Breach Science Publishers, New York-London 1963 xiii+291 pp.

[2] V.S. Buslaev, V.N. Fomin, *An Inverse Scattering Problem for the One-Dimensional Schrödinger Equation on the Entire Axis.* Vestnik Leningrad. Univ., **17** (1962), 56–64.

120   I. Alvarez-Romero and Yu.I. Lyubarskii

[3] R. Carlson, *Inverse eigenvalue problems on directed graphs*. Trans. Amer. Math. Soc., **351** (1999), no. 10, 4069–4088.

[4] A. Cohen, T. Kappeler, *Scattering and inverse scattering for steplike potentials in the Schrödinger equation*. Indiana Univ. Math. J., **34** (1985), no. 1, 127–180.

[5] I. Egorova, *The scattering problem for step-like Jacobi operator*. Matematicheskaya fizika, analiz, geometriya, **9** (2002), no. 2, 188-205.

[6] I. Egorova, J. Michor, G. Teschl, *Scattering theory for Jacobi operators with general steplike quasi-periodic background*, Zh. Mat. Fiz. Anal. Geom. **4-1**, 33–62 (2008).

[7] N.I. Gerasimenko, B.S. Pavlov *A scattering problem on noncompact graphs* (Russian), Teoret. Mat. Fiz. 74 (1988), no. 3, 345–359; translation in Theoret. and Math. Phys., **74** (1988), no. 3, 230–240.

[8] F. Gesztesy, R. Nowell, and W. Pötz, *One-Dimensional Scattering Theory for Quantum Systems with Nontrivial Spatial Asymptotics*, Differential and Integral Equations, **10** (1997), 521–546.

[9] B. Gutkin, U. Smilansky, *Can one hear the shape of a graph?* J. Phys. A, **34:31** (2001), 6061–6068.

[10] P. Kuchment, *Quantum graphs. II. Some spectral properties of quantum and combinatorial graphs*. J. Phys. A, **38** (2005), no. 22, 4887–4900.

[11] P. Kurasov, M. Nowaczyk, *Inverse spectral problem for quantum graphs*, J. Phys. A **38** (2005), 4901–4915.

[12] V. Kostrykin, R. Schrader, *The inverse scattering problem for metric graphs and the traveling salesman problem*, `arXiv:math-ph/0603010`, pp. 1–68, 2006.

[13] Yu.I. Lyubarskii, V.A. Marchenko, *Direct and inverse problems of multichannel scattering* (Russian), Funktsional. Anal. i Prilozhen. **41** (2007), 58–77, (Translation in Funct. Anal. Appl. **41** (2007), no. 2, 126–142)

[14] Yu.I. Lyubarskii, V.A. Marchenko, *Inverse problem for small oscillations*, Spectral analysis, differential equations and mathematical physics: a festschrift in honor of Fritz Gesztesy's 60th birthday, 263–290, Proc. Sympos. Pure Math., 87, Amer. Math. Soc., Providence, RI, 2013.

[15] V.A. Marchenko, *Introduction to the theory of inverse problems of spectral analysis*, (Russian), Kharkov, 2005, pp. 1–135.

[16] V.A. Marchenko, V.V. Slavin, *Inverse problems it theory of small oscillations*, (Russian), Kiev, Naukova Dumka, 2015, 219 pp.

[17] R.G. Newton, *Scattering theory of waves and particles*, Dover Publications, Inc., Mineola, NY, 2002, pp. xx+745

[18] B.S. Pavlov, M.D. Faddeev, *A model of free electrons and the scattering problem*. (Russian), Teoret. Mat. Fiz. **55** (1983), no. 2, 257–268.

[19] G. Teschl, *Jacobi operators and completely integrable nonlinear lattices*, American Mathematical Society, Providence, RI, 2000, xvii+351 pp.

[20] S. Venakides, P. Deift, and R. Oba, *The Toda shock problem*, Comm. Pure. Appl. Math., **44** (1991), no. 8, 1171–1242.

Isaac Alvarez-Romero and Yurii Lyubarskii
Department of Mathematical Sciences
Norwegian University of Science and Technology
NO–7491 Trondheim, Norway
e-mail: `isaac.romero@math.ntnu.no`, `isaacalrom@gmail.com`, `yura@math.ntnu.no`

Operator Theory:
Advances and Applications, Vol. 261, 121–139
© Springer International Publishing AG, part of Springer Nature 2018

Note on the Resonance Method for the Riemann Zeta Function

Andriy Bondarenko and Kristian Seip

To the memory of Victor Havin

Abstract. We improve Montgomery's Ω-results for $|\zeta(\sigma + it)|$ in the strip $1/2 < \sigma < 1$ and give in particular lower bounds for the maximum of $|\zeta(\sigma+it)|$ on $\sqrt{T} \le t \le T$ that are uniform in σ. We give similar lower bounds for the maximum of $|\sum_{n \le x} n^{-1/2-it}|$ on intervals of length much larger than x. We rely on our recent work on lower bounds for maxima of $|\zeta(1/2 + it)|$ on long intervals, as well as work of Soundararajan, Gál, and others. The paper aims at displaying and clarifying the conceptually different combinatorial arguments that show up in various parts of the proofs.

Mathematics Subject Classification (2010). Primary 11M06. Secondary 11C20.

Keywords. Resonance method, Riemann zeta function, Hardy spaces, Dirichlet series.

1. Introduction

Soundararajan [22] and Hilberdink [13] presented independently slightly different versions of a technique, known as the resonance method, for detecting large values of the Riemann zeta function $\zeta(s)$. In our recent paper [6], we used Soundararajan's version of this method and the construction of a special multiplicative function to show that

$$\max_{\sqrt{T} \le t \le T} \left| \zeta\left(\frac{1}{2} + it\right) \right| \ge \exp\left(\left(\frac{1}{\sqrt{2}} + o(1)\right) \sqrt{\frac{\log T \log \log \log T}{\log \log T}} \right) \qquad (1.1)$$

when $T \to \infty$. This gave an improvement by a power of $\sqrt{\log \log \log T}$ compared with previously known estimates [3, 22].

Research supported in part by Grant 227768 of the Research Council of Norway.

In this note, we will apply the resonance method to two closely related problems, namely to find large values of respectively $\zeta(\sigma + it)$ for $1/2 < \sigma < 1$ and the partial sum $\sum_{n \leq M} n^{-1/2-it}$ on certain long intervals (depending on M). We will find uniform lower bounds on the maximum in the strip $1/2 < \sigma < 1$ and show in particular that the bound on the right-hand side of (1.1) (with $1/\sqrt{2}$ replaced by a different constant) holds as far as $1/\log \log T$ to the right of the critical line.

Before proceeding to the details of these new results, we would like to comment on the relationship between our subject and Hardy spaces, the presumed topic of the present volume. As outlined in [21], our construction of resonators originates in Bohr's several complex variables perspective of Dirichlet series and our recent work on Hardy spaces of Dirichlet series. The present paper can thus be viewed as an outgrowth of the remarkably rich subject of Hardy spaces and more specifically of a branch of it that interacts with number theory. Moreover, one may interpret our theorem on partial sums (Theorem 2 below) as dealing with a well-known type of problem in complex analysis, namely how small the maximal size of an analytic function can be on a set of uniqueness that in some sense is "small". For further information about Hardy spaces of Dirichlet series and connections with number theory, we refer to the survey paper [21] and the monograph [18].

2. Statement of main results

A less precise version of the following result was stated without proof in [6].

Theorem 1. *There exists a positive and continuous function $\nu(\sigma)$ on $(1/2, 1)$, bounded below by $1/(2 - 2\sigma)$, with the asymptotic behavior*

$$\nu(\sigma) = \begin{cases} (1 - \sigma)^{-1} + O(|\log(1 - \sigma)|), & \sigma \nearrow 1 \\ (1/\sqrt{2} + o(1))\sqrt{|\log(2\sigma - 1)|}, & \sigma \searrow 1/2, \end{cases}$$

and such that the following holds. For $1/2 + 1/\log \log T \leq \sigma \leq 3/4$, if T is sufficiently large, then

$$\max_{t \in [\sqrt{T}, T]} \left| \zeta(\sigma + it) \right| \geq \exp\left(\nu(\sigma) \frac{(\log T)^{1-\sigma}}{(\log \log T)^\sigma} \right) \tag{2.1}$$

and for $3/4 \leq \sigma \leq 1 - 1/\log \log T$,

$$\max_{t \in [T/2, T]} \left| \zeta(\sigma + it) \right| \geq \log \log T \exp\left(c + \nu(\sigma) \frac{(\log T)^{1-\sigma}}{(\log \log T)^\sigma} \right), \tag{2.2}$$

with c an absolute constant independent of T.

To place this result in context, we recall Levinson's classical estimate[1] [15]

$$\max_{1 \leq t \leq T} |\zeta(1 + it)| \geq e^\gamma \log \log T + O(1), \tag{2.3}$$

---

[1] This result was later improved by Granville and Soundararajan [12] who managed to add a positive term of size $\log \log \log T$ on the right-hand side of (2.3).

where γ is the Euler–Mascheroni constant. We now observe that Theorem 1 gives a "smooth" transition between the two endpoint cases (2.3) and (1.1). The factor $\log\log T$ on the right-hand side of (2.2) is only needed for σ close to the right endpoint $1 - 1/\log\log T$, to get the transition to Levinson's estimate. Theorem 1 gives a notable improvement of a classical estimate of Montgomery [17] for the range $1/2 < \sigma < 1$. See [19] and the discussion in [6] for the best estimates known previously.

The choice of intermediate abscissa $\sigma = 3/4$ is somewhat arbitrary (any fixed σ_0, $1/2 < \sigma_0 < 1$ would do), and we could have shortened the interval in (2.1) (depending on σ). Indeed, the precise statement of Theorem 1 is a tradeoff between conveying the main point of the transition between the two endpoint cases and keeping the technicalities reasonably simple.

We have refrained from making a precise statement about sharp estimates in the short intervals $[1/2, 1/2 + 1/\log\log T]$ and $[1 - 1/\log\log T, 1]$, although our method would certainly allow us to do it. The main point is that the order of magnitude of the respective endpoint estimates persists in these intervals. It may seem surprising that these intervals are as long as $1/\log\log T$ on either side. We will see in the course of the proof that this can be attributed to the resonance method's selection of smooth numbers[2] in the construction of resonating Dirichlet polynomials.

In our proof of Theorem 1, we use the approximate formula

$$\zeta(\sigma + it) = \sum_{n \leq x} n^{-\sigma - it} - \frac{x^{1-\sigma-it}}{1 - \sigma - it} + O(x^{-\sigma}), \tag{2.4}$$

which holds uniformly in the range $\sigma \geq \sigma_0 > 0$, $|t| \leq x$ (see [23, Theorem 4.11]). This means that detecting large values of $\zeta(\sigma + it)$ for $1/2 \leq \sigma \leq 1$ and $|t| \leq T$ is mainly a question about finding large values of the Dirichlet polynomial $\sum_{n \leq T} n^{-\sigma - it}$ for $|t| \leq T$.

We find it to be of interest to see what we get when we look for large values of just the partial sum itself on longer intervals. Thus we remove the a priori restriction on the length of the interval forced upon us by the approximate formula (2.4). We will only consider $\sigma = 1/2$ and introduce the notation

$$D_M(t) = \sum_{n \leq M} n^{-1/2 - it}.$$

Theorem 2. *Suppose that c, $0 < c < 1/2$, is given. If T is sufficiently large and $M \geq \exp\left(e\sqrt{\log T \log\log T \log\log\log T/2}\right)$, then*

$$\max_{t \in [\sqrt{T}, T]} |D_M(t)| \geq \exp\left(c\sqrt{\frac{\log T \log\log\log T}{\log\log T}}\right).$$

---

[2] The smoothness (sometimes called the friability) of a positive integer n is measured by the largest prime p dividing n. The smaller this prime is, the smoother the number is.

This theorem gives information about the precision of the resonance method as well as its limitations. We notice that the global maximum satisfies

$$\|D_M\|_\infty := \max_t |D_M(t)| \sim \sqrt{M},$$

and hence we see that our method gives us that when M takes the minimal value $\exp\left(e\sqrt{\log T \log\log T \log\log\log T}/2\right)$, the maximum on $[\sqrt{T}, T]$ is at least $\|D_M\|_\infty^{\eta/\log\log M}$ for some positive number η. This means that the value of the maximum "predicted" by the resonance method is at most a power of order $1/\log\log T$ (or equivalently of order $1/\log\log M$) off the true maximum (whatever it is). On the other hand, while we know that $|D_M(t)|$ "eventually" will come arbitrarily close to the absolute maximum, the interval $[\sqrt{T}, T]$ is of course far too short for us to guarantee, by general considerations, that we get anywhere near $\|D_M\|_\infty$. Hence the resonance method could in fact be considerably more precise than what we can safely conclude that it is in this case.

A word on notation, before we turn to a general discussion of the resonance method and the proofs of our theorems: we will use the shorthand notation $\log_2 x := \log\log x$ and $\log_3 x := \log\log\log x$.

3. The resonance method – general considerations

The basic idea of either version of the resonance method is to identify a special Dirichlet polynomial

$$R(t) = \sum_{m \in \mathcal{N}} r(m) m^{-it}$$

that "resonates well" with the object at hand, which in our case is the partial sum $\sum_{n \le x} n^{-\sigma - it}$ on a given interval. The precise meaning of this is that the integral of $|R(t)|^2$ (mollified by multiplication by a suitable smooth bump function) times $\sum_{n \le x} n^{-\sigma - it}$ is as large as possible, given that the coefficients $r(m)$ have a fixed square sum and also subject to whatever a priori restrictions we need to put on the set of integers $\mathcal{N}$. The method will not only produce large values, but also give information about which of the terms in $\sum_{n \le x} n^{-\sigma - it}$ do contribute in an "essential" way.

The technicalities will differ considerably depending on σ, for reasons that will become clear below. In our study of $\zeta(\sigma + it)$ in the range $3/4 \le \sigma \le 1$, we will use Soundararajan's original method. This means that we choose a smooth function Ψ compactly supported in $[1/2, 1]$, taking values in the interval $[0, 1]$ with $\Psi(t) = 1$ for $5/8 \le t \le 7/8$. We define

$$M_1(R, T) := \int_{-\infty}^{\infty} |R(t)|^2 \Psi\left(\frac{t}{T}\right) dt, \tag{3.1}$$

$$M_2(R, T) := \int_{-\infty}^{\infty} \zeta(\sigma + it) |R(t)|^2 \Psi\left(\frac{t}{T}\right) dt. \tag{3.2}$$

Then

$$\max_{T/2 \le t \le T} |\zeta(\sigma + it)| \ge \frac{|M_2(R,T)|}{M_1(R,T)}, \tag{3.3}$$

and the goal is therefore to maximize the ratio on the right-hand side of (3.3). We require that $\max \mathcal{N} \le T^{1-\varepsilon}$ for some fixed ε, $0 < \varepsilon < 1$, and get by straightforward computations (see [22, pp. 471–472]) that

$$M_1(R,T) = T\hat{\Psi}(0)\big(1 + O(T^{-1})\big) \sum_{n \in \mathcal{N}} |r(n)|^2 \tag{3.4}$$

and

$$M_2(R,T) = T\hat{\Psi}(0) \sum_{n \in \mathcal{N}, mk=n} \frac{r(m)\overline{r(n)}}{k^\sigma} + O(T^{1-\sigma} \log T) \sum_{n \in \mathcal{N}} |r(n)|^2. \tag{3.5}$$

Hence the problem of estimating the right-hand side of (3.3) boils down to finding out how large the ratio

$$\sum_{n \in \mathcal{N}, mk=n} \frac{r(m)\overline{r(n)}}{k^\sigma} \Big/ \sum_{n \in \mathcal{N}} |r(n)|^2 \tag{3.6}$$

can be under the a priori restriction that $\max \mathcal{N} \le T^{1-\varepsilon}$. This problem was solved in [22, Theorem 2.1] for $\sigma = 1/2$.

As shown in [6], we can do better when $\sigma = 1/2$ by removing the a priori restriction that $\max \mathcal{N} \le T^{1-\varepsilon}$, and this is also true when σ is not too close to 1. If we again manage to reduce the problem to that of maximizing a ratio like the one in (3.6), then we clearly are in a better position. However, arriving at such an optimization problem is less straightforward, mainly because more terms will contribute in either of the sums representing respectively $M_1(R,T)$ and $M_2(R,T)$. An additional problem is that sets of integers $\mathcal{N}$ involved in making expressions like (3.6) large typically enjoy a multiplicative structure, while estimating sums like those representing $M_1(R,T)$ and $M_2(R,T)$ requires some "additive control". We will now present the remedies, introduced in [6], for getting around these problems.

We begin with the problem of "additive control". We go "backwards" and start with the problem of maximizing

$$\sum_{n \in \mathcal{M}, mk=n} \frac{f(m)\overline{f(n)}}{k^\sigma} \Big/ \sum_{n \in \mathcal{M}} |f(n)|^2 \tag{3.7}$$

for a suitable set $\mathcal{M}$ and arithmetic function $f(n)$ under the condition that $|\mathcal{M}| \le N$. We then extract the resonating Dirichlet polynomial from a solution (or approximate solution) to this problem as follows, assuming as we may that $f(n)$ is nonnegative. Following an idea from [1], we let $\mathcal{J}$ be the set of integers j such that

$$[(1 + T^{-1})^j, (1 + T^{-1})^{j+1}) \cap \mathcal{M} \ne \emptyset,$$

and let m_j be the minimum of $[(1 + T^{-1})^j, (1 + T^{-1})^{j+1}) \cap \mathcal{M}$ for j in $\mathcal{J}$. Then set

$$\mathcal{N} := \{m_j : j \in \mathcal{J}\}$$

and

$$r(m) := \left(\sum_{n \in \mathcal{M}, 1 - T^{-1}(\log T)^2 \leq n/m \leq 1 + T^{-1}(\log T)^2} f(n)^2 \right)^{1/2} \tag{3.8}$$

for every m in $\mathcal{N}$. Note that plainly $|\mathcal{N}| \leq |\mathcal{M}| \leq N$. By taking the local ℓ^2 average as in (3.8), we get a precise relation between $f(n)$ and $r(n)$, while at the same time we get the desired "additive control" because each of the intervals $[(1 + T^{-1})^j, (1 + T^{-1})^{j+1})$ contains at most one integer from $\mathcal{N}$.

We turn next to the counterparts to (3.1) and (3.2). We consider now a longer interval of the form $[T^\beta, T]$ with $0 < \beta < 1$; it will be convenient for us to fix once and for all $\beta = 1/2$. Moreover, we use the Gaussian $\Phi(t) := e^{-t^2/2}$ as a mollifier. Our replacements for (3.1) and (3.2) are then, respectively,

$$\widetilde{M}_1(R, T) := \int_{\sqrt{T} \leq |t| \leq T} |R(t)|^2 \Phi\left(\frac{\log T}{T} t \right) dt,$$

$$\widetilde{M}_2(R, T) := \int_{\sqrt{T} \leq |t| \leq T} \zeta(\sigma + it) |R(t)|^2 \Phi\left(\frac{\log T}{T} t \right) dt, \tag{3.9}$$

and we get the inequality

$$\max_{\sqrt{T} \leq t \leq T} |\zeta(\sigma + it)| \geq \frac{|\widetilde{M}_2(R, T)|}{\widetilde{M}_1(R, T)}. \tag{3.10}$$

We state first the estimate for $\widetilde{M}_1(R, T)$ obtained in [6, Formula (22)]. This is a matter of direct computation based on the definitions given above.

Lemma 3. *There exists an absolute constant C such that*

$$\widetilde{M}_1(R, T) \leq CT(\log T)^3 \sum_{n \in \mathcal{M}} f(n)^2. \tag{3.11}$$

To estimate (3.9), we extend the integral to the whole real line, so that we can take advantage of the fact that the Fourier transform $\widehat{\Phi}$ of Φ is positive. We chose the larger set $\sqrt{T} \leq |t| \leq T$ and a different dilation factor of the mollifier $((\log T)/T$ instead of $1/T)$, because these choices allow us to get the control we need of the integral over the complementary set. Indeed, the estimation for $|t| \geq T$ is trivial because of the rapid decay of the Gaussian and our choice of dilation factor, while the following estimate takes care of the interval $|t| \leq \sqrt{T}$: for arbitrary numbers $\lambda > 0, 0 < \beta < 1$, and $0 < \sigma < 1$, we have

$$\left| \sum_{1 \leq n \leq M} n^{-\sigma} \int_{-T^\beta}^{T^\beta} \left(\frac{\lambda}{n} \right)^{it} \Phi\left(\frac{\log T}{T} t \right) dt \right| \leq C \max\left(T^\beta, M^{1-\sigma} \log M \right), \tag{3.12}$$

where the constant C is independent of λ, β, σ. This is proved by making a minor adjustment of the proof of [6, Lemma 4], which deals only with the case $\sigma = 1/2$. Doing the same computations as in [6], relying crucially on the positivity of $\widehat{\Phi}$, we arrive at the following lemma (see formula (14) in [6]).

Lemma 4. *Suppose* $1/2 \leq \sigma \leq 1$ *and* $|\mathcal{M}| \leq \sqrt{T}$. *There exist absolute positive constants* c, C *such that*

$$\widetilde{M}_2(R,T) \geq c \left(\frac{T}{\log T} \sum_{n \in \mathcal{M}, mk=n, k \leq T} \frac{f(n)f(m)}{k^\sigma} - CT(\log T)^4 \sum_{n \in \mathcal{M}} f(n)^2 \right).$$

The powers of $\log T$ are harmless if $\sigma \leq \sigma_0 < 1$ for some fixed σ_0, but they make this lemma useless when σ is close to 1 since the lower bound in (2.3) is of order $\log_2 T$. This is why we need both versions of the resonance method when considering the whole range $1/2 + 1/\log_2 T \leq \sigma \leq 1 - 1/\log_2 T$.

The resonance method for the partial sum problem yields the same bounds, up to an obvious modification of the indices in the summation in Lemma 4. Indeed, defining

$$\widetilde{\widetilde{M}}_2(R,T) := \int_{\sqrt{T} \leq |t| \leq T} D_M(t)|R(t)|^2 \Phi\left(\frac{\log T}{T} t\right) dt,$$

we get:

Lemma 5. *Suppose that* $|\mathcal{M}| \leq \sqrt{T}$. *There exist absolute positive constants* c, C *such that*

$$\widetilde{\widetilde{M}}_2(R,T) \geq c \left(\frac{T}{\log T} \sum_{n \in \mathcal{M}, mk=n, k \leq M} \frac{f(n)f(m)}{\sqrt{k}} - CT(\log T)^4 \sum_{n \in \mathcal{M}} f(n)^2 \right).$$

We are now left with the problem of making the first sum on the right-hand side large; the problem of making the right-hand side of (3.7) large is just the special case when $M = [T]$. In the next session, we will show how to deal with this problem for a wide range of values of M.

4. Gál-type extremal problems and the proof of Theorem 1

4.1. Background on Gál-type extremal problems

Before turning to the extremal problems arrived at in the previous section, we would like to place them in context by describing briefly a line of research that has been instrumental for our approach. This is the study of greatest common divisor (GCD) sums of the form

$$\sum_{m,n \in \mathcal{M}} \frac{(m,n)^{2\sigma}}{(mn)^\sigma} \tag{4.1}$$

and the associated (normalized) quadratic forms

$$\sum_{m,n \in \mathcal{M}} f(m)f(n) \frac{(m,n)^{2\sigma}}{(mn)^\sigma} \Big/ \sum_{n \in \mathcal{M}} f(n)^2, \tag{4.2}$$

where $\mathcal{M}$ is as above and we again assume that $f(n)$ is nonnegative and does not vanish on $\mathcal{M}$. We observe that (3.7) is smaller than (4.2) because the former is obtained from the latter by restricting the sum in the nominator to a subset of $\mathcal{M} \times \mathcal{M}$. In most cases of interest when $1/2 \leq \sigma < 1$, we may obtain a reverse inequality so that the two expressions are of the same order of magnitude. In general, it is clear that if (4.2) is large, then also (3.7) will be large.

The problem is to decide how large either of the two expressions (4.1) or (4.2) can be under the assumption that $|\mathcal{M}| = N$, and more specifically we are interested in the asymptotics when $N \to \infty$ and σ is fixed with $0 < \sigma \leq 1$. We refer to (4.1) and (4.2) as Gál-type sums because the topic begins with a sharp bound of Gál [11] (of order $CN(\log \log N)^2$) for the growth of (4.1) when $\sigma = 1$. Dyer and Harman [9] obtained the first nontrivial estimates for the range $1/2 \leq \sigma < 1$, and during the past few years, we have reached an essentially complete understanding for the full range $0 < \sigma \leq 1$, thanks to the papers [2, 4, 5, 16]. The techniques used for different values of σ differ considerably, and the problem is particularly delicate for $\sigma = 1/2$ at which an interesting "phase transition" occurs. We refer to [21] for an overview of these results and to [2, 16] for information about the many different applications of such asymptotic estimates.

It is the insight accumulated in this research that has led to the constructions given below. More specifically, we will follow Gál [11] when σ is close to 1 and [6] when σ is close to $1/2$. The reader will notice that our set $\mathcal{M}$ will contain very smooth numbers when σ is close to 1 in contrast to what happens near $\sigma = 1/2$. Our treatment of the latter case shows that more and more primes are needed when σ decreases; the simplest possible choice (made by Aistleitner in [1]) of taking r to be of size $\log N / \log 2$ and the n_j to be the divisors of the square-free number $p_1 \cdots p_r$ will be nearly optimal only when σ is "far" from the endpoints 1 and $1/2$. Translating this philosophy to Soundararajan's method, we find that the terms picked out in the approximating sum $\sum_{n \leq T} n^{-\sigma - it}$ correspond to decreasingly smooth numbers when σ goes from 1 to $1/2$.

4.2. Levinson's case $\sigma = 1$ revisited

It is instructive to consider first the endpoint case $\sigma = 1$. We will now show that

$$\max_{T/2 \leq t \leq T} |\zeta(1 + it)| \geq e^\gamma \log_2 T + O(\log_3 T). \tag{4.3}$$

This estimate is slightly worse than (2.3) and the best known result of Granville and Soundararajan [12], but the benefit is the simplicity of the proof and also that the interval has been shortened. We notice at this point that Hilberdink got the estimate (2.3) by his version of the resonance method.

We fix a positive number x and an integer ℓ (to be determined later) and let $\mathcal{M}$ be the set of divisors of the number

$$K = K(x, \ell) := \prod_{p \leq x} p^{\ell - 1}.$$

We require that $K \leq [\sqrt{T}]$ and choose $r(n)$ to be the characteristic function of $\mathcal{M}$. A computation shows that

$$\sum_{n\in\mathcal{M},mk=n} \frac{1}{k^\sigma} = \prod_{p\leq x}\left(\ell + \sum_{\nu=1}^{\ell-1} \frac{\ell - \nu}{p^{\nu\sigma}}\right).$$

Hence

$$\sum_{mk=n} \frac{r(m)r(n)}{k^\sigma} \Big/ \sum_{n\in\mathcal{M}} r(n)^2 = \prod_{p\leq x}\left(1 + \sum_{\nu=1}^{\ell-1}\left(1 - \frac{\nu}{\ell}\right)p^{-\nu\sigma}\right). \qquad (4.4)$$

We now set $\sigma = 1$ and find that

$$\sum_{mk=n} \frac{r(m)r(n)}{k} \Big/ \sum_{n\in\mathcal{M}} r(n)^2 = \prod_{p\leq x}\left((1-p^{-1})^{-1} + \sum_{\nu=2}^{\ell-1}\frac{\nu}{\ell}p^{-\nu} + O(p^{-\ell})\right)$$

$$= (1 + O(\ell^{-1})) \prod_{p\leq x}(1-p^{-1})^{-1}$$

$$= \left(1 + O(\ell^{-1}) + O\left(\frac{1}{\sqrt{x}\log x}\right)\right) e^\gamma \log x,$$

where at the last step we used Mertens' third theorem (see [8] for a precise analysis of the error term). By the prime number theorem, we may choose $x = (\log T)/(2\log_2 T)$ and $\ell = [\log_2 T]$ if T is large enough. Taking into account (3.4) and (3.5), we obtain the desired result (4.3).

4.3. The case $3/4 \leq \sigma \leq 1 - 1/\log_2 T$

We follow the argument of the preceding subsection up to (4.4), from which we deduce that

$$\sum_{mk=n} \frac{r(m)\overline{r(n)}}{k^\sigma} \Big/ \sum_{n\in\mathcal{M}} |r(n)|^2 \geq \prod_{p\leq x}\left(1 + \left(1 - \frac{1}{\ell}\right)p^{-\sigma}\right)$$

$$\geq \prod_{p\leq x}\left(1 + p^{-\sigma}\right)^{1-\frac{1}{\ell}}; \qquad (4.5)$$

here we have used Bernoulli's inequality at the last step. We will use the following lemma to estimate the latter expression.

Lemma 6. *There exists an absolute constant C such that*

$$\sum_{p\leq x} p^{-\sigma} \geq \sigma \log_2 x + C + \frac{x^{1-\sigma}}{(1-\sigma)\log x}$$

whenever $(1-\sigma)\log x \geq 1/2$.

Proof. By Abel summation and the inequality $\pi(x) > x/\log x$ which is valid for $x \geq 17$ (see [20]), we find that

$$\sum_{p\leq x} p^{-\sigma} \geq \frac{x^{1-\sigma}}{\log x} + \sigma \int_2^x \frac{dy}{y^\sigma \log y} + C', \qquad (4.6)$$

where C' is an absolute constant. Making the change of variables $u = \log_2 y$, we see that

$$\int_2^x \frac{dy}{y^\sigma \log y} = \int_{\log_2 2}^{\log_2 x} e^{(1-\sigma)e^u} du$$

$$= \log_2 x - \log_2 2 + \sum_{j=1}^\infty \frac{(1-\sigma)^j (\log^j x - \log^j 2)}{j \cdot j!}$$

$$= \log_2 x - \log_2 2 + \int_{(1-\sigma)\log 2}^{(1-\sigma)\log x} \frac{e^y - 1}{y} dy.$$

Now using the trivial bound

$$\int_0^a \frac{e^y - 1}{y} dy \geq \frac{e^a}{a} - \frac{a+1}{a}$$

and returning to (4.6), we obtain the desired estimate. $\square$

We are now prepared to give the first part of the proof of Theorem 1.

Proof of Theorem 1 – part 1. Making the same choices $x = (\log T)/(2\log_2 T)$ and $\ell = [\log_2 T]$ as in the case $\sigma = 1$ and returning to (4.5), we see that Lemma 6 gives

$$\sum_{mk=n} \frac{r(m)\overline{r(n)}}{k^\sigma} \bigg/ \sum_{n\in M} |r(n)|^2 \geq \exp\left(\sigma \log_3 T + \frac{2^{\sigma-1}(\log T)^{1-\sigma}}{(1-\sigma)(\log_2 T)^\sigma} - E(T,\sigma)\right),$$

where

$$E(T,\sigma) \leq C + \frac{(1+\delta)\log_3 T (\log T)^{1-\sigma}}{(1-\sigma)(\log_2 T)^{\sigma+1}}$$

for arbitrary $\delta > 0$ when T is sufficiently large. Returning to (3.3), (3.4), and (3.5), we now obtain (2.2) and the desired asymptotic behavior of $\nu(\sigma)$ when $\sigma \nearrow 1$ because

$$\frac{\log_3 T}{\log_2 T} \leq (1-\sigma)|\log(1-\sigma)|$$

when $\log_2 T \geq e$, by our a priori assumption that $\sigma \leq 1 - 1/\log_2 T$. We also get the uniform lower bound $\nu(\sigma) \geq 1/(2-2\sigma)$ because $2^{\sigma-1} > 1/2$ when $\sigma \geq 1/2$ and $\log_3 T/\log_2 T \to 0$ when $T \to \infty$. $\square$

4.4. The case $1/2 + 1/\log_2 T \leq \sigma \leq 3/4$

In view of the preceding section, we already know that (2.2) holds for large T when we choose $\nu(\sigma)$ to be an appropriate function bounded below by $1/(2-2\sigma)$. This is just because the interval $[T/2, T]$ is shorter than $[\sqrt{T}, T]$ when $T \geq 4$. What remains is therefore to prove that $\nu(\sigma)$ can be chosen such that it also has the desired asymptotic behavior when $\sigma \searrow 1/2$, while (2.2) still holds for large T.

We will use a construction from [6, Section 3] which one should understand as a "reversion" of an application of the Cauchy–Schwarz inequality in [5]. This key step in bounding Gál-type sums from above when $\sigma = 1/2$ relies on the existence of so-called divisor closed extremal sets of square-free numbers and a certain completeness property enjoyed by such sets. The interested reader is advised to consult [5] to see the close connection between our construction and the proof given in that paper.

We recall that, in view of Lemma 5, our goal is to find a multiplicative function $f(n)$ (depending on σ) and an associated set of integers $\mathcal{M}$ with $|\mathcal{M}| \leq \sqrt{T}$ such that

$$\sum_{n \in \mathcal{M}, mk=n, k \leq M} \frac{f(n)f(m)}{k^{\sigma}} \geq W(T, \sigma) \times \sum_{n \in \mathcal{M}} f(n)^2 \qquad (4.7)$$

for suitable values of M, depending on T, where

$$W(T, \sigma) = \begin{cases} \exp\left(c\sqrt{\frac{\log T \log_3 T}{\log_2 T}}\right), & \sigma = 1/2 \\ \exp\left(\nu(\sigma)\frac{(\log T)^{1-\sigma}}{(\log_2 T)^{\sigma}}\right), & 1/2 + 1/\log_2 T \leq \sigma \leq 3/4 \end{cases}$$

and $0 < c < 1/2$. This was done for $\sigma = 1/2$ and $M \geq N^{\varepsilon}$ for every $\varepsilon > 0$ in [6, Section 3]. We will now extend this construction to the range $1/2 + 1/\log_2 T \leq \sigma \leq 3/4$, and we will show that we can allow much smaller values of M. Since we already obtained the lower bound $\nu(\sigma) \geq 1/(2 - 2\sigma)$ in the preceding subsection, we will mainly be interested in estimates for $\nu(\sigma)$ when σ is sufficiently close to $1/2$.

We begin with the construction of $f(n)$ and $\mathcal{M}$ when $1/2 + 1/\log_2 T \leq \sigma \leq 3/4$. We follow the scheme in [6, Section 3] word for word, the only essential difference being that we let P be the set of all primes p such that

$$e \log N \log_2 N < p \leq \log N \exp((2\sigma - 1)^{-\alpha}) \log_2 N$$

for a suitable α, $0 < \alpha < 1$, and set

$$f(p) := \frac{(\log N \log_2 N)^{1-\sigma}}{\sqrt{|\log(2\sigma - 1)|}} \cdot \frac{1}{p^{1-\sigma}(\log p - \log_2 N - \log_3 N)},$$

where $N = [\sqrt{T}]$. This defines a multiplicative function $f(n)$, if we require it to be supported on the square-free numbers with prime factors in P. Arguing as in [6, Section 3], we are now led to consider the quantity

$$A(N, \sigma) := \frac{1}{\sum_{j \in \mathbb{N}} f(j)^2} \sum_{n \in \mathbb{N}} \frac{f(n)}{n^{\sigma}} \sum_{d \mid n} f(d)d^{\sigma} = \prod_{p \in P} \frac{1 + f(p)^2 + f(p)p^{-\sigma}}{1 + f(p)^2}.$$

The following estimate, which is a counterpart to [6, Lemma 1] for the case $\sigma = 1/2$, is of basic importance.

Lemma 7. *We have*

$$A(N, \sigma) \geq \exp\left((\alpha + o(1))\frac{|\log(2\sigma - 1)|^{3/2}}{1 + |\log(2\sigma - 1)|} \frac{(\log N)^{1-\sigma}}{(\log_2 N)^{\sigma}}\right) \qquad (4.8)$$

uniformly for $1/2 + 1/\log_2 T \leq \sigma \leq 3/4$ when $T \to \infty$.

Proof. Since $f(p) < 1/\sqrt{|\log(2\sigma - 1)|}$ for every p in P, we find that

$$A(N, \sigma) \geq \exp\left((1 + o(1))\frac{|\log(2\sigma - 1)|}{1 + |\log(2\sigma - 1)|}\sum_{p \in P} f(p)p^{-\sigma}\right). \qquad (4.9)$$

Here and in what follows, the error term goes to 0 when $T \to \infty$ uniformly for $1/2 + 1/\log_2 T \leq \sigma \leq 3/4$. Now we obtain

$$\sum_{p \in P} f(p)p^{-\sigma} = \frac{1}{\sqrt{|\log(2\sigma - 1)|}} \cdot (\log N \log_2 N)^{1-\sigma}$$

$$\times \sum_{p \in P} \frac{1}{p(\log p - \log_2 N - \log_3 N)},$$

and

$$\sum_{p \in P} \frac{1}{p(\log p - \log_2 N - \log_3 N)}$$

$$= (1 + o(1))\int_{e \log N \log_2 N}^{\log N \exp((2\sigma-1)^{-\alpha}) \log_2 N} \frac{dx}{x \log x(\log x - \log_2 N - \log_3 N)}$$

$$= (1 + o(1))\int_{1 + \log_2 N + \log_3 N}^{\log_2 N + (2\sigma-1)^{-\alpha} + \log_3 N} \frac{dt}{t(t - \log_2 N - \log_3 N)}$$

$$= (\alpha + o(1))\frac{|\log(2\sigma - 1)|}{\log_2 N}.$$

Returning to (4.9), we obtain the desired estimate (4.8). $\square$

We proceed next to choose our set $\mathcal{M}$. To this end, we let P_k be the set of all primes p such that $e^k \log N \log_2 N < p \leq e^{k+1} \log N \log_2 N$ for $k = 1, \ldots, [(2\sigma - 1)^{-\alpha}]$. Fix $1 < a < 1/\alpha$. Then let M_k be the set of integers that have at least $\frac{a \log N}{k^2 |\log(2\sigma - 1)|}$ prime divisors in P_k, and let M'_k be the set of integers from M_k that have prime divisors only in P_k. Finally, set

$$\mathcal{M} := \mathrm{supp}(f) \setminus \bigcup_{k=1}^{[(2\sigma-1)^{-\alpha}]} M_k.$$

We need to show that we have the bound $|\mathcal{M}| \leq N$ if N is large and α and a have been chosen appropriately. As in [6, Section 3], we start with the facts that

$$\binom{m}{n} \leq \exp\left(n(\log m - \log n) + n + \log m\right) \qquad (4.10)$$

holds when $n \leq m$ and m is large enough and that

$$\frac{\binom{m}{n}}{\binom{m}{n-1}} = \frac{m-n+1}{n} \geq 2 \qquad (4.11)$$

when $m \geq 3n - 1$. By the prime number theorem, the cardinality of each P_k is at most $e^{k+1} \log N$, and we therefore see, using first (4.11) and then (4.10), that

$$|\mathcal{M}| \leq \prod_{k=1}^{[(2\sigma-1)^{-\alpha}]} \sum_{j=0}^{\left[\frac{a \log N}{k^2 |\log(2\sigma-1)|}\right]} \binom{e^{k+1} \log N}{j} \leq \prod_{k=1}^{[(2\sigma-1)^{-\alpha}]} 2 \binom{e^{k+1} \log N}{\left[\frac{a \log N}{k^2 |\log(2\sigma-1)|}\right]}$$

$$\leq \exp\left(\sum_{k=1}^{[(2\sigma-1)^{-\alpha}]} \left(1 + \frac{a \log N \left(k+2+\log|\log(2\sigma-1)|+2\log k\right)}{k^2 |\log(2\sigma-1)|} + k + 1 + \log_2 N\right)\right).$$

Hence, choosing α and a suitably, depending on σ, we obtain $|\mathcal{M}| \leq N$ for all N large enough. In fact, we notice that the closer σ is to $1/2$, the closer to 1 we can choose α and hence also a.

We have now identified the desired set $\mathcal{M}$. The proof of the next lemma shows that we can choose α and a such that $f(n)$ is essentially concentrated on this set. Here we use the following terminology: a set of positive integers $\mathcal{M}$ is said to be divisor closed if d is in $\mathcal{M}$ whenever m is in $\mathcal{M}$ and d divides m. Note that this lemma is a counterpart to [6, Lemma 2] for the case $\sigma = 1/2$.

Lemma 8. *We can choose α depending on σ, with $\alpha \nearrow 1$ when $\sigma \searrow 1/2$, such that there exists a divisor closed set of integers $\mathcal{M}$ of cardinality at most N and the following estimate holds:*

$$\frac{1}{\sum_{j \in \mathbb{N}} f(j)^2} \sum_{n \in \mathbb{N}, n \notin \mathcal{M}} \frac{f(n)}{n^\sigma} \sum_{d|n} f(d) d^\sigma = o(A(N,\sigma)), \quad N \to \infty. \qquad (4.12)$$

This estimate is uniform in σ for $1/2 + 1/\log_2 T \leq \sigma \leq 3/4$.

Proof. We use the set $\mathcal{M}$ constructed above. We have already seen that we can in a suitable way let $\alpha \nearrow 1$ when $\sigma \searrow 1/2$. To prove the desired estimate (4.12), we begin by noting that

$$\frac{1}{A(N,\sigma) \sum_{j \in \mathbb{N}} f(j)^2} \sum_{n \in \mathbb{N}, n \notin \mathcal{M}} \frac{f(n)}{n^\sigma} \sum_{d|n} f(d) d^\sigma$$

$$\leq \frac{1}{A(N,\sigma) \sum_{j \in \mathbb{N}} f(j)^2} \sum_{k=1}^{[(2\sigma-1)^\alpha]} \sum_{n \in M_k} \frac{f(n)}{n^\sigma} \sum_{d|n} f(d) d^\sigma. \qquad (4.13)$$

Now for each $k = 1, \ldots, [(2\sigma - 1)^{-\alpha}]$, we have

$$\frac{1}{A(N,\sigma)\sum_{j\in\mathbb{N}}f(j)^2}\sum_{n\in M_k}\frac{f(n)}{n^\sigma}\sum_{d|n}f(d)d^\sigma$$

$$= \frac{1}{\prod_{p\in P_k}(1+f(p)^2+f(p)p^{-\sigma})}\sum_{n\in M'_k}\frac{f(n)}{n^\sigma}\sum_{d|n}f(d)d^\sigma$$

$$\leq \frac{1}{\prod_{p\in P_k}(1+f(p)^2)}\sum_{n\in M'_k}f(n)^2\prod_{p\in P_k}\left(1+\frac{1}{f(p)p^\sigma}\right). \qquad (4.14)$$

To deal with the product to the right in (4.14), we make the following computation:

$$\prod_{p\in P_k}\left(1+\frac{1}{f(p)p^\sigma}\right)$$

$$= \prod_{p\in P_k}\left(1+\frac{\sqrt{|\log(2\sigma-1)|}}{(\log N\log_2 N)^{1-\sigma}}p^{1-2\sigma}(\log p-\log_2 N-\log_3 N)\right)$$

$$\leq \left(1+(k+1)e^{k(1-2\sigma)}\sqrt{|\log(2\sigma-1)|}(\log N\log_2 N)^{-\sigma}\right)^{e^{k+1}\log N}$$

$$\leq \exp\left((k+1)e^{k(2-2\sigma)+1}\sqrt{|\log(2\sigma-1)|}(\log N)^{1-\sigma}(\log_2 N)^{-\sigma}\right)$$

$$= \exp\left(o\left(\frac{\log N}{|\log(2\sigma-1)|}\right)\frac{1}{k^2}\right), \qquad (4.15)$$

where the last relation holds simply because $k \leq (2\sigma-1)^{-\alpha}$. Since every number in M'_k has at least $\frac{a\log N}{k^2|\log(2\sigma-1)|}$ prime divisors and $f(n)$ is a multiplicative function, it follows that

$$\sum_{n\in M'_k}f(n)^2 \leq b^{-a\frac{\log N}{k^2|\log(2\sigma-1)|}}\prod_{p\in P_k}(1+bf(p)^2)$$

whenever $b > 1$ and hence

$$\frac{\sum_{n\in M'_k}f(n)^2}{\prod_{p\in P_k}(1+f(p)^2)} \leq b^{-a\frac{\log N}{k^2|\log(2\sigma-1)|}}\exp\left(\sum_{p\in P_k}(b-1)f(p)^2\right). \qquad (4.16)$$

Finally,

$$\sum_{p\in P_k}f(p)^2 = \frac{(\log N\log_2 N)^{(2-2\sigma)}}{|\log(2\sigma-1)|}\sum_{p\in P_k}\frac{1}{p^{2-2\sigma}(\log p-\log_2 N-\log_3 N)^2}$$

$$\leq (1+o(1))\frac{(\log N\log_2 N)^{(2-2\sigma)}}{|\log(2\sigma-1)|}\int_{e^k\log N\log_2 N}^{e^{k+1}\log N\log_2 N}\frac{dx}{k^2x^{2-2\sigma}\log x}$$

$$\leq (1+o(1))\frac{\log N}{k^2|\log(2\sigma-1)|}\frac{e^{2\sigma-1}-1}{2\sigma-1}e^{k(2\sigma-1)}.$$

Combining the last inequality with (4.16) and (4.15), we deduce that (4.14) is at most

$$\exp\left(\left(\frac{e^{2\sigma-1}-1}{2\sigma-1}e^{k(2\sigma-1)}(b-1)-a\log b+o(1)\right)\frac{\log N}{k^2|\log(2\sigma-1)|}\right).$$

We see that when σ is close to $1/2$, the factor in front of $b-1$ is close to 1. In this case, we may therefore choose both α and a close to 1 and then b close to 1, to arrange it so that

$$(b-1)\frac{e^{2\sigma-1}-1}{2\sigma-1}e^{k(2\sigma-1)}-a\log b<0;$$

the latter inequality can of course be obtained trivially for all values of σ if we choose α, a, b appropriately. Returning to (4.13), we therefore see that the desired relation (4.12) has been established, as well as the asymptotic relation between σ and α.   □

It remains to make the additional restriction $k \leq M$ in (4.7). The next lemma addresses this point and proves a result which is much stronger than what we need to finish the proof of Theorem 1. In fact, this lemma is what we would need to prove a counterpart to Theorem 2 for $1/2 + 1/\log_2 T < \sigma \leq 3/4$.

Lemma 9. *Let $\mathcal{M}$ be as defined above. Then*

$$\frac{1}{\sum_{j\in\mathbb{N}}f(j)^2}\sum_{n\in\mathcal{M}}\frac{f(n)}{n^\sigma}\sum_{d|n,\,d\leq n/M}f(d)d^\sigma = o(A(N,\sigma)),\quad N\to\infty \qquad (4.17)$$

uniformly for $1/2 + 1/\log_2 T \leq \sigma \leq 3/4$, where

$$M := \exp\left(e(\sqrt{|\log(2\sigma-1)|}+3)(\log N\log_2 N)^{1-\sigma}\right).$$

We notice that here we only need that $1 < a < 1/\alpha$. It may also be observed that with some extra effort one may replace e by a somewhat smaller constant in the definition of M.

Proof of Lemma 9. To begin with, we observe that

$$\sum_{n\in\mathcal{M}}\frac{f(n)}{n^\sigma}\sum_{d|n,\,d\leq n/M}f(d)d^\sigma = \sum_{n\in\mathcal{M}}f(n)^2\sum_{k|n,\,k\geq M}\frac{1}{f(k)k^\sigma}.$$

It is therefore enough to show that for each n in $\mathcal{M}$ we have

$$\sum_{k|n,\,k\geq M}\frac{1}{f(k)k^\sigma} = o(A(N,\sigma)),\quad N\to\infty.$$

Multiplying and dividing the kth term by $k^{-\delta}$ and using that $f(k)$ is a multiplicative function[3], we deduce that

$$\sum_{k|n,\,k\geq M}\frac{1}{f(k)k^{\sigma}} \leq M^{-\delta}\prod_{p|n}\left(1+\frac{1}{p^{\sigma-\delta}f(p)}\right)$$

$$\leq M^{-\delta}\exp\left((1+o(1))\max_{p|n}p^{\delta}\sum_{p|n}\frac{1}{p^{\sigma}f(p)}\right)$$

whenever $\delta > 0$. We find that by the definition of $\mathcal{M}$,

$$\sum_{p|n}\frac{1}{p^{\sigma}f(p)}=\sum_{p|n}\frac{\sqrt{|\log(2\sigma-1)|}}{(\log N\log_2 N)^{1-\sigma}}p^{1-2\sigma}(\log p-\log_2 N-\log_3 N)$$

$$\leq \sum_{k=1}^{[(2\sigma-1)^{-\alpha}]}\frac{a\log N}{k^2|\log(2\sigma-1)|}\frac{\sqrt{|\log(2\sigma-1)|}}{(\log N\log_2 N)^{1-\sigma}}(e^{k+1}\log N\log_2 N)^{1-2\sigma}(k+1)$$

$$\leq a\alpha(\sqrt{|\log(2\sigma-1)|}+3)\frac{(\log N)^{1-\sigma}}{(\log_2 N)^{\sigma}}.$$

We now set $\delta=1/\log_2 N$ and obtain

$$\sum_{k|n,\,k\geq M}\frac{1}{f(k)\sqrt{k}}\leq \exp\left((ae\alpha-e)(\sqrt{|\log(2\sigma-1)|}+3)\frac{(\log N)^{1-\sigma}}{(\log_2 N)^{\sigma}}\right)<1,$$

provided that $a\alpha < 1$.   □

We are now finally prepared to finish the proof of Theorem 1.

Proof of Theorem 1 – part 2. We recall that $N=[\sqrt{T}]$. This means that we need to prove that (4.7) holds for $M=T\geq N^2$ and suitable choices of the parameters α and a, ensuring the desired asymptotic behavior

$$\nu(\sigma)=(1/\sqrt{2}+o(1))\sqrt{|\log(2\sigma-1)|}$$

when $\sigma \searrow 1/2$. We conclude by observing that this follows from the three preceding lemmas.   □

5. Proof of Theorem 2

All the work needed for the proof of Theorem 2 has now been done. Indeed, we may use the construction in [6, Section 3] and the estimates established there, corresponding to Lemma 7 and Lemma 8. We only need a minor modification of Lemma 9, which is as follows.

---

[3] This estimation technique is known as Rankin's trick.

Lemma 10. *Let $f(n)$ and $\mathcal{M}$ be as defined above in the case $\sigma = 1/2 + 1/\log_2 T$. Then*

$$\frac{1}{\sum_{j \in \mathbb{N}} f(j)^2} \sum_{n \in \mathcal{M}} \frac{f(n)}{\sqrt{n}} \sum_{d|n,\, d \leq n/M} f(d)\sqrt{d} = o(A(N, 1/2)), \quad N \to \infty,$$

where

$$M := \exp\left(e(\sqrt{\log N \log_2 N \log_3 N}\right).$$

The proof of Lemma 10 is word for word the same as that of Lemma 9 and is therefore omitted. The desired estimate for $\widetilde{M}_2(R, T)$ (see Lemma 5) is now obtained in exactly the same way as in the preceding case when $\sigma = 1/2 + 1/\log_2 T$.

6. Concluding remarks

To obtain a more precise estimate in the range $\sigma_0 \leq \sigma \leq 1 - 1/\log_2 T$ for a suitable σ_0, $1/2 < \sigma_0 < 1$, we could combine the two constructions in the range in which the powers of $\log T$ in Lemma 3 and Lemma 4 are insignificant, say when $\sigma_0 \leq \sigma \leq 1 - 1/\sqrt{\log_2 T}$. This can be done as follows. Let ℓ and x be two positive integers such that $N := \ell^{\pi(x)}$ satisfies the inequality $N \leq \sqrt{T}$; here $\pi(x)$ is as usual the number of primes not exceeding x. We let again $\mathcal{M}$ be the set of divisors of the number $\prod_{p \leq x} p^{\ell-1}$ and choose $f(n)$ to be the characteristic function of $\mathcal{M}$. We observe that the only difference from Subsection 4.3 is that we have replaced the condition $\max \mathcal{M} \leq \sqrt{T}$ by the less severe requirement that $|\mathcal{M}| \leq \sqrt{T}$. The computation is precisely as in Subsection 4.3, but we are now free to choose a larger x and consequently a smaller ℓ. From Lemma 6, we see that this should be done so that we make

$$\frac{x^{1-\sigma}}{(1-\sigma)\log x}$$

as large as possible. A calculus argument shows that ℓ should be of order $1/(1-\sigma)$ and consequently x of order $(1 - \sigma)\log_3 T$. We see again the phenomenon that more and more primes are used when σ decreases.

Our final remark is about what we might expect the true growth of $|\zeta(\sigma + it)|$ to be. Farmer, Gonek, and Hughes [10] conjectured, appealing to random matrix theory, that

$$\max_{1 \leq t \leq T} |\zeta(1/2 + it)| = \exp\left(\left(\frac{1}{\sqrt{2}} + o(1)\right)\sqrt{\log T \log_2 T}\right)$$

and in [14, Remark 2], it was suggested that for example

$$\max_{T/2 \leq t \leq T} |\zeta(1/2 + 1/\log_2 T + it)| = \exp\left(\left(c + o(1)\right)\sqrt{\log T \log_2 T}\right)$$

for some $c < 1/\sqrt{2}$. Hence the asymptotics of our estimates when $\sigma \searrow 1/2$ is an order of magnitude smaller than this prediction. On the other hand, it is expected that the true growth of $\max_{1 \leq t \leq T} |\zeta(\sigma + it)|$ is of order $\exp((c +$

$o(1))(\log T)^{1-\sigma}(\log_2 T)^{-\sigma})$ for some constant c when $1/2 < \sigma < 1$, and the asymptotics of Theorem 1 when $\sigma \nearrow 1$ is indeed consistent with what is predicted in [14, Remark 2].

The predictions of [10] and [14] are however very far from the known upper bounds for the growth of $\zeta(1/2 + it)$. We refer to Bourgain's recent paper [7] for the best result when $\sigma = 1/2$:

$$|\zeta(1/2 + it)| \le C_\varepsilon |t|^{13/84+\varepsilon}$$

for every $\varepsilon > 0$.

Acknowledgment

We are grateful to Maksym Radziwiłł for helpful remarks regarding Lamzouri's paper [14].

References

[1] C. Aistleitner, *Lower bounds for the maximum of the Riemann zeta function along vertical lines*, Math. Ann. **365** (2016), 473–496.

[2] C. Aistleitner, I. Berkes, and K. Seip, *GCD sums from Poisson integrals and systems of dilated functions*, J. Eur. Math. Soc. **17** (2015), 1517–1546.

[3] R. Balasubramanian and K. Ramachandra, *On the frequency of Titchmarsh's phenomenon for $\zeta(s)$. III*, Proc. Indian Acad. Sci. Sect. A **86** (1977), 341–351.

[4] A. Bondarenko, T. Hilberdink, and K. Seip, *Gál-type GCD sums beyond the critical line*, J. Number Theory **166** (2016), 93–104.

[5] A. Bondarenko and K. Seip, *GCD sums and complete sets of square-free numbers*, Bull. London Math. Soc. **47** (2015), 29–41.

[6] A. Bondarenko and K. Seip, *Large greatest common divisor sums and extreme values of the Riemann zeta function*, Duke Math. J., **166** (2017), 1685–1701.

[7] J. Bourgain, *Decoupling, exponential sums and the Riemann zeta function*, J. Amer. Math. Soc. **30** (2017), 205–224.

[8] H.G. Diamond and J. Pintz, *Oscillation of Mertens' product formula* (English, J. Théor. Nombres Bordeaux **21** (2009), 523–533.

[9] T. Dyer and G. Harman, *Sums involving common divisors*, J. London Math. Soc. **34** (1986), 1–11.

[10] D.W. Farmer, S.M. Gonek, and C.P. Hughes, *The maximum size of L-functions*, J. Reine Angew. Math. **609** (2007), 215–236.

[11] I.S. Gál, *A theorem concerning Diophantine approximations*, Nieuw Arch. Wiskunde **23** (1949), 13–38.

[12] A. Granville and K. Soundararajan, *Extreme values of $|\zeta(1+it)|$*, "The Riemann Zeta Function and Related Themes: Papers in Honour of Professor K. Ramachandra", pp. 65–80, Ramanujan Math. Soc. Lect. Notes Ser., 2, Ramanujan Math. Soc., Mysore, 2006.

[13] T. Hilberdink, *An arithmetical mapping and applications to Ω-results for the Riemann zeta function*, Acta Arith. **139** (2009), 341–367.

[14] Y. Lamzouri, *On the distribution of extreme values of zeta and L-functions in the strip $1/2 < \sigma < 1$*, Int. Math. Res. Not. IMRN **2011**, 5449–5503.

[15] N. Levinson, *Ω-theorems for the Riemann zeta-function*, Acta Arith. **20** (1972), 317–330.

[16] M. Lewko and M. Radziwiłł, *Refinements of Gál's theorem and applications*, Adv. Math. **305** (2017), 280–297.

[17] H.L. Montgomery, *Extreme values of the Riemann zeta function*, Comment. Math. Helv. **52** (1977), 511–518.

[18] H. Queffélec and M. Queffélec, *Diophantine Approximation and Dirichlet series*. HRI Lecture Notes Series – 2, Hindustan Book Agency, New Delhi, 2013.

[19] K. Ramachandra and A. Sankaranarayanan, *Note on a paper by H.L. Montgomery (Omega theorems for the Riemann zeta-function)*, Publ. Inst. Math. (Beograd) (N.S.) **50 (64)** (1991), 51–59.

[20] J.B. Rosser and L. Schoenfeld, *Approximate formulas for some functions of prime numbers*, Illinois J. Math. **6** (1962), 64–97.

[21] E. Saksman and K. Seip, *Some open questions in analysis for Dirichlet series*, in "Recent Progress on Operator Theory and Approximation in Spaces of Analytic Functions", pp. 179–193, Contemp. Math. **679**, Amer. Math. Soc., Providence RI, 2016.

[22] K. Soundararajan, *Extreme values of zeta and L-functions*, Math. Ann. **342** (2008), 467–486.

[23] E.C. Titchmarsh, *The Theory of the Riemann Zeta-Function*, 2nd Edition, Oxford University Press, New York, 1986.

Andriy Bondarenko and Kristian Seip
Department of Mathematical Sciences
Norwegian University of Science and Technology
NO-7491 Trondheim, Norway
e-mail: andriybond@gmail.com
       kristian.seip@ntnu.no.

Operator Theory:
Advances and Applications, Vol. 261, 141–157
© Springer International Publishing AG, part of Springer Nature 2018

Spectra of Stationary Processes on $\mathbb{Z}$

Alexander Borichev, Mikhail Sodin and Benjamin Weiss

Dedicated to the memory of V.P. Havin

Abstract. We will discuss a somewhat striking spectral property of finitely valued stationary processes on $\mathbb{Z}$ that says that if the spectral measure of the process has a gap then the process is periodic. We will give some extensions of this result and raise several related questions.

Mathematics Subject Classification (2010). Primary 60G10. Secondary 30D99.

Keywords. Spectral measure, spectrum, finitely valued stationary process, Carleman spectrum, spectral set.

1. Introduction

One of the basic features of a complex-valued stationary random process $\xi \colon \mathbb{Z} \to \mathbb{C}$ is its *spectral measure ρ* and its closed support, $\mathrm{spt}(\rho)$, called *the spectrum* of the process ξ. As is well known, the spectral measure is obtained in the following way. Assuming for simplicity of notation that ξ has zero mean, one first forms the covariance function $r(m) = \mathbb{E}[\xi(0)\bar{\xi}(m)]$, and then observing that it is positive definite, defines a measure ρ on the unit circle $\mathbb{T}$ by $r(m) = \hat{\rho}(m)$, i.e., r is the Fourier transform of ρ. Any positive measure on $\mathbb{T}$ is the spectral measure of a stationary process obtained, for instance, by a familiar Gaussian construction. The case of finitely valued stationary processes is strikingly different. We will show that *if ξ is a finitely valued stationary process on $\mathbb{Z}$ and the spectrum of ξ is not all of $\mathbb{T}$, then ξ is periodic and therefore its spectrum is contained in $\{t \colon t^N = 1\}$, where N is a period of ξ.*

This result bears a close resemblance to a theorem of Szegő. Szegő's original version [6, Section 83] says that if a sequence of Taylor coefficients f_n attains finitely many values and the sum of the Taylor series $f(z) = \sum_{n \geq 0} f_n z^n$ can be analytically continued through an arc of the unit circle, then the sequence (f_n)

The second author was supported in part by ERC Advanced Grant 692616 and ISF Grant 382/15.

is eventually periodic and f is a rational function with poles at roots of unity. Helson's harmonic analysis version of Szegő's theorem [9, Section 6.4] says that if a sequence $(f_n)_{n \in \mathbb{Z}}$ attains finitely many values and its Beurling's spectrum (defined below in Section 2.4) is not all of $\mathbb{T}$, then the sequence (f_n) is periodic.

We will give two proofs of our result. They are short, not too far from each other, both use some ideas of Szegő and Helson, and each of these proofs will give us more than we have stated above.

For our first proof we will show the following general result. If ξ is a finite-valued stationary ergodic process, the Beurling spectrum of almost every realization $\xi(n)$ coincides with the support of the spectral measure of the process. In addition we will give a new proof of Helson's result that gives an effective estimate for the size of the period in terms of the set of values of the process and the size of the gap in the spectrum.

The second proof exploits a relation to the polynomial approximation problem in the space $L^2(\rho)$. It allows us to replace the assumption that $\mathrm{spt}(\rho) \neq \mathbb{T}$ by a much weaker condition, which in particular forbids ρ to have sufficiently deep exponential zeroes.

It would be very interesting to reveal more restrictions on the spectral measures of finitely valued stationary sequences. In this connection we mention that McMillan [11] described covariances of stationary processes that attain the values ± 1, the proofs appeared later in the work of Shepp [14]. Unfortunately, it seems to be very difficult to apply this description. On the other hand, in the book by M. Queffélec [13] one can find a Zoo of various spectral measures of finitely valued stationary processes of dynamical origin.

In the last section of this note we will turn to unimodular stationary processes on $\mathbb{Z}$. Using a result of Eremenko and Ostrovskii [7], we will show that if ξ is an ergodic process that takes the values from the unit circle and if the spectrum of ξ is contained in an arc of length less than π, then $\xi(n) = ts^n$, $n \in \mathbb{Z}$, with a unimodular constant s and a random variable t which has either a uniform distribution on the circle if s is not a root of unity, or a uniform distribution on a cyclic group if s is a root of unity.

*   *   *

We sadly dedicate this note to the memory of Victor Havin, a wonderful person, teacher, and mathematician. The interplay between harmonic and complex analysis and probability theory presented here, probably, would be close to his heart.

We thank Fedor Nazarov for several illuminating discussions of the material presented here, and especially for a suggestion to use monic polynomials with small $L^2(\rho)$ norms.

2. Spectra of stationary random sequences

Some of our results deal with wide-sense stationary processes. Recall that the random process $\xi \colon \mathbb{Z} \to \mathbb{C}$ is called *wide-sense stationary* (a.k.a. weak stationary or second-order stationary) if $\mathbb{E}|\xi(n)|^2 < \infty$ for every n and $\mathbb{E}\,\xi(n)$ and $\mathbb{E}[\xi(n)\bar{\xi}(n+m)]$ do not depend on n. Then the spectral measure ρ of ξ is defined in the same way as above so that the covariance function of ξ is the Fourier transform of ρ. As above, we will call the closed support of ρ, $\mathrm{spt}(\rho)$, *the spectrum* of ξ.

2.1. Spectrum of a single realization

Now, we define the spectrum of a single realization of a wide-sense stationary process ξ. First, we note that by the Borel–Cantelli lemma, almost surely, we have

$$|\xi(n)| = o(|n|^\alpha), \quad n \to \infty,$$

with any $\alpha > \frac{1}{2}$. Hence, the sequence $(\xi(n))_{n \in \mathbb{Z}}$ generates the random distribution (or the random functional) F_ξ on $C^\infty(\mathbb{T})$ acting as follows:

$$F_\xi(\varphi) = \sum_{n \in \mathbb{Z}} \xi(n)\widehat{\varphi}(-n), \quad \varphi \in C^\infty(\mathbb{T}).$$

We denote by $\sigma(\xi) \subset \mathbb{T}$ the support of this distribution, that is, the complement to the largest open set $O \subset \mathbb{T}$ on which F_ξ vanishes (which means that $F_\xi(\varphi) = 0$ for any smooth function φ whose closed support is contained in O).

We consider the Borel structure on the family of the compact subsets of the unit circle corresponding to the Hausdorff distance. Then the map $\xi \mapsto \sigma(\xi)$ is measurable. Indeed, we can take a countable collection of smooth functions $\{\varphi_k\}$ such that the complement of the support of $\sigma(\xi)$ is determined by the set of k such that $F_\xi(\varphi_k) = 0$. With a natural Borel structure on the set of sequences that grow polynomially this is a measurable set for a fixed k. Finally, one can characterize the Hausdorff distance between two compact subsets of the unit circle in terms of the open arcs that constitute theirs complements.

Note that the random compact set $\sigma(\xi)$ is invariant with respect to the translations of ξ, so if the translations act ergodically on ξ, the spectrum $\sigma(\xi)$ is not random. We also note that $\sigma(\xi)$ is reflected in the real line under the flip $\xi(n) \mapsto \xi(-n)$.

2.2. Spectrum of the process and spectra of its realizations

Theorem 1. *Suppose $\xi \colon \mathbb{Z} \to \mathbb{C}$ is a wide-sense stationary process with zero mean. Let ρ be the spectral measure of ξ. Then,*

(A) *almost surely, $\sigma(\xi) \subseteq \mathrm{spt}(\rho)$;*
(B) *if $O \subsetneq \mathbb{T}$ is an open set such that, almost surely, $O \cap \sigma(\xi) = \varnothing$, then $O \cap \mathrm{spt}(\rho) = \varnothing$.*

Corollary 2. *Let ξ be a stationary square integrable ergodic process on $\mathbb{Z}$. Then, almost surely, $\sigma(\xi) = \mathrm{spt}(\rho)$.*

2.2.1. Proof of Part (A). Suppose that $\mathrm{spt}(\rho) \neq \mathbb{T}$ and take an arbitrary function $\varphi \in C_0^\infty(\mathbb{T} \setminus \mathrm{spt}(\rho))$. Then

$$\int_{\mathbb{T}} \left| \sum_{n \in \mathbb{Z}} \widehat{\varphi}(n) t^n \right|^2 \mathrm{d}\rho(t) = 0 \,. \tag{1}$$

Recall that there exists a linear isomorphism between the closure of the linear span of $(\xi(n))_{n \in \mathbb{Z}}$ in $L^2(\mathbb{P})$ and the space $L^2(\rho)$ that maps $\xi(n)$ to t^n, $n \in \mathbb{Z}$. Therefore, almost surely,

$$\sum_{n \in \mathbb{Z}} \xi(n) \widehat{\varphi}(n) = 0 \,, \tag{2}$$

that is, $F_\xi(\bar{\varphi}) = 0$. Since this holds for *any* φ as above, we conclude that almost surely $\sigma(\xi) \subseteq \mathrm{spt}(\rho)$. □

2.2.2. Proof of Part (B). One simply needs to read the same lines in the reverse direction. Suppose that $I \subsetneq \mathbb{T}$ is an open arc such that almost surely $I \cap \sigma(\xi) = \varnothing$. We take an arbitrary function $\varphi \in C_0^\infty(I)$. Then, almost surely, $F_\xi(\bar{\varphi}) = 0$, that is, (2) holds. By the same linear isomorphism, this yields (1), i.e., $\varphi = 0$ in $L^2(\rho)$. Since φ was arbitrary, we conclude that $\mathrm{spt}(\rho) \cap I = \varnothing$. □

2.3. Carleman spectrum

In some instances, it is helpful to combine Theorem 1 with another definition of the spectrum which is based on analytic continuation and goes back at least to Carleman.

Let $\xi \colon \mathbb{Z} \to \mathbb{C}$ be a sequence such that $\limsup\limits_{|n| \to \infty} |\xi(n)|^{1/|n|} = 1$. Put

$$F_\xi^+(z) = \sum_{n \geq 0} \xi(n) z^n, \qquad \text{analytic in } \{|z| < 1\},$$

$$F_\xi^-(z) = - \sum_{n \leq -1} \xi(n) z^n, \qquad \text{analytic in } \{|z| > 1\} \cup \{\infty\}.$$

Then the Carleman spectrum $\sigma_C(\xi)$ is the minimal compact set $\sigma \subseteq \mathbb{T}$ such that the function

$$F_\xi(z) = \begin{cases} F_\xi^+(z), & |z| < 1 \\ F_\xi^-(z), & |z| > 1 \end{cases}$$

is analytic on $\widehat{\mathbb{C}} \setminus \sigma$.

It is a classical fact of harmonic analysis that if the sequence ξ has at most polynomial growth (which is almost surely the case for realizations of a wide-sense stationary process), then $\sigma_C(\xi) = \sigma(\xi)$, i.e., both definitions of the spectrum of ξ coincide (see, for example, [10, VI.8], where the proof is given for bounded functions on $\mathbb{R}$).

2.4. Spectral sets of bounded sequences

It is worth to recall here that for bounded sequences ξ there is another equivalent definition of the spectrum that goes back to Beurling. The spectral set (a.k.a. Beurling's spectrum or weak-star spectrum), $\sigma_B(\xi)$, of a bounded sequence ξ is the set of $t \in \mathbb{T}$ such that the sequence $(t^n)_{n\in\mathbb{Z}}$ belongs to the closure of the linear span of translates of ξ in the weak-star topology in $\ell^\infty(\mathbb{Z})$ as a dual to $\ell^1(\mathbb{Z})$.

The Hahn–Banach theorem easily gives an equivalent definition: $\sigma_B(\xi)$ is the set of $t \in \mathbb{T}$ such that whenever $\varphi \in \ell^1(\mathbb{Z})$ and $\varphi * \xi = 0$ identically, necessarily $\sum_{n\in\mathbb{Z}} \varphi(n)\bar{t}^n = 0$.

It is not difficult to show that for any bounded sequence ξ, $\sigma_B(\xi) = \sigma(\xi)$ (see for instance [10, VI.6], where the proof is given for bounded functions on $\mathbb{R}$).

The advantage of the definition of the spectral set is that it can be transferred to bounded functions on locally compact Abelian groups. This allows one without extra work to extend Theorem 1 to bounded stationary processes on arbitrary locally compact Abelian groups with a countable base of open sets.

3. A random variation on the theme of Szegő and Helson

We call the wide-sense stationary random process ξ *periodic* if there exists a positive integer N so that almost every realization $(\xi(n))_{n\in\mathbb{Z}}$ is N-periodic. In this case the spectrum of ξ is contained in the set of roots of unity of order N.

Theorem 3. *Let $X \subset \mathbb{C}$ be a finite set, and let $\xi \colon \mathbb{Z} \to X$ be a wide-sense stationary process with the spectral measure ρ. Suppose that $\mathrm{spt}(\rho) \neq \mathbb{T}$. Then the process ξ is periodic.*

Theorem 3 is an immediate corollary to the following theorem combined with Theorem 1.

3.1. Helson's theorem

In the following theorem which is essentially due to Helson [9, Section 5.4] we have added the claim that the period N does not depend on the sequence ξ but only on the spectrum and the set X. Helson's proof used a compactness argument which does not give this and we have found a new proof of his theorem which does.

Theorem 4 (Helson). *Let $X \subset \mathbb{C}$ be a finite set. Then any sequence $\xi \colon \mathbb{Z} \to X$ with $\sigma(\xi) \neq \mathbb{T}$ is N-periodic with $N \leq \Psi(\ell, X)$, where ℓ is the length of the largest gap in $\sigma(\xi)$, and $\ell \mapsto \Psi(\ell, X)$ is a non-increasing function.*

3.2. Two lemmas on polynomials on an arc of the circle

The proof of Theorem 4 uses the following lemma.

Lemma 1. *Given a closed arc $J \subsetneq \mathbb{T}$ and $\delta > 0$, there exists a polynomial P such that $P(0) = 1$ and $\|P\|_{C(J)} < \delta$.*

Proof: By Runge's theorem, there exists a polynomial S so that $\left\| \frac{1}{z} - S \right\|_{C(J)} < \delta$. Put $P(z) = 1 - zS(z)$. $\qquad\square$

Next, we will need a version of the Bernstein inequality for polynomials on an arc which is due to V.S. Videnskii [3, Section 5.1.E19.c].

Lemma 2. *Let $J \subsetneq \mathbb{T}$ be a closed arc with $mJ \geq 1/2$, where m is the normalized Lebesgue measure on $\mathbb{T}$. There exists a constant K such that for any polynomial P of degree n, we have $\|P'\|_{C(J)} \leq Kn^2 \|P\|_{C(J)}$.*

Note that any bound polynomial (or even subexponential) in n would suffice for our purposes.

3.3. The δ-prediction lemma

The next lemma (which can be traced to Szegő) is the main ingredient in the proof of Theorem 4. By $\|\xi\|_p$ we denote the weighted norm of ξ defined as

$$\|\xi\|_p^2 = \sum_{m \in \mathbb{Z}} \left(\frac{|\xi(m)|}{1 + |m|^p} \right)^2.$$

Lemma 3. *Given $\delta, p, M > 0$ and given an open arc J, $\bar{J} \subsetneq \mathbb{T}$, $mJ = 1 - \ell \leq 1/2$, there exist $n = n(p, \ell, M, \delta) \in \mathbb{N}$ and $q_0, \ldots, q_{n-1} \in \mathbb{C}$ such that for any sequence $\xi \colon \mathbb{Z} \to \mathbb{C}$ with $\|\xi\|_p \leq M$ and $\sigma(\xi) \subset J$, we have*

$$\left| \xi(n) + \sum_{k=0}^{n-1} q_k \xi(k) \right| < \delta.$$

Moreover, the function $\ell \mapsto n(p, \ell, M, \delta)$ does not increase.

Proof: It will be convenient to assume that $J = \{e^{i\theta} \colon |\theta| < \pi(1 - \ell)\}$. Put $J' = \{e^{i\theta} \colon |\theta| < \pi(1 - \ell/2)\}$ and let $\varphi \colon \mathbb{T} \to [0, 1]$ be a C^∞-function such that $\varphi = 1$ on J and $\varphi = 0$ on $\mathbb{T} \setminus J'$.

By Lemma 1, there exists a polynomial P such that $P(0) = 1$ and $\|P\|_{C(J')} \leq \frac{1}{2}$. Let $b \in \mathbb{N}$ be a sufficiently large number to be chosen later, $n = b \cdot \deg P$, and put

$$Q(t) = t^{-n} P(t)^b = t^{-n} + \sum_{k=0}^{n-1} q_k t^{-k}.$$

Then $\|Q\|_{C(J')} \leq e^{-cn}$, whence by Lemma 2, $\|Q^{(j)}\|_{C(J')} \leq \left(Kn^2 \right)^j e^{-cn}$.

Next, we note that since the function $1 - \varphi$ vanishes on an open neighbourhood of the spectrum of ξ, we have $F_\xi(Q) = F_\xi(Q\varphi)$, and therefore,

$$\xi(n) + \sum_{k=0}^{n-1} q_k \xi(k) = F_\xi(Q)$$

$$= F_\xi(Q\varphi) = \sum_{m \in \mathbb{Z}} \xi(m) \widehat{Q\varphi}(-m).$$

Hence, by the Cauchy–Schwarz inequality, we have

$$\left| \xi(n) + \sum_{k=0}^{n-1} q_k \xi(k) \right| \leq \|\xi\|_p \Big(\sum_{m \in \mathbb{Z}} (1 + |m|^p)^2 |\widehat{Q\varphi}(-m)|^2 \Big)^{1/2}$$

$$\lesssim_p M \sum_{j=0}^{p} \|(Q\varphi)^{(j)}\|_{L^2(\mathbb{T})}$$

$$\lesssim_p M \sum_{j=0}^{p} \|(Q\varphi)^{(j)}\|_{C(J')}$$

$$\lesssim_p M \ell^{-p} n^{2p} e^{-cn}$$

$$\leq \delta \qquad \text{for } n \geq n_0(p, \ell, M, \delta),$$

with a non-increasing function $\ell \mapsto n_0(p, \ell, M, \delta)$ which completes the proof. $\qquad \square$

3.4. Proof of Theorem 4

We fix a finite set $X \subset \mathbb{C}$ and put $\delta_X = \inf\{|z - w|: z, w \in X, z \neq w\}$. Applying the previous lemma with $\delta < \frac{1}{2}\delta_X$, we see that the n-tuple $\langle \xi(0), \ldots, \xi(n-1) \rangle$ uniquely defines the value of $\xi(n)$. Repeating this argument, we proceed with $\xi(n+1), \xi(n+2), \ldots$. We proceed in the same way with $\xi(-1), \xi(-2), \ldots$. Hence, the n-tuple $\langle \xi(0), \ldots, \xi(n-1) \rangle$ uniquely determines the whole sequence ξ.

Observe that among the $|X|^n + 1$ collections

$$\langle \xi(k), \xi(k+1), \ldots, \xi(k+n-1) \rangle, \quad 0 \leq k \leq |X|^n,$$

at least two must coincide. If $0 \leq k_1 < k_2 \leq |X|^n$ are the corresponding indices, after a minute reflection we conclude that the sequence ξ is periodic with period $N = k_2 - k_1 \leq |X|^n$. $\qquad \square$

4. More random variations on Szegő's theme

Here we will give another version of our main result, Theorem 3. Now we will assume that the random process $\xi \colon \mathbb{Z} \to X$ is stationary in the usual sense and replace our previous assumption that the spectrum of ξ is not all of $\mathbb{T}$ by a much weaker condition imposed on the spectral measure ρ of ξ. We will also allow X to be a uniformly discrete subset of $\mathbb{C}$.

4.1. Condition (Θ)

Given a positive measure ρ on $\mathbb{T}$, we put

$$e_n(\rho) = \mathrm{dist}_{L^2(\rho)}\big(\mathbb{1}, \mathcal{P}_n^0\big),$$

where $\mathcal{P}_n^0 \subset \mathbb{C}[z]$ is the set of algebraic polynomials of degree at most n vanishing at the origin. We say that *the measure ρ satisfies condition (Θ)* if

$$\sum_{n \geq 1} e_n(\rho)^2 < \infty.$$

Recall that by the classical Szegő theorem,

$$\lim_{n\to\infty} e_n(\rho) = 0 \iff \int_{\mathbb{T}} \log \rho'\, dm = -\infty \,,$$

where m is the Lebesgue measure on $\mathbb{T}$ and $\rho' = \dfrac{d\rho}{dm}$. Note that by Lemma 1, $e_n(\rho) = O(e^{-cn})$ provided that $\mathrm{spt}(\rho) \neq \mathbb{T}$, that is, condition (Θ) is essentially weaker than the condition that the support of ρ is not all of $\mathbb{T}$. In Section 5 we will show that condition (Θ) forbids ρ to have certain exponentially deep zeroes.

In spite of the omnipresence of Szegő's theorem [8] (see also [1, 5, 15] for recent development) we are not aware of any results that in the case when the logarithmic integral of ρ' is divergent would relate the rate of decay of the sequence $(e_n(\rho))$ with the properties of the measure ρ (the only exceptions are several results that relate the gaps in $\mathrm{spt}(\rho)$ with the exponential rate of decay of $e_n(\rho)$'s).

4.2. Stationary processes on $\mathbb{Z}$ with uniformly discrete set of values

Recall that the set $X \subset \mathbb{C}$ is called *uniformly discrete* if

$$\delta_X \stackrel{\mathrm{def}}{=} \inf\{|z - w|\colon z, w \in X, z \neq w\} > 0 \,.$$

Theorem 5. *Suppose that $X \subset \mathbb{C}$ is a uniformly discrete set and $\xi\colon \mathbb{Z} \to X$ is a stationary process with the spectral measure ρ satisfying condition (Θ). Then almost every realization $(\xi(n))_{n\in\mathbb{Z}}$ of the process ξ is periodic.*

Corollary 6. *Suppose that the stationary ergodic process ξ on $\mathbb{Z}$ attains values in a uniformly discrete subset of $\mathbb{C}$ and that its spectral measure ρ satisfies condition (Θ). Then the process ξ is periodic.*

This was proven in [2] under a stronger assumption that $\mathrm{spt}(\rho) \neq \mathbb{T}$. The proof given in [2] used a somewhat different approach.

4.3. Proof of Theorem 5

Applying the linear isomorphism between the closure of the linear span of $(\xi(n))_{n\in\mathbb{Z}}$ in $L^2(\mathbb{P})$ and the space $L^2(\rho)$ that maps $\xi(n)$ to t^n, we see that, given $N \in \mathbb{N}$, there exist $q_0, \ldots, q_{N-1} \in \mathbb{C}$ so that

$$\mathbb{E}\left[\left|\xi(N) + \sum_{k=0}^{N-1} q_k\xi(k)\right|^2\right] = e_N(\rho)^2 \,,$$

whence

$$\mathbb{P}\left\{\left|\xi(N) + \sum_{k=0}^{N-1} q_k\xi(k)\right| \geq \frac{1}{2}\delta_X\right\} \leq \frac{4}{\delta_X^2} e_N(\rho)^2 \,.$$

Thus, with probability at least $1 - \frac{4}{\delta_X^2}e_N(\rho)^2$, the term $\xi(N)$ is determined by the values of the terms $\xi(0), \ldots, \xi(N-1)$. Proceeding to $\xi(N+1), \xi(N+2), \ldots$ and then to $\xi(-1), \xi(-2), \ldots$, we see that the whole sequence $(\xi(n))$ is determined by the values $\xi(0), \ldots, \xi(N-1)$ with probability at least

$$1 - \frac{8}{\delta_X^2} \sum_{n \geq N} e_n(\rho)^2 \, .$$

Recalling that the series of $e_n(\rho)^2$ converges (and that the set X is countable), we conclude that, given $\varepsilon > 0$, there is a countable part Ω' of our probability space Ω with $\mathbb{P}\{\Omega'\} > 1 - \varepsilon$. Hence, we may assume that the whole probability space Ω is countable.

Since the process ξ is a stationary one, there exists a measure preserving transformation τ of $(\Omega, \mathbb{P})$. The only possibility for this is that $\Omega = \bigcup_j \Omega_j$, each Ω_j consists of finitely many atoms of equal probability and is τ-invariant. That is, τ acts as a permutation of elements of Ω_j. Therefore, for each j, the restriction $\tau|_{\Omega_j}$ is periodic. $\qquad\square$

4.4. Integer-valued stationary processes

Here is another variation on the same theme taken from [2].

Theorem 7. *Suppose $\xi \colon \mathbb{Z} \to \mathbb{Z}$ is a stationary process with $\mathrm{spt}(\rho) \neq \mathbb{T}$. Then the process ξ is periodic.*

4.5. Proof of Theorem 7

By Theorem 5, almost every realization $\xi(n)$ is periodic with some period N. We need to show that the period N is non-random. For this, it suffices to see that it is bounded.

We have

$$F_\xi^+(z) \stackrel{\text{def}}{=} \sum_{n \geq 0} \xi(n) z^n$$

$$= (1 - z^N)^{-1} \sum_{n=0}^{N-1} \xi(n) z^n$$

$$= \frac{T(z)}{1 - z^N} \, ,$$

where T is a polynomial with integral coefficients of degree at most $N-1$. Since the ring $\mathbb{Z}[z]$ of polynomials over the integers is a unique factorization domain, we can write

$$F_\xi^+(z) = \frac{P(z)}{Q(z)} \, ,$$

where $P, Q \in \mathbb{Z}[z]$, have no common factors, and Q is monic and divides $z^N - 1$.

First, we claim that P and Q have no common zeroes. Indeed, we decompose $P = P_1 \cdot \ldots \cdot P_\ell$ and $Q = Q_1 \cdot \ldots \cdot Q_m$ in the product of monic polynomials irreducible on $\mathbb{Z}$. By the Gauss lemma, these polynomials are also irreducible on $\mathbb{Q}$. Since $\mathbb{Q}$

is a field, $\mathbb{Q}[z]$ is a principal ideal domain, that is, if $P_i \neq Q_j$ are irreducible on $\mathbb{Q}$, then there are $R, S \in \mathbb{Q}[z]$ so that $P_i R + Q_j S = 1$. Thus, P_i and Q_j have no common zeroes. Hence, P and Q have no common zeroes.

Next, let

$$\Phi_n(z) = \prod_{k:\, \gcd(k,n)=1} (z - e(k/n))$$

be the cyclotomic polynomial of degree n. It belongs to $\mathbb{Z}[z]$ and is irreducible [16, § 8.4]. Therefore, we can factor

$$Q = \prod_{1 \leq k \leq r} \Phi_{n(k)} .$$

Since the rational function F_ξ^+ does not have singularities on a fixed arc of $\mathbb{T}$, there is a (non-random) integer n^* so that $n(k) \leq n^*$. Thus, there is a (non-random) integer N^* so that

$$\text{zeroes}(Q) \subset \{t \colon t^{N^*} = 1\} .$$

That is, the period of the realization $(\xi(n))_{n \in \mathbb{Z}}$ is bounded by N^*. $\qquad \square$

5. Measures satisfying condition (Θ)

The following theorem shows that if a positive measure ρ on $\mathbb{T}$ has a deep exponential root, then it must satisfy condition (Θ), and if the exponential root is not deep enough, then ρ may not satisfy condition (Θ).

As before, we denote by m the normalized Lebesgue measure on $\mathbb{T}$.

Theorem 8. *Let ρ be a positive measure on $\mathbb{T}$ and let β be a positive parameter.*

(A) *There exists a positive numerical constant K such that if*

$$\int_{-\pi}^{\pi} \exp\left(\frac{\beta}{|\theta|}\right) d\rho(e^{i\theta}) < \infty ,$$

then $e_n(\rho) \lesssim n^{-K\beta}$.

(B) *Suppose that $d\rho \gtrsim \exp\left(-\frac{\beta}{|\theta|}\right) dm$ everywhere on $\mathbb{T}$. Then $e_n(\rho) \gtrsim_\beta n^{-\beta/(2\pi)}$.*

We will deduce Theorem 8 from a more general estimate which in spite of its crudeness might be of an independent interest. We let $W \colon \mathbb{T} \to [1, +\infty]$ be an arbitrary measurable function, and for $A > 0$ we put $W_A = \min(W, e^A)$. By

$$P_r(t) = \frac{1 - r^2}{|1 - rt|^2}, \quad 0 \leq r < 1,\ t \in \mathbb{T},$$

we denote the Poisson kernel for the unit disk.

Theorem 9. *Let ρ be a positive measure on $\mathbb{T}$, let β, M be positive parameters, and let $A \geq A_0(\beta, M) > 1$.*

(A) *Suppose that*

$$\int_{\mathbb{T}} W^\beta \, d\rho \leq C < \infty, \tag{3}$$

and that

$$(\log W_A) * P_{1-A^{-1}} \leq M \log W \tag{4}$$

everywhere on $\mathbb{T}$. Then, for $n \geq \dfrac{A^2\beta}{2M}$,

$$e_n(\rho) \leq \sqrt{2C + \tfrac{1}{2}\rho(\mathbb{T})} \, \exp\!\left(-\frac{\beta}{2M} \int_{\mathbb{T}} \log W_A \, dm\right).$$

(B) *Suppose that $d\rho \gtrsim W^{-\beta} \, dm$ everywhere on $\mathbb{T}$. Then, for $n \cdot m\{W > e^A\} \lesssim 1$,*

$$e_n(\rho) \gtrsim \exp\!\left(-\frac{\beta}{2} \int_{\mathbb{T}} \log W_A \, dm\right).$$

5.1. Proof of Theorem 9, Part (A)

Let F_A be an outer function in the unit disk with the boundary values $|F_A|^2 = W_A^{\beta/M}$ on $\mathbb{T}$ (this means that

$$\log|F_A(rt)| = \left(\frac{\beta}{2M} \log W_A * P_r\right)(t)$$

everywhere in the unit disk $\mathbb{D}$). Then, by (4),

$$\int_{\mathbb{T}} \left|F_A\!\left((1-A^{-1})t\right)\right|^2 d\rho(t) \leq \int_{\mathbb{T}} W^\beta \, d\rho \leq C.$$

Furthermore, $|F_A| \leq e^{A\beta/(2M)}$ everywhere in $\mathbb{D}$, and

$$|F_A(0)| = \exp\!\left(\int_{\mathbb{T}} \log|F_A| \, dm\right)$$

$$= \exp\!\left(\frac{\beta}{2M} \int_{\mathbb{T}} \log W_A \, dm\right).$$

Let

$$F_A\!\left((1-A^{-1})z\right) = \sum_{k \geq 0} f_k z^k.$$

Then, by Cauchy's inequalities,

$$|f_k| \leq \left(1 - A^{-1}\right)^k e^{A\beta/(2M)},$$

and for $n \geq A^2\beta/M$ we get

$$\sum_{k>n} |f_k| \leq \left(1 - A^{-1}\right)^n e^{A\beta/(2M)} \cdot A$$

$$\leq A \exp\left(\frac{A\beta}{2M} - \frac{n}{A}\right)$$

$$\leq A \exp\left(-\frac{A\beta}{2M}\right)$$

$$< \frac{1}{2}, \quad \text{provided that } A \geq A_0(\beta, M).$$

Put

$$P_A(z) = \sum_{k=0}^{n} f_k z^k.$$

Then

$$\int_{\mathbb{T}} |P_A|^2 \, d\rho \leq 2 \int_{\mathbb{T}} |F_A|^2 d\rho + \frac{1}{2}\rho(\mathbb{T}) \leq 2C + \frac{1}{2}\rho(\mathbb{T}),$$

while

$$|P_A(0)| = |F_A(0)| \geq \exp\left(\frac{\beta}{2M} \int_{\mathbb{T}} \log W_A \, dm\right).$$

Therefore,

$$e_n(\rho) \leq \sqrt{2C + \tfrac{1}{2}\rho(\mathbb{T})} \, \exp\left(-\frac{\beta}{2M} \int_{\mathbb{T}} \log W_A \, dm\right),$$

proving part (A). $\qquad\square$

5.2. Proof of Theorem 9, Part (B)

To simplify notation, we denote here by C, c various positive numerical constant whose values are not important for our purposes. The values of these constants may vary from line to line.

Let P be a polynomial of degree n such that $P(0) = 1$ and

$$\int_{\mathbb{T}} |P|^2 \, d\rho = e_n(\rho)^2.$$

Then

$$0 = \log |P(0)| \leq \int_{\mathbb{T}} \log |P| \, dm$$

$$= \frac{1}{2} \int_{\mathbb{T}} \log(|P|^2 W_A^{-\beta}) \, dm + \frac{\beta}{2} \int_{\mathbb{T}} \log W_A \, dm.$$

Furthermore,

$$\int_{\mathbb{T}} |P|^2 W_A^{-\beta} \, dm = \int_{\{W \leq e^A\}} |P|^2 W_A^{-\beta} \, dm + e^{-\beta A} \int_{\{W > e^A\}} |P|^2 \, dm$$

$$\leq \int_{\{W \leq e^A\}} |P|^2 W_A^{-\beta} \, dm + e^{-\beta A} \int_{\mathbb{T}} |P|^2 \, dm,$$

and by Nazarov's version of Turán's lemma [12] (for the reader's convenience, we will recall it after the proof):

$$\int_{\mathbb{T}} |P|^2 \, dm \leq e^{Cn \cdot m\{W > e^A\}} \int_{\{W \leq e^A\}} |P|^2 \, dm$$

$$\leq C \int_{\{W \leq e^A\}} |P|^2 \, dm \,,$$

provided that $n \cdot m\{W > e^A\} \leq C$. Thus,

$$\int_{\mathbb{T}} |P|^2 W_A^{-\beta} \, dm \leq \int_{\{W \leq e^A\}} |P|^2 W_A^{-\beta} \, dm + C e^{-\beta A} \int_{\{W \leq e^A\}} |P|^2 \, dm$$

$$\lesssim \int_{\mathbb{T}} |P|^2 W^{-\beta} \, dm$$

$$\lesssim \int_{\mathbb{T}} |P|^2 \, d\rho$$

$$= e_n(\rho)^2 \quad \text{by the choice of the polynomial } P.$$

Therefore,

$$0 \leq \frac{1}{2} \log\big(e_n(\rho)^2\big) + C + \frac{\beta}{2} \int_{\mathbb{T}} \log W_A \, dm \,,$$

whence,

$$e_n(\rho) \gtrsim \exp\Big(-\frac{\beta}{2} \int_{\mathbb{T}} \log W_A \, dm\Big)$$

completing the proof.   □

Here is the result of Nazarov used in the proof:

Theorem 10 (Nazarov). *Let $n \geq 1$. There exists a positive numerical constant A such that for any set $\Lambda \subset \mathbb{Z}$ of cardinality $n + 1$, any trigonometric polynomial*

$$p(t) = \sum_{\lambda \in \Lambda} c_\lambda t^\lambda \,, \quad (c_\lambda)_{\lambda \in \Lambda} \subset \mathbb{C}, \ t \in \mathbb{T},$$

and any measurable set $E \subset \mathbb{T}$ of Lebesgue measure $m(E) \geq \frac{1}{3}$,

$$\int_{\mathbb{T}} |p|^2 \, dm \leq e^{An \cdot m(\mathbb{T} \backslash E)} \int_E |p|^2 \, dm \,.$$

We have used this theorem with $\Lambda = \{0, 1, \ldots, n\}$.

5.3. Proof of Theorem 8

We apply Theorem 9 with

$$W(e^{i\theta}) = \exp\Big(\frac{1}{|\theta|}\Big), \quad -\pi \leq \theta \leq \pi.$$

Then

$$\Big| \int_{\mathbb{T}} \log W_A \, dm - \frac{1}{\pi} \log A \Big| \leq C,$$

and $m\{W > e^A\} = (\pi A)^{-1}$. Thus, choosing $A \simeq n$, we get

$$e_n(\rho) \gtrsim \exp\left(-\frac{\beta}{2\pi} \log A\right) \gtrsim_\beta n^{-\beta/(2\pi)},$$

proving the lower bound.

To get the upper bound we need to check condition (4). This is straightforward but somewhat long. To simplify our writing, we put

$$w_A(\theta) = \min\left(\frac{1}{|\theta|}, A\right) \quad \text{and} \quad p_r(\theta) = P_r(e^{i\theta}), \quad -\pi \le \theta \le \pi.$$

In this notation, we will show that

$$\int_{-\pi}^{\pi} w_A(\theta - \varphi) p_{1-A^{-1}}(\varphi)\,\mathrm{d}\varphi \lesssim w_A(\theta). \tag{5}$$

Since the function w_A is even, we may assume that $0 \le \theta \le \pi$.

Using the estimate of the Poisson kernel

$$p_r(\varphi) \lesssim (1-r)\min\left(\varphi^{-2}, (1-r)^{-2}\right),$$

we get

$$\int_{-\pi}^{\pi} w_A(\theta - \varphi) p_{1-A^{-1}}(\varphi)\,\mathrm{d}\varphi$$

$$\lesssim \frac{1}{A} \int_{-\pi}^{\pi} w_A(\theta - \varphi) \min\left(\varphi^{-2}, A^2\right)\,\mathrm{d}\varphi$$

$$= \frac{1}{A} \int_{0}^{\pi} \left(w_A(\theta - \varphi) + w_A(\theta + \varphi)\right) \min\left(\varphi^{-2}, A^2\right)\,\mathrm{d}\varphi$$

$$\le \frac{2}{A} \int_{0}^{\pi} w_A(\theta - \varphi) \min\left(\varphi^{-2}, A^2\right)\,\mathrm{d}\varphi$$

$$\lesssim A \int_{0}^{1/A} w_A(\theta - \varphi)\,\mathrm{d}\varphi + \frac{1}{A} \sum_{j \ge 0} \int_{2^j/A}^{2^{j+1}/A} w_A(\theta - \varphi)\varphi^{-2}\,\mathrm{d}\varphi.$$

It remains to estimate the integrals. We have

$$A \int_{0}^{1/A} w_A(\theta - \varphi)\,\mathrm{d}\varphi = A \int_{0}^{1/A} \left(w_A(\theta) + O(1)\min(\theta^{-2}, A^2)\varphi\right)\,\mathrm{d}\varphi$$

$$= w_A(\theta) + O(1)A^{-1}\min(\theta^{-2}, A^2)$$

$$\lesssim w_A(\theta).$$

Next,

$$\frac{1}{A} \sum_{j \ge 0} \int_{2^j/A}^{2^{j+1}/A} w_A(\theta - \varphi)\varphi^{-2}\,\mathrm{d}\varphi \le A \sum_{j \ge 0} 2^{-2j} \int_{2^j/A}^{2^{j+1}/A} w_A(\theta - \varphi)\,\mathrm{d}\varphi.$$

Let k be the minimal non-negative integer such that $\theta \leq 2^k/A$. Then

$$A \sum_{j \geq k+1} 2^{-2j} \int_{2^j/A}^{2^{j+1}/A} w_A(\theta - \varphi) \, d\varphi$$

$$= A \sum_{j \geq k+1} 2^{-2j} \int_{2^j/A}^{2^{j+1}/A} \frac{d\varphi}{\varphi - \theta} \lesssim A \sum_{j \geq k+1} 2^{-2j} \int_{2^j/A}^{2^{j+1}/A} \frac{d\varphi}{\varphi}$$

$$\lesssim A \sum_{j \geq k+1} 2^{-2j} \lesssim 2^{-k} w_A(\theta),$$

and, for $k \geq 2$,

$$A \sum_{0 \leq j \leq k-2} 2^{-2j} \int_{2^j/A}^{2^{j+1}/A} w_A(\theta - \varphi) \, d\varphi$$

$$= A \sum_{0 \leq j \leq k-2} 2^{-2j} \int_{2^j/A}^{2^{j+1}/A} \frac{d\varphi}{\theta - \varphi} \lesssim \frac{1}{\theta} A \sum_{0 \leq j \leq k-2} 2^{-2j} \frac{2^j}{A} \lesssim \frac{1}{\theta}.$$

At last, we are left with two terms corresponding to $j = k$ and $j = k - 1$ (the second term disappears when $k = 0$). Each of them is bounded by

$$\frac{A}{2^{2k}} \cdot A \cdot \frac{2^k}{A} = \frac{A}{2^k} \lesssim w_A(\theta).$$

This completes the proof of estimate (5), and hence of Theorem 8. □

6. Unimodular processes on $\mathbb{Z}$

Here we are interested in spectral properties of unimodular stationary processes $\xi \colon \mathbb{Z} \to \mathbb{T}$.

6.1. An example

We start with a simple example that shows that *any* probability measure on $\mathbb{T}$ can be realized as the spectral measure of a unimodular stationary process on $\mathbb{Z}$.

Consider the product of two circles $\mathbb{T} \times \mathbb{T}$, the first equipped with a probability measure ρ and the second with the normalized Lebesgue measure, and let P be the product of these measures on $\mathbb{T} \times \mathbb{T}$. Let τ be the measure preserving transformation which is the identity on the first circle, and rotates the second by the first, i.e., $\tau(s,t) = (s, st)$, $(s,t) \in \mathbb{T} \times \mathbb{T}$. Consider the process ξ defined by the function $f(s,t) = t$ and the transformation τ, i.e., put $\xi(n) = f(\tau^n) = t \cdot s^n$, $n \in \mathbb{Z}$. It remains to note that the covariance function of this process is the Fourier transform of the measure ρ:

$$r(m) = \mathbb{E}[\xi(0)\bar{\xi}(m)] = \int_{\mathbb{T}} \bar{s}^m \, d\rho(s) = \hat{\rho}(m), \qquad m \in \mathbb{Z}.$$

6.2. A theorem

Curiously enough, it turns out that if the spectrum of the unimodular stationary process is contained in an arc of length less than π, then the only possibility is given by the example we gave.

Theorem 11. *Let $\xi \colon \mathbb{Z} \to \mathbb{T}$ be a unimodular wide-sense stationary process with the spectral measure ρ. Suppose that $\mathrm{spt}(\rho)$ is contained in an arc of length less than π. Then almost every realization of ξ has the form*

$$\xi(n) = t \cdot s^n, \qquad n \in \mathbb{Z}, \tag{6}$$

with some (random) $t, s \in \mathbb{T}$.

Corollary 12. *The only unimodular stationary ergodic process ξ on $\mathbb{Z}$ with the support of the spectral measure contained in an arc of length less than π is the process defined by (6) for some constant $s \in \mathbb{T}$ and for $t \in \mathbb{T}$ which is a random variable with either a uniform distribution on the circle if s is not a root of unity, or a uniform distribution on a cyclic group if s is a root of unity.*

Indeed, the ergodicity of ξ yields that $s \in \mathbb{T}$ is a constant. Furthermore, the distribution of $t \in \mathbb{T}$ is invariant under multiplication by s and ergodic. $\qquad\square$

It would be very interesting to reveal other restrictions on the spectral measures of unimodular ergodic processes on $\mathbb{Z}$.

6.2.1. Eremenko–Ostrovskii theorem. The proof of Theorem 11 is based on the following deterministic result of Eremenko and Ostrovskii [7, Theorem 1′]:

Theorem 13 (Eremenko–Ostrovskii). *Suppose that the Taylor series*

$$F(z) = \sum_{n \geq 0} f_n z^n$$

with unimodular coefficients $(f_n)_{n \geq 0} \subset \mathbb{T}$ admits analytic continuation to $\bar{\mathbb{C}} \setminus J$, where $J \subset \mathbb{T}$ is an arc of length less than π. Then $f_n = t \cdot s^n$, $n \geq 0$, with $t, s \in \mathbb{T}$.

6.2.2. Proof of Theorem 11. Let $(\xi(n))_{n \in \mathbb{Z}}$ be a realization of the process ξ. By Theorem 1, the Fourier–Carleman transform F_ξ is analytic on $\bar{\mathbb{C}} \setminus J$ where $J \subset \mathbb{T}$ is an arc of length less than π. Then, applying Theorem 13 first to F_ξ^+ and then to F_ξ^-, we conclude that (6) holds with some $t, s \in \mathbb{T}$. $\qquad\square$

References

[1] N.H. Bingham, *Szegő's theorem and its probabilistic descendants.* Probab. Surv. **9** (2012), 287–324.

[2] A. Borichev, A. Nishry, M. Sodin, *Entire functions of exponential type represented by pseudo-random and random Taylor series.* J. d'Analyse Math. **133** (2017), 361–396.

[3] P. Borwein, T. Erdélyi, *Polynomials and polynomial inequalities.* Springer-Verlag, New York, 1995.

[4] J. Breuer, B. Simon, *Natural boundaries and spectral theory.* Adv. Math. **226** (2011), 4902–4920.

[5] P. Deift, A. Its, I. Krasovsky, *Toeplitz matrices and Toeplitz determinants under the impetus of the Ising model: some history and some recent results.* Comm. Pure Appl. Math. **66** (2013), 1360–1438.

[6] P. Dienes, *The Taylor series: an introduction to the theory of functions of a complex variable.* Dover Publications, New York, 1957.

[7] A. Eremenko, I.V. Ostrovskii, *On the 'pits effect' of Littlewood and Offord.* Bull. Lond. Math. Soc. **39** (2007), 929–939.

[8] U. Grenander, G. Szegő, *Toeplitz forms and their applications.* Second edition. Chelsea Publishing Co., New York, 1984

[9] H. Helson, *Harmonic analysis.* Addison-Wesley Publishing Company, Reading, MA, 1983.

[10] Y. Katznelson, *An Introduction to Harmonic Analysis.* 3rd edition. Cambridge University Press, Cambridge, 2004.

[11] B. McMillan, *History of a problem.* J. Soc. Indust. Appl. Math. **3** (1955), 119–128.

[12] F. Nazarov, *Complete version of Turán's lemma for trigonometric polynomials on the unit circumference.* Complex analysis, operators, and related topics, 239–246, Oper. Theory Adv. Appl., 113, Birkhäuser, Basel, 2000.

[13] M. Queffelec, *Substitution dynamical systems – spectral analysis.* 2nd edition. Lecture Notes in Mathematics, 1294. Springer-Verlag, Berlin, 2010.

[14] L. Shepp, *Covariance of unit processes.* Proc. Working Conf. on Stochastic Processes, Santa Barbara, Ca, 1967, 205–218.

[15] B. Simon, *Szegő's theorem and its descendants.* Spectral theory for L^2 perturbations of orthogonal polynomials. Princeton University Press, Princeton, NJ, 2011.

[16] B.L. van der Waerden, *Algebra.* Vol. I. Based in part on lectures by E. Artin and E. Noether. Translation from the 7th German edition. Springer-Verlag, New York, 1991.

Alexander Borichev
Aix-Marseille Université, CNRS
Centrale Marseille, I2M
F-13453 Marseille, France
e-mail: alexander.borichev@math.cnrs.fr

Mikhail Sodin
School of Mathematics
Tel Aviv University
Tel Aviv, 69978 Israel
e-mail: sodin@post.tau.ac.il

Benjamin Weiss
Institute of Mathematics
Hebrew University of Jerusalem
Jerusalem, 91904 Israel
e-mail: weiss@math.huji.ac.il

Operator Theory:
Advances and Applications, Vol. 261, 159–171
© Springer International Publishing AG, part of Springer Nature 2018

Index Formulas for Toeplitz Operators, Approximate Identities, and the Wolf–Havin Theorem

Albrecht Böttcher

In memory of Viktor Petrovich Havin

Abstract. This is a review of a development in the 1970s and 1980s that was concerned with index formulas for block Toeplitz operators with discontinuous symbols. We cite the classical results by Douglas and Sarason, outline Silbermann's approach to index formulas via algebraization, and embark on the replacement of the Abel–Poisson means by arbitrary approximate identities. One question caused by this development was whether the Abel–Poisson means are the only approximative identities that are asymptotically multiplicative on (H^∞, H^∞), and the review closes with Wolf and Havin's theorem, which gives an affirmative answer to this question.

Mathematics Subject Classification (2010). Primary 47B35. Secondary 30H05, 42A10, 46J15.

Keywords. Toeplitz operator, index formula, approximate identity, asymptotic multiplicativity, BMO, VMO, Viktor Havin.

1. A personal note

It was in 1986 when I was one week in Leningrad, now again Saint Petersburg, following an invitation of Nikolai Nikolski. I knew many of the young mathematicians of his group from their publications and was overwhelmed to meet all of them in person, including Khrushchov, Kisliakov, Peller, Tolokonnikov, Treil, Vasyunin, Volberg, Yakubovich. My first encounter with Viktor Havin dates back to this visit, too: I was invited to give a talk at the weekly seminar directed by Nikolai Nikolski and Viktor Havin. This seminar was a regular and big event. It took place in a proud hall and the number of listeners was impressive. The imposing round desk in the seminar room was about hundred years old and it was in fact the desk of Andrei Markov, one of my great mathematical heroes [6]. The speaker assigned to that day was Nikolai Makarov, so that actually the date was already occupied. However, I was told not to worry about that. And indeed, Nikolai Makarov first gave a stirring talk of about one hour on his fresh results on harmonic measures and after him I was given another hour for my talk, and all the time I felt admirable interest and unbroken attention in the audience. The privilege to give that talk and the warm reception I received are unforgettable in my mind, and I am happy about the opportunity to remind of that event here.

In the years after that visit in Leningrad, I talked and corresponded with Viktor Havin on several occasions, and it was always a great pleasure to benefit from his outstanding personality and his insights into mathematics. In the late 1980s, my student Hartmut Wolf spent one year in Leningrad under the guidance of Viktor Havin, and one result of their collaboration was the paper [23], which will enter this review later. The intention of this review is in fact to tell the entire pre-history to this paper.

2. Continuous symbols

We think of $\mathbf{C}^N$ as a column space and denote by L_N^2 and H_N^2 the usual $\mathbf{C}^N$-valued Lebesgue and Hardy spaces on the complex unit circle $\mathbf{T}$. For $A \subset L^\infty := L^\infty(\mathbf{T})$, we define $A_{N \times N}$ as the set of all functions $a : \mathbf{T} \to \mathbf{C}^{N \times N}$ whose entries are in A. The Toeplitz operator $T(a)$ generated by a function $a \in L_{N \times N}^\infty$ is the bounded linear operator that acts on H_N^2 by the rule $T(a)f = P(af)$, where P is the orthogonal projection of L_N^2 onto H_N^2. The function a is referred to as the symbol of the operator $T(a)$. In the scalar case, that is, for $N = 1$, we omit the subscript N.

If $a \in C_{N \times N} := [C(\mathbf{T})]_{N \times N}$, then $T(a)$ is known to be a Fredholm operator, that is, to be invertible modulo compact operators, if and only if the function $\det a$ has no zeros on $\mathbf{T}$, which is equivalent to saying that $0 \notin (\det a)(\mathbf{T})$. In that case the index $\operatorname{Ind} T(a)$ of $T(a)$, which is defined as the difference of the kernel and co-kernel dimensions of $T(a)$, equals $-\operatorname{wind} \det a$, that is, minus the winding number of $\det a$ about the origin.

For symbols with absolutely convergent Fourier series, the index formula is due to Gohberg and Krein, who proved it via Wiener–Hopf factorization and partial indices. In the case of continuous symbols, the index formula was proved independently by Atiyah [1] and Douglas [8, 9] using homotopy arguments. Another proof, the one given in [7], is based on a result of Markus and Feldman [14] which says that if the entries of an operator matrix commute pairwise up to trace class operators, then the index of the operator matrix equals the index of its determinant. One more very elegant proof was recently given in [22].

3. Symbols in the Douglas algebra

Let H^∞ be the algebra of (non-tangential limits of) bounded analytic functions in the unit disk. In 1966, Sarason discovered that the smallest closed subalgebra of L^∞ that contains $C \cup H^\infty$, is just the sum $C + H^\infty := \{f + g : f \in C, g \in H^\infty\}$.

See, e.g., Theorem 2.53 of [7] or Corollary 2.3.1 of [15] or Sections 2.2.9 and 2.3.8 of [16]. In 1968, Douglas showed that if $a \in (C + H^\infty)_{N \times N}$, then $T(a)$ is Fredholm if and only if a is invertible in $(C + H^\infty)_{N \times N}$. Nowadays, at least in parts of the community, $C + H^\infty$ is referred to as "the" Douglas algebra. (The definite article is ambiguous, because $C + H^\infty$ is actually the smallest within a whole family of algebras which are also called Douglas algebras.) To say more, Douglas considered the harmonic extension of a into the unit disk. If a has the Fourier series $\sum_{n=-\infty}^{\infty} a_n t^n$ ($t = e^{i\theta} \in \mathbf{T}$), the harmonic extension $\widehat{a}$ of a at the point rt with $r \in [0,1)$ and $t \in \mathbf{T}$ is

$$\widehat{a}(rt) = (h_r a)(t) := \sum_{n=-\infty}^{\infty} a_n r^{|n|} t^n.$$

The function $h_r a$ is also referred to as the rth Abel–Poisson mean of a. The family $\{h_r a\} := \{h_r a\}_{r \in [0,1)}$ is a family of functions in $C_{N \times N}$. This family is said to be stable if there is an $r_0 \in (0,1)$ such that $\det h_r a$ has no zeros on $\mathbf{T}$ for $r \in [r_0, 1)$ and

$$\sup_{r \in [r_0,1)} \|(h_r a)^{-1}\|_\infty < \infty, \tag{1}$$

where $\|\cdot\|_\infty$ is any norm in $L^\infty_{N \times N}$, for example, the C^*-norm. Note that replacing $(h_r a)^{-1}$ by $(\det h_r a)^{-1}$ in (1) would result in an equivalent definition. If $\{h_r a\}$ is a stable family, then the limit $\lim_{r \to 1}$ wind $\det h_r a$ exists and is an integer. Here is the result of [8].

Theorem 1 (Douglas). *Let $a \in (C + H^\infty)_{N \times N}$. Then the following are equivalent:* (i) *the operator $T(a)$ is Fredholm,* (ii) *the function a is invertible in $(C+H^\infty)_{N \times N}$,* (iii) *the family $\{h_r a\}$ is stable. If one of these conditions is satisfied, then*

$$\operatorname{Ind} T(a) = -\lim_{r \to 1} \text{wind} \det h_r a. \tag{2}$$

A key ingredient to the proof is the following theorem, which, incidentally, implies that in (2) we could replace $\det h_r a$ by $h_r(\det a)$.

Theorem 2 (Douglas). *If $a, b \in (C + H^\infty)_{N \times N}$, then*

$$\|h_r(ab) - (h_r a)(h_r b)\|_\infty \to 0 \ \ as \ r \to 1. \tag{3}$$

The proof is actually simple: it is not difficult to show that (3) holds if $a \in C$ and $b \in L^\infty$, while if both a and b are in H^∞, we have $h_r(ab) = (h_r a)(h_r b)$.

For proofs of Theorem 1 itself, see the original sources [8, 9] or Theorem 2.94 of [7] or Section 4.5 of [15] or Section 3.4.3 of [16]. Note that in [7] the matrix case is again reduced to the scalar case by invoking the Markus–Feldman theorem [14].

A proof of the index formula in Theorem 1 different from the ones just mentioned was given by Silbermann [19]. His argument is as follows. Let $\mathcal{B}$ be the C^*-algebra of all bounded linear operators on H^2_N and let $\mathcal{A}$ denote the set of all functions $[0,1) \to \mathcal{B}$, $r \mapsto A_r$ with the property that there is an operator $A \in \mathcal{B}$ such that $A_r \to A$ strongly and $A_r^* \to A^*$ strongly as $r \to 1$. Clearly, we may think of the elements of $\mathcal{A}$ as families $\{A_r\} := \{A_r\}_{r \in [0,1)}$ of operators $A_r \in \mathcal{B}$. With the operations $\{A_r\} + \{B_r\} = \{A_r + B_r\}$, $\alpha\{A_r\} = \{\alpha A_r\}$, $\{A_r\}\{B_r\} = \{A_r B_r\}$, $\{A_r\}^* = \{A_r^*\}$ and the norm $\|\{A_r\}\| = \sup_{r \in [0,1)} \|A_r\|$, the set $\mathcal{A}$ becomes a C^*-algebra. The set $\mathcal{J}$ of all $\{A_r\}$ of the form $A_r = R + C_r$ where R is compact and $\|C_r\| \to 0$ as $r \to 1$, is a closed two-sided ideal of $\mathcal{A}$. For $\{A_r\} \in \mathcal{A}$, we denote the coset $\{A_r\} + \mathcal{J} \in \mathcal{A}/\mathcal{J}$ by $\{A_r\}^\pi$. $A \in \mathcal{A}$ if $A_r \to A$ strongly and if there invertible for $r \in [r_0, 1)$ and Herewith the result of [19], which, in a more general form, can also be found with a full proof as Theorem 4.28 of [7].

Theorem 3 (Silbermann). *Let $a \in L_{N \times N}^\infty$ and suppose $\{T(h_r a)\}^\pi$ is invertible in $\mathcal{A}/\mathcal{J}$. Then $T(a)$ is Fredholm, $\{h_r a\}$ is a stable family, and*

$$\operatorname{Ind} T(a) = -\lim_{r \to 1} \operatorname{wind} \det h_r a. \tag{4}$$

Silbermann's derivation of the index formula of Douglas is now as follows. Let $b := a^{-1} \in (C + H^\infty)_{N \times N}$ and write $b = f + g$ with $f \in C_{N \times N}$ and $g \in H_{N \times N}^\infty$. With $Q = I - P$, we have

$$T(h_r a)T(h_r b) = T((h_r a)(h_r b)) - Ph_r a Q h_r b P = T((h_r a)(h_r b)) - Ph_r a Q h_r f P,$$

the second equality resulting from the fact that $Qh_r g P = 0$. Since f is continuous, we get $\|h_r f - f\|_\infty \to 0$ and thus $Qh_r f P = Qf P + C_r'$ with $\|C_r'\| \to 0$ as $r \to 1$. Consequently,

$$Ph_r a Q h_r f P = Ph_r a Q f P + Ph_r a Q C_r' = Ph_r a Q f P + C_r'' \text{ with } \|C_r''\| \to 0$$

as $r \to 1$. As $Ph_r a Q \to PaQ$ strongly (for every $a \in L_{N \times N}^\infty$) and the Hankel operator QfP is compact due to Hartman's theorem, we conclude that $Ph_r a Q f P = PaQfP + C_r'''$ with $\|C_r'''\| \to 0$ as $r \to 1$. In summary,

$$T(h_r a)T(h_r b) = T((h_r a)(h_r b)) - PaQfP - C_r''' - C_r'',$$

and since $PaQfP$ is compact, we obtain the identity

$$\{T(h_r a)\}^\pi \{T(h_r b)\}^\pi = \{T((h_r a)(h_r b))\}^\pi.$$

Finally, Theorem 2 implies that $\|T((h_r a)(h_r b)) - T(h_r(ab))\| \to 0$ as $r \to 1$, which gives that $\{T((h_r a)(h_r b))\}^\pi = \{T(h_r(ab))\}^\pi = \{I\}^\pi$. Thus, $\{T(h_r b)\}^\pi$ is a right inverse of $\{T(h_r a)\}^\pi$. It can be shown in a completely analogous manner that $\{T(h_r b)\}^\pi$ is also a left inverse of $\{T(h_r a)\}^\pi$. Theorem 3 completes the proof.

4. Locally sectorial symbols

The actual power of Theorem 3 comes into play when we consider locally sectorial symbols. Namely, as already mentioned after Theorem 2, it is easy to show that if $\varphi \in C$ and $a \in L^\infty$, then $\|h_r(\varphi a) - (h_r \varphi)(h_r a)\|_\infty \to 0$ as $r \to 1$. This implies that $\{T(h_r \varphi)\}^\pi$ and $\{T(h_r a)\}^\pi$ commute and thus offers application of the so-called local principles, that is, of theorems on the localization over central subalgebras.

A function $a : S \to \mathbf{C}^{N \times N}$ defined on some set S is said to be sectorial on S if there exist invertible matrices $b, c \in \mathbf{C}^{N \times N}$ and a number $\varepsilon > 0$ such that $\operatorname{Re}(ba(s)cz, z) \geq \varepsilon \|z\|^2$ for all $s \in S$ and all $z \in \mathbf{C}^N$, where $(\cdot, \cdot)$ is the usual inner product on $\mathbf{C}^N$. If a is defined only almost everywhere on S, then the "for all $s \in S$" has to be replaced by "for almost all $s \in S$". A function $a \in L^\infty_{N \times N}$ is called locally sectorial over C (= the continuous functions) if for every point $\tau \in \mathbf{T}$ there exists an open arc $U_\tau \subset \mathbf{T}$ containing τ such that a is sectorial on U_τ. Douglas and Widom [11] ($N = 1$) and Simonenko [21] ($N \geq 1$) showed that $T(a)$ is Fredholm whenever $a \in L^\infty_{N \times N}$ is locally sectorial over C and expressed the index of $T(a)$ in terms of what I call the sectorial cloud moving around with a. See also [2] and the beginning of Chapter 3 of [7] for some subtleties in the case $N > 1$. Silbermann used a local principle to prove that $\{T(h_r a)\}^\pi$ is invertible if $a \in L^\infty_{N \times N}$ is locally sectorial over C and then employed his Theorem 3 to deduce that $T(a)$ is Fredholm with the index given by (4).

Since Sarason's paper [18] and Douglas' paper [10], localization over the algebra QC of quasicontinuous functions was a hot topic of the 1970s and 1980s. The fascination of Silbermann and myself for the topic is reflected in [7, 20]. The algebra QC is defined as $QC = (C + H^\infty) \cap (C + \overline{H^\infty})$; see also Section 2.80 of [7] or Section 2.3.2 of [15]. A key result of Sarason [18] is as follows.

Theorem 4 (Sarason). *If $\varphi \in QC$ and $a \in L^\infty$, then*

$$\|h_r(\varphi a) - (h_r \varphi)(h_r a)\|_\infty \to 0 \text{ as } r \to 1.$$

Here is a simple proof. Let $P_r(x) = (1 - r^2)/(2\pi(1 - 2r\cos x + r^2))$ be the periodic Poisson kernel. We have

$$|h_r(\varphi a)(e^{ix}) - (h_r \varphi)(e^{ix})(h_r a)(e^{ix})|$$

$$= \left| \int_0^{2\pi} [\varphi(e^{i\theta}) - (h_r \varphi)(e^{ix})]a(e^{i\theta})P_r(x - \theta)d\theta \right|$$

$$\leq \|a\|_\infty \int_0^{2\pi} |\varphi(e^{i\theta}) - (h_r \varphi)(e^{ix})|P_r(x - \theta)d\theta$$

$$\leq \|a\|_\infty \left(\int_0^{2\pi} |\varphi(e^{i\theta}) - (h_r \varphi)(e^{ix})|^2 P_r(x - \theta)d\theta \right)^{1/2}$$

$$= \|a\|_\infty \left((h_r(\varphi\overline{\varphi}))(e^{ix}) - (h_r\varphi)(e^{ix})(h_r\overline{\varphi})(e^{ix}) \right)^{1/2},$$

and since $\varphi \in C + H^\infty$ and $\overline{\varphi} \in C + H^\infty$, we get $\|h_r(\varphi\overline{\varphi}) - (h_r\varphi)(h_r\overline{\varphi})\|_\infty \to 0$ by virtue of Theorem 2.

Theorem 4 implies that if $\varphi \in QC$ and $a \in L^\infty$, then $\{T(h_r\varphi)\}^\pi$ and $\{T(h_r a)\}^\pi$ commute, and using his Theorem 3, Silbermann [20] was so able to prove the following result, which, with a full proof, is also contained in Corollary 4.30 of [7]. The conclusion that $T(a)$ is Fredholm is already in [10].

Theorem 5 (Silbermann). *If $a \in L^\infty_{N \times N}$ is locally sectorial over QC, then $T(a)$ is Fredholm, $\{h_r a\}$ is a stable family, and the index of $T(a)$ is given by (4).*

So what is local sectoriality over QC? Let X denote the maximal ideal space of L^∞ and think of $L^\infty_{N \times N}$ as $C_{N \times N}(X)$. The C^*-subalgebra $QC \subset L^\infty$ induces a fibration of X into fibers over the maximal ideal space $M(QC)$ of QC, and a function in $L_{N\times N}^\infty = C_{N\times N}(X)$ is said to be locally over QC if it is sectorial on each of the fibers of the fibration. Analogously one can define local sectoriality over an arbitrary C^*-subalgebra B of L^∞. For $B = C$, one gets just local sectoriality in the sense introduced above. A function $g \in L_{N\times N}^\infty$ is called globally sectorial if it is sectorial on all of $\mathbf{T}$ (equivalently, on all of X). Finally, given a unital Banach algebra A, we denote by GA the group of its invertible elements. The following theorem demystifies local sectoriality. It was established in [3] for $N = 1$ and then, using a result of Roch [17], in [5] for general N. It can also be found as Theorem 3.8 in [7].

Theorem 6. *A function $a \in L_{N\times N}^\infty$ is locally sectorial over a C^*-algebra B between $\mathbf{C}$ and L^∞ if and only if $a = \varphi g$ with a function $\varphi \in GB_{N\times N}$ and a globally sectorial function $g \in GL_{N\times N}^\infty$.*

Clearly, this theorem eases index computation significantly: if $B \subset QC$, we have $T(a) = T(\varphi)T(g) +$ a compact operator, so

$$\operatorname{Ind} T(a) = \operatorname{Ind} T(\varphi) + \operatorname{Ind} T(g),$$

the index of $T(g)$ is zero, because globally sectorial functions generate invertible Toeplitz operators, from Theorem 1 we infer that the index of the operator $T(\varphi)$ is $-\lim \operatorname{wind} \det h_r \varphi$, and using Theorem 4, we eventually obtain that

$$\lim \operatorname{wind} \det h_r \varphi = \lim \operatorname{wind} \det h_r \varphi + \lim \operatorname{wind} \det h_r g = \lim \operatorname{wind} \det h_r a.$$

5. Approximate identities

The index formula (4) expresses the index of a Toeplitz operator with a discontinuous symbol via the winding numbers of approximations of the symbol by smooth functions. In the mid 1980s, it came to my mind that it could be advantageous to replace these smooth approximations even with trigonometric polynomials, for example, the Fejér–Cèsaro means

$$(\sigma_n a)(t) := \sum_{|j|\leq n}\left(1 - \frac{|j|}{n+1}\right)a_j t^j \quad (t \in \mathbf{T}) \quad \text{for} \quad a(t) = \sum_{j=-\infty}^{\infty} a_j t^j,$$

because determining the winding number of $\sigma_n a$ simply amounts to counting the zeros of $z^n(\sigma_n a)(z)$ inside the unit circle. The integral representations through the Poisson and Fejér kernels,

$$(h_r a)(e^{ix}) = \int_{\mathbf{R}} \lambda K(\lambda(x - \theta))a(e^{i\theta})d\theta \text{ with } \lambda = -\frac{1}{\log r}, \ K(x) = \frac{1}{\pi(1 + x^2)},$$

$$(\sigma_n a)(e^{ix}) = \int_{\mathbf{R}} \lambda K(\lambda(x - \theta))a(e^{i\theta})d\theta \text{ with } K(x) = \frac{2}{\pi}\frac{\sin^2(x/2)}{x^2},$$

actually led me to the consideration of more general approximate identities.

Thus, let K be a function in $L^1(\mathbf{R})$ with the following properties: $K(x) \geq 0$ and $K(x) = K(-x)$ for all $x \in \mathbf{R}$, $\int_{\mathbf{R}} K(x)dx = 1$, the essential infimum of K on

$(-\pi, \pi)$ is (strictly) positive, the essential supremum of K on $(-\pi, \pi)$ is finite, and $|K(x)| = O(1/x^2)$ as $|x| \to \infty$. Let further Λ be an unbounded subset of $(0, \infty)$. Given a function a in $L^1_{N \times N}$ on the unit circle $\mathbf{T}$ and $\lambda \in \Lambda$, we define

$$(k_\lambda a)(e^{ix}) = \int_{\mathbf{R}} \lambda K(\lambda(x - \theta)) a(e^{i\theta}) d\theta. \tag{5}$$

If K is the Poisson kernel, then $\Lambda = (0, \infty)$ and $k_\lambda a$ is $h_r a$ at $r = e^{-1/\lambda}$. If K is the Fejér kernel, we could take $\Lambda = \mathbf{N}$ to get $k_\lambda a = \sigma_\lambda a$ for a natural number λ, but nothing prevents us from alternatively taking $\Lambda = (0, \infty)$ and thus from working with $\sigma_\lambda a = k_\lambda a$ also in the case where λ is a positive real number. In what follows, the notation $\lambda \to \infty$ means that $\lambda \to \infty, \lambda \in \Lambda$.

Let $a \in L^1_{N \times N}$. Our assumptions on K guarantee that the functions $k_\lambda a$ belong to $C_{N \times N}$. It is also easy to show that

$$\|k_\lambda(\varphi a) - (k_\lambda \varphi)(k_\lambda a)\|_\infty \to 0 \text{ as } \lambda \to \infty \text{ whenever } \varphi \in C_{N \times N}. \tag{6}$$

We call the family $\{k_\lambda a\}$ stable if there is a $\lambda_0 \in \Lambda$ such that $k_\lambda a \in GC_{N \times N}$ for $\lambda \in \Lambda \cap (\lambda_0, \infty)$ and the supremum of $\|(k_\lambda a)^{-1}\|_\infty$ over $\lambda \in \Lambda \cap (\lambda_0, \infty)$ is finite. If the family $\{k_\lambda a\}$ is stable and Λ is a connected set, then the limit of the winding numbers of $\det(k_\lambda a)$ exists and is finite as $\lambda \to \infty$.

It is clear how to transfer Silbermann's construction of the C^*-algebra $\mathcal{A}/\mathcal{J}$ given above to the present setting. We now define $\mathcal{A}_\Lambda$ as the set of functions $a : \Lambda \to \mathcal{B}, \lambda \mapsto A_\lambda$ such that there is an operator $A \in \mathcal{B}$ with $A_\lambda \to A$ strongly and $A^*_\lambda \to A^*$ strongly as $\lambda \to \infty$, we consider the elements of $\mathcal{A}_\Lambda$ as families $\{A_\lambda\} := \{A_\lambda\}_{\lambda \in \Lambda}$, and we equip $\mathcal{A}_\Lambda$ with algebraic operations and a C^*-norm in an obvious fashion. The ideal $\mathcal{J}_\Lambda$ is the set of all families $\{R + C_\lambda\}$ with R compact and $\|C_\lambda\| \to 0$ as $\lambda \to \infty$. Finally, for $\{A_\lambda\} \in \mathcal{A}_\Lambda$, we define $\{A_\lambda\}^\pi_\Lambda \in \mathcal{A}_\Lambda/\mathcal{J}_\Lambda$ as the coset $\{A_\lambda\} + \mathcal{J}_\Lambda$. Proceeding as in [19], we get the following analogue to Theorem 3.

Theorem 7. *Let $a \in L^\infty_{N \times N}$ and suppose $\{T(k_\lambda a)\}^\pi$ is invertible in $\mathcal{A}_\Lambda/\mathcal{J}_\Lambda$. Then $T(a)$ is Fredholm and $\{k_\lambda a\}$ is a stable family. If, in addition, Λ is connected, we have*

$$\operatorname{Ind} T(a) = -\lim_{\lambda \to \infty} \operatorname{wind} \det k_\lambda a. \tag{7}$$

If $a \in L^\infty_{N \times N}$ is locally sectorial over C, then, by employing (6) in conjunction with either a local principle or Theorem 6, one can show that $\{T(k_\lambda a)\}^\pi$ is invertible in $\mathcal{A}_{\operatorname{conv} \Lambda}/\mathcal{J}_{\operatorname{conv} \Lambda}$, where $\operatorname{conv} \Lambda$ is the convex hull of Λ. Since this convex hull is connected, formula (7) is true with $\lambda \to \infty$ and $\lambda \in \operatorname{conv} \Lambda$ and hence also true for every sequence $\lambda_n \to \infty, \lambda_n \in \Lambda$. In particular, the theorem implies that if $a \in L^\infty_{N \times N}$ is locally sectorial over C, then

$$\operatorname{Ind} T(a) = -\lim_{n \to \infty} \operatorname{wind} \det \sigma_n a.$$

However, as already indicated above, in the 1980s, a kind of a grail in the theory of Toeplitz operators was the passage from localization over C to localization over QC. We turn to this topic in Section 7 after a deviation in the following Section 6.

6. Partial indices cannot be detected by approximate identities

Let W be the Wiener algebra of all functions on the unit circle with absolutely convergent Fourier series. Put $W^+ = W \cap H^\infty$ and $W^- = W \cap \overline{H^\infty}$. As shown by Gohberg and Krein, every $a \in GW_{N\times N}$ admits a so-called (right) Wiener–Hopf factorization $a = a_- d a_+$ with $a_\pm \in GW^\pm_{N\times N}$ and $d(t) = \mathrm{diag}(t^{\kappa_1}, \ldots, t^{\kappa_N})$ $(t \in \mathbf{T})$ where $\kappa_1 \geq \cdots \geq \kappa_N$ are integers. These integers are uniquely determined by a and are called the (right) partial indices of a. Having a Wiener–Hopf factorization, we can write $T(a) = T(a_-)T(d)T(a_+)$, and the operators $T(a_\pm)$ are invertible, the inverses being $T(a_\pm^{-1})$. It follows that the kernel and cokernel dimensions of $T(a)$ are given by

$$\dim \mathrm{Ker}\, T(a) = \sum_{\kappa_j < 0} |\kappa_j|, \quad \dim \mathrm{Ker}\,[T(a)]^* = \sum_{\kappa_j > 0} \kappa_j,$$

and, clearly, $\mathrm{Ind}\, T(a) = -(\kappa_1 + \cdots + \kappa_N)$.

Since a function $a \in GW_{N\times N}$ is locally sectorial over C, we deduce from what was said after Theorem 7 that the sum of its partial indices is equal to the sum of the partial indices of $k_\lambda a$ for all sufficiently large λ. In [4], it was shown that, for $N \geq 2$, this is in general no longer true for the individual partial indices if the approximate identity is produced by the Poisson or the Fejér kernels. The following result shows that the situation is the same for arbitrary approximate identities.

Theorem 8. *There exist $a \in GW_{2\times 2}$ such that the partial indices of $k_\lambda a$ are zero for all $\lambda \in \Lambda$ but the partial indices of a itself are not all equal to zero.*

Proof. Krupnik and Feldman [13] proved that if $c(t) = \sum_{j=-n+1}^{n-1} c_j t^j$, c_j in $\mathbf{C}$, then the Toeplitz matrix $T_n(c) := (c_{j-\ell})_{j,\ell=1}^n$ is invertible if and only if the partial indices of $a(t) = \left(\begin{smallmatrix} t^{-n} & 0 \\ c(t) & t^n \end{smallmatrix}\right)$ are zero. Thus, if we let $c(t) = t^{-1} + 1 + t$ and $a(t) = \left(\begin{smallmatrix} t^{-2} & 0 \\ c(t) & t^2 \end{smallmatrix}\right)$, then $T_2(c) = \left(\begin{smallmatrix} 1 & 1 \\ 1 & 1 \end{smallmatrix}\right)$ is not invertible, which implies that the two partial indices of a are not equal to zero. Let $\widehat{K}(\xi) := \int_{\mathbf{R}} K(x)e^{i\xi x}dx$ be the Fourier transform of the kernel K. We have

$$(k_\lambda t)(e^{ix}) = \int_{\mathbf{R}} \lambda K(\lambda(x-\theta))e^{i\theta}d\theta = \int_{\mathbf{R}} K(s)e^{ix}e^{-is/\lambda}ds = e^{ix}\widehat{K}(-1/\lambda),$$

and analogously, $(k_\lambda t^{-1})(e^{ix}) = e^{-ix}\widehat{K}(1/\lambda)$ and $(k_\lambda 1)(e^{ix}) = \widehat{K}(0) = 1$. Consequently, $T_2(k_\lambda c) = \left(\begin{smallmatrix} 1 & \widehat{K}(1/\lambda) \\ \widehat{K}(-1/\lambda) & 1 \end{smallmatrix}\right)$. But if $\xi \neq 0$, then $-1 < \widehat{K}(\xi) < 1$ because K is even and hence

$$1 - \widehat{K}(\xi) = 2\int_0^\infty K(x)dx - 2\int_0^\infty K(x)\cos(\xi x)dx$$

$$= 2\int_0^\infty K(x)(1 - \cos(\xi x))dx > 0,$$

and analogously, $\widehat{K}(\xi) + 1 > 0$. It follows that $T_2(k_\lambda c)$ is invertible for all $\lambda \in \Lambda$, which tells us that the partial indices of $k_\lambda a$ are zero for all $\lambda \in \Lambda$. $\square$

In fact, the Wiener–Hopf factorization of the matrix a introduced in the previous proof is

$$\begin{pmatrix} t^{-2} & 0 \\ t^{-1}+1+t & t^2 \end{pmatrix} = \begin{pmatrix} 1 & -1+t^{-1} \\ 0 & 1 \end{pmatrix} \begin{pmatrix} t & 0 \\ 0 & t^{-1} \end{pmatrix} \begin{pmatrix} 1 & -1+t \\ 1+t+t^3 & 1 \end{pmatrix},$$

which shows that the partial indices are $1, -1$. It is a well-known result that if $a \in GW_{N \times N}$ has two partial indices κ_j and κ_ℓ such that $\kappa_j - \kappa_\ell \geq 2$, then every ball $\{b \in GW_{N \times N} : \|b - a\|_\infty < \varepsilon\}$ contains a b whose partial indices differ from those of a. Theorem 8 says that even every "curve" $\{k_\lambda a : \lambda \in \Lambda\}$ may entirely consist of such b.

7. Asymptotic multiplicativity

The kernel K or the approximate identity k_λ induced by it are said to be asymptotically multiplicative on a pair (A, B) of subsets $A, B \subset L^\infty$ if

$$\|k_\lambda(ab) - (k_\lambda a)(k_\lambda b)\|_\infty \to 0 \quad \text{as } \lambda \to \infty \text{ for all } a \in A, \, b \in B.$$

Thus, Theorems 2 and 4 say that the Poisson kernel is asymptotically multiplicative on $(C + H^\infty, C + H^\infty)$ and (QC, L^∞), respectively, and (6) tells us that every approximate identity is asymptotically multiplicative on (C, L^∞). Thus, asymptotic multiplicativity on $(C + H^\infty, C + H^\infty)$ is in fact equivalent to asymptotic multiplicativity on (H^∞, H^∞) only.

The derivation of Theorem 5 from Theorem 3 makes essential use of the asymptotic multiplicativity of the Poisson kernel on (QC, L^∞). With Theorem 7, an analogue of Theorem 3 for general approximate identities is at our disposal. An analogue of Theorem 5 can therefore be established for all approximate identities which are asymptotically multiplicative on (QC, L^∞). The following theorem from [5], which entered [7] as Theorem 3.23, shows that fortunately all approximate identities enjoy this property.

Theorem 9. *Every approximate identity is asymptotically multiplicative on the pair* (QC, L^∞).

When writing [5], I was not able to prove that arbitrary approximate identities are asymptotically multiplicative on (H^∞, H^∞) and so I could not transfer the proof of Theorem 4 presented above to the general setting. I therefore proceeded differently and used the equality $QC = VMO \cap L^\infty$, which was established by Sarason in 1975, and ideas from Garnett's book [12], especially of Section VI.1 of that book. (In those times, access to such books was difficult on the eastern side of the curtain until the books were translated into Russian and distributed by Soviet publishing houses. That I had this book on my desk in 1985 is due to the fact that E. M. Dyn'kin translated it into Russian under the editorship of Viktor Havin.) Here is an outline of the proof of Theorem 9.

In what follows we write $k_{\lambda,t} a$ for $(k_\lambda a)(t)$. The moving average m_λ, also called the Steklov mean, is the approximate identity that is generated by the kernel

$K(x)$ that is $1/(2\pi)$ for $x \in (-\pi, \pi)$ and zero for $x \notin (-\pi, \pi)$. The corresponding Λ is $(1, \infty)$. Thus, for $\lambda > 1$,

$$m_{\lambda,t}a = (m_\lambda a)(t) = \frac{\lambda}{2\pi} \int_{x-\pi/\lambda}^{x+\pi/\lambda} a(e^{i\theta})d\theta \quad \text{with} \quad t = e^{ix}.$$

The space BMO is defined as the space of all $a \in L^1$ for which

$$\|a\|_* := \sup_{\lambda \in (1,\infty)} \sup_{t \in \mathbf{T}} m_{\lambda,t}(|a - m_{\lambda,t}a|) < \infty.$$

A well-known equivalent definition of BMO is through the so-called Garsia norm

$$\|a\|_G := \sup_{r \in (0,1)} \sup_{t \in \mathbf{T}} h_{r,t}(|a - h_{r,t}a|),$$

where, of course, $h_{r,t}a = (h_r a)(t)$ is the rth Abel–Poisson mean of a at t; see, e.g., Section 1.6.2(b) of [15]. Thus, the following result of [5], which is also cited as Theorem 3.21 in [7], should not come as too much a surprise.

Theorem 10. *Let k_λ be an arbitrary approximate identity with the generating kernel K and put*

$$\|a\|_K := \sup_{\lambda \in \Lambda} \sup_{t \in \mathbf{T}} k_{\lambda,t}(|a - k_{\lambda,t}a|).$$

Then a function $a \in L^1$ belongs to BMO if and only if $\|a\|_K < \infty$. Moreover, there exist constants $0 < c_1 \le c_2 < \infty$ (depending on K and Λ) such that, for $a \in L^1$,

$$c_1\|a\|_K \le \|a\|_* \le c_2\|a\|_K.$$

Armed with this theorem one can show that $\|\varphi - k_\lambda\varphi\|_K \to 0$ as $\lambda \to \infty$ whenever $\varphi \in VMO \cap L^\infty$. This in conjunction with the estimate

$$\sup_{t \in \mathbf{T}} |k_\lambda(\varphi a)(t) - (k_\lambda\varphi)(t)(k_\lambda a)(t)| \le \|a\|_\infty \sup_{t \in \mathbf{T}} k_{\lambda,t}(|\varphi - k_{\lambda,t}\varphi|)$$

and an $\varepsilon/3$ argument then completes the proof of Theorem 9. Combining Theorems 7 and 9 with a local principle or with Theorem 6, we finally arrive at the following theorem, which is the actual goal of our journey.

Theorem 11. *Let k_λ be an arbitrary approximate identity. If $a \in L^\infty_{N \times N}$ is locally sectorial over QC, then $T(a)$ is Fredholm, $\{k_\lambda a\}$ is a stable family, and the index of $T(a)$ is given by (7).*

This theorem was established in [5]. It is, with a full proof, Corollary 4.30 in [7].

8. The Wolf–Havin theorem

As said, asymptotic multiplicativity on (QC, L^∞) is crucial for getting Theorem 11. The asymptotic multiplicativity of the Poisson kernel on (QC, L^∞) (Theorem 4) was an easy consequence of Theorem 2, which in turn readily follows from the (exact) multiplicativity of the Poisson kernel on (H^∞, H^∞). When writing [5], I first tried to prove the asymptotic multiplicativity of arbitrary approximate identities on (QC, L^∞) in an analogous fashion, but I had to give up, and eventually I took the tour through Theorem 10. The problem whether the Poisson kernel is the only approximate identity that is asymptotically multiplicative on (H^∞, H^∞) was nevertheless interesting, but it remained open. In 1988 or so, my student Hartmut Wolf was about to go one year to Viktor Havin in Leningrad, and I gave him this problem to take along. In 1989, Wolf and Havin solved it [23], and this section is devoted to their solution.

First, if $K(x)$ generates an asymptotically multiplicative approximate identity, then so also does $K^\mu(x) := \mu K(\mu x)$ for every $\mu > 0$ because $(k_\lambda^\mu a)(e^{ix}) = (k_{\lambda\mu} a)(e^{ix})$. For the Poisson kernel $K(x) = 1/(\pi(1 + x^2))$, the kernels $K^\mu(x)$ are

$$\frac{\mu}{\pi(1 + \mu^2 x^2)}. \tag{8}$$

The result of Wolf and Havin is that the kernels (8) with $\mu > 0$ are the only kernels that generate approximate identities that are asymptotically multiplicative on (H^∞, H^∞). In fact, Wolf and Havin proved a little more. Namely, so far we have always supposed that $K(x)$ is even and non-negative, that its restriction to $(-\pi, \pi)$ is in $GL^\infty(-\pi, \pi)$, and that $K(x) = O(1/x^2)$ as $|x| \to \infty$. All these assumptions were dropped by Wolf and Havin. Here is their result of [23].

Theorem 12 (Wolf and Havin). *Let $K(x)$ be a real-valued function in $L^1(\mathbf{R})$ with $\int_{\mathbf{R}} K(x)dx = 1$ and define $k_\lambda a$ for λ in an unbounded subset $\Lambda \subset (0, \infty)$ by (5). The approximate identity k_λ is asymptotically multiplicative on (H^∞, H^∞) if and only if there are $\mu > 0$ and $h \in \mathbf{R}$ such that*

$$K(x) = \frac{\mu}{\pi(1 + \mu^2(x - h)^2)}. \tag{9}$$

In other words, shifts of the functions (8) and only these induce approximate identities that are asymptotically multiplicative on (H^∞, H^∞).

The basic ideas of the proof in [23] are as follows. Wolf and Havin first replace (5) by

$$(k_\lambda a)(x) = \lambda \int_{\mathbf{R}} K(\lambda(t - x))a(t)dt$$

to move from $\mathbf{T}$ to $\mathbf{R}$ and to have a convenient switch of the sign, and then their plan is to prove that if k_λ is asymptotically multiplicative on $(H^\infty(\mathbf{R}), H^\infty(\mathbf{R}))$, then $K(x)$ is of the form (9). Here $H^\infty(\mathbf{R})$ stands for the non-tangential limits of functions in $H^\infty(\mathbf{C}_+)$, the bounded analytic functions in the upper half-plane. With $\arg z \in (0, \pi)$ and for $x \in \mathbf{R}$, the function g_x defined as $g_x(z) = z^{ix}$ belongs to $H^\infty(\mathbf{C}_+)$. Let g_x also denote the boundary function on $\mathbf{R}$. Thus, $g_x \in H^\infty(\mathbf{R})$. Then

$$(k_\lambda g_x)(0) = \lambda \int_{\mathbf{R}} K(\lambda t)t^{ix}dt = \lambda^{-ix} \int_{\mathbf{R}} K(t)t^{ix}dt =: \lambda^{-ix}(\mathcal{M}K)(x).$$

Notice that $\mathcal{M}K$ is nothing but the Mellin transform of K. If k_λ is asymptotically multiplicative on $(H^\infty(\mathbf{R}), H^\infty(\mathbf{R}))$, we have in particular

$$(k_\lambda g_x g_y)(0) - (k_\lambda g_x)(0)(k_\lambda g_y)(0) \to 0,$$

and since $g_x g_y = g_{x+y}$, it follows that

$$\lambda^{-ix}\lambda^{-iy}(\mathcal{M}K)(x+y) - \lambda^{-ix}(\mathcal{M}K)(x)\lambda^{-iy}(\mathcal{M}K)(y) \to 0,$$

which – what a beautiful step in the proof! – implies that the continuous function $\mathcal{M}K$ satisfies the functional equation

$$(\mathcal{M}K)(x+y) = (\mathcal{M}K)(x)(\mathcal{M}K)(y) \tag{10}$$

and thus must be of the form $(\mathcal{M}K)(x) = Ce^{\alpha x}$. In addition, $(\mathcal{M}K)(0) = 1$, and some more analysis gives $(\mathcal{M}K)(+\infty) = 0$ and $(\mathcal{M}K)(x) = o(e^{-\pi x})$ as $x \to -\infty$. Consequently, $C = 1$ and $-\pi < \operatorname{Re}\alpha < 0$. Each such number α is of the form $\alpha = i\log(h + i/\mu)$ with $\mu > 0$. Wolf and Havin show that if $K(x)$ is given by (9), then $(\mathcal{M}K)(x) = e^{\alpha x}$ with $\alpha = i\log(h + i/\mu)$. They also prove that if $\mathcal{M}K_1 = \mathcal{M}K_2$, then $K_1 = K_2$. This implies that the K they started with must be of the form (9). Thus, asymptotic multiplicativity on $(H^\infty(\mathbf{R}), H^\infty(\mathbf{R}))$ is settled.

Finally, to perform the passage from $\mathbf{R}$ to $\mathbf{T}$, they use a sophisticated and nontrivial argument based on the test functions $G_x \in H^\infty = H^\infty(\mathbf{T})$ given by the formula $G_x(z) := g_x(i(1 - e^{iz}))$ to show that if the k_λ defined by (5) is asymptotically multiplicative on (H^∞, H^∞), then $\mathcal{M}K$ again satisfies the functional equation (10), so that the circle case can be disposed of by repeating the above reasoning.

References

[1] M.F. Atiyah, *Algebraic topology and operators in Hilbert space.* In: Lectures in Modern Analysis and Applications I, pp. 101–121, Springer-Verlag, Berlin, 1969.

[2] E. Azoff and K.F. Clancey, *Toeplitz operators with sectorial matrix-valued symbol.* Indiana Univ. Math. J. 28, 975–983 (1979).

[3] A. Böttcher, *Scalar Toeplitz operators, distance estimates, and localization over subalgebras of $C + H^\infty$.* In: Seminar Analysis 1985/86, pp. 1–17, Akad. Wiss. DDR, Berlin, 1986.

[4] A. Böttcher, *A remark on the relation between the partial indices of a matrix function and its harmonic extension.* In: Seminar Analysis 1985/86, pp. 19–22, Akad. Wiss. DDR, Berlin, 1986.

[5] A. Böttcher, *Analysis lokal sektorieller Matrixfunktionen und geometrische Spektraltheorie von Toeplitzoperatoren.* Habilitationsschrift, Tech. Univ. Karl-Marx-Stadt, 1987.

[6] A. Böttcher, *Best constants for Markov type inequalities in Hilbert space norms.* In: Recent Trends in Analysis, Proceedings of the Conference in Honor of Nikolai Nikolski, Bordeaux, 2011, pp. 73–83, Theta, Bucharest, 2013.

[7] A. Böttcher and B. Silbermann, *Analysis of Toeplitz Operators.* Second edition, Springer-Verlag, Berlin, 2006.

[8] R.G. Douglas, *Toeplitz and Wiener–Hopf operators in $H^\infty + C$*. Bull. Amer. Math. Soc. 74, 895–899 (1968).

[9] R.G. Douglas, *Banach Algebra Techniques in the Theory of Toeplitz Operators*. CBMS Lecture Notes 15, Amer. Math. Soc., Providence, 1973.

[10] R.G. Douglas, *Local Toeplitz operators*. Proc. London Math. Soc. (3) 36, 243–272 (1978).

[11] R.G. Douglas and H. Widom, *Toeplitz operators with locally sectorial symbols*. Indiana Univ. Math. J. 20, 385–388 (1970).

[12] J.B. Garnett, *Bounded Analytic Functions*. Academic Press, New York, 1981.

[13] N.Ya. Krupnik and I.A. Feldman, *On the relation between factorization and inversion of finite Toeplitz matrices*. Izv. Akad. Nauk Mold. SSR, Ser. Fiz.Tekh. Mat. Nauk, No. 3, 20–26 (1985) [Russian].

[14] A.S. Markus and I.A. Feldman, *On the index of an operator matrix*. Funkts. Anal. Prilozh. 11, 83–84 (1977) [Russian]; Engl. transl. in Funct. Anal. Appl. 11, 149–151 (1977).

[15] N.K. Nikolski, *Operators, Functions, and Systems: An Easy Reading. Vol. 1. Hardy, Hankel, and Toeplitz*. Amer. Math. Soc., Providence, 2002.

[16] N.K. Nikolski, *Matrices et Opérateurs de Toeplitz*. Calvage & Mounet, Paris, 2017.

[17] S. Roch, *Locally strongly elliptic singular integral operators*. Wiss. Z. Tech. Univ. Karl-Marx-Stadt 29, 224–229 (1987).

[18] D. Sarason, *Toeplitz operators with piecewise quasicontinuous symbols*. Indiana Univ. Math. J. 26, 817–838 (1977).

[19] B. Silbermann, *Harmonic approximation of Toeplitz operators and index computation*. Integral Equations Operator Theory 8, 842–853 (1985).

[20] B. Silbermann, *Local objects in the theory of Toeplitz operators*. Integral Equations Operator Theory 9, 706–738 (1986).

[21] I.B. Simonenko, *The Riemann boundary value problem for n pairs of functions with measurable coefficients and its application to the investigation of singular integrals in the spaces L^p with weight*. Izv. Akad. Nauk SSSR, Ser. Mat., 28, 277–306 (1964) [Russian].

[22] T. Tradler, S.O. Wilson, and M. Zeinalian, *One more proof of the index formula for block Toeplitz operators*. J. Oper. Theory 76, 101–104 (2016).

[23] H. Wolf [Kh. Vol'f] and V.P. Havin [V.P. Khavin], *The Poisson kernel is the only approximate identity that is asymptotically multiplicative on H^∞*. Zap. Nauchn. Sem. Leningrad. Otdel. Mat. Inst. Steklov. (LOMI) 170, 82–89 (1989) [Russian]; Engl. transl. in J. Soviet Math. 63, 159–163 (1993).

Albrecht Böttcher
Fakultät für Mathematik
Technische Universität Chemnitz
D-09107 Chemnitz, Germany
e-mail: `aboettch@mathematik.tu-chemnitz.de`

Operator Theory:
Advances and Applications, Vol. 261, 173–190
© Springer International Publishing AG, part of Springer Nature 2018

Bounded Point Derivations on Certain Function Algebras

J.E. Brennan

Dedicated to the memory of V.P. Havin

Abstract. Let X be a compact nowhere dense subset of the complex plane $\mathbb{C}$, let $C(X)$ be the linear space of all continuous functions on X endowed with the uniform norm, and let dA denote two-dimensional Lebesgue (or area) measure in $\mathbb{C}$. Denote by $R(X)$ the closure in $C(X)$ of the set of all rational functions having no poles on X. It is well known that if X is sufficiently massive, then the functions in $R(X)$ can inherit many of the properties usually associated with the analytic functions, such as unlimited degrees of differentiability and even the uniqueness property itself. Here we shall examine the extent to which some of those properties are inherited by the larger algebra $H^\infty(X)$, which by definition is the weak-* closure of $R(X)$ in $L^\infty(X) = L^\infty(X, dA)$.

Mathematics Subject Classification (2010). Primary 30H50.

Keywords. Point derivation, monogeneity, Swiss cheese, peak point, analytic capacity, Wang's theorem.

1. Introduction

Let X be a compact nowhere dense subset of the complex plain $\mathbb{C}$, and let dA stand for area, that is, two-dimensional Lebesgue measure in $\mathbb{C}$. Denote by $C(X)$ the linear space of all continuous functions on X endowed with the uniform norm, and let $R(X)$ be the closure in $C(X)$ of the set $R_0(X)$ consisting of all rational functions whose poles lie outside of X. By a *point derivation* on $R(X)$ at a point $x_0 \in X$ we shall mean a nontrivial linear functional D (that is, $D \neq 0$) with the property that

$$D(fg) = f(x_0)Dg + g(x_0)Df$$

for every pair of functions $f, g \in R(X)$. If, however, D happens to be bounded as a linear functional on $R(X)$, then it is said to be a *bounded point derivation* at x_0.

Moreover, in this context, it is easy to see that there is a bounded point derivation D at a fixed point $x_0 \in X$ if and only if there exists some constant k such that

$$|f'(x_0)| \leq k\|f\| \tag{1.1}$$

for all $f \in R_0(X)$, where $\|f\|$ is understood to designate the norm of f in $C(X)$. Evidently, whenever (1.1) is satisfied, the map D defined on $R_0(X)$ by setting $Df = f'(x_0)$ admits a unique extension to a bounded point derivation on $R(X)$. By reversing the process for the extended map, still denoted by D, and defining $f'(x_0) = Df$ for all $f \in R(X)$, we can arrange that $f'(x_0)$ exists in a weak sense for every nonrational function $f \in R(X)$. But, the question then arises as to whether each $f \in R(X)$ is at least *monogenic* at x_0 in the sense that

$$f'(x_0) = Df = \lim_{x \to x_0, x \in E} \frac{f(x) - f(x_0)}{x - x_0}, \tag{1.2}$$

where the limit is taken through some large subset $E \subset X$ having x_0 as a limit point. The idea that certain properties of the analytic functions such as *monogeneity*, or equivalently *pointwise differentiability*, can be inherited by larger classes of functions defined on closed planar sets having no interior points originated with Borel during the last decade of the 19th century, and was finally brought to fruition almost a quarter of a century later in [2] (cf. also [3] and [13]).

Since it is known that the algebra $C(X)$ admits no bounded, or even unbounded, point derivations (cf. [28]), we shall assume from the outset that $R(X) \neq C(X)$. On the other hand, Roth [26] and Mergeljan [21] have identified an entire class of compact nowhere dense sets for which that is indeed the case. To construct one such example simply remove from the interior of the closed unit disk $\bar{B}$ countably many open disks B_j, $j = 1, 2, 3, \ldots$, having mutually disjoint closures, the sum of whose radii is finite, and so that the resulting set X (referred to colloquially as a *Swiss cheese*) has no interior. The set $\partial^* X = \partial B \cup (\cup_{j=1}^\infty \partial B_j)$ is known as the *reduced boundary* of X, and when $\partial^* X$ is suitably oriented $\int_{\partial^* X} f\,dz = 0$ for all $f \in R(X)$. As dz is then a nontrivial measure of finite total variation on X which annihilates $R(X)$, it follows that $R(X) \neq C(X)$. Here, we may also conclude, as a consequence of a theorem of Hartogs and Rosenthal, that $|X| > 0$ (cf. [8, p. 163] and [14, p. 47]).

By definition, a point $x \in X$ is a *peak point* for $R(X)$ if there exists a function $f \in R(X)$ such that $f(x) = 1$, but $|f(y)| < 1$ whenever $y \in X$ and $y \neq x$. Bishop [1] has shown that the set P of peak points is a G_δ, and that $R(X) = C(X)$ if and only if dA almost every point of X is a peak point for $R(X)$. It follows that if the set of non-peak points $Q = X \setminus P$ is non-empty, then $|Q| > 0$. More details on the connection between the existence of peak points and rational approximation can be found in Gamelin [14] and Zalcman [33]. On the other hand, Browder [7] (cf. also [8, p. 178]) has also shown that there exists a non-zero point derivation (not necessarily bounded) on $R(X)$ at a fixed point $x_0 \in X$ if and only if x_0 is not a peak point, and so the set of points corresponding to point derivations for $R(X)$ has positive area, and even more surprisingly contains no isolated points. He also obtained a sufficient condition (cf. [31, p. 34]) in order that there exist a bounded point derivation at a given point $x_0 \in X$, and at the same time, conjectured that some Swiss cheeses admit no bounded point derivations for $R(X)$ at all. Subsequently, Wermer [31] confirmed Browder's conjecture by actually constructing a Swiss cheese having the required property. The situation was further clarified by Hallstrom [16] when he obtained necessary and sufficient criteria (expressed in terms of analytic capacity) for $R(X)$ to admit a bounded point derivation of a given integral order s at a fixed point $x_0 \in X$, where by analogy with (1.1), such a bounded point derivation is said to exist at x_0 whenever the inequality

$$|f^{(s)}(x_0)| \leq k\|f\|$$

is satisfied for all $f \in R_0(X)$, and some constant k. These criteria then enabled Hallstrom to construct a Swiss cheese X for which $R(X)$ admits bounded point derivations of all orders at a.e.-dA point of X. A similar example, not depending on analytic capacity, but more directly on the notion of monogeneity, can be found in [3].

Wang [30] seems to have been the first to deal explicitly with the question raised here in (1.2), insofar as it concerns the algebra $R(X)$. In fact, he was able to prove that if D is a bounded point derivation at x_0, then there exists a set E having *full area density* at x_0 such that (1.2) is satisfied for every $f \in R(X)$. If, however, X also happens to be a nowhere dense continuum, such as a Swiss cheese, Dolzhenko [12] (cf. also [13, p. 95]) has shown that there must always be a function $f \in R(X)$ which is nowhere differentiable on X. Hence, given that situation, there evidently exists an *exceptional set* having x_0 as a point of accumulation, and through which the limit in (1.2) cannot exist for at least some functions in $R(X)$.

One of our principle objectives here is to extend the notion of a bounded point derivation of order s from $R(X)$ to a larger subalgebra of $L^\infty(X, dA)$, to obtain integral formulas for the corresponding weak derivatives, and to determine the extent to which Wang's theorem remains valid in the extended context. In general, we shall suppress explicit mention of the underlying measure dA, and simply write $L^\infty(X)$ in place of $L^\infty(X, dA)$. Whenever possible the page numbers on individual citations to Russian articles will refer to the English translation.

2. Non-peak points and weak-∗ limits

Let $H^\infty(X)$ be the weak-∗ closure of $R(X)$ in $L^\infty(X)$, and denote by $H^\infty(X)^\perp$ the closed subspace of $L^1(X)$ orthogonal to $H^\infty(X)$. In this way, $H^\infty(X)$ can be regarded as the dual of a Banach space, while the dual of $H^\infty(X)$ can be identified with the quotient

$$L^1(X)/[L^1(X) \cap H^\infty(X)^\perp],$$

and so every weak-∗ continuous linear functional L on $H^\infty(X)$ has the form $L(f) = \int fk \, dA$ for some, possibly non-unique, $k \in L^1(X)$ (cf. [34, p. 23] and [27, pp. 55–66]). In order to achieve our objectives we first need to establish that for any

$x_0 \in Q$ the linear functional

$$L : f \to f(x_0)$$

can be extended from $R(X)$ to a weak-$*$ continuous functional on $H^\infty(X)$, and then to verify that the same assertion is valid for the map $L : f \to f^{(s)}(x_0)$, whenever $R(X)$ has a bounded point derivation of order s at x_0. The initial step is to show that each $f \in H^\infty(X)$ has a precise representative f^* such that $f^*(x_0)$ is unambiguously defined at each point $x_0 \in Q$. The key in this regard is the following theorem of Davie [11] (cf. also [15]). Let B_R denote the linear subspace consisting of all $f \in L^\infty(X)$ for which there exists a sequence $f_n \in R(X)$ such that

   (i) $\sup \|f_n\|_\infty < \infty$;
   (ii) $f_n \to f$ weak-$*$ in $L^\infty(X)$.

Theorem 2.1 (Davie). *If $f \in B_R$ there exists a sequence $f_n \in R(X)$ such that $f_n \to f$ weak-$*$ in $L^\infty(X)$ and $\| f_n \|_\infty \leq \| f \|_\infty$ for all $n = 1, 2, \ldots$*

It is a consequence of Davie's result and the Krein–Shmulian theorem in a form due to Hoffman and Rossi [17] that B_R is weak-$*$ closed in $L^\infty(X)$. Therefore, $B_R = H^\infty(X)$, since $R(X) \subset B_R \subseteq H^\infty(X)$. That is, the weak-$*$ and bounded pointwise closures of $R(X)$ in $L^\infty(X)$ are identical. From that fact alone it can be deduced that every $f \in H^\infty(X)$ is precisely defined on the set of non-peak points Q in the sense that

$$f(x_0) = \lim_{j \to \infty} f_j(x_0)$$

for any point $x_0 \in Q$, and any sequence of rational functions $f_j \to f$ pointwise and boundedly a.e.-dA on X. Note, moreover, that the value $f(x_0)$ depends only on f, and not on the approximating sequence f_j, as can be seen by interlacing two different sequences of rational functions converging pointwise and boundedly to f. For additional details see [4, p. 1137] and [11, p. 132]. If $R(X)$ has a bounded point derivation at x_0, then x_0 is not a peak point as Theorems (3.2) and (3.3) of Mel'nikov and Hallstrom in the next section clearly indicate. Since we ultimately hope to prove that, even if $f \in H^\infty(X)$, then $f'(x_0)$ is well defined and

$$f'(x_0) = \lim_{x \to x_0, x \in E} \frac{f(x) - f(x_0)}{x - x_0} \tag{2.1}$$

for some subset E having full area density at x_0, we can at least be reassured here that the difference quotient on the right side of (2.1) is well defined for every $f \in H^\infty(X)$ and every $x \in Q$. Our task is to verify that Q has full area density at x_0, and to select an appropriate subset $E \subset Q$ for which (2.1) is satisfied.

In the particular case of a Swiss cheese X it is easy to see that the set of non-peak points Q is dense in X, since $\int_{\partial^* X} |z - x|^{-1}|dz| < \infty$ for a.e.-dA point $x \in X$, and no such point can be a peak point for $R(X)$. Thus, if $f \in H^\infty(X)$ and a sequence of rational functions $f_j \to f$ pointwise and boundedly a.e.-dA on X, and hence everywhere on Q, it follows that

$$\|f_j\|_{\partial^* X} = \|f_j\|_Q \leq C, \tag{2.2}$$

for some constant C, and all $j = 1, 2, \ldots$ We can, therefore, choose an additional subsequence, which we may assume to be the original f_j, $j = 1, 2, \ldots$, so that $f_j \to f^*$ weak-$*$ in $L^\infty(\partial^* X, |dz|)$, and it can then be shown that

$$f(x_0) = \lim_{\epsilon \to 0} \frac{1}{2\pi i} \int_{\partial^* X \cap (|z - x_0| \geq \epsilon)} \frac{f^*(z)}{(z - x_0)} \, dz, \qquad (2.3)$$

at any point $x_0 \in Q$ (cf. [4, p. 1139]). Note, it can happen that $x_0 \in Q$, but nevertheless the Cauchy integral $\int_{\partial^* X} |z - x|^{-1} |dz| = \infty$ (cf. [4, p. 1142]). For that reason, in order to obtain a representation formula of this kind and valid in this setting for all $x_0 \in Q$, the integral in (2.3) must be interpreted as a principal value integral at points of Q where the Cauchy integral fails to converge absolutely. Later, we shall indicate how to obtain a corresponding formula for $f^{(s)}(x_0)$, whenever $R(X)$ has a bounded point derivation of order s at x_0, the function f belongs to $H^\infty(X)$, and the derivative $f^{(s)}(x_0)$ has been suitably defined.

3. Bounded point derivations and analytic capacity

A first step in addressing the pointwise differentiability, or monogeneity, question for functions in $H^\infty(X)$ is to establish effective criteria for determining whether $H^\infty(X)$ has a bounded point derivation at a given point in the more abstract sense of (1.1). In the case of $R(X)$ this has been carried out by Hallstrom [16] and is best described in terms of analytic capacity. The *analytic capacity* of a compact set K, denoted by $\gamma(K)$, is defined as

$$\gamma(K) = \sup |f'(\infty)|,$$

the supremum being taken over all functions f analytic in $\Omega(K)$, the unbounded component of $\widehat{\mathbb{C}} \setminus K$, where $\widehat{\mathbb{C}}$ designates the extended complex plane, $\|f\|_\infty = \sup_{\Omega(K)} |f| \leq 1$ and $f(\infty) = 0$. For a general set E, we define $\gamma(E) = \sup \gamma(K)$ where this supremum is taken over all compact sets $K \subset E$. An extensive survey of the properties of analytic capacity and its application to problems of approximation theory can be found in [14] and [33], where proofs of the following basic properties are available:

(1) $\gamma(B_r) = r$ for every disk B_r of radius r;
(2) Area$(E) \leq 4\pi \gamma(E)^2$ for every bounded measurable subset $E \subset \mathbb{C}$.

Much deeper, however, is the *semi-additivity* of analytic capacity; that is,

(3) If $E_1, E_2, \ldots$ are Borel sets, then $\gamma\left(\bigcup_n E_n\right) \leq C \sum_n \gamma(E_n)$,

where C is an absolute constant, independent of the particular sets $E_1, E_2, \ldots$. The semi-additivity property was long conjectured, but only relatively recently confirmed in full generality by Tolsa [29].

Mel'nikov [19] has obtained a necessary and sufficient condition in terms of analytic capacity in order for a point $x_0 \in X$ to be a peak point for $R(X)$. A key step in his argument consists in establishing the following estimate, which will be critical at every stage of this investigation. Suppose that $0 < r < R$ and let

$\mathcal{A}(x_0) = \{z : r \le |z - x_0| \le R\}$. If f is continuous on $\mathcal{A}(x_0)$ and analytic in $\mathcal{A}(x_0) \setminus K$, where K is compact, then we have the lemma.

Lemma 3.1. *There exists a constant $C = C(r/R)$, depending only on the ratio r/R and bounded away from ∞ as long as r/R is bounded away from 1, such that*

$$\left| \int_{|z-x_0|=R} f(z)\,dz - \int_{|z-x_0|=r} f(z)\,dz \right| \le C\|f\|_\infty \gamma(K). \tag{3.1}$$

From this estimate the *necessity* of the peak point criterion is a, more or less, straightforward consequence (cf. [14, p. 234]). For each $x_0 \in X$, and each $n = 1, 2, \ldots$, let

$$\mathcal{A}_n(x_0) = \left\{ z : 2^{-(n+1)} \le |z - x_0| \le 2^{-n} \right\}.$$

Theorem 3.2 (Mel'nikov). *A point $x_0 \in X$ is a peak point for $R(X)$ if and only if*

$$\sum_{n=1}^{\infty} 2^n \gamma(\mathcal{A}_n(x_0) \setminus X) = \infty. \tag{3.2}$$

Because of the connection between point derivations (not necessarily bounded) and peak points established by Browder [7], Mel'nikov's criterion also provides a necessary and sufficient condition for the existence of point derivations on $R(X)$. The corresponding criteria for the existence of *bounded* point derivations were obtained by Hallstrom [16].

Curtis [10] (cf. [14, p. 204]) has also given, independently, a proof of the sufficiency of Mel'nikov's peak point criterion (3.2). The same methods also allowed him to establish the following sufficient condition which is more restrictive than Mel'nikov's, but somewhat easier to apply in practice.

Theorem 3.3 (Curtis). *Fix a point $x_0 \in X$, and let $B_r(x_0) = \{z : |z - x_0| < r\}$. If*

$$\limsup_{r \to 0} \frac{\gamma(B_r(x_0) \setminus X)}{r} > 0, \tag{3.4}$$

then x_0 is a peak point for $R(X)$.

Since $\text{Area}(E) \le 4\pi\gamma(E)^2$ for every bounded measurable subset $E \subset \mathbb{C}$, it follows immediately from Curtis' theorem that the set X has full area density in any neighborhood of a non-peak point, or bounded point derivation. We shall see in due course that the set Q of non-peak points also has full area density at such points.

Theorem 3.4 (Hallstrom). *$R(X)$ has a bounded point derivation of order s at $x_0 \in X$ if and only if*

$$\sum_{n=1}^{\infty} (2^{s+1})^n \gamma(\mathcal{A}_n(x_0) \setminus X) < \infty. \tag{3.5}$$

When $s = 0$ the sum in (3.5) reduces to Mel'nikov's *peak point criterion* for $R(X)$. In that case, x_0, is a peak point for $R(X)$ if, and only if, the sum diverges. Whatever the value of s, however, the proof of Theorem 3.4 also depends in an essential way on Mel'nikov's estimate in Lemma 2.1.

In order to extend the notion of a bounded point derivation to $H^\infty(X)$ Hallstrom's argument must be modified. To that end fix a point $x_0 \in X$. Assume for the moment that $s = 0$, and that the corresponding sum

$$\sum_{n=1}^{\infty} 2^n \gamma(A_n(x_0) \setminus X) < \infty.$$

In particular, x_0 is a non-peak point for $R(X)$; that is, $x_0 \in Q$. It can be shown that at any non-peak point x_0 the map

$$L : f \to f(x_0)$$

is a well-defined weak-$*$ continuous linear functional on $H^\infty(X)$. Although the details can be found in [4], certain estimates in the proof will play an important role in our treatment of bounded point derivations for $H^\infty(X)$. For that reason we shall first review that portion of the argument related to point evaluations, and then indicate how it can be extended to cover point derivations.

Given a function $f \in H^\infty(X)$, choose a sequence of rational functions f_j, $j = 1, 2, \ldots$, so that $f_j \to f$ pointwise and boundedly a.e.-dA on X. Set each $f_j = 0$ outside a neighborhood of X in such a way that the modified functions, still denoted by f_j, are continuous on $\mathbb{C}$ with $f_j \to 0$ on $\mathbb{C} \setminus X$ and

$$\|f_j\|_\infty \leq 2\|f_j\|_X.$$

Next, fix an $r > 0$ and choose a smooth function g supported inside the open disk $B_r = B_r(x_0)$ in such a way that

(1) $0 \leq g \leq 1$,
(2) $g = 1$ in a neighborhood of x_0,
(3) $\|\frac{\partial g}{\partial \bar{z}}\|_\infty \leq \frac{4}{r}$.

For the construction of a function g with these properties see [14, p. 211]. Now form a new function by setting

$$F_j(\zeta) = g(\zeta) f_j(\zeta) + \frac{1}{\pi} \int_{B_r} \frac{f_j(z)}{(z - \zeta)} \frac{\partial g}{\partial \bar{z}} dA_z \qquad (3.6)$$

By construction F_j is continuous on $\widehat{\mathbb{C}}$, $F_j(\infty) = 0$, and moreover

(4) F_j is analytic wherever f_j is analytic,
(5) $F_j - f_j$ is analytic in a neighborhood of x_0,
(6) $\|F_j\|_\infty \leq 16\|f_j\|_\infty$.

We may assume that $r = 2^{-n_0}$ for some positive integer n_0. The telescoping series

$$\int_{|z-x_0|=2^{-N}} \frac{F_j(z)}{(z - x_0)} dz = \int_{|z-x_0|=r} \frac{F_j(z)}{(z - x_0)} dz - \sum_{n=n_0}^{N-1} \int_{\partial A_n(x_0)} \frac{F_j(z)}{(z - x_0)} dz \quad (3.7)$$

then converges to $2\pi i F_j(x_0)$ as $N \to \infty$. Because F_j is analytic outside B_r and $F_j(\infty) = 0$, the first term on the right vanishes, and since we are assuming that x_0 is not a peak point, it follows first from (3.1) and then from Mel'nikov's theorem that, given any $\epsilon > 0$,

$$|F_j(x_0)| \leq C\|F_j\|_\infty \sum_{\mathcal{A}_n(x_0) \subset B_r} 2^n \gamma(\mathcal{A}_n(x_0) \setminus X) < \epsilon \qquad (3.8)$$

independent of j, provided r is sufficiently small. Likewise, $|F_j(x_0) - F_m(x_0)| < 2\epsilon$ for all j and m if r is small.

On the other hand, it follows from (3.6) that

$$|(F_j - F_m)(x_0) - (f_j - f_m)(x_0)| \to 0$$

as $j, m \to \infty$, since $f_j - f_m \to 0$ pointwise and boundedly a.e.-dA, and therefore

$$\limsup_{j,m\to\infty} |f_j(x_0) - f_m(x_0)| \leq \limsup_{j,m\to\infty} |F_j(x_0) - F_m(x_0)| < 2\epsilon.$$

Since the left side is independent of ϵ, we conclude that $f_j(x_0)$, $j = 1, 2, \ldots$, is a convergent Cauchy sequence, and $\lim f_j(x_0)$ depends only on f, and not on the approximating sequence f_j, as can be seen by interlacing two different sequences of rational functions converging pointwise and boundedly to f.

Equivalently, from (3.6), (3.7), and (3.8) we conclude that

$$|f_j(x_0) - f_m(x_0)|$$
$$\leq C \sum_{\mathcal{A}_n(x_0) \subset B_r} 2^n \gamma(\mathcal{A}_n(x_0) \setminus X) + \left| \frac{1}{\pi} \int_{B_r} \frac{(f_j(z) - f_m(z))}{(z - x_0)} \frac{\partial g}{\partial \bar{z}} \, dA_z \right|, \qquad (3.9)$$

where the first term on the right of the inequality is independent of j and m, depending only on r. Since we are assuming that x_0 is not a peak point, given $\epsilon > 0$, we can choose r sufficiently small to ensure that

$$|f_j(x_0) - f_m(x_0)| \leq \epsilon + \left| \frac{1}{\pi} \int_{B_r} \frac{(f_j(z) - f_m(z))}{(z - x_0)} \frac{\partial g}{\partial \bar{z}} \, dA_z \right|,$$

the inequality holding for all j and m. If we now assume that $R(X)$ has a bounded point derivation of order $s \geq 1$ at x_0 so that the series $\sum (2^{s+1})^n \gamma(\mathcal{A}_n(x_0) \setminus X)$ converges, then by replacing (3.6) with the corresponding telescoping series arising from Cauchy's formula for derivatives and arguing as above, we obtain the following inequality for derivatives analogous to that in (3.9)

$$|f_j^{(s)}(x_0) - f_m^{(s)}(x_0)|$$
$$\leq Cs! \sum_{\mathcal{A}_n(x_0) \subset B_r} (2^{s+1})^n \gamma(\mathcal{A}_n(x_0) \setminus X) + \left| \frac{s!}{\pi} \int_{B_r} \frac{(f_j(z) - f_m(z))}{(z - x_0)^{s+1}} \frac{\partial g}{\partial \bar{z}} \, dA_z \right|,$$

from which we conclude that the sequence of derivatives $f_j^{(s)}(x_0)$, $j = 1, 2, \ldots$, converges. Having begun with a function $f \in H^\infty(X)$ and a sequence of rational functions $f_j \to f$ pointwise and boundedly a.e.-dA, we may define $f^{(s)}(x_0) = \lim f_j^{(s)}(x_0)$. And, in so doing it can then be inferred that

$$|f^{(s)}(x_0)| \leq k\|f\|_\infty \tag{3.10}$$

for some constant k and all $f \in H^\infty(X)$. Whenever the map $f \to f^{(s)}(x_0)$ can be extended from $R(X)$ to $H^\infty(X)$ in this way with the inequality (3.10) being preserved we will say that $H^\infty(X)$ has a bounded point derivation of order s at the point x_0. In particular, if $R(X)$ has a bounded point derivation of order s at x_0 then so also does $H^\infty(X)$, and conversely. Thus, $H^\infty(X)$ has a bounded point derivation of order s at x_0 if, and only if, Hallstrom's capacitary criterion (3.5) is satisfied (cf. also [23, p. 552]).

4. Bounded point derivations and representing measures

Our main objective in this section is to prove the following theorem.

Theorem 4.1. *For each point $x_0 \in X$ at which $H^\infty(X)$ has a bounded point derivation there exists a corresponding function $k \in L^1(X)$, depending only on x_0, such that*

$$f'(x_0) = \int fk \, dA_Q \tag{4.1}$$

for all $f \in H^\infty(X)$, where dA_Q denotes the restriction of dA to Q.

Since, by assumption, the map $f \to f'(x_0)$ is also a bounded linear functional on $R(X)$, there is evidentally a measure μ with the property that $f'(x_0) = \int_X f \, d\mu$ for all $f \in R(X)$, and hence for all $f \in H^\infty(X)$. The interesting feature here is the fact that in each case the measure μ can be taken to be absolutely continuous with respect to area dA restricted to the set of non-peak points. If there also happens to be a bounded point derivation of order s at x_0, then a similar formula $f^{(s)}(x_0) = \int fk \, dA_Q$ is valid for all $f \in H^\infty(X)$ with the measure $k \, dA$ now depending on s as well. This includes a portion of Wilken's main theorem [32, p. 372], but with additional information on the nature of the corresponding representing measures. We shall be content here, however, to limit the discussion to the case $s = 1$, since any modifications required to treat higher-order point derivations are completely evident.

Proof of Theorem 4.1. Fix a point $x_0 \in X$ at which $H^\infty(X)$ has a bounded point derivation, and consider the bounded linear functional $L : f \to f'(x_0)$. Let f_j be any sequence in $H^\infty(X)$ such that $f_j \to f$ pointwise and boundedly a.e.-dA on X; $\|f_j\|_\infty \leq M$, say. Suppose, moreover, that $f_j \in \ker L$, or equivalently that $f_j'(x_0) = 0$ for all $j = 1, 2, \ldots$. If we can show that $f'(x_0) = 0$, then $\Delta \cap \ker f$ is weak-$*$ closed for every bounded norm closed ball $\Delta = \{h \in L^\infty : \|h\|_\infty \leq M\}$. Since $\ker L$ is convex, it will follow from the Krein–Shmulian theorem that $\ker L$ is weak-$*$ closed, and that L is therefore weak-$*$ continuous (cf. [9, p. 161]). From remarks in Section 2, each f_j is the pointwise bounded limit of a sequence of *rational functions*

h_{jn}, and by Davie's theorem we can arrange that $\|h_{jn}\|_\infty \le \|f_j\|_\infty \le M$ for all n. Passing to a subsequence if necessary we can, according to Egorov's theorem and the discussion at the end of Section 3 concerning the pointwise convergence of the sequence $f_j'(x_0)$, $j = 1, 2, \ldots$, select rational functions h_j and subsets $E_j \subseteq X$ with the property that $\|h_j\|_\infty \le M$, $j = 1, 2, \ldots$, and moreover

(i) $|E_j| < 2^{-j}$;

(ii) $|f_j - h_j| < 2^{-j}$ on $X \setminus E_j$;

(iii) $|f_j'(x_0) - h_j'(x_0)| < 1/j$.

It follows from (i) and (ii) that f is the bounded pointwise limit a.e.-dA on X of a sequence of rational functions h_j, and since x_0 is a bounded point derivation for $H^\infty(X)$ we conclude that $\lim h_j'(x_0) = f'(x_0)$. By assumption, however, $f_j'(x_0) = 0$ for all $j = 1, 2, \ldots$, and so (iii) implies that $f'(x_0) = 0$, from which we conclude that L is weak-$*$ continuous on $H^\infty(X)$.

At this point it can be inferred that for any bounded point derivation x_0 for $H^\infty(X)$ there exists some $k \in L^1(X)$ depending on x_0 with the property that

$$f'(x_0) = \int_X fk\, dA \tag{4.2}$$

for all $f \in H^\infty(X)$. Thus, the measure $(z - x_0)^2 k\, dA$ is orthogonal to $H^\infty(X)$, and so also to $R(X)$. On the other hand, by a theorem of Øksendal [25, p. 334] an absolutely continuous annihilating measure for $R(X)$ can place no mass on the set of peak points. Hence, the integration in (4.2) is carried out only over Q, which completes the proof of (4.1).   □

For sets of finite perimeter, and in particular for a Swiss cheese X, Khavinson [18] has shown that if $f \in H^\infty(X)$ there always exists a unique boundary function $f^* \in L^\infty(\partial^* X, |dz|)$ such that

$$f(x_0) = \frac{1}{2\pi i} \int_{\partial^* X} \frac{f^*(z)}{(z - x_0)}\, dz \tag{4.3}$$

at any point $x_0 \in X$ where $\int_{\partial^* X} |z - x_0|^{-1} |dz| < \infty$. The formula (4.3) is evidentally valid for all $f \in R(X)$ with $f = f^*$, and the difficulty here consists in passing from $R(X)$ to $H^\infty(X)$. Although the integrability requirement at x_0 ensures that $x_0 \in Q$, it can happen that $x_0 \in Q$ but nevertheless $\int_{\partial^* X} |z - x_0|^{-1} |dz| = \infty$. Even at such a point, however, (4.3) is still valid, provided the integral is interpreted in a principal value sense (cf. [4, p. 1139]). More generally, we have the following theorem.

Theorem 4.2. *Let X be a Swiss cheese, and assume that $H^\infty(X)$ has a bounded point derivation of order s at a fixed point $x_0 \in X$. Then there exists a unique boundary function $f^* \in L^\infty(\partial^* X, |dz|)$ such that*

$$f^{(n)}(x_0) = \lim_{\epsilon \to 0} \frac{n!}{2\pi i} \int_{\partial^* X \cap (|z - x_0| \ge \epsilon)} \frac{f(z)}{(z - x_0)^{n+1}}\, dz, \tag{4.4}$$

for all $f \in H^\infty(X)$, and all $n = 0, 1, 2, \ldots, s$.

Implicit in the statement of the theorem is the assertion that if $H^\infty(X)$ has a
bounded point derivation of a given order at a fixed point x_0, then it has bounded
point derivations of all lower orders at the same point. This is a straightforward
consequence of Hallstrom's theorem, and the fact that $H^\infty(X)$ has a bounded point
derivation of order s at x_0 if, and only if, $\sum_{n=1}^\infty (2^{(s+1)})^n \gamma(\mathcal{A}_n(x_0) \setminus X) < \infty$. The
latter, of course, was noted earlier by O'Farrell (cf. [23, p. 552]), but in a somewhat
different context.

Proof of Theorem 4.2. Fix $f \in H^\infty(X)$, and choose a sequence of rational functions $f_j \to f$ pointwise and boundedly a.e.-dA on X, and hence everywhere on Q.
Since Q is dense in X, it follows as in (2.2) that

$$\|f_j\|_{\partial^* X} \le \|f_j\|_Q \le C,$$

for some constant C, and all $j = 1, 2, \ldots$. We can, therefore, choose an additional
subsequence, which we may take to be the original f_j, $j = 1, 2, \ldots$, so that $f_j \to f^*$
weak-$*$ in $L^\infty(\partial^* X, |dz|)$. By the Cauchy integral formula for derivatives applied
to rational functions having no poles on X it follows that

$$f_j^{(s)}(x_0) = \frac{s!}{2\pi i} \int_{\partial^* X \cap (|z-x_0| \ge r)} + \frac{s!}{2\pi i} \int_{\partial^* X \cap (|z-x_0| < r)} \frac{f_j(z)}{(z-x_0)^{s+1}} \, dz. \qquad (4.5)$$

Here and throughout this discussion $\partial^* X$ is assumed to be oriented in a manner
consistent with a positive orientation for the boundary of each approximating
domain in the construction of the Swiss cheese X. Our first task is to find a bound
for the second integral in (4.5) that depends on r, but is independent of j. To
accomplish that task set each $f_j = 0$ in a neighborhood of its singularities such
that the modified function, still denoted by f_j, is continuous on $\widehat{\mathbb{C}}$, vanishes in a
neighborhood of ∞, and satisfies the inequality

$$\|f_j\|_\infty \le 2\|f_j\|_X.$$

Assume initially that $r = 2^{-n_0}$ for some positive integer n_0. If $\partial^* X$ does not
meet the circle $|z - x_0| = r$, then according to Cauchy's integral theorem

$$\frac{-1}{2\pi i} \int_{\partial^* X \cap (|z-x_0|<r)} \frac{f_j(z)}{(z-x_0)^{s+1}} \, dz = \sum_{n=n_0}^\infty \frac{1}{2\pi i} \int_{\partial \mathcal{A}_n(x_0)} \frac{f_j(z)}{(z-x_0)^{s+1}} \, dz,$$

and it follows from Mel'nikov's estimate (3.1) and Hallstrom's criterion (3.5) that,
given $\epsilon > 0$,

$$\left| \frac{1}{2\pi i} \int_{\partial^* X \cap (|z-x_0|<r)} \frac{f_j(z)}{(z-x_0)^{s+1}} \, dz \right| \le C \sum_{n=n_0}^\infty (2^{s+1})^n \gamma(\mathcal{A}_n(x_0) \setminus X) < \epsilon$$

independent of j, provided r is sufficiently small, since there is a bounded point
derivation of order s at x_0.

If, however, $\partial^* X$ intersects the circle $|z - x_0| = r$, then more care needs to be
taken in order to obtain a similar estimate. For each $r > 0$ let $\Delta_r = (|z - x_0| \le r)$.
It is known that we can select a family of functions F_δ each of which is analytic

wherever f_j is, as well as in a neighborhood of $\partial\Delta_r$, in such a way that $F_\delta \to f_j$ uniformly on Δ_r as $\delta \to 0$ (cf. [4, p. 1140] and [14, p. 230]). In particular, we can also choose $t < r$ so that each F_δ is analytic in a region abutting $\partial\Delta_r$, and containing the annulus $\Delta_r \setminus \Delta_t$. Next, let D_k, $k = 1, 2, \ldots$, denote the collection of all complementary components of X that intersect $\partial\Delta_r$, and for each k set $D_k^t = D_k \cap \Delta_t$. In this way, some disks near $\partial\Delta_r$ will be modified, but others further away will be unaffected. Finally, consider the set $X_t = \Delta_r \setminus \bigcup_k D_k^t$.

Since F_δ is analytic near $\partial\Delta_r$, as well as in a neighborhood of x_0, we are back in the situation described earlier. It follows that, given $\epsilon > 0$,

$$\left| \frac{1}{2\pi i} \int_{\partial^* X_t \cap (|z-x_0|<r)} \frac{F_\delta(z)}{(z-x_0)^{s+1}}\, dz \right| \leq C \sum_{n=n_0}^{\infty} (2^{s+1})^n \gamma(\mathcal{A}_n(x_0) \setminus X) < \epsilon$$

provided r is sufficiently small, since $H^\infty(X)$ is assumed to have a bounded point derivation of order s at x_0. Letting $t \to r$ we can infer that

$$\left| \frac{1}{2\pi i} \int_{\partial^* X \cap (|z-x_0|<r)} + \frac{1}{2\pi i} \int_I \frac{F_\delta(z)}{(z-x_0)^{s+1}}\, dz \right|$$
$$\leq C \sum_{n=n_0}^{\infty} (2^{s+1})^n \gamma(\mathcal{A}_n(x_0) \setminus X) < \epsilon,$$

where the set I consists of the collection of intervals on $\partial\Delta_r$ corresponding to $\partial\Delta_r \cap (\bigcup D_k)$. Because $F_\delta \to f_j$ uniformly on Δ_r as $\delta \to 0$,

$$\left| \frac{1}{2\pi i} \int_{\partial^* X \cap (|z-x_0|<r)} + \frac{1}{2\pi i} \int_I \frac{f_j(z)}{(z-x_0)^{s+1}}\, dz \right| \tag{4.6}$$
$$\leq C \sum_{n=n_0}^{\infty} (2^{s+1})^n \gamma(\mathcal{A}_n(x_0) \setminus X) < \epsilon.$$

On the other hand, we are free from the outset to set each $f_j = 0$ on an arbitrarily large portion of any D_k without affecting its subsequent behavior on X, and so the contribution coming from integration over I in (4.6) can be made arbitrarily small, from which it follows that

$$\left| \frac{1}{2\pi i} \int_{\partial^* X \cap (|z-x_0|<r)} \frac{f_j(z)}{(z-x_0)^{s+1}}\, dz \right| \leq C \sum_{n=n_0}^{\infty} (2^{s+1})^n \gamma(\mathcal{A}_n(x_0) \setminus X) < \epsilon, \tag{4.7}$$

independent of j, provided $r = 2^{-n_0}$ is sufficiently small. Should $2^{-(n_0+1)} < r < 2^{-n_0}$ consider the annulus $W_n(x_0) = \{z : 2^{-(n_0+2)} \leq |z - x_0| \leq r\}$. Repeating the above argument, we conclude first from Lemma 3.1, and then from the semi-additivity of analytic capacity in a weak form due to Mel'nikov [19](cf. also [33, pp. 38–39] and [29]), that the right side of the inequality (4.7) can be estimated by simply observing that

$$\left((2^{s+1})^{(n_0+2)}\gamma(W_n(x_0)\setminus X) + \sum_{n=n_0+2}^{\infty}(2^{s+1})^n\gamma(\mathcal{A}_n(x_0)\setminus X)\right)$$
$$\leq C\sum_{n=n_0}^{\infty}(2^{s+1})^n\gamma(\mathcal{A}_n(x_0)\setminus X),$$

where the constant C depends only on the fact that the conformal ratio between the inside and outside radii of $W_n(x_0)$ is bounded away from 1 for all $n = 1, 2, \ldots$, and so in this case C can be regarded as absolute. In particular, the integral in (4.7) can be made arbitrarily small for all j by choosing r sufficiently small. Letting $j \to \infty$, and recalling that $f_j \to f^*$ weak-$*$ in $L^\infty(\partial^* X, |dz|)$ while $f_j^{(s)}(x_0) \to f^{(s)}(x_0)$, it follows from (4.5) and (4.7) that

$$\left|f^{(s)}(x_0) - \frac{s!}{2\pi i}\int_{\partial^* X \cap (|z-x_0|\geq r)} \frac{f^*(z)}{(z-x_0)}\,dz\right| < \epsilon,$$

whenever r is sufficiently small. Therefore, we are left to conclude that

$$f^{(s)}(x_0) = \lim_{r\to 0}\frac{s!}{2\pi i}\int_{\partial^* X \cap (|z-x_0|\geq r)} \frac{f^*(z)}{(z-x_0)^{s+1}}\,dz,$$

the formula holding for all $f \in H^\infty(X)$ at every point x_0 where there is a bounded point derivation of order s. $\qquad\square$

5. Bounded point derivations and monogeneity

In this section we take up the following question raised in the introduction: If $R(X)$ has a bounded point derivation at a point $x_0 \in X$ does there exist a set E so that

$$f'(x_0) = \lim_{x\to x_0, x\in E}\frac{f(x) - f(x_0)}{x - x_0} \tag{5.1}$$

for every $f \in R(X)$, or equivalently for every $f \in H^\infty(X)$? On the other hand, O'Farrell [22] has given an example of a Swiss cheese X for which $R(X)$ has bounded point derivations of all orders at a single isolated point $x_0 \in X$, and no others. In this case $f^{(s)}(x_0)$ is well defined at x_0 for all $s = 1, 2, \ldots$, but is not defined at any other point when $s \geq 1$. We cannot, therefore, expect higher-order derivatives to be represented in general by the limit of difference quotients as in (5.1) when $s \geq 2$. For first-order derivations we have the following theorem which is an extension of known results (cf. [23] and [30]) from $R(X)$ to the weak-$*$ algebra $H^\infty(X)$.

Theorem 5.1. *If $x_0 \in X$ is a point at which $H^\infty(X)$ has a bounded point derivation, then there exists a set E having full area density at x_0 such that*

$$f'(x_0) = \lim_{x\to x_0, x\in E}\frac{f(x) - f(x_0)}{x - x_0} \tag{5.2}$$

for all $f \in H^\infty(X)$.

Evidently, the set $E \subset Q$ since it is only at non-peak points that every $f \in H^\infty(X)$ is precisely defined, and for which the difference quotient in the statement of the theorem is therefore guaranteed to be well defined. For an arbitrary $f \in H^\infty(X)$ it is understood that

$$f'(x_0) = \lim_{j \to \infty} f'_j(x_0)$$

for any sequence of rational functions $f_j \to f$ pointwise and boundedly a.e.-dA.

Proof of Theorem 5.1. Every function $f \in H^\infty(X)$ can be expressed in the form

$$f(x) = f(x_0) + f'(x_0)(x - x_0) + F(x) \tag{5.3}$$

where $F \in H^\infty(X)$, while $\|F\|_\infty \leq C\|f\|_\infty$ with C depending only on the norm of the point derivation at x_0. To establish the theorem we need only exhibit a subset $E \subset Q$ having full area density at x_0 such that

$$\lim_{x \to x_0, x \in E} \frac{F(x)}{(x - x_0)} = 0. \tag{5.4}$$

With that in mind we shall adopt a technique of Browder [7] which allows us to compare a representing measure for $H^\infty(X)$ based at the point x_0 with certain associated representing measures at nearby points.

Let μ be any representing measure for x_0; that is, $f(x_0) = \int_X f d\mu$ for all $f \in H^\infty(X)$. Assume, moreover, that μ places no mass at x_0. Since $\nu = (z - x_0)\mu$ is then an annihilating measure for $H^\infty(X)$ it follows from an argument of Bishop [1] that

$$f(x) = \frac{1}{\widehat{\nu}(x)} \int_X \frac{f(z)}{(z - x)} d\nu_z$$

for all $f \in H^\infty(X)$ at any point $x \in X$ where the Cauchy integral $\widehat{\nu}(x) = \int (z - x)^{-1} d\nu$ is defined and not equal to 0. In particular, $\widehat{\nu}(x)$ is well defined at any point x where

$$\widetilde{\nu}(x) = \int_X \frac{1}{|z - x|} d|\nu_z| < \infty.$$

In terms of the original measure μ this yields the reproducing formula

$$f(x) = c(x)^{-1} \int_X f(z) \frac{(z - x_0)}{(z - x)} d\mu_z, \tag{5.5}$$

where $c(x) = \widehat{\nu}(x) = \int_X (z - x_0)(z - x)^{-1} d\mu = 1 + (x - x_0)\widehat{\mu}(x)$. Hence, whenever $\widehat{\mu}(x)$ is well defined and $|(x - x_0)\widetilde{\mu}(x)| < \delta < 1$ it follows that (5.5) is valid for all $f \in H^\infty(X)$.

Let us recall here that the map $f \to f'(x_0)$ is weak-$*$ continuous, and so by Theorem 4.1 there exists some $k \in L^1(X)$ with the property that

$$f'(x_0) = \int_X f k \, dA_Q$$

for all $f \in H^\infty(X)$. We may assume, therefore, that $k = 0$ a.e.-dA on $X \setminus Q$, and it is easily checked that $\mu = (z - x_0)k \, dA$ is a representing measure for $H^\infty(X)$ at x_0.

Note that in this context $\widehat{\mu}(z) = (z - x_0)\widehat{k}(z)$, where $\widehat{k}$ is by definition the Cauchy transform of the measure $k \, dA$. Formula (5.5) is therefore valid for all $f \in H^\infty(X)$, whenever $\widetilde{k}(x) < \infty$ and

$$|(x - x_0)\widetilde{k}(x)| < \delta < 1. \tag{5.6}$$

In order to complete the proof of the theorem, we shall make use of a result of Browder (cf. [7, p. 22] and [8, p. 157]) to the effect that if μ is any compactly supported measure in the plane, then

$$\lim_{r \to 0} \frac{1}{|B_r(x_0)|} \int_{B_r(x_0)} |z - x_0| \widetilde{\mu}(z) \, dA_z = |\mu\{x_0\}|. \tag{5.7}$$

In the particular case of interest here the absolutely continuous measure $k \, dA$ places no mass at the point x_0, and we may infer from (5.7) that

$$\lim_{r \to 0} \frac{1}{|B_r(x_0)|} \int_{B_r(x_0)} |z - x_0| \widetilde{k}(z) \, dA_z = 0. \tag{5.8}$$

For any $\delta, 0 < \delta < 1$, it follows immediately that the set $E = \{x : |(x - x_0)\widetilde{k}(x)| < \delta\}$ has *full area density* at x_0. We shall, in the end, conclude that E is a set through which the derivative $f'(x_0)$ in (5.2) can be realized as a limit of the corresponding difference quotient.

Since $(z - x_0)^2 k(z) \, dA$ is an annihilating measure for $R(X)$, it follows from (5.5) that

$$\begin{aligned}
F(x) &= c(x)^{-1} \int F(z) \frac{(z - x_0)}{(z - x)} (z - x_0)^2 k(z) \, dA_z \\
&= c(x)^{-1} \int F(z) \frac{(x - x_0)}{(z - x)} (z - x_0)^2 k(z) \, dA_z \\
&= c(x)^{-1}(x - x_0) \int F(z) \left(1 + \frac{x - x_0}{z - x} \right) (z - x_0) k(z) \, dA_z.
\end{aligned}$$

Then, because $\int F(z)(z - x_0) k \, dA = F(x_0) = 0$, we easily obtain the estimate

$$|F(x)| \le C\|F\|_\infty |x - x_0|^2 \widetilde{k}(x). \tag{5.9}$$

Since $R(X)$ evidently admits a representing measure at each point $x \in E$ which places no mass at x, it is clear that $E \subset Q$ (cf. [8, p. 172] and [14, p. 54]). If $x \in E$, then by (5.9)

$$\left| \frac{F(x)}{(x - x_0)} \right| \le C\|F\|_\infty.$$

To complete the proof of the theorem we shall argue as in Wang [30, p. 353]. Whenever $x \in E$ the linear functional

$$L_x(f) = \frac{F(x)}{|x - x_0|}$$

is well defined for all $f \in H^\infty(X)$, since $x \in Q$. The two functions F and f are of course connected to one another through the relation (5.3). By virtue of that connection

$$|L_x(f)| \leq C\|f\|_\infty$$

for all $x \in E$, all $f \in H^\infty(X)$, and some absolute constant C; that is, as linear functionals on $H^\infty(X)$, the norms $\|L_x\| \leq C$ for all $x \in E$. Moreover, since $H^\infty(X)$ is a weak-* closed *separable* subspace of $L^\infty(X)$, we can select from E a subsequence $x_j \to x_0$ in such a way that the corresponding bounded linear functionals L_{x_j} converge weakly to another bounded linear functional L on $H^\infty(X)$; in particular,

   (i) $L(f) = \lim_{j\to\infty} L_{x_j}(f)$ for all $f \in H^\infty(X)$,
  (ii) $\|L\| \leq C$.

On the other hand, it is clear that $L_{x_j}(f) \to 0$ for all $f \in R_0(X)$ whenever $x_j \to x_0$ through the set E. Since, however, $R_0(X)$ is dense in $H^\infty(X)$, we may conclude that $L_{x_j}(f) \to 0$ for all $f \in H^\infty(X)$. Otherwise, L would not be continuous on $H^\infty(X)$, and (ii) would be violated. Finally, it is clear from the above argument that

$$\lim_{x \to x_0, x \in E} L_x(f) = 0$$

as $x \to x_0$ over any sequence of points in E, and this establishes (5.4), from which the theorem evidently follows.   □

There is a wide range of compact nowhere dense sets $X \subset \mathbb{C}$ for which $R(X)$ and $H^\infty(X)$ inherit certain properties usually associated with the notion of analyticity. At one extreme O'Farrell [22] has constructed a set X with the property that $R(X)$ admits a bounded point derivation of *every order* at a fixed point $x_0 \in X$, while having no other bounded point derivations. At the other end of the spectrum there is the following theorem, as yet unpublished (cf. [5]).

Theorem 5.2. *There exists a compact nowhere dense set $X \subset \mathbb{C}$ such that every function $f \in H^\infty(X)$ is uniquely determined by its values on any subset of X having positive one-dimensional Hausdorff measure.*

As indicated in the introduction, the idea of extending the uniqueness property of the analytic functions to larger classes of functions defined on closed planar sets having no interior points originated with Borel during the last decade of the 19th century, and was eventually brought to fruition almost a quarter of a century later in [2]. The first nontrivial example of the transfer of the uniqueness property in this way to $R(X)$ was obtained by Keldysh, apparently around 1940 (cf. [13, p. 95], [20, pp. 745–748], and [21, pp. 317–318]). In a precursor to Theorem 5.2, Gonchar subsequently showed that there exists a Swiss cheese X such that every function $f \in R(X)$ is uniquely determined by its values on any subset of X having positive one-dimensional Hausdorff measure (cf. [20, pp. 746–748] and [6, pp. 225–228]). In either case, however, the uniqueness property depends on the fact that each function f belonging to the particular space in question has *approximate derivatives* $f^{(n)}$ of all orders $n = 0, 1, 2, \ldots$ in the sense of Theorem 5.1 on a large dense subset $E \subset X$, and that these derivatives satisfy a growth restriction as n increases which is consistent with the Denjoy–Carleman criteria for *quasi-analyticity*. Paradoxically, even when $R(X)$ or $H^\infty(X)$ enjoys the uniqueness property associated with the analytic functions each of those spaces nevertheless contains functions nowhere differentiable on X (cf. [12] and [13, p. 95]). In a recent paper [24] O'Farrell has studied many of the problems considered here in connection with the existence of point derivations at boundary points for analytic Lipschitz functions.

References

[1] E. Bishop, *A minimal boundary for function algebras*, Pacific J. Math. **9** (1959), 629–642.

[2] É. Borel, *Leçons sur les Fonctions Monogènes Uniformes d'une Variable Complexe*, Gauthier-Villars, Paris, 1917.

[3] J.E. Brennan, *Approximation in the mean and quasianalyticity*, J. Funct. Anal. **12** (1973), 307–320.

[4] J.E. Brennan, *Absolutely continuous representing measures for $R(X)$*, Bull. London Math. Soc. **46** (2014), 1133–1144.

[5] J.E. Brennan, *The uniqueness property of the analytic functions on closed sets without interior points*, (preprint).

[6] J.E. Brennan and C.N. Mattingly, *Approximation by rational functions on compact nowhere dense subsets of the complex plane*, Anal. Math. Phys. **3** (2013), 201–234.

[7] A. Browder, *Point derivations on function algebras*, J. Funct. Anal. **1** (1967), 22–27.

[8] A. Browder, *Introduction to Function Algebras*, W.A. Benjamin, New York, 1969.

[9] J.B. Conway, *A Course in Functional Analysis*, 2nd ed., Springer, New York, 1990.

[10] P.C. Curtis, *Peak points for algebras of analytic functions*, J. Funct. Anal. **3** (1969), 35–47.

[11] A.M. Davie, *Bounded limits of analytic functions*, Proc. Amer. Math. Soc. **32** (1972), 127–133.

[12] E.P. Dolzhenko, *The construction on a nowhere-dense continuum of a nowhere differentiable function which can be expanded into a series of rational functions*, Dokl. Akad. Nauk SSSR **125** (1959), 970–973 (Russian).

[13] E.P. Dolzhenko, *The work of D.E. Men'shov in the theory of analytic functions and the present state of the theory of monogeneity*, Uspekhi Mat. Nauk **47**, no. 5, (1992), 67–96 (Russian); Russian Math. Surveys **47**, no. 5, (1992), 71–102 (English).

[14] T.W. Gamelin, *Uniform Algebras*, Prentice-Hall, Englewood Cliffs, NJ, 1969.

[15] I. Glicksberg. *Recent results on function algebras*, Conference Board of the Mathematical Sciences Regional Conference Series in Mathematics, No. 11, Amer. Math. Soc., Providence, RI, 1972.

[16] A.P. Hallstrom, *On bounded point derivations and analytic capacity*, J. Funct. Anal. **4** (1969), 153–165.

190   J.E. Brennan

[17] K. Hoffman and H. Rossi, *Extensions of positive weak-∗ continuous functionals*, Duke Math. J. **34** (1967), 453–466.

[18] D. Khavinson, *The Cauchy–Green formula and rational approximation on sets with finite perimeter in the complex plane*, J. Funct. Anal. **64**, no. 1, (1985), 112–123.

[19] M.S. Mel'nikov, *A bound for the Cauchy integral along an analytic curve*, Mat. Sb. **71** (1966), 503–515 (Russian); Amer. Math. Soc. Transl. **80** (1969), 243–255 (English).

[20] M.S. Mel'nikov and S.O. Sinanjan, *Questions in the theory of approximation of functions of one complex variable*, Itogi Nauki i Tekhniki, Sovrem. Probl. Mat., vol. **4**, 143–250, VINITI, Moscow 1975 (Russian); J. Soviet Math. **5** (1976), 688–752 (English).

[21] S.N. Mergeljan, *Uniform approximations to functions of a complex variable*, Uspekhi Mat. Nauk (N.S.) **7** (1952), 31–122 (Russian); Amer. Math. Soc. Transl. **101**, (1954), 294–391 (English).

[22] A.G. O'Farrell, *An isolated bounded point derivation*, Proc. Amer. Math. Soc. **39** (1973), 559–562.

[23] A.G. O'Farrell, *Equiconvergence of derivations*, Pacific J. Math. **53**, no. 2, (1974), 539–554.

[24] A.G. O'Farrell, *Derivatives at the boundary for analytic Lipschitz functions*, Mat. Sb. **207**, no. 10, (2016), 119–140 (Russian); Sb. Math. **207**, no. 10, (2016), 1471–1490 (English)

[25] B.K. Øksendal, *Null sets for measures orthogonal to R(X)*, Amer. J. Math. **94**, no. 2, (1972), 331–342.

[26] A. Roth, *Approximationseigenschaften und Strahlengrenzwerte meromorpher und ganzer Funktionen*, Comment. Math. Helv. **11**, no. 1 (1938), 477–507.

[27] W. Rudin, *Functional Analysis*, McGraw-Hill, New York, 1973.

[28] I.M. Singer and J. Wermer, *Derivations on commutative normed algebras*, Math. Ann. **129** (1955), 260–264.

[29] X. Tolsa, *Painlevé's problem and the semiadditivity of analytic capacity*, Acta Math. **190** (2003), 105–149.

[30] J.L.-M. Wang, *An approximate Taylor's theorem for R(X)*, Math. Scand. **33** (1973), 343–358.

[31] J. Wermer, *Bounded point derivations on certain Banach algebras*, J. Funct. Anal. **1** (1967), 28–36.

[32] D.R. Wilken, *Bounded point derivations and representing measures for R(X)*, Proc. Amer. Math. Soc. **24** (1970), 371–373.

[33] L. Zalcman, *Analytic Capacity and Rational Approximation*, Lecture Notes in Math. **50**, Springer, 1968.

[34] R.J. Zimmer, *Essential Results of Functional Analysis*, Chicago Lectures in Mathematics, Univ. Chicago Press 1990.

J.E. Brennan
Department of Mathematics
University of Kentucky
Lexington, KY 40506, USA
e-mail: james.brennan@uky.edu

Operator Theory:
Advances and Applications, Vol. 261, 191–227

Three Problems in Function Theory

J.E. Brennan

Dedicated to the memory of V.P. Havin

Abstract. The three problems mentioned in the title are as follows: L^p-approximation by polynomials and rational functions; the uniqueness property of the analytic functions; and the integrability of the derivative in conformal mapping. The emphasis in this exposition is on potential theoretic ideas underlying many results presented.

Mathematics Subject Classification (2010). Primary 30-02. Secondary 30E10, 30C35, 30C85, 30D60.

Keywords. Approximation in the mean by polynomials, crescent, subnormal operator, harmonic measure, conformal map, Dyn'kin extension theorem, asymptotically holomorphic functions, capacity.

1. Introduction

On the occasion of this tribute to the founder of the St. Petersburg school of mathematical analysis I am happy to recall a number of interests that I have shared over many years with Victor Havin, as well as with a number of others in the St. Petersburg mathematical community. A common feature of our individual efforts has been the application of potential theoretic ideas to certain problems having to do with approximation by polynomials or rational functions on, more or less arbitrary, bounded domains or compact subsets of the complex plane. For the purpose of this discussion our themes can be roughly categorized as follows;

1. L^p-approximation by polynomials and rational functions,
2. The uniqueness property of the analytic functions,
3. The integrability of the derivative in conformal mapping.

It is my intention to summarize some of what has been achieved in the first two areas with a particular emphasis on the contributions of Victor Havin, and to mention two problems that initially arose in connection with L^p-approximation by polynomials, have been widely studied, and have remained open for decades.

In the process I will draw on the work of Beurling and Vol'berg on *general quasi-analyticity*, the work of Maz'ja and Havin on *nonlinear potential theory*, as well as the more recent work of Mel'nikov and Tolsa on *analytic capacity*.

When I entered Brown University in 1961 as a graduate student in mathematics Mergeljan's theorem concerning uniform polynomial approximation on compact planar sets was still the subject of much discussion, and here my initial exposure to approximation theory is uncannily similar to that reported by Havin [35] in his tribute to Maz'ja. Although the statement of Mergeljan's theorem is quite simple, the original proof was rather complicated (cf. [56]). It was in this atmosphere that I attended my first advanced seminar at which John Wermer presented a simplified proof of Mergeljan's theorem, based on joint work with Irving Glicksberg (cf. [72]). Perhaps the shortest and most self contained proof now available is due to Carleson [19]. At the time, however, I had just begun working on the invariant subspace problem for subnormal operators on a Hilbert space, which with the help of the spectral theorem, reduces to describing the closure of the polynomials in $L^2(\mu)$ for an arbitrary compactly supported positive measure μ. But, it soon occurred to me that I should first simplify the problem and consider weighted polynomial approximation in $L^2(dA)$, or more generally in $L^p(dA)$, where dA is two-dimensional Lebesgue (or area) measure. Having made that decision, I began to read Mergeljan's paper [57] which is devoted to a myriad of problems associated with approximation in the mean by polynomials. It was here that I encountered for the first time the subtle differences between uniform and mean approximation by polynomials that would occupy my attention for many years, and eventually bring me into contact with Victor Havin.

Throughout this discussion X will be a compact subset of the complex plane $\mathbb{C}$, and $C(X)$ will denote the linear space of all continuous functions on X endowed with the uniform norm. By $P(X)$ and $R(X)$ we shall mean the closure in $C(X)$ of the polynomials and rational functions having no poles on X, respectively. And, $A(X)$ will stand for the closed subspace of $C(X)$ consisting of all functions that are continuous on X and analytic in its interior X°. Thus,

$$P(X) \subseteq R(X) \subseteq A(X) \subseteq C(X).$$

The problem for uniform polynomial approximation is to determine for which X is $P(X) = C(X)$ or $P(X) = A(X)$, and here in both cases there is a purely topological answer.

Theorem (Lavrentiev, 1936). $P(X) = C(X)$ *if and only if* $X^\circ = \emptyset$, *and* $\mathbb{C} \setminus X$ *is connected.*

Theorem (Mergeljan, 1951). $P(X) = A(X)$ *if and only if* $\mathbb{C} \setminus X$ *is connected.*

As a precursor to Mergeljan's theorem Keldysh was able to show in 1945 that $P(X) = A(X)$ if X happens to be the closure of a simply connected domain. The problem of L^p-approximation by polynomials is of a much different nature, however, and the fact that no purely topological characterization is possible in that context is what both Havin and I found so intriguing.

In general we shall suppress explicit reference to the underlying measure dA, and simply write $L^p(X)$ in place of $L^p(X, dA)$. If, on the other hand, we happen to be dealing with a weighted measure $w\, dA$ that fact will be reflected in the accompanying notation. Whenever possible the page numbers of individual citations to Russian articles will refer to the English translation; otherwise, they will correspond to the original article.

2. Approximation in the mean by polynomials

In this setting Ω will be a bounded simply connected domain in $\mathbb{C}$, and Ω_∞ will denote the unbounded component of $\mathbb{C} \setminus \bar\Omega$. By definition Ω is a *Carathéodory domain* if $\partial\Omega = \partial\Omega_\infty$. Although the closure of a Carathéodory domain need not have a connected complement, such domains are nevertheless the natural analogues in connection with L^p-approximation to sets with connected complements in the case of uniform polynomial approximation. Under these circumstances two spaces each defined for $1 \le p < \infty$ will be considered. We shall designate by $H^p(\Omega) = H^p(\Omega, dA)$ the closure of the polynomials in $L^p(\Omega)$, and $L_a^p(\Omega)$ will stand for the set of functions in $L^p(\Omega)$ that are also analytic in Ω. Evidently,

$$H^p(\Omega) \subseteq L_a^p(\Omega) \text{ for all } p, 1 \le p < \infty.$$

As early as 1934 A.I. Markushevich and O.J. Farrell (cf. [57, p. 112]) established independently the fact that if Ω is a Carathéodory domain then $H^p(\Omega) = L_a^p(\Omega)$ whenever $1 \le p < \infty$. In case $H^p(\Omega) = L_a^p(\Omega)$ the polynomials are then said to be *complete* in $L_a^p(\Omega)$. The L^p analogue of Mergeljan's theorem for uniform polynomial approximation on a compact subset $X \subset \mathbb{C}$ was finally obtained by S.O. Sinanjan [65] in 1966, and is this:

Theorem (Sinanjan, 1966). $H^p(X, dA) = L_a^p(X, dA)$ *whenever X is a closed Carathéodory set; that is whenever $\partial X = \partial X_\infty$.*

Initially, we shall be principally concerned with approximation on domains, and there are two, more or less, typical non-Carathéodory domains that we will consider:

(i) A *crescent*; that is, a region which is topologically equivalent to one bounded by two internally tangent circles.

(ii) A Jordan domain with a cut or incision in the form of a simple arc from an interior point to a boundary point.

For a domain Ω with a single boundary cut it is clear that $H^p(\Omega) \ne L_a^p(\Omega)$. In that case, to achieve completeness we need to introduce a weighted measure $w\, dA$, where w is allowed to approach zero rapidly at each point of the cut. For the present we shall defer consideration of approximation in the presence of boundary cuts, and focus on the completeness problem in the context of a crescent. In contrast to the situation presented in case (ii), Keldysh [44] discovered in 1939 that $L^2(dA)$-completeness for a crescent Ω may or may not occur depending on the *thickness*

of Ω near the multiple-boundary point (cf. also [57, p. 116]). In any case, it was
now clear that there can be no topological characterization of polynomial completeness in the $L^p(dA)$-norm. The article of Mel'nikov and Sinanjan [55] contains
an excellent summary of results obtained between the publication of Mergeljan's
survey [57] and the appearance of their subsequent survey in 1976, including contributions of Hedberg, Maz'ja and Havin, as well as those of the author, and many
others.

(i) The crescent

For the purpose of this discussion it will be convenient to adopt a somewhat less
restrictive notion of a crescent domain. To that end let X be a compact set with
connected complement, and let U be a Jordan domain with $U \subset X^\circ$. The set
$\Omega = (X \setminus U)^\circ$ will again be referred to as a *crescent* provided $\partial\Omega_\infty \cap \partial U \neq$
$\emptyset$, and our subsequent use of that term will be in this generalized sense. The
first broad result concerning $L^p(dA)$-approximation on a crescent was obtained
by M.M. Dzhrbashjan and A.L. Shaginjan in 1948 under rather sever regularity
restrictions on $\partial\Omega_\infty$ and ∂U, while assuming that

$$\partial\Omega_\infty \cap \partial U = \{p_0\}$$

is a singleton, and that $l(r) = \text{length}(\Omega \cap (|z - p_0| = r))$ is subject to a strong
convexity requirement, they proved the following theorem [57, p. 158]:

Theorem (Dzhrbashjan and Shaginjan). $H^2(\Omega) = L_a^2(\Omega)$ *if and only if*

$$\int_0 \log l(r)\, dr = -\infty.$$

In [50] and [52] Maz'ja and Havin took an approach to the L^p-completeness problem based on the theory of *Sobolev spaces*, and its associated *nonlinear
potential theory* (cf. [51]). Retaining the notation introduced above they established
a stronger result valid not only for $p = 2$, but for all p, $1 < p < \infty$ (cf. [50, p. 236]).
Their result is still valid, for $p = 1$, but in that case it needs to be modified along
the lines found in Hedberg [37, p. 168], making use of an artifice associated with
what is known as the *Ahlfors mollifier*:

Theorem (Maz'ja and Havin, 1968). *For any p, $1 \leq p < \infty$,*

$$\int_0 \log l(r)\, dr = -\infty \text{ implies that } H^p(\Omega) = L_a^p(\Omega),$$

where now

(i) $\partial\Omega_\infty$ *can be arbitrary, and is no longer required to be smooth,*
(ii) ∂U *is still required to be C^2 smooth,*
(iii) $\partial\Omega_\infty \cap \partial U$ *can be arbitrary, and is no longer required to be a singleton.*

This, of course, left open the question as to whether, under the stated hypotheses, the sufficient condition $\int_0 \log l(r)\, dr = -\infty$ is also necessary for L^p-
completeness, as it is in the theorem of Dzhrbashjan and Shaginjan. At the time,

however, I had been unknowingly working on the same problem, and obtained the following improvement (cf. [10, p. 182]).

Theorem (1973). *If ∂U is a C^1-curve with a Lipschitz normal and*

$$\delta(z) = \mathrm{dist}(z, \partial\Omega_\infty),$$

then

$$H^p(\Omega) = L^p_a(\Omega) \text{ if and only if } \int_{\partial U} \log \delta(z)\, |dz| = -\infty \qquad ((2.1))$$

for all p, $1 \le p < \infty$.

Thus, it suddenly became clear, at least in this situation, that

$$H^1(\Omega) = L^1_a(\Omega) \text{ if and only if } H^p(\Omega) = L^p_a(\Omega) \text{ for all } p,\ 1 \le p < \infty. \qquad (2.2)$$

But, even today, it remains an open question as to whether (2.2) is valid for the most general crescent.

As for the proof of the theorem itself, the demonstration of *necessity* is based on an idea of Shaginjan (cf. [57, p. 121]). The regularity conditions imposed on ∂U ensure that there exists a small tubular neighborhood T around ∂U such that whenever $x \in T$ there is a unique point of ∂U nearest to x. The set of all points x having a unique nearest point in ∂U is denoted $\mathrm{Unp}(\partial U)$, and those points are said to lie within the *reach* of ∂U, the terminology being due to Federer. It is then possible, by pushing out along a *smoothed nearly normal vector field* to ∂U, to construct a class C^1 Jordan curve Γ lying in Ω, and moreover having the following properties:

(i) $\bar U$ lies in the region bounded by Γ;

(ii) If $z \in \Gamma$, then $\rho(z) = \mathrm{dist}(z, \partial U)$ and $\delta(z) = \mathrm{dist}(z, \partial\Omega_\infty)$ are *comparable*.

Under the assumption that $\int_{\partial U} \log \delta(z)|dz| > -\infty$ it can be shown that

$$\int_\Gamma \log \delta(z)|dz| > -\infty,$$

and that as a consequence any sequence of polynomials bounded in the $L^p(\Omega, dA)$-norm is in fact a normal family in the region bounded by Γ. Hence, every function in $H^p(\Omega)$ admits an analytic continuation to U, and the polynomials cannot therefore be complete in $L^p_a(\Omega)$.

The proof of *sufficiency*, on the other hand, depends heavily on the theory of Sobolev spaces and the Calderón–Zygmund theory of *singular integrals*, first employed in this context by Havin and Maz'ja (cf. [50] and [52]). The initial step is to fix p and to let k be any function in $L^q(\Omega)$, $q = p/(p-1)$, with the property that $\int fk\, dA = 0$ for all $f \in H^p(\Omega)$. To prove that $H^p(\Omega) = L^p_a(\Omega)$ it suffices, in view of the *fine continuity* of functions in $W_1^q(\mathbb{R}^2)$ and an idea of Havin [33] (cf. also [15, p. 770]), to show that the Cauchy integral

$$\widehat{k}(z) = \int \frac{k(\zeta)}{(z-\zeta)}\, dA_\zeta = 0$$

everywhere outside of $\bar{\Omega}$. Since k is orthogonal to the polynomials, and thus $\widehat{k}$ automatically vanishes identically in the unbounded complementary component of $\bar{\Omega}$, we need only verify that $\widehat{k} = 0$ in U. The idea is to multiply and divide $|\widehat{k}(z)|^q$ by $\delta(z)^\epsilon$, which yields the identity

$$\int_{\partial U} \log |\widehat{k}(z)|^q \, |dz| = \int_{\partial U} \log \delta(z)^\epsilon \, |dz| + \int_{\partial U} \log\Big(\frac{|\widehat{k}(z)|^q}{\delta(z)^\epsilon}\Big) \, |dz|, \qquad (2.3)$$

and then to prove that we can choose $\epsilon > 0$ in such a way that the second term on the right side of the identity in (2.3) is finite. That would imply $\int_{\partial U} \log |\widehat{k}(z)| \, |dz| = -\infty$, from which it can be inferred that $\widehat{k}(z) \equiv 0$ in U, provided integration over ∂U is well defined. Herein lies the difficulty.

If $1 \leq p < 2$ then $k \in L^q(\Omega, dA)$, $q > 2$, and therefore $\widehat{k}$ satisfies a Hölder condition with exponent $\mu = (q - 2)/q$ (cf. [18, p. 205]). In particular, $|\widehat{k}(z)| \leq C \, \delta(z)^\mu$ for all $z \in \partial U$, since $\widehat{k}$ vanishes identically in Ω_∞, and

$$\int_{\partial U} \log |\widehat{k}(z)| \, |dz| \leq C' + \mu \int_{\partial U} \log \delta(z) \, |dz| = -\infty.$$

In this case there is no difficulty integrating over ∂U, and it follows that $\widehat{k} \equiv 0$ in U.

If $2 \leq p < \infty$ then $1 < q \leq 2$ and the situation is much different. Although $\widehat{k}$ is analytic outside of $\bar{\Omega}$, the most that can inferred globally is that $\widehat{k}$ belongs to the Sobolev space $W_1^q(\mathbb{R}^2)$, and so integration over ∂U becomes an issue. To avoid that difficulty fix t_0, $0 < t_0 < \text{reach}(\partial U)$, and whenever $0 < t < t_0$ let $U_t = \{z \in U : \rho(z) \geq t\}$. Let us begin by replacing (2.3) with a corresponding identity in which integration is now carried out over ∂U_t, a curve lying in the region where $\widehat{k}$ is analytic; more specifically,

$$\int_{\partial U_t} \log |\widehat{k}(z)|^q \, |dz| = \int_{\partial U_t} \log \delta(z)^\epsilon \, |dz| + \int_{\partial U_t} \log\Big(\frac{|\widehat{k}(z)|^q}{\delta(z)^\epsilon}\Big) \, |dz|. \qquad (2.4)$$

To procure a bound for the second term on the right that is independent of $t < t_0$ we can simply remove the logarithm from the integrand in question, and apply the divergence theorem to the resulting integral, thereby obtaining the estimate (cf. [10, p. 186]):

$$\int_{\partial U_t} \frac{|\widehat{k}(z)|^q}{\delta(z)^\epsilon} \, |dz| \leq C_1 \int_{U_t} \frac{|\widehat{k}(z)|^q}{\delta(z)^{1+\epsilon}} \, dA + C_2 \int_{U_t} \frac{|\widehat{k}(z)|^{q-1}}{\delta(z)^\epsilon} \, |\,\text{grad}\,|\widehat{k}(z)|| \, dA. \qquad (2.5)$$

If $1 + \epsilon < q$ it can be shown, with the aid of an inequality due to Hedberg [37, p. 168] involving q-capacity and the Hardy–Littlewood maximal function, that the first term on the right in (2.5) is bounded by a constant that does not depend on t, $0 < t < t_0$. By Hölder's inequality the second term does not exceed

$$C_2 \left(\int_{\Omega_t} \frac{|\widehat{k}(z)|^q}{\delta(z)^{p\epsilon}}\right)^{1/p} \left(\int_{\Omega_t} |\,\text{grad}\,|\widehat{k}(z)||^q \, dA\right)^{1/q}. \qquad (2.6)$$

If $p\epsilon < q$ it follows again from Hedberg's lemma that the first factor in (2.6) admits a bound independent of t. By the Calderón–Zygmund theory of singular integrals the second factor does not exceed $C_3\|k\|_q$. Therefore, if ϵ is chosen so that $\epsilon < q - 1 = q/p$, then

$$\int_{\partial U_t} \log\left(\frac{|\widehat{k}(z)|^q}{\delta(z)^{1+\epsilon}}\right)|dz| \le M,$$

where M is a constant independent of t, $0 < t < t_0$. In light of the identity in (2.4), we can now infer that

$$\lim_{t\to 0}\int_{\partial U_t} \log|\widehat{k}(z)|\,|dz| = -\infty,$$

which implies that $\widehat{k} \equiv 0$ in the bounded complementary region U, and from which the theorem follows.   □

In the same year that my 1973 theorem was published Havin and Maz'ja [52, p. 562] also established the following sufficient criterion for L^p-completeness of the polynomials in a rather general crescent domain.

Theorem (Maz'ja and Havin, 1973). *Let $\Omega = (X \setminus U)^\circ$ be a crescent as described above, and let μ be harmonic measure on ∂U with respect to any fixed point $x_0 \in U$. If ∂U is an arbitrary Jordan curve, then $H^p(\Omega, dA) = L_a^p(\Omega, dA)$ whenever $1 \le p < 2$ and*

$$\int_{\partial U} \log|\psi|\,d\mu = -\infty \tag{2.7}$$

for every $\psi \in \mathring{W}_1^q(X)$ that is q-precisely defined in $W_1^q(\mathbb{R}^2)$. If ∂U is a smooth curve, then the conclusion of the theorem is valid for all p, $1 \le p < \infty$.

As Havin [35, p. 102] has noted, the divergence of the integral in (2.7) expresses the thinness of the crescent Ω near a multiple-boundary point in a *rather implicit manner*. My recollection here is that I began to think about how harmonic measure might be used to describe thinness in a more direct fashion. Building on the ideas in my 1973 theorem, I was soon able to prove that if ∂U is an arbitrary Jordan curve and μ is harmonic measure on ∂U, then $H^p(\Omega, dA) = L_a^p(\Omega, dA)$ for $1 \le p < 3$ whenever $\int_{\partial U}\log\delta(z)\,d\mu = -\infty$ (cf. [11, p. 143]). But, initially it was not clear whether $p = 3$ represented a real impediment beyond which L^p-completeness could not be expected to occur in this setting. Within a year, however, I obtained the following theorem.

Theorem 1 (1978). *Let $\Omega = (X \setminus U)^\circ$ be a crescent with X and U as above, and let μ be harmonic measure relative to any fixed point in U. There exists an absolute constant $\tau > 0$ such that, if*

$$\int_{\partial U} \log\delta(z)\,d\mu(z) = -\infty, \tag{2.8}$$

then $H^p(\Omega) = L_a^p(\Omega, dA)$ whenever $1 \le p < 3 + \tau$.

Here the proof proceeds generally along the same lines as that of my 1973 theorem outlined above, but with one significant difference. To prove that $H^p(\Omega) = L^p_a(\Omega)$ we again assume that $k \in L^q(\Omega, dA)$ and that $\int Q k \, dA = 0$ for every polynomial Q, which implies that $\widehat{k} \equiv 0$ in the unbounded complementary component of $\overline{\Omega}$, and we have only to prove that $\widehat{k} \equiv 0$ in U. To accomplish that task we let φ be a conformal map of U onto the open unit disk with $\varphi(x_0) = 0$. Next choose a sequence $r_j \uparrow 1$, and let U_j be the region bounded by the level set $|\varphi| = r_j$. The proof that $\widehat{k} \equiv 0$ in U is based on the identity

$$\int_{\partial U_j} \log |\widehat{k}(z)|^q \, d\omega_j = \int_{\partial U_j} \log \delta(z)^\epsilon \, d\omega_j + \int_{\partial U_j} \log\left(\frac{|\widehat{k}(z)|^q}{\delta(z)^\epsilon}\right) d\omega_j,$$

where $d\omega_j$ is the harmonic measure for x_0 on ∂U_j. By assumption the first integral on the right approaches $-\infty$ as $j \to \infty$. The proof is then completed by proving that the second integral can be bounded by some power of $\int |\varphi'|^{p+\tau} dA$, and then appealing to the following theorem also established in the same paper:

Theorem 2 (1978). *There exists a constant $\tau > 0$ which does not depend on U such that*

$$\int_U |\varphi'|^p \, dA < \infty \text{ whenever } \frac{4}{3} < p < 3 + \tau. \tag{2.9}$$

Concerning Theorem 2 it is known that $4/3$ is a sharp lower bound and that 4 is an upper bound. Moreover, it is conjectured that the correct range of integrability is $4/3 < p < 4$ in all cases. Details on the proofs of both Theorems 1 & 2 can be found in [12]. They do, however, leave several questions unanswered, and raise other problems as well; for example,

1. *Does (2.8) imply that $H^p(\Omega) = L^p_a(\Omega)$ for $1 \le p < \infty$, or is $p = 4$ an upper bound?*
2. *Is $\int_U |\varphi'|^p \, dA < \infty$ whenever $\frac{4}{3} < p < 4$, for every simply connected domain U?*
3. *If Ω is a crescent and $H^p(\Omega) \ne L^p_a(\Omega)$, give a complete description of all $f \in H^p(\Omega)$.*

The first two questions have remained open for decades, while the second continues to be the subject of several ongoing investigations. The third problem, on the other hand, was solved by Havin [34] in 1968 with certain regularity restrictions being placed on $\partial\Omega$. Because problem 2 occupies a central position in geometric function theory, and is related to a number of open questions, it is especially appropriate to include a few remarks on that topic before moving on to Havin's treatment of Problem 3.

It is known that $4/3 < p < 4$ is the correct interval of convergence for the integral (2.9) in a number of special cases, that together suggest where the essential difficulties may lie. These regions include starlike and close-to-convex domains, as well as any domain whose boundary is locally the graph of a function, which seems to suggest that the extremal situation is somehow associated with the presence of twist points in ∂U. By definition a point $\xi \in \partial U$ is a *twist point* if

$$\limsup_{\substack{z \to \xi \\ z \in U}} \arg(z - \xi) = +\infty \quad \text{and} \quad \liminf_{\substack{z \to \xi \\ z \in U}} \arg(z - \xi) = -\infty$$

simultaneously. These are points near which boundary sets experience their greatest distortion under a conformal mapping (cf. [20] and [60]). The original proof of Theorem 2 as presented in [12] depends on the notion of extremal length and a probabilistic method of estimating harmonic measure introduced by Carleson in [20]. Subsequently, Carleson and Makarov [21] used a strengthened version of the ideas in [20] to obtain a reformulation of the integrability conjecture itself. In his thesis (apparently never published) Bertilsson [3] extended many of the ideas in [21], and as a byproduct he was able to show that for an arbitrary simply connected domain the integral in (2.9) converges for $4/3 < p < 3.421$. Later Shimorin [61] established the conjecture for $p < 3.7858$, and that was further improved by Hedenmalm and Shimorin [41].

In order to test the integrability conjecture on domains with a fractal boundary one is led to consider complex dynamics as a source of such domains. For a fixed $c \in \mathbb{C}$ let $F(z) = F_c(z) = z^2 + c$. By definition $F^n = F \circ \cdots \circ F$ is the n-fold iterate of F with itself and

$$\Omega_c = \{z \in \mathbb{C} : F^n(z) \to \infty \text{ as } n \to \infty\}.$$

Thus $\infty \in \Omega_c$, and in some cases Ω_c is simply connected. By definition

$$\mathcal{M} = \{c \in \mathbb{C} : \Omega_c \text{ is simply connected}\} = \{c \in \mathbb{C} : J = \partial\Omega_c \text{ is connected}\}$$

and is known as the *Mandelbrot set*. Here, $J = \partial\Omega_c$ is the Julia set corresponding to F, and is most often a fractal curve. Baranski, Vol'berg and Zdunik [2] have shown that if $c \in \mathcal{M}$ and φ is a conformal map of Ω_c onto the open unit disk, then $\int_{\Omega_c} |\varphi'|^p \, dA < \infty$ whenever $4/3 < p < 4$. Pommerenke has also remarked that some aspects of Makarov's results on the distortion of harmonic measure would follow more easily if the integrability conjecture was first known to be valid, thereby providing further evidence that the conjecture is likely to be correct in general.

In order to present Havin's solution in [34] to Problem 3 we must first describe a class of functions $E^p(\Omega)$ defined on a simply connected domain Ω, introduced and studied by V.I. Smirnov. By definition a function f analytic in Ω belongs to $E^p(\Omega)$ if there exists a sequence of rectifiable curves C_j, $j = 1, 2, \ldots$, in Ω such that

(i) C_j bounds a region Ω_j, with $\Omega_j \subset \Omega_{j+1}$ for all $j = 1, 2, \ldots$,
(ii) $C_j \to \partial\Omega$ as $j \to \infty$,
(iii) $\sup_j \int_{C_j} |f(z)|^p \, |dz| < \infty$.

The answer to Problem 3 can be found in [34], and was brought up again years later by Nikolskii and Havin [59, p. 118] in connection with the subsequent development of Smirnov's ideas in complex analysis.

Assume that $\Omega = (X \setminus U)^\circ$ is a crescent and that ∂U is a class $C^{1+\epsilon}$ curve. By my 1973 theorem, if $H^p(\Omega) \neq L^p_a(\Omega)$ then

$$\int_{\partial U} \log \delta(z) \, |dz| > -\infty,$$

and this is sufficient to ensure the existence of a function Q which is analytic in U, continuous on $\bar{U}$ and has the additional property that $|Q(z)| = \delta(z)$ for all $z \in \partial U$. Havin's solution to Problem 3 is expressed in terms to the function Q.

Theorem (Havin, 1968). *Let $\Omega = (X \setminus U)^\circ$ be a crescent such that $\partial\Omega_\infty$ and ∂U are each class C^3 Jordan curves, and assume that*

$$\int_{\partial U} \log \delta(z) \, |dz| > -\infty$$

so that $H^p(\Omega) \neq L^p_a(\Omega)$. If $f \in L^p_a(\Omega, dA)$ the following are equivalent:

(1) *$f \in H^p(\Omega)$;*

(2) *f extends to a function $\widetilde{f}$ analytic in U such that $\widetilde{f}Q^{1/p} \in E^p(U)$.*

Almost a decade later in 1977 (cf. [11, p. 149]) I was able to show that Havin's theorem remains valid if it is only assumed that $\partial\Omega_\infty$ and ∂U are curves of class $C^{2+\epsilon}$, $\epsilon > 0$. My proof depends on the following lemma.

Lemma. *If F is an outer function defined in the open unit disk $\mathbb{D}$, and if $|F| \in \Lambda_\alpha(\partial\mathbb{D})$, that is if $|F|$ satisfies a Lipschitz condition of order α on $\partial\mathbb{D}$, then*

(1) *$F \in \Lambda_{\alpha/2}(\bar{\mathbb{D}})$ if $0 < \alpha < 2$,*

(2) *$F \in \Lambda_1(\bar{\mathbb{D}})$ if $\alpha > 2$.*

Ironically, part (1) of the Lemma was obtained independently by Havin and Shamojan [36] and Jacobs [43, p. 29]. My contribution was to extend the Lemma to cover the case $\alpha = 2 + \epsilon$, which was needed in order to relax the boundary restrictions in Havin's theorem. Later Shirokov ([62, 63]) extended it to cover all $\alpha < \infty$.

(ii) Domains with boundary cuts

Let Ω be a simply connected region, obtained from a Jordan domain by introducing cuts or slits in the form of simple (but not necessarily smooth) arcs extending from the interior outward to the boundary, and let $w(z) > 0$ be a bounded continuous function defined on Ω. We will assume throughout that $w(z) \to 0$ at each point of $\partial\Omega$. In this section we shall be concerned with the following two spaces of functions:

(i) $H^p = H^p(\Omega, w \, dA)$, the closure of the polynomials in $L^p(\Omega, w \, dA)$, $1 \leq p < \infty$,

(ii) $L^p_a = L^p_a(\Omega, w \, dA)$, the set of functions in $L^p(\Omega, w \, dA)$ that are analytic in Ω.

Evidentally, $H^p \subseteq L^p_a$ and we can ask, for which weights w does $H^p = L^p_a$? In a sense, this is a two-dimensional version of the classic *Bernstein problem*. By duality, the problem is equivalent to the question: If $k \in L^q(\Omega, w \, dA)$, $q = p/(p-1)$, and if the integral

$$f(z) = \int_\Omega \frac{kw(\zeta)}{(\zeta - z)} \, dA_\zeta$$

vanishes identically in Ω_∞, must $f = 0$ a.e. with respect to harmonic measure on $\partial\Omega$? Our task then is to prove that $f = 0$ on the cuts from the knowledge that $f = 0$ on $\partial\Omega_\infty$. To accomplish this we shall make extensive use of the work of Beurling and Vol'berg on *general quasi-analyticity* (cf. [4, 69], and [71]) and that of Maz'ja and Havin on *nonlinear potential theory* [51], suitably adapted to contend with the difficulties presented by a general boundary.

We shall assume as a standing requirement that $w = w(g)$ depends only on Green's function g. In particular, we may assume that w can be written in the form

$$w(z) = e^{-h(\log 1/|\varphi(z)|)},$$

where φ is a conformal map of Ω onto the open unit disk $\mathbb{D}$. With one additional assumption that $y\,h(y) \uparrow +\infty$ as $y \downarrow 0$, we have the following theorem (cf. [15, p. 769].

Theorem 1 (1994). *The polynomials are dense in $L^2(\Omega, w\,dA)$ (i.e., $H^2 = L_a^2$) whenever*

$$\int_0 \log h(y)\,dy = +\infty, \qquad (2.10)$$

or, equivalently, whenever $\int_0 \log\log \frac{1}{w(y)}\,dy = +\infty$.

Roughly speaking, $H^2 = L_a^2$ if the weight $w(z) \to 0$ sufficiently rapidly as $z \to \partial\Omega$. The theorem gives a quantitative description of that dependence, and also generalizes earlier work of the author [14], Beurling [5, p. 413], Dzhrbashjan and Shaginjan (cf. [57, §8]), and Keldysh [45] (cf. also [57, §7]). It will become evident as we proceed that our line of reasoning can be extended to cover all of the cases $1 \leq p \leq 2$. The theorem, however, is also true when $p > 2$, and in that context follows easily from the author's results in [13] on bounded point evaluations, a topic that will be discussed later after outlining the proof of the $p = 2$ case.

Proof of the Theorem (Outline). Let $k \in L^2(w\,dA)$ be any function with the property that $\int_\Omega Qkw\,dA = 0$ for all polynomials Q, and once again form the Cauchy integral

$$f(z) = \int_\Omega \frac{kw(\zeta)}{(\zeta - z)}\,dA_\zeta.$$

In this case $f \in W_1^2(\mathbb{R}^2)$, while $f \equiv 0$ in Ω_∞. Because of the rate at which $w(z) \to 0$ as $z \to \partial\Omega$, it follows that f is continuous on the entire boundary $\partial\Omega$. Hence, by the fine continuity of functions in $W_1^2(\mathbb{R}^2)$, we also know that $f \equiv 0$ on $\partial\Omega_\infty$. The essential difficulty is to prove that $f \equiv 0$ on all of $\partial\Omega$.

Let us assume for the moment that this has been established. Then, according to the work of Havin and Deny (cf. [33] and [15, p. 770]) we may choose a sequence $\eta_j \in C_0^\infty(\Omega)$ such that

$$\int_\Omega |\bar\partial\eta_j - kw|^2\,dA \to 0;$$

in other words $f \in \mathring{W}_1^2(\Omega)$. If, now, $\sigma \in L_a^2(\Omega, w\,dA) \cap L^2(\Omega, dA)$ we see, after integration by parts, that

$$\int_\Omega \sigma \bar{\partial} \eta_j \, dA = - \int_\Omega \eta_j \bar{\partial} \sigma \, dA = 0$$

for all $j = 1, 2, \ldots$. Then, because

$$\lim_{j \to \infty} \int_\Omega \sigma \bar{\partial} \eta_j \, dA = \int_\Omega f k w \, dA,$$

we see that $\int f k w \, dA = 0$ and, therefore, $\sigma \in H^2(\Omega, w\,dA)$. Since we can let $\sigma = \varphi^n(\varphi')$ for all $n = 0, 1, 2, \ldots$, and these functions are known to span $L_a^2(\Omega, w\,dA)$ (cf. [57, p. 136]), it follows that $H^2(\Omega, w\,dA) = L_a^2(\Omega, w\,dA)$. Here is precisely where the requirement that $w = w(g)$ is needed.

To complete the proof of the theorem it remains to show that $f \equiv 0$ on $\partial\Omega \setminus \partial\Omega_\infty$; that is on the cuts. This is where the hypothesis that

$$\int_0 \log h(y) \, dy = +\infty$$

comes into play. In essence, it provides a quantitative description of the rate at which the weight $w(z)$ must decay as $z \to \partial\Omega$. In this portion of the proof a key role is played by works of E.M. Dynkin and A.L. Vol'berg, each of whom, at one time or another, took part in the Seminar founded by Victor Havin.

The first step is to transfer the problem to the open unit disk by means of the conformal mapping $\varphi : \Omega \to \mathbb{D}$. Setting $F = Tf = f(\varphi^{-1})$ we obtain a mapping

$$T : W_1^2(\Omega) \to W_1^2(\mathbb{D}), \tag{2.11}$$

and we note that $p = 2$ is essential here (cf. [15]). Although the function F belongs to $W_1^2(\mathbb{D})$, and as such has boundary values $F(e^{i\theta})$ a.e.-$d\theta$ on $\partial\mathbb{D}$, it need not be smooth or even bounded in $\mathbb{D}$. That apparent obstacle, however, can be removed by making use of the Dyn'kin *extension theorem* [25] (cf. also [46, pp. 338–343]) to replace F by a new Sobolev function $\widetilde{F}$ having the same boundary values as F. This step is central to our entire argument, and may be useful in other problems as well. In particular, the function $\widetilde{F} \in W_1^2(\mathbb{D})$, and by construction it has boundary values $F(e^{i\theta})$ a.e.-$d\theta$ on $\partial\mathbb{D}$. Our task is to prove that $F(e^{i\theta}) = 0$ a.e.-$d\theta$ on $\partial\mathbb{D}$. Here we shall merely provide a brief glimpse of the proof, sufficient only to indicate at what points the individual contributions of Beurling, Vol'berg, and Dyn'kin enter the argument. More details and additional references can be found in [15]. The Dyn'kin extension theorem is this (cf. [15, pp. 770–771]).

Theorem (Dyn'kin, 1972). *$F(e^{i\theta})$ coincides with the boundary values of a new Sobolev function $\widetilde{F}$ such that*

(1) *$\widetilde{F}$ is defined and continuous in $\mathbb{D}$,*

(2) *$|\bar{\partial}\widetilde{F}(z)| \leq C\,W(z)$, where $W = w(\varphi^{-1})$ and C is a fixed constant.*

In particular, $\bar{\partial}\widetilde{F}$ can be expressed in the form $\bar{\partial}\widetilde{F} = -\rho W$, $\rho \in L^\infty(\mathbb{D})$.

Having secured the function $\widetilde{F}$, let $E = \{z \in \mathbb{D} : |\widetilde{F}(z)| \leq W(z)\}$, and set $U = \mathbb{D} \setminus E$. Since $\widetilde{F}$ is continuous in $\mathbb{D}$, we can be sure that U is an open set. We may assume with no loss of generality that E is the union of countably many disjoint, smoothly bounded, Jordan regions only finitely many of which meet any compact subset of $\mathbb{D}$. Should any connected component of E abut $\partial \mathbb{D}$ we are in the situation covered by case (A) below, and may proceed along the lines suggested there. Otherwise we can surround E by a family of smooth curves lying in the region where $|\widetilde{F}(z)| < 2W(z)$, thereby obtaining a new set $E' \supset E$ and having the properties mentioned above. The important point is that $|\widetilde{F}(z)| > W(z)$ in the complement of E', and that $|\widetilde{F}(z)| \leq 2W(z)$ on $\partial E'$. We may also assume, for purposes of argument, that ∂U contains all of $\partial \mathbb{D}$. Otherwise, $\widetilde{F}(e^{i\theta}) = 0$ on some subarc, and hence $\widetilde{F}(e^{i\theta}) = F(e^{i\theta}) \equiv 0$ by a theorem of Levinson [48, pp. 19–21].

There are two possibilities: Either $\partial U \cap \mathbb{D}$ is so sparse that harmonic measure for U, denoted $d\omega_U$, is boundedly equivalent to arc length $d\theta$ along $\partial \mathbb{D}$, or it is not. The situation can be conveniently expressed in the following way. Let $\lambda_0 = e^{i\theta_0}$ be a fixed point in $\partial \mathbb{D}$, and for each $\epsilon \leq 1/2$ form the small sectorial box

$$B_\epsilon = \left\{ re^{i\theta} : 1 - \epsilon \leq r \leq 1,\ |\theta - \theta_0| \leq \frac{1}{2} \log \frac{1}{(1 - \epsilon)} \right\}.$$

The mapping $\chi(z) = -i \log z$ carries B_ϵ onto a square, and the point $\lambda_\epsilon = \sqrt{1 - \epsilon}\, e^{i\theta}$ corresponds to the center of the image; see [15, p. 772] for a diagram. Whenever $\lambda_\epsilon \in U$, we let $d\omega_\epsilon$ be harmonic measure for λ_ϵ relative to the region $B_\epsilon \cap U$. We may assume that $B_\epsilon \cap U$ is connected, which can always be arranged. Then either

(A) $\displaystyle \int_{\partial U \cap B_\epsilon} \frac{d\omega_\epsilon}{(1 - |z|)} \geq c > 0$ for all ϵ (at some $\lambda_0 \in \partial \mathbb{D}$), or

(B) $\displaystyle \int_{\partial U \cap B_\epsilon} \frac{d\omega_\epsilon}{(1 - |z|)} \to 0$ for some sequence of $\epsilon \to 0$ (at every λ_0).

In the situation represented by case (A) there is a high concentration of harmonic measure on ∂U near a single point $\lambda_0 \in \partial \mathbb{D}$. On the other hand, case (B) cannot be summarily dismissed. It could happen that $\partial \Omega_\infty$ corresponds to a single *prime end*, as would be the case if Ω is a region obtained by removing from the open unit disk $\mathbb{D}$ a spiral that accumulates at every point of $\partial \mathbb{D}$, in which case we could not be sure, *a priori*, that $\widetilde{F}$ vanishes at more that a single point on $\partial \mathbb{D}$.

Although F will not in general be analytic in $\mathbb{D}$, the restrictions imposed on the weight w ensure that it is nevertheless almost analytic in the sense that its boundary values $F(e^{i\theta})$ have negative Fourier coefficients a_{-n}, $n = 1, 2, \ldots$, which decay sufficiently rapidly that the function $F^-(e^{i\theta}) = \sum_{-\infty}^{-1} a_n e^{in\theta}$ is given by an absolutely convergent series, is therefore continuous on $\partial \mathbb{D}$, and clearly satisfies the requirements of Dyn'kin's theorem (cf. [15, pp. 770–771]). Hence, it admits an extension to the interior of $\mathbb{D}$, still denoted F^-, which is explicitly given in the form

$$F^-(z) = \frac{1}{\pi} \int_{\mathbb{D}} \frac{\rho(\zeta)W(\zeta)}{(\zeta - z)} \, dA_\zeta,$$

where $\rho \in L^\infty(\mathbb{D})$, and $|\bar\partial F^-(z)| \le C\,W(z)$ on $\mathbb{D}$. In this way we obtain a function with the property that $(F - F^-) \in W_1^2(\mathbb{D})$, and has vanishing negative Fourier coefficients, so that

$$0 = \frac{1}{2\pi i} \int_{\partial\mathbb{D}} (F - F^-) z^{n-1} \, dz = \frac{1}{\pi} \int_{\mathbb{D}} \bar\partial(F - F^-) z^{n-1} \, dA$$

for $n = 1, 2, \ldots$. That is, the derivative $\bar\partial(F - F^-)$ is orthogonal to the polynomials over $\mathbb{D}$, and so by Weyl's lemma there is an analytic $Q \in W_1^2(\mathbb{D})$ such that

$$F(z) - F^-(z) = Q(z) - \frac{1}{\pi} \int_{\mathbb{D}} \frac{\bar\partial(F - F^-)}{(\zeta - z)} \, dA_\zeta$$

almost everywhere and, in keeping with our remarks above and the discussion in [15, §2], the integral vanishes a.e. on $\partial\mathbb{D}$ in the Sobolev sense. Setting $\widetilde{F} = Q + F^-$ on $\bar{\mathbb{D}}$, it follows that $F = \widetilde{F}$ on $\partial\mathbb{D}$, or equivalently that $(F - \widetilde{F}) \in \mathring{W}_1^2(\mathbb{D})$. Our goal, therefore, is to prove that $\widetilde{F} = 0$ a.e. on $\partial\mathbb{D}$, and here the cases (A) and (B) need to be treated separately.

To get at least an idea of where the hypothesis (2.10) enters this part of the proof, we define for each $\epsilon < 1$ a function

$$\widetilde{F}_\epsilon(z) = Q(z) + \frac{1}{\pi} \int_{D_\epsilon} \frac{\rho(\zeta)W(\zeta)}{(\zeta - z)} \, dA_\zeta,$$

where $D_\epsilon = \{z : |z| < 1 - \epsilon\}$, and Q is the analytic part of $\widetilde{F}$. Evidently, $\widetilde{F}_\epsilon$ is analytic for $1 - \epsilon < |z| < 1$, and it can be shown that

(1) $|\widetilde{F}(z) - \widetilde{F}_\epsilon| \le e^{-ch(\epsilon)}$
(2) $|\widetilde{F}_\epsilon(z)| \le K,$

where the constants c and K are independent of ϵ. Here, (1) is clear and (2) is actually a consequence of the argument in case (B).

To deal with the difficulties inherent in case (A) it is best to transfer the problem to the real line $\mathbb{R}$ by means of the mapping $\chi(z) = -i \log z$. In this way we obtain functions H and H_ϵ corresponding to $\widetilde{F}$ and $\widetilde{F}_\epsilon$, respectively, such that

(3) H and H_ϵ are defined and periodic on $\mathbb{R}$;
(4) H_ϵ has an analytic extension to a horizontal strip S_ϵ of height $\log \frac{1}{1-\epsilon}$

To prove that $H \equiv 0$ on $\mathbb{R}$, and therefore that $\widetilde{F} \equiv 0$ on $\partial\mathbb{D}$, we choose an interval $[a, b]$ containing a full period of H and form the Fourier transform

$$\widehat{H}(z) = \int_a^b H(x) e^{izx} \, dx.$$

Since $\widehat{H}$ is entire and satisfies an inequality of the form $|\widehat{H}(x+iy)| \leq Ke^{c|y|}$, it suffices to verify that

$$\int_1^\infty \log|\widehat{H}(t)| \frac{dt}{t^2} = -\infty, \tag{2.11}$$

from which it follows that $\widehat{H} \equiv 0$ on $\mathbb{R}$, and hence $H \equiv 0$ (cf. [46, pp. 50–51]).

To carry out that procedure we argue as in Beurling [4]. Let R_ϵ be the rectangle with base $[a, b]$ and top lying along the opposite side of S_ϵ. We can arrange that a and b correspond to the distinguished point $\lambda_0 \in \partial\mathbb{D}$ under the conformal map χ. If we integrate over $\Gamma_\epsilon = \partial R_\epsilon \setminus [a, b]$ it follows from inequality (1) that

$$|\widehat{H}(t)| \leq e^{-ch(\epsilon)} + \left| \int_{\Gamma_\epsilon} \widehat{H}_\epsilon(z) e^{itz} \, dz \right|,$$

from which, after replacing h by its Legendre transform, it can be deduced along the lines employed by Beurling in [4] that the integral in (2.11) diverges (cf. [15, p. 773]). Therefore, $\widehat{H}(z) \equiv 0$, which establishes the theorem in case (A).

To deal with the situation in case (B) we follow Vol'berg and introduce an auxiliary function $\Phi = \widetilde{F}e^R$, where

$$R(z) = -\frac{1}{\pi} \int_U \frac{1}{\overline{F}(\zeta)} \frac{\rho(\zeta)W(\zeta)}{(\zeta - z)} dA_\zeta.$$

Because $\bar{\partial}\Phi = 0$ and because the integrand is bounded,

(1) Φ is analytic in U, and

(2) $C_1|\widetilde{F}(z)| \leq |\Phi(z)| \leq C_2|\widetilde{F}(z)|$ with $C_1, C_2 > 0$.

The idea of modifying a given function in this way to obtain an analytic Φ has its origins in the 1930s, was used by Vol'berg [69], and is widely known as the *similarity principle* (cf. [15, 774]). To determine the behavior of Φ in U, fix a point $z_0 \in U$. For each ϵ sufficiently small let $D_\epsilon = \{z : |z| < 1 - \epsilon\}$ and denote by U_ϵ the component of $D_\epsilon \setminus E$ containing z_0. Next, let $d\omega_{U_\epsilon}$ be harmonic measure for z_0 relative to U_ϵ and integrate $\log|\Phi|$ over ∂U_ϵ to obtain the identity

$$\int_{\partial U_\epsilon} \log|\Phi| \, d\omega_{U_\epsilon} = \int_{\partial U_\epsilon} \log|\widetilde{F}| \, d\omega_{U_\epsilon} + \int_{\partial U_\epsilon} \text{Re}(R) \, d\omega_{U_\epsilon}$$

It can be shown that $\int \log|\widetilde{F}| \, d\omega_{U_\epsilon} \to -\infty$ for some sequence of $\epsilon \to 0$. Since the second term on the right is bounded, it then follows

$$\int_{\partial U_\epsilon} \log|\Phi| \, d\omega_{U_\epsilon} \to -\infty, \tag{2.12}$$

from which we may infer that $\Phi \equiv 0$ in U. Hence, $\widetilde{F}(e^{i\theta}) = F(e^{i\theta}) = 0$ on $\partial\mathbb{D}$, since $\widetilde{F} \equiv 0$ in U and $\widetilde{F}(z) \to 0$ as $z \to \partial\mathbb{D}$ through $\mathbb{D} \setminus U$, by construction. Verification of the divergence of the integral in (2.12) requires a number of delicate estimates concerning harmonic measure on ∂U_ϵ (cf. [15, pp. 774–777]). $\qquad\square$

Although to this point we have been concerned primarily with approximation in the $L^2(\Omega, w\,dA)$ norm, it is clear that Theorem 1 is actually valid whenever $1 \leq p \leq 2$, and in that context is implied by the argument for $p = 2$. In order to treat the case $p > 2$, however, it is convenient to introduce the notion of a bounded point evaluation. By definition, $H^p(\Omega, w\,dA)$ is said to have a *bounded point evaluation* (abbreviated bpe) at a point ξ_0 if the map $Q \to Q(\xi_0)$ can be extended from the polynomials to a bounded linear functional on $H^p(\Omega, w\,dA)$; that is, if the inequality

$$|Q(\xi_0)| \leq C\,\|Q\|_{L^p(\Omega, w\,dA)}$$

is valid for all polynomials Q and some absolute constant C. The following theorem was obtained by the author [13, pp. 416–418], and provides a complete solution to the weighted approximation problem for polynomials in terms of bpe's, confirming an earlier conjecture of Mergeljan [58] (cf. also [17]) in a more general setting than that in which it was originally posed. The proof, which will not be presented here does, however, rely heavily on the work of Havin [33] and the subsequent work of Maz'ja and Havin [51] concerning L^q-capacities, and the fine continuity properties of the associated potentials.

Theorem (1979). *Let Ω be a bounded simply connected domain, and let $w \in L^\infty$ be a weight depending only on Green's function. In order that $H^p(\Omega, w\,dA) = L^p_a(\Omega, w\,dA)$ for any p it is necessary and sufficient that H^p have no bpe's on $\partial\Omega$. Moreover, if H^p has a bpe at a point $\xi_0 \in \partial\Omega$, then every function in H^p admits an analytic continuation to some fixed open set U containing ξ_0.*

Let us assume, as above, that $w(z) = e^{-h(\log 1/|\varphi(z)|)}$ is a weight depending only on Green's function, and that it has the two properties:

(i) $yh(y) \uparrow +\infty$ as $y \downarrow 0$,
(ii) $\int_0 \log h(y)\,dy = +\infty$.

It follows from the area mean value property and the Koebe distortion theorem that

$$\sup_\Omega |Q|\tilde{w} \leq C\,\|Q\|_{L^1(\Omega, w\,dA)} \tag{2.13}$$

for all polynomials Q with $\tilde{w}(y) = y^4 w(y/2)$. Since (i) is assumed, we can replace $w(z)$ by $\tilde{w}(z) = e^{-2h(\frac{1}{2}\log 1/|\varphi(z)|)}$, and properties (i) and (ii) will be preserved in passing from w to $\tilde{w}$, and conversely.

Suppose now that $H^p(\Omega, \tilde{w}dA) \neq L^p_a$ for some p. By the above theorem $H^p(\Omega, \tilde{w}dA)$ must have a bpe at some point $\xi_0 \in \partial\Omega$, and so by our remarks in connection with (2.13), there is another weight w having the same properties as $\tilde{w}$ such that

$$|Q(\xi_0)| \leq C_1\,\|Q\|_{L^p(\Omega), \tilde{w}dA} \leq C_2\,\|Q\|_{L^1(\Omega, w\,dA)}$$

for all polynomials Q. But, this contradicts the fact that $H^1(\Omega, w\,dA) = L^1_a$ as has already been established. We conclude, therefore:

Theorem 2 (1994). *For any* p, $1 \leq p < \infty$, *the polynomials are dense in* $L_a^p(\Omega, w\, dA)$ *that is,* $H^p(\Omega, w\, dA) = L_a^p$ *whenever* w *is a weight satisfying* (i) *and* (ii).

Much of what has transpired has a natural interpretation in the context of uniform weighted approximation. Beurling [5], for example, has considered the following generalization of the classic *Bernstein problem*. With Ω and w as above, let $C_w(\Omega)$ be the Banach space of all complex-valued functions f such that the product $f(z)w(z)$ is continuous on $\bar{\Omega}$ and vanishes on $\partial\Omega$, with the norm being defined by

$$\|f\|_w = \sup_{\Omega} |f|w.$$

Evidentally, the collection of functions

$$A_w(\Omega) = \{f \in C_w(\Omega) : f \text{ is analytic in } \Omega\}$$

is a closed linear subspace of $C_w(\Omega)$. The problem is to determine whether the polynomials are dense in $A_w(\Omega)$.

Theorem 3 (1994). *If* (i) *and* (ii) *are satisfied, then the polynomials are dense in* $A_w(\Omega)$.

As stated, Theorem 3 represents an extension of a theorem of Beurling [5, p. 413] to arbitrary simply connected domains. Beurling assumes, as part of the hypotheses, that the conformal map $\varphi^{-1} : \mathbb{D} \to \Omega$ has a continuous extension to $\bar{\mathbb{D}}$, and that unnecessarily restricts the scope of his result. The proof of Theorem 3, in outline, is as follows. Fix $f \in A_w(\Omega)$, and set $F = f(\varphi^{-1})$. For each $r < 1$, let $f_r = F(r\varphi)$ so that $f_r \in A_w(\Omega)$. By the monotonicity of w with respect to Green's function,

$$\|f_r - f\|_w \to 0 \text{ as } r \to 1.$$

Because f_r is also bounded it can be approximated arbitrarily closely by a sequence of polynomials in $L^1(\Omega, \widetilde{w}\, dA)$ for any weight $\widetilde{w}$ dominating w in the sense of (2.13). Then same sequence, therefore, also converges to f_r in the norm of $C_w(\Omega)$, and the theorem follows. (Note, in the application of (2.13) the roles of w and $\widetilde{w}$ are reversed here.) $\qquad\square$

In order to address the difficulties inherent in the approximation problem we have made extensive use of the theory of *asymptotically holomorphic* functions in a form that grew out of the work of Dyn'kin [25] and Vol'berg [69]. Roughly speaking, a function $F(e^{i\theta})$ defined initially on $\partial\mathbb{D}$ is asymptotically holomorphic if it can be realized as the boundary values of a corresponding Sobolev function $F(z)$ defined on $\mathbb{D}$ and having the property that

$$|\bar{\partial}F(z)| \to 0 \text{ rapidly as } z \to \partial\mathbb{D}.$$

The rate at which $|\bar{\partial}F|$ decays at $\partial\mathbb{D}$ can be taken as a measure (or asymptotic estimate) of the extent to which $F(e^{i\theta})$ deviates from the boundary values of an actual analytic function. If the drop off is sufficiently rapid and the consequent deviation is below a certain critical level, then $F(e^{i\theta})$ will inherit many of the properties usually associated with analyticity. Most of what was used here can be found in a more precise form in Koosis' book [46]. One of the more striking applications of the theory of asymptotically holomorphic functions is to the problem of weighted polynomial approximation as presented above, and was duly noted by Vol'berg [70, pp. 965–967] in his address to the International Congress of Mathematicians in Kyoto.

3. Approximation on compact nowhere dense sets and the uniqueness property

It is well known that if a sequence of analytic functions f_n, $n = 1, 2, \ldots$, converges uniformly, or even pointwise and boundedly on a region $\Omega \subseteq \mathbb{C}$, then the limit function f is also analytic on Ω. At his thesis defense in 1894, at which Poincaré was the rapporteur, Émile Borel suggested that it should be possible to extend the theory of analytic functions to larger classes of functions defined on planar sets without interior points in such a way that the distinctive property of unique continuation is preserved. Poincaré, having constructed certain examples with the aim of disproving the possibility of such an extension, took a negative point of view. At the time, however, Borel was not yet able to exhibit the kind of extension he envisioned, and so the matter remained unresolved for many years. But, eventually Borel was able to put his ideas on a more solid foundation, and to fully realize his stated goal. To that end he introduced the notion of a monogenic function. By definition a function f defined on a set E is *monogenic* at a point $x_0 \in E$ if it has a derivative at x_0 in the sense that

$$\lim_{x \to x_0, x \in E} \frac{f(x) - f(x_0)}{x - x_0} \tag{3.1}$$

exists through points of E. Thus, if Ω is an open set then a function f is monogenic at each point of Ω if and only if it is analytic in Ω. Borel's main result (cf. [7]), published almost a quarter of a century after his confrontation with Poincaré, was to construct a compact set X having no interior and containing a large dense subset E such that every function f continuous on X and monogenic on E is actually infinitely differentiable on E, and is uniquely determined by its value and the values of all its derivatives at any fixed point of E.

  In the intervening years it has become clear that certain large classes of functions defined on a compact nowhere dense subset X of the complex plane, and obtained as limits of analytic functions in various metrics, can sometimes inherit the uniqueness property. The first nontrivial example of the transfer of the uniqueness property in this way to $R(X)$, the space of functions that can be uniformly approximated on X by a sequence of rational functions whose poles lie outside of X, was obtained by Keldysh, apparently around 1940, but never published. He constructed a compact nowhere dense planar set X of positive area such that every function $f \in R(X)$ is uniquely determined by its values on any relatively open subset of X. Later, Sinanjan [64] produced similar examples illustrating the transfer of the uniqueness property in certain cases to $R^p(X)$, $p \geq 2$, where $R^p(X)$ denotes the evidently larger space of functions obtained as limits of rational functions in the $L^p(dA)$-norm on X. Here again, if X is sufficiently massive then every function $f \in R^p(X)$ is uniquely determined as an element of $L^p(X)$ by its values on any relatively open subset of X. Eventually, these results were further refined by Gonchar (cf. [55]) and the author [9] by describing more accurately in each case the critical size of any determining subset of X. An extensive survey of the early results in this area and their subsequent development can be found in the survey articles of Bilodeau [6], Dolzhenko [24] and Fuglede [28, 30, 31, 29].

My goal in this section is to take a look at the problem of approximation by polynomials and rational functions on compact nowhere dense subsets of the complex plane, both in the uniform and $L^p(dA)$ metrics, and to provide a glimpse of the manner in which the uniqueness property can be transferred to large classes of functions defined on sets without interior points. In the case of L^p-approximation the nonlinear L^q-capacities and corresponding potential theory introduced by Maz'ja and Havin (cf. [51] and [52]) will play a decisive role. An analogous role is played by *analytic capacity* in the case of uniform approximation (cf. [68]).

(iii) Approximation on nowhere dense sets

As indicated in the introduction, my initial interest in questions having to do with approximation by polynomials in $L^2(\mu)$ for an arbitrary compactly supported measure μ was ignited by its connection to the invariant subspace problem for subnormal operators on a Hilbert space. By definition, an operator $T : H \to H$ mapping a Hilbert space H into itself is *subnormal* if it has a normal extension to a larger Hilbert space. Insofar as the invariant subspace problem is concerned, it can always be assumed that the operator T has a cyclic vector; that is, a vector $x \in H$ such that the closed linear span of $x, Tx, T^2x, \ldots$ coincides with H, otherwise invariant subspaces abound and the question is settled in the affirmative. If, on the other hand, T is a subnormal operator with a cyclic vector, then there exists a regular Borel measure on the spectrum of T with the property that H is isometrically isomorphic to $H^2(\mu)$, the closure in $L^2(\mu)$ of the complex polynomials in z, and the action of T in H corresponds to multiplication by z in $H^2(\mu)$. By analogy with our remarks at the end of the preceding section $H^2(\mu)$ is said to admit a *bounded point evaluation* (bpe) at a point $\xi_0 \in \mathbb{C}$ if the inequality

$$|f(\xi_0)| \leq C \|f\|_{L^2(\mu)} \tag{3.2}$$

is satisfied for all polynomials f and some absolute constant C. If either $H^2(\mu)$ has a bpe or $H^2(\mu) = L^2(\mu)$, then evidently $H^2(\mu)$ admits a nontrivial closed subspace invariant under multiplication by z. At this point in 1971 the following question arose, and remained open for another twenty years when it was eventually settled by Thomson [66]:

1. *If μ is not equal to a point mass, does $H^2(\mu) = L^2$ if and only if $H^2(\mu)$ has no bpe?*

Prior to 1971 the answer to this question was not known, even in the simplest case
when $d\mu$ is the restriction of dA to a compact nowhere dense subset X of positive
area in $\mathbb{C}$. The same question can, of course, be asked of $H^p(\mu)$, $1 \le p < \infty$, where
$H^p(\mu)$ is by definition the closure of the polynomials in $L^p(\mu)$, and bounded point
evaluations are defined by analogy with (3.2). At times, it will be instructive to
consider all p, $1 \le p < \infty$, simultaneously.

Suppose, for example, that X is a compact nowhere dense subset of $\mathbb{C}$, and we
wish to determine whether $H^p(X, dA) = L^p(X, dA)$. As before, let $k \in L^q(X, dA)$,
$q = p/(p-1)$, be any function for which $\int_X fk \, dA = 0$ for all polynomials f. If
$H^p(XdA)$ has no bpe in $\mathbb{C} \setminus X$ it is easy to see that

$$\widehat{k}(z) = \int_X \frac{k(\zeta)}{(\zeta - z)} dA_\zeta \equiv 0 \text{ in } \mathbb{C} \setminus X. \tag{3.3}$$

In particular, $H^p(X, dA)$ contains every rational function with poles off X, and
so $H^p(X, dA) = R^p(X, dA)$. If, on the other hand, $p < 2$ then $q > 2$, and the
function $\widehat{k}(z)$ in (3.3) is continuous in the entire complex plane. Hence, $\widehat{k} \equiv 0$, and
so $\bar{\partial}\widehat{k} = -\pi k = 0$ a.e.-dA, from which it follows that $H^p(X, dA) = L^p(X, dA)$.
A corollary here, of course, is that $R^p(X, dA) = L^p(X, dA)$ whenever X has no
interior and $p < 2$. This much can be found in the papers of Sinanjan [64] and the
author [8].

At this point it was clear that for approximation by rational functions in
the $L^p(X, dA)$-norm the case $p = 2$ is exceptional. In fact, it was known at the
time that if $p \ne 2$, then $R^p(X, dA) = L^P(X, dA)$ if and only if $R^p(X, dA)$ has no
bounded point evaluations. By 1975, however, Fernstrom [26] had constructed a set
X for which $R^2(X, dA)$ has no bpe's, but nevertheless $R^2(X, dA) \ne L^2(X, dA)$.
That, of course, left open the corresponding question for $H^2(X, dA)$, which is
where the matter stood prior to 1979. In that year, however, I was able to prove
the following theorem (cf. [13, p. 410]).

Theorem 1 (1979). *Let $w \ge 0$ be a function of compact support and fix p, $1 \le
p < \infty$. If $w \in L^{1+\epsilon}(dA)$ for some $\epsilon > 0$ then, either $H^p(w \, dA)$ has at least one
bounded point evaluation or $H^p(w \, dA) = L^p(w \, dA)$.*

Although this theorem did not establish a fundamental connection between
the nonexistence of bounded point evaluations and the density of the polynomials
in $L^p(d\mu)$ for an arbitrary measure μ, it did indicate a method by which to approach the general question. Broadly speaking, that was the scheme followed by
Thomson in [66], albeit by overcoming significant technical difficulties. Moreover,
shortly after receiving a preprint of my paper S.V. Hruschev [42] noticed that the
theorem remains valid for any measure of *finite entropy*; that is, for any measure
$w \, dA$ with the property that $\int w(\log w)^p dA < \infty$. I subsequently included a proof
of Hruschev's theorem in my paper [13, p. 413], the proof of which is more in keeping with my original argument. Due to its implications for identifying a suitable
scheme in the most general situation, an outline of the proof is included here.

First, however, it is appropriate to say a few words about the underlying potential theory due to Maz'ja and Havin [51], and upon which the proof rests. To address accurately the exceptional sets associated with Sobolev functions we need to introduce the notion of capacity. For our purpose it can be assumed that $p \geq 2$, or equivalently that the dual exponent $q \leq 2$. If $1 \leq q < 2$ the q-capacity of a compact set $X \subset \mathbb{C}$ is defined by

$$\Gamma_q(E) = \inf_u \int |\operatorname{grad} u|^q \, dA,$$

where the infimum is taken over all functions $u \in C_0^\infty$ such that $u \equiv 1$ on X. If $q = 2$ each function u is required to have its support in some fixed disk. For an arbitrary set E the q-capacity is defined by setting $\Gamma_q(E) = \sup \Gamma_q(X)$, the supremum being taken over all compact sets $X \subset E$. A property is said to hold q-quasi-everywhere if the set where it fails has q-capacity zero. It is known that each function $u \in W_1^q(\mathbb{R}^2)$ can be redefined on a set of measure zero in such a way that the resulting function, still denoted as u, is then q-finely continuous at quasi-every point. In particular, at q-quasi-every point x_0 there exists a set E which is thin at x_0 in a potential theoretic sense, and for which

$$u(x_0) = \lim_{x \to x_0, x \notin E} u(x).$$

For our purpose here it will be sufficient to know that if $1 < q < 2$ a set E is thick (i.e., not thin) at a point x_0 if

$$\limsup_{r \to 0} \frac{\Gamma_q(B_r(x_0) \cap E))}{r^{2-q}} > 0. \tag{3.4}$$

Here, $B_r(x_0)$ is the disk with center x_0 and radius r; when no specific center is implied we shall simply write B_r. Moreover, $\Gamma_q(B_r) \approx \Gamma_q(\operatorname{diam} B_r) \approx r^{2-q}$ if $1 < q < 2$, and $\Gamma_2(B_r) \approx \left(\log \frac{1}{r}\right)^{-1}$. Thus, the thickness of a set E in the neighborhood of a fixed point x_0 is in this instance reflective of a capacitary density condition. Additional information on these various capacities can be found in [1] (cf. also [39]).

Proof of Theorem 1 (Outline). We shall suppose that $H^p(w \, dA)$ has no bpe's, and argue that $H^p(w \, dA) = L^p(w \, dA)$. And, for purposes of exposition we may assume that $p > 2$, and that therefore $1 < q < 2$. With this as our goal let $k \in L^q(w \, dA)$ be any function such that $\int Qk \, w \, dA = 0$ for all polynomials Q, and form the Cauchy transform

$$\widehat{kw}(z) = \int \frac{k(\zeta)w(\zeta)}{(\zeta - z)} \, dA_\zeta.$$

Since $k \in L^q(w \, dA)$ and $w \in L^{1+\epsilon}(dA)$, it follows from Hölder's inequality that $kw \in L^\alpha(dA)$ for some $\alpha > 1$, and because $\widehat{kw}$ has compact support $\widehat{kw} \in W_1^\alpha$. For each $\lambda > 0$ consider $E_\lambda = \{z : |\widehat{kw}(z)| < \lambda\}$. Having fixed a point x_0, we shall establish two things:

(1) Almost every circle $|z - x_0| = r$ meets E_λ in a set of positive linear measure.
(2) E_λ is thick at x_0 with respect to each of the capacities Γ_α, $\alpha > 1$.

Once these facts are known it can be inferred from (3.4) that E_λ is thick at x_0, since α-capacity decreases modulo a fixed multiplicative constant under a contraction, in this case under circular projection onto a line segment (cf. [1, p. 140]). Thus, if x_0 is a point of α-fine continuity for $\widehat{kw}$, then $|\widehat{kw}(x_0)| < \lambda$. Since this holds for every $\lambda > 0$ we must conclude that $\widehat{kw}(x_0) = 0$.

It remains then to establish property (1), which is where the absence of bpe's is needed. If we assume that assertion (1) is false, then there exists a set E of nonnegative real numbers, having positive linear measure, such that $|\widehat{kw}| \geq \lambda$ almost everywhere on $|z - x_0| = r$ for every $r \in E$. Setting E^* equal to the union of all circles $|z - x_0| = r$ corresponding to these values of r, by the mean value theorem

$$Q(x_0) = C \int_{E^*} Q(\zeta) \, dA_\zeta$$

for a suitable constant C. Since $\int Qk \, w \, dA = 0$ for every polynomial Q, it follows from an argument due essentially to Cauchy that

$$Q(\zeta) = \frac{1}{\widehat{kw}(\zeta)} \int Q(z) \frac{kw(z)}{(z - \zeta)} \, dA_z$$

at every point ζ where $\widehat{kw}(\zeta)$ is defined and not equal to zero. In particular, this is valid for almost every $\zeta \in (\mathbb{C} \setminus E_\lambda)$, and therefore

$$Q(x_0) = C \int_{E^*} \frac{1}{\widehat{kw}(\zeta)} \int Q(z) \frac{kw(z)}{(z - \zeta)} \, dA_z \, dA_\zeta$$

$$= C \int Q(z) \, kw(z) \left(\int_{E^*} \frac{1}{\widehat{kw}(\zeta)} \frac{1}{(z - \zeta)} \, dA_\zeta \right) dA_z.$$

Because $|\widehat{kw}| \geq \lambda$ almost everywhere on E^* and $(z - \zeta)^{-1}$ is locally integrable,

$$|Q(x_0)| \leq C \int |Q| \, |k| w \, dA$$

and hence $|Q(x_0)| \leq \|Q\|_{L^p(w \, dA)}$ for every polynomial Q. This, of course, violates our assumption that $H^p(w \, dA)$ has no bpe's, and so we are forced to accept assertion (1). $\square$

We have included a certain amount of detail on this latter point because it is an essential part of Thomson's theorem, where analytic capacity γ replaces q-capacity. It is by no means clear, for example, as to what might replace assertion (1), since Vitushkin has shown that analytic capacity need not decrease under a contraction (cf. [18, p. 215]). At this point, it was not even clear how to treat the bpe problem for an absolutely continuous measure $w \, dA$ if it is only assumed that $w \in L^1(dA)$. Ironically, however, this turned out to be the essential difficulty. Throughout this discussion μ will be a compactly supported measure in $\mathbb{C}$, and $g \in L^2(d\mu)$ will be an annihilator of $H^2(d\mu)$; that is $\int Qg \, d\mu = 0$ for all polynomials Q. By convention, $\nu = g\mu$. The existence of bounded point evaluations (bpe's) for

$H^2(d\mu)$ is inextricably linked to the analogous question for $L^1(|\hat{\nu}|\,dA)$, where $\hat{\nu}$ is the Cauchy transform of the measure ν. Treatment of the general problem is based on the following elementary fact.

Lemma 1. *If there is a bpe for the polynomials in the $L^1(|\hat{\nu}|\,dA)$ norm at a point x_0, then $H^2(d\mu)$ also has a bpe at x_0.*

Proof. By assumption there exists a function $h \in L^\infty(dA)$ such that $Q(x_0) = \int Qh\,|\hat{\nu}|\,dA$ for all polynomials Q. Setting $k = h\frac{|\hat{\nu}|}{\hat{\nu}}$ when $\hat{\nu} \neq 0$ we have $\|k\|_\infty = \|h\|_\infty$, and by an interchange in the order of integration we obtain

$$Q(x_0) = \int Qk\hat{\nu}\,dA = -\int \widehat{Qk}\,d\nu.$$

On the other hand, by Weyl's lemma $\widehat{Qk} = Q\hat{k} + F$, where F is a entire function. Since ν annihilates the polynomials and so also F (that is, $\int F\,d\nu = 0$) we see that

$$Q(x_0) = -\int Q\hat{k}\,d\nu = -\int Q\hat{k}g\,d\mu.$$

By an application of Schwarz' inequality it follows that $|Q(x_0)| \leq C\|Q\|_{L^2(d\mu)}$ for all polynomials Q, and some constant C. Thus, $H^2(d\mu)$ has a bpe at x_0 as claimed. $\qquad\square$

To facilitate a discussion of the bpe problem for $L^2(d\mu)$ a few words concerning analytic capacity are in order. The *analytic capacity* of a compact set X, denoted $\gamma(X)$, is defined as follows:

$$\gamma(X) = \sup |f'(\infty)|,$$

where the supremum is taken over all functions f analytic in $\widehat{\mathbb{C}}\backslash X$ and normalized so that

(a) $\|f\|_\infty = \sup_{\widehat{\mathbb{C}}\backslash X} |f| \leq 1$,
(b) $f(\infty) = 0$.

For an arbitrary planar set E we let $\gamma(E) = \sup\gamma(X)$, where the supremum is over all compacts $X \subset E$. It is known that

(i) $\gamma(B_r) = r$ for every disk B_r of radius r;
(ii) $\gamma(X) \approx \operatorname{diam}(X)$ if X is compact and connected; in fact, $\gamma(X) \leq \operatorname{diam}(X) \leq 4\gamma(X)$.

Although these represent similarities between analytic capacity and the potential-theoretic capacities considered above, there are also some significant differences. Two that initially made it impossible to directly extend our earlier reasoning on the bpe problem to the general situation are these:

(iii) γ is *not* known to be subadditive, as is q-capacity;
(iv) if Φ is a contraction it may not happen that $\gamma(\Phi E) \leq C\,\gamma(E)$, for any $C > 0$.

Even though γ may not be subadditive, Tolsa [67] has established countable semi-additivity.

Theorem (Tolsa 2003). *There exists an absolute constant C such that for any collection of Borel sets E_n, $n = 1, 2, 3, \ldots$,*

$$\gamma\left(\bigcup_{n=1}^{\infty} E_n\right) \leq C \sum_{n=1}^{\infty} \gamma(E_n).$$

This, together with the following lemma (cf. [16, pp. 224–225]), will allow us to give a brief outline of the proof of Thomson's theorem. The lemma itself was suggested by an idea of Carleson [19, Lemma 1] used to give a short proof of Mergeljan's theorem on uniform polynomial approximation. In addition to its application in that regard, it evidently implies Lavrentiev's characterization of those compact sets for which $P(X) = C(X)$, as well as Vitushkin's condition ensuring that $R(X) = C(X)$ (cf. [16, pp. 235–236]).

Lemma 2 (2005). *Let μ be a finite, compactly supported Borel measure in $\mathbb{C}$, and let x_0 be any point where the Newtonian potential $\widetilde{\mu}(x_0) < \infty$. Assume that E is a set with the property that for every $r > 0$ there is a relatively large subset $E_r \subseteq (E \cap B_r(x_0))$ on which $\widetilde{\mu}$ is bounded; that is,*

(1) $\widetilde{\mu} \leq M < \infty$ *on* E_r,
(2) $\gamma(E_r) \geq \epsilon \gamma(E \cap B_r)$ *for some absolute constant ϵ.*

If, moreover, E is thick at x_0 in the sense that

$$\limsup_{r \to 0} \frac{\gamma(E \cap B_r)}{r} > 0,$$

then $|\widehat{\mu}(x_0)| \leq \limsup_{z \to x_0, z \in E} |\widehat{\mu}(z)|$.

The main theorem on bpe's for $L^2(\mu)$ can now be established along lines similar to those presented earlier under conditions which allowed the use of purely potential-theoretic methods. The proof that we will present here in outline is not entirely that originally published by Thomson, although it clearly contains critical elements of the original argument. The use of Lemma 2 and Tolsa's theorem on the semi-additivity of analytic capacity allow for a more accurate description of the points where $H^2(\mu)$ can actually have a bpe (cf. [16] for details).

Theorem 2 (Thomson 1991). *If μ is a positive measure of compact support in $\mathbb{C}$, and not a point mass, then $H^2(\mu) = L^2(\mu)$ if and only if $H^2(\mu)$ admits no bpe's.*

First consider an application of the Vitushkin approximation scheme due to Mel'nikov. For each positive integer n form a grid in the plane consisting of lines parallel to the coordinate axes, intersecting at those points whose coordinates are both integral multiples of 2^{-n}. The resulting collection of squares $\mathcal{G} = \{S_{nj}\}_{j=1}^{\infty}$ of side lengths 2^{-n} is an edge-to-edge tiling of the entire plane; its members will be referred to as squares of the nth generation. Beginning with a fixed positive integer k, let Π_k be a polygon formed by taking the union of finitely many squares from $\mathcal{G}_k$, and denote its boundary by Γ_k (not to be confused with Sobolev capacity). Next, adjoin to Π_k finitely many additional squares from the next generation $\mathcal{G}_{k+1}$

extending outward from Π_k and separating Γ_k from ∞ to obtain a second polygon Π_{k+1} bounded by Γ_{k+1}. Continuing in this way we construct a finite sequence of polygons

$$\Pi_k \subseteq \Pi_{k+1} \subseteq \cdots \subseteq \Pi_{k+l}, \qquad (3.5)$$

and we shall assume that at the final stage Π_{k+l} extends to ∞ in all directions. Enlarging each square S_r occurring in the chain (3.5) by a factor of 5/4 we obtain an open covering of the plane such that no point $z \in \mathbb{C}$ lies in more than four of the corresponding enlarged squares. Except for minor modifications the next lemma is due to Mel'nikov [54, p. 262] (cf. [16, pp. 226–229]). Throughout this discussion the area of a set E will be denoted $|E|$.

Lemma 3. *Suppose that a nested sequence of polygons*

$$\Pi_k \subseteq \Pi_{k+1} \subseteq \cdots \subseteq \Pi_{k+l}$$

has been constructed in the manner indicated above. Let $K = \bigcup K_n$ be a compact set such that for each n,

(1) $K_n = \bigcup S_{nj}$ *is a finite union of closed squares S_{nj} in $\Pi_n \setminus \Pi_{n-1}$, and*
(2) $n^2\, 2^{-n} \le \mathrm{dist}(K_{n+1}, \Pi_n) \le 3n^2\, 2^{-n}$.

If $E \subseteq K$ is a measurable set with the property that $|E \cap S_{nj}| > \epsilon\, |S_{nj}|$ for some fixed constant $\epsilon > 0$ and all $S_{nj} \subseteq K$, then

$$\gamma(E) \ge C\epsilon\, \gamma(K),$$

where C is an absolute constant.

Proof of Theorem 2 (Outline). Let $g \in L^2(\mu)$ be an annihilator of $H^2(\mu)$; that is, $\int Pg\, d\mu = 0$ for all polynomials P. Let $\nu = g\mu$ and consider the sets $E_\lambda = \{z : |\widehat{\nu}(z)| < \lambda\}$. Select a point $x_0 \in \mathbb{C}$ where

$$\widetilde{\nu}(x_0) = \int \frac{d|\nu|(\zeta)}{|\zeta - x_0|} < \infty,$$

and fix $\lambda > 0$. Beginning with a particular generation, the nth say, choose a square $S^* \in \mathcal{G}_n$ with $x_0 \in S^*$. Denote by $\mathcal{G}_n^\lambda$ the collection of all squares in $\mathcal{G}_n$ for which

$$|E_\lambda \cap S| > \frac{1}{100}\,|S|. \qquad (3.6)$$

Let K_n stand for the union of those squares in $\mathcal{G}_n^\lambda$ that can be joined to S^* by a finite chain of squares lying in $\mathcal{G}_n^\lambda$. Should K_n be bounded, or perhaps empty, there exists a *corridor*, or *barrier*, $Q_n = \bigcup_j S_{nj}$ consisting of squares from the nth generation abutting $S^* \cup K_n$, separating the latter from ∞, adjacent along their sides, and such that

$$|E_\lambda \cap S_{nj}| \le \frac{1}{100}\,|S_{nj}| \qquad (3.7)$$

for each j. The polynomially convex hull of Q_n is a polygon Π_n with its boundary Γ_n lying along the sides of squares for which (3.7) is satisfied. In particular, $|\widehat{\nu}| \ge \lambda$ on a large portion of every square S_{nj} meeting Γ_n.

Next, construct a polygon Π_n^* with boundary Γ_n^* in such a way that

  (i) $\Pi_n^* \supseteq \Pi_n$, and
  (ii) $n^2 2^{-n} \leq \mathrm{dist}(\Gamma_n^*, \Gamma_n) \leq 3n^2 2^{-n}$

This can be done by simply adjoining to Π_n additional squares from $\mathcal{G}_n$. Now, let K_{n+1} denote the union of all squares in $\mathcal{G}_{n+1}^\lambda$ that can be joined to Π_n^* by a finite chain of squares in $\mathcal{G}_{n+1}^\lambda$. If K_{n+1} is bounded, or empty, there is a second barrier Q_{n+1} abutting $\Pi_n^* \cup K_{n+1}$ with the property that

$$|E_\lambda \cap S| \leq \frac{1}{100}\,|S|$$

for every square $S \in Q_{n+1}$. The polygon Π_{n+1} is then the polynomial convex hull of Q_{n+1}, its boundary is Γ_{n+1}, and the process continues.

The result is a nested sequence of polygons

$$\Pi_n \subseteq \Pi_{n+1} \subseteq \cdots \subseteq \Pi_{n+l} \subseteq \cdots \tag{3.8}$$

and compact sets $K_j \subseteq \Pi_j$, $j \geq n$, for which the hypotheses of Lemma 3 are satisfied:

  (i) $K_j \subseteq \Pi_j \setminus \Pi_{j-1}$,
  (ii) $j^2 2^{-j} \leq \mathrm{dist}(K_{j+1}, \Pi_j) \leq 3j^2 2^{-j}$

There are two mutually exclusive possibilities; either the sequence (3.8)

  (A) terminates after l steps and $\infty \in \Pi_{n+l}$, or
  (B) it continues indefinitely, and $\infty \notin \Pi_j$ for any j.

In the second case (B) there exists an *infinite sequence* of barriers Q_j bounded by polygonal curves Γ_j extending outward from x_0, accumulating in a finite portion of the plane. Moreover, $|\hat{\nu}| \geq \lambda$ on a sufficiently large percentage of each barrier Q_j to ensure that $L^1(|\hat{\nu}|\, dA)$ has a bounded point evaluation at x_0 (cf. [16, pp. 230–232]). According to Lemma 1, therefore, $H^2(\mu)$ must also have a bpe at x_0. In this context then the barriers arising from the situation in case (2) take the place of the circles used to obtain a bpe for $H^p(w\, dA)$ in the proof of Theorem 1.

Let us now assume that $H^2(\mu)$ has no bounded point evaluation. Then there can be no infinite sequence of barriers associated with $\mathbb{C} \setminus E_\lambda$ for any λ. Hence, beginning with an arbitrary generation, the nth say, there exists a finite sequence of polygons

$$\Pi_n \subseteq \Pi_{n+1} \subseteq \cdots \subseteq \Pi_{n+l}$$

constructed in the manner of (3.8) with $x_0 \in \Pi_n$ and $\infty \in \Pi_{n+l}$. In particular, if $n < j < n+l$ there exist compact sets $K_j \subset \Pi_j \setminus \Pi_{j-1}$, some of which may be empty, such that if $K_j \neq \emptyset$

  (i) K_j is the union of squares in $\mathcal{G}_j$ and connects Γ_{j-1}^* to Q_j,
  (ii) $|E_\lambda \cap S| > \frac{1}{100}|S|$ for each square $S \subseteq K_j$,
  (iii) $\mathrm{dist}(K_j, \Gamma_j^*) \leq \mathrm{dist}(K_j, \Gamma_j) + \mathrm{dist}(\Gamma_j, \Gamma_j^*) < 2^{-j} + 3j^2 2^{-j} < 4j^2 2^{-j}$.

As in the earlier description of the chain formation, let S^* be the nth generation square containing x_0, and form a new chain of squares leading from S^* to ∞ as follows. Join S^* to the first nonempty K_j by a narrow rectangle R_0 so that

$\mathrm{diam}(R_0) \approx \mathrm{dist}(S^*, \Gamma^*_{j-1})$. Next, choose a connected component of K_j meeting R_0, and so obtain a chain from S^* to the barrier Q_j. By property (iii) this chain can then be joined to Γ^*_j by another narrow rectangle R_j with $\mathrm{diam}(R_j) < 4j^2\, 2^{-j}$. The resulting chain either meets K_{j+1} at Γ^*_j, or it does not. If it does we adjoin a connected component of K_{j+1} extending the chain to the next barrier Q_{j+1}; if not, adjoin a narrow rectangle R_{j+1} extending the chain to Γ^*_{j+1}, being sure to arrange that $\mathrm{diam}(R_{j+1}) < 4(j+1)^2\, 2^{-(j+1)}$. We continue in this way until the chain of squares from S^* escapes to ∞. The result is a connected set X joining S^* to ∞, which is composed of squares satisfying property (ii) and certain narrow rectangles.

Given $r > 0$, let $B_r = B_r(x_0)$ be the disk with center at x_0 and radius r. Fix n, choose $S^* \in \mathcal{G}_n$ with $x_0 \in S^*$, and let the set X be formed as above. By discarding certain superfluous pieces we can assume that $(X \cap B_r)$ is connected and joins S^* to ∂B_r. Therefore,

$$\gamma(X \cap B_r) \geq \frac{1}{4}\mathrm{diam}(X \cap B_r) \geq \frac{r}{8}.$$

Denote by K the collection of all squares in $(X \cap B_r)$ that lie entirely inside B_r. By construction $K = \bigcup K_j$, where each K_j is the union of squares from $\mathcal{G}_j$ that satisfy

(1) $|E_\lambda \cap S| > \frac{1}{100}|S|$ for each $S \subseteq K$,
(2) $j^2\, 2^{-j} \leq \mathrm{dist}(K_j, \Gamma^*_j) < 4j^2\, 2^{-j}$.

It follows from these remarks and the semiadditivity of analytic capacity that

$$\frac{r}{16} \leq \gamma(X \cap B_{r/2}) \leq C\left[\gamma(K) + \sum_{j=n}^{\infty} j^2\, 2^{-j}\right], \tag{3.9}$$

where C is an absolute constant. Since construction of the set X can begin with an arbitrary generation we are free to choose n as large as we please, and (3.9) remains valid with the same constant C. In particular, we can choose n sufficiently large that the infinite sum on the right of (3.9) becomes negligible, from which it follows that $\gamma(K) \geq Cr$ for some other constant C. Properties (1) and (2) together with Lemma 3 then imply that

$$\gamma(E_\lambda \cap B_r) \geq C\epsilon\,\gamma(K) \geq C\epsilon\, r$$

for some constant ϵ independent of r. Hence, E_λ is thick at x_0 in the sense that

$$\limsup_{r \to 0} \frac{(\gamma(E_\lambda \cap B_r)}{r} > 0.$$

Since $|\tilde{\nu}| < \infty$ a.e.-dA, we can find a subset $\Omega_r \subseteq (E_\lambda \cap B_r)$ on which $|\tilde{\nu}| \leq M$ and for which $|\Omega_r \cap S)| > \frac{1}{100}|S|$. By the same reasoning presented above in support of the lower estimate for $\gamma(E_\lambda \cap B_r)$ we may conclude that

$$\gamma(\Omega_r) \geq C\epsilon\,\gamma(E\lambda \cap B_r)$$

The hypotheses of Lemma 2 are therefore satisfied and we may conclude that $|\widehat{\nu}(x_0)| \leq \limsup_{z\to x_0, z\in E_\lambda} |\widehat{\nu}(z)| \leq \lambda$. Since the inequality is valid for all $\lambda > 0$, we can infer that $\widehat{\nu}(x_0) = 0$. Hence, $\widehat{\nu} = 0$ a.e.-dA and so $\nu = gu = 0$ as a measure. Therefore, $H^2(d\mu) = L^2(d\mu)$ □

(iv) The uniqueness property and rational approximation

Let X be a compact subset of $\mathbb{C}$. We shall say that $R(X)$ has a *bounded point derivation of order n* at a point $x_0 \in X$ if there exists a constant $C > 0$ such that

$$|f^{(n)}(x_0)| \leq C \|f\|_{L^\infty}$$

for every rational function f having no poles on X. Bounded point derivations for $R^p(X)$ are defined similarly. Evidentally, both $R(X)$ and $R^p(X)$ have bounded point derivations of all orders at each interior point of X. Here, however, we shall adopt as a standing assumption the requirement that X has no interior. The study of point derivations for $R(X)$, relative to the uniform norm, was initiated by Wermer in 1967. Included among his results is an example of a *Swiss cheese* X with the property that $R(X)$ admits a bounded point derivation of order 1 at almost every point of X (cf. [18] for additional references). Hallstrom [32] subsequently generalized this by exhibiting a Swiss cheese X with the added feature that $R(X)$ has bounded point derivations of all orders at almost every point of X. To accomplish that he first obtained a necessary and sufficient condition in terms of analytic capacity for $R(X)$ to have a bounded point derivation of a given order at a fixed point of X. That condition can be viewed as a generalization of Mel'nikov's peak point criterion (cf. [53]). Later Hedberg [38] obtained analogous criteria in terms of the Sobolev capacity Γ_q governing the possible existence of bounded point derivations on $R^p(X)$ when $p \geq 2$. Recall that $R^p(X) = L^p(X)$ if $1 \leq p < 2$ and so $R^p(X)$ can have no bounded point derivations for that range of p.

Meantime, in 1965 Sinanjan [64] was able to construct a compact nowhere dense set X with the property that whenever two functions in $R^p(X)$, $p \geq 2$, agree on a *relatively open* subset of X, they agree almost everywhere. Of course, in an open set, if two analytic functions agree only on a subset with a limit point, they agree everywhere. My initial contribution to this circle of problems was to construct a set X such that the determining subsets are essentially minimal. Here is that result (cf. [9]), which incidentally also includes Hallstrom's example, but does not involve analytic capacity.

Theorem 3 (1973). *There exists a compact nowhere dense set X of positive dA-measure such that whenever two functions in $R^P(X)$, $p \geq 2$ coincide on any subset of positive dA-measure in X they coincide almost everywhere.*

Before outlining the proof there are two results that need to be mentioned. The first is due to W.K. Allard, and is the result of a lunchtime conversation we had when we were both graduate students at Brown University. Allard's result furnishes additional information on the geometric nature of a Swiss cheese $E = \bar{\mathbb{D}} \setminus \bigcup_{j=1}^\infty D_j$ formed by deleting from the open unit disk $\mathbb{D}$ countably many open disks D_j,

$j = 1, 2, \ldots$, having mutually disjoint closures, the sum of whose radii is finite, and so that no interior is left. The second item required is a theorem of Denjoy and Carleman on quasianalyticity from the early 1920s.

Lemma (Allard). *Let E be a Swiss cheese, and for each $x \in [-1, 1]$ let*

$$E_x = \{z \in E : \operatorname{Re} z = x\}.$$

For almost every $x \in [-1, 1]$ the set E_x is the union of finitely many disjoint non-degenerate intervals.

It follows from the Lemma that almost every pair of points in a Swiss cheese E can be joined by finitely many line segments in E and finitely many subarcs of $\partial \mathbb{D}$ and ∂D_j, $j = 1, 2, \ldots$. This property will play a key role in the proof of the uniqueness theorem. Since the proof is quite short, and is related to another important idea from geometric measure theory, namely to the *Banach indicatrix theorem*, we include it here in its entirety.

Proof of the lemma. Set $D_0 = \mathbb{D}$, and for each $x \in [-1, 1]$ let $N_j(x)$, $j = 0, 1, 2, \ldots$, denote the number of points in $E_x \cap \partial D_j$. Thus, $N_j(x) = 0$, 1 or 2, and it follows from the monotone convergence theorem for integrals that

$$\int_{-1}^{1} \sum_{j=0}^{\infty} N_j(x)\, dx = \sum_{j=0}^{\infty} \int_{-1}^{1} N_j(x)\, dx \le \sum_{j=0}^{\infty} \operatorname{length}(\partial D_j) < \infty.$$

Hence, $\sum_{j=0}^{\infty} N_j(x) < \infty$ for almost every $x \in [-1, 1]$. For any such x all but finitely many $N_j(x) = 0$, and the lemma follows. $\qquad\square$

Theorem (Denjoy–Carleman). *Let f be a function of one real variable defined and having derivatives of all orders on a closed interval $[a, b]$. Assume that*

$$\sup_{[a,b]} |f^{(n)}| \le M_n, \quad n = 0, 1, 2 \ldots.$$

Then, f is uniquely determined by its value and the values of all its derivatives at any fixed point $x_0 \in [a, b]$ whenever

$$\sum_{n=0}^{\infty} \left(\frac{1}{M_n} \right)^{1/n} = \infty. \tag{3.10}$$

Proof of the Uniqueness Theorem (Outline). We may assume that $p = 2$. The idea is to construct a Swiss cheese X in such a way that it contains an increasing sequence compact sets $E_1 \subset E_2 \subset \cdots \subset E_k \subset \cdots$ having the following properties:

(1) each E_k is a Swiss cheese whose complement is bounded by polygonal arcs;
(2) $\operatorname{meas}(X \setminus E_k) \to 0$ as $k \to \infty$;
(3) for each k there are positive constants A_n, $n = 1, 2, \ldots$ such that

$$|f^{(n)}(\zeta)| \le A_n \, \|f\|_{L^2(X)}$$

for all rational functions f having no poles on X, and all $\zeta \in E_k$;
(4) $\sum_{n=1}^{\infty} (1/A_n)^{1/n} = \infty$.

The crucial properties (3) and (4) are easily arranged by taking advantage of the fact that in the process of constructing a Swiss cheese the removal of additional *small disks* does not disturb the already established norms of point derivations at points far away by more than a negligible amount.

Suppose now that X has been constructed as indicated, and that $f \in R^2(X)$ vanishes on a set of positive dA-measure in X. By assumption f is the limit in the $L^2(X)$-norm of a sequence of rational functions f_j, $j = 1, 2, \ldots$, having no poles on X, and so by (3) the sequence of derivatives $f_j^{(n)}$, $j = 1, 2, \ldots$, converges uniformly on each E_k, $k = 1, 2, \ldots$, for each n. Thus, the restriction of f to any line segment belonging to E_k can be viewed as a C^∞ function of one real variable. Moreover, since $f = 0$ on a set of positive dA-measure in X, it follows from (2) that $f = 0$ on a set of positive dA-measure in E_k if k is sufficiently large. For such a k it is a consequence of Fubini's theorem and Allard's lemma that $f = 0$ on a set of positive linear measure on some line segment l lying in E_k. Hence the limit function f has directional derivatives of all orders along any line segment l lying in E_k. On the other hand,

$$\sup_l |f^{(n)}| \leq A_n \|f\|_{L^2(X)}$$

and so by property (4), f satisfies the conditions of the Denjoy–Carleman Theorem with $M_n = A_n \|f\|_{L^2(X)}$. If x_0 is a point of linear measure density in l for the zero set of f, it follows that $f^{(n)}(x_0) = 0$, $n = 0, 1, 2, \ldots$, where differentiation is along l. By quasianalyticity, $f \equiv 0$ on l. But, by Allard's Lemma, we may assume that x_0 can be joined to almost every other point of E_k by a polygonal arc in E_k whose initial segment belongs to l. Again, by quasianalyticity, $f \equiv 0$ on any such arc, since at each vertex its derivatives coincide in the appropriate directions. Therefore, $f = 0$ almost everywhere on E_k for all k, $k = 1, 2, \ldots$, and so by property (2) almost everywhere on X.   □

Several years later in 1976 the following theorem of Gonchar appeared in the survey article of Mel'nikov and Sinanjan [55], and Mattingly and I [18] subsequently gave another proof of Gonchar's theorem which is more transparent in certain of its technical aspects.

Theorem (Gonchar). *There exists a compact nowhere dense set X with the property that each function in $R(X)$ is completely determined by its values on any subset of positive one-dimensional Hausdorff measure.*

It can be shown that the algebra $H^\infty(X)$ defined as the weak-$*$ closure of the rational functions in $L^\infty(X)$, or equivalently the bounded pointwise limits of rational functions, also inherits the same uniqueness property as in Gonchar's theorem if the set X is sufficiently dense, although that has not yet been published. Paradoxically, however, even when $R(X)$ or $H^\infty(X)$ enjoys the uniqueness property associated with the analytic functions each of those spaces nevertheless contains functions nowhere differentiable on X (cf. [23] and [24, p. 95]). The existence of nowhere differentiable functions in this setting was established by Dolzhenko [23] in his first published paper, and that was what first brought him to the attention of the mathematical community at large. For some reason, however, the existence of nowhere differentiable functions in $R(X)$ when the latter enjoys the uniqueness property received little attention in subsequent years.

(v) Finely holomorphic functions

In 1968 Havin [33] studied the problem of approximation by analytic functions in $L^2(X)$ for an arbitrary compact set $X \subset \mathbb{C}$, and he obtained a necessary and sufficient condition in terms of the *Cartan fine topology* of classical potential theory in order that $R^2(X) = L_a^2(X)$. In 1938, however, Keldysh had considered the following question concerning uniform approximation by harmonic functions on compact subsets of $\mathbb{C}$. Denote by $H(X)$ the closure in $C(X)$ of the set of all functions harmonic in a neighborhood of X, and by $C_h(X)$ the closed subspace of $C(X)$ consisting of those functions harmonic in the interior of X. Evidentally.

$$H(X) \subseteq C_h(X) \subseteq C(X).$$

The problem here is to determine for which X is $H(X) = C_h(X)$, or $H(X) = C(X)$ in case X has no interior, and it was completely solved by Keldysh. Moreover, it was soon realized that the necessary and sufficient conditions obtained by Havin and Keldysh are equivalent (cf. [40, p. 194]), and therefore

Theorem. *If* $X \subset \mathbb{C}$ *is compact, then* $H(X) = C_h(X)$ *if and only if* $R(X) = L_a^2(X)$.

At this point it was rapidly becoming clear that the fine topology is associated in a fundamental way to certain questions having to do with approximation by analytic functions, which drew attention to the possibility of its also being connected to the transferal of the uniqueness property to large classes of functions defined on sets without interior points. The fine topology has its origin in the notion of thin sets (cf. [47, Ch. V, §3] and [1, p. 176]). To make the concept of thinness more precise in this instance consider the *logarithmic potential* $U^\mu(x)$ defined by

$$U^\mu(x) = \int \log \frac{1}{|z - x|} \, d\mu_z.$$

A set E is by definition 2-*thin* at a point $x_0 \in \mathbb{C}$ if either $x_0 \notin \bar{E}$, or there exists a positive measure μ supported on E such that

$$U^\mu(x_0) < \liminf_{z \to x_0, z \in E} U^\mu(z),$$

where $z \neq x_0$. The *Wiener capacity* (i.e., 2-*capacity*) of a set $E \subset \mathbb{C}$, which we may assume to be contained in the unit disk, is denoted $C_2(E)$ and is by definition

$$C_2(E) = \sup \mu(E),$$

with the supremum being taken over all positive measures μ concentrated on E for which $\sup_z U^\mu(z) \leq 1$. In addition, it can be shown [37, p. 160] that $\Gamma_2(E) = 2\pi \, C_2(E)$, where Γ_2 is the *Sobolev capacity* introduced in Section 3(iii).

The associated *logarithmic capacity* of a set E denoted by $C_l(E)$ is defined by the relation

$$C_l(E) = e^{-C_2(E)^{-1}}.$$

Moreover, a set E is 2-thin at a point x_0 if and only if

$$\int_0 \frac{C_2(E \cap B_r(x_0))}{r} \, dr < \infty \tag{3.11}$$

(cf. [1, §6.3] or [39, p. 316]), which by virtue of the subadditivity of 2-capacity condition (3.11) is equivalent to the condition

$$\sum_{n=1}^{\infty} n\, C_2(E \cap \mathcal{A}_n(x_0)) < \infty, \tag{3.12}$$

where $\mathcal{A}_n(x_0) = \{z : 2^{-(n+1)} \leq |z - x_0| < 2^{-n}\}$. Setting $\gamma(E) = \log(1/C_l(E))$ the series in (3.12) becomes the familiar Wiener series

$$\sum_{n=1}^{\infty} \frac{n}{\gamma(E \cap \mathcal{A}_n(x_0))} = \sum_{n=1}^{\infty} \frac{n}{\log\left[C_l(E \cap \mathcal{A}_n(x_0))\right]^{-1}} < \infty. \tag{3.13}$$

The weakest topology on $\mathbb{C}$ making all superharmonic functions continuous, or equivalently making all potentials $U^\mu(x)$ of measures μ continuous, was introduced by Cartan in 1940, is therefore known as the *Cartan fine topology*, and it is strictly stronger than the standard Euclidean topology on $\mathbb{C}$. A set V is by definition a *fine neighborhood* of x if $\mathbb{C} \setminus V$ is thin at x, and such sets constitute a *neighborhood basis* for the fine topology. A *fine domain* $\Omega \subseteq \mathbb{C}$ is understood to be a connected open set in the fine topology. A function $f : \Omega \to \mathbb{C}$ is finely differentiable at a point $z_0 \in \Omega$ if the fine limit

$$f'(z_0) = \text{fine} \lim_{z \to z_0} \frac{f(z) - f(z_0)}{z - z_0} \tag{3.14}$$

exists. This is to be understood in the following way, namely that there exists a set E which is 2-thin at x_0 such that the limit

$$f'(z_0) = \lim_{z \to z_0, z \notin E} \frac{f(z) - f(z_0)}{z - z_0} \tag{3.15}$$

exists in the usual sense; that is, as $z \to z_0$ through the fine neighborhood $V = \mathbb{C} \setminus E$. A function $f : \Omega \to \mathbb{C}$ is said to be *finely holomorphic* in Ω if it is finely differentiable at every point of Ω and if the fine derivative $f' : \Omega \to \mathbb{C}$ is itself finely continuous. A good introduction to the basic properties of finely holomorphic functions can be found in the articles [28, 30, 31] by Fuglede.

The following theorem [22, p. 300] has a particular relevance in connection with Theorem 3 of this section, and to the uniqueness question generally. Let us recall that in Theorem 3 we constructed a Swiss cheese X with the property that X contains an increasing sequence of compact sets $E_1 \subset E_2 \subset E_3 \subset \cdots$, such that $\text{meas}(X \setminus E_k) \to 0$ as $k \to \infty$, and $R(X)$ has bounded point derivations of all orders on each E_k with norms depending only on the set E_k.

Theorem 4 (Debiard and Gaveau, 1974). *If $X \subset \mathbb{C}$ is compact, then $R(X)$ admits bounded point derivations of all orders at each point in the fine interior of X.*

Proof from [22]. If x_0 lies in the fine interior of X, then according to (3.13)

$$\sum_{n=1}^{\infty} \frac{n}{\log\left[C_l\big(\mathcal{A}_n(x_0) \setminus X\big)\right]^{-1}} < \infty.$$

Since the terms of the series tend to zero as $n \to 0$, given any positive integer k there exists a corresponding integer N_k such that

$$\frac{n}{\log\left[C_l\big(\mathcal{A}_n(x_0) \setminus X\big)\right]^{-1}} < \frac{1}{(k+2)\log 2} \quad \text{whenever } n \geq N_k.$$

Exponentiating we obtain the inequality $C_l\big(\mathcal{A}_n(x_0) \setminus X\big) < 2^{-(k+2)n}$ for $n \geq N_k$, and since logarithmic capacity C_l dominates analytic capacity γ we are able to conclude that

$$\sum_{n \geq N_k} 2^{(k+1)n} \gamma(\mathcal{A}_n(x_0) \setminus X) \leq \sum_{n \geq N_k} 2^{(k+1)n} C_l\big(\mathcal{A}_n(x_0) \setminus X\big) < \sum_{n \geq N_k} 2^{-n} < \infty.$$

Since $k \geq 1$ is arbitrary, it follows from Hallstrom's theorem [32, p. 155] that $R(X)$ has bounded point derivations of all orders at x_0. $\qquad\square$

It follows from remarks of Fuglede (cf. [31, §10]) that each of the Swiss cheeses E_k in the uniqueness set $X = \bigcup_k E_k$ is a fine domain, less a polar set. Fuglede has also shown that every finely holomorphic function f on a fine domain has at most countably many zeros, unless $f \equiv 0$. This is easily seen to be the case in Theorem 3 for each $f \in R(X)$, and the proof is the same as that presented there. Gonchar's theorem and its generalization to $H^{\infty}(X)$ present additional difficulties, since in those situations uniqueness subsets of X involve boundary values of functions from $R(X)$ or $H^{\infty}(X)$ on $\partial^* X$, the perimeter or reduced boundary of X in the measure theoretic sense of DeGiorgi.

In Theorem 3 use was made of the fact that as sets of finite perimeter, almost every pair of points in each *fine domain* E_k can be joined by a polygonal arc of finite length lying completely in E_k. This could be taken as a precursor to a theorem of Lyons who subsequently showed that any two points in a *fine domain* Ω can likewise be joined by a suitable path in Ω, by first establishing a *probabilistic version* of the indicatrix theorem (cf. [49, p. 16]. Later Fuglede [27, pp. 107–113] gave a simplified proof of Lyons' theorem along more classical lines. In addition, Fuglede [29] has extended Havin's theorem [33] on L^2-approximation by analytic functions to the context of finely holomorphic functions.

References

[1] D.R. Adams and L.I. Hedberg, *Function Spaces and Potential Theory*, Grundlehren Math. Wiss., vol. **314**, Springer-Verlag, Berlin, 1996.

[2] K. Baranski, A. Vol'berg and A. Zdunik, *Brennan's conjecture and the Mandelbrot set*, Internat. Math. Res. Notices, no. 12, 589–600, (1998).

[3] D. Bertilsson, *On Brennan's conjecture in conformal mapping*, Doctoral Thesis, Royal Institute of Technology, Stockholm, Sweden 1999.

[4] A. Beurling, *On quasianalyticity and general distributions*, Multilithed Lecture Notes, Summer School, Stanford Univ., Stanford, CA 1961.

[5] A. Beurling, *Collected Works*, vol. **1**, Birkhäuser, Boston 1989.

[6] G.G. Bilodeau, *The origin and early development of non-analytic infinitely differentiable functions*, Arch. Hist. Exact Sci. **27** (1982), 115–135.

[7] É. Borel, *Leçons sur les Fonctions Monogènes Uniformes d'une Variable Complexe*, Gauthier-Villars, Paris, 1917.

[8] J.E. Brennan, *Invariant subspaces and rational approximation*, J. Funct. Anal. **7** (1971), 285–310.

[9] J.E. Brennan, *Approximation in the mean and quasianalyticity*, J. Funct. Anal. **12** (1973), 307–320.

[10] J.E. Brennan, *Invariant subspaces and weighted polynomial approximation*, Ark. Mat. **11** (1973), 167–189.

[11] J.E. Brennan, *Approximation in the mean by polynomials on non-Carathéodory domains*, Ark. Mat. **15** (1977), 117–168.

[12] J.E. Brennan, *The integrability of the derivative in conformal mapping*, J. London Math. Soc. **18** (1978), 261–272.

[13] J.E. Brennan, *Point evaluations, invariant subspaces, and approximation in the mean by polynomials*, J. Funct. Anal. **34** (1979), 407–420.

[14] J.E. Brennan, *Weighted polynomial approximation, quasianalyticity and analytic continuation*, J. Reine Angew. Math. **357** (1985), 23–50.

[15] J.E. Brennan, *Weighted polynomial approximation and quasianalyticity for general sets*, Algebra i analiz **6** (1994), 69–89; St. Petersburg Math. J. **6** (1995), 763–779.

[16] J.E. Brennan, *Thomson's theorem on mean-square polynomial approximation*, Algebra i analiz **17**, no. 2, (2005), 1–32 (Russian); St. Petersburg Math. J. **17**, no. 2, (2006), 217–238 (English).

[17] J.E. Brennan, *On a conjecture of Mergelyan*, Izv. Nats. Akad. Armenii Mat. **43**, no. 6, (2008), 341–352; J. Contemp. Math. Anal. **43**, no. 6, (2008), 341–352.

[18] J.E. Brennan and C.N. Mattingly, *Approximation by rational functions on compact nowhere dense subsets of the complex plane*, Anal. Math. Phys. **3** (2013), 201–234.

[19] L. Carleson, *Mergelyan's theorem on uniform polynomial approximation*, Math. Scand. **15** (1964), 167–175.

[20] L. Carleson, *On the distortion of sets on a Jordan curve under a conformal mapping*, Duke Math. J. **40** (1973), 547–559.

[21] L. Carleson and N.G. Makarov, *Some results related to Brennan's conjecture*, Ark. Mat. **32**, (1994), 33–62.

[22] A. Debiard and B. Gaveau, *Potentiel fin et algèbres de fonctions analytiques I*, J. Funct. Anal. **16** (1974), 289–304.

[23] E.P. Dolzhenko, *The construction on a nowhere-dense continuum of a nowhere differentiable function which can be expanded into a series of rational functions*, Dokl. Akad. Nauk SSSR **125** (1959), 970–973 (Russian).

[24] E.P. Dolzhenko, *The work of D.E. Men'shov in the theory of analytic functions and the present state of the theory of monogeneity*, Uspekhi Mat. Nauk **47**, no. 5, (1992), 67–96 (Russian); Russian Math. Surveys **47**, no. 5, (1992), 71–102 (English).

[25] E.M. Dyn'kin, *Functions with a given estimate for $\partial f/\partial \bar{z}$ and a theorem of N. Levinson*, Mat. Sb. **89**, no. 2., (1972), 182–190 (Russian); Math. USSR-Sb. **18** (1972), 181–189 (English).

[26] C. Fernström, *Bounded point evaluations and approximation in L^p by analytic functions*, Spaces of analytic functions (Sem. Functional Anal. and Function Theory, Kristiansand, 1975); Lecture Notes in Math., vol. **512**, Springer, Berlin, 1976, 65–68.

[27] B. Fuglede, *Asympotic paths for subharmonic functions and polygonal connectedness of fine domains*, Lecture Notes in Math. **814**, Springer, 1980, 97–116.

[28] B. Fuglede, *Fine topology and finely holomorphic functions*, Proc. 18th Scandinavian Cong. Math. (Aarhus 1980), Prog. Math. **11**, Birkhäuser, Boston, MA, 1981, 22–28.

[29] B. Fuglede, *Complements to Havin's theorem on L^2-approximation by analytic functions*, Ann. Acad. Sci. Fenn. Math. **10** (1985), 187–201.

[30] B. Fuglede, *Finely holomorphic functions: a survey*, Rev. Roumaine Math. Pures Appl. **33** (1988), 283–295.

[31] B. Fuglede, *Fine potential theory*, Lecture Notes in Math. **1344**, Springer 1988, 81–97.

[32] A.P. Hallstrom, *On bounded point derivations and analytic capacity*, J. Funct. Anal. **4** (1969), 153–165.

[33] V.P. Havin, *Approximation in the mean by analytic functions*, Dokl. Akad. Nauk SSSR **178**, no. 5, (1968), 1025–1028 (Russian); Soviet Math. Dokl. **9**, no. 1, (1968), 245–248 (English).

[34] V.P. Havin, *Approximation in the mean by polynomials on certain non-Carathéodory domains I and II*, Izv. Vysh. Uchebn. Zaved. Mat. **76**, 86–93; **77**, 87–94, (1968) (Russian).

[35] V.P. Havin, *On some potential theoretic themes in function theory*, in The Maz'ya Anniversary Collection vol. 1, Operator Theory Advances and Applications **109**, 99–110, Birkhäuser (1999).

[36] V.P. Havin and F.A. Shamojan, *Analytic functions with a Lipschitzian modulus of boundary values*, Zap. Naučn. Sem. Leningrad Otdel. Mat. Inst. Steklov (LOMI) **19**, (1970), 237–239 (Russian).

[37] L.I. Hedberg, *Approximation in the mean by analytic functions*, Trans. Amer. Math. Soc. **163** (1972), 157–171.

[38] L.I. Hedberg, *Bounded point evaluations and capacity*, J. Funct. Anal. **10** (1972), 269–280.

[39] L.I. Hedberg, *Non-linear potentials and approximation in the mean by analytic functions*, Math. Z. **129** (1972), 299–319.

226   J.E. Brennan

[40] L.I. Hedberg, *Approximation by harmonic functions and stability of the Dirichlet problem*, Exposition. Math. **11** (1993), 193–259.

[41] H. Hedenmalm and S. Shimorin, *Weighted Bergman spaces and the integral means spectrum of conformal mappings*, Duke Math. J. **127**, no. 2, (2005), 341–393.

[42] S.V. Hruscev, *The Brennan alternative for measures with finite entropy*, Izv. Akad. Nauk Armyan. SSR Ser. Mat., **14**, no. 3, (1979), 184–191 (Russian).

[43] S. Jacobs, *Ett extremalproblem för analytiska funktioner i multipelt sammanhängande områden*, Uppsala Univ. Math. Dept. Report **2**, 1969 (Swedish).

[44] M.V. Keldysh, *Sur l'approximation en moyenne quadratique des fonctions analytiques*, Mat. Sb. (N. S.) **5(47)**, no. 2, (1939), 391–401.

[45] M.V. Keldysh, *Sur l'approximation en moyenne par polynômes des fonctions d'une variable complexe*, Mat. Sb. (N. S.) **16 (58)**, no. 1, (1945), 1–20.

[46] P. Koosis, *The Logarithmic Integral*, vol. **1**, Cambridge Univ. Press, Cambridge, UK (1988).

[47] N.S. Landkof, *Foundations of Modern Potential Theory*, Nauka, Moscow, 1966 (Russian); Grundlehren Math. Wiss., vol. **180**, Springer, 1972 (English).

[48] N. Levinson, *Gap and Density Theorems*, Amer. Math. Soc. Colloq. Publ., vol. **26**, Amer. Math. Soc., New York, 1940.

[49] T.J. Lyons, *Finely holomorphic functions*, J. Funct. Anal. **37** (1980), 1–18.

[50] V.G. Maz'ja and V.P. Havin, *On approximation in the mean by analytic functions*, Vestnik Leningrad Univ. **23**, no. 13, (1968), 62–74 (Russian); Vestnik Leningrad Univ. Math **1**, no. 3, (1974), 231–245 (English).

[51] V.G. Maz'ja and V.P. Havin, *Non-linear potential theory*, Uspekhi Mat. Nauk **27**, no. 6, (1972), 67–138 (Russian); Russian Math. Surveys **27**, no. 6, (1972), 71–148 (English).

[52] V.G. Maz'ja and V.P. Havin, *Applications of (p, l)-capacity to some problems in the theory of exceptional sets*, Mat. Sb. **90**, no. 4, (1973), 558–591 (Russian); Math. USSR-Sb., no. 4, **19** (1973), 547–579 (English).

[53] M.S. Mel'nikov, *A bound for the Cauchy integral along an analytic curve*, Mat. Sb. **71** (1966), 503–515 (Russian); Amer. Math. Soc. Transl. **80** (1969), 243–255 (English).

[54] M.S. Mel'nikov, *On the Gleason parts of the algebra $R(X)$*, Mat. Sb. **101**, no. 2, (1976), 293–300 (Russian); Math. USSR-Sb. **30** (1976), 261–268 (English).

[55] M.S. Mel'nikov and S.O. Sinanjan, *Questions in the theory of approximation of functions of one complex variable*, Itogi Nauki i Tekhniki, Sovrem. Probl. Mat. **4**, VINITI, Moscow (1975), 143–250 (Russian); J. Soviet Math. **5** (1976), 688–752 (English).

[56] S.N. Mergeljan, *Uniform approximations to functions of a complex variable*, Uspekhi Mat. Nauk (N.S.) **7** (1952), 31–122 (Russian); Amer. Math. Soc. Transl. **101**, (1954), 294–391 (English).

[57] S.N. Mergeljan, *On the completeness of systems of analytic functions*, Uspekhi Mat. Nauk (N.S.) **8**, no. 4, (1953), 3–63 (Russian); Amer. Math. Soc. Transl. **19** (1962), 109–166 (English).

[58] S.N. Mergeljan, *General metric criteria for the completeness of systems of polynomials*, Dokl. Akad. Nauk SSSR **105**, no. 5, (1955), 901–904 (Russian).

[59] N.K. Nikolskii and V.P. Havin, *Results of V.I. Smirnov in complex analysis and their subsequent develpoment*, in V.I. Smirnov Selected Works: Complex Analysis, Mathematical theory of Diffraction, Publ. Leningrad Univ. (1988), 111–145 (Russian).

[60] Ch. Pommerenke, *Boundary Behavior of Conformal Maps*, Grundlehren Math. Wiss., vol. **299**, Springer, Berlin 1992.

[61] S. Shimorin, *A multiplier estimate of the Schwarzian derivative of univalent functions*, Internat. Math. Res. Notices **30** (2003), 1623–1633.

[62] N.A. Shirokov, *Properties of the moduli of analytic functions that are smooth up to the boundary*, Dokl. Akad. Nauk SSSR **269**, no. 6, 1320–1323, (1983), (Russian); Soviet Math. Dokl. **27**, no. 2, 507–510, (1983) (English).

[63] N.A. Shirokov, *Analytic Functions Smooth up to the Boundary*, Lecture Notes in Math. **1312**, Springer 1988.

[64] S.O. Sinanjan, *The uniqueness property of the analytic functions on closed sets without interior points*, Sibirsk. Mat. Zh. **6**, no. 6, (1965), 1365–1381 (Russian).

[65] S.O. Sinanjan, *Approximation by polynomials and analytic functions in the areal mean*, Mat. Sb. **69** (1966), 546–578 (Russian); Amer. Math. Soc. Transl. **74** (1968), 91–124 (English).

[66] J.E. Thomson, *Approximation in the mean by polynomials*, Ann. of Math. **133** (1991), 477–507.

[67] X. Tolsa, *Painlevé's problem and the semiadditivity of analytic capacity*, Acta Math. **190** (2003), 105–149.

[68] A.G. Vitushkin, *Analytic capacity of sets in problems of approximation theory*, Uspekhi Mat. Nauk **22** (1967), 141–199 (Russian); Russian Math. Surveys **22** (1967), 139–200 (English).

[69] A.L. Vol'berg, *The logarithm of an almost analytic function is summable*, Dokl. Akad. Nauk SSSR **265** (1982), 1297–1301 (Russian); Soviet Math. Dokl. **26** (1982), 238–243 (English).

[70] A.L. Vol'berg, *Asympotically holomorphic functions and certain of their applications*, Proc. International Congress of Mathematicians, Kyoto, Japan, 659–667, 1990.

[71] A.L. Vol'berg and B. Jöricke, *The summability of the logarithm of an almost analytic function and a generalization of a theorem of Levinson–Cartwright*, Mat. Sb. **130** (1986), 335–348 (Russian); Math. USSR-Sb. **58** (1987), 337–349 (English).

[72] J. Wermer, *Seminar über Funktionen-Algebren*, Lecture Notes in Math. **1**, Springer 1964.

J.E. Brennan
Department of Mathematics
University of Kentucky
Lexington, KY 40506, USA
e-mail: james.brennan@uky.edu

Operator Theory:
Advances and Applications, Vol. 261, 229–255
© Springer International Publishing AG, part of Springer Nature 2018

Various Sharp Estimates for Semi-discrete Riesz Transforms of the Second Order

K. Domelevo, A. Osękowski and S. Petermichl

Dedicated to the memory of V.P. Havin

Abstract. We give several sharp estimates for a class of combinations of second-order Riesz transforms on Lie groups $\mathbb{G} = \mathbb{G}_x \times \mathbb{G}_y$ that are multiply connected, composed of a discrete Abelian component $\mathbb{G}_x$ and a connected component $\mathbb{G}_y$ endowed with a biinvariant measure. These estimates include new sharp L^p estimates via Choi type constants, depending upon the multipliers of the operator. They also include weak-type, logarithmic and exponential estimates. We give an optimal $L^q \to L^p$ estimate as well.

It was shown recently by Arcozzi–Domelevo–Petermichl that such second-order Riesz transforms applied to a function may be written as conditional expectation of a simple transformation of a stochastic integral associated with the function.

The proofs of our theorems combine this stochastic integral representation with a number of deep estimates for pairs of martingales under strong differential subordination by Choi, Banuelos and Osękowski.

When two continuous directions are available, sharpness is shown via the laminates technique. We show that sharpness is preserved in the discrete case using Lax–Richtmyer theorem.

Mathematics Subject Classification (2010). Primary 42B20. Secondary 60G46.

Keywords. Riesz transforms, martingales, jump process, strong subordination, L^p.

Adam Osekowski is supported by Narodowe Centrum Nauki Poland (NCN), grant DEC-2014/14/E/ST1/00532
Stefanie Petermichl is supported by ERC project CHRiSHarMa DLV-862402.

1. Introduction

Sharp, classical L^p norm inequalities for pairs of differentially subordinate martingales date back to the celebrated work of Burkholder [15] in 1984 where the optimal constant is exhibited. See also from the same author [17, 18]. The relation between differentially subordinate martingales and CZ (i.e., Calderón–Zygmund) operators is known at least since Gundy–Varopoulos [32]. Banuelos–Wang [12] were the first to exploit this connection to prove new sharp inequalities for singular integrals. This intersection of probability theory with classical questions in harmonic analysis has lead to much interest and a vast literature has been accumulating on this line of research.

In this article we state a number of sharp estimates that hold in the very recent, new direction concerning the semi-discrete setting, applying it to a family of second-order Riesz transforms on multiply-connected Lie groups. We recall their representation through stochastic integrals using jump processes on multiply-connected Lie groups from [3]. In this representation formula jump processes play a role, but the strong differential subordination holds between the martingales representing the test function and the operator applied to the test function.

The usual procedure for obtaining (sharp) inequalities for operators of Calderón–Zygmund type from inequalities for martingales is the following. Starting with a test function f, martingales are built using Brownian motion or background noise and harmonic functions in the upper half-space $\mathbb{R}^+ \times \mathbb{R}^n$. Through the use of Itø formula, it is shown that the martingale arising in this way from Rf, where R is a Riesz transform in $\mathbb{R}^n$, is a martingale transform of the martingale arising from f. The two form a pair of martingales that have differential subordination and (in case of Hilbert or Riesz transforms) orthogonality. One then derives sharp martingale inequalities under hypotheses of strong differential subordination (and orthogonality) relations.

In the case of Riesz transforms of the second order, the use of heat extensions in the upper half-space instead of Poisson extensions originated in the context of a weighted estimate in Petermichl–Volberg [44] and was used to prove L^p estimates for the second order-Riesz transforms based on the results of Burkholder in Nazarov–Volberg [51] as part of their best-at-time estimate for the Beurling–Ahlfors operator, whose real and imaginary parts themselves are second-order Riesz transforms. We mention the recent version on discrete Abelian groups Domelevo–Petermichl [24] also using a type of heat flow. These proofs are deterministic. The technique of Bellman functions was used. This deterministic strategy does well when no orthogonality is present and when strong subordination is the only important property. Stochastic proofs (aside from giving better estimates in some situations) also have the advantage that once the integral representation is known, the proofs are a very concise consequence of the respective statements on martingales.

In [3] the authors proved sharp L^p estimates for semi-discrete second-order Riesz transforms R_α^2 using stochastic integrals. There is an array of Riesz transforms of the second order that are treated, indexed my a matrix index α (see below for precisions on acceptable α). The following representation formula of semi-discrete second-order Riesz transforms R_α^2 à la Gundy–Varopoulos (see [32]) is instrumental.

Theorem (Arcozzi–Domelevo–Petermichl, 2016). *The second-order Riesz transform $R_\alpha^2 f$ of a function $f \in L^2(\mathbb{G})$ as defined in* (1.1) *can be written as the conditional expectation*

$$\mathbb{E}(M_0^{\alpha,f}|\mathcal{Z}_0 = z).$$

Here $M_t^{\alpha,f}$ is a suitable martingale transform of a martingale M_t^f associated to f, and $\mathcal{Z}_t$ is a suitable random walk on $\mathbb{G}$

We remark that the L^p estimates of the discrete Hilbert transform on the integers are still open. It is a famous conjecture that this operator has the same norm as its continuous counterpart.

These known L^p norm inequalities use special functions found in the results of Pichorides [45], Verbitsky [50], Essén [28], Banuelos–Wang [12] when orthogonality is present in addition to differential subordination or Burkholder [15, 16, 17], Wang [52] when differential subordination is the only hypothesis.

The aim of the present paper is to establish new estimates for semi-discrete Riesz transforms by using the martingale representation above together with recent martingale inequalities found in the literature.

Here is a brief description of the new results in this paper.

- In the case where the function f is real-valued, we can obtain better estimates for R_α^2 than in the general case. These estimates depend upon the make of the matrix index α. The precise statement is found in Theorem 1.2.
- We prove a refined sharp weak type estimate using a weak type norm defined just before the statement of Theorem 1.3.
- We prove logarithmic and exponential estimates, in a sense limiting (in p) cases of the classical sharp L^p estimate. See Theorem 1.4.
- We consider the norm estimates of the $R_\alpha^2 : L^q \to L^p$, spaces of different exponent. The statement is found in Theorem 1.5.

1.1. Differential operators and Riesz transforms

First-order derivatives and tangent planes. We will consider Lie groups $\mathbb{G} := \mathbb{G}_x \times \mathbb{G}_y$, where $\mathbb{G}_x$ is a discrete Abelian group with a fixed set G of m generators, and their reciprocals, and $\mathbb{G}_y$ is a connected, Lie group of dimension n endowed with a biinvariant metric. The choice of the set G of generators in $\mathbb{G}_x$ corresponds to the choice of a biinvariant metric structure on $\mathbb{G}_x$. We will use on $\mathbb{G}_x$ the multiplicative notation for the group operation. We will define a product metric structure on $\mathbb{G}$, which agrees with the Riemannian structure on the first factor, and with the discrete "word distance" on the second. We will at the same time define a "tangent space" $T_z\mathbb{G}$ for $\mathbb{G}$ at a point $z = (x,y) \in (\mathbb{G}_x \times \mathbb{G}_y) = \mathbb{G}$. We will do this in three steps.

First, since $\mathbb{G}_y$ is an n-dimensional connected Lie group with Lie algebra $\mathfrak{G}_y$, we can identify each left-invariant vector field Y in $\mathfrak{G}_y$ with its value at the identity e, $\mathfrak{G}_y \equiv T_e\mathbb{G}_y$. Since $\mathbb{G}$ is compact, it admits a biinvariant Riemannian metric, which is unique up to a multiplicative factor. We normalize it so that the measure μ_y associated with the metric satisfies $\mu_y(\mathbb{G}_y) = 1$. The measure μ_y is also the normalized Haar measure of the group. We denote by $\langle\,\cdot\,,\,\cdot\,\rangle_y$ be the corresponding inner product on $T_y\mathbb{G}_y$ and by $\nabla_{\mathbf{y}}f(y)$ the gradient at $y \in \mathbb{G}_y$ of a smooth function $f : \mathbb{G}_y \to \mathbb{R}$. Let $Y_1, \ldots, Y_n$ be an orthonormal basis for $\mathfrak{G}_y$. The gradient of f can be written in the form $\nabla_{\mathbf{y}}f = Y_1(f)Y_1 + \cdots + Y_n(f)Y_n$.

Second, in the discrete component $\mathbb{G}_x$, let $\mathfrak{G}_x = (g_i)_{i=1,\ldots,m}$ be a set of generators for $\mathbb{G}_x$ such that for $i \neq j$ and $\sigma = \pm 1$ we have $g_i \neq g_j^\sigma$. The choice of a particular set of generators induces a word metric, hence, a geometry, on $\mathbb{G}_x$. Any two sets of generators induce bi-Lipschitz equivalent metrics.

At any point $x \in \mathbb{G}_x$, and given a direction $i \in \{1, \ldots, m\}$, we can define the right and the left derivative at x in the direction i:

$$(\partial^+ f/\partial x_i)(x,y) := f(x + g_i, y) - f(x,y) := (\partial_i^+ f)(x,y)$$
$$(\partial^- f/\partial x_i)(x,y) := f(x,y) - f(x - g_i, y) := (\partial_i^- f)(x,y).$$

Comparing with the continuous component, this suggests that the tangent plane $\widehat{T}_x\mathbb{G}_x$ at a point x of the discrete group $\mathbb{G}_x$ might actually be split into a "right" tangent plane $T_x^+\mathbb{G}_x$ and a "left" tangent plane $T_x^-\mathbb{G}_x$, according to the direction with respect to which discrete differences are computed. We consequently define the **augmented** discrete gradient $\widehat{\nabla}_x f(x)$, with a *hat*, as the $2m$-vector of $\widehat{T}_x\mathbb{G}_x :=$ $T_x^+\mathbb{G}_x \oplus T_x^-\mathbb{G}_x$ accounting for all the local variations of the function f in the direct vicinity of x; that is, the $2m$-column-vector

$$\widehat{\nabla}_x f(x) := (X_1^+ f, X_2^+ f, \ldots, X_1^- f, X_2^- f, \ldots)(x) = \sum_{i=1}^m \sum_{\tau=\pm} X_i^\tau f(x)$$

with $X_i^\tau \in \widehat{T}_x\mathbb{G}_x$, where we noted the discrete derivatives $X_i^\pm f := \partial_i^\pm f$ and introduced the discrete $2m$-vectors $X_i^\pm$ as the column vectors of $\mathbb{Z}^{2m}$

$$X_i^+ = (0, \ldots, 1, \ldots, 0) \times \mathbf{0}_m, \quad X_i^- = \mathbf{0}_m \times (0, \ldots, 1, \ldots, 0).$$

Here the 1's in $X_i^\pm$ are located at the ith position of respectively the first or the second m-tuple. Notice that those vectors are independent of the point x. The scalar product on $\widehat{T}_x\mathbb{G}_x := T_x^+\mathbb{G}_x \oplus T_x^-\mathbb{G}_x$ is defined as

$$(U, V)_{\widehat{T}_x\mathbb{G}_x} := \frac{1}{2} \sum_{i=1}^m \sum_{\tau=\pm} U_i^\tau V_i^\tau.$$

We chose to put a factor $\frac{1}{2}$ in front of the scalar product to compensate for the fact that we consider both left and right differences.

Finally, for a function f defined on the Cartesian product $\mathbb{G} := \mathbb{G}_x \times \mathbb{G}_y$, the (augmented) gradient $\widehat{\nabla}_z f(z)$ at the point $z = (x, y)$ is an element of the tangent plane $\widehat{T}_z\mathbb{G} := \widehat{T}_x\mathbb{G}_x \oplus T_y\mathbb{G}_y$, that is a $(2m+n)$-column-vector

$$\widehat{\boldsymbol{\nabla}}_z f(z) := \sum_{i=1}^m \sum_{\tau=\pm} X_i^\tau f(z)\widehat{X}_i^\tau + \sum_{j=1}^n Y_j f(z)\widehat{Y}_j(z)$$
$$= (X_1^+ f, X_2^+ f, \ldots, X_1^- f, X_2^- f, \ldots, Y_1 f, Y_2 f, \ldots)(z)$$

where $\widehat{X}_i^\tau$ and $\widehat{Y}_j(z)$ can be identified with column vectors of size $(2m+n)$ with obvious definitions and scalar product $(\,\cdot\,,\,\cdot\,)_{\widehat{T}_z\mathbb{G}_z}$.

Let $d\mu_z := d\mu_x\, d\mu_y$, $d\mu_x$ be the counting measure on $\mathbb{G}_x$ and $d\mu_y$ the Haar measure on $\mathbb{G}_y$. The inner product of φ, ψ in $L^2(\mathbb{G})$ is

$$(\varphi,\psi)_{L^2(\mathbb{G})} := \int_{\mathbb{G}} \varphi(z)\psi(z)d\mu_z(z).$$

Finally, we adopt the following hypotheses.

Hypothesis. We assume everywhere in the sequel:
1) the discrete component $\mathbb{G}_x$ of the Lie group $\mathbb{G}$ is an Abelian group;
2) the connected component $\mathbb{G}_y$ of the Lie group $\mathbb{G}$ is a Lie group that can be endowed with a biinvariant Riemannian metric, so that the family $(Y_j)_{j=1,\ldots,n}$ commutes with Δ_y.

Notice that this includes compact Lie groups $\mathbb{G}_y$ since those can be endowed with a biinvariant metric. It also includes the usual Euclidian spaces since those are commutative.

Riesz transforms. Following [1, 2], recall first that for a compact Riemannian manifold $\mathbb{M}$ without boundary, one denotes by $\boldsymbol{\nabla}_{\mathbb{M}}$, $\mathrm{div}_{\mathbb{M}}$ and $\Delta_{\mathbb{M}} := \mathrm{div}_{\mathbb{M}} \boldsymbol{\nabla}_{\mathbb{M}}$ respectively the gradient, the divergence and the Laplacian associated with $\mathbb{M}$. Then $-\Delta_{\mathbb{M}}$ is a positive operator and the vector Riesz transform is defined as the linear operator

$$\mathbf{R}_{\mathbb{M}} := \boldsymbol{\nabla}_{\mathbb{M}} \circ (-\Delta_{\mathbb{M}})^{-1/2}$$

acting on $L_0^2(\mathbb{M})$ (L^2 functions with vanishing mean). It follows that if f is a function defined on $\mathbb{M}$ and $y \in \mathbb{M}$ then $\mathbf{R}_{\mathbb{M}} f(y)$ is a vector of the tangent plane $T_y\mathbb{M}$.

Similarly on $\mathbb{M} = \mathbb{G}$, we define $\boldsymbol{\nabla}_{\mathbb{G}} := \widehat{\boldsymbol{\nabla}}_z$ as before, and then we define the divergence operator as its formal adjoint, that is $-\mathrm{div}_{\mathbb{G}} = -\widehat{\mathrm{div}}_z := \widehat{\boldsymbol{\nabla}}_z^*$, with respect to the natural L^2 inner product of vector fields:

$$(U,V)_{L^2(\widehat{T}\mathbb{G})} := \int_{\mathbb{G}} (U(z), V(z))_{\widehat{T}_z\mathbb{G}}\, d\mu_z(z)$$

We have the L^2-adjoints $(X_i^\pm)^* = -X_i^\mp$ and $Y_j^* = -Y_j$. If $U \in \widehat{T}\mathbb{G}$ is defined by

$$U(z) = \sum_{i=1}^m \sum_{\tau=\pm} U_i^\tau(z)\widehat{X}_i^\tau + \sum_{j=1}^n U_j(z)\widehat{Y}_j,$$

we define its divergence $\widehat{\boldsymbol{\nabla}}_z^* U$ as

$$\widehat{\boldsymbol{\nabla}}_z^* U(z) := -\frac{1}{2} \sum_{i=1}^m \sum_{\tau=\pm} X_i^{-\tau} U_i^\tau(z) - \sum_{j=1}^n Y_j U_j(z).$$

The Laplacian $\Delta_{\mathbb{G}}$ is as one might expect:

$$\begin{aligned}
\Delta_z f(z) &:= -\widehat{\boldsymbol{\nabla}}_z^* \widehat{\boldsymbol{\nabla}}_z f(z) = -\widehat{\boldsymbol{\nabla}}_x^* \widehat{\boldsymbol{\nabla}}_x f(z) - \widehat{\boldsymbol{\nabla}}_y^* \widehat{\boldsymbol{\nabla}}_y f(z) \\
&= \sum_{i=1}^m X_i^- X_i^+ f(z) + \sum_{j=1}^n Y_j^2 f(z) \\
&= \sum_{i=1}^m X_i^2 f(z) + \sum_{j=1}^n Y_j^2 f(z) \\
&=: \Delta_x f(z) + \Delta_y f(z)
\end{aligned}$$

where we have denoted $X_i^2 := X_i^+ X_i^- = X_i^- X_i^+$. We have chosen signs so that $-\Delta_{\mathbb{G}} \geq 0$ as an operator. The Riesz vector $(\widehat{\mathbf{R}}_z f)(z)$ is the $(2m+n)$-column-vector of the tangent plane $\widehat{T}_z \mathbb{G}$ defined as the linear operator

$$\widehat{\mathbf{R}}_z f := \left(\widehat{\boldsymbol{\nabla}}_z f\right) \circ (-\Delta_z f)^{-1/2}.$$

We also define transforms along the coordinate directions:

$$R_i^\pm = X_i^\pm \circ (-\Delta_z)^{-1/2} \quad \text{and} \quad R_j = Y_j \circ (-\Delta_z)^{-1/2}.$$

Plan of the paper. In the next two sections, we present successively the main results of the paper and recall the weak formulations involving second-order Riesz transforms and semi-discrete heat extensions. Section 2 introduces the stochastic setting for our problems. This includes in Subsection 2.1 semi-discrete random walks, martingale transforms and quadratic covariations. Subsection 2.2 presents a set of martingale inequalities already known in the literature. Finally, in Section 3 we give the proof of the main results.

1.2. Main results

In this text, we are concerned with second-order Riesz transforms and combinations thereof. We first define the square Riesz transform in the (discrete) direction i to be

$$R_i^2 := R_i^+ R_i^- = R_i^- R_i^+.$$

Then, given $\alpha := ((\alpha_i^x)_{i=1...m}, (\alpha_{jk}^y)_{j,k=1...n}) \in \mathbb{C}^m \times \mathbb{C}^{n \times n}$, we define R_α^2 to be the following combination of second-order Riesz transforms:

$$R_\alpha^2 := \sum_{i=1}^m \alpha_i^x R_i^2 + \sum_{j,k=1}^n \alpha_{jk}^y R_j R_k, \tag{1.1}$$

where the first sum involves squares of discrete Riesz transforms as defined above, and the second sum involves products of continuous Riesz transforms. This combination is written in a condensed manner as the quadratic form

$$R_\alpha^2 = (\widehat{\mathbf{R}}_z, \mathbf{A}_\alpha \widehat{\mathbf{R}}_z)$$

where $\mathbf{A}_\alpha$ is the $(2m+n) \times (2m+n)$ block matrix

$$\mathbf{A}_\alpha := \begin{pmatrix} \mathbf{A}_\alpha^x & \mathbf{0} \\ \mathbf{0} & \mathbf{A}_\alpha^y \end{pmatrix} \tag{1.2}$$

with

$$\mathbf{A}_\alpha^x = \mathrm{diag}(\alpha_1^x, \ldots, \alpha_m^x, \alpha_1^x, \ldots, \alpha_m^x) \in \mathbb{C}^{2m \times 2m}, \mathbf{A}_\alpha^y = (\alpha_{jk}^y)_{j,k=1\ldots n} \in \mathbb{C}^{n \times n}.$$

In the theorems below, we assume that $\mathbb{G}$ is a Lie group and R_α^2 is a combination of second-order Riesz transforms as defined above. The first application of the stochastic integral formula, Theorem 1.1 was done in [3], while the other applications, Theorems 1.2, 1.3, 1.4 and 1.5 are new.

Theorem 1.1 (Arcozzi–Domelevo–Petermichl, 2016). *For any $1 < p < \infty$ we have*

$$\|R_\alpha^2\|_p \leq \|\mathbf{A}_\alpha\|_2 (p^* - 1),$$

where, as previously, $p^* = \max\{p, p/(p-1)\}$.

Above, we have set:

$$\|\mathbf{A}_\alpha\|_2 = \max(\|\mathbf{A}_\alpha^x\|_2, \|\mathbf{A}_\alpha^y\|_2) = \max(|\alpha_1^x|, \ldots, |\alpha_m^x|, \|\mathbf{A}_\alpha^y\|_2).$$

In the case where $\mathbb{G} = \mathbb{G}_x$ only consists of the discrete component, this was proved in [25, 24] using the deterministic Bellman function technique. In the case where $\mathbb{G} = \mathbb{G}_y$ is a connected compact Lie group, this was proved in [8] using Brownian motions defined on manifolds and projections of martingale transforms.

In the case where the function f is real-valued, we can obtain better estimates. For any real numbers $a < b$ and any $1 < p < \infty$, let $\mathfrak{C}_{a,b,p}$ be the constants introduced in Bañuelos and Osękowski [10].

Theorem 1.2. *Assume that $a\mathbf{I} \leq \mathbf{A}_\alpha \leq b\mathbf{I}$ in the sense of quadratic forms. Then $R_\alpha^2 : L^p(\mathbb{G}, \mathbb{R}) \to L^p(\mathbb{G}, \mathbb{R})$ enjoys the norm estimate $\|R_\alpha^2\|_p \leq \mathfrak{C}_{a,b,p}$.*

We should point out here that the constants $\mathfrak{C}_{a,b,p}$ appear in earlier works of Burkholder [15] (for $a = -b$: then $\mathfrak{C}_{a,b,p} = b(p^* - 1)$), and in the paper [19] by Choi (in the case when one of a, b is zero). The Choi constants are not explicit; an approximation of $\mathfrak{C}_{0,1,p}$ is known and writes as

$$\mathfrak{C}_{0,1,p} = \frac{p}{2} + \frac{1}{2}\log\left(\frac{1+e^{-2}}{2}\right) + \frac{\beta_2}{p} + \cdots,$$

with $\beta_2 = \log^2\left(\frac{1+e^{-2}}{2}\right) + \frac{1}{2}\log\left(\frac{1+e^{-2}}{2}\right) - 2\left(\frac{e^{-2}}{1+e^{-2}}\right)^2$.

Coming back to complex-valued functions, we will also establish the following weak-type bounds. We consider the norms

$$\|f\|_{L^{p,\infty}(\mathbb{G},\mathbb{C})} = \sup\left\{\mu_z(E)^{1/p-1} \int_E f \, d\mu_z\right\},$$

where the supremum is taken over the class of all measurable subsets E of $\mathbb{G}$ of positive measure.

Theorem 1.3. *For any $1 < p < \infty$ we have*

$$\|R_\alpha^2\|_{L^p(\mathbb{G},\mathbb{C}) \to L^{p,\infty}(\mathbb{G},\mathbb{C})} \leq \|\mathbf{A}_\alpha\|_2 \cdot \begin{cases} \left(\frac{1}{2}\Gamma\left(\frac{2p-1}{p-1}\right)\right)^{1-1/p} & \text{if } 1 < p \leq 2, \\ \left(\frac{p^{p-1}}{2}\right)^{1/p} & \text{if } p \geq 2. \end{cases}$$

We will also prove the following logarithmic and exponential estimates, which can be regarded as versions of Theorem 1.1 for $p = 1$ and $p = \infty$. Consider the Young functions $\Phi, \Psi : [0, \infty) \to [0, \infty)$, given by $\Phi(t) = e^t - 1 - t$ and $\Psi(t) = (t+1)\log(t+1) - t$.

Theorem 1.4. *Let $K > 1$ be fixed.*

(i) *For any measurable subset E of $\mathbb{G}$ and any f on $\mathbb{G}$ we have*

$$\int_E |R_\alpha^2 f| d\mu_z \leq \|\mathbf{A}_\alpha\|_2 \cdot \left(K \int_{\mathbb{G}} \Psi(|f|) d\mu_z + \frac{\mu_z(E)}{2(K-1)}\right).$$

(ii) *For any $f : \mathbb{G} \to \mathbb{C}$ bounded by 1,*

$$\int_{\mathbb{G}} \Phi\left(\frac{|R_\alpha^2 f|}{K\|\mathbf{A}_\alpha\|_2}\right) d\mu_z \leq \frac{\|f\|_{L^1(\mathbb{G},\mathbb{C})}}{2K(K-1)}.$$

Our final result concerns another extension of Theorem 1.1, which studies the action of R_α^2 between two different L^p spaces. For $1 \leq p < q < \infty$, let $C_{p,q}$ be the constant defined by Osękowski in [40].

Theorem 1.5. *For any $1 \leq p < q < \infty$, any measurable subset E of $\mathbb{G}$ and any $f \in L^q(\mathbb{G})$ we have*

$$\|R_\alpha^2 f\|_{L^p(E,\mathbb{C})} \leq C_{p,q} \|\mathbf{A}_\alpha\|_2 \|f\|_{L^q(\mathbb{G},\mathbb{C})} \mu_z(E)^{1/p-1/q}.$$

An interesting feature is that all the estimates in the five theorems above are sharp when $\mathbb{G} = \mathbb{G}_x \times \mathbb{G}_y$ and $\dim(\mathbb{G}_y) + \dim^\infty(\mathbb{G}_x) \geq 2$, where $\dim^\infty(\mathbb{G}_x)$ denotes the number of infinite components of $\mathbb{G}_x$.

1.3. Weak formulations

Let $f : \mathbb{G} \to \mathbb{C}$ be given. The heat extension $\tilde{f}(t)$ of f is defined as $\tilde{f}(t) := e^{t\Delta_z} f =: P_t f$. We have therefore $\tilde{f}(0) = f$. Our aim in this section is to derive weak formulations for second-order Riesz transforms. We start with the weak formulation of the identity operator $\mathcal{I}$, that is obtained by using semi-discrete heat extensions (see [3] for details).

Assume f in $L^2(\mathbb{G})$ and g in $L^2(\mathbb{G})$. Let $\bar{f}$ be the average of f on $\mathbb{G}$ if $\mathbb{G}$ has finite measure and zero otherwise. Then

$$(\mathcal{I}f, g) = (f, g)_{L^2(\mathbb{G})}$$

$$= \bar{f}\bar{g} + 2\int_0^\infty \left(\widehat{\boldsymbol{\nabla}}_z P_t f, \widehat{\boldsymbol{\nabla}}_z P_t g\right)_{L^2(\widehat{T}\mathbb{G})} dt$$

$$= \bar{f}\bar{g} + 2\int_0^\infty \int_{z \in \mathbb{G}} \left\{\frac{1}{2}\sum_{i=1}^m \sum_{\tau = \pm} (X_i^\tau P_t f)(z)(X_i^\tau P_t g)(z)\right.$$

$$\left. + \sum_{j=1}^n (Y_j P_t f)(z)(Y_j P_t g)(z)\right\} d\mu_z(z)\, dt$$

and the sums and integrals that arise converge absolutely.

In order to pass to the weak formulation for the squares of Riesz transforms, we first observe that the following commutation relations hold

$$Y_j \circ \Delta_z = \Delta_z \circ Y_j$$
$$X_i^\tau \circ \Delta_z = \Delta_z \circ X_i^\tau, \tau \in \{+, -\}$$

This is an easy consequence of the hypothesis made on the Lie group. In accordance with [3], the following weak formulation for second-order Riesz transforms holds.

Assume f in $L^2(\mathbb{G})$ and g in $L^2(\mathbb{G})$, then

$$(R_\alpha^2 f, g)_{L^2(\mathbb{G})} = -2\int_0^\infty \left(\mathbf{A}_\alpha \widehat{\boldsymbol{\nabla}}_z P_t f, \widehat{\boldsymbol{\nabla}}_z P_t g\right)_{L^2(\widehat{T}\mathbb{G})} dt$$

$$= -2\int_0^\infty \int_{z \in \mathbb{G}} \left\{\frac{1}{2}\sum_{i=1}^m \sum_{\tau = \pm} \alpha_i^x (X_i^\tau P_t f)(z)(X_i^\tau P_t g)(z)\right.$$

$$\left. + \sum_{j,k=1}^n \alpha_{jk}^y (Y_j P_t f)(z)(Y_k P_t g)(z)\right\} d\mu_z(z)\, dt$$

and the sums and integrals that arise converge absolutely.

2. Stochastic integrals and martingale transforms

In what follows, we assume that we have a complete probability space $(\Omega, \mathcal{F}, \mathbb{P})$ with a càdlàg (i.e., right continuous left limit) filtration $(\mathcal{F}_t)_{t \geq 0}$ of sub-σ-algebras of $\mathcal{F}$. We assume as usual that $\mathcal{F}_0$ contains all events of probability zero. All random walks and martingales are adapted to this filtration.

We define below a continuous-time random process $\mathcal{Z}$ with values in $\mathbb{G}$, $\mathcal{Z}_t := (\mathcal{X}_t, \mathcal{Y}_t) \in \mathbb{G}_x \times \mathbb{G}_y$, having infinitesimal generator $L = \Delta_z$. The pure-jump component $\mathcal{X}_t$ is a compound Poisson jump process on the discrete set $\mathbb{G}_x$, whereas the continuous component $\mathcal{Y}_t$ is a standard Brownian motion on the manifold $\mathbb{G}_y$. Then, Itô's formula ensures that semi-discrete "harmonic" functions

$f : \mathbb{R}^+ \times \mathbb{G} \to \mathbb{C}$ solving the backward heat equation $(\partial_t + \Delta_z)f = 0$ give rise to martingales $M_t^f := f(t, \mathcal{Z}_t)$ for which we define a class of martingale transforms.

2.1. Stochastic integrals, martingale transforms and quadratic covariations

Stochastic integrals on Riemannian manifolds and Itô integral. Following Emery [26, 27], see also Arcozzi [1, 2], we define the Brownian motion $\mathcal{Y}_t$ on $\mathbb{G}_y$, a compact Riemannian manifold, as the process $\mathcal{Y}_t : \Omega \to (0, T) \times \mathbb{G}_y$ such that for all smooth functions $f : \mathbb{G}_y \to \mathbb{R}$, the quantity

$$f(\mathcal{Y}_t) - f(\mathcal{Y}_0) - \frac{1}{2} \int_0^t (\Delta_y f)(\mathcal{Y}_s) \, \mathrm{d}s =: (I_{\mathrm{d}_y f})_t \qquad (2.1)$$

is an $\mathbb{R}$-valued continuous martingale. For any adapted continuous process Ψ with values in the cotangent space $T^* \mathbb{G}_y$ of $\mathbb{G}_y$, if $\Psi_t(\omega) \in T^*_{Y_t(\omega)} \mathbb{G}_y$ for all $t \geq 0$ and $\omega \in \Omega$, then one can define the *continuous* Itô integral I_Ψ of Ψ as

$$(I_\Psi)_t := \int_0^t \langle \Psi_s, \mathrm{d}\mathcal{Y}_s \rangle$$

so that in particular

$$(I_{\mathrm{d}_y f})_t := \int_0^t \langle \mathrm{d}_y f(\mathcal{Y}_s), \mathrm{d}\mathcal{Y}_s \rangle$$

The integrand above involves the 1-form of $T^*_y \mathbb{G}_y$

$$\mathrm{d}_y f(y) := \sum_j (Y_j f)(y) \, Y_j^*.$$

A pure jump process on $\mathbb{G}_x$. We will now define the *discrete* m-dimensional process $\mathcal{X}_t$ on the *discrete* Abelian group $\mathbb{G}_x$ as a generalized compound Poisson process. In order to do this we need a number of independent variables and processes:

First, for any given $1 \leq i \leq m$, let $\mathcal{N}_t^i$ be a càdlàg Poisson process of parameter λ, that is

$$\forall t, \mathbb{P}(\mathcal{N}_t^i = n) = \frac{(\lambda t)^n}{n!} e^{-\lambda t}.$$

The sequence of instants where the jumps of the $\mathcal{N}_t^i$ occur is denoted by $(T_k^i)_{k \in \mathbb{N}}$, with the convention $T_0^i = 0$.

Second, we set

$$\mathcal{N}_t = \sum_{i=1}^m \mathcal{N}_t^i$$

Almost surely, for any two distinct i and j, we have $\{T_k^i\}_{k \in \mathbb{N}} \cap \{T_k^j\}_{k \in \mathbb{N}} = \emptyset$. Let therefore $\{T_k\}_{k \in \mathbb{N}} = \bigcup_{i=1}^m \{T_k^i\}_{k \in \mathbb{N}}$ be the ordered sequence of instants of jumps of $\mathcal{N}_t$ and let $i_t \equiv i_t(\omega)$ be the index of the coordinate where the jump occurs at

time t. We set $i_t = 0$ if no jump occurs. The random variables i_t are measurable:
$i_t = (\mathcal{N}_t^1 - \mathcal{N}_{t-}^1, \mathcal{N}_t^2 - \mathcal{N}_{t-}^2, \ldots, \mathcal{N}_t^m - \mathcal{N}_{t-}^m) \cdot (1, 2, \ldots, m)$. In a differential form,

$$\mathrm{d}\mathcal{N}_t = \sum_{i=1}^{m} \mathrm{d}\mathcal{N}_t^i = \mathrm{d}\mathcal{N}_t^{i_t}.$$

Third, we denote by $(\tau_k)_{k \in \mathbb{N}}$ a sequence of independent Bernoulli variables

$$\forall k, \mathbb{P}(\tau_k = 1) = \mathbb{P}(\tau_k = -1) = 1/2.$$

Finally, the random walk $\mathcal{X}_t$ started at $\mathcal{X}_0 \in \mathbb{G}_x$ is the càdlàg compound Poisson process (see, e.g., Protter [48], Privault [46, 47]) defined as

$$\mathcal{X}_t := \mathcal{X}_0 + \sum_{k=1}^{\mathcal{N}_t} G_{i_k}^{\tau_k},$$

where $G_i^\tau = (0, \ldots, 0, \tau g_i, 0, \ldots, 0)$ when $i \neq 0$ and $(0, \ldots, 0)$ when $i = 0$.

Stochastic integrals on discrete groups. We recall for the convenience of the reader the derivation of stochastic integrals for jump processes. We will emphasize the fact that Itô's corresponding formula involves the action of a discrete 1-form written in a well-chosen local coordinate system of the discrete *augmented* cotangent plane (see details below). Let $1 \leq k < \mathcal{N}_t$ and let (T_k, i_k, τ_k) be respectively the instant, the axis and the direction of the kth jump. We set $T_0 = 0$. Let $f := f(t, x)$, $t \in \mathbb{R}^+$, $x \in \mathbb{G}_x$ be a function defined on $\mathbb{R}^+ \times \mathbb{G}_x$. Then

$$f(t, \mathcal{X}_t) - f(0, \mathcal{X}_0) = \int_0^t (\partial_t f)(s, \mathcal{X}_s)\, \mathrm{d}s + \sum_{i=1}^{m} \int_0^t (f(s, \mathcal{X}_s) - f(s, \mathcal{X}_{s-}))\, \mathrm{d}\mathcal{N}_s^i.$$

At an instant s, the integrand in the last term writes as

$$\begin{aligned}
(f(s, \mathcal{X}_s) - f(s, \mathcal{X}_{s-}))\, \mathrm{d}\mathcal{N}_s^i &= (f(s, \mathcal{X}_{s-} + G_i^{\tau_{\mathcal{N}_s}}) - f(s, \mathcal{X}_{s-}))\, \mathrm{d}\mathcal{N}_s^i \\
&= (X_i^{\tau_{\mathcal{N}_s}} f)(s, \mathcal{X}_{s-})\tau_{\mathcal{N}_s}\, \mathrm{d}\mathcal{N}_s^i \\
&= \frac{1}{2}\{(X_i^2 f)(s, \mathcal{X}_{s-}) + \tau_{\mathcal{N}_s}(X_i^0 f)(s, \mathcal{X}_{s-})\}\, \mathrm{d}\mathcal{N}_s^i
\end{aligned}$$

where we have introduced, for all $1 \leq i \leq m$,

$$X_i^0 := X_i^+ + X_i^-$$
$$X_i^2 := X_i^+ - X_i^-.$$

Notice that, for any given $1 \leq i \leq m$, up to a normalisation factor, the system of coordinates (X_i^2, X_i^0) is obtained thanks to a *rotation* through $\pi/4$ of the canonical system of coordinates (X_i^+, X_i^-). Finally,

$$
\begin{aligned}
f(t, \mathcal{X}_t) - f(0, \mathcal{X}_0) \\
= \int_0^t \left\{ (\partial_t f)(s, \mathcal{X}_s) + \frac{\lambda}{2}(\Delta_x f)(s, \mathcal{X}_s) \right\} \mathrm{d}s + \int_0^t \langle \widehat{\mathbf{d}} f(s, \mathcal{X}_{s_-}), \mathrm{d}\widehat{\mathcal{W}}_s \rangle \\
=: \int_0^t \left\{ (\partial_t f)(s, \mathcal{X}_s) + \frac{\lambda}{2}(\Delta_x f)(s, \mathcal{X}_s) \right\} \mathrm{d}s + \left(I_{\widehat{\mathbf{d}}_x f} \right)_t.
\end{aligned}
$$

where we set $\mathrm{d}\mathcal{X}_s^i = \tau_{\mathcal{N}_s} \mathrm{d}\mathcal{N}_s^i$. It is easy to see that $\mathrm{d}\mathcal{X}_s^i$ is the stochastic differential of a martingale. Here and in the sequel, we take $\lambda = 2$.

Discrete Itô integral. The stochastic integral above shows that the Itô formula (2.1) for continuous processes has a discrete counterpart involving stochastic integrals for jump processes, namely we have the *discrete* Itô integral

$$
\left(I_{\widehat{\mathbf{d}}_x f} \right)_t := \frac{1}{2} \sum_{i=1}^m \int_0^t (X_i^2 f)(s, \mathcal{X}_{s_-}) \, \mathrm{d}(\mathcal{N}_s^i - \lambda s) + (X_i^0 f)(s, \mathcal{X}_{s_-}) \, \mathrm{d}\mathcal{X}_s
$$

This has a more intrinsic expression similar to the continuous Itô integral (2.1). If we regard the discrete component $\mathbb{G}_x$ as a "discrete Riemannian" manifold, then this discrete Itô integral involves discrete vectors (respectively, 1-forms) defined on the *augmented* discrete tangent (respectively, cotangent) space $\widehat{T}_x \mathbb{G}_x$ (respectively, $\widehat{T}_x^* \mathbb{G}_x$) of dimension $2m$ defined as

$$
\begin{aligned}
\widehat{T}_x \mathbb{G}_x &= \operatorname{span}\{ X_1^+, X_2^+, \ldots, X_1^-, X_2^-, \ldots \} \\
&= \operatorname{span}\{ X_1^2, X_2^2, \ldots, X_1^0, X_2^0, \ldots \} \\
\widehat{T}_x^* \mathbb{G}_x &= \operatorname{span}\{ (X_1^+)^*, (X_2^+)^*, \ldots, (X_1^-)^*, (X_2^-)^*, \ldots \} \\
&= \operatorname{span}\{ (X_1^2)^*, (X_2^2)^*, \ldots, (X_1^0)^*, (X_2^0)^*, \ldots \}.
\end{aligned}
$$

Let $\mathrm{d}\widehat{\mathcal{W}}_s \in \widehat{T}_{\mathcal{X}_s} \mathbb{G}_x$ be the vector and $\widehat{\mathbf{d}} f \in \widehat{T}_{\mathcal{X}_s}^* \mathbb{G}_x$ the 1-form respectively defined as:

$$
\begin{aligned}
\mathrm{d}\widehat{\mathcal{W}}_s &= \mathrm{d}(\mathcal{N}_s^1 - \lambda s) X_1^2 + \cdots + \mathrm{d}(\mathcal{N}_s^m - \lambda s) X_m^2 + \mathrm{d}\mathcal{X}_s^1 X_1^0 + \cdots + \mathrm{d}\mathcal{X}_s^m X_m^0 \\
\widehat{\mathbf{d}}_x f &= X_1^2 f (X_1^2)^* + \cdots + X_m^2 f (X_m^2)^* + X_1^0 f (X_1^0)^* + \cdots + X_m^0 f (X_m^0)^*
\end{aligned}
$$

With these notation, we have

$$
\left(I_{\widehat{\mathbf{d}}_x f} \right)_t := \langle \widehat{\mathbf{d}}_x f, \mathrm{d}\widehat{\mathcal{W}}_s \rangle_{\widehat{T}_x^* \mathbb{G}_x \times \widehat{T}_x \mathbb{G}_x}
$$

where the factor $1/2$ is included in the pairing $\langle \cdot, \cdot \rangle_{\widehat{T}_x^* \mathbb{G}_x \times \widehat{T}_x \mathbb{G}_x}$.

Semi-discrete stochastic integrals. Let finally $\mathcal{Z}_t = (\mathcal{X}_t, \mathcal{Y}_t)$ be a semi-discrete random walk on the Cartesian product $\mathbb{G} = \mathbb{G}_x \times \mathbb{G}_y$, where $\mathcal{X}_t$ is the random walk defined above on $\mathbb{G}_x$ with generator Δ_x and where $\mathcal{Y}_t$ is the Brownian motion defined on $\mathbb{G}_y$ with generator Δ_y. For $f := f(t, z) = f(t, x, y)$ defined from

$\mathbb{R}^+ \times \mathbb{G}$ onto $\mathbb{C}$, we have easily the stochastic integral involving both discrete and continuous parts:

$$f(t, \mathcal{Z}_t) = \int_0^t \{(\partial_t f)(s, \mathcal{Z}_s) + (\Delta_z f)(s, \mathcal{Z}_s)\}\, ds + (I_{\hat{\mathbf{d}}_z f})_t$$

where the *semi-discrete* Itô integral writes as

$$
\begin{aligned}
(I_{\hat{\mathbf{d}}_z f})_t &:= (I_{\hat{\mathbf{d}}_x f})_t + (I_{\mathbf{d}_y f})_t \\
&:= \int_0^t \langle \hat{\mathbf{d}}_x f(s, \mathcal{Z}_{s_-}), d\widehat{\mathcal{W}}_s \rangle_{\widehat{T}^*_{\mathcal{X}_s}\mathbb{G}_x \times \widehat{T}_{\mathcal{X}_s}\mathbb{G}_x} \\
&\quad + \int_0^t \langle \mathbf{d}_y f(s, \mathcal{Z}_{s_-}), d\mathcal{Y}_s \rangle_{\widehat{T}^*_{\mathcal{Y}_s}\mathbb{G}_y \times \widehat{T}_{\mathcal{Y}_s}\mathbb{G}_y}.
\end{aligned}
$$

Martingale transforms. We are interested in martingale transforms allowing us to represent second-order Riesz transforms. Let $f(t, z)$ be a solution to the heat equation $\partial_t - \Delta_z = 0$. Fix $T > 0$ and $\mathcal{Z}_0 \in \mathbb{G}$. Then define

$$\forall 0 \le t \le T, M_t^{f, T, \mathcal{Z}_0} = f(T - t, \mathcal{Z}_t).$$

This is a martingale since $f(T-t)$ solves the backward heat equation $\partial_t + \Delta_z = 0$, and we have in terms of stochastic integrals

$$M_t^{f, T, \mathcal{Z}_0} = f(T - t, \mathcal{Z}_t) = f(T, \mathcal{Z}_0) + \int_0^t \langle \hat{\mathbf{d}}_z f(T - s, \mathcal{Z}_{s_-}), d\mathcal{Z}_s \rangle$$

Given $\mathbf{A}_\alpha$ the $\mathbb{C}^{(2m+n) \times (2m+n)}$ matrix defined earlier, we denote by $M_t^{\alpha, f, T, \mathcal{Z}_0}$ the martingale transform $\mathbf{A}_\alpha * M_t^{f, T, \mathcal{Z}_0}$ defined as

$$
\begin{aligned}
M_t^{\alpha, f, T, \mathcal{Z}_0} &:= f(T, \mathcal{Z}_0) + \int_0^t (\mathbf{A}_\alpha \widehat{\boldsymbol{\nabla}}_z f(s, \mathcal{Z}_{s_-}), d\mathcal{Z}_s) \\
&= f(T, \mathcal{Z}_0) + \int_0^t \langle \hat{\mathbf{d}}_z f(T - s, \mathcal{Z}_{s_-}) \mathbf{A}_\alpha^*, d\mathcal{Z}_s \rangle
\end{aligned}
$$

where the first integral involves the L^2 scalar product on $\widehat{T}_z\mathbb{G} \times \widehat{T}_z\mathbb{G}$ and the second integral involves the duality $\widehat{T}^*_z\mathbb{G} \times \widehat{T}_z\mathbb{G}$. In a differential form:

$$
\begin{aligned}
dM_t^{\alpha, f, T, \mathcal{Z}_0} &= (\mathbf{A}_\alpha \widehat{\boldsymbol{\nabla}}_z f(s, \mathcal{Z}_{s_-}), d\mathcal{Z}_s) \\
&= \sum_{i=1}^m \alpha_i^x \{(X_i^2 f)(T - t, \mathcal{Z}_{t_-})\, d(\mathcal{N}_t^i - \lambda t) + (X_i^0 f)(t, \mathcal{Z}_{t_-})\, d\mathcal{X}_t^i\} \\
&\quad + \sum_{j=1}^n \alpha_{j,k}^y (X_j f)(T - t, \mathcal{Z}_{t_-})\, d\mathcal{Y}_t^k
\end{aligned}
$$

Quadratic covariation and subordination. We have the quadratic covariations (see Protter [48], Dellacherie–Meyer [22], or Privault [46, 47]). Since

$$d[\mathcal{N}^i - \lambda t, \mathcal{N}^i - \lambda t]_t = d\mathcal{N}^i_t$$
$$d[\mathcal{N}^i - \lambda t, \mathcal{X}^i]_t = \tau_{\mathcal{N}_t} \, d\mathcal{N}^i_t$$
$$d[\mathcal{X}^i, \mathcal{X}^i]_t = d\mathcal{N}^i_t$$
$$d[\mathcal{Y}^j, \mathcal{Y}^j]_t = dt,$$

it follows that

$$d[M^f, M^g]_t = \sum_{i=1}^{m} \sum_{\tau=\pm} (X_i^\tau f)(X_i^\tau g)(T - t, \mathcal{Z}_{t_-}) \mathbb{1}(\tau_{\mathcal{N}_t} = \tau 1) \, d\mathcal{N}^i_t$$
$$+ (\boldsymbol{\nabla}_y f, \boldsymbol{\nabla}_y g)(T - t, \mathcal{Z}_{t_-}) \, dt.$$

Differential subordination. Following Wang [52], given two adapted càdlàg Hilbert space-valued martingales X_t and Y_t, we say that Y_t is differentially subordinate by quadratic variation to X_t if $|Y_0|_{\mathbb{H}} \le |X_0|_{\mathbb{H}}$ and $[Y, Y]_t - [X, X]_t$ is nondecreasing nonnegative for all t. In our case, we have

$$d[M^{\alpha, f}, M^{\alpha, f}]_t = \sum_{i=1}^{m} |\alpha_i^x|^2 \big\{ (X_i^+ f)^2 (T - t, \mathcal{Z}_{t_-}) \mathbb{1}(\tau_{\mathcal{N}_t} = 1)$$
$$+ (X_i^- f)^2 (T - t, \mathcal{Z}_{t_-}) \mathbb{1}(\tau_{\mathcal{N}_t} = -1) \big\} d\mathcal{N}^i_t$$
$$+ (\mathbf{A}_\alpha^y \boldsymbol{\nabla}_y f, \mathbf{A}_\alpha^y \boldsymbol{\nabla}_y f)(T - t, \mathcal{Z}_{t_-}) \, dt.$$

Hence

$$d[M^{\alpha, f}, M^{\alpha, f}]_t \le \|\mathbf{A}_\alpha\|_2^2 \, d[M^f, M^f]_t. \tag{2.2}$$

This means that $M_t^{\alpha, f}$ is differentially subordinate to $\|\mathbf{A}_\alpha\|_2 M_t^f$.

2.2. Martingale inequalities under differential subordination

In the final part of the section we discuss a number of sharp martingale inequalities which hold under the assumption of the differential subordination imposed on the processes. Our starting point is the following celebrated L^p bound.

Theorem 2.1 (Wang, 1995). *Suppose that X and Y are martingales taking values in a Hilbert space $\mathbb{H}$ such that Y is differentially subordinate to X. Then for any $1 < p < \infty$ we have*

$$\|Y\|_p \le (p^* - 1)\|X\|_p$$

and the constant $p^* - 1$ is the best possible, even if $\mathbb{H} = \mathbb{R}$.

This result was first proved by Burkholder in [15] in the following discrete-time setting. Suppose that $(X_n)_{n\ge0}$ is an $\mathbb{H}$-valued martingale and $(\alpha_n)_{n\ge0}$ is a predictable sequence with values in $[-1, 1]$. Let $Y := \alpha * X$ be the martingale transform of X defined for almost all $\omega \in \Omega$ by

$$Y_0(\omega) = \alpha_0 X_0(\omega) \quad \text{and} \quad (Y_{n+1} - Y_n)(\omega) = \alpha_n (X_{n+1} - X_n)(\omega).$$

Then the above L^p bound holds true and the constant $p^* - 1$ is optimal. The general continuous-time version formulated above is due to Wang [52]. To see that the preceding discrete-time version is indeed a special case, treat a discrete-time martingale $(X_n)_{n \geq 0}$ and its transform $(Y_n)_{n \geq 0}$ as continuous-time processes via $X_t = X_{\lfloor t \rfloor}$, $Y_t = Y_{\lfloor t \rfloor}$ for $t \geq 0$; then Y is differentially subordinate to X.

In 1992, Choi [19] established the following non-symmetric, discrete-time version of the L^p estimate.

Theorem 2.2 (Choi, 1992). *Suppose that $(X_n)_{n \geq 0}$ is a real-valued discrete time martingale and let $(Y_n)_{n \geq 0}$ be its transform by a predictable sequence $(\alpha_n)_{n \geq 0}$ taking values in $[0, 1]$. Then there exists a constant $\mathfrak{C}_p$ depending only on p such that $\|Y\|_p \leq \mathfrak{C}_p \|X\|_p$ and the estimate is best possible.*

This result can be regarded as a non-symmetric version of the previous theorem, since the transforming sequence $(\alpha_n)_{n \geq 0}$ takes values in a non-symmetric interval $[0, 1]$. There is a natural question whether the estimate can be extended to the continuous-time setting; in particular, this gives rise to the problem of defining an appropriate notion of non-symmetric differential subordination. The following statement obtained by Bañuelos and Osękowski addresses both these questions. For any real numbers $a < b$ and any $1 < p < \infty$, let $\mathfrak{C}_{a,b,p}$ be the constant introduced in [10].

Theorem 2.3 (Banuelos–Osękowski, 2012). *Let $(X_t)_{t \geq 0}$ and $(Y_t)_{t \geq 0}$ be two real-valued martingales satisfying*

$$d \left[Y - \frac{a+b}{2} X, Y - \frac{a+b}{2} X \right]_t \leq d \left[\frac{b-a}{2} X, \frac{b-a}{2} X \right]_t \qquad (2.3)$$

for all $t \geq 0$. Then for all $1 < p < \infty$, we have $\|Y\|_p \leq \mathfrak{C}_{a,b,p} \|X\|_p$.

The condition (2.3) is a continuous counterpart of the condition that the transforming sequence $(\alpha_n)_{n \geq 0}$ takes values in the interval $[a, b]$. Thus, in particular, Choi's constant $\mathfrak{C}_p$ is, in the terminology of the above theorem, equal to $\mathfrak{C}_{p,0,1}$.

We return to the context of the "classical" differential subordination introduced in the preceding subsection and study other types of martingale inequalities. The following statements, obtained by Bañuelos–Osękowski, [11] will allow us to deduce sharp weak-type and logarithmic estimates for Riesz transforms, respectively.

Theorem 2.4 (Banuelos–Osękowski, 2015). *Suppose that X and Y are martingales taking values in a Hilbert space $\mathbb{H}$ such that Y is differentially subordinate to X.*

(i) *Let $1 < p < 2$. Then for any $t \geq 0$,*

$$\mathbb{E} \max \left\{ |Y_t| - \frac{p^{-1/(p-1)}}{2} \Gamma \left(\frac{p}{p-1} \right), 0 \right\} \leq \mathbb{E} |X_t|^p.$$

(ii) *Suppose that* $2 < p < \infty$. *Then for any* $t \geq 0$,

$$\mathbb{E} \max \left\{ |Y_t| - 1 + p^{-1}, 0 \right\} \leq \frac{p^{p-2}}{2} \mathbb{E}|X_t|^p.$$

Both estimates are sharp: for each p, *the numbers* $\frac{p^{-1/(p-1)}}{2}\Gamma\left(\frac{p}{p-1}\right)$ *and* $1-p^{-1}$ *cannot be decreased.*

Recall that $\Phi, \Psi : [0, \infty) \to [0, \infty)$ are conjugate Young functions given by $\Phi(t) = e^t - 1 - t$ and $\Psi(t) = (t+1)\log(t+1) - t$.

Theorem 2.5 (Banuelos–Osękowski, 2015). *Suppose that* X *and* Y *are martingales taking values in a Hilbert space* $\mathbb{H}$ *such that* Y *is differentially subordinate to* X. *Then for any* $K > 1$ *and any* $t \geq 0$ *we have*

$$\mathbb{E} \left\{ |Y_t| - (2(K-1))^{-1}, 0 \right\} \leq K\mathbb{E}\Psi(|X_t|).$$

For each K, *the constant* $(2(K-1))^{-1}$ *appearing on the left, is the best possible (it cannot be replaced by any smaller number).*

The following exponential estimate, established by Osękowski in [41], can be regarded as a dual statement to the above logarithmic bound.

Theorem 2.6 (Osękowski, 2013). *Assume that* X, Y *are* $\mathbb{H}$*-valued martingales such that* $\|X\|_\infty \leq 1$ *and* Y *is differentially subordinate to* X. *Then for any* $K > 1$ *and any* $t \geq 0$ *we have*

$$\mathbb{E}\Phi(|Y_t|/K) \leq \frac{1}{2K(K-1)} \mathbb{E}|X_t|. \tag{2.4}$$

Finally, we will need the following sharp $L^q \to L^p$ estimate, established by Osękowski in [43], which will allow us to deduce the corresponding estimate for Riesz transforms.

Theorem 2.7 (Osękowski, 2014). *Assume that* X, Y *are* $\mathbb{H}$*-valued martingales such that* Y *is differentially subordinate to* X. *Then for any* $1 \leq p < q < \infty$ *there is a constant* $L_{p,q}$ *such that*

$$\mathbb{E} \max\{|Y_t|^p - L_{p,q}, 0\} \leq \mathbb{E}|X_t|^q. \tag{2.5}$$

Actually, the paper [43] identifies, for any p and q as above, the optimal (i.e., the least) value of the constant $L_{p,q}$ in the estimate above. As the description of this constant is a little complicated (and will not be needed in our considerations below), we refer the reader to that paper for the formal definition of $L_{p,q}$.

Let us conclude with the observation which will be crucial in the proofs of our main results. Namely, all the martingale inequalities presented above are of the form $\mathbb{E}\zeta(|Y_t|) \leq \mathbb{E}\xi(|X_t|)$, $t \geq 0$, where ζ, ξ are certain convex functions. This will allow us to successfully apply a conditional version of Jensen's inequality.

3. Proofs of the main results

We turn our attention to the proofs of the estimates for R_α^2 formulated in the introductory section. We will focus on Theorems 1.1, 1.2 and 1.3 only; the remaining statements are established by similar arguments. Also, we postpone the proof of the sharpness of these estimates to the next section.

3.1. Proof of Theorem 1.1

Recall that the subordination estimate (2.2) shows that the martingale transform $Y_t := M_t^\alpha$ is differentially subordinate to the martingale $X_t := \|\mathbf{A}_\alpha\|_2 M_t^f$. Therefore, by Theorem 2.1, we immediately deduce that

$$\|M_t^{\alpha,f}\|_p \le \|\mathbf{A}_\alpha\|_2 (p^* - 1)\|M_t^f\|_p$$

for all $t \ge 0$. Since the operator $\mathcal{T}^\alpha$ is a conditional expectation of $M_t^{\alpha,f}$, an application of Jensen's inequality proves the estimate $\|\mathcal{T}^\alpha\|_p \le \|\mathbf{A}_\alpha\|_2 (p^* - 1)$, which is the desired bound.

3.2. Proof of Theorem 1.2

The argument is the same as above and exploits the fine-tuned L^p estimate of Theorem 2.3 applied to $X_t = M_t^f$ and $Y_t = M_t^{\alpha,f}$. It is not difficult to prove that the difference of quadratic variations above writes in terms of a jump part and a continuous part as

$$
\left[Y - \frac{a+b}{2}X, Y - \frac{a+b}{2}X \right]_t - \mathrm{d}\left[\frac{b-a}{2}X, \frac{b-a}{2}X \right]_t
$$
$$
= \sum_{i=1}^m \sum_{\pm} (\alpha_i^x - a)(\alpha_i^x - b)(X_i^\pm f)^2(\mathcal{B}_t)\mathbb{1}(\tau_{N_t} = \pm 1)\,\mathrm{d}\mathcal{N}_t^i
$$
$$
+ \langle (\mathbf{A}_\alpha^y - a\mathbf{I})(\mathbf{A}_\alpha^y - b\mathbf{I})\nabla_y f(\mathcal{B}_t), \nabla_y f(\mathcal{B}_t) \rangle \,\mathrm{d}t,
$$

which is nonpositive since we assumed precisely $a\mathbf{I} \le \mathbf{A}_\alpha \le b\mathbf{I}$. Thus, the estimate of Theorem 1.2 follows. The sharpness is established in a similar manner.   □

3.3. Proof of Theorem 1.3

We will focus on the case $1 < p < 2$; for remaining values of p the argument is similar. An application of Theorem 2.4 to the processes $X_t = \|\mathbf{A}_\alpha\|_2 M_t^f$ and $Y_t = M_t^{\alpha,f}$ yields

$$
\mathbb{E}\max\left\{ |M_t^{\alpha,f}| - \frac{p^{-1/(p-1)}}{2}\Gamma\left(\frac{p}{p-1} \right), 0 \right\} \le \|\mathbf{A}_\alpha\|_2^p \mathbb{E}|M_t^f|^p
$$

and hence, by Jensen's inequality, we obtain

$$
\int_{\mathbb{G}} \max\left\{ |R_\alpha^2 f| - \frac{p^{-1/(p-1)}}{2}\Gamma\left(\frac{p}{p-1} \right), 0 \right\}\,d\mu_z \le \|\mathbf{A}_\alpha\|_2^p \|f\|_{L^p(\mathbb{G})}^p.
$$

K. Domelevo, A. Osękowski and S. Petermichl

Therefore, if E is an arbitrary measurable subset of $\mathbb{G}$, we get

$$\int_E |R_\alpha^2 f| d\mu_z \leq \int_E \left(|R_\alpha^2 f| - \frac{p^{-1/(p-1)}}{2} \Gamma\left(\frac{p}{p-1}\right) \right) d\mu_z$$

$$+ \frac{p^{-1/(p-1)}}{2} \Gamma\left(\frac{p}{p-1}\right) \mu_z(E)$$

$$\leq \|f\|_{L^p(\mathbb{G})}^p + \frac{p^{-1/(p-1)}}{2} \Gamma\left(\frac{p}{p-1}\right) \mu_z(E).$$

Apply this bound to λf, where λ is a nonnegative parameter, then divide both sides by λ and optimize the right-hand side over λ to get the desired assertion.

4. Sharpness

The proof of the sharpness of different results is made in several steps. In some cases the sharpness for certain second-order Riesz transform estimates in the continuous setting (such as in Theorem 1.1) is already known. In these cases we prove below the sharpness for the discrete (or semidiscrete) case by using sequences of finite difference approximates of continuous functions and their finite difference second-order Riesz transforms. In other cases, we need to prove first sharpness for certain continuous second-order Riesz transforms. The key point here is to transfer the sharp result for zigzag martingales into a sharp result for certain continuous second-order Riesz transforms by the laminate technique. We will illustrate this for the weak-type estimate of Theorem 1.3 and establish the following statement.

Theorem 4.1. *Let $\Theta : [0, \infty) \to [0, \infty)$ be a given function and let $\lambda > 0$ be a fixed number. Assume further that there is a pair (F, G) of finite martingales starting from $(0,0)$ such that G is a ± 1-transform of F and*

$$\mathbb{E}(|G_\infty| - \lambda)_+ > \mathbb{E}\Theta(|F_\infty|).$$

Then there is a function $f : \mathbb{R}^2 \to \mathbb{R}$ supported on the unit disc $\mathbb{D}$ of $\mathbb{R}^2$ such that

$$\int_{\mathbb{R}^2} \left(|(R_1^2 - R_2^2)f| - \lambda \right)_+ dx > \int_{\mathbb{D}} \Theta(|f|) dx.$$

We will prove this statement with the use of laminates, important family of probability measures on matrices. It is convenient to split this section into several separate parts. For the sake of convenience, and to make this section as self-contained as possible, we recall the preliminaries on laminates and their connections to martingales from [14] and [39], Section 4.2.

4.1. Laminates

Assume that $\mathbb{R}^{m \times n}$ stands for the space of all real matrices of dimension $m \times n$ and $\mathbb{R}^{n \times n}_{\text{sym}}$ denotes the subclass of $\mathbb{R}^{n \times n}$ which consists of all symmetric matrices of dimension $n \times n$.

Definition 4.2. A function $f : \mathbb{R}^{m \times n} \to \mathbb{R}$ is said to be *rank-one convex*, if for all $A, B \in \mathbb{R}^{m \times n}$ with rank $B = 1$, the function $t \mapsto f(A + tB)$ is convex.

For other equivalent definitions of rank-one convexity, see [21, p. 100]. Suppose that $\mathcal{P} = \mathcal{P}(\mathbb{R}^{m \times n})$ is the class of all compactly supported probability measures on $\mathbb{R}^{m \times n}$. For a measure $\nu \in \mathcal{P}$, we define

$$\bar{\nu} = \int_{\mathbb{R}^{m \times n}} X \, d\nu(X),$$

the associated *center of mass* or *barycenter* of ν.

Definition 4.3. We say that a measure $\nu \in \mathcal{P}$ is a *laminate* (and write $\nu \in \mathcal{L}$), if

$$f(\bar{\nu}) \leq \int_{\mathbb{R}^{m \times n}} f \, d\nu$$

for all rank-one convex functions f. The set of laminates with barycenter 0 is denoted by $\mathcal{L}_0(\mathbb{R}^{m \times n})$.

Laminates can be used to obtain lower bounds for solutions of certain PDEs, as observed by Faraco in [30]. In addition, laminates appear naturally in the context of convex integration, where they lead to interesting counterexamples, see, e.g., [5, 20, 34, 37, 49]. For our results here we will be interested in the case of 2×2 symmetric matrices. The key observation is that laminates can be regarded as probability measures that record the distribution of the gradients of smooth maps: see Corollary 4.7 below. We briefly explain this and refer the reader to the works [33, 37, 49] for full details.

Definition 4.4. Let U be a subset of $\mathbb{R}^{2 \times 2}$ and let $\mathcal{PL}(U)$ denote the smallest class of probability measures on U which

(i) contains all measures of the form $\lambda \delta_A + (1-\lambda)\delta_B$ with $\lambda \in [0, 1]$ and satisfying rank$(A - B) = 1$;

(ii) is closed under splitting in the following sense: if $\lambda \delta_A + (1 - \lambda)\nu$ belongs to $\mathcal{PL}(U)$ for some $\nu \in \mathcal{P}(\mathbb{R}^{2 \times 2})$ and μ also belongs to $\mathcal{PL}(U)$ with $\bar{\mu} = A$, then also $\lambda\mu + (1 - \lambda)\nu$ belongs to $\mathcal{PL}(U)$.

The class $\mathcal{PL}(U)$ is called the *prelaminates* in U.

It follows immediately from the definition that the class $\mathcal{PL}(U)$ only contains atomic measures. Also, by a successive application of Jensen's inequality, we have the inclusion $\mathcal{PL} \subset \mathcal{L}$. The following are two well-known lemmas in the theory of laminates; see [5, 33, 37, 49].

Lemma 4.5. *Let* $\nu = \sum_{i=1}^{N} \lambda_i \delta_{A_i} \in \mathcal{PL}(\mathbb{R}^{2 \times 2}_{\text{sym}})$ *with* $\bar{\nu} = 0$. *Moreover, let* $0 < r < \frac{1}{2} \min |A_i - A_j|$ *and* $\delta > 0$. *For any bounded domain* $\mathcal{B} \subset \mathbb{R}^2$ *there exists* $u \in W_0^{2,\infty}(\mathcal{B})$ *such that* $\|u\|_{C^1} < \delta$ *and for all* $i = 1 \dots N$

$$\left\| \{x \in \mathcal{B} : |D^2 u(x) - A_i| < r\} \right\| = \lambda_i |\mathcal{B}|.$$

Lemma 4.6. *Let $K \subset \mathbb{R}_{\mathrm{sym}}^{2\times2}$ be a compact convex set and suppose that $\nu \in \mathcal{L}(\mathbb{R}_{\mathrm{sym}}^{2\times2})$ satisfies $\operatorname{supp}\nu \subset K$. For any relatively open set $U \subset \mathbb{R}_{\mathrm{sym}}^{2\times2}$ with $K \subset U$, there exists a sequence $\nu_j \in \mathcal{PL}(U)$ of prelaminates with $\bar{\nu}_j = \bar{\nu}$ and $\nu_j \overset{*}{\rightharpoonup} \nu$, where $\overset{*}{\rightharpoonup}$ denotes weak convergence of measures.*

Combining these two lemmas and using a simple mollification, we obtain the following statement, proved by Boros, Shékelyhidi Jr. and Volberg [14]. It exhibits the connection between laminates supported on symmetric matrices and second derivatives of functions. It will be our main tool in the proof of the sharpness. Recall that $\mathbb{D}$ denotes the unit disc of $\mathbb{C}$.

Corollary 4.7. *Let $\nu \in \mathcal{L}_0(\mathbb{R}_{\mathrm{sym}}^{2\times2})$. Then there exists a sequence $u_j \in C_0^\infty(\mathbb{D})$ with uniformly bounded second derivatives, such that*

$$\frac{1}{|\mathbb{D}|} \int_{\mathbb{D}} \phi(D^2 u_j(x)) \, dx \to \int_{\mathbb{R}_{\mathrm{sym}}^{2\times2}} \phi \, d\nu$$

for all continuous $\phi : \mathbb{R}_{\mathrm{sym}}^{2\times2} \to \mathbb{R}$.

4.2. Biconvex functions and a special laminate

The next step in our analysis is devoted to the introduction of a certain special laminate. We need some additional notation. A function $\zeta : \mathbb{R} \times \mathbb{R} \to \mathbb{R}$ is said to be *biconvex* if for any fixed $z \in \mathbb{R}$, the functions $x \mapsto \zeta(x, z)$ and $y \mapsto \zeta(z, y)$ are convex. Now, take the martingales F and G appearing in the statement of Theorem 4.1. Then the martingale pair

$$(\mathsf{F}, \mathsf{G}) := \left(\frac{F + G}{2}, \frac{F - G}{2} \right)$$

is finite, starts from $(0,0)$ and has the following *zigzag* property: for any $n \geq 0$ we have $\mathsf{F}_n = \mathsf{F}_{n+1}$ with probability 1 or $\mathsf{G}_n = \mathsf{G}_{n+1}$ almost surely; that is, in each step (F, G) moves either vertically, or horizontally. Indeed, this follows directly from the assumption that G is a ±1-transform of F. This property combines nicely with biconvex functions: if ζ is such a function, then a successive application of Jensen's inequality gives

$$\mathbb{E}\zeta(\mathsf{F}_n, \mathsf{G}_n) \geq \mathbb{E}\zeta(\mathsf{F}_{n-1}, \mathsf{G}_{n-1}) \geq \cdots \geq \mathbb{E}\zeta(\mathsf{F}_0, \mathsf{G}_0) = \zeta(0, 0). \tag{4.1}$$

The distribution of the terminal variable $(\mathsf{F}_\infty, \mathsf{G}_\infty)$ gives rise to a probability measure ν on $\mathbb{R}_{\mathrm{sym}}^{2\times2}$: put

$$\nu\left(\operatorname{diag}(x, y)\right) = \mathbb{P}\left((\mathsf{F}_\infty, \mathsf{G}_\infty) = (x, y)\right), \qquad (x, y) \in \mathbb{R}^2,$$

where $\operatorname{diag}(x, y)$ stands for the diagonal matrix $\left(\begin{smallmatrix} x & 0 \\ 0 & y \end{smallmatrix} \right)$. Observe that ν is a laminate of barycenter 0. Indeed, if $\psi : \mathbb{R}^{2\times2} \to \mathbb{R}$ is a rank-one convex, then $(x, y) \mapsto \psi(\operatorname{diag}(x, y))$ is biconvex and thus, by (4.1),

$$\int_{\mathbb{R}^{2\times2}} \psi \, d\nu = \mathbb{E}\psi(\operatorname{diag}(\mathsf{F}_\infty, \mathsf{G}_\infty)) \geq \psi(\operatorname{diag}(0, 0)) = \psi(\bar{\nu}).$$

Here we have used the fact that (F, G) is finite, so $(\mathsf{F}_\infty, \mathsf{G}_\infty) = (\mathsf{F}_n, \mathsf{G}_n)$ for some n.

4.3. A proof of Theorem 4.1

Consider a continuous function $\phi : \mathbb{R}^{2\times2}_{\text{sym}} \to \mathbb{R}$ given by

$$\phi(A) = (|A_{11} - A_{22}| - \lambda)_+ - \Theta(|A_{11} + A_{22}|).$$

By Corollary 4.7, there is a functional sequence $(u_j)_{j\geq1} \subset C_0^\infty(\mathbb{D})$ such that

$$\frac{1}{|\mathbb{D}|} \int_{\mathbb{R}^2} \phi(D^2 u_j)\, dx = \frac{1}{|\mathbb{D}|} \int_{\mathbb{D}} \phi(D^2 u_j)\, dx$$

$$\xrightarrow{j\to\infty} \int_{\mathbb{R}^{2\times2}_{\text{sym}}} \phi\, d\nu = \mathbb{E}(|G_\infty| - \lambda)_+ - \mathbb{E}\Theta(|F_\infty|) > 0.$$

Therefore, for sufficiently large j, we have

$$\int_{\mathbb{R}^2} \left(\left\| \frac{\partial^2 u_j}{\partial x^2} - \frac{\partial^2 u_j}{\partial y^2} \right\| - \lambda \right)_+ dx\, dy > \int_{\mathbb{R}^2} \Theta(|\Delta u_j|)\, dx\, dy.$$

Setting $f = \Delta u_j$, we obtain the desired assertion.

In the remaining part of this subsection, let us briefly explain how Theorem 4.1 yields the sharpness of weak-type and logarithmic estimates for second-order Riesz transforms (in the classical setting). We will focus on the weak-type bounds for $1 < p < 2$ – the remaining estimates can be treated analogously. Suppose that λ_p is the best constant in the estimate

$$\mathbb{E}(|G_\infty| - \lambda_p)_+ \leq \mathbb{E}|F_\infty|^p,$$

valid for all pairs (F, G) of finite martingales starting with 0 such that G is a ±1-transform of F. The value of λ_p appears in the statement of Theorem 9 above, the fact that it is already the best for martingale transforms follows from the examples exhibited in [38]. For any $\varepsilon > 0$, Theorem 4.1 yields the existence of $f : \mathbb{R}^2 \to \mathbb{R}$, supported on the unit disc, such that

$$\int_{\mathbb{R}^2} \left(|(R_1^2 - R_2^2)f| - \lambda_p + \varepsilon \right)_+ dx\, dy > \int_{\mathbb{R}^2} |f|^p\, dx\, dy.$$

That is, if we set $A = \{|(R_1^2 - R_2^2)f| \geq \lambda_p - \varepsilon\}$, we get

$$\int_A |(R_1^2 - R_2^2)f|\, dx\, dy > \int_{\mathbb{R}^2} |f|^p\, dx\, dy + (\lambda_p - \varepsilon)|A|. \qquad (4.2)$$

However, if the weak-type estimate holds with a constant c_p, Young's inequality implies

$$\int_A |(R_1^2 - R_2^2)f|\, dx\, dy \leq \int_{\mathbb{R}^2} |f|^p\, dx\, dy + \frac{(p-1)c_p^{p/(p-1)}}{p^{p/(p-1)}} |A|.$$

Therefore, inequality (4.2) enforces that

$$\frac{(p-1)c_p^{p/(p-1)}}{p^{p/(p-1)}} \geq \lambda_p$$

(since ε was arbitrary). This estimate is equivalent to

$$c_p \geq \left(\frac{1}{2}\Gamma\left(\frac{2p-1}{p-1}\right)\right)^{1-1/p},$$

which is the desired sharpness.

4.4. From continuous to discrete sharp estimates

We claim that the sharp bounds found for the continuous second-order Riesz transforms also hold in the case of purely discrete groups. Groups of mixed type would be treated in the same manner. We illustrate those results only for the sharpness in Theorem 1.1 and in Theorem 1.3 since other results follow the same lines. Precisely, we show that the sharpness in the discrete case is inherited from the sharpness of the continuous case through the use of the so-called fundamental theorem of finite difference methods from Lax and Richtmyer [35] (see also [36]). This result states that **stability** and **consistency** of the finite difference scheme implies **convergence** of the approximate finite difference solution towards the continuous solution, in a sense that we detail below.

Finite difference Riesz transforms. Let $u = R_i^2 f$ be the ith second-order Riesz transform in $\Omega := \mathbb{R}^N$ of a function $f \in L^p$. The function u is a unique solution to the Poisson problem in $\mathbb{R}^N$, $\Delta u = \partial_i^2 f$ in $\mathbb{R}^N$ (see [29]). This is a problem of the form $Au = Bf$, where $A = \Delta$ and $B = \partial_{ij}^2$. Introduce now a finite difference grid of step-size $h > 0$, that is the grid $\Omega_h := h\mathbb{Z}^N$. The functions v_h defined on Ω_h are equipped with the L_h^p norm defined as

$$\|v_h\|_{L_h^p}^p := \sum_{x \in \Omega_h} |v_h(x)|^p h^N.$$

It is common to identify a finite difference function v_h defined on the grid Ω_h with the piecewise constant function (also denoted) $v_h : \mathbb{R}^N \to \mathbb{C}$ such that $v_h(x) = v_h(y)$ for all x's in the open cube $\Omega(y)$ of volume h^N centered around the grid point $y \in \Omega_h$. With this notation, we might write finite difference integrals in the form

$$\|v_h\|_{L_h^p} = \int_{x \in \mathbb{R}^N} |v_h(x)| \, \mathrm{d}\mu_h(x).$$

The finite difference second-order Riesz transform $u_h = R_i^2 f_h$ of f_h is the solution to the problem $A_h u_h = B_h^2 f_h$, where $A_h := \Delta_h$ is the finite difference Laplacian and $B_h := \partial_{i,h}^2$ the 3-point finite difference second-order derivative. Precisely, for any $x \in \Omega_h$, any $v_h : \Omega_h \to \mathbb{R}$,

$$(\partial_{i,h}^2 v_h)(x) := \frac{v_h(x + he_i) - 2v_h(x) + v_h(x - he_i)}{h^2}$$

$$(\Delta_h v_h)(x) := \sum_{i=1}^{N} (\partial_{i,h}^2 v_h)(x).$$

It is classical that we have the **consistency** of the discrete problem with respect to the continuous problem, that is for given smooth functions u and f we have

$\Delta_h u = \Delta u + \mathcal{O}(h)$ and $\partial^2_{i,h} f = \partial^2_i f + \mathcal{O}(h)$, where the coefficients in $\mathcal{O}(h)$ include as a factor up to fourth-order derivatives of u or f. This implies in particular that $B_h f = Bf + \mathcal{O}(h)$ in L^p_h for any given smooth function f with compact support. It is also classical that $(-\Delta_h)^{-1}$ is bounded in L^p_h uniformly with respect to h. This is the L^p **stability** of the finite difference scheme. The fundamental theorem of finite difference methods implies the L^p **convergence** of the sequence of discrete second-order Riesz transforms u_h towards the continuous second-order Riesz transform u.

Discrete Riesz transforms on a Lie group. Observe that the finite difference Riesz transform $u_h = R^2_{i,h} f_h$ defined on the grid Ω_h, also gives rise to a Riesz transform on the Lie group $\Omega_1 = \mathbb{Z}^N$. This is a consequence of the homogeneity of order zero of the Riesz transforms. Indeed, the equation $\Delta_h u_h = \partial^2_{i,h} f_h$ rewrites as $\Delta_1 u_1 = \partial^2_{i,1} f_1$, where $u_1(y) := u_h(y/h)$, $f_1(y) := f_h(y/h)$ for all $y \in \mathbb{Z}^N$, and where Δ_1 and $\partial^2_{i,1}$ are the discrete differential operators defined on $\mathbb{Z}^N$. We have also $\|u_h\|_{L^p_h} = h^{N/p} \|u_1\|_{L^p_1}$ and $\|f_h\|_{L^p_h} = h^{N/p} \|f_1\|_{L^p_1}$. Notice that for all h, this ensures that $\|u_h\|_{L^p_h} / \|f_h\|_{L^p_h} = \|u_1\|_{L^p_1} / \|f_1\|_{L^p_1}$.

Sharpness for Theorem 1.1 in the discrete setting. In the continuous setting, the sharpness was proved in [31] based on the combination $R^2_\alpha = R^2_1 - R^2_2$ of second-order Riesz transforms.

Let $u^{(k)} = R^2_\alpha f^{(k)}$ be a sequence of second-order Riesz transforms yielding the sharp constant C_p in the estimate, that is $\|u^{(k)}\|_p / \|f^{(k)}\|_p \to C_p$ as k goes to infinity. For each $k \in \mathbb{N}$ and $h > 0$, we introduce the finite difference approximation $f^{(k)}_h$ of $f^{(k)}$ and the corresponding finite difference Riesz transform. Thanks to the convergence of the finite difference scheme, we can extract a subsequence $f^{(k)}_{h_k}$ such that $\|u^{(k)}_1\|_p / \|f^{(k)}_1\|_p = \|u^{(k)}_{h_k}\|_p / \|f^{(k)}_{h_k}\|_p \to C_p$. Therefore C_p is also the sharp constant for the second-order Riesz transforms in $\mathbb{Z}^N$.

Sharpness for Theorem 1.3 in the discrete setting. Recall that we have a bound of the form

$$\|R^2_\alpha f\|_{L^{p,\infty}(\mathbb{G},\mathbb{C})} := \sup_E \left\{ \mu_z(E)^{1/p-1} \int_E |R^2_\alpha f| \, d\mu_z \right\} \le C_p \|f\|_{L^p}$$

for a certain constant C_p that is known to be sharp in the case of continuous second-order Riesz transforms. In order to prove sharpness when the Lie group $\mathbb{G}$ does not have enough continuous components, it suffices again to approximate a sequence of continuous extremizers by a sequence of finite difference approximations. Take $\mathbb{G} = \mathbb{R}^N$. For any $\varepsilon > 0$, let f, $u := R^2_\alpha f$, and E with finite measure be chosen so that

$$\mu_z(E)^{1/p-1} \|u\|_{L^1(E)} / \|f\|_{L^p} \ge C_p - \varepsilon.$$

We can assume without loss of generality that f is a smooth function with compact support. Let f_h a finite difference approximation of f defined as its L^2 projection on the grid, and u_h its discrete second-order Riesz transform both defined on $\Omega_h := h\mathbb{Z}^N$. Since $\mu_z(E)$ is the finite N-dimensional Lebesgue measure of E, we

use outer measure approximations of E followed by approximations from below
by a finite number of sufficiently small cubes of size h centered around the grid
points of Ω_h, to define a "finite difference" approximation E_h of E such that

$$\mu_h(E_h) := \sum_{x \in E_h} h^N \to \mu_z(E)$$

when h goes to zero. Since the discrete Riesz transforms are stable in L^2, the Lax–
Richtmyer theorem ensures that $\|u_h\|_{L^2_h} \to \|u\|_{L^2}$ which implies $\|u_h\|_{L^1_h(E)} \to$
$\|u\|_{L^1(E)}$ and also $\|u_h\|_{L^1_h(E_h)} \to \|u\|_{L^1(E)}$. Therefore for h sufficiently small, we
have

$$\mu_h(E_h)^{1/p-1} \|u_h\|_{L^1_h(E_h)} / \|f_h\|_{L^p} \geq C_p - 2\varepsilon.$$

Let as before $u_1(y) := u_h(y/h)$, $f_1(y) := f_h(y/h)$ for all $y \in \Omega_1 := \mathbb{Z}^N$, and
$E_1 := E/h$. We have successively $\mu_h(E) = h^N \mu_1(E_1)$, $\|u_h\|_{L^1_h(E_h)} = h^N \|u_1\|_{L^1_h(E_1)}$
and $\|f_h\|_{L^p} = h^{N/p} \|f_1\|_{L^p}$. This yields immediately

$$\mu_1(E_1)^{1/p-1} \|u_1\|_{L^1_h(E_1)} / \|f_1\|_{L^p} \geq C_p - 2\varepsilon,$$

allowing us to prove sharpness for the class of discrete groups we are interested in.

References

[1] Nicola Arcozzi, *Riesz transforms on spheres and compact Lie groups.* ProQuest LLC,
Ann Arbor, MI, 1995. Thesis (Ph.D.) – Washington University in St. Louis.

[2] Nicola Arcozzi, *Riesz transforms on compact Lie groups, spheres and Gauss space.*
Ark. Mat., 36(2):201–231, 1998.

[3] N. Arcozzi, K. Domelevo, and S. Petermichl, *Second Order Riesz Transforms
on Multiply-Connected Lie Groups and Processes with Jumps.* Potential Anal.,
45(4):777–794, 2016.

[4] K. Astala, *Area distortion of quasiconformal mappings*, Acta. Math. **173** (1994), 37–
60.

[5] K. Astala, D. Faraco, L. Székelyhidi, Jr., *Convex integration and the L^p theory of
elliptic equations.* Ann. Sc. Norm. Super. Pisa Cl. Sci. (5), Vol. 7 (2008), pp. 1–50.

[6] R. Bañuelos and P.J. Méndez-Hernández, *Space-time Brownian motion and the
Beurling–Ahlfors transform.* Indiana Univ. Math. J., 52(4):981–990, 2003.

[7] Rodrigo Bañuelos, *Martingale transforms and related singular integrals.* Trans. Amer.
Math. Soc., 293(2):547–563, 1986.

[8] Rodrigo Bañuelos and Fabrice Baudoin, *Martingale Transforms and Their Projection
Operators on Manifolds.* Potential Anal., 38(4):1071–1089, 2013.

[9] Rodrigo Bañuelos and Prabhu Janakiraman, *L^p-bounds for the Beurling–Ahlfors
transform.* Trans. Amer. Math. Soc., 360(7):3603–3612, 2008.

[10] Rodrigo Banuelos and Adam Osękowski, *Martingales and sharp bounds for Fourier
multipliers.* Ann. Acad. Sci. Fenn., Math., 37(1):251–263, 2012.

[11] Rodrigo Banuelos and Adam Osękowski, *Sharp martingale inequalities and applications to Riesz transforms on manifolds, Lie groups and Gauss space.* J. Funct. Anal.,
269(6):1652–1713, 2015.

[12] Rodrigo Bañuelos and Gang Wang, *Sharp inequalities for martingales with applications to the Beurling–Ahlfors and Riesz transforms.* Duke Math. J., 80(3):575–600, 1995.

[13] Alexander Borichev, Prabhu Janakiraman, and Alexander Volberg, *Subordination by conformal martingales in L^p and zeros of Laguerre polynomials.* Duke Math. J., 162(5):889–924, 2013.

[14] N. Boros, L. Székelyhidi Jr. and A. Volberg, *Laminates meet Burkholder functions,* Journal de Mathématiques Pures et Appliquées, to appear.

[15] D.L. Burkholder, *Boundary value problems and sharp inequalities for martingale transforms.* Ann. Probab., 12(3):647–702, 1984.

[16] D.L. Burkholder, *A sharp and strict L^p-inequality for stochastic integrals.* Ann. Probab., 15(1):268–273, 1987.

[17] D.L. Burkholder, *Sharp inequalities for martingales and stochastic integrals.* Astérisque, (157–158):75–94, 1988. Colloque Paul Lévy sur les Processus Stochastiques (Palaiseau, 1987).

[18] D.L. Burkholder, *Explorations in martingale theory and its applications.* In École d'Été de Probabilités de Saint-Flour XIX – 1989, vol. 1464 of Lecture Notes in Math., pages 1–66. Springer, Berlin, 1991.

[19] K.P. Choi, *A sharp inequality for martingale transforms and the unconditional basis constant of a monotone basis in $L^p(0,1)$.* Trans. Amer. Math. Soc., 330(2):509–529, 1992.

[20] S. Conti, D. Faraco, F. Maggi, *A new approach to counterexamples to L^1 estimates: Korn's inequality, geometric rigidity, and regularity for gradients of separately convex functions,* Arch. Rat. Mech. Anal. 175 no. 2 (2005), pp. 287–300.

[21] B. Dacoronga, *Direct Methods in the Calculus of Variations.* Springer, 1989.

[22] Claude Dellacherie and Paul-André Meyer. *Probabilities and potential. B,* volume 72 of North-Holland Mathematics Studies. North-Holland Publishing Co., Amsterdam, 1982. Theory of martingales, translated from the French by J.P. Wilson.

[23] Komla Domelevo and Stefanie Petermichl, *Bilinear embeddings on graphs.* preprint, 2014.

[24] Komla Domelevo and Stefanie Petermichl, *Sharp L^p estimates for discrete second order Riesz transforms.* Adv. Math., 262:932–952, 2014.

[25] Komla Domelevo and Stefanie Petermichl, *Sharp L^p estimates for discrete second-order Riesz transforms.* C. R. Math. Acad. Sci. Paris, 352(6):503–506, 2014.

[26] Michel Emery, *Martingales continues dans les variétés différentiables.* In Lectures on probability theory and statistics (Saint-Flour, 1998), volume 1738 of Lecture Notes in Math., pages 1–84. Springer, Berlin, 2000.

[27] Michel Emery, *An Invitation to Second-Order Stochastic Differential Geometry.* https://hal.archives-ouvertes.fr/hal-00145073, 2005. 42 pages.

[28] Matts Essén, *A superharmonic proof of the M. Riesz conjugate function theorem.* Ark. Mat., 22(2):241–249, 1984.

[29] Lawrence C. Evans, *Partial Differential Equations.* Amer. Math. Soc., 1998.

[30] D. Faraco, *Milton's conjecture on the regularity of solutions to isotropic equations.* Ann. Inst. H. Poincaré Anal. Non Linéaire, **20** (2003), pp. 889–909.

[31] Stefan Geiss, Stephen Montgomery-Smith, and Eero Saksman, *On singular integral and martingale transforms*. Trans. Amer. Math. Soc., 362(2):553–575, 2010.

[32] Richard F. Gundy and Nicolas Th. Varopoulos, *Les transformations de Riesz et les intégrales stochastiques*. C. R. Acad. Sci. Paris Sér. A-B, 289(1):A13–A16, 1979.

[33] B. Kirchheim, *Rigidity and Geometry of Microstructures*. Habilitation Thesis, University of Leipzig (2003), http://www.mis.mpg.de/publications/other-series/ln/lecturenote-1603.html

[34] B. Kirchheim, S. Müller, V. Šverák, *Studying nonlinear pde by geometry in matrix space*. Geometric Analysis and nonlinear partial differential equations, Springer (2003), pp. 347–395.

[35] P.D. Lax and R.D. Richtmyer, *Survey of the stability of linear finite difference equations*. Comm. Pure Appl. Math., 9:267–293, 1956.

[36] Randall LeVeque, *Finite Difference Methods for Ordinary and Partial Differential Equations: Steady-State and Time-Dependent Problems (Classics in Applied Mathematics Classics in Applied Mathemat.)*. Society for Industrial and Applied Mathematics, Philadelphia, PA, USA, 2007.

[37] S. Müller, V. Šverák, *Convex integration for Lipschitz mappings and counterexamples to regularity*. Ann. of Math. (2), 157 no. 3 (2003), pp. 715–742.

[38] A. Osękowski, *Weak-type inequalities for Fourier multipliers with applications to the Beurling–Ahlfors transform*, J. Math. Soc. Japan 66 (2014), no. 3, 745–764.

[39] A. Osękowski, *On restricted weak-type constants of Fourier multipliers*. Publ. Mat. 58 (2014), no. 2, 415–443.

[40] A. Osękowski, *Sharp moment inequalities for differentially subordinated martingales*. Studia Math. 201:103–131, 2010.

[41] A. Osękowski, *Logarithmic inequalities for second-order Riesz transforms and related Fourier multipliers*. Colloq. Math., 130(1):103–126, 2013.

[42] A. Osękowski, *Sharp weak-type inequalities for Fourier multipliers and second-order Riesz transforms*. Cent. Eur. J. Math., 12(8):1198–1213, 2014.

[43] A. Osękowski, *Sharp localized inequalities for Fourier multipliers*. Canadian J. Math., 66: 1358–1381, 2014.

[44] Stefanie Petermichl and Alexander Volberg, *Heating of the Ahlfors–Beurling operator: weakly quasiregular maps on the plane are quasiregular*. Duke Math. J., 112(2):281–305, 2002.

[45] S.K. Pichorides, *On the best values of the constants in the theorems of M. Riesz, Zygmund and Kolmogorov*. Studia Math., 44:165–179 (errata insert), 1972. Collection of articles honoring the completion by Antoni Zygmund of 50 years of scientific activity, II.

[46] Nicolas Privault, *Stochastic analysis in discrete and continuous settings with normal martingales*. vol. 1982 of Lecture Notes in Mathematics. Springer-Verlag, Berlin, 2009.

[47] Nicolas Privault, *Stochastic finance*. Chapman & Hall/CRC Financial Mathematics Series. CRC Press, Boca Raton, FL, 2014. An introduction with market examples.

[48] Philip E. Protter, *Stochastic integration and differential equations*. Volume 21 of Stochastic Modelling and Applied Probability. Springer-Verlag, Berlin, 2005. Second edition. Version 2.1, Corrected third printing.

[49] L. Székelyhidi, Jr., *Counterexamples to elliptic regularity and convex integration*. Contemp. Math. 424 (2007), pp. 227–245.

[50] Igor E. Verbitsky, *Estimate of the norm of a function in a Hardy space in terms of the norms of its real and imaginary parts*. Mat. Issled., (54):16–20, 164–165, 1980. English transl.: Amer. Math. Soc. Transl. (2) 124 (1984), 11–15.

[51] A. Volberg and F. Nazarov, *Heat extension of the Beurling operator and estimates for its norm*. St. Petersburg Math J., 15(4):563–573, 2004.

[52] Gang Wang, *Differential subordination and strong differential subordination for continuous-time martingales and related sharp inequalities*. Ann. Probab., 23(2):522–551, 1995.

K. Domelevo and S. Petermichl
IMT, Université Paul Sabatier
118 Route de Narbonne
F-31062 Toulouse, France
e-mail: komla.domelevo@math.univ-toulouse.fr
         stefanie.petermichl@math.univ-toulouse.fr

A. Osękowski
Faculty of Mathematics Informatics and Mechanics
University of Warsaw
Banacha 2
PL-02-097 Warsaw, Poland
e-mail: ados@mimuw.edu.pl

Operator Theory:
Advances and Applications, Vol. 261, 257–266
© Springer International Publishing AG, part of Springer Nature 2018

On the Maximum Principle for the Riesz Transform

Vladimir Eiderman and Fedor Nazarov

Dedicated to the memory of Victor Petrovich Havin, a remarkable mathematician and personality

Abstract. Let μ be a measure in $\mathbb{R}^d$ with compact support and continuous density, and let

$$R^s\mu(x) = \int \frac{y-x}{|y-x|^{s+1}}\, d\mu(y), \quad x, y \in \mathbb{R}^d, \ 0 < s < d.$$

We consider the following conjecture:

$$\sup_{x \in \mathbb{R}^d} |R^s\mu(x)| \le C \sup_{x \in \operatorname{supp}\mu} |R^s\mu(x)|, \quad C = C(d,s).$$

This relation was known for $d - 1 \le s < d$, and is still an open problem in the general case. We prove the maximum principle for $0 < s < 1$, and also for $0 < s < d$ in the case of radial measure. Moreover, we show that this conjecture is incorrect for non-positive measures.

Mathematics Subject Classification (2010). Primary 42B20. Secondary 31B05, 31B15.

Keywords. Riesz transform, maximum principle, reflectionless measure.

1. Introduction

Let μ be a non-negative finite Borel measure with compact support in $\mathbb{R}^d$, and let $0 < s < d$. The truncated Riesz operator $R^s_{\mu,\varepsilon}$ is defined by the equality

$$R^s_{\mu,\varepsilon} f(x) = \int_{|y-x|>\varepsilon} \frac{y-x}{|y-x|^{s+1}} f(y)\, d\mu(y), \quad x, y \in \mathbb{R}^d, \ f \in L^2(\mu), \ \varepsilon > 0.$$

For every $\varepsilon > 0$ the operator $R_{\mu,\varepsilon}^s$ is bounded on $L^2(\mu)$. By R_μ^s we denote a linear operator on $L^2(\mu)$ such that

$$R_\mu^s f(x) = \int \frac{y-x}{|y-x|^{s+1}} f(y) \, d\mu(y),$$

whenever the integral exists in the sense of the principal value. We say that R_μ^s is bounded on $L^2(\mu)$ if

$$\|R_\mu^s\| := \sup_{\varepsilon>0} \|R_{\mu,\varepsilon}^s\|_{L^2(\mu)\to L^2(\mu)} < \infty.$$

In the case $f \equiv 1$ the function $R_\mu^s 1(x)$ is said to be *the s-Riesz transform (potential)* of μ and is denoted by $R^s \mu(x)$. If μ has continuous density with respect to the Lebesgue measure m_d in $\mathbb{R}^d$, that is if $d\mu(x) = \rho(x) \, dm_d(x)$ with $\rho(x) \in C(\mathbb{R}^d)$, then $R^s \mu(x)$ exists for every $x \in \mathbb{R}^d$.

By C, c, possibly with indexes, we denote various constants which may depend only on d and s.

We consider the following well-known conjecture.

Conjecture 1.1. *Let μ be a nonnegative finite Borel measure with compact support and continuous density with respect to the Lebesgue measure in $\mathbb{R}^d$. There is a constant C such that*

$$\sup_{x\in\mathbb{R}^d} |R^s\mu(x)| \leq C \sup_{x\in\mathrm{supp}\,\mu} |R^s\mu(x)|. \tag{1.1}$$

For $s = d - 1$ the proof is simple. Obviously,

$$R^s\mu(x) = \nabla U_\mu^s(x), \tag{1.2}$$

where

$$U_\mu^s(x) = \frac{1}{s-1} \int \frac{d\mu(y)}{|y-x|^{s-1}}, \quad s \neq 1, \quad U_\mu^1(x) = -\int \log|y-x| \, d\mu(y).$$

Thus each component of the vector function $R^s\mu(x)$, $s = d - 1$, is harmonic in $\mathbb{R}^d \setminus \mathrm{supp}\,\mu$. Applying the maximum principle for harmonic functions we get (1.1).

For $d - 1 < s < d$, the relation (1.1) was established in [2] under the stronger assumption that $\rho \in C^\infty(\mathbb{R}^d)$. In fact it was proved that (1.1) holds for each component of $R^s\mu$ with $C = 1$ as in the case $s = d - 1$. The proof is based on the formula which recovers a density ρ from U_μ^s. But this method does not work for $s < d - 1$.

The problem under consideration has a very strong motivation and also is of independent interest. In [2] it is an important ingredient of the proof of the following theorem. By $\mathcal{H}^s$ we denote the s-dimensional Hausdorff measure.

Theorem 1.2 ([2]). *Let $d - 1 < s < d$, and let μ be a positive finite Borel measure such that $\mathcal{H}^s(\mathrm{supp}\,\mu) < \infty$. Then $\|R^s\mu\|_{L^\infty(m_d)} = \infty$ (equivalently, $\|R_\mu^s\| = \infty$).*

If s is an integer, the conclusion of Theorem 1.2 is incorrect. For $0 < s < 1$ Theorem 1.2 was proved by Prat [10] using a different approach. The obstacle for extension of this result to all nonintegers s between 1 and $d - 1$ is the lack of the maximum principle. The same issue concerns the quantitative version of Theorem 1.2 obtained by Jaye, Nazarov, and Volberg [3].

The maximum principle is important for other problems on the connection between geometric properties of a measure and the boundedness of the operator R_μ^s on $L^2(\mu)$ – see for example [3, 5, 6, 7]. All these results are established for $d - 1 < s < d$ or $s = d - 1$.

The problem of the lower estimate for $\|R_\mu^s\|$ in terms of the Wolff energy (a far reaching development of Theorem 1.2) which is considered in [3, 5], was known for $0 < s < 1$. And the results in [6, 7] are $(d-1)$-dimensional analogs of classical facts known for $s = 1$ (in particular, [7] contains the proof of an analog of the famous Vitushkin conjecture in higher dimensions). For $0 < s \leq 1$, the proofs essentially use the Melnikov curvature techniques and do not require the maximum principle. But this tool is absent for $s > 1$.

At the same time the validity of the maximum principle itself remained open even for $0 < s < 1$. This is especially interesting because no analog of (1.1) is available for each component of $R^s \mu$ when $0 < s < d - 1$ unlike the case $d - 1 \leq s < d$ – see Proposition 2.1 below.

We prove Conjecture 1.1 for $0 < s < 1$ in Section 2 (Theorem 2.3). The proof is completely different from the proof in the case $d - 1 \leq s < d$. In Section 3 we prove Conjecture 1.1 in the special case of a radial density of μ (that is when $d\mu = h(|x|)\, dm_d(x)$), but for all $s \in (0, d)$. Section 4 contains an example showing that Conjecture 1.1 is incorrect for non-positive measures, even for radial measures with C^∞-density (note that in [14, Conjecture 7.3] Conjecture 1.1 was formulated for all finite signed measures with compact support and C^∞-density).

2. The case $0 < s < 1$

We start with a statement showing that the maximum principle fails for every component of $R^s \mu$ if $0 < s < d - 1$.

Proposition 2.1. *For any $d \geq 2$, $0 < s < d - 1$, and any $M > 0$, there is a positive measure μ in $\mathbb{R}^d$ with C^∞-density such that*

$$\sup_{x \in \mathbb{R}^d} |R_1^s \mu(x)| > M \sup_{x \in \operatorname{supp} \mu} |R_1^s \mu(x)|, \tag{2.1}$$

where $R_1^s \mu$ is the first component of $R^s \mu$.

Proof. Let $E = \{(x_1, \ldots, x_d) \in \mathbb{R}^d : x_1 = 0,\ x_2^2 + \cdots + x_d^2 \leq 1\}$, and let E_δ, $\delta > 0$, be the δ-neighborhood of E in $\mathbb{R}^d$. Let $\mu = \mu_\delta$ be a positive measure supported on $\overline{E_\delta}$ with $\mu(\overline{E_\delta}) = 1$ and with C^∞-density $\rho(x)$ such that $\rho(x) < 2/\operatorname{vol}(E_\delta) \leq C_d/\delta$. Then

$$|R_1^s \mu(x')| > A_d, \text{ where } x' = (1, 0, \ldots, 0),\ \ 0 < \delta < 1/2.$$

On the other hand, for $x \in \operatorname{supp} \mu$ integration by parts yields

$$|R_1^s \mu(x)| < \int_{|y-x|<\delta} \frac{1}{|y-x|^s}\, d\mu(y) + \int_{|y-x|\geq\delta} \frac{\delta}{|y-x|^{s+1}}\, d\mu(y)$$

$$= \frac{\mu(B(x,\delta))}{\delta^s} + s \int_0^\delta \frac{\mu(B(x,r))}{r^{s+1}} \, dr + \delta(s+1) \int_\delta^\infty \frac{\mu(B(x,r))}{r^{s+2}} \, dr$$

$$< C \frac{C_d}{\delta} \frac{\delta^d}{\delta^s} + \frac{Cs}{\delta} \int_0^\delta \frac{r^d}{r^{s+1}} \, dr + \frac{C\delta(s+1)}{\delta} \int_\delta^2 \frac{r^{d-1}\delta}{r^{s+2}} \, dr + C\delta.$$

Here by C we denote different constants depending only on d, and $B(x,r) := \{y \in \mathbb{R}^d : |y - x| < r\}$. We have

$$\delta \int_\delta^2 \frac{r^{d-1}}{r^{s+2}} \, dr = \begin{cases} \delta \ln \frac{2}{\delta}, & s = d-2, \\ \frac{1}{d-s-2}(2^{d-s-2}\delta - \delta^{d-s-1}), & s \neq d-2. \end{cases}$$

Thus, all terms on the right-hand side of the estimate for $|R_1^s \mu(x)|$ tend to 0 as $\delta \to 0$, and we may choose δ and a corresponding measure μ satisfying (2.1). $\square$

We need the following lemma. The notation $A \approx B$ means that $cA < B < CB$ with constants c, C which may depend only on d and s.

Lemma 2.2. *Let μ be a non-negative measure in $\mathbb{R}^d$ with continuous density and compact support. Let $0 < s < d - 1$. Then for every ball $B = B(x_0, r)$,*

$$\left| \int_{\partial B} (R^s \mu \cdot \mathbf{n}) \, d\sigma \right| \approx r^{d-s-1} \mu(B) + r^d \int_r^\infty \frac{d\mu(B(x_0, t))}{t^{s+1}}, \tag{2.2}$$

where $\mathbf{n}$ is the outer normal vector to B and σ is the surface measure on ∂B.

Proof. We will use the Ostrogradsky–Gauss theorem and differentiation under the integral sign. To justify these operations and make an integrand sufficiently smooth, we approximate $K(x) = x/|x|^{s+1}$ by a smooth kernel K_ε in the following standard way. Let $\phi(t)$, $t \geq 0$, be a C^∞-function such that $\phi(t) = 0$ as $0 \leq t \leq 1$, $\phi(t) = 1$ as $t \geq 2$, and $0 \leq \phi'(t) \leq 2$, $t > 0$. Let $\phi_\varepsilon(t) := \phi(\frac{t}{\varepsilon})$, $K_\varepsilon(x) := \phi_\varepsilon(|x|)K(x)$, and $\widetilde{R}_\varepsilon^s \mu := K_\varepsilon * \mu$. We have

$$\int_{\partial B} (\widetilde{R}_\varepsilon^s \mu \cdot \mathbf{n}) \, d\sigma = \int_B \nabla \cdot \widetilde{R}_\varepsilon^s \mu(x) \, dm_d(x)$$

$$= \int_B \left[\int_{\mathbb{R}^d} \nabla \cdot \phi_\varepsilon(|y-x|) \frac{y-x}{|y-x|^{s+1}} \, d\mu(y) \right] dm_d(x).$$

The inner integral is equal to

$$\int_{|y-x| \leq 2\varepsilon} \nabla \cdot \phi_\varepsilon(|y-x|) \frac{y-x}{|y-x|^{s+1}} \, d\mu(y) + \int_{|y-x| > 2\varepsilon} \nabla \cdot \frac{y-x}{|y-x|^{s+1}} \, d\mu(y)$$

$$=: I_1(x) + I_2(x).$$

One can easily see that

$$\left| \frac{\partial}{\partial x_i} \left[\phi_\varepsilon(|y-x|) \frac{y-x}{|y-x|^{s+1}} \right] \right| < C \left[\frac{1}{\varepsilon} \frac{1}{|y-x|^s} + \frac{1}{|y-x|^{s+1}} \right] < \frac{C}{|y-x|^{s+1}},$$
$$|y-x| \leq 2\varepsilon.$$

Hence,

$$|I_1(x)| < C \int_{|y-x|\leq 2\varepsilon} \frac{1}{|y-x|^{s+1}} \, d\mu(y) < C \int_0^{2\varepsilon} \frac{1}{t^{s+1}} \, d\mu(B(x,t))$$

$$\approx \frac{\mu(B(x,2\varepsilon))}{(2\varepsilon)^{s+1}} + \int_0^{2\varepsilon} \frac{\mu(B(x,t))}{t^{s+2}} \, dt.$$

Since μ has a continuous density with respect to m_d, we have $\mu(B(x,t)) < A_{\mu,B} t^d$ as $t \leq 2\varepsilon < 1$, $x \in B$. Taking into account that $s < d-1$, we obtain the relation $\int_B I_1(x) \, dm_d(x) \to 0$ as $\varepsilon \to 0$.

To estimate the integral of $I_2(x)$ we use the equality $\nabla \cdot \frac{x}{|x|^{s+1}} = \frac{d-s-1}{|x|^{s+1}}$. Thus,

$$\left| \int_B I_2(x) \, dm_d(x) \right| = C \int_B \left[\int_{|y-x|>2\varepsilon} \frac{d\mu(y)}{|y-x|^{s+1}} \right] dm_d(x)$$

$$= C \left(\int_{B(x_0, r+2\varepsilon)} \left[\int_{B \cap \{|y-x|>2\varepsilon\}} \frac{dm_d(x)}{|y-x|^{s+1}} \right] d\mu(y) \right.$$

$$\left. + \int_{\mathbb{R}^d \setminus B(x_0, r+2\varepsilon)} \left[\int_B \frac{dm_d(x)}{|y-x|^{s+1}} \right] d\mu(y) \right) =: C(J_1 + J_2).$$

Obviously,

$$\int_B \frac{dm_d(x)}{|y-x|^{s+1}} \approx \begin{cases} \int_0^r \frac{t^{d-1} \, dt}{t^{s+1}} \approx r^{d-s-1}, & |y-x_0| \leq r, \\[2mm] \dfrac{r^d}{|y-x_0|^{s+1}}, & |y-x_0| > r. \end{cases}$$

In order to estimate J_1 we note that for sufficiently small ε,

$$\int_{B \cap \{|y-x|>2\varepsilon\}} \frac{dm_d(x)}{|y-x|^{s+1}} \approx \int_B \frac{dm_d(x)}{|y-x|^{s+1}} \approx r^{d-s-1}, \quad y \in B(x_0, r+2\varepsilon).$$

Hence, $J_1 \approx r^{d-s-1} \mu(B(x_0, r+2\varepsilon))$. Moreover,

$$J_2 \approx \int_{\mathbb{R}^d \setminus B(x_0, r+2\varepsilon)} \frac{r^d}{|y-x_0|^{s+1}} \, d\mu(y) = r^d \int_{r+2\varepsilon}^\infty \frac{d\mu(B(x_0,t))}{t^{s+1}}.$$

Passing to the limit as $\varepsilon \to 0$, we get (2.2) □

Now we are ready to prove our main result.

Theorem 2.3. *Let μ be a non-negative measure in $\mathbb{R}^d$ with continuous density and compact support. Let $0 < s < 1$. Then (1.1) holds with a constant C depending only on d and s.*

Proof. Let us sketch the idea of the proof. Let a measure μ be such that $\mu(B(y,t)) \leq Ct^s$, $y \in \mathbb{R}^d$, $t > 0$. For Lipschitz continuous compactly supported functions φ, ψ, define the form $\langle R^s(\psi\mu), \varphi \rangle_\mu$ by the equality

$$\langle R^s(\psi\mu), \varphi \rangle_\mu = \frac{1}{2} \iint_{\mathbb{R}^d \times \mathbb{R}^d} \frac{y-x}{|y-x|^{s+1}} (\psi(y)\varphi(x) - \psi(x)\varphi(y)) \, d\mu(y) \, d\mu(x);$$

the double integral exists since $|\psi(y)\varphi(x) - \psi(x)\varphi(y)| \le C_{\psi,\varphi}|x - y|$. If we assume in addition that $\int \psi \, d\mu = 0$, we may define $\langle R^s(\psi\mu), \varphi \rangle_\mu$ for any (not necessarily compactly supported) bounded Lipschitz continuous function φ on $\mathbb{R}^d$; here we follow [4]. Let supp $\psi \in B(0, R)$. For $|x| > 2R$ we have

$$\left| \int_{\mathbb{R}^d} \frac{y - x}{|y - x|^{s+1}} \psi(y) \, d\mu(y) \right| = \left| \int_{\mathbb{R}^d} \left[\frac{y - x}{|y - x|^{s+1}} + \frac{x}{|x|^{s+1}} \right] \psi(y) \, d\mu(y) \right|$$

$$\le \frac{C}{|x|^{s+1}} \int_{\mathbb{R}^d} |y\psi(y)| \, d\mu(y) = \frac{C_\psi}{|x|^{s+1}}.$$

Choose a Lipschitz continuous compactly supported function ξ which is identically 1 on $B(0, 2R)$. Then we may define the form $\langle R^s(\psi\mu), \varphi \rangle_\mu$ as

$$\langle R^s(\psi\mu), \varphi \rangle_\mu$$

$$= \langle R^s(\psi\mu), \xi\varphi \rangle_\mu + \int_{\mathbb{R}^d} \left[\int_{\mathbb{R}^d} \frac{y - x}{|y - x|^{s+1}} \psi(y) \, d\mu(y) \right] (1 - \xi(x))\varphi(x) \, d\mu(x).$$

The repeated integral is well defined because

$$\int_{|x|>2R} \frac{d\mu(x)}{|x|^{s+1}} \le C \int_{2R}^\infty \frac{\mu(B(0,t))}{t^{s+2}} \, dt \le C \int_{2R}^\infty \frac{1}{t^2} \, dt.$$

Assuming that Theorem 2.3 is incorrect and using the Cotlar inequality we establish the existence of a positive measure ν such that ν has no point masses, the operator R_ν^s is bounded on $L^2(\nu)$, and $\langle R^s(\psi\nu), 1 \rangle_\nu = 0$ for every Lipschitz continuous function ψ with $\int \psi \, d\mu = 0$. This means that ν is a reflectionless measure, that is a measure without point masses with the following properties: R_ν^s is bounded on $L^2(\nu)$, and $\langle R^s(\psi\nu), 1 \rangle_\nu = 0$ for every Lipschitz continuous compactly supported function ψ such that $\int \psi \, d\mu = 0$. But according to the recent result by Prat and Tolsa [11] such measures do not exist for $0 < s < 1$. We remark that the proof of this result contains estimates of an analog of Melnikov's curvature of a measure. This is the obstacle to extent the result to $s \ge 1$. We now turn to the details.

Suppose that C satisfying (1.1) does not exists. Then for every $n \ge 1$ there is a positive measure μ_n such that

$$\sup_{x \in \mathbb{R}^d} |R^s \mu_n(x)| = 1, \qquad \sup_{x \in \text{supp}\,\mu_n} |R^s \mu_n(x)| \le 1/n.$$

Let

$$\theta_\mu(x, r) := \mu(B(x, r))/\, r^s, \qquad \theta_\mu := \sup_{x,r} \theta_\mu(x, r).$$

We prove that

$$0 < c < \theta_{\mu_n} < C. \tag{2.3}$$

The estimate from above is a direct consequence of Lemma 2.2. Indeed, for any ball $B(x, r)$ (2.2) implies the estimate

$$c_d r^{d-1} \ge \left| \int_{\partial B} (R^s \mu_n \cdot \mathbf{n}) \, d\sigma \right| \ge C r^{d-s-1} \mu_n(B),$$

which implies the desired inequality.

The estimate from below follows immediately from a Cotlar-type inequality

$$\sup_{x \in \mathbb{R}^d} |R^s \mu_n(x)| \leq C \Big[\sup_{x \in \text{supp}\, \mu_n} |R^s \mu_n(x)| + \theta_{\mu_n} \Big]$$

(see [8, Theorem 7.1] for a more general result).

Let $B(x_n, r_n)$ be a ball such that $\theta_{\mu_n}(x_n, r_n) > c = c(s, d)$, and let $\nu_n(\cdot) = r_n^{-s} \mu_n(x_n + r_n \cdot)$. Then

$$R^s \mu_n(x) = R^s \nu_n \left(\frac{x - x_n}{r_n} \right), \quad \theta_{\nu_n}(y, t) = \theta_{\mu_n}(r_n y + x_n, r_n t).$$

In particular, $\nu_n(B(0,1)) = \theta_{\mu_n}(x_n, r_n) > c$. Choosing a weakly converging subsequence of $\{\nu_n\}$, we obtain a positive measure ν. If we prove that

(a) $\nu(B(y,t)) \leq Ct^s$,
(b) $\langle R^s \nu, \psi \rangle_\nu = 0$ for every Lipschitz continuous compactly supported function ψ with $\int \psi \, d\nu = 0$,
(c) the operator R_ν^s is bounded on $L^2(\nu)$,

then ν is reflectionless, and we arrive at a contradiction with Theorem 1.1 in [11] mentioned above. Thus, the proof would be completed.

Property (a) follows directly from (2.3). For weakly converging measures ν_n with $\theta_{\mu_n} < C$ we may apply Lemma 8.4 in [4] which yields (b). To establish (c) we use the inequality

$$R^{s,*} \mu(x) := \sup_{\varepsilon > 0} |R_{\mu,\varepsilon}^s 1(x)| \leq \|R^s \mu\|_{L^\infty(m_d)} + C, \quad x \in \mathbb{R}^d, \ C = C(s),$$

for any positive Borel measure μ such that $\mu(B(x,r)) \leq r^s$, $x \in \mathbb{R}^d$, $r > 0$, – see [12, Lemma 2] or [13, p. 47], [1, Lemma 5.1] for a more general setting. Thus, $R_\varepsilon^s \nu_n(x) := R_{\nu_n,\varepsilon}^s 1(x) \leq C$ for every $\varepsilon > 0$. Hence, $R_\varepsilon^s \nu(x) \leq C$ for $\varepsilon > 0$, $x \in \mathbb{R}^d$, and the non-homogeneous $T1$-theorem [9] implies the boundedness of R_ν^s on $L^2(\nu)$. $\qquad \square$

3. The case of a radial density

Lemma 2.2 allows us to prove the maximum principle for all $s \in (0, d)$ in the special case of a radial density.

Proposition 3.1. *Let $d\mu(x) = h(|x|) \, dm_d(x)$, where $h(t)$ is a continuous function on $[0, \infty)$, and let $s \in (0, d - 1)$. Then (1.1) holds with a constant C depending only on d and s.*

We remind the reader that for $s \in [d - 1, d)$ Conjecture 1.1 is proved in [2] for any compactly supported measure with C^∞ density. Thus, for compactly supported radial measures with C^∞ density (1.1) holds for all $s \in (0, d)$.

Proof. Because μ is radial, by (2.2) we have

$$c_d r^{d-1} |R^s \mu(x)| = \left| \int_{\partial B(0,r)} (R^s \mu \cdot \mathbf{n}) \, d\sigma \right|$$

$$\approx r^{d-s-1} \mu(B(0,r)) + r^d \int_r^\infty \frac{d\mu(B(0,t))}{t^{s+1}}, \qquad r = |x|.$$

Thus,

$$|R^s \mu(x)| \approx \frac{\mu(B(0,r))}{r^s} + r \int_r^\infty \frac{d\mu(B(0,t))}{t^{s+1}}.$$

Fix $w \notin \operatorname{supp} \mu$, and let $r = |w|$. If

$$\frac{\mu(B(0,r))}{r^s} \geq r \int_r^\infty \frac{d\mu(B(0,t))}{t^{s+1}},$$

then there is $r_1 \in (0,r)$ such that $\{y : |y| = r_1\} \subset \operatorname{supp} \mu$ and $\mu(B(0,r)) = \mu(B(0,r_1))$. Hence,

$$|R^s \mu(w)| \approx \frac{\mu(B(0,r))}{r^s} < \frac{\mu(B(0,r_1))}{r_1^s} \leq C |R^s \mu(x_1)|, \quad |x_1| = r_1.$$

If

$$\frac{\mu(B(0,r))}{r^s} < r \int_r^\infty \frac{d\mu(B(0,t))}{t^{s+1}},$$

then there is $r_2 > r$ such that $\{y : |y| = r_2\} \subset \operatorname{supp} \mu$ and $\mu(B(0,r)) = \mu(B(0,r_2))$. Hence,

$$\frac{\mu(B(0,r_2))}{r_2^s} < \frac{\mu(B(0,r))}{r^s} < r \int_{r_2}^\infty \frac{d\mu(B(0,t))}{t^{s+1}},$$

and we have

$$|R^s \mu(w)| \approx r \int_{r_2}^\infty \frac{d\mu(B(0,t))}{t^{s+1}} \leq C |R^s \mu(x_2)|, \quad |x_2| = r_2. \qquad \square$$

4. Counterexample

Given $\varepsilon > 0$, we construct a signed measure $\nu = \nu(\varepsilon)$ in $\mathbb{R}^5$ with the following properties:

(a) ν is a radial signed measure with C^∞-density;
(b) $\operatorname{supp} \nu \in D_\varepsilon := \{1 - \varepsilon \leq |x| \leq 1 + \varepsilon\}$;
(c) $|R^2 \nu(x)| < \varepsilon$ for $x \in \operatorname{supp} \nu$; $|R^2 \nu(x)| > a > 0$ for $|x| = 2$, where a is an absolute constant. Here $R^2 \nu$ means $R^s \nu$ with $s = 2$.

Let $\Delta^2 := \Delta \circ \Delta$, and let

$$u(x) = \begin{cases} 2/3, & |x| \leq 1, \\ \dfrac{1}{|x|} - \dfrac{1}{3|x|^3}, & |x| > 1. \end{cases}$$

Note that $\Delta\left(\frac{1}{|x|^3}\right) = 0$ and $\Delta^2\left(\frac{1}{|x|}\right) = 0$ in $\mathbb{R}^5 \setminus \{0\}$. Hence, $\Delta^2 u(x) = 0, |x| \neq 1$. Moreover, ∇u is continuous in $\mathbb{R}^5$ and $\nabla u(x) = 0, |x| = 1$.

For $\delta \in (0, \varepsilon)$, let $\varphi_\delta(x)$ be a C^∞-function in $\mathbb{R}^5$ such that $\varphi_\delta > 0$, $\operatorname{supp} \varphi_\delta = \{x \in \mathbb{R}^5 : |x| \leq \delta\}$, and $\int \varphi_\delta(x)\, dm_5(x) = 1$ (for example, a bell-like function on $|x| \leq \delta$). Let $U_\delta := u * \varphi_\delta$. Then $\Delta^2 U_\delta(x) = 0$ as $x \notin D_\delta$. Also, $\Delta U_\delta(x) \to 0$ as $|x| \to \infty$. Hence, the function ΔU_δ can be represented in the form $\Delta U_\delta = c\left(\frac{1}{|x|^3} * \Delta^2 U_\delta\right)$ (here and in the sequel by c we denote various absolute constants). Set $d\nu_\delta = \Delta^2 U_\delta\, dm_5$. Then $\operatorname{supp} \nu_\delta \in D_\delta$, and (b) is satisfied. Since $\Delta\left(\frac{1}{|x|}\right) = \frac{c}{|x|^3}$, we have $\Delta U_\delta = c\Delta\left(\frac{1}{|x|} * \Delta^2 U_\delta\right)$, that is $U_\delta = c\left(\frac{1}{|x|} * \Delta^2 U_\delta\right) + h$, where h is a harmonic function in $\mathbb{R}^5$. Since both U_δ and $\frac{1}{|x|} * \Delta^2 U_\delta$ tend to 0 as $x \to \infty$, we have

$$U_\delta = c\left(\frac{1}{|x|} * \Delta^2 U_\delta\right).$$

Thus,

$$R^2\nu(x) = c \int \frac{y - x}{|y - x|^3}\, d\nu_\delta(y) = c\nabla U_\delta(x).$$

Obviously, $\nabla U_\delta = \nabla(u * \varphi_\delta) = (\nabla u) * \varphi_\delta$, and hence $\max_{x \in D_\delta} |\nabla U_\delta(x)| \to 0$ as $\delta \to 0$. On the other hand, for fixed x with $|x| > 1$ (say, $|x| = 2$) we have $\lim_{\delta \to 0} |\nabla u * \varphi_\delta| = |\nabla u(x)| > 0$. Thus, (c) is satisfied if δ is chosen sufficiently small.

Remark. It is well known that the maximum principle (with a constant C) holds for potentials $\int K(|x - y|)\, d\mu(y)$ with non-negative kernels $K(t)$ decreasing on $(0, \infty)$, and non-negative finite Borel measures μ. Our arguments show that for non-positive measures the analog of (1.1) fails even for potentials with positive Riesz kernels. In fact we have proved that *for every $\varepsilon > 0$, there exists a signed measure $\eta = \eta(\varepsilon)$ in $\mathbb{R}^5$ with C^∞-density and such that $\operatorname{supp} \eta \in D_\varepsilon := \{1 - \varepsilon \leq |x| \leq 1 + \varepsilon\}$, $|u_\eta(x)| < \varepsilon$ for $x \in \operatorname{supp} \eta$, but $|u_\eta((2, 0, \ldots, 0))| > b > 0$*, where $u_\eta(x) := \int \frac{d\eta(y)}{|y - x|}$, and b is an absolute constant.

Indeed, for the first component $R_1^2\nu$ of $R^2\nu$ we have

$$R_1^2\nu = c\frac{\partial}{\partial x_1} U_\delta = c\left(\frac{1}{|x|} * \frac{\partial}{\partial x_1}(\Delta^2 U_\delta)\right) = cu_\eta, \quad \text{where } d\eta = \frac{\partial}{\partial x_1}(\Delta^2 U_\delta)\, dm_5.$$

References

[1] D.R. Adams, V.Ya. Eiderman, *Singular operators with antisymmetric kernels, related capacities, and Wolff potentials*, Internat. Math. Res. Notices 2012, no. 24, 5554–5584; arXiv:1012.2877, 22 pp.

[2] V. Eiderman, F. Nazarov, A. Volberg, *The s-Riesz transform of an s-dimensional measure in $\mathbb{R}^2$ is unbounded for $1 < s < 2$*, J. Anal. Math. **122** (2014), 1–23; arXiv:1109.2260.

[3] B. Jaye, F. Nazarov, A. Volberg, *The fractional Riesz transform and an exponential potential*, Algebra i Analiz **24** (2012), no. 6, 77–123; reprinted in St. Petersburg Math. J. **24** (2013), no. 6, 903–938; arXiv:1204.2135.

[4] B. Jaye, F. Nazarov, *Reflectionless measures for Calderón–Zygmund operators I: general theory*, to appear in J. Anal. Math., arXiv:1409.8556.

[5] B. Jaye, F. Nazarov, M.C. Reguera, X. Tolsa, The Riesz transform of codimension smaller than one and the Wolff energy, arXiv:1602.02821.

[6] F. Nazarov, X. Tolsa, A. Volberg, *On the uniform rectifiability of AD regular measures with bounded Riesz transform operator: the case of codimension 1*, Acta Math. **213** (2014), no. 2, 237–321; arXiv:1212.5229, 88 p.

[7] F. Nazarov, X. Tolsa, A. Volberg, *The Riesz transform, rectifiability, and removability for Lipschitz harmonic functions*, Publ. Mat. **58** (2014), no. 2, 517–532, arXiv:1212.5431, 15 p.

[8] F. Nazarov, S. Treil, and A. Volberg, *Weak type estimates and Cotlar inequalities for Calderón–Zygmund operators on nonhomogeneous spaces*, Internat. Math. Res. Notices 1998, no. 9, 463–487.

[9] F. Nazarov, S. Treil, and A. Volberg, *The Tb-theorem on non-homogeneous spaces*, Acta Math. **190** (2003), no. 2, 151–239.

[10] L. Prat, *Potential theory of signed Riesz kernels: capacity and Hausdorff measure*, Internat. Math. Res. Notices, 2004, no. 19, 937–981.

[11] L. Prat, X. Tolsa, Non-existence of reflectionless measures for the s-Riesz transform when $0 < s < 1$, Ann. Acad. Sci. Fenn. Math. 40 (2015), no. 2, 957–968.

[12] M. Vihtilä, *The boundedness of Riesz s-transforms of measures in* $\mathbb{R}^n$, Proc. Amer. Math. Soc. **124** (1996), no. 12, 3797–3804.

[13] A. Volberg, *Calderón–Zygmund capacities and operators on nonhomogeneous spaces*. CBMS Regional Conference Series in Mathematics, **100**. AMS, Providence, RI, 2003.

[14] A.L. Volberg, V.Ya. Eiderman, *Nonhomogeneous harmonic analysis: 16 years of development*, Uspekhi Mat. Nauk **68** (2013), no. 6(414), 3–58 (Russian); translation in Russian Math. Surveys **68** (2013), no. 6, 973–1026.

Vladimir Eiderman
Department of Mathematics
Indiana University, Bloomington, IN
e-mail: veiderma@indiana.edu

Fedor Nazarov
Department of Mathematics
Kent State University, Kent, OH
e-mail: nazarov@math.kent.edu

Operator Theory:
Advances and Applications, Vol. 261, 267–279
© Springer International Publishing AG, part of Springer Nature 2018

Submultiplicative Operators on C^k-Spaces

Dmitry Faifman, Hermann König and Vitali Milman

Dedicated to the memory of the great analyst Victor Havin, F-K-M.
Expressing my warmest gratitude for fifty years of our friendship, V.M.

Abstract. Let $I \subset \mathbb{R}$ be open, let $k \in \mathbb{N}_0$ and $T : C^k(I) \to C^k(I)$ be bijective, pointwise continuous and submultiplicative, i.e.,

$$T(f \cdot g) \leq Tf \cdot Tg; \quad f, g \in C^k(I).$$

Suppose also that $T(-\mathbb{1}) < 0$ and that T and T^{-1} are positivity preserving. We show that there is a homeomorphism $u : I \to I$ and that there are continuous functions $p, A \in C(I)$ with $p > 0$, $A \geq 1$ such that

$$(Tf)(u(x)) = \begin{cases} f(x)^{p(x)} & f(x) \geq 0, \\ -A(x)|f(x)|^{p(x)} & f(x) < 0. \end{cases}$$

For $k \in \mathbb{N}$, we have $A = p = 1$, and u is a C^k-diffeomorphismus so that

$$Tf(u(x)) = f(x),$$

i.e., T is multiplicative for $k \in \mathbb{N}$. This rigidity property also holds for supermultiplicative operators.

Mathematics Subject Classification (2010). Primary 39B62. Secondary 46E25, 46J10.

Keywords. Submultiplicative operators, approximate indicator, positivity preserving diffeomorphism.

1. Introduction and results

In 1949 Milgram [Mi] characterized bijective multiplicative operators $T : C(M) \to C(M)$ on real-valued spaces of continuous functions $C(M)$ on real manifolds M. They have the form

$$Tf(u(x)) = |f(x)|^{p(x)} \operatorname{sgn} f(x)$$

Dmitry Faifman is supported by NSERC Discovery grant 500549.
Hermann König is supported in part by Minerva.
Vitali Milman is supported in part by the Alexander von Humboldt Foundation, by Minerva, by ISF grant 826/13 and by BSF grant 0361-4561.

where $u : M \to M$ is a homeomorphism and $p : M \to \mathbb{R}_+$ is a continuous function. This result was generalized to spaces of smooth functions $C^k(M)$ on a differentiable manifold M by Mrčun [M] and Mrčun, Šemrl [MS]. It was extended to $C^\infty(M)$-functions and other function algebras like the Schwartz space $\mathcal{S}(\mathbb{R}^n)$ and to spaces of complex-valued functions by Artstein-Avidan, Faifman, Milman [AFM], cf. also the earlier paper by Alesker, Artstein-Avidan, Faifman, Milman [AAFM]. In the case of $C^k(M)$ with $k \geq 1$, we have $p = 1$ and T has the simpler form $Tf(u(x)) = f(x)$, where $u : M \to M$ is a C^k-diffeomorphism.

The corresponding result for $\mathcal{S}(\mathbb{R}^n)$ is useful to characterize the Fourier transform as a bijective map on $S(\mathbb{R}^n)$ which maps convolutions to products and *vice versa*; this is a characterization up to diffeomorphisms and complex conjugation, cf. [AFM], [AAFM].

In this paper, we study stability properties of bijective multiplicative maps $T : C(I) \to C(I)$ on $C(M)$-spaces where $M = I \subset \mathbb{R}$ is an open set, by replacing the equality by an inequality. We consider bijective submultiplicative (or supermultiplicative) operators $T : C^k(I) \to C^k(I)$ and give the general form of such maps under weak additional assumptions on T. In the case of differentiable functions ($k \geq 1$), the solutions are the same as in the bijective multiplicative case, i.e., the problem is very rigid. In the case of continuous functions ($k = 0$), the negative part of Tf has a slightly more general form. This was already established for continuous bijective submultiplicative functions $f : \mathbb{R} \to \mathbb{R}$ which have the form

$$f(t) = \begin{cases} t^p & t \geq 0, \\ -A|t|^p & t < 0. \end{cases}$$

for a suitable $p > 0$ and $A \geq 1$, as shown in König, Milman [KM].

The proof of the main result consists firstly of a localization step which modifies a corresponding proof in [AFM] and is based on ideas of [MS], and secondly of an analysis of the representing function F, $Tf(x) = F(x, f(x), \ldots, f^{(k)}(x))$, where the result of [KM] for bijective submultiplicative functions $f : \mathbb{R} \to \mathbb{R}$ is used. To formulate our main result, we need the following notation and definitions.

Definition. Let $k \in \mathbb{N}_0$ and $I \subset \mathbb{R}$ be open. We denote $C^k(I) := \{f : I \to \mathbb{R} \mid f$ is k times continuously differentiable$\}$, with $C(I) = C^0(I)$ being the continuous functions on I. By $C_c^k(I)$ we denote the functions in $C^k(I)$ with compact support in I.

Definition. An operator $T : C^k(I) \to C^k(I)$ is *pointwise continuous* provided that for any functions $(f_n)_{n \in \mathbb{N}}$ and f in $C^k(I)$ such that the sequence $(f_n)^{(j)}$ converges to $f^{(j)}$ uniformly on all compact sets in I for all $0 \leq j \leq k$, the limit $\lim_{n \to \infty} T(f_n)(x)$ exists for all $x \in I$ and is equal to $T(f)(x)$.

Definition. For $f \in C^k(I)$ we write $f > 0$ ($f \geq 0$) if $f(x) > 0$ ($f(x) \geq 0$) for all $x \in I$. An operator is *positivity preserving* if for every $f \in C^k(I)$, $f \geq 0$ implies $Tf \geq 0$. Furthermore, $f \leq 0$ is defined similarly and T is *negativity preserving* if $f \leq 0$ always implies $Tf \leq 0$.

Theorem 1. *Let $I \subset \mathbb{R}$ be open and $k \in \mathbb{N}_0$. Suppose that $T : C^k(I) \to C^k(I)$ is bijective, pointwise continuous and submultiplicative, i.e.,*

$$T(f \cdot g)(x) \leq Tf(x) \cdot Tg(x); \quad f, g \in C^k(I), \ x \in I. \tag{1}$$

Assume also that $T(-\mathbb{1}) < 0$ and that T and T^{-1} are positivity preserving. Then there exists a homeomorphism $u : I \to I$ and two continuous functions $p, A \in C(I)$ with $A \geq 1$, $p > 0$ such that

$$(Tf)(u(x)) = \begin{cases} f(x)^{p(x)} & f(x) \geq 0, \\ -A(x)|f(x)|^{p(x)} & f(x) < 0. \end{cases}$$

If $k \in \mathbb{N}$, then $A = p = 1$ and u is a C^k-diffeomorphism, so that

$$Tf(u(x)) = f(x).$$

Remark. Hence for $k \in \mathbb{N}$, under the conditions of the theorem, T is even *multiplicative*, which is the case considered by Mrčun, Šemrl [MS] and Artstein-Avidan, Faifman, Milman [AFM] in a more general context. The result for *multiplicative* bijective maps in $C(I)$ goes back to Milgram [Mi]. A similar result holds when I is the unit circle. In this case the assumption that T^{-1} is positivity preserving is not needed. It is used in Proposition 4 to prove the compactness of the support of certain functions, which is automatically true in the case of S^1. The analogue for supermultiplicative functions is

Theorem 2. *Let $I \subset \mathbb{R}$ be open and let $k \in \mathbb{N}_0$. Suppose that $T : C^k(I) \to C^k(I)$ is bijective, pointwise continuous and supermultiplicative, i.e.,*

$$T(f \cdot g)(x) \geq Tf(x) \cdot Tg(x); \quad f, g \in C^k(I), \ x \in I.$$

Assume also that $T(-\mathbb{1}) < 0$ and that T and T^{-1} are negativity preserving. Then there exists a homeomorphism $u : I \to I$ and two continuous functions $p, B \in C(I)$ with $0 < B \leq 1$, $p > 0$ such that

$$(Tf)(u(x)) = \begin{cases} f(x)^{p(x)} & f(x) \geq 0, \\ -B(x)|f(x)|^{p(x)} & f(x) < 0. \end{cases}$$

If $k \in \mathbb{N}$, then $B = p = 1$ and u is a C^k-diffeomorphism, so that

$$Tf(u(x)) = f(x).$$

The formulas also show that, e.g., in the case of Theorem 1 when T is submultiplicative, the inverse T^{-1} is supermultiplicative.

2. Localization

To prove Theorem 1, we first show that T is locally defined, i.e., that $Tf(y)$ is determined by the values and derivatives of f at a single point x, first at least for a dense set of points y. Let $k \in \mathbb{N}_0$, and let $I \subset \mathbb{R}$ be an open interval or $I = S^1$. A function $f \in C^k(I)$ is an *approximate indicator* at $x \in I$ if $f \geq 0$ and there are open neighborhoods $x \in J_1 \subset J_2$ of x such that $f|_{J_1} = \mathbb{1}$ and $f|_{I \setminus J_2} = 0$. For $x \in I$, let

$$AI_x := \{f \in C^k(I) \mid f \text{ is an approximate indicator at } x\}.$$

Denote by $\mathcal{F}(I)$ the closed subsets of I and define a map $u : I \to \mathcal{F}(I)$ by $u(x) := \cap_{f \in AI_x} \text{supp}(Tf)$. Here $\text{supp}(Tf)$ denotes the support of Tf.

Proposition 3. *Suppose the assumptions of Theorem 1 hold. Then*

(i) *For $f \in AI_x$, $Tf|_{u(x)} = \mathbb{1}$. More generally, if the germ of f at x is either 1 or 0, the same is true of Tf at any $y \in u(x)$.*

(ii) *For $f, g \in C^k(I)$, $g \geq 0$ with $fg = g$ (respectively, $fg = 0$) we have $TfTg = Tg$ (respectively $TfTg = 0$).*

(iii) *For $f \in C^k(I)$, $\text{sgn}\, Tf|_{u(x)} = \text{sgn}\, f(x)$.*

(iv) *For $x \in I$, $u(x)$ consists of at most one point. For all $x, y \in I$ with $x \neq y$, $u(x) \cap u(y) = \emptyset$.*

Proof. We partially use arguments from Mrčun [M], cf. also [MS], [AFM].

(a) For any $f \in C^k(I)$ and $x \in I$, we have $Tf(x) \leq T\mathbb{1}(x)Tf(x)$. Since T is surjective, $Tf(x)$ may be chosen either positive or negative which yields the relation $T\mathbb{1} = \mathbb{1}$. Then $\mathbb{1} = T\mathbb{1} \leq T(-\mathbb{1})^2$ and $T(-\mathbb{1}) < 0$ implies $T(-\mathbb{1}) \leq -\mathbb{1}$. Similarly, $T0(x) \leq T0(x)Tf(x)$ implies that $T0 = 0$. If $f \leq 0$ then $Tf \leq 0$ since in this case $-f \geq 0$ and $Tf \leq T(-f)T(-\mathbb{1}) \leq 0$.

(b) Let $f, g \in C^k(I)$ with $g \geq 0$ and $fg = g$ (respectively $fg = 0$). We claim that then $TfTg = Tg$ (respectively $TfTg = 0$).

First note that $\text{supp}(Tg) \subset \text{supp}(T(-g))$ since $T(-g) \leq T(-\mathbb{1})Tg \leq 0$. If $fg = g$ and $g \geq 0$, then $0 \leq Tg \leq TfTg$ since T is positivity preserving and submultiplicative and hence $Tf|_{\text{supp}(Tg)} \geq 1$. But also $T(-g) \leq T(-g)Tf$, yielding $Tf|_{\text{supp}(T(-g))} \leq 1$. Hence $Tf|_{\text{supp}(Tg)} = \mathbb{1}$ and $TfTg = Tg$.

If $fg = 0$ and $g \geq 0$, then $0 \leq TfTg$, so $Tf|_{\text{supp}(Tg)} \geq 0$. But also $0 \leq T(-g)Tf$, hence $Tf|_{\text{supp}(T(-g))} \leq 0$. Therefore $Tf|_{\text{supp}(Tg)} = 0$, so that $TfTg = 0$. This shows (ii).

(c) For $f \in AI_x$ or if the germ of f at x is 1, we may choose $g \in AI_x$ such that $fg = g$. By (b), $Tf|_{\text{supp}(Tg)} = \mathbb{1}$, hence $Tf|_{u(x)} = \mathbb{1}$. If the germ of f at x is 0, choose $g \in AI_x$ with $fg = 0$. By (b), $Tf|_{\text{supp}(Tg)} = 0$ and $Tf|_{u(x)} = 0$, proving (i).

(d) We claim that $f(x) > 0$ implies $Tf|_{u(x)} > 0$. Namely, if $f(x) > 0$, we may choose a compactly supported $g \geq 0$ such that $fg \in AI_x$. Then by (i), $T(fg)|_{u(x)} = \mathbb{1}$ and hence $Tf|_{u(x)}Tg|_{u(x)} \geq 1$. Since $g \geq 0$, also $Tg \geq 0$. Hence $Tf|_{u(x)} > 0$. Similarly $f(x) < 0$ implies $Tf|_{u(x)} < 0$.

To finish the proof of (iii), it remains to show that $f(x) = 0$ implies $Tf|_{u(x)} = 0$. Choose a positive sequence $\varepsilon_n \to 0$ such that $f_n := f + \varepsilon_n$ satisfies $f_n|_{[x-\frac{1}{n}, x+\frac{1}{n}]} > 0$. It follows that $Tf_n|_{u(x)} > 0$ and $Tf|_{u(x)} \geq 0$ by the pointwise continuity of T. Repeating this for $f_n := f - \varepsilon_n$ with $f_n|_{[x-\frac{1}{n}, x+\frac{1}{n}]} < 0$, we deduce that $Tf|_{u(x)} = 0$.

(e) To prove (iv), assume first that $u(x)$ contains at least two points, say $y \neq z$. Choose $g \in C^k(I)$ with $g(y) = 0$ and $g(z) > 0$. By the surjectivity of T, we find $f \in C^k(I)$ with $Tf = g$. By (iii), $f(x)$ must be both 0 and strictly positive, a contradiction. Now suppose that $x \neq y$. Choose $f \in AI_x$ and $g \in AI_y$ with $fg = 0$. By (ii), $TfTg = 0$, $u(x) \cap u(y) \subset \mathrm{supp}(Tf) \cap \mathrm{supp}(Tg) = \emptyset$. Hence $u(x) \cap u(y) = \emptyset$. $\qquad\square$

To determine whether $u(x) \neq \emptyset$, we distinguish between "good" and "bad" points. Let

$$G := \{x \in I \mid \text{ there is } f \in AI_x \text{ such that } \mathrm{supp}(Tf) \text{ is compact } \}$$

be the set of "good" points. The complement $B := I \setminus G$ is the set of "bad" points.

Proposition 4. *Suppose the assumptions of Theorem 1 hold. Then $u(x) \neq \emptyset$ for all $x \in G$. Hence $u(x)$ consists of a single point and thus $u|_G$ can be regarded as a map $u|_G : G \to I$. The set G is open and $G \subset I$ and $u(G) \subset I$ are both dense in I. The map $u|_G : G \to I$ is injective, continuous and has dense open range.*

Proof. (a) By definition $u(x) = \cap_{f \in AI_x} \mathrm{supp}(Tf)$. By the injectivity of T, $\mathrm{supp}(Tf) \neq \emptyset$ for any $f \in AI_x$. Also any finite intersection of such supports is non-void. Indeed, take $f_1, \dots, f_m \in AI_x$. Then $f := \prod_{j=1}^m f_j \in AI_x$ and $0 \leq Tf \leq \prod_{j=1}^m T(f_j)$ so that $\mathrm{supp}(Tf) \subset \mathrm{supp}(T(f_1)) \cap \cdots \cap \mathrm{supp}(T(f_m))$ and $\mathrm{supp}(Tf) \neq \emptyset$ again by the injectivity of T. For "good" points $x \in G$, at least one of the sets in the intersection is compact, and we may replace all intersecting sets by the intersections with this subset, which are compact. It follows that $u(x) \neq \emptyset$ for all $x \in G$.

(b) For $x \in G$ and $f \in AI_x$ such that $\mathrm{supp}(Tf)$ is compact, the same f belongs to AI_y for y sufficiently close to x. Hence G is open. We now show that $u(G) \subset I$ is dense in I. Take any bounded open set $J_1 \subset I$, $J_1 \neq I$, and any closed set $J_2 \subset J_1$ with non-empty interior. Choose $g \in C^k(I)$, $g \geq 0$ with $g|_{J_2} = 0$ and $g|_{I \setminus J_1} = 1$. Let w be an approximate indicator supported on J_2. By the surjectivity of T, we may find f and v with $Tf = g$ and $Tv = w$. Since T^{-1} is positivity preserving, we have $f \geq 0$ and $v \geq 0$. Since $0 = gw = TfTv \geq T(fv)$ with $fv \geq 0$, we conclude that $fv = 0$. In particular, since $v \neq 0$, there is an open interval I_f on which f vanishes identically. Taking $x \in I_f$ and $f_x \in AI_x$ supported on I_f, we see that $f f_x = 0$ and hence by Proposition 3 (ii), $0 = TfT(f_x) = gT(f_x)$. Hence $\mathrm{supp}(T(f_x)) \subset J_1$ is compact and x is good with $u(x) \subset J_1$.

(c) To prove that also G is dense in I, suppose to the contrary that there is an open interval $J \subset I$ consisting of "bad" points. Choose $x \in J$ and $f \in AI_x$ with $\mathrm{supp}(f) \subset J$. For any $y \in G$ there is $g \in AI_y$ with $fg = 0$. Applying again Proposition 3 (ii), we conclude that $\mathrm{supp}(Tf) \cap \{x \in I \mid Tg(x) = 1\} = \emptyset$ and therefore $u(G)$ is disjoint from $\mathrm{supp}(Tf)$, contradicting the density of $u(G)$ in I.

(d) By Proposition 3 (iv), $u|_G$ is injective. Suppose $u|_G : G \to I$ were not continuous. Then there would be $x_n, x \in G$, $x_n \to x$ such that $u(x_n)$ would not converge to $u(x)$. Passing to a subsequence, we may find $\varepsilon > 0$ such that $|u(x_n) - u(x)| \geq \varepsilon$ for any $n \in \mathbb{N}$. Hence there exists $g \in C^k(I)$ with $g(u(x)) = 1$ and $g(u(x_n)) = -1$ for all $n \in \mathbb{N}$. By surjectivity, $g = Tf$ for some $f \in C^k(I)$. By Proposition 3 (iii), $f(x_n) < 0$ and $f(x) > 0$, contradicting the continuity of f.

Thus $u|_G$ is continuous. Since G is open, it can be written as a countable disjoint union of open intervals, $G = \cup I_n$. Since u is injective and continuous, $u(I_n)$ is an open interval and any two such intervals are disjoint. Therefore $u(G) = \cup u(I_n)$ is open and dense. $\qquad\square$

Proposition 5. *Under the assumptions of Theorem 1 we have: if $J \subset I$ is open and $f \in C^k(I)$ is a function with $f|_J = -\mathbb{1}$, then $Tf|_{u(J)} = T(-\mathbb{1})|_{u(J)}$. Here $u(J) = \bigcup_{x \in J} u(x)$.*

Proof. Let $J \subset I$ be open and let $f \in C^k(I)$ be a function with $f|_J = -\mathbb{1}$. Then $T(f) \leq T(-f)T(-\mathbb{1})$. Since $-f|_J = \mathbb{1}$, we obtain $Tf|_{u(J)} \leq T(-\mathbb{1})|_{u(J)}$ by Proposition 3 (i).

Now assume first that f with $f|_J = -\mathbb{1}$ is invertible, i.e., never zero. Put $g := \frac{1}{f}$. Then $T(-\mathbb{1}) \leq T(f)T(-g)$ with $-g|_J = \mathbb{1}$, so again by Proposition 3 (i), $T(-\mathbb{1})|_{u(J)} \leq T(f)|_{u(J)}$, i.e., $T(-\mathbb{1})|_{u(J)} = T(f)|_{u(J)}$.

For a general $f \in C^k(I)$ with $f|_J = -\mathbb{1}$ note that $T(-f^2) \leq T(-f)T(f)$, so again by Proposition 3 (i) $T(-f^2)|_{u(J)} \leq T(f)|_{u(J)}$ and hence

$$T(-f^2)|_{u(J)} \leq T(f)|_{u(J)} \leq T(-\mathbb{1})|_{u(J)}.$$

It now suffices to show that $T(-\mathbb{1})|_{u(J)} = T(-f^2)|_{u(J)}$. Let (ε_n) be a positive sequence with $\varepsilon_n \to 0$ and let $g_n := \frac{-f^2 - \varepsilon_n}{1 + \varepsilon_n}$. Then $g_n < 0$ and $g_n|_J = -\mathbb{1}$. For these invertible functions g_n we know by the above that $T(-\mathbb{1})|_{u(J)} = T(g_n)|_{u(J)}$. Since $(g_n)^{(j)} \to (-f^2)^{(j)}$ uniformly on all compacta for all $0 \leq j \leq k$, the pointwise continuity of T implies that $T(-\mathbb{1})|_{u(J)} = T(-f^2)|_{u(J)}$. $\qquad\square$

Proposition 6. *Suppose the assumptions of Theorem 1 hold. Then we have: if $J \subset G$ is open and $f_1, f_2 \in C^k(I)$ are functions with $f_1|_J = f_2|_J$, then $Tf_1|_{u(J)} = Tf_2|_{u(J)}$.*

Proof. (a) Let $J \subset G$ be open with $f_1|_J = f_2|_J > 0$. Let $x \in J$. We may assume that $f_1|_J = f_2|_J \geq c > 0$, if necessary by restricting f_1, f_2 to an open subset $J_1 \subset J$ with $x \in J_1$. Let $f_3 \in C^k(I)$ be an invertible extension of $f_1^{-1}|_J = f_2^{-1}|_J$ to I, i.e., $f_3 > 0$ on I with $\mathbb{1} = f_1|_J f_3|_J = f_2|_J f_3|_J$. By (1) and Proposition 3

$$\mathbb{1} = T(\mathbb{1})|_{u(J)} = T(f_1 f_3)|_{u(J)} \leq T(f_1)|_{u(J)} T(f_3)|_{u(J)}. \qquad (2)$$

Since $f_3 \geq 0$, we also have $T(f_3) \geq 0$. This yields $T(f_3)|_{u(J)} > 0$ and $T(f_1)|_{u(J)} > 0$. Similarly, $h := -f_1 f_3$ satisfies $h|_J = -\mathbb{1} = -f_2|_J f_3|_J$ and by Proposition 5

$$T(-\mathbb{1})|_{u(J)} = T(h)|_{u(J)} = T(-f_1 f_3)|_{u(J)}$$
$$\leq T(-f_1)|_{u(J)} T(f_3)|_{u(J)} \leq T(-\mathbb{1})|_{u(J)} T(f_1)|_{u(J)} T(f_3)|_{u(J)}.$$

Since $T(-\mathbb{1}) < 0$, this yields $T(f_1)|_{u(J)} T(f_3)|_{u(J)} \leq 1$ and together with (2) we find

$$\mathbb{1} = T(f_1)|_{u(J)} T(f_3)|_{u(J)} = T(f_2)|_{u(J)} T(f_3)|_{u(J)}.$$

In particular, $T(f_1)|_{u(J)} = T(f_2)|_{u(J)}$.

(b) If $f_1|_J = f_2|_J \leq -c < 0$ and $f_3 < 0$ is an invertible extension of $f_1^{-1}|_J = f_2^{-1}|_J < 0$ to I, we have $f_1(-f_3)|_J = -\mathbb{1}$, $(-f_1)(-f_3)|_J = \mathbb{1}$. Since $(-f_1)|_J > 0$, $(-f_3)|_J > 0$, we see that $T(-f_1)|_{u(J)} > 0$, $T(-f_3)|_{u(J)} > 0$ and by Proposition 5

$$T(-\mathbb{1})|_{u(J)} = T(f_1(-f_3))|_{u(J)} \leq T(-f_3)|_{u(J)}T(f_1)|_{u(J)}$$
$$\leq T(-f_3)|_{u(J)}T(-\mathbb{1})|_{u(J)}T(-f_1)|_{u(J)}. \tag{3}$$

By part (a) with $-f_1|_J > 0$, $-f_3|_J > 0$, $1 = T(-f_1)|_{u(J)}T(-f_3)|_{u(J)}$. Hence in (3) there is equality everywhere and we find that

$$T(f_1)|_{u(J)}T(-f_3)|_{u(J)} = T(f_2)|_{u(J)}T(-f_3)|_{u(J)} = T(-\mathbb{1})|_{u(J)}.$$

In particular, $T(f_1)|_{u(J)} = T(f_2)|_{u(J)}$.

(c) Let $f_1|_J = f_2|_J$. Suppose J is open with $\overline{J} \subset G$ and $x \in J$ with $f_1(x) = 0$. If f_1 is identically zero in a neighborhood of x, $Tf_1(x) = Tf_2(x) = 0$ by Proposition 3 (i). If not, choose open intervals J_n with $x \in J_{n+1} \subset J_n \subset J$ and $|J_n| \to 0$. Let $\epsilon_n := 2\max_{J_n}|f|$. Then $|f + \epsilon_n||_{J_n} \geq \epsilon_n$ and hence by (a) or (b), $T(f_1 + \epsilon_n)|_{u(J_n)} = T(f_2 + \epsilon_n)|_{u(J_n)}$. The pointwise continuity of T then implies $Tf_1(u(x)) = Tf_2(u(x))$ for $n \to \infty$. □

Proposition 7. *Suppose the assumptions of Theorem 1 hold. Let G and $u|_G : G \to I$ be as in Proposition 4. Then there exists a function $F : G \times \mathbb{R}^{k+1} \to \mathbb{R}$ such that*

$$(Tf)(u(x)) = F(x, f(x), \ldots, f^{(k)}(x))$$

for all $f \in C^k(I)$ and all $x \in G$.

Proof. Let $x_0 \in G$ be arbitrary but fixed, and let $f \in C^k(I)$. Let $J \subset G$ be open with $x_0 \in J$. Denote by $P_k f \in C^k(I)$ the kth order Taylor polynomial of f at x_0 and define $g : I \to \mathbb{R}$ by

$$g(y) = \begin{cases} f(y) & y \in I, y \leq x_0, \\ P_k f(y) & y \in I, y > x_0. \end{cases}$$

Then $g \in C^k(I)$ and $g|_{J_1} = f|_{J_1}$, $g|_{J_2} = P_k f|_{J_2}$ where $J_1 := J \cap (-\infty, x_0)$, $J_2 := J \cap (x_0, \infty)$. By Proposition 6,

$$Tf|_{u(J_1)} = Tg|_{u(J_1)}, \quad Tg|_{u(J_2)} = T(P_k f)|_{u(J_2)}.$$

Now $u|_G : G \to I$ is continuous and $u(G)$ is dense in I. Thus $\overline{u(J_1)} \cap \overline{u(J_2)} = \{u(x_0)\}$. Since Tf, Tg and $T(P_k f)$ are continuous functions, we find at $u(x_0)$

$$(Tf)(u(x_0)) = (Tg)(u(x_0)) = T(P_k f)(u(x_0)).$$

However, $T(P_k f)$ only depends on $(x_0, f(x_0), \ldots, f^{(k)}(x_0))$, because these values determine the Taylor polynomial $P_k f$ at x_0 completely. Therefore $(Tf)(u(x_0))$ is a function of these parameters: there is a function $F : G \times \mathbb{R}^{k+1} \to \mathbb{R}$ such that

$$(Tf)(u(x_0)) = F(x_0, f(x_0), \ldots, f^{(k)}(x_0))$$

for all $f \in C^k(I)$ and all $x_0 \in G$. □

3. Analysis of the representing function

Under the assumptions of Theorem 1 we know by Proposition 7 that T has the form

$$(Tf)(u(x)) = F(x, f(x), f'(x), \ldots, f^{(k)}(x)), \tag{4}$$

$f \in C^k(I)$, $x \in G$, $k \in \mathbb{N}_0$, where $G \subset I$ is the dense set studied in Proposition 4. We now show for $k \in \mathbb{N}$ that F does not depend on the derivative variables.

Proposition 8. *Under the assumptions of Theorem 1, the function $F : G \times \mathbb{R}^{k+1} \to \mathbb{R}$ with (4) does not depend on the derivative variables. We have*

$$F(x, \alpha_0, \alpha_1, \ldots, \alpha_k) = F(x, \alpha_0, 0, \ldots, 0) = \begin{cases} \alpha_0^{p(x)} & \alpha_0 \geq 0, \\ -A(x)|\alpha_0|^{p(x)} & \alpha_0 < 0 \end{cases}$$

for all $x \in G$, $(\alpha_0, \ldots, \alpha_k) \in \mathbb{R}^{k+1}$. Here $p : G \to \mathbb{R}$, $A : G \to \mathbb{R}$ are continuous functions determined by T, with $A \geq 1$, $p > 0$.

In the proof, we need the following lemma which can be found in Hille, Phillips [HP, Chap. VII].

Lemma 9. *Let $f : \mathbb{R} \to \mathbb{R}$ be measurable and subaddiditve, i.e.,*

$$f(s+t) \leq f(s) + f(t); \quad s, t \in \mathbb{R}$$

and define $r := \sup_{t<0} \frac{f(t)}{t}$, $s := \inf_{t>0} \frac{f(t)}{t}$. Then f is bounded on compact sets and $-\infty < r \leq s < \infty$. Moreover the limits $\lim_{t \to -\infty} \frac{f(t)}{t}$ and $\lim_{t \to \infty} \frac{f(t)}{t}$ exist and $r = \lim_{t \to -\infty} \frac{f(t)}{t}$ and $s = \lim_{t \to \infty} \frac{f(t)}{t}$. If $f(0) = 0$ and f is continuous at 0, f is continuous on $\mathbb{R}$.

As a consequence of Lemma 9, we have with $-\infty < r \leq s < \infty$

$$f(t) = \begin{cases} rt + a(t) & t < 0, \\ st + a(t) & t > 0. \end{cases} \tag{5}$$

where $a : \mathbb{R} \to \mathbb{R}$ is a non-negative function with $\lim_{|t| \to \infty} \frac{a(t)}{t} = 0$. We also need

Lemma 10. *Let $g : \mathbb{R} \to \mathbb{R}$ be measurable and submultiplicative,*

$$g(\alpha\beta) \leq g(\alpha)g(\beta); \quad \alpha, \beta \in \mathbb{R}$$

with $g(-1) < 0 < g(1)$. If g is continuous at 0 and at 1, there is $p > 0$ and $A \geq 1$ such that

$$g(\alpha) = \begin{cases} \alpha^p & \alpha > 0, \\ -A|\alpha|^p & \alpha < 0. \end{cases}$$

This is Theorem 2 in König, Milman [KM].

Proof of Proposition 8. (a) Assume $k \in \mathbb{N}$. Writing

$$F_x(\alpha_0, \ldots, \alpha_k) = F(x, \alpha_0, \ldots, \alpha_k),$$

we will show that F_x does not depend on the last variable α_k, i.e.,

$$F_x(\alpha_0, \ldots, \alpha_{k-1}, \alpha_k) = F_x(\alpha_0, \ldots, \alpha_{k-1}, 0)$$

for all $(\alpha_0, \ldots, \alpha_{k-1}, \alpha_k) \in \mathbb{R}^{k+1}$, $x \in G$. The argument may then be repeated to prove independence of $(\alpha_1, \ldots, \alpha_{k-1})$ as well.

Let $x \in G$. We know that $T(\mathbb{1}) = \mathbb{1}$, $T(-\mathbb{1}) =: -A \leq -\mathbb{1}$ and $T(0) = 0$ so that

$$1 = F_x(1, 0, \ldots, 0) > 0 > F_x(-1, 0, \ldots, 0), F_x(0, \ldots, 0) = 0.$$

For all $(\alpha_0, \ldots, \alpha_k), (\beta_0, \ldots, \beta_k) \in \mathbb{R}^{k+1}$ and $x \in G$ there are functions $f, g \in C^k(I)$ with $f^{(j)}(x) = \alpha_j$, $g^{(j)}(x) = \beta_j$ for all $j \in \{0, \ldots, k\}$. By Leibniz' formula we have

$$(f \cdot g)^{(l)} = \sum_{j=0}^{l} \binom{l}{j} \alpha_j \beta_{l-j}.$$

Using (4), the submultiplicativity of T, $T(f \cdot g)(u(x)) \leq Tf(u(x)) \cdot Tg(u(x))$ translates into the functional inequality for F

$$F_x\left(\alpha_0\beta_0, \alpha_0\beta_1 + \alpha_1\beta_0, \ldots, \sum_{j=0}^{k} \binom{k}{j} \alpha_j \beta_{k-j}\right)$$

$$\leq F_x(\alpha_0, \alpha_1, \ldots, \alpha_k) \cdot F_x(\beta_0, \beta_1, \ldots, \beta_k). \tag{6}$$

There is extensive knowledge about functional equations, cf. Aczél [A], but less on functional inequalities. In our case, however, it is not too difficult to determine all solutions of (6).

For $f_n(y) := \sum_{j=0}^{k} \frac{\alpha_j(n)}{j!}(y-x)^j$ with $\alpha_j(n) \to \alpha_j$ as $n \to \infty$ for all $j \in \{0, \ldots, k\}$, we put $f(y) := \sum_{j=0}^{k} \frac{\alpha_j}{j!}(y-x)^j$. Then the functions $f_n^{(j)}$ converge to $f^{(j)}$ uniformly on compact sets for all $j \in \{0, \ldots, k\}$. By the assumption on T, we see that $Tf_n(x) \to Tf(x)$ which means

$$F_x(\alpha_0(n), \ldots, \alpha_k(n)) \to F_x(\alpha_0, \ldots, \alpha_k),$$

i.e., F_x is continuous in $\mathbb{R}^{k+1}$.

(b) **Case 1.** Assume first that $F_x(1, 0, \ldots, 0, \beta_k) \geq 1$ for all $\beta_k \in \mathbb{R}$. By Proposition 3 (iii), $\operatorname{sgn} Tf(u(x)) = \operatorname{sgn} f(x)$ for all $f \in C^k(I)$. By (4), this means that for all $(\gamma_0, \ldots, \gamma_k) \in \mathbb{R}^{k+1}$ we have

$$\operatorname{sgn} F_x(\gamma_0, \gamma_1, \ldots, \gamma_k) = \operatorname{sgn} \gamma_0. \tag{7}$$

Using (7), we see that $F_x(1, 0, \ldots, 0, \beta_k) > 0$. Let $\alpha_0 > 0$ and $\alpha_1, \ldots, \alpha_k, \tilde{\alpha}_k \in \mathbb{R}$ be arbitrary. Put $\beta_0 = 1, \beta_1 = \cdots = \beta_{k-1} = 0$ and $\beta_k := \frac{\tilde{\alpha}_k - \alpha_k}{\alpha_0}$ in (6) to find

$$F_x(-\alpha_0, \alpha_1, \ldots, \alpha_{k-1}, \alpha_k) \leq F_x(-\alpha_0, \alpha_1, \ldots, \alpha_{k-1}, \tilde{\alpha}_k) F_x(1, 0, \ldots, 0, \beta_k).$$

Again by (7), $F_x(-\alpha_0, \alpha_1, \ldots, \alpha_{k-1}, \alpha_k) < 0$ and $F_x(-\alpha_0, \alpha_1, \ldots, \alpha_{k-1}, \tilde{\alpha}_k) < 0$.

Therefore

$$|F_x(-\alpha_0, \alpha_1, \ldots, \alpha_{k-1}, \alpha_k)| \geq |F_x(-\alpha_0, \alpha_1, \ldots, \alpha_{k-1}, \tilde{\alpha}_k)| F_x(1, 0, \ldots, 0, \beta_k)$$
$$\geq |F_x(-\alpha_0, \alpha_1, \ldots, \alpha_{k-1}, \tilde{\alpha}_k)|.$$

Since this holds for all $\alpha_k, \tilde{\alpha}_k$, we deduce that $F_x(-\alpha_0, \alpha_1, \ldots, \alpha_{k-1}, \cdot)$ is constant in the last variable:

$$F_x(-\alpha_0, \alpha_1, \ldots, \alpha_{k-1}, \alpha_k) = F_x(-\alpha_0, \alpha_1, \ldots, \alpha_{k-1}, 0).$$

This also implies that $F_x(1, 0, \ldots, 0, \beta_k) = 1$ for all $\beta_k \in \mathbb{R}$. Therefore

$$F_x(\alpha_0, \alpha_1, \ldots, \alpha_{k-1}, \alpha_k) \leq F_x(\alpha_0, \alpha_1, \ldots, \alpha_{k-1}, \tilde{\alpha}_k) F_x(1, 0, \ldots, 0, \tilde{\beta}_k)$$
$$= F_x(\alpha_0, \alpha_1, \ldots, \alpha_{k-1}, \tilde{\alpha}_k)$$

with $\tilde{\beta}_k := \frac{\alpha_k - \tilde{\alpha}_k}{\alpha_0}$. Since this is true for all $\alpha_k, \tilde{\alpha}_k$, also

$$F_x(\alpha_0, \alpha_1, \ldots, \alpha_{k-1}, \alpha_k) = F_x(\alpha_0, \alpha_1, \ldots, \alpha_{k-1}, 0).$$

Therefore F_x is independent of the last variable α_k.

(c) **Case 2.** Assume now that there is $\beta_k \in \mathbb{R}$ with $0 < F(1, 0, \ldots, 0, \beta_k) < 1$. We will show that this case is impossible. By (6), for all $\alpha_k, \tilde{\alpha}_k \in \mathbb{R}$

$$F_x(1, 0, \ldots, 0, \alpha_k + \tilde{\alpha}_k) \leq F_x(1, 0, \ldots, 0, \alpha_k) \cdot F_x(1, 0, \ldots, 0, \tilde{\alpha}_k).$$

Define $f := \mathbb{R} \to \mathbb{R}$ by $f(\alpha) := \ln F_x(1, 0, \ldots, 0, \alpha)$. Then f is continuous and subadditive, $f(\alpha + \tilde{\alpha}) \leq f(\alpha) + f(\tilde{\alpha})$, with $f(0) = 0$ and $f(\beta_k) < 0$. By Lemma 9 and (5) there are r, s with $-\infty < r \leq s < \infty$ such that

$$f(\alpha) = \begin{cases} r\alpha + a(\alpha) & \alpha < 0, \\ s\alpha + a(\alpha) & \alpha > 0 \end{cases}$$

where $a : \mathbb{R} \to \mathbb{R}_{\geq 0}$ is nonnegative with $\lim_{|\alpha| \to \infty} \frac{a(\alpha)}{|\alpha|} = 0$. Also, $f(n\beta_k) \leq nf(\beta_k) \to -\infty$ as $n \to \infty$.

Assume first that $\beta_k < 0$. Then $r > 0$ since $f(\alpha) \geq r\alpha$ for all $\alpha < 0$ and $\lim_{\alpha \to -\infty} f(\alpha) = -\infty$. Using $\lim_{\alpha \to -\infty} \frac{a(\alpha)}{|\alpha|} = 0$, we find $\bar{\alpha} < 0$ such that for all $\alpha < \bar{\alpha}$ we have $f(\alpha) \leq \frac{r}{2}\alpha$ and $F_x(1, 0, \ldots, 0, \alpha) \leq \exp(\frac{r}{2}\alpha)$. Therefore for all $\alpha_0 > 0$ with $-\frac{1}{\alpha_0} \leq \bar{\alpha}$, again using (6), we obtain

$$F_x(\alpha_0, 0, \ldots, 0, 0) \leq F_x(\alpha_0, 0, \ldots, 0, 1) F_x\left(1, 0, \ldots, 0, -\frac{1}{\alpha_0}\right)$$
$$\leq F_x(\alpha_0, 0, \ldots, 0, 1) \exp\left(-\frac{r}{2}\frac{1}{\alpha_0}\right).$$

Note that by (6), $F_x(\alpha_0\beta_0, 0, \ldots, 0) \leq F_x(\alpha_0, 0, \ldots, 0) \cdot F_x(\beta_0, 0, \ldots, 0)$. Therefore by Lemma 10, there is $p = p(x) > 0$ such that $F_x(\alpha_0, 0, \ldots, 0) = \alpha_0^p$ for all $\alpha_0 > 0$. We find

$$\alpha_0^p \exp\left(\frac{r}{2}\frac{1}{\alpha_0}\right) \leq F_x(\alpha_0, 0, \ldots, 0, 1).$$

Let $f(t) := \frac{1}{n!}(t-x)^n$. By Proposition 3 (iii), $F_x(0,0,\ldots,0,1) = Tf(u(x)) = 0$. The continuity of F_x implies $\lim_{\alpha_0 \to 0} F_x(\alpha_0, 0, \ldots, 0, 1) = F_x(0,0,\ldots,0,1) = 0$. However, the left side tends to infinity as $\alpha_0 \to 0$, a contradiction.

Assume now that $\beta_k > 0$. Then $\lim_{\alpha \to \infty} f(\alpha) = -\infty$. Since $f(\alpha) \geq s\alpha$ for all $\alpha > 0$, we have $s < 0$. Then there is $\bar\alpha > 0$ such that $f(\alpha) \leq \frac{s}{2}\alpha$ and $F_x(1, 0, \ldots, 0, \alpha) \leq \exp(\frac{s}{2}\alpha)$ for all $\alpha > \bar\alpha$. For $\alpha_0 > 0$ with $\frac{1}{\alpha_0} \geq \bar\alpha$, using (6), we obtain

$$F_x(\alpha_0, 0, \ldots, 0, 0) \leq F_x(\alpha_0, 0, \ldots, 0, -1) F_x\left(1, 0, \ldots, 0, \frac{1}{\alpha_0}\right)$$

$$\leq F_x(\alpha_0, 0, \ldots, 0, -1)\exp\left(\frac{s}{2}\frac{1}{\alpha_0}\right), \quad s = -|s|,$$

$$\alpha_0^p \exp\left(\frac{|s|}{2}\frac{1}{\alpha_0}\right) \leq F_x(\alpha_0, 0, \ldots, 0, -1).$$

Again we arrive at a contradiction as $\alpha_0 \to 0$ since $F_x(0,0,\ldots,0,-1) = 0$.

(d) Therefore only Case 1 is possible, and iterating this for $k-1, \ldots, 1$, we find

$$F_x(\alpha_0, \alpha_1, \ldots, \alpha_k) = F_x(\alpha_0, 0, \ldots, 0)$$

for all $(\alpha_0, \ldots, \alpha_k) \in \mathbb{R}^{k+1}$. Since

$$F_x(\alpha_0\beta_0, 0, \ldots, 0) \leq F_x(\alpha_0, 0, \ldots, 0)F_x(\beta_0, 0, \ldots, 0)$$

by (6), Lemma 10 yields the existence of $p(x) > 0$ and $A(x) \geq 1$ such that

$$F_x(\alpha_0, \alpha_1, \ldots, \alpha_k) = \begin{cases} \alpha_0^{p(x)} & \alpha \geq 0, \\ -A(x)|\alpha_0|^{p(x)} & \alpha < 0. \end{cases}$$

Hence, using (4), we arrive at

$$Tf(u(x)) = \begin{cases} f(x)^{p(x)} & f(x) \geq 0, \\ -A(x)|f(x)|^{p(x)} & f(x) < 0. \end{cases}$$

Applying this to the constant function $f = 2$, we deduce that p is continuous on G since Tf is continuous. Then applying it to $f = -2$, we conclude that A is also continuous on G. $\square$

Proof of Theorem 1. By Propositions 7 and 8, T has the form

$$Tf(u(x)) = \begin{cases} f(x)^{p(x)} & f(x) \geq 0, \\ -A(x)|f(x)|^{p(x)} & f(x) < 0 \end{cases} \tag{8}$$

with continuous functions $p > 0$ and $A \geq \mathbb{1}$ on G, for any $f \in C^k(I)$, $x \in G$. Recall that by Proposition 4, G and $u(G)$ are dense in I. We now show that $u : G \to u(G)$ extends to a C^k-diffeomorphism $u : I \to I$. By surjectivity, there is $f \in C^k(I)$ with $Tf = 2$. By Proposition 3 (iii), $f > 0$. Hence for $x \in G$ we have $2 = Tf(u(x)) = f(x)^{p(x)}$, $2^{1/p(x)} = f(x)$. Let $z \in I$. Since G is dense in I, we may choose $x_j \in G$ with $x_j \to z$. Then by the continuity of f, we see that $f(x_j) \to f(z)$.

Hence $p(x_j)$ has a well-defined limit, denoted $p(z)$, which is independent of the particular sequence (x_j) in G tending to z. This yields a continuous extension $p : I \to I$. Similarly, choose $f \in C^k(I)$ with $Tf = -2$. Again by Proposition 3 (iii), $f < 0$. This implies that A can be extended continuously to $A : I \to I$. Hence for any $f \in C^k(I)$ the quantity

$$Tf(u(x_j)) = \begin{cases} f(x_j)^{p(x_j)} & f(x_j) \geq 0, \\ -A(x_j)|f(x_j)|^{p(x_j)} & f(x_j) < 0 \end{cases}$$

has a well-defined limit $\lim_{j \to \infty} Tf(u(x_j))$ as $x_j \in G \to z \in I$. Since T is bijective, $u(x_j)$ must have a finite limit, denoted by $u(z)$. By the bijectivity of T, $u : I \to I$ is a homeomorphism and (8) holds for all $x \in I$ and $f \in C^k(I)$. This formula then yields the following expression for the inverse of T, where $g \in C^k(I)$ and $x \in I$:

$$T^{-1}g(u^{-1}(x)) = \begin{cases} g(x)^{1/p(x)} & g(x) \geq 0, \\ -A(x)^{-1/p(x)}|g(x)|^{1/p(x)} & g(x) < 0. \end{cases}$$

For $k \in \mathbb{N}$, T and T^{-1} map differentiable functions into differentiable functions. Again there are $f, g \in C^k(I)$ such that $Tf = 2$ and $Tg = -2$, and we have $f > 0 > g$. Therefore $Tf(u(x)) = f(x)^{p(x)} = 2$ which shows that $p \in C^k(I)$. Then the relation $Tg(u(x)) = -A(x)|g(x)|^{p(x)}$ implies that A is also in $C^k(I)$. For positive functions $h \in C^k(I)$, we have $Th(u(x)) = h(x)^{p(x)}$, $\ln Th(u(x)) = p(x) \ln h(x)$ which implies that u is in $C^k(I)$ as well. Let $f(y) = y - x_0$. The derivatives of Tf at $u(x_0)$ and of $T^{-1}f$ at $u^{-1}(x_0)$ for $y > x_0$ are given by

$$(Tf)'(u(y)) = p(y)(y - x_0)^{p(y)-1} + p'(y)(y - x_0)^{p(y)} \ln(y - x_0),$$

$$(T^{-1}f)'(u^{-1}(y)) = 1/p(y)(y - x_0)^{1/p(y)-1} - p'(y)/p(y)^2(y - x_0)^{1/p(y)} \ln(y - x_0).$$

The boundedness of these functions for $y \to x_0$ requires $p(x_0) = 1$. Therefore $p = \mathbb{1}$ on G, and the differentiability of Tf at $u(x_0)$ requires $A(x_0) = 1$, $A = \mathbb{1}$ on I. Hence $Tf(u(x)) = f(x)$ if $k \in \mathbb{N}$. This proves Theorem 1. $\square$

The proof of Theorem 2 requires only minor modification. Again the relations $T(\mathbb{1}) = \mathbb{1}$ and $T(0) = 0$ are seen immediately. In the proof of the result corresponding to Proposition 3 we now have $\mathrm{supp}(T(-f)) \subset \mathrm{supp}(T(f))$. The analogues of Lemmas 9 and 10 in the supermultiplicative case are found in [HP] and [KM].

In the proof of the result corresponding to Proposition 8, one has to distinguish Cases 1 and 2 where Case 1 now reads: assume that $F_x(1, 0, \ldots, 0, \beta_k) \leq 1$ for all $\beta_k \in \mathbb{R}$, and Case 2 is the negation of it which again is shown to be impossible.

References

[A] J. Aczél, *Lectures on functional equations and their applications.* Academic Press, 1966.

[AAFM] S. Alesker, S. Artstein-Avidan, D. Faifman, V. Milman, *A characterization of product preserving maps with applications to a characterization of the Fourier transform.* Illinois J. Math. **54** (2010), 1115–1132.

[AFM] S. Artstein-Avidan, D. Faifman, V. Milman, *On multiplicative maps of continuous and smooth functions.* Geometric aspects of functional analysis, Lecture Notes in Math. 2050, 35–59, Springer, Heidelberg, 2012.

[HP] E. Hille, R. Phillips, *Functional Analysis and Semi-Groups.* Amer. Math. Soc., Providence, 1968.

[KM] H. König, V. Milman, *Submultiplicative functions and operator inequalities.* Studia Math. **223** (2014), 217–231.

[M] J. Mrčun, *On isomorphisms of algebras of smooth functions.* Proc. Amer. Math. Soc. **133**, 2005, 3109–3113.

[Mi] A. N. Milgram, *Multiplicative semigroups of continuous functions.* Duke Math. J. **16** (1940), 377–383.

[MS] J. Mrčun, P. Šemrl, *Multplicative bijections between algebras of differentiable functions.* Ann. Acad. Sci. Fenn. Math. **32** (2007), 471–480.

Dmitry Faifman
Department of Mathematics
University of Toronto
Toronto, Ontario, Canada M5S 2E4
e-mail: `dfaifman@gmail.com`

Hermann König
Mathematisches Seminar
Universität Kiel
D-24098 Kiel, Germany
e-mail: `hkoenig@math.uni-kiel.de`

Vitali Milman
School of Mathematical Sciences
Tel Aviv University
Ramat Aviv, Tel Aviv 69978, Israel
e-mail: `milman@post.tau.ac.il`

Operator Theory:
Advances and Applications, Vol. 261, 281–305
© Springer International Publishing AG, part of Springer Nature 2018

Isoperimetric Functional Inequalities via the Maximum Principle: The Exterior Differential Systems Approach

Paata Ivanisvili and Alexander Volberg

In memory of V.P. Havin

Abstract. Our goal in this note is to give a unified approach to the solutions of a class of isoperimetric problems by relating them to the exterior differential systems studied by R. Bryant and P. Griffiths.

Mathematics Subject Classification (2010). Primary 42B20. Secondary 42B35, 47A30.

Keywords. Log-Sobolev inequality, Poincaré inequality, Bobkov's inequality, Gaussian isoperimetry, semigroups, maximum principle, Monge–Ampère equations, exterior differential systems, backward heat equation, (B) theorem.

1. Introduction: a function and its gradient

In this note we list several classical by now isoperimetric inequalities which can be proved in a unified way. This unified approach reduces them to the so-called exterior differential systems studied by Robert Bryant and Phillip Griffiths. To the best of our knowledge, this is the first article where this link is found.

Let $d\gamma(x)$ be the standard n-dimensional Gaussian measure

$$d\gamma(x) = \frac{1}{\sqrt{(2\pi)^n}} e^{-\frac{|x|^2}{2}} dx.$$

Set $\Omega \subset \mathbb{R}$ to be a bounded closed interval and let $\mathbb{R}^+ := \{x \in \mathbb{R} : x \geq 0\}$. By the symbol $C^\infty(\mathbb{R}^n; \Omega)$ we denote the smooth functions on $\mathbb{R}^n$ with values in Ω. We prove the following theorem.

PI is partially supported by the Hausdorff Institute for Mathematics, Bonn, Germany. AV is partially supported by the NSF grants DMS-1265549, DMS-1600065 and by the Hausdorff Institute for Mathematics, Bonn, Germany.

Theorem 1.1. *If a real-valued function $M(x,y)$ is such that $M(x, \sqrt{y}) \in C^2(\Omega \times \mathbb{R}_+)$ and it satisfies the differential inequality*

$$\begin{pmatrix} M_{xx} + \frac{M_y}{y} & M_{xy} \\ M_{xy} & M_{yy} \end{pmatrix} \leq 0 \tag{1.1}$$

then

$$\int_{\mathbb{R}^n} M(f, \|\nabla f\|) d\gamma \leq M\left(\int_{\mathbb{R}^n} f d\gamma, 0\right) \quad \text{for all} \quad f \in C^\infty(\mathbb{R}^n; \Omega). \tag{1.2}$$

One can obtain a similar result for uniformly log-concave probability measures, and the short way to see this is based on the mass transportation argument. In fact, let $d\mu = e^{-U(x)}dx$ be a probability measure such that $U(x)$ is smooth and $\text{Hess}\, U \geq R \cdot Id$ for some $R > 0$. By the result of Caffarelli (see [16]) there exists a Brenier map $T = \nabla\phi$ for some convex function ϕ such that T pushes forward $d\gamma$ to $d\mu$, moreover $0 \leq \text{Hess}\,\phi \leq \frac{1}{\sqrt{R}} \cdot Id$. We apply (1.1) to $f(x) = g(\nabla\phi(x))$ and use the fact $M_y \leq 0$ which follows from (1.1). Since $\|\nabla f(x)\| = \|\text{Hess}\,\phi(x)\,\nabla g(\nabla\phi)\| \leq \frac{1}{\sqrt{R}}\|\nabla g(\nabla\phi)\|$ we obtain:

Corollary 1.2. *If $M(x, \sqrt{y})$ satisfies $M(x, \sqrt{y}) \in C^2(\Omega \times \mathbb{R}_+)$ and (1.1) then for any $g \in C^\infty(\mathbb{R}^n; \Omega)$ we have*

$$\int_{\mathbb{R}^n} M\left(g, \frac{\|\nabla g\|}{\sqrt{R}}\right) d\mu \leq M\left(\int_{\mathbb{R}^n} g d\mu, 0\right), \tag{1.3}$$

where $d\mu = e^{-U(x)}dx$ is a probability measure such that $\text{Hess}\, U(x) \geq R \cdot Id$.

In Section 1.1 we present applications of the functional inequality (1.3). In Section 2 we prove a theorem about equivalence of some functional inequalities and partial differential inequalities. Corollary 1.2 is merely a consequence of this result. We will notice that our proof of Corollary 1.2 for a general log-concave measure will not differ from the case of Gaussian measures and it will be completely self-contained (it will not need the mass transportation argument).

In Section 3 we describe the solutions of (1.1) (in the case important for us when the determinant of the matrix in (1.1) is zero) by reducing it to the exterior differential system (EDS) studied by R. Bryant and P. Griffiths.

This allows us to linearize the underlying non-linear PDE that appeared by the requirement of the determinant of the matrix in (1.1) to vanish. In Section 4 we investigate the one-dimensional case of the results obtained in Section 2, and in Section 5 we present further applications. In particular, we sharpen the Beckner–Sobolev inequality (already sharp of course), and we show other examples of new isoperimetric inequalities, which one obtains through the EDS method.

1.1. A unified approach to classical inequalities via one and the same PDE

In this section we list classical isoperimetric inequalities that can be obtained by choosing different solution of one and the same PDE

$$y(M_{xx}M_{yy} - M_{xy}^2) + M_y M_{yy} = 0 \tag{1.4}$$

corresponding to different initial values at $y = 0$. In the next sections we will show how exterior differential systems (EDS) method allows us to reduce it to a linear PDE, and thus match this classical isoperimetric inequalities with interesting solution of a linear PDE that happened to be none other than a reverse heat equation.

Then later, starting with Subsection 1.1.4, we show that one can choose other interesting solutions of (1.4), and, in its turn, this translates to new isoperimetric inequalities. In particular, we will show an instance when the Beckner–Sobolev inequality can be further sharpened in an ultimate way.

1.1.1. Log-Sobolev inequalities: entropy estimates. The log-Sobolev inequality of Gross (see [20]) states that

$$\int_{\mathbb{R}^n} |f|^2 \ln |f|^2 d\gamma - \left(\int_{\mathbb{R}^n} |f|^2 d\gamma \right) \ln \left(\int_{\mathbb{R}^n} |f|^2 d\gamma \right) \le 2 \int_{\mathbb{R}^n} \|\nabla f\|^2 d\gamma \qquad (1.5)$$

whenever the right-hand side of (1.5) is well defined and finite for complex-valued f. This implies that if f and $\|\nabla f\|$ are in $L^2(d\gamma)$ then f is in the Orlicz space $L^2 \ln L$. The proof of Gross involves *a two-point inequality* which by the central limit theorem establishes the hypercontractivity of the Ornstein–Uhlenbeck semigroup $\|e^{t(\Delta - x \cdot \nabla)}\|_{L^p(d\gamma) \to L^q(d\gamma)} \le 1$ for all $t \ge 0$ such that $e^{-2t} \le \frac{p-1}{q-1}$. Then as a corollary, differentiating this estimate at the point $t = 0$ for $q = 2$ one obtains (1.5). Earlier than Gross, a similar *two-point inequality* was proved by Aline Bonami (see [9, 10]). For more on *two-point inequalities* we refer the reader to [33]. For a simple proof of the hypercontractivity of the Ornstein–Uhlenbeck semigroup we refer the reader to [25, 29], and also to the earlier works [22, 31]. Bakry and Emery [2] extended the inequality to log-concave measures. Namely the inequality

$$\int_{\mathbb{R}^n} f^2 \ln f^2 d\mu - \left(\int_{\mathbb{R}^n} f^2 d\mu \right) \ln \left(\int_{\mathbb{R}^n} f^2 d\mu \right) \le \frac{2}{R} \int_{\mathbb{R}^n} \|\nabla f\|^2 d\mu \qquad (1.6)$$

holds for a bounded real-valued $f \in C^1$ and a log-concave probability measure $d\mu = e^{-U(x)} dx$ such that $\operatorname{Hess} U(x) \ge R \cdot \operatorname{Id}$. For further remarks we refer the reader to [3].

Proof of (1.6). Take

$$M(x, y) = x \ln x - \frac{y^2}{2x}, \qquad x > 0 \quad \text{and} \quad y \ge 0. \qquad (1.7)$$

The matrix in (1.1) takes the form

$$\begin{pmatrix} -\frac{y^2}{x^3} & \frac{y}{x^2} \\ \frac{y}{x^2} & -\frac{1}{x} \end{pmatrix} \le 0. \qquad (1.8)$$

By Corollary 1.3 we obtain

$$\int_{\mathbb{R}^n} \left(g \ln g - \frac{1}{2R} \frac{\|\nabla g\|^2}{g} \right) d\mu \le \left(\int_{\mathbb{R}^n} g \, d\mu \right) \ln \left(\int_{\mathbb{R}^n} g \, d\mu \right). \qquad (1.9)$$

Taking $g = f^2$ for positive f and rearranging terms in (1.9) we arrive at (1.6). $\quad \square$

Remark 1.3. The proof we have presented meets an obstruction: $M(x, \sqrt{y}) \notin C^2(\mathbb{R}_+ \times \mathbb{R}_+)$. In order to avoid this obstruction one has to consider $M^\varepsilon(x, y) := M(x + \varepsilon, y)$ for some $\varepsilon > 0$. Then surely $M^\varepsilon(x, y)$ will satisfy (1.1), what is more $M^\varepsilon(x, \sqrt{y}) \in C^2(\mathbb{R}_+ \times \mathbb{R}_+)$ and we can repeat the same proof as above for $M^\varepsilon(x, y)$. Finally, we simply send $\varepsilon \to 0$ assuming that $\int f^2 d\mu \neq 0$ and we obtain the desired estimate. We should use the same idea in the applications presented below.

1.1.2. Bobkov's inequality: Gaussian isoperimetry. In [7] Bobkov obtained the following functional version of the Gaussian isoperimetry. Let

$$\Phi(x) = \frac{1}{\sqrt{2\pi}} \int_{-\infty}^{x} e^{-x^2/2} dx,$$

and let $\Phi'(x)$ be the derivative of Φ. Set $I(x) := \Phi'(\Phi^{-1}(x))$. Then for any locally Lipschitz function $f : \mathbb{R}^n \to [0, 1]$, we have

$$I\left(\int_{\mathbb{R}^n} f d\mu\right) \leq \int_{\mathbb{R}^n} \sqrt{I^2(f) + \frac{\|\nabla f\|^2}{R}} d\mu \tag{1.10}$$

where $d\mu = e^{-U(x)} dx$ is a log-concave probability measure such that $\operatorname{Hess} U \geq R \cdot Id$. Bobkov's proof involves a *two-point inequality*: for all $0 \leq a, b \leq 1$ we have

$$I\left(\frac{a+b}{2}\right) \leq \frac{1}{2}\sqrt{I^2(a) + \left|\frac{a-b}{2}\right|^2} + \frac{1}{2}\sqrt{I^2(b) + \left|\frac{a-b}{2}\right|^2}. \tag{1.11}$$

Iterating (1.11) appropriately and using the central limit theorem Bobkov obtained (1.10) for the Gaussian measures. By a mass transportation argument one immediately obtains (1.10) for uniformly log-concave probability measures. Notice that $I(0) = I(1) = 0$. Testing (1.10) for $d\mu = d\gamma$ and $f(x) = \mathbb{1}_A$ where A is a Borel subset of $\mathbb{R}^n$ one obtains Gaussian isoperimetry: for any Borel measurable set $A \subset \mathbb{R}^n$

$$\gamma^+(A) \geq \Phi'(\Phi^{-1}(\gamma(A))) \quad \text{where} \quad \gamma^+(A) := \liminf_{\varepsilon \to 0} \frac{\gamma(A_\varepsilon) - \gamma(A)}{\varepsilon} \tag{1.12}$$

denotes the Gaussian perimeter of A, here $A_\varepsilon = \{x \in \mathbb{R}^n : \operatorname{dist}_{\mathbb{R}^n}(A, x) < \varepsilon\}$. For further remarks on (1.10) see [4]. The Gaussian isoperimetry (1.12) can be derived also from Ehrhard's inequality (see for example [25]).

Proof of (1.10). Take

$$M(x, y) = -\sqrt{I^2(x) + y^2} \quad \text{where} \quad x \in [0, 1], \quad y \geq 0. \tag{1.13}$$

The matrix in (1.1) takes the form

$$\begin{pmatrix} -\dfrac{(I'(x))^2 y^2}{(I^2(x)+y^2)^{3/2}} - \dfrac{I(x)I''(x)+1}{\sqrt{I^2(x)+y^2}} & y\dfrac{I(x)I'(x)}{(I^2(x)+y^2)^{3/2}} \\ y\dfrac{I(x)I'(x)}{(I^2(x)+y^2)^{3/2}} & -\dfrac{I^2(x)}{(I^2(x)+y^2)^{3/2}} \end{pmatrix}. \tag{1.14}$$

Notice that $I''(x)I(x) + 1 = 0$ therefore (1.14) is negative semidefinite. So by Corollary 1.2 we obtain

$$\int_{\mathbb{R}^n} -\sqrt{I^2(f) + \frac{\|\nabla f\|^2}{R}}\, d\mu \leq -I\left(\int_{\mathbb{R}^n} f d\mu\right);\qquad(1.15)$$

rearranging terms in (1.15) we obtain (1.10) for differentiable $f : \mathbb{R}^n \to [0,1]$. Notice that (1.15) still holds if $I''(x)I(x) + 1 \geq 0$ for arbitrary smooth $I(x)$. $\qquad\square$

1.1.3. Poincaré inequality and the spectral gap. The classical Poincaré inequality for the Gaussian measure obtained by J. Nash [32] (see p. 941) states that

$$\int_{\mathbb{R}^n} f^2 d\gamma - \left(\int_{\mathbb{R}^n} f d\gamma\right)^2 \leq \int_{\mathbb{R}^n} \|\nabla f\|^2 d\gamma.\qquad(1.16)$$

The inequality also says that the spectral gap, i.e., the first nontrivial eigenvalue of the self-adjoint positive operator $L = -\Delta + x \cdot \nabla$ in $L^2(\mathbb{R}^n, d\gamma)$ is bounded from below by 1. If $d\mu = e^{-U(x)}dx$ is a probability measure such that $\operatorname{Hess} U \geq R \cdot Id$ then we have

$$\int_{\mathbb{R}^n} g^2 d\mu - \left(\int_{\mathbb{R}^n} g d\mu\right)^2 \leq \frac{1}{R}\int_{\mathbb{R}^n} \|\nabla g\|^2 d\mu.\qquad(1.17)$$

It is folklore that inequality (1.17), besides of mass transportation argument, follows from the log-Sobolev inequality (1.6): apply (1.6) to the function $f(x) = 1 + \varepsilon g(x)$ where $\int g d\mu = 0$, and send $\varepsilon \to 0$. Then the left-hand side of (1.6) is $2\varepsilon^2 \int g^2 d\mu + o(\varepsilon^2)$ whereas the right-hand side of (1.6) is $\frac{2\varepsilon^2}{R}\int \|\nabla g\|^2 d\mu$. This gives (1.17). In [11] Brascamp and Lieb obtained an improvement of (1.17): instead of $\frac{\|\nabla g\|^2}{R}$ one can put $\langle(\operatorname{Hess}U)^{-1}\nabla g, \nabla g\rangle$ on the right-hand side of (1.17), where we only assume that $\operatorname{Hess} U$ is positive. For a simple proof of this improvement we refer the reader to [17] (see also [8] by using the Prékopa–Leindler inequality). A subtler result of Bobkov [6] in this direction says that for any log-concave probability measure $d\mu = e^{-U(x)}dx$ one can put $K\|x - \int x d\mu\|_{L^2(d\mu)}^2\|\nabla g\|^2$ instead of $\frac{\|\nabla g\|^2}{R}$ for some universal constant $K > 0$. This implies that the nonnegative operator $L = -\Delta + \nabla U \cdot \nabla$ has a spectral gap.

In [5] Beckner found an inequality which interpolates in a sharp way between the Poincaré inequality and the log-Sobolev inequality. The inequality was obtained for Gaussian measures but, again, by a mass transportation argument it can easily be transferred to a log-concave probability measure. Beckner–Sobolev inequality states that for $f \in L^2(d\mu)$ and $1 \leq p \leq 2$ we have

$$\int |f|^2 d\mu - \left(\int |f|^p\right)^{2/p} \leq \frac{(2-p)}{R}\int_{\mathbb{R}^n} \|\nabla f\|^2 d\mu\qquad(1.18)$$

where $d\mu = e^{-U(x)}dx$ is a probability measure such that $\operatorname{Hess} U \geq R \cdot Id$. The case of $p = 1$ gives the Poincaré inequality (1.17) and the case where $p \to 2$ after dividing (1.18) by $2 - p$ gives (1.6). The Beckner–Sobolev inequality was studied for various measures in [28].

Proof of (1.18). Take

$$M(x,y) = x^{\frac{2}{p}} - \frac{2-p}{p^2} x^{\frac{2}{p}-2} y^2 \quad \text{where} \quad x, y \geq 0 \quad 1 \leq p \leq 2. \tag{1.19}$$

The matrix in (1.1) takes the form

$$\begin{pmatrix} -\dfrac{2(2-p)(1-p)(2-3p)x^{\frac{2}{p}-4}y^2}{p^4} & -\dfrac{4(2-p)(1-p)x^{\frac{2}{p}-3}y}{p^3} \\ -\dfrac{4(2-p)(1-p)x^{\frac{2}{p}-3}y}{p^3} & -\dfrac{4(2-p)x^{\frac{2}{p}-2}}{p^2} \end{pmatrix}, \tag{1.20}$$

and it is negative semidefinite. By Corollary 1.2 we have

$$\int_{\mathbb{R}^n} g^{\frac{2}{p}} - \frac{2-p}{p^2} g^{\frac{2}{p}-2} \frac{\|\nabla g\|^2}{R} \, d\mu \leq \left(\int_{\mathbb{R}^n} g d\mu \right)^{\frac{2}{p}} \tag{1.21}$$

for positive (in fact nonnegative) functions g. Now set $g = |f|^p$, and notice that $\|\nabla |f|\| \leq \|\nabla f\|$. After rearranging terms in (1.21) we obtain (1.18). $\qquad\square$

1.1.4. 3/2 function. Beckner's inequality (1.18) can be rewritten in an equivalent form

$$\int_{\mathbb{R}^n} f^p d\gamma - \left(\int_{\mathbb{R}^n} f d\gamma \right)^p \leq \frac{p(p-1)}{2} \int_{\mathbb{R}^n} f^{p-2} \|\nabla f\|^2 d\gamma, \quad p \in [1,2]. \tag{1.22}$$

In fact inequality (1.22) can be improved essentially for $p \in (1,2)$. We will illustrate the improvement in the case of $p = 3/2$ and for the general case we should refer the reader to our recent paper [24] which is based on application of Theorem 1.1.

The following inequality valid for all smooth bounded nonnegative f was proved in our recent paper [24]:

$$\int_{\mathbb{R}^n} f^{3/2} d\gamma - \left(\int_{\mathbb{R}^n} f d\gamma \right)^{3/2}$$
$$\leq \int_{\mathbb{R}^n} \left(f^{3/2} - \frac{1}{\sqrt{2}} (2f - \sqrt{f^2 + \|\nabla f\|^2}) \sqrt{f + \sqrt{f^2 + \|\nabla f\|^2}} \right) d\gamma. \tag{1.23}$$

Inequality (1.23) improves Beckner's bound (1.22) for $p = 3/2$. Indeed, notice that for $x, y \geq 0$ we have the following *pointwise* inequality

$$x^{3/2} - \frac{1}{\sqrt{2}} \left(2x - \sqrt{x^2 + y^2} \right) \sqrt{x + \sqrt{x^2 + y^2}} \leq \frac{3}{8} x^{-1/2} y^2, \tag{1.24}$$

which follows from homogeneity, i.e., take $x = 1$. By plugging f for x, $|\nabla f|$ for y and integrating we see that (1.23) improves up to (1.22).

Inequality (1.24) is always strict except when $y = 0$. Also notice that, as $y \to +\infty$, the right-hand side of (1.24) increases as y^2 whereas the left-hand side of (1.24) increases as $y^{3/2}$. It should be mentioned as well that, as $x \to 0$, the difference in (1.24) goes to infinity. The only place where the quantities in (1.24) are comparable is when $y/x \to 0$. It should be noted that since the left-hand side of (1.24) is a monotone decreasing function in x (see [24]), and when $x = 0$ it becomes $\frac{y^{3/2}}{\sqrt{2}}$, from (1.23) it follows that

$$\int_{\mathbb{R}^n} f^{3/2} d\gamma - \left(\int_{\mathbb{R}^n} f d\gamma\right)^{3/2} \leq \frac{1}{\sqrt{2}} \int_{\mathbb{R}^n} \|\nabla f\|^{3/2} d\gamma. \tag{1.25}$$

Inequality (1.25) gives some information about the measure concentration of γ.

Proof of (1.23). Take

$$M(x,y) = \frac{1}{\sqrt{2}} \left(2x - \sqrt{x^2 + y^2}\right) \sqrt{x + \sqrt{x^2 + y^2}} \quad \text{where} \quad x, y \geq 0. \tag{1.26}$$

The matrix in (1.1) takes the form

$$\frac{3\sqrt{2}}{8\sqrt{x^2 + y^2}} \begin{pmatrix} -\frac{y^2}{(x+\sqrt{x^2+y^2})^{3/2}} & \frac{y}{\sqrt{x+\sqrt{x^2+y^2}}} \\ \frac{y}{\sqrt{x+\sqrt{x^2+y^2}}} & -\sqrt{x + \sqrt{x^2 + y^2}} \end{pmatrix}. \tag{1.27}$$

Clearly (1.27) is negative semidefinite. So by Corollary 1.2 we obtain

$$\int_{\mathbb{R}^n} \frac{1}{\sqrt{2}} \left(2f - \sqrt{f^2 + \frac{\|\nabla f\|^2}{R}}\right) \sqrt{f + \sqrt{f^2 + \frac{\|\nabla f\|^2}{R}}} d\mu \leq \left(\int_{\mathbb{R}^n} f d\mu\right)^{3/2}. \tag{1.28}$$

This is of course (1.23) for the Gaussian measure γ: by taking $R = 1$ and rearranging terms in (1.28) we obtain (1.23). □

1.1.5. Banaszczyk's problem: the (B) Theorem. A problem proposed by W. Banaszczyk (see for example [27]) says that, given a symmetric convex body $K \subset \mathbb{R}^n$, the function $\phi(t) = \gamma(e^t K)$ is log-concave on $\mathbb{R}$. The problem was solved in [19]: clearly one only needs to check log-concavity at one point: $(\ln \phi(t))''|_{t=0} \leq 0$. This is the same as

$$\int_{\mathbb{R}^n} \|x\|^4 d\gamma_k - \left(\int_{\mathbb{R}^n} \|x\|^2 d\gamma_K\right)^2 \leq 2 \int_{\mathbb{R}^n} \|x\|^2 d\gamma_K, \tag{1.29}$$

where

$$d\gamma_K = \frac{\mathbb{1}_K(x) e^{-\|x\|^2/2} dx}{\int_K e^{-\|y\|^2/2} dy} = e^{-\|x\|^2/2 - \psi(x)} dx,$$

where ψ is a convex function constant on K and equal to $+\infty$ outside of K. In other words one can assume that $d\gamma_K = e^{-U(x)} dx$ is a probability measure where $U(x)$ is even and such that $\text{Hess}\, U \geq Id$. Setting $f(x) = \|x\|^2$, we can rewrite inequality (1.29) as follows:

$$\int_{\mathbb{R}^n} f^2 d\mu - \left(\int_{\mathbb{R}^n} f d\mu\right)^2 \leq \frac{1}{2} \int_{\mathbb{R}^n} \|\nabla f\|^2 d\mu, \tag{1.30}$$

which is better than the Poincaré inequality (1.17). This is a key ingredient in the (B) Theorem, and it was proved by Cordero-Erausquin–Fradelizi–Maurey in [19] that (1.30) holds provided that $\int_{\mathbb{R}^n} \nabla f d\mu = 0$, and $d\mu = e^{-U(x)} dx$ is a probability measure such that $\operatorname{Hess} U \geq Id$ (which is true for $f(x) = \|x\|^2$).

If one tries to apply Corollary 1.2 then the right choice of the function M must be

$$M(x, y) = x^2 - \frac{y^2}{2} \qquad (1.31)$$

but unfortunately this function does not satisfy (1.1). However, we want (1.30) to hold only for the functions such that $\int \nabla f d\mu = 0$ therefore one can slightly modify the proof of Theorem 1.1 in order to obtain (1.30). In Section 4 we will show how this works and we will present a different proof of (1.30) with the additional conditions that f is even and $d\mu$ is even (which definitely suffices for the (B) Theorem).

1.1.6. Φ-entropy. Let $\Phi : \Omega \to \mathbb{R}$ be a convex function. Given a probability measure $d\mu$ on $\mathbb{R}^n$, define Φ-entropy (see [18]) as follows:

$$\mathbf{Ent}_{\mu}^{\Phi}(f) \overset{\text{def}}{=} \int_{\mathbb{R}^n} \Phi(f) d\mu - \Phi\left(\int_{\mathbb{R}^n} f d\mu\right).$$

Corollary 1.2 provides us with a systematic approach to finding the bounds of Φ-entropy for uniformly log-concave probability measures $d\mu$. Indeed, let us illustrate this by the example of the Gaussian measure. Given a convex function Φ on $\Omega \subset \mathbb{R}$, let $M(x, y)$ be such that $M(x, 0) = \Phi(x)$, $M(x, \sqrt{y}) \in C^2(\Omega \times \mathbb{R}_+)$ and M satisfies (1.1). Then by Theorem 1.1 we obtain

$$\int_{\mathbb{R}^n} \Phi(f(x)) d\gamma - \Phi\left(\int_{\mathbb{R}^n} f d\gamma\right) \leq \int_{\mathbb{R}^n} [M(f, 0) - M(f, \|\nabla f\|)] d\gamma.$$

In our recent paper [24], we did find the bounds for the Φ-entropy as an application of Theorem 1.1 for the following fundamental examples

$$\Phi(x) = x^p \qquad \text{for} \quad p \in \mathbb{R} \setminus [0, 1]; \qquad (1.32)$$

$$\Phi(x) = -x^p \quad \text{for} \quad p \in (0, 1); \qquad (1.33)$$

$$\Phi(x) = e^x; \qquad (1.34)$$

$$\Phi(x) = -\ln x. \qquad (1.35)$$

Finding the best possible M is based on solving a PDE problem (1.37) with boundary conditions (1.32, 1.33, 1.34, 1.35) (see Subsection 1.1.8, Section 3 and [24]).

1.1.7. Yet another isoperimetric inequality obtained by the EDS method. In Section 3 we consider a peculiar example (see Subsection 3.1.5) of the elliptic solution of the PDE (1.4) with initial data

$$M(x, 0) = x \arccos(-x) + \sqrt{1 - x^2} \quad \text{for} \quad x \in [-1, 1],$$

which is not related to the applications that we have discussed before, but which gives yet another example of a new isoperimetric inequality. It looks like a useful one, in particular because the Poincaré inequality is its consequence.

1.1.8. Concluding remarks. As we shall notice, application of Theorem 1.1 to functional (and thereby isoperimetric) inequalities meets a difficulty: one has to find a right function $M(x, y)$, for example such as (1.7), (1.19), (1.13), (1.26), (1.31) and functions M mentioned in Subsection 1.1.6 (see [24]).

If one knows what inequality should be proved then one can try to guess what function $M(x, y)$ should be chosen: in the integrand one needs to set $g = x$ and $\|\nabla g\| = y$ and then the integrand in terms of x and y will be $M(x, y)$.

In general finding $M(x, y)$ will be based purely on solving PDEs. Let us recall the discussions of Subsection 1.1.6. First suppose that given, for example, a convex function $\Phi : \Omega \to \mathbb{R}$ one wants to find an optimal *error term* in Jensen's inequality (the Φ-entropy, see [18])

$$0 \le \int_{\mathbb{R}^n} \Phi(f(x)) d\gamma - \Phi\left(\int_{\mathbb{R}^n} f d\gamma\right) \le \int_{\mathbb{R}^n} \mathrm{Error}(f, \|\nabla f\|) d\gamma$$

for all $f \in C^\infty(\mathbb{R}^n; \Omega)$. If we find $M(x, y) \in C^2(\Omega \times \mathbb{R}_+)$ such that $M(x, 0) = \Phi(x)$ and $M(x, y)$ satisfies (1.1) then by Theorem 1.1 we can find a possible error term as follows

$$\int_{\mathbb{R}^n} \Phi(f(x)) d\gamma - \Phi\left(\int_{\mathbb{R}^n} f d\gamma\right) \le \int_{\mathbb{R}^n} [M(f, 0) - M(f, \|\nabla f\|)] d\gamma. \qquad (1.36)$$

In fact we would like to minimize the error term which corresponds to maximizing $M(x, y)$ under the constraints (1.1) and $M(x, 0) = \Phi(x)$. This suggests that the partial differential inequality (1.1) should degenerate. Indeed, if $\lambda_1(x, y)$ and $\lambda_2(x, y)$ denote eigenvalues of the matrix in (1.1) then condition (1.1) becomes $\lambda_1 + \lambda_2 \le 0$ and $\lambda_1 \cdot \lambda_2 \ge 0$. If we have strict inequality $\lambda_1 \cdot \lambda_2 > 0$ then $\lambda_1 + \lambda_2 < 0$. In this case we can slightly perturb M at a point (x, y) so as to make $M(x, y)$ larger but still keep the inequality $\lambda_1 \cdot \lambda_2 > 0$. Clearly the condition $\lambda_1 + \lambda_2 < 0$ still holds. We can continue perturbing M until (1.1) degenerates. Therefore we will seek $M(x, y)$ among those functions which, together with (1.1) also satisfy a *degenerate elliptic Monge–Ampère equation of general type*:

$$\det \begin{pmatrix} M_{xx} + \frac{M_y}{y} & M_{xy} \\ M_{xy} & M_{yy} \end{pmatrix} = M_{xx} M_{yy} - M_{xy}^2 + \frac{M_y M_{yy}}{y} = 0 \qquad (1.37)$$

for $(x, y) \in \Omega \times \mathbb{R}_+$.

For example in the log-Sobolev inequality (1.6) and in Bobkov's inequality (1.10) the determinants of the matrices (1.8) and (1.14) are zero. In the Beckner–Sobolev inequality (1.18) the determinant of (1.20) is zero if and only if $p = 1, 2$. Notice that these are exactly the cases when the Beckner–Sobolev inequality interpolates the Poincaré and log-Sovbolev inequalities. Moreover, since the determinant in the Beckner–Sobolev inequality is not zero for $p \in (1, 2)$, this indicates that one should improve the inequality, and this is exactly what was done in (1.23). We refer the reader to our recent paper [24] where we do improve the

Beckner–Sobolev inequality by solving the elliptic Monge–Ampère equation (1.37) with the boundary condition $M(x,0) = x^p$, where $p \in \mathbb{R}$.

   In Section 3 we will show that thanks to the exterior differential systems studied by R. Bryant and P. Griffiths (see [12, 13, 14]), the nonlinear equation (1.37) can be reduced (after a suitable change of variables) to the linear backward heat equation. In Section 3.1 we will illustrate this by the examples

$$M(x,0) = x \ln x, \ M(x,0) = x^2, \ M(x,0) = -I(x) \quad \text{and} \quad M(x,0) = x^{3/2}$$

which correspond to the log-Sobolev, Poincaré, Bobkov, and 3/2 inequalities.

   To justify the claim that Section 3 makes the approach to bounds of Φ-entropy systematic, we do consider a peculiar example (see Subsection 3.1.5)

$$M(x,0) = x \arccos(-x) + \sqrt{1-x^2} \quad \text{for} \quad x \in [-1,1]$$

which is not related to the applications that we have discussed before.

2. Function of the variables $D^\alpha f$. Proof of Theorem 1.1

We prove here Theorem 1.1 and even the more general Theorem 2.1. The reader who *a priori* believes in Theorem 1.1 can skip to the next Section 3 devoted to the exterior differential systems (EDS) method of finding the elliptic solutions of the PDE (1.4) (by elliptic solutions we mean the solutions M satisfying condition (1.1) on M).

   Let $d\mu = e^{-U(x)}dx$ be a log-concave probability measure such that U is smooth and $\text{Hess}\, U \geq R \cdot Id$. Set $L = \Delta - \nabla U \cdot \nabla$. Then $-L$ is a selfadjoint positive operator in $L^2(\mathbb{R}^n, d\mu)$, moreover by (1.17) it has a spectral gap. Let $P_t := e^{tL}$ be the corresponding semigroup generated by L. Let $\alpha = (\alpha_0, \ldots, \alpha_m)$, where $\alpha_j = (\alpha_j^1, \ldots, \alpha_j^n)$ is a multi-index of size n and $\alpha_j^i \in \mathbb{N} \cup \{0\}$ for each $j = 0, \ldots, m$ and $i = 1, \ldots, n$. Let $|\alpha_j|$ be the length of the multi-index, i.e., $|\alpha_j| = \alpha_j^1 + \cdots + \alpha_j^n$. By D^{α_j} we denote the differential operator

$$D^{\alpha_j} = \frac{\partial^{|\alpha_j|}}{\partial x_1^{\alpha_j^1} \cdots \partial x_n^{\alpha_j^n}}. \tag{2.1}$$

Further we fix some *multi-multi*-index $\alpha = (\alpha_0, \ldots, \alpha_m)$ where each α_j is a multi-index of size n as above.

Test functions $C^\infty(\mathbb{R}^n; \Lambda)$

Let Λ be a closed convex subset of $\mathbb{R}^m$. By $C^\infty(\mathbb{R}^n; \Lambda)$ we denote the set of test functions $\mathbf{f} = (f_0, \ldots, f_m) : \mathbb{R}^n \to \Lambda$, i.e., smooth bounded vector functions with values in Λ. Let

$$D^\alpha \mathbf{f} = (D^{\alpha_0} f_0, \ldots, D^{\alpha_m} f_m) \quad \text{and} \quad P_t \mathbf{f} := (P_t f_0, \ldots, P_t f_m).$$

We require that $C^\infty(\mathbb{R}^n; \Lambda)$ be closed under taking D^α, i.e., $D^\alpha C^\infty(\mathbb{R}^n; \Lambda) \subset C^\infty(\mathbb{R}^n; \Lambda)$. The linearity and positivity of P_t imply that $P_t \mathbf{f}, P_t D^\alpha \mathbf{f} \in \Lambda$ for any $f \in C^\infty(\mathbb{R}^n; \Lambda)$.

Let $B(u_1, \ldots, u_m) : \Lambda \to \mathbb{R}$ be a smooth (at least C^2) function such that $P_t B(D^\alpha \mathbf{f})$ is well defined for all $t \geq 0$. Set

$$[L, D^\alpha]\mathbf{f} \overset{\text{def}}{=} ([L, D^{\alpha_0}]f_0, \ldots, [L, D^{\alpha_m}]f_m) \quad \text{and}$$

$$\Gamma(D^\alpha \mathbf{f}) \overset{\text{def}}{=} \{\langle \nabla D^{\alpha_i} f_i, \nabla D^{\alpha_j} f_j \rangle\}_{i,j=0}^m$$

where $\Gamma(D^\alpha \mathbf{f})$ denotes a matrix of size $(m+1) \times (m+1)$, and $[A, B] = AB - BA$ denotes the commutator of A and B.

Theorem 2.1. *The following conditions are equivalent:*

(i) $\nabla B(D^\alpha \mathbf{f}) \cdot [L, D^\alpha]\mathbf{f} + \text{Tr}[\text{Hess}\, B(D^\alpha \mathbf{f}) \Gamma(D^\alpha \mathbf{f})] \leq 0$ *for all* $f \in C^\infty(\mathbb{R}^n; \Lambda)$;

(ii) $P_t[B(D^\alpha \mathbf{f})](x) \leq B(D^\alpha[P_t \mathbf{f}](x))$ *for all* $t \geq 0$, $x \in \mathbb{R}^n$ *and* $f \in C^\infty(\mathbb{R}^n; \Lambda)$.

Proof. (i) implies (ii): let $V(x,t) = P_t[B(D^\alpha \mathbf{f})](x) - B(D^\alpha[P_t \mathbf{f}](x))$. Notice that

$$(\partial_t - L)V(x,t) = (L - \partial_t)B(D^\alpha[P_t \mathbf{f}](x))$$

$$= \sum_j \frac{\partial B}{\partial u_j} L D^{\alpha_j} P_t f_j + \sum_{i,j} \frac{\partial^2 B}{\partial u_i \partial u_j} \nabla D^{\alpha_i} P_t f_i \cdot \nabla D^{\alpha_j} P_t f_j$$

$$- \sum_j \frac{\partial B}{\partial u_j} D^{\alpha_j} L P_t f_j$$

$$= \nabla B(D^\alpha P_t \mathbf{f}) \cdot [L, D^\alpha]P_t \mathbf{f} + \text{Tr}(\text{Hess}\, B(D^\alpha P_t \mathbf{f})\, \Gamma(D^\alpha P_t \mathbf{f})) \leq 0$$

The last inequality follows from (i) and the fact that $P_t \mathbf{f}(x) \in \Lambda$. Indeed, we can find a function $\mathbf{g} \in C^\infty(\mathbb{R}^n; \Lambda)$ such that $\mathbf{g} = P_t \mathbf{f}$ in a neighborhood of x and we can apply (i) to $\mathbf{g}$.

By the maximum principle we obtain $V(x,t) \leq \sup_x V(x,0) = 0$. Another way (without maximum principle) is that

$$V(x,t) = \int_0^t \frac{\partial}{\partial s} P_s B(D^\alpha P_{t-s} \mathbf{f}) ds = \int_0^t P_s \left[\left(L - \frac{\partial}{\partial t} \right) B(D^\alpha P_{t-s} \mathbf{f}) \right] ds, \quad (2.3)$$

and the integrand in (2.3) is non-positive by (2.2).

(ii) impies (i): for all $\mathbf{f} \in C^\infty(\mathbb{R}^n; \Lambda)$ we have

$$0 \geq \lim_{t \to 0} \frac{V(x,t)}{t} = \lim_{t \to 0} \frac{V(x,t) - V(x,0)}{t} = \frac{\partial}{\partial t} V(x,t)|_{t=0}$$

$$= \nabla B(D^\alpha \mathbf{f}) \cdot [L, D^\alpha]\mathbf{f} + \text{Tr}(\text{Hess}\, B(D^\alpha \mathbf{f})\, \Gamma(D^\alpha \mathbf{f})). \qquad \square$$

Remark 2.2. We notice that if one considers diffusion semigroups generated by

$$L = \sum_{ij} a_{ij}(x) \frac{\partial^2}{\partial x_i \partial x_j} + \sum_j b_j(x) \frac{\partial}{\partial x_j}$$

where $A = \{a_{ij}\}_{i,j=1}^n$ is positive then absolutely nothing changes in Theorem 2.1 except the matrix $\Gamma(D^\alpha \mathbf{f})$ takes the form

$$\Gamma(D^\alpha \mathbf{f}) = \{\nabla D^{\alpha_i} f_i\, A\, (\nabla D^{\alpha_j} f_j)^T\}_{i,j=0}^m.$$

2.1. Proof of Theorem 1.1

Consider the special case when $n = m$, $\mathbf{f} = \underbrace{(f, \ldots, f)}_{n+1}$, $\alpha_0 = \underbrace{(0, \ldots, 0)}_{n}$, $\alpha_1 = (1, 0, \ldots, 0), \ldots$, and $\alpha_n = (0, \ldots, 0, 1)$. Then $D^\alpha \mathbf{f} = (f, \nabla f)$, and, given the relation $L = \Delta - \nabla U \cdot \nabla$, we obtain

$$\nabla B(D^\alpha \mathbf{f}) \cdot [L, D^\alpha] \mathbf{f} = \nabla_{1, \ldots, n} B \, (\mathrm{Hess}\, U)(\nabla f)^T.$$

Here $\nabla_{1, \ldots, n} B$ is the gradient of $B(u_0, \ldots, u_n)$ taken with respect to the variables $u_1, \ldots, u_n$. Assume that f takes values in a closed convex set $\Omega \subset \mathbb{R}$. Take

$$B(u_0, \ldots, u_n) = M\left(u_0, \sqrt{\frac{u_1^2 + \cdots + u_n^2}{R}}\right), \tag{2.4}$$

where $M(x, \sqrt{y}) \in C^2(\Omega \times \mathbb{R}_+)$ satisfies (1.1). Notice that $M_y \leq 0$. Indeed, if we multiply the first diagonal entry of (1.1) by y and send $y \to 0$ we obtain $M_y(x, 0) \leq 0$. On the other hand since the second diagonal entry of (1.1) is nonpositive we obtain $M_y(x, y) \leq 0$ for all y.

Next we notice that

$$\nabla_{1, \ldots, n} B(D^\alpha \mathbf{f}) = \frac{M_y}{\|\nabla f\| \sqrt{R}} \nabla f.$$

Since $M_y \leq 0$ and $\mathrm{Hess}\, U \geq R \cdot Id$, we have

$$\nabla B(D^\alpha \mathbf{f}) \cdot [L, D^\alpha] \mathbf{f} = \frac{M_y}{\|\nabla f\| \sqrt{R}} \nabla f (\mathrm{Hess}\, U)(\nabla f)^T \leq \sqrt{R} \|\nabla f\| M_y.$$

Therefore

$$\nabla B(D^\alpha \mathbf{f}) \cdot [L, D^\alpha] \mathbf{f} + \mathrm{Tr}(\mathrm{Hess}\, B(D^\alpha \mathbf{f})) \, \Gamma(D^\alpha \mathbf{f})) \leq \mathrm{Tr}(W \, \Gamma(D^\alpha \mathbf{f}))$$

where

$$W = \begin{bmatrix} \partial_{00}^2 B + \frac{\sqrt{R} \cdot M_y}{\|\nabla f\|} & \partial_{01}^2 B & \cdots & \partial_{0n}^2 B \\ \partial_{10}^2 B & \partial_{11}^2 B & \cdots & \partial_{1n}^2 B \\ \cdots\cdots\cdots\cdots\cdots\cdots\cdots\cdots\cdots\cdots\cdots \\ \partial_{n0}^2 B & \partial_{n1}^2 B & \cdots & \partial_{nn}^2 B \end{bmatrix}$$

where $\partial_{ij}^2 B = \frac{\partial^2 B}{\partial u_i \partial u_j}$.

We will show that $W \leq 0$, and then we will obtain $\mathrm{Tr}(W \, \Gamma(D^\alpha \mathbf{f})) \leq 0$ because $\Gamma(D^\alpha \mathbf{f}) \geq 0$.

We have $\partial_{00}^2 B = M_{xx}$, $\partial_{0j}^2 B = \frac{M_{xy}}{\|\nabla f\| \sqrt{R}} f_{x_j}$ for all $j \geq 1$ and

$$\partial_{ij}^2 B = \frac{M_{yy}}{\|\nabla f\|^2 R} f_{x_i} f_{x_j} - \frac{M_y}{\|\nabla f\|^3 \sqrt{R}} f_{x_i} f_{x_j} + \frac{M_y \delta_{ij}}{\|\nabla f\| \sqrt{R}} \quad \text{for} \quad i, j \geq 1$$

where δ_{ij} is the Kronecker symbol.

Notice that since $M(x, \sqrt{y}) \in C^2(\Omega \times \mathbb{R}_+)$, we have $B \in C^2(\Omega \times \mathbb{R}^n)$. If $\nabla f = 0$ then there is nothing to prove because W becomes a diagonal matrix with negative entries on the diagonal. Further assume that $\|\nabla f\| \neq 0$.

Now notice that

$$W = S \left(W_1 + \frac{M_y \sqrt{R}}{\|\nabla f\|} W_2 \right) S$$

where S is the diagonal matrix with diagonal $(1, \frac{\nabla f}{\|\nabla f\| \sqrt{R}})$, and

$$W_1 = \begin{bmatrix} M_{xx} + \frac{\sqrt{R} \cdot M_y}{\|\nabla f\|} & M_{xy} & \cdots & M_{xy} \\ M_{xy} & M_{yy} & \cdots & M_{yy} \\ \cdots\cdots\cdots\cdots\cdots\cdots\cdots\cdots \\ M_{xy} & M_{yy} & \cdots & M_{yy} \end{bmatrix} \quad \text{and}$$

$$W_2 = \begin{bmatrix} 0 & 0 & 0 & \cdots & 0 \\ 0 & \frac{\|\nabla f\|^2}{(f_{x_1})^2} - 1 & -1 & \cdots & -1 \\ 0 & -1 & \frac{\|\nabla f\|^2}{(f_{x_2})^2} - 1 & \cdots & -1 \\ \cdots\cdots\cdots\cdots\cdots\cdots\cdots\cdots\cdots\cdots\cdots \\ 0 & -1 & \cdots & -1 & \frac{\|\nabla f\|^2}{(f_{x_n})^2} - 1 \end{bmatrix}$$

It is clear that $W_1 \le 0$ because M satisfies (1.1) at the point x and $\frac{y}{\sqrt{R}}$.

For W_2, first notice that if $f_{x_j} \ne 0$ for all $j \ge 1$ then W_2 is well defined and $W_2 \ge 0$. Otherwise if $f_{x_j} = 0$ for some j, then consider the initial expression SW_2S and notice that $SW_2S = S\tilde{W}_2S + D$, where $\tilde{W}_2$ is the same as W_2 except that the jth column and row are replaced by zeros, and D is the zero matrix except that the element (j, j) is equal to $\frac{1}{R}$. We again see that $SW_2S \ge 0$. Hence $M_y SW_2S \le 0$ as soon as (1.1) holds.

Thus we have proved that if $M(x, \sqrt{y}) \in C^2(\Omega \times \mathbb{R}_+)$, M satisfies (1.1) then by Theorem 2.1 we have

$$P_t M(f, \|\nabla f\|) \le M(P_t f, \|\nabla P_t f\|) \quad \text{for all} \quad f \in C^\infty(\mathbb{R}^n; \Omega). \qquad (2.5)$$

We send $t \to \infty$ and because of the fact that $\|\nabla P_t f\| \le e^{-tR} P_t \|\nabla f\|$ (see [3]) we obtain

$$\int_{\mathbb{R}^n} M(f, \|\nabla f\|) d\mu \le M \left(\int_{\mathbb{R}^n} f d\mu, 0 \right),$$

where $d\mu = e^{-U(x)} dx$ is a probability measure.

Remark 2.3. It is worth mentioning but not necessary for our purposes that (2.5) also implies (1.1) in the case of a Gaussian measure. This follows from the fact that the matrix $\lambda \Gamma(D^\alpha \mathbf{f})$ can be an arbitrary positive definite matrix where $\lambda > 0$ and $\mathbf{f} \in C^\infty(\mathbb{R}^n; \Lambda)$. Then the condition $\mathrm{Tr}(W\Gamma(D^\alpha \mathbf{f})) \le 0$ implies that $W \le 0$ and this gives us condition (1.1).

2.2. Relationship with the stochastic calculus and Γ-calculus approach

Inequality (2.5) implies that the map

$$t \to \int_{\mathbb{R}^n} M(P_t f, \|\nabla P_t f\|) d\gamma \qquad (2.6)$$

is monotone provided that M satisfies (1.1). Indeed, by sending $t \to 0$ in (2.5) we obtain its infinitesimal form $LM(f, \|\nabla f\|) \leq \frac{d}{ds} M(P_s f, \|\nabla P_s f\|)\big|_{s=0}$. Finally, if the last inequality is true for any f then it is true for any f of the form $P_t f$. This implies that

$$LM(P_t f, \|\nabla P_t f\|) \leq \frac{d}{ds} M(P_s f, \|\nabla P_s f\|)\bigg|_{s=t},$$

and it gives the monotonicity of (2.6).

The interpolation (2.5) (or even the monotonicity (2.6)) plays a fundamental role in functional inequalities and it was known before for some particular functions $M(x, y)$ as a consequence of their special properties and some linear algebraic manipulations (see [3], [1]). The purpose of Theorem 1.1 was to exclude the linear algebra involved in the interpolation (2.5) and to show that in fact (2.5) boils down (actually it is equivalent) to the fact that M satisfies an *elliptic Monge–Ampère equation of a general form* (1.1). The Monge–Ampère equation (1.1), apparently, was not noticed before or it was hidden in the literature from the wide audience. Equations of Monge–Ampère type are of course among the most important fully nonlinear partial differential equations (see [35], [36]).

Next we will show that in fact (1.1) gives monotonicity of the type (2.6) in different settings as well.

2.2.1. Stochastic calculus approach.

Proposition 2.4. *For $t \geq 0$, let W_t, N_t be $\mathcal{F}_t$ real-valued martingales with $W_t = W_0 + \int_0^t w_s dB_s$, $N_t = N_0 + \int_0^t n_s dB_s$, and let $A_t = A_0 + \int_0^t a_s ds$ where $A_0, a_s \geq 0$. Assume that A_t is bounded, $a_t |N_t|^2 \geq |w_t|^2$ and $W_t \in \Omega$ for $t \geq 0$. Assume that $M(x, \sqrt{y}) \in C^2(\Omega \times \mathbb{R})$ satisfies (1.1). Then*

$$z_t = M(W_t, |N_t| \sqrt{A_t})$$

is a supermartingale for $t \geq 0$.

Proof. The proof of the proposition proceeds absolutely in the same way as in [1], which treated the case of a *particular* function M involved in Bobkov's inequality. In fact, it is (1.1) which makes the drift $\Delta(t)$ nonpositive where $dz_t = u_t dB_t + \Delta(t)dt$. □

One may obtain another proof of Theorem 1.1 for the case of $n = 1$ by using Proposition 2.4 for the special case when $W_t = \mathbb{E}[f(B_1)|\mathcal{F}_t]$, $A_t = t$ and $N_t = \mathbb{E}[\nabla f(B_1)|\mathcal{F}_t]$ where $0 \leq t \leq 1$, $B_0 = 0$, and f is a real-valued smooth bounded function. Indeed, in this case, by the optional stopping theorem one obtains

$$M(\mathbb{E}[f(B_1)|\mathcal{F}_0], 0) = z_0 \geq \mathbb{E}z_1 = \mathbb{E}M(f(B_1), |\nabla f(B_1)|).$$

2.2.2. Γ-calculus approach. One may obtain Theorem 1.1 using the remarkable Γ-calculus (see [3]). In fact, setting $\Gamma(f,g) = \nabla f \cdot \nabla g$ one can easily show that if $M(x, \sqrt{y}) \in C^2(\Omega \times \mathbb{R}_+)$ satisfies (1.1) then the following map

$$s \to P_s B(P_{t-s}f, \Gamma(P_{t-s}f, P_{t-s}f))$$

is monotone for $0 \le s \le t$ for any given $t > 0$, where $B(x, y^2) = M(x, y)$. Initially this was the way we obtained Theorem 1.1 (see, for example, [23]). Later it became clear to us that one does not have to be limited by the symbols $\Gamma, \Gamma_2, \Gamma_3$, etc. in order to enjoy interpolations of the form (2.5). In fact, one can directly work with an arbitrary differential operator D^α (2.1), and Theorem 1.1 is simply a consequence of Theorem 2.1 for an appropriate choice of the *Bellman function* (2.4). In support of the classical notation, we should say that it is not clear to us how Γ-calculus can be used in proving Theorem 2.1 which is a simple statement if we work with the classical notation of differential operators D^α.

3. Reduction to the exterior differential systems and backward heat equation

As we already mentioned in Subsection 1.1.8 (and it also follows from the proof of Theorem 1.1) in order that inequality (1.2) to be *sharp* we need to assume that (1.1) degenerates, i.e.,

$$\det \begin{pmatrix} M_{xx} + \frac{M_y}{y} & M_{xy} \\ M_{xy} & M_{yy} \end{pmatrix} = M_{xx}M_{yy} - M_{xy}^2 + \frac{M_y M_{yy}}{y} = 0. \tag{3.1}$$

Let us make the following observation: consider the 1-graph of $M(x, y)$, i.e.,

$$(x, y, p, q) = (x, y, M_x(x, y), M_y(x, y))$$

in $xypq$-space. This is a simply connected surface Σ in 4-space on which $\Upsilon = dx \wedge dy$ is non-vanishing but to which the two 2-forms

$$\Upsilon_1 = dp \wedge dx + dq \wedge dy \quad \text{and} \quad \Upsilon_2 = (y dp + q dx) \wedge dq$$

pull back to be zero.

Conversely, suppose we are given a simply connected surface Σ in $xypq$-space (with $y > 0$) on which Υ is non-vanishing but to which Υ_1 and Υ_2 pull back to be zero. The 1-form $p dx + q dy$ pulls back to Σ to be closed (since Υ_1 vanishes on Σ) and hence exact, and therefore there exists a function $m : \Sigma \to \mathbb{R}$ such that $dm = p dx + q dy$ on Σ. We then have (at least locally), $m = M(x, y)$ on Σ and, by its definition, we have $p = M_x(x, y)$ and $q = M_y(x, y)$ on the surface. Then the fact that Υ_2 vanishes when pulled back to Σ implies that $M(x, y)$ satisfies the desired equation.

Thus, we have encoded the given PDE as an exterior differential system on $\mathbb{R}^4$. Note that we can make a change of variables on the open set where $q < 0$:

set $y = qr$ and let $t = \frac{1}{2}q^2$. Then, using these new coordinates on this domain, we have

$$\Upsilon_1 = dp \wedge dx + dt \wedge dr \quad \text{and} \quad \Upsilon_2 = (rdp + dx) \wedge dt.$$

Now, when we take an integral surface Σ on these 2-forms on which $dp \wedge dt$ is vanishing, it can be written locally as a graph of the form

$$(p, t, x, r) = (p, t, u_p(p, t), u_t(p, t))$$

(since Σ is an integral of Υ_1), where $u(p, t)$ satisfies $u_t + u_{pp} = 0$ (since Σ is an integral of Υ_2). Thus, "generically" our PDE is equivalent to the backward heat equation, up to a change of variables. Thus the function $M(x, y)$ can be parametrized as follows

$$x = u_p\left(p, \frac{1}{2}q^2\right); \quad y = qu_t\left(p, \frac{1}{2}q^2\right);$$

$$M(x, y) = pu_p\left(p, \frac{1}{2}q^2\right) + q^2 u_t\left(p, \frac{1}{2}q^2\right) - u\left(p, \frac{1}{2}q^2\right). \tag{3.2}$$

Note that $y \geq 0$, $q = M_y \leq 0$ then $u_t\left(p, \frac{1}{2}q^2\right) \leq 0$. Let us rewrite the conditions $M_{yy} \leq 0$ and $M_{xx} + \frac{M_y}{y} \leq 0$ in terms of $u(p, t)$. In other words we want q_y and $p_x + \frac{q}{y}$ be non-positive. We have

$$0 = u_{pp}p_y + u_{pt}qq_y \quad \text{and} \quad 1 = q_y u_t + qp_y u_{tp} + q^2 q_y u_{tt}.$$

Then

$$1 = q_y u_t + q^2 q_y \frac{u_{pt}^2}{u_{pp}} + q^2 q_y u_{tt} \quad \text{and} \quad M_{yy} = q_y = \frac{u_t}{u_t^2 - 2t(u_{tt}u_{pp} - u_{pt}^2)}.$$

Thus the negative definiteness of the matrix (1.1) (if it is known that its determinant vanishes) is equivalent to

$$u_t^2 - 2t \det(\text{Hess } u) \geq 0. \tag{3.3}$$

Let us show that the function $u(p, t)$ must satisfy a boundary condition:

$$u(f'(x), 0) = xf'(x) - f(x) \quad \text{for} \quad x \in \Omega \quad \text{where} \quad f(x) = M(x, 0). \tag{3.4}$$

Indeed, we know that $M(x, \sqrt{y}) \in C^2(\Omega \times \mathbb{R}_+)$ therefore $M_y(x, 0) = 0$. By choosing $y = 0$ in (3.2), we have $q = 0$, and we obtain the desired boundary condition:

$$M(x, 0) = xM_x(x, 0) - u(M_x(x, 0), 0).$$

Now it is clear how to find the function $M(x, y)$ provided that $M(x, 0)$ is given.

First we try to find a function $u(p, t)$ such that

$$u_{pp} + u_t = 0, \quad u_t \leq 0, \tag{3.5}$$

$$u(M_x(x, 0), 0) = xM_x(x, 0) - M(x, 0) \quad x \in \Omega, \tag{3.6}$$

$$u_t^2 - 2t \det(\text{Hess } u) \geq 0. \tag{3.7}$$

Then a candidate for $M(x,y)$ will be given by (3.2). We should mention that if $M(x,0)$ is convex then (3.6) simply means that $u_p(p,0)$ is the Legendre transform of $M_x(x,0)$. Indeed, if we take the derivative in (3.6) with respect to x we obtain $u_p(M_x(x,0),0) = x$.

3.1. Back to the applications, old and new. Revisiting Section 1.1 with our new tool

Further we assume that we know the expression $M(x,0)$ and we would like to restore the function $M(x,y)$ which satisfies conditions of Theorem 1.1, the PDE (3.1) and hence it gives us inequality (1.3), or the error term in Jensen's inequality (see Subsection 1.1.8 for the explanations).

3.1.1. Gross function. In this case we have $M(x,0) = x \ln x$. Condition (3.6) can be rewritten as follows: $u(p,0) = e^{p-1}$ for all $p \in \mathbb{R}$. If we set $D = \frac{\partial^2}{\partial p^2}$ then

$$u(p,t) = e^{-tD} e^{p-1} = \sum_{k=0}^{\infty} \frac{(-t)^k}{k!} e^{p-1} = e^{p-t-1} \quad \text{for all} \quad t \geq 0.$$

Clearly $u(p,t)$ satisfies (3.7) because $\det(\operatorname{Hess} u) = 0$. Notice that we have $u_t < 0$,

$$\begin{cases} x = e^{p - \frac{q^2}{2} - 1}; \\ y = -q e^{p - \frac{q^2}{2} - 1}; \end{cases} \quad \text{then} \quad \begin{cases} q = -\frac{y}{x}; \\ p = \ln x + \frac{y^2}{2x^2} + 1. \end{cases}$$

Therefore we obtain

$$M(x,y) = xp + qy - u\left(p, \frac{1}{2}q^2\right) = x \ln x + \frac{y^2}{2x} + x - \frac{y^2}{x} - x = x \ln x - \frac{y^2}{2x}.$$

3.1.2. Nash's function. In this case we have $M(x,0) = x^2$. Condition (3.6) takes the form $u(p,0) = \frac{p^2}{4}$ for all $p \in \mathbb{R}$. Then

$$u(p,t) = e^{-tD} \frac{p^2}{4} = (1 - tD)\frac{p^2}{4} = \frac{p^2}{4} - \frac{t}{2} \quad t \geq 0.$$

The function $u(p,t)$ satisfies (3.7) because $\det(\operatorname{Hess} u) = 0$. We have $u_t < 0$

$$\begin{cases} x = \frac{p}{2}; \\ y = -\frac{q}{2}; \end{cases} \quad \text{then} \quad \begin{cases} p = 2x; \\ q = -2y. \end{cases}$$

We obtain

$$M(x,y) = 2x^2 - 2y^2 - (x^2 - y^2) = x^2 - y^2.$$

3.1.3. Bobkov's function. It is not clear at all where the function $M(x,y) = -\sqrt{I(x)^2 + y^2}$ comes from. Apparently it was a pretty good guess.

Let us show how easily it can be restored by solving the Monge–Ampère equation (1.37). In this case we have $M(x,0) = -I(x)$. Condition (3.6) takes the form

$$u(p,0) = p\Phi(p) + \Phi'(p) \quad \text{for all} \quad p \in \mathbb{R}. \tag{3.8}$$

Now we will try to find the usual heat extension of $u(p,0)$ (call it $\tilde{u}(p,t)$) which satisfies $\tilde{u}_{pp} = \tilde{u}_t$, and then we try to consider the formal candidate $u(p,t) := \tilde{u}(p,-t)$.

It is easier to find the heat extension of $\tilde{u}_p(p,0)$ and then take the antiderivative in p. Indeed, notice that (3.8) implies $u_p(p,0) = \Phi(p)$. The heat extension of $\Phi(p)$ is $\Phi\left(\frac{p}{\sqrt{1+2t}}\right)$. Indeed, the heat extension of the function $\mathbb{1}_{(-\infty,0]}(p)$ at time $t = 1/2$ is $\Phi(p)$. Then by the semigroup property, the heat extension of $\Phi(p)$ at time t will be the heat extension of $\mathbb{1}_{(-\infty,0]}(p)$ at time $1/2 + t$ which equals $\Phi\left(\frac{p}{\sqrt{1+2t}}\right)$. Thus $\tilde{u}_p(p,t) = \Phi\left(\frac{p}{\sqrt{1+2t}}\right)$. Taking the antiderivative in p and using (3.8) if necessary we obtain

$$\tilde{u}(p,t) = \sqrt{1+2t}\,\Phi'\left(\frac{p}{\sqrt{1+2t}}\right) + p\Phi\left(\frac{p}{\sqrt{1+2t}}\right).$$

This expression is well defined even for $t \in (-1/2,0)$. Therefore if we set

$$u(p,t) = \tilde{u}(p,-t) = \sqrt{1-2t}\,\Phi'\left(\frac{p}{\sqrt{1-2t}}\right) + p\Phi\left(\frac{p}{\sqrt{1-2t}}\right) \quad \text{for} \quad p \in \mathbb{R}$$

and $t \in \left[0,\frac{1}{2}\right)$, direct computations show that $u(p,t)$ satisfies (3.5), (3.8) and (3.7) because $\det(\text{Hess}\,u) = -\left(\frac{\Phi'\left(\frac{p}{\sqrt{1-2t}}\right)}{1-2t}\right)^2 < 0$. We have $u_t = -\frac{\Phi'\left(\frac{p}{\sqrt{1-2t}}\right)}{\sqrt{1-2t}} < 0$ and $u_p = \Phi\left(\frac{p}{\sqrt{1-2t}}\right)$. Therefore,

$$\begin{cases} x = \Phi\left(\frac{p}{\sqrt{1-q^2}}\right); \\ y = \frac{-q}{\sqrt{1-q^2}}\Phi'\left(\frac{p}{\sqrt{1-q^2}}\right); \end{cases} \quad \text{then} \quad \begin{cases} \Phi^{-1}(x) = \frac{p}{\sqrt{1-q^2}}; \\ y = \frac{-q}{\sqrt{1-q^2}}\Phi'(\Phi^{-1}(x)). \end{cases}$$

From the last equalities we obtain $M_y = q = -\frac{y}{\sqrt{I^2(x)+y^2}}$ and $M_x = p = \frac{I(x)\Phi^{-1}(x)}{\sqrt{I^2(x)+y^2}}$ where we remind the reader that $I(x) = \Phi'(\Phi^{-1}(x))$. It follows that

$$M(x,y) = -\sqrt{I^2(x) + y^2}.$$

3.1.4. Function 3/2. In this case we have $M(x,0) = x^{3/2}$ for $x \geq 0$. It follows from (3.6) that $u(p,0) = \frac{4}{27}p^3$ for $p \geq 0$. The solution of the backward heat equation is the Hermite polynomial, i.e., we have $u(p,t) = \frac{4}{27}(p^3 - 6tp)$. The function $u(p,t)$

satisfies (3.7) because Hess $u < 0$. Since $p \geq 0$ we have $u_t \leq 0$. Next we obtain

$$\begin{cases} x = \frac{4}{9}(p^2 - q^2), \\ y = -\frac{8}{9}pq; \end{cases} \quad \text{then} \quad \begin{cases} p = \frac{3}{4}\sqrt{2x + 2\sqrt{x^2 + y^2}}, \\ q = -\frac{3}{4}\sqrt{-2x + 2\sqrt{x^2 + y^2}}. \end{cases}$$

Finally

$$M(x, y) = xp + qy - u\left(p, \frac{1}{2}q^2\right) = \frac{1}{\sqrt{2}}(2x - \sqrt{x^2 + y^2})\sqrt{x + \sqrt{x^2 + y^2}}.$$

3.1.5. Function arccos(x). Consider the increasing convex function

$$M(x, 0) = x\arccos(-x) + \sqrt{1 - x^2}.$$

It follows from (3.6) that $u(p, 0) = -\sin(p)$ for $p \in [0, \pi]$. The solution of the backward heat equation (3.5) becomes $u(p, t) = -e^t \sin(p)$. Notice that $u_t \leq 0$ for $p \in [0, \pi]$, and

$$u_t^2 - 2t \det(\text{Hess}\, u) = e^{2t}(2t + \sin^2(x)) \geq 0.$$

Conditions (3.2) can be rewritten as follows

$$x = -e^{q^2} \cos(p);$$

$$y = -qe^{q^2/2} \sin(p);$$

$$M(x, y) = px + qy + e^{q^2/2} \sin(p) = px + qy - \frac{y}{q}, \quad x \in [-1, 1], \quad y \geq 0.$$

It follows that the negative number q satisfies the equation

$$-q\sqrt{e^{q^2} - x^2} = y, \tag{3.9}$$

and $p = \arccos(-xe^{-q^2/2})$. Thus we obtain

$$M(x, y) = x\arccos(-xe^{-q^2/2}) + (1 - q^2)\sqrt{e^{q^2} - x^2}$$

where a negative number q is a unique solution of (3.9). Consequently, we arrive at

$$\int_{\mathbb{R}^n} f\arccos(-fe^{-r/2}) + (1 - r)\sqrt{e^r - f^2}\, d\gamma \leq \tag{3.10}$$

$$\left(\int_{\mathbb{R}^n} f d\gamma\right)\arccos\left(-\int_{\mathbb{R}^n} f\right) + \sqrt{1 - \left(\int_{\mathbb{R}^n} f d\gamma\right)^2}$$

for any smooth bounded $f : \mathbb{R}^n \to (-1, 1)$ where $r > 0$ solves the equation

$$\|\nabla f\|^2 = r(e^r - f^2).$$

One can obtain the Poincaré inequality from (3.10). Indeed, take $f_\varepsilon = \varepsilon f$ and send $\varepsilon \to 0$. Notice that

$$r = \varepsilon^2 \|\nabla f\|^2 + O(\varepsilon^2);$$

$$M\left(\int_{\mathbb{R}^n} f_\varepsilon d\gamma, 0\right) = 1 + \frac{\pi}{2}\varepsilon \int_{\mathbb{R}^n} f d\gamma + \frac{1}{2}\left(\int_{\mathbb{R}^n} f d\gamma\right)^2 \varepsilon^2 + O(\varepsilon^2);$$

$$M(f_\varepsilon, \|\nabla f_\varepsilon\|) = 1 + \frac{\pi}{2}f\varepsilon + \frac{1}{2}\left(f^2 - \|\nabla f\|^2\right)\varepsilon^2 + O(\varepsilon^2).$$

Substituting these expressions in (3.10) and sending $\varepsilon \to 0$ we obtain the Poincaré inequality.

4. One-dimensional case

Let $n = 1$, and set $\alpha = (\alpha_0, \ldots, \alpha_m)$ where $\alpha_0 = 0$, $\alpha_1 = 1, \ldots, \alpha_n = n$. Take $\mathbf{f} = \underbrace{(f, \ldots, f)}_{m+1}$, where $f \in C_0^\infty(\mathbb{R}; \Omega)$. Then $D^\alpha \mathbf{f} = (f, f', f'', \ldots, f^{(m)})$. Given a log-concave probability measure $e^{-U(x)}dx$ such that $U''(x) \geq R > 0$, the associated semigroup P_t has the generator $L = d^2 x - U'(x)dx$. Let $\mathbf{u} = (u_0, \tilde{\mathbf{u}})$ where $\tilde{\mathbf{u}} = (u_1, \ldots, u_m) \in \mathbb{R}^m$ is arbitrary and $u_0 \in \Omega$. Let the function $B(u_0, \ldots, u_m)$ belong to $C^2(\Omega \times \mathbb{R}^m)$. Let $B_j := \frac{\partial B}{\partial u_j}$ and $B_{ij} := \frac{\partial^2 B}{\partial u_i \partial u_j}$. Set

$$L_j(\mathbf{u}, y) = \sum_{k=j+1}^{m} \binom{k}{j} B_k(\mathbf{u})U^{(k-j+1)}(y) \quad \text{for} \quad j = 0, \ldots, m-1$$

Remark 4.1. Notice that if $e^{-U(x)}dx = \frac{1}{\sqrt{2\pi}}e^{-x^2/2}dx$ then

$$L_j(\mathbf{u}, y) = (j+1)B_{j+1}(\mathbf{u}).$$

Further we assume that $B_{mm} \neq 0$. Theorem 2.1 implies the following corollary.

Corollary 4.2. *The following conditions are equivalent:*

(i) *for all $\mathbf{u} \in \Omega \times \mathbb{R}^m$ we have*

$$B_{mm} \leq 0,$$

$$\tilde{\mathbf{u}}\{B_{mj}(\mathbf{u})B_{mi}(\mathbf{u}) - B_{mm}(\mathbf{u})B_{ij}(\mathbf{u}) - \delta_{i-j}\frac{B_{mm}}{u_{j+1}}L_j(\mathbf{u}, y)\}_{i,j=0}^{m-1}\tilde{\mathbf{u}}^T \leq 0;$$

(ii) *for all $f \in C_0^\infty(\mathbb{R}^m; \Omega)$ and $t \geq 0$ we have*

$$P_t B(f, f', \ldots, f^{(m)}) \leq B(P_t f, P_t f', \ldots, P_t f^{(m)}).$$

Remark 4.3. If we send $t \to \infty$ then (ii) in the corollary implies the inequality

$$\int_{\mathbb{R}} B(f, f', \ldots, f^{(m)})d\mu(x) \leq B\left(\int_{\mathbb{R}} f d\mu, 0, \ldots, 0\right) \quad \text{for all } f \in C_0^\infty(\mathbb{R}^m; \Omega).$$

Proof. It suffices to show that (i) in Corollary 4.2 is the same as (i) in Theorem 2.1. Notice that

$$[L, D^{\alpha_0}] = 0 \quad \text{and} \quad [L, D^{\alpha_k}] = \sum_{\ell=1}^{k} \binom{k}{\ell} U^{(\ell+1)}(x) d^{k+1-\ell} x \quad \text{for } 1 \le k \le m.$$

Thus

$$\nabla B[L, D^{\alpha}]\mathbf{f} = \sum_{k=1}^{n} \frac{\partial B}{\partial u_k} \left(\sum_{\ell=1}^{k} \binom{k}{\ell} U^{(\ell+1)}(x)(P_t f)^{k+1-\ell} \right)$$

and

$$\Gamma(D^{\alpha} f) = \begin{bmatrix} g' \cdot g' & g' \cdot g'' & \cdots & g' \cdot g^{(m+1)} \\ g'' \cdot g' & g'' \cdot g'' & \cdots & g'' \cdot g^{(m+1)} \\ \cdots \cdots \cdots \cdots \cdots \cdots \cdots \cdots \cdots \cdots \cdots \cdots \\ g^{(m+1)} \cdot g' & g'' \cdot g^{(m+1)} & \cdots & g^{(m+1)} \cdot g^{(m+1)} \end{bmatrix}$$

Therefore the quantity (i) in Theorem 2.1 takes the form

$$\sum_{k=1}^{m} B_k(\mathbf{u}) \left[\sum_{\ell=1}^{k} \binom{k}{\ell} U^{(\ell+1)}(y) u_{k+1-\ell} \right] + \sum_{i,j=0}^{m} B_{ij}(\mathbf{u}) u_{i+1} u_{j+1}$$

where $u_1, \ldots, u_{n+1}, y$ are arbitrary real numbers and u_0 takes values in Ω. Notice that the above expression can be rewritten as follows:

$$B_{mm} u_{m+1}^2 + 2u_{m+1} \left(\sum_{j=0}^{m-1} B_{mj} u_{j+1} \right) + \sum_{i,j=0}^{m-1} B_{ij} u_{i+1} u_{j+1}$$

$$+ \sum_{k=1}^{m} B_k(\mathbf{u}) \left[\sum_{\ell=1}^{k} \binom{k}{\ell} U^{(\ell+1)}(y) u_{k+1-\ell} \right]$$

This expression is nonpositive if and only if condition (i) of Corollary 4.2 holds.   □

5. Further applications

Houdré–Kagan [21] obtained an extension of the classical Poincaré inequality:

$$\sum_{k=1}^{2d} \frac{(-1)^{k+1}}{k!} \int_{\mathbb{R}^n} \|\nabla^k f\|^2 d\gamma \le \int_{\mathbb{R}^n} f^2 d\gamma - \left(\int_{\mathbb{R}^n} f \right)^2$$

$$\le \sum_{k=1}^{2d-1} \frac{(-1)^{k+1}}{k!} \int_{\mathbb{R}^n} \|\nabla^k f\|^2 d\gamma \qquad (5.1)$$

for all compactly supported functions f on $\mathbb{R}^n$, and any $d \ge 1$. Here by the symbol $\|\nabla^k f\|$ we denote

$$\|\nabla^k f\|^2 = \sum_{|\alpha|=k} (D^{\alpha} f)^2.$$

We refer the reader to [34] for further remarks on (5.1) in the one-dimensional case $n = 1$. A remarkable paper [30] explains (5.1) via integration by parts.

We will illustrate the use of Corollary 4.2 by (5.1) in case $n = 1$.

Proof of (5.1) *in case* $n = 1$. Consider

$$B(u_0, u_1, \ldots, u_m) = \sum_{k=0}^{m} \frac{(-1)^k}{k!} u_k^2,$$

and $d\mu = \frac{1}{\sqrt{2\pi}} e^{-x^2/2} dx$. If m is odd then $B_{mm} \leq 0$ and condition (i) of Corollary 4.2 holds. Indeed, in this case $L_j(\mathbf{u}, y) = B_{j+1}(\mathbf{u})(j+1) = u_{j+1}(-1)^{j+1}\frac{2}{j!}$, and

$$\tilde{\mathbf{u}}\{B_{mj}(\mathbf{u})B_{mi}(\mathbf{u}) - B_{mm}(\mathbf{u})B_{ij}(\mathbf{u}) - \delta_{i-j}\frac{B_{mm}}{u_{j+1}}L_j(\mathbf{u}, y)\}_{i,j=0}^{m-1}\tilde{\mathbf{u}}^T$$

$$= -B_{mm}\tilde{\mathbf{u}}\left\{B_{jj} + B_{j+1}\frac{j+1}{u_{j+1}}\right\}_{i,j=0}^{m-1}\tilde{\mathbf{u}}^T = 0.$$

Thus by (ii) of Corollary 4.2 we deduce that

$$\int_{\mathbb{R}} \sum_{k=0}^{m} \frac{(-1)^k}{k!}[f^{(k)}(x)]^2 d\mu \leq \left(\int_{\mathbb{R}} f(x) d\mu\right)^2$$

for all $f \in C_0^\infty(\mathbb{R})$ and for odd m, and similarly we obtain the opposite inequality for even m. □

Proof of (1.30) (*Banaszczyk conjecture*). We will show that *if f is even and $d\mu = e^{-U(x)}dx$ is an even log-concave measure such that* $\operatorname{Hess} U \geq Id$ *then*

$$\left(\int_{\mathbb{R}^n} f^2 d\mu\right)^2 - \left(\int_{\mathbb{R}^n} f d\mu\right)^2 \leq \frac{1}{2}\int_{\mathbb{R}^n} \|\nabla f\|^2 d\mu. \tag{5.2}$$

Indeed, take $M(x, y)$ as in (1.31), i.e.,

$$M(x, y) = x^2 - \frac{y^2}{2} \quad \text{for} \quad x \in \mathbb{R}, \quad y \geq 0.$$

Unfortunately $M(x, y)$ does not satisfy (1.1) (because $M_{xx} + M_y/y = 1 > 0$) therefore we cannot directly apply Theorem 1.1.

Let P_t be the semigroup associated with $d\mu$ and let L be its generator. Consider the function $V(x, t) = P_t M(f, \|\nabla f\|) - M(P_t f, \|\nabla P_t f\|)$ as in the proof of Theorem 2.1. Then

$$(\partial_t - L)V(x, t) = -\nabla P_t f(\operatorname{Hess} U)(\nabla P_t f)^T + 2\|\nabla P_t f\|^2 - \|\nabla^2 P_t f\|^2$$

$$\leq \|\nabla P_t f(x)\|^2 - \|\nabla^2 P_t f\|^2.$$

Clearly it is not true that the above expression is non-positive pointwise, i.e., for all $x \in \mathbb{R}^n$ (consider $t = 0$). Therefore we cannot directly apply the maximum principle as in the proof of Theorem 2.1 in order to get the pointwise bound

$V(x,t) \leq 0$. Actually, we do not need the pointwise estimate $V(x,t) \leq 0$ in order to get (5.2), for example $\int_{\mathbb{R}^n} V(x,t)d\mu \leq 0$ will suffice. Notice that

$$\int_{\mathbb{R}^n} V(x,T)d\mu = \int_0^T \int_{\mathbb{R}^n} (\partial_t - L)V(x,t)d\mu dt$$

$$\leq \int_0^T \int_{\mathbb{R}^n} \|\nabla P_t f(x)\|^2 - \|\nabla^2 P_t f\|^2 d\mu ds \leq 0$$

for all $T \geq 0$. The last inequality follows by application of the Poincaré inequality (1.17) to the functions $\partial_{x_j} P_t f(x)$ for all $j = 1, \ldots, n$, and the fact that $\int_{\mathbb{R}^n} \partial_{x_j} P_t f(x)d\mu = 0$ because $P_t f(x)$ is an even function. Thus we obtain

$$\int_{\mathbb{R}^n} M(f, \|\nabla f\|)d\mu \leq \int_{\mathbb{R}^n} M(P_T f, \|\nabla P_T f\|)d\mu \quad \text{for all} \quad T \geq 0.$$

By sending $T \to \infty$ we arrive at (5.2) because $\lim_{T \to \infty} \|\nabla P_T f\| = 0$. $\quad\square$

In the end we should mention that even though the current paper is self-contained it should be viewed as a continuation of the ideas developed in our recent papers [25, 26] where certain PDEs similar to (1.1) happen to rule some functional inequalities.

Acknowledgment

We are very grateful to Robert Bryant from whom we learned how to solve a non-linear PDE important for our goals (see [15]). In Section 3.1 this allowed us to explain how one could find the right functions $M(x,y)$ for all the applications mentioned in Section 1.1, and how to find new functions M each responsible for a particular isoperimetric inequality.

References

[1] F. Barthe, B. Maurey, *Some remarks on isoperimetry of Gaussian type.* Ann. Inst. H. Poincaré Probab. Statist., 36 (4):419–434, 2000

[2] D. Bakry, M. Emery, *Hypercontractivité de semi-groups de diffusion*, C. R. Acad. Sci. Paris Sér I Math. **299**, 775–778 (1984).

[3] D. Bakry, I. Gentil, M. Ledoux, *Analysis and Geometry of Markov Diffusion Operators*, Grundlehren der Mathematischen Wissenschaften **348**. Springer, Cham.

[4] D. Bakry, M. Ledoux, *Lévy–Gromov's isoperimetric inequality for an infinity dimensional diffusion generator.* Invent. math. 123, 259–281 (1996)

[5] W. Beckner, *A generalized Poincaré inequality for Gaussian measures*, Proceedings of the American Mathematical Society **105**, no. 2, 397–400 (1989).

[6] S.G. Bobkov, *Isoperimetric and analytic inequalities for log-concave probability measures*, Ann. Probab. **27**, no. 4, 1903–1921 (1999).

[7] S.G. Bobkov, *An isoperimetric inequality on the discrete cube, and an elementary proof of the isoperimetric inequality in Gauss space* The Annals of Probability **25**, no. 1, 206–214 (1997).

[8] S.G. Bobkov, M. Ledoux *From Brunn–Minkowski to Brascamp–Lieb and to logarithmic Sobolev inequalities* Geom. Funct. Anal., 10(5) (2000), 1028–1052.

[9] A. Bonami, *Ensembles* $\Lambda(p)$ *dans le dual de* D^∞. Ann. Inst. Fourier, 18 (2) 193–204 (1968).

[10] A. Bonami, *Étude des coefficients de Fourier des fonctions de* $L^p(G)$. Ann. Inst. Fourier, 20: 335–402 (1970).

[11] H.J. Brascamp, E.H. Lieb, *On extensions of the Brunn–Minkowski and Prékopa–Leindler theorems, including inequalities for log-concave functions, and with an application to the diffusion equation*, J. Funct. Anal. **22**, 366–389 (1976).

[12] R. Bryant et al., *Exterior differential systems*.

[13] R. Bryant, P. Griffiths *Characteristic cohomology of differential systems II: Conservation laws for a class of parabolic equations*, Duke Math. J. **78**, no. 3 (1995), 531–676.

[14] R. Bryant, P. Griffiths, *Characteristic cohomology of differential systems I: General theory*, J. Amer. Math. Soc. **8** (1995), 507–596.

[15] R. Bryant, *Monge–Ampère with drift*, URL (version 2015-06-29):
http://mathoverflow.net/q/210127

[16] L. Caffarelli, *Monotonicity properties of optimal transportation and the FKG and related inequalities*, Comm. Math. Phys. **214**, (2000), 547–563.

[17] E.A. Carlen, D. Cordero-Erausquin, E.H. Lieb, *Asymmetric covariance estimates of Brascamp–Lieb type and related inequalities for log-concave measures*, Ann. Inst. Henri Poincaré Probab. Stat. **49** (2013), 1–12.

[18] D. Chafai, *On* Φ-*entropies and* Φ-*Sobolev inequalities*, preprint 2002.

[19] D. Cordero-Erausquin, M. Fradelizi, B. Maurey, *The* (B) *conjecture for the Gaussian measure of dilates of symmetric convex sets and related problems*, J. Funct. Anal., **214** (2004), no. 2, 410–427.

[20] L. Gross, *Logarithmic Sobolev inequalities*. Amer. J. Math. **97** (1975), no. 4, 1061–1083.

[21] C. Houdré, A. Kagan, *Variance inequalities for functions of Gaussian variables*, J. Theoret. Probab. **8** (1995), no. 1, 23–30

[22] Y. Hu, *A unified approach to several inequalities for Gaussian and diffusion measures*. Séminare de Probabilités XXXIV. Lecture Notes in Math. 1729, 329–335 (2000). Springer.

[23] P. Ivanisvili, *Geometric aspects of exact solutions of Bellman equations of harmonic analysis problems*, PhD thesis, 2015.

[24] P. Ivanisvili, A. Volberg, *Improving Beckner's bound via Hermite polynomials*, preprint, arXiv:1606.08500.

[25] P. Ivanisvili, A. Volberg, *Bellman partial differential equation and the hill property for classical isoperimetric problems*, preprint (2015) arXiv:1506.03409, pp. 1–30.

[26] P. Ivanisvili, A. Volberg, *Hessian of Bellman functions and uniqueness of the Brascamp–Lieb inequality*, J. London Math. Soc (2015) doi:10.1112/ jlms/jdv040.

[27] R. Latała *On some inequalities for Gaussian measures*. Proceedings of the International Congress of Mathematicians, Vol. II. (Beijing, 2002), 813–822, Higher Ed. Press, Beijing 2002.

[28] R. Latała, K. Oleszkiewicz, *Between Sobolev and Poincaré*, Geometric Aspects of Functional Analysis. Lect. Notes Math., 1745: 147–168, 2000

[29] M. Ledoux, *Remarks on Gaussian noise stability, Brascamp–Lieb and Slepian inequalities*, Geometric Aspects of Functional Analysis, 309–333, Lecture Notes in Math., 2116, Springer (2014).

[30] M. Ledoux, *L'algèbre de Lie des gradients itérés d'un générateur markovien – développements de moyennes et entropies*, Annales scientifiques de l'É.N.S. 4e série, tome 28, n° 4 (1995), p. 435–460.

[31] E. Mossel, J. Neeman, *Robust optimality of Gaussian noise stability* (2012).

[32] J. Nash, *Continuity of solutions of parabolic and elliptic equations*, Amer. J. Math. **88** (1958), 931–954.

[33] R. O'Donell, *Analysis of Boolean Functions.* (2013).

[34] I. Popescu, *A refinement of the Brascamp–Lieb–Poincaré inequality in one dimension*, Comptes Rendus Mathématique, **352**, Issue 1 (2014), 55–58

[35] G. Philippis, A. Figalli, *The Monge–Ampère equation and its link to optimal transportation*, Bulletin of the AMS, Vo. 5, no. 4, pp. 527-580.

[36] N.S. Trudinger, X. Wang, *The Monge–Ampère equation and its geometric applications*, Handbook of Geometric Analysis, International Press, 2008, Vol I, pp. 467–524.

Paata Ivanisvili
Department of Mathematics
Kent State University
Kent, OH 44240, USA
e-mail: ivanishvili.paata@gmail.com

Alexander Volberg
Department of Mathematics
Michigan State University
East Lansing, MI 48824, USA
e-mail: volberg@math.msu.edu

Operator Theory:
Advances and Applications, Vol. 261, 307–315
© Springer International Publishing AG, part of Springer Nature 2018

Fundamental Groups, Slalom Curves and Extremal Length

Burglind Jöricke

To the memory of my teacher and collaborator Viktor Havin,
his enthusiasm and his ability to convey a great feeling
of the beauty of mathematics

Abstract. We define two versions of the extremal length of elements of the fundamental group of the twice punctured complex plane and give upper and lower bounds for two versions of this invariant. The bounds differ by a multiplicative constant. The main motivation comes from 3-braid invariants and their application.

Mathematics Subject Classification (2010). Primary 30Cxx; Secondary 20F34, 20F36, 57Mxx.

Keywords. Fundamental group, extremal length, conformal module, 3-braids.

In this paper we will describe two versions of a conformal invariant for the elements of the fundamental group $\pi_1(\mathbb{C} \setminus \{-1, 1\}, 0)$ of the twice punctured complex plane with base point 0 and give upper and lower bounds for the versions of this invariant. The group $\pi_1(\mathbb{C} \setminus \{-1, 1\}, 0)$ is a free group with two generators. We choose generators a_1 and a_2 so that a_1 is represented by a simple closed curve α_1 with base point 0 which surrounds the point -1 counterclockwise such that the image of the curve except the point 0 is contained in the open left half-plane. Respectively, a standard representative α_2 of the generator a_2 surrounds the point 1 counterclockwise and the image of the curve except the point 0 is contained in the open right half-plane.

The fundamental group $\pi_1 \overset{def}{=} \pi_1(\mathbb{C} \setminus \{-1, 1\}, 0)$ is isomorphic to the relative fundamental group $\pi_1^{tr} \overset{def}{=} \pi_1(\mathbb{C} \setminus \{-1, 1\}, (-1, 1))$ whose elements are homotopy classes of curves in $\mathbb{C} \setminus \{-1, 1\}$ with endpoints on the interval $(-1, 1)$. We refer to π_1^{tr} as the fundamental group with totally real boundary values (*tr*-boundary values for short). To establish the isomorphism one has to use that the set $(-1, 1)$ is connected and simply connected and contains 0. In the same way π_1 is isomorphic to the relative fundamental group with perpendicular bisector boundary values $\pi_1^{pb} \overset{def}{=} \pi_1(\mathbb{C} \setminus \{-1, 1\}, i\mathbb{R})$ whose elements are homotopy classes of curves in $\mathbb{C} \setminus \{-1, 1\}$ with endpoints on the imaginary axis $i\mathbb{R}$. For an element $w \in \pi_1$ we denote by w_{tr} and by w_{pb}, respectively, the elements in π_1^{tr} and in π_1^{pb}, respectively, corresponding to w.

Consider an open rectangle $\mathcal{R}$ with sides parallel to the axes and with length of the horizontal sides equal to b and length of the vertical sides equal to a. Recall that according to Ahlfors's definition [1] the extremal length $\lambda(\mathcal{R})$ of such a rectangle is equal to $\frac{\mathsf{a}}{\mathsf{b}}$ and its conformal module $m(\mathcal{R})$ equals $\frac{\mathsf{b}}{\mathsf{a}}$. A continuous mapping of the rectangle $\mathcal{R}$ into $\mathbb{C} \setminus \{-1, 1\}$ is said to represent w_{tr} if it has a continuous extension to the closure of $\mathcal{R}$ which maps open horizontal sides to the interval $(-1, 1)$ and whose restriction to each closed vertical side represents w_{tr}. We make the respective convention for w_{pb} instead of w_{tr}.

We are now in a position to define the versions of the extremal length of elements of the relative fundamental groups.

Definition 1. For an element w of the fundamental group $\pi_1(\mathbb{C} \setminus \{-1, 1\}, 0)$ the extremal length of w with perpendicular bisector boundary values (*pb*-boundary values for short) is defined as

$$\lambda_{pb}(w) \overset{def}{=} \inf\{\lambda(\mathcal{R}) : \mathcal{R} \text{ admits a holomorphic map}$$
$$\text{to } \mathbb{C} \setminus \{-1, 1\} \text{ that represents } w_{pb}\}.$$

An analogous definition can be given for λ_{tr}.

We will give upper and lower bounds for λ_{pb} and λ_{tr} differing by a multiplicative constant. This is of independent interest for the fundamental group of the twice punctured plane, but the main motivation was to give estimates of conformal invariants of braids. Recall that a pure geometric n-braid with base point is a continuous mapping of the unit interval $[0, 1]$ into the n-dimensional configuration space $C_n(\mathbb{C}) = \{(z_1, \ldots, z_n) : z_j \neq z_k \text{ for } j \neq k\}$ whose values at the endpoints are equal to a given base point in $C_n(\mathbb{C})$. More geometrically, a pure geometric n-braid consists of n pairwise disjoint curves in the cylinder $[0, 1] \times \mathbb{C}$, each joining a point in the top $\{1\} \times \mathbb{C}$ of the cylinder with its copy in the bottom so that for each curve the canonical projection to the interval $[0, 1]$ is a homeomorphism. A pure n-braid with base point is an isotopy class of pure geometric n-braids with fixed base point.

Consider a pure geometric 3-braid. Associate to it a curve in $\mathbb{C} \setminus \{-1, 1\}$ as follows. For a point $z = (z_1, z_2, z_3) \in C_3(\mathbb{C})$ we denote by M_z the Möbius transformation that maps z_1 to 0, z_3 to 1 and fixes ∞. Then $M_z(z_2)$ omits 0, 1, and ∞. Notice that z_2 is equal to the cross-ratio $(z_2, z_3; z_1, \infty) = \frac{z_2 - z_1}{z_3 - z_1} \cdot \frac{z_3 - \infty}{z_2 - \infty} = \frac{z_2 - z_1}{z_3 - z_1}$.

Let $\gamma(t) = (\gamma_1(t), \gamma_2(t), \gamma_3(t))$, $t \in [0, 1]$, be a curve in $C_3(\mathbb{C})$. Associate to it the curve $\mathfrak{C}(\gamma)(t) \overset{def}{=} 2 \frac{\gamma_2(t) - \gamma_1(t)}{\gamma_3(t) - \gamma_1(t)} - 1$, $t \in [0, 1]$, in $\mathbb{C}$ which omits the points -1 and 1. If γ is a loop with base point $\gamma(0) = (-1, 0, 1)$, then $\mathfrak{C}(\gamma)$ is a loop with base point $\mathfrak{C}(\gamma)(0) = 0$. The homotopy class of $\mathfrak{C}(\gamma)$ in $\mathbb{C} \setminus \{-1, 1\}$ with base point 0 depends only on the homotopy class of γ in the configuration space $C_3(\mathbb{C})$ with base point $(-1,0,1)$. We obtain a surjective homomorphism $\mathfrak{C}_*$ from the fundamental group of $C_3(\mathbb{C})$ with base point $(-1,0,1)$ to the fundamental group of $\mathbb{C}\setminus\{-1,1\}$ with base point 0. The kernel of $\mathfrak{C}_*$ equals $\langle\Delta_3^2\rangle$, the subgroup of $\mathcal{B}_3$ generated by the full twist obtained by twisting the cylinder keeping the bottom fixed and turning the top by the angle 2π. Respective facts hold for loops in $C_3(\mathbb{C})$ with specified boundary values instead of loops with a base point. With the natural definition of the extremal length with totally real boundary values of a pure 3-braid b this extremal length is equal to $\lambda_{tr}(\mathfrak{C}_*(b))$. The respective fact holds for perpendicular bisector boundary values. The obtained invariants are invariants of 3-braids rather than invariants of conjugacy classes of 3-braids. In particular, they are finer than a popular invariant of braids, the entropy. Our estimates imply estimates of the entropy of pure 3-braids b in terms of the representing word of the image $\mathfrak{C}_*(b)$. Notice that the names "totally real" and "perpendicular bisector" are motivated by the definition in the case of braids. Details will be given in a later paper. For an introduction to braids see, e.g., [2]. For more information on the conformal module, the extremal length and entropy of braids, or of conjugacy classes of braids, respectively, see also [3] and [4].

We will now lift the elements of π_1^{pb} to the logarithmic covering $U_{\log}$ of $\mathbb{C}\setminus\{-1,1\}$ and identify the lifts with homotopy slalom curves. This geometric interpretation will suggest how to estimate the extremal length with perpendicular bisector boundary values.

The logarithmic covering of $\mathbb{C}\setminus\{-1,1\}$ is the universal covering of the twice punctured Riemann sphere $\mathbb{P}^1\setminus\{-1,1\}$ with all preimages of ∞ under the covering map removed. Geometrically the universal covering of $\mathbb{P}^1\setminus\{-1,1\}$ can be described as follows. Take copies of $\mathbb{P}^1\setminus(-1,1)$ labeled by the set $\mathbb{Z}$ of integer numbers. Attach to each copy of $\mathbb{P}^1\setminus(-1,1)$ two copies of $(-1,1)$, the $+$-edge (the accumulation set of points of the upper half-plane) and the $-$-edge (the accumulation set of points of the lower half-plane). For each $k\in\mathbb{Z}$ we glue the $+$-edge of the kth copy to the $-$-edge of the $(k+1)$st copy (using the identity mapping on $(-1,1)$ to identify points on different edges). Denote by $U_{\log}$ the set obtained from the described covering of $\mathbb{P}^1\setminus(-1,1)$ by removing all preimages of ∞.

The following proposition holds.

Proposition 1. *The set $U_{\log}$ is conformally equivalent to $\mathbb{C}\setminus i\mathbb{Z}$. The mapping $f_1\circ f_2$, $f_2(z)=\frac{e^{\pi z}-1}{e^{\pi z}+1}$, $z\in\mathbb{C}\setminus i\mathbb{Z}$, $f_1(w)=\frac{1}{2}(w+\frac{1}{w})$, $w\in\mathbb{C}\setminus\{0\}$, is a covering map from $\mathbb{C}\setminus i\mathbb{Z}$ to $\mathbb{C}\setminus\{-1,1\}$.*

The lift of α_1 with initial point $\frac{-i}{2}+ik$ is a curve which joins $\frac{-i}{2}+ik$ with $\frac{-i}{2}+i(k+1)$ and is contained in the closed left half-plane. The only points on the imaginary axis are the endpoints.

The lift of α_2 with initial point $\frac{-i}{2}+ik$ is a curve which joins $\frac{-i}{2}+ik$ with $\frac{-i}{2}+i(k-1)$ and is contained in the closed right half-plane. The only points on the imaginary axis are the endpoints.

Figure 1 shows the curves α_1 and α_2 which represent the generators of the fundamental group $\pi_1(\mathbb{C} \setminus \{-1,1\}, 0)$ and their lifts under the covering maps f_1 and $f_2 \circ f_1$. For $j = 1, 2$ the curves α'_j and α''_j are the two lifts of α_j under the double branched covering $f_1 : \mathbb{C} \setminus \{0\} \to \mathbb{C} \setminus \{0\}$ with branch points 1 and -1. The curve $\tilde{\alpha}'_1$ is the lift of α'_1 under the mapping f_2 with initial point $\frac{-i}{2}$, the curve $\tilde{\alpha}'_2$ lifts α'_2 and has initial point $\frac{i}{2}$.

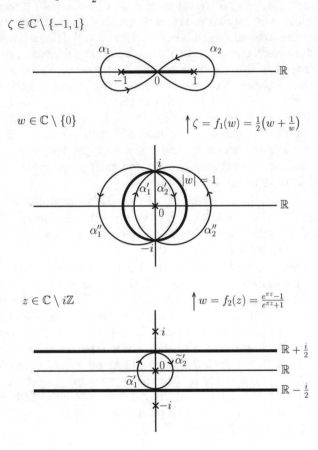

FIGURE 1

Consider the curve α_1^n, $n \subset \mathbb{Z} \setminus \{0\}$. It runs n times along the curve α_1 if $n > 0$ and $|n|$ times along the curve which is inverse to α_1 if $n < 0$. For each $k \in \mathbb{Z}$, the curve α_1^n lifts to a curve with initial point $\frac{-i}{2} + ik$ and terminating point $\frac{-i}{2} + ik + in$ which is contained in the closed left half-plane and omits the points in $i\mathbb{Z}$. Respectively, α_2^n, $n \in \mathbb{Z} \setminus \{0\}$, lifts to a curve with initial point $\frac{+i}{2} + ik$ and terminating point $\frac{+i}{2} + ik - in$ which is contained in the closed right half-plane and omits the points in $i\mathbb{Z}$. The mentioned lifts are homotopic through curves in $\mathbb{C} \setminus i\mathbb{Z}$ with endpoints on $i\mathbb{R} \setminus i\mathbb{Z}$ to curves with interior contained in the open (left, respectively, right) half-plane. We have the following definition where we identify a curve with its image, ignoring orientation.

Definition 2. A simple arc in $\mathbb{C} \setminus i\mathbb{Z}$ with endpoints on different connected components of $i\mathbb{R} \setminus i\mathbb{Z}$ is called an elementary slalom curve if its interior (i.e., the complement of its endpoints) is contained in one of the open half-planes $\mathbb{C}_r \overset{def}{=} \{z \in \mathbb{C} : \operatorname{Re} z > 0\}$ or $\mathbb{C}_\ell \overset{def}{=} \{z \in \mathbb{C} : \operatorname{Re} z < 0\}$.

A curve in $\mathbb{C} \setminus i\mathbb{Z}$ is called an elementary half slalom curve if one of the endpoints is contained in the horizontal line $\left\{ \operatorname{Im} z = k + \frac{1}{2} \right\}$ for an integer k and the union of the curve with its mirror reflection in the line $\left\{ \operatorname{Im} z = k + \frac{1}{2} \right\}$ is an elementary slalom curve.

A slalom curve in $\mathbb{C} \setminus i\mathbb{Z}$ is a curve which can be divided into a finite number of elementary slalom curves so that consecutive elementary slalom curves are contained in different half-planes.

A curve which is homotopic to a slalom curve in $\mathbb{C} \setminus i\mathbb{Z}$ through curves with endpoints in $\mathbb{R} \setminus i\mathbb{Z}$ is called a homotopy slalom curve.

Figure 2 below shows a slalom curve which represents a lift of the element $a_2^{-1} a_1^2 a_2^{-3} a_1^{-1} a_2^{-1} a_1^{-1} a_2 a_1^{-1}$ with perpendicular bisector boundary values.

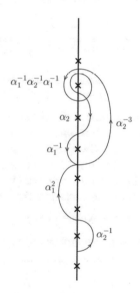

FIGURE 2

Elementary slalom curves and elementary half slalom curves will serve as building blocks. Note that each curve in $\mathbb{C} \setminus i\mathbb{Z}$ with endpoints in $i\mathbb{R} \setminus i\mathbb{Z}$ is a homotopy slalom curve or is homotopic to the identity in $\mathbb{C} \setminus i\mathbb{Z}$ with endpoints in $i\mathbb{R} \setminus i\mathbb{Z}$. Proposition 1 implies that each lift of a curve in $\mathbb{C} \setminus \{-1, 1\}$ with endpoints on the imaginary axis is of such type. The extremal length of slalom curves (more precisely, of homotopy classes of slalom curves) can be defined in the same way as the respective object for elements of the fundamental group π_1^{pb}. The extremal length with perpendicular bisector boundary values of an element of π_1 is equal to that of its lift.

Consider the extremal length of an elementary slalom curve which corresponds to the word $a^n \in \pi_1$ where a equals either a_1 or a_2. Without loss of generality we may assume that $a = a_1$, hence the curve is contained in the closed left half-plane. After a translation the endpoints of the curve are contained in the intervals $(-i(M+1), -iM)$ and $(iM, i(M+1))$, respectively, with $M = \frac{|n|-1}{2}$. For $M = 0$ (thus, for $|n| = 1$) the extremal length equals 0. In this case we call the original curve a trivial elementary slalom curve. Let M be positive. The curve is represented by a conformal mapping of an open rectangle $\mathcal{R}^M$ onto the left half-plane whose extension to the boundary maps the horizontal sides onto $[-i(M+1), -iM]$, and $[iM, i(M+1)]$, respectively. Hence its extremal length is bounded from above by the extremal length of the rectangle $\mathcal{R}^M$.

The conformal mapping of a rectangle onto the left half-plane $\mathbb{C}_\ell$ whose extension to the boundary maps the horizontal sides onto $[-i(M+1), -iM]$, and $[iM, i(M+1)]$, respectively, is related to elliptic integrals. With a suitable normalization of the rectangle, the inverse of the mapping is equal to the elliptic integral

$$
\begin{aligned}
\mathcal{F}_M(z) &= \int_0^z \frac{d\zeta}{\sqrt{(\zeta^2 - (iM)^2)(\zeta^2 - (i(M+1))^2)}} \\
&= \frac{i}{M} \int_0^{\frac{z}{iM}} \frac{dw}{\sqrt{(1-w^2)((1+\frac{1}{M})^2 - w^2)}}, \quad z \in \mathbb{C}_\ell.
\end{aligned}
\tag{1}
$$

We use the branch of the square root which is positive on the positive real axis. The function $\mathcal{F}_M$ extends continuously to the imaginary axis (the integral converges). The extended map takes the closed left half-plane to a closed rectangle. The points $-i(M+1)$, $-iM$, iM, and $i(M+1)$ are mapped to the vertices of the rectangle. It is known and follows from formula (1) for the elliptic integral that for $M \geq \frac{1}{2}$ the extremal length of the rectangle $\mathcal{R}^M$ satisfies the inequalities

$$
c \log(1+M) \leq \lambda(\mathcal{R}^M) \leq C \log(1+M)
\tag{2}
$$

for positive constants c and C not depending on M.

Equation (2) suggests the following proposition.

Proposition 2. *The extremal length $\lambda_{k,\ell}$ of an elementary slalom curve with endpoints in the intervals $(ik, i(k+1))$ and $(i\ell, i(\ell+1))$, respectively, with $|k - \ell| \geq 2$, satisfies the inequalities*

$$
c' \log\left(1 + \frac{|k-\ell|-1}{2}\right) \leq \lambda_{k,\ell} \leq C' \log\left(1 + \frac{|k-\ell|-1}{2}\right)
\tag{3}
$$

for positive constants c' and C' not depending on k and ℓ.

There are explicit estimates for the constants c' and C'.

The proof will be given elsewhere. Here we have already discussed the estimate from above. The estimate from below is subtler. The first difficulty is that the representing mappings for an elementary slalom curve are not necessarily conformal mappings, they are merely holomorphic. The second difficulty is that the image of the rectangle is not necessarily contained in the half-plane. We can only say about the mapping that it lifts to a holomorphic mapping into the universal covering of $\mathbb{C} \setminus i\mathbb{Z}$ with specified boundary values. The universal covering is a half-plane, but the horizontal sides of the rectangle are not mapped any more into boundary intervals of the half-plane but into some curves in the half-plane. One tool for dealing with these difficulties is an analog of the following lemma, which is of independent interest.

Lemma 1. *Let $\mathcal{R}_1$ and $\mathcal{R}_2$ be rectangles with sides parallel to the axes. Suppose S_2 is a vertical strip bounded by the two vertical lines which are prolongations of the vertical sides of the rectangle $\mathcal{R}_2$. Let $f : \mathcal{R}_1 \to S_2$ be a holomorphic map whose extension to the closure maps the two horizontal sides of $\mathcal{R}_1$ into different horizontal sides of $\mathcal{R}_2$. Then $\lambda(\mathcal{R}_1) \geq \lambda(\mathcal{R}_2)$.*
Equality occurs if and only if the mapping is a surjective conformal map from $\mathcal{R}_1$ to $\mathcal{R}_2$.

The proof of the lemma is based on the Cauchy–Riemann equations.

To estimate the extremal length of an arbitrary element of the fundamental group $\pi_1 = \pi_1(\mathbb{C} \setminus \{-1, 1\}, \{0\})$, we represent the element as a word in the generators (and identify it with the word). A word is in reduced form (or a reduced word) if it is written as a product of powers of generators where consecutive terms correspond to different generators. Consider the reduced word

$$w = a_1^{n_1} \cdot a_2^{n_2} \cdot \ldots, \tag{4}$$

where the n_j are integers. (Here $a_j^0 \overset{def}{=} \mathrm{id}$, we allow $n_1 = 0$.) We are interested first in the extremal length with perpendicular bisector boundary conditions. One can show that any curve which represents this element with perpendicular boundary values can be represented as composition of curves $\alpha_1^{n_1}$, $\alpha_2^{n_2}$, ..., which represent $a_1^{n_1}$, $a_2^{n_2}$, ..., with perpendicular boundary values. Together with Theorems 2 and 4 of [1], this implies the following estimate from below

$$\lambda_{pb}(a_1^{n_1} a_2^{n_2} \ldots) \geq \lambda_{pb}(a_1^{n_1}) + \lambda_{pb}(a_2^{n_2}) + \cdots.$$

This gives a good lower bound if all terms $a_j^{n_j}$ of the reduced word enter with power of absolute value at least 2. It does not give a good lower bound, for example, for the word $(a_1 a_2^{-1})^n$ with $n \geq 1$, or for the word $a_1^{n_1}(a_2 a_1)^{n_2}$ for integers n_1 and n_2 larger than 1 and n_2 much bigger than n_1. In the first example the reason is the following. Each representing curve for $a_1 a_2^{-1}$ with pb-boundary values can be written as composition of the following two curves: a curve α_1 with pb-boundary values on the left and tr-boundary values on the right representing a_1, and a curve α_2^{-1} with tr-boundary values on the left and pb-boundary values on the right representing a_2^{-1}. The lift of each of the two curves is a non-trivial half slalom curve. Hence the extremal length with pb-boundary values of the element $(a_1 a_2^{-1})^n$ is proportional to n.

For the second example one can show that each representing curve with pb-boundary values contains a piece corresponding to $(a_2 a_1)^{n_2}$ with mixed boundary values. A different choice of a lift gives a half slalom curve which shows that the extremal length of this piece is proportional to $\log\left(\frac{n_2-1}{2}\right)$.

The discussion suggests that the extremal length of a general element of π_1^{pb} can be given in terms of a syllable decomposition of the representing reduced word.

We describe now the syllable decomposition of the word (4).

(1) Any term $a_j^{n_j}$ of the reduced word with $|n_j| \geq 2$ is a syllable.
(2) Any maximal sequence of at least two consecutive terms of the reduced word which have equal power equal to either $+1$ or -1 is a syllable.
(3) Each remaining term of the reduced word is characterized by the following properties. It enters with power $+1$ or -1 and the neighbouring term on the right (if there is one) and also the neighbouring term on the left (if there is one) has power different from that of the given one. Each term of this type is a syllable, called a singleton.

Define the degree of a syllable $\deg(\text{syllable})$ to be the sum of the absolute values of the powers of terms entering the syllable.

For example, the syllables of the word $a_2^{-1} a_1^2 a_2^{-3} a_1^{-1} a_2^{-1} a_1^{-1} a_2 a_1^{-1}$ (see Figure 2) from left to right are the singleton a_2^{-1}, the syllable a_1^2 of degree 2, the syllable a_2^{-3} of degree 3, the syllable $a_1^{-1} a_2^{-1} a_1^{-1}$ of degree 3, the singleton a_2 and the singleton a_1^{-1}.

Put $\Lambda(w) \overset{def}{=} \sum_{\text{syllables of } w} \log(1 + \deg(\text{syllable}))$.

The following theorem holds.

Theorem 1. *There are absolute positive constants C_+ and C_- such that the following holds. Let w be the word representing an element of $\pi_1 = \pi_1(\mathbb{C} \setminus \{-1, 1\}, \{0\})$. Then*

(1) $C_- \cdot \Lambda(w) \leq \lambda_{tr}(w) \leq C_+ \cdot \Lambda(w)$, *except in the following cases: $w = a_1^n$ or $w = a_2^n$ for an integer n. In these cases $\lambda_{tr}(b) = 0$.*
(2) $C_- \cdot \Lambda(w) \leq \lambda_{pb}(w) \leq C_+ \cdot \Lambda(w)$, *except in the following case: each term in the reduced word w has the same power, which equals either $+1$ or -1. In these cases $\lambda_{pb}(w) = 0$.*

Corollary 1. *For an element $w \in \pi_1$ which is not one of the exceptional cases of Theorem 1, the two versions of the extremal length are comparable:*

$$C_1 \lambda_{tr}(w) \leq \lambda_{pb}(w) \leq C_2 \lambda_{tr}(w)$$

for positive constants C_1 and C_2 which do not depend on w.

Corollary 2. *There are positive constants C'_- and C'_+ such that for each element $w \in \pi_1$ which is not a singleton the estimate*

$$C'_- \cdot \Lambda(w) \leq \lambda_{tr}(w) + \lambda_{pb}(w) \leq C'_+ \cdot \Lambda(w)$$

holds.

The extremal length of elements of the fundamental group of the complex plane with an arbitrary number of punctures will be treated in a forthcoming paper. The case of n-braids with arbitrary n is subtler.

During the work on the present paper, the author was supported by the SFB "Space-Time-Matter" at Humboldt-University Berlin. She is grateful to K. Mshagskiy and A. Khrabrov for the professional drawing of the figures.

References

[1] L. Ahlfors, *Lecture on Quasiconformal Mappings*, Van Nostrand, Princeton (1966).

[2] C. Kassel, V. Turaev, *Braid Groups*, Graduate Texts in Mathematics **247**, Spinger (2008).

[3] B. Jöricke, *Braids, Conformal Module and Entropy* (145 p.), `arXiv:1412.7000`.

[4] B. Jöricke, *Braids, Conformal Module and Entropy*, C. R. Acad. Sci. Paris, Ser.I, **351** (2013) 289–293.

Burglind Jöricke
Institut für Mathematik
Humboldt-University Berlin
Unter den Linden 6
D-10099 Berlin, Germany
e-mail: `joericke@googlemail.com`

Operator Theory:
Advances and Applications, Vol. 261, 317–332
© Springer International Publishing AG, part of Springer Nature 2018

Sparse Bounds for Random Discrete Carleson Theorems

Ben Krause and Michael T. Lacey

Dedicated to the memory of V.P. Havin

Abstract. We study discrete random variants of the Carleson maximal operator. Intriguingly, these questions remain subtle and difficult, even in this setting. Let $\{X_m\}$ be an independent sequence of $\{0,1\}$ random variables with expectations

$$\mathbb{E}X_m = \sigma_m = m^{-\alpha}, \ 0 < \alpha < 1/2,$$

and $S_m = \sum_{k=1}^{m} X_k$. Then the maximal operator below almost surely is bounded from ℓ^p to ℓ^p, provided the Minkowski dimension of $\Lambda \subset [-1/2, 1/2]$ is strictly less than $1 - \alpha$.

$$\sup_{\lambda \in \Lambda} \left| \sum_{m \neq 0} X_{|m|} \frac{e(\lambda m)}{\operatorname{sgn}(m) S_{|m|}} f(x - m) \right|.$$

This operator also satisfies a sparse type bound. The form of the sparse bound immediately implies weighted estimates in all ℓ^2, which are novel in this setting. Variants and extensions are also considered.

Mathematics Subject Classification (2010). Primary 42B20. Secondary 42A45.

Keywords. Carleson maximal operator, arithmetic Minkowski dimension, sparse operator, arithmetic ergodic theorems, strong law of large numbers.

1. Introduction

The Carleson maximal operator [6] controls the pointwise convergence of Fourier series. In the discrete setting, this estimate is as follows.

B. K. is an NSF Postdoctoral Research Fellow. Research of M. L. supported in part by grant NSF-DMS 1265570 and NSF-DMS-1600693.

Theorem 1.1. *The discrete Carleson maximal operator*

$$Cf(x) := \sup_{0 \le \lambda \le 1} \left| \sum_{m \ne 0} f(n - m) \frac{e(\lambda m)}{m} \right| \tag{1.2}$$

is bounded on $\ell^p(\mathbb{Z})$, $1 < p < \infty$. Here and throughout, $e(t) := e^{2\pi i t}$.

The original article of Carleson addressed the theorem above with the integers replaced by the circle group, in the case of $p = 2$, with its extension to L^p due to Hunt [12]. It was transferred to the real line by Kenig and Tomas [15], but the variant for the integers was not noticed for several years. We are aware of two independent references for the theorem above, that of Campbell and Petersen [5, Lemma 2] and Stein and Wainger [33].

In its much more well-known version on the real line, this theorem has several variants and extensions. For instance, the polynomial variant of Stein and Wainger [35], and its deep extension by Victor Lie [22, 23]. Pierce and Yung [31] have recently established certain Radon transform versions of Carleson's theorem. These are powerful and deep facts.

The discrete versions of these results has only recently been investigated. To give the flavor of results that are under consideration, we recall this conjecture of Lillian Pierce [29]. Below, and throughout this paper we set $e(t) = e^{2\pi i t}$, and identify the fundamental domain for $\mathbb{T} = \mathbb{R}/\mathbb{Z}$ as $[-1/2, 1/2]$.

Conjecture 1.3. *The following inequality holds on $\ell^2(\mathbb{Z})$.*

$$\left\| \sup_{-\frac{1}{2} \le \lambda \le \frac{1}{2}} \left| \sum_{n \ne 0} f(x - n) \frac{e(\lambda n^2)}{n} \right| \right\|_{\ell^2} \lesssim \|f\|_{\ell^2}.$$

A recent paper of the authors [16] supplied sufficient conditions on Λ so that if one forms a restricted supremum over $\lambda \in \Lambda$, the maximal function above would be bounded on ℓ^2. Even under this restriction, in which we require sufficiently small *arithmetic Minkowski dimension*, our proof is difficult, even for examples of Λ being a sequence that converge very rapidly to the origin. The interested reader is referred to [7, 16] for more background (including the definition of arithmetic Minkowski dimension), and related results.

It is therefore of some interest to study random versions of these questions, in which we expect some of the severe obstacles in the arithmetic case to be of an easier nature. This is so, but even still, we will not be able to prove the most natural conjectures, and indeed even find that the random versions still have remnants of the arithmetic difficulties of the non-random versions.

We consider two examples of random Carleson operators. From now on $\{X_n : n \in \mathbb{Z}\}$ will denote a sequence of independent $\{0, 1\}$ random variables (on a probability space Ω) with expectations

$$\mathbb{E}X_m := \sigma_m = m^{-\alpha}, \ 0 < \alpha < 1. \tag{1.4}$$

Also define the partial sums by

$$S_n = \begin{cases} \sum_{m=1}^n X_m & n > 0 \\ -S_{-n} & n < 0 \end{cases}. \tag{1.5}$$

By the Law of Large Numbers, S_n is approximately $c_\alpha n^{1-\alpha}$.

In the first random examples, the analogy to the (linear) Carleson theorem is stronger, since the frequency modulation and shift parameters agree.

$$\mathcal{T}^\omega_{\alpha,\Lambda} f(x) := \sup_{\lambda \in \Lambda} \left| \sum_{m \neq 0} X_{|m|} \frac{e(\lambda m)}{S_m} f(x - m) \right|. \tag{1.6}$$

We consider arbitrary $0 < \alpha < 1$ for the above operator, but in the second example below, we only consider $\alpha = 1 - \frac{1}{d}$ where $d \geq 3$ is an integer, and have distinct frequency modulations and shift parameters.

$$\mathcal{C}^\omega_{\alpha,\Lambda} f(x) := \sup_{\lambda \in \Lambda} \left| \sum_{m \neq 0} X_m \frac{e(\lambda m)}{S_m} f(x - S_m) \right|. \tag{1.7}$$

Note that $|S_m| \approx m^d$ above is a random version of a monomial power. In both definitions, we are using the definition (1.5) to define S_m for negative m.

We will not be able to control the unrestricted supremum of λ, using *Minkowski dimension* as a sufficient condition for the boundedness of our maximal operators. Given $\Lambda \subset [-1/2, 1/2]$, and $0 < \delta < 1$, let $N(\delta) = N_\Lambda(\delta)$ be the fewest number of intervals $I_1, \ldots, I_N$ required to cover Λ, subject to the condition at $|I_n| < \delta$ for all $1 \leq n \leq N$. We say that Λ has *Minkowski dimension d* if

$$C_d := \sup_{0 < \delta \leq 1} N(\delta) \delta^d < \infty. \tag{1.8}$$

The points of interest in the next theorem are that we (a) allow arbitrary $0 < \alpha < 1$, (b) have an explicit assumption on the Minkowski dimension of Λ, and (c) obtain *a sparse bound* for the operator.

Theorem 1.9. *Suppose*

$$\mathbb{E}X_m = \sigma_m = m^{-\alpha}, \qquad 0 < \alpha < 1,$$

and let $\Lambda \subset [0,1]$ have upper Minkowski dimension δ strictly less than $1 - \alpha$. Then almost surely, these two properties hold.

1. *For all $1 < p < \infty$, we have $\|\mathcal{T}^\omega_{\alpha,\Lambda} : \ell^p \mapsto \ell^p\| < \infty$.*
2. *There is an $r = r(\alpha, \delta)$ with $1 < r < 2$ so that for finitely supported functions f and g, there is a sparse operator $\Pi_{S,r}$ so that*

$$|\langle \mathcal{T}^\omega_{\alpha,\Lambda} f, g \rangle| \lesssim \Pi_{S,r}(f, g).$$

In particular, there holds almost surely, for all weights $w \in A_{2/r} \cap RH_{r/(2-r)}$,

$$\|\mathcal{T}^\omega_{\alpha,\Lambda} : \ell^2(\mathbb{Z}, w) \mapsto \ell^2(\mathbb{Z}, w)\| < \infty. \tag{1.10}$$

In the second conclusion, we are using the notation of §2.3, specifically see (2.6). It implies the weighted inequalities (1.10), as is explained in that section. In particular, there is a slightly wider class of inequalities that are true, as specified in (2.10). We are *not aware* of any prior weighted inequality for a discrete variant of the Carleson operators (except the Carleson operator itself). (For discrete random Hilbert transforms, see [19].) We remark that we could keep track of the dependence of the sparse index r as a function of α and δ, but we don't do so.

Theorem 1.11. *Suppose $d \geq 3$ is an integer and*

$$\mathbb{E}X_m = \sigma_m = m^{-\alpha}, \quad \alpha = 1 - 1/d.$$

Let $\Lambda \subset [-1/2, 1/2]$ have upper Minkowski dimension δ strictly less than $1/d$, and $\Lambda \cap (-\epsilon, \epsilon) = \emptyset$ for some $0 < \epsilon < 1/4$. Then almost surely, these two conclusions hold.

1. *For all $1 < p < \infty$, we have $\|\mathcal{C}^\omega_{\alpha,\Lambda} : \ell^p \mapsto \ell^p\| < \infty$.*
2. *There is a $1 < r = r(d, \delta) < 2$ so that for finitely supported functions f and g, there is a sparse operator $\Pi_{\mathcal{S},r}$ so that*

$$|\langle \mathcal{C}^\omega_{\alpha,\Lambda} f, g \rangle| \lesssim \Pi_{\mathcal{S},r}(f, g).$$

In particular, the inequality (1.10) also holds for $\mathcal{C}^\omega_{\alpha,\Lambda}$.

Our assumption that the set Λ is bounded away from the origin is rather severe. But, interestingly, removing this assumption would entail many extra subtleties, which we comment on at the end of the paper.

We are inspired by the arithmetic ergodic theorems of Bourgain [3, 4]. To explore the underlying complexity of these theorems, Bourgain studied the pointwise ergodic theorem formed from randomly selected subsets of the integers. In our notation, this lead to the study of maximal functions

$$\sup_n \left| \frac{1}{S_n} \sum_{m=1}^n X_m f(x - m) \right|,$$

for non-negative $f \in \ell^p(\mathbb{Z})$. This theme was studied by several authors [20, 24, 32], and we point in particular to the definitive results in the 'lacunary' case [1]. Our theorems are also closely related to the Wiener–Wintner theorem [36], which itself continues to have powerful and deep connections to ergodic theory [5, 11] and harmonic analysis [10, 28]. One can compare the results here with that of say [17], which obtains much stronger theorems, but for averages, as opposed to the singular sums of this paper.

Our subject is also connected to the discrete harmonic analysis also inspired by Bourgain's arithmetic ergodic theorems, and promoted by Stein and Wainger [33, 34]. This area remains quite active. Besides these older papers [13, 14], the reader should also reference these very recent papers for interesting new developments [7, 16, 25–27, 30].

The sparse bounds have been a recent and quite active topic in continuous harmonic analysis, see [8, 9, 18] and references therein for a guide to this subject. Their appearance in the discrete settings is new. In particular, the weighted inequalities that are corollaries to our main theorem have very few precedents in the literature.

The techniques of our proofs straddle (discrete) harmonic analysis and probability theory. We will use standard facts about maximal functions, and the Carleson theorem itself. On the probability side, we reference standard large deviation inequalities for iid random variables, and martingales, to control random Fourier series. The method to obtain the sparse bounds is illustrated, in a simpler way, in the argument of [19], which also addresses random discrete inequalities.

2. Preliminaries

2.1. Notation

With X_m as in (1.4), we let $Y_m = X_m - \sigma_m$, and

$$W_m = \sum_{n=1}^{m} \sigma_m.$$

By the integral test, we note that $W_m = \frac{1}{1-\alpha}m^{1-\alpha} + O(1)$. Moreover $\mathrm{Var}(S_n) \leq W_n$. And, it is well known that, for any $\epsilon > 0$,

$$|S_m - W_m| \lesssim m^{\epsilon + \frac{1-\alpha}{2}} \tag{2.1}$$

We will make use of the modified Vinogradov notation. We use $X \lesssim Y$, or $Y \gtrsim X$ to denote the estimate $X \leq CY$ for an absolute constant C. We use $X \approx Y$ as shorthand for $Y \lesssim X \lesssim Y$. We also make use of big-O notation: we let $O(Y)$ denote a quantity that is $\lesssim Y$.

Since we will be concerned with establishing *a priori* $\ell^p(\mathbb{Z})$-estimates in this paper, we will restrict every function considered to be a member of a "nice" dense subclass: each function on the integers will be assumed to have finite support.

2.2. Fourier transform

As previously mentioned, we let $e(t) := e^{2\pi i t}$. The Fourier transform on $f \in \ell^2(\mathbb{Z})$ is defined by

$$\mathcal{F}f(\beta) = \sum_n f(n)e(-\beta n).$$

This is a unitary map from $\ell^2(\mathbb{Z})$ to $L^2(\mathbb{T})$. In particular, we have for convolution

$$\mathcal{F}(f * g) = \mathcal{F}f \cdot \mathcal{F}g.$$

In particular, all of our theorems can be understood as maximal theorems over convolutions. It will be convenient to study the corresponding Fourier multipliers.

Indeed, the following technical lemma exhibits the way that small Minkowski dimension is used. (It is so to speak a variant of Sobolev embedding, for sets of small Minkowski dimension.)

Lemma 2.2 (Lemma 2.4 of [16]). *Suppose* $\Lambda \subset [0,1]$ *has upper Minkowski dimension at most* $0 < d < 1$, *as given in (1.8). Suppose that* $\{T_\lambda : \lambda \in [0,1]\}$ *is a family of operators so that for each* $f \in \ell^2(\mathbb{Z})$, $T_\lambda f(x)$ *is differentiable in* $\lambda \in [0,1]$. *Set*

$$a := \sup_{\lambda \in [0,1]} \|T_\lambda\|_{\ell^2(\mathbb{Z}) \to \ell^2(\mathbb{Z})}, \quad \text{and} \tag{2.3}$$

$$A := \sup_{\lambda \in [0,1]} \|\partial_\lambda T_\lambda\|_{\ell^2(\mathbb{Z}) \to \ell^2(\mathbb{Z})}. \tag{2.4}$$

Then we have the maximal inequality below

$$\|\sup_\Lambda |T_\lambda f|\|_{\ell^2(\mathbb{Z})} \lesssim C_d^{1/2}(a + a^{1-d/2}A^{d/2})\|f\|_{\ell^2(\mathbb{Z})}. \tag{2.5}$$

In application, the quantities in (2.3) and (2.4) are estimated on the Fourier side. This will be used in settings where $a \ll 1$ and $1 \ll A \ll a^{-m}$, for some large integer m. Then, for $0 < d < 1/m$ sufficiently small, the right side of (2.5) will be small.

2.3. Sparse operators

A *sparse* collection of intervals $\mathcal{S}$ satisfies this essential condition: There is a collection of pairwise disjoint sets of the integers $\{E(I) : I \in \mathcal{S}\}$ so that $|E(I)| > \frac{1}{10}|I|$ for all $I \in \mathcal{S}$. A *sparse bilinear form* is defined in terms of a choice of an index $1 \leq r < \infty$, and a sparse collection of intervals $\mathcal{S}$. Define

$$\Pi_{\mathcal{S},r}(f,g) = \sum_{I \in \mathcal{S}} \langle f \rangle_{I,r} \langle g \rangle_{I,r} |I|$$

$$\langle f \rangle_{I,r} := \left[|I|^{-1} \sum_{n \in I} |f(n)|^r \right]^{1/r}. \tag{2.6}$$

If the role of the sparse collection is not essential, it will be suppressed in the notation.

Sparse bounds are known for some operators T, taking this form: For all f, g finitely supported on $\mathbb{Z}$, there is a choice of sparse operator Π_r so that

$$|\langle Tf, g \rangle| \lesssim \Pi_r(f,g), \tag{2.7}$$

where the implied constant is independent of f, g. We refer to this as *the sparse property of index* r, and write $T \in \text{Sparse}_r$.

Theorem 2.8. *We have these sparse bounds.*

1. *For the maximal function* M, *we have* $M \in \text{Sparse}_1$.
2. *For the Carleson operator* C *of (1.2), we have* $C \in \text{Sparse}_r$ *for all* $1 < r < 2$.

The bound for the maximal function is very easy, and not that sharp. The bound for the Carleson operator follows for instance from [21, Theorem 4.6]. One of the fascinating things about the sparse bounds is that they easily imply weighted inequalities.

For non-negative function w, we define the Muckenhoupt A_p and *reverse Hölder* characteristics by

$$[w]_{A_p} = \sup_Q \left[\frac{w^{\frac{1}{1-p}}(Q)}{|Q|}\right]^{p-1} \frac{w(Q)}{|Q|} < \infty$$

$$[w]_{RH_p} = \sup_Q \frac{\langle w^p \rangle_Q^{1/p}}{\langle w \rangle_Q} < \infty.$$

Above, we are conflating w as a measure and a density, thus

$$w^{\frac{1}{1-p}}(Q) = \int_Q w(x)^{\frac{1}{1-p}}\,dx.$$

And, we are stating the definition as if it were on Euclidean space, but the theory transfers to the integers in a straightforward way. We have these estimates, a corollary to [2, Prop 6.4].

Theorem 2.9. *For all $1 < r < 2$, $r < p < r'$ and weights w there holds*

$$\Pi_r(f,g) \le C([w]_{A_{p/r}}, [w]_{RH_{r/(r-p(r-1))}})\|f\|_{\ell^p(w)}\|g\|_{\ell^p(w)}. \tag{2.10}$$

The sharp bound for the constant on the right is computed in [2]. There is little doubt that the results of this paper can be improved, so we don't track that constant.

3. The Proof of Theorem 1.9

Theorem 1.9 concerns the maximal function in (1.6). For fixed λ the summands in (1.6) consist in part of

$$X_{|m|}\frac{e(\lambda m)}{S_{|m|}} = c_\alpha \frac{e(\lambda m)}{|m|} \tag{3.1}$$

$$+ \left[\frac{\sigma_{|m|}}{W_{|m|}} - \frac{c_\alpha}{|m|}\right]e(\lambda m) \tag{3.2}$$

$$+ Y_{|m|}\frac{e(\lambda m)}{W_{|m|}} \tag{3.3}$$

$$+ X_{|m|}e(\lambda m)\left[\frac{1}{S_{|m|}} - \frac{1}{W_{|m|}}\right]. \tag{3.4}$$

This leads to the decomposition of the maximal operator in (1.6), upon multiplying each term by $\text{sgn}(m)$. We will address them in order, with the restriction on Minkowski dimension arising from only the term in (3.3).

The first and most significant term is associated with (3.1), which is entirely deterministic, and in fact the associated maximal function is exactly the Carleson Theorem 1.1, hence we have the sparse bound from Theorem 2.8. The relevant sparse bound is $C \in \mathrm{Sparse}_r$, for all $1 < r < 2$. The second term (3.2) is entirely trivial. As follows from (2.1), we have almost surely

$$\left| \frac{\sigma_m}{W_m} - \frac{c_\alpha}{m} \right| = \frac{|m^{1-\alpha} - c_\alpha W_m|}{W_m \cdot m} \lesssim m^{-2}.$$

Convolution with $\frac{1}{m^2}$ is easily seen to satisfy a sparse bound, with $r = 1$.

The third term (3.3) is the one that imposes a condition on Λ, the set that defines the maximal operator. We have this important lemma, which controls a relevant maximal function in ℓ^2-norm. Define

$$P_k f := \sup_{\lambda \in \Lambda} \left| \sum_{m \,:\, 2^k \le |m| < 2^{k+1}} Y_m \frac{e(\lambda m)}{W_m} f(x - m) \right|. \tag{3.5}$$

Lemma 3.6. *Assume that Λ has Minkowski dimension strictly less than $1 - \alpha$. Then there is a positive choice of $\eta = \eta(\alpha) > 0$, so for all integers $k \in \mathbb{N}$, we have almost surely*

$$\sup_{k \in \mathbb{N}} \| P_k \,:\, \ell^1 \to \ell^1 \| + \| P_k \,:\, \ell^\infty \to \ell^\infty \| < \infty, \tag{3.7}$$

$$\sup_{k \in \mathbb{N}} 2^{\eta k} \| P_k \,:\, \ell^2 \to \ell^2 \| < \infty. \tag{3.8}$$

Proof. The first claim is a consequence of the Strong Law of Large Numbers. Note that

$$\| P_k \,:\, \ell^1 \to \ell^1 \| \lesssim \sum_{m \,:\, 2^k \le |m| < 2^{k+1}} \frac{|Y_m|}{W_m}.$$

And, the latter is uniformly bounded almost surely.

The ℓ^2 estimate, which has a gain in operator norm, is a consequence of Lemma 2.2, which bounds supremums like those in (3.8) in terms of the ℓ^∞ norm of the multipliers, and the derivatives of the multipliers. Due to the form of the sums, the multipliers are translations by $\lambda \in \Lambda$ of the functions of θ below. Now, Lemma 2.2 requires two estimates, the first is the $L^\infty(d\theta)$ estimate, for which we have

$$\mathbb{P}\left(\left\| \sum_{m \,:\, 2^k \le |m| < 2^{k+1}} Y_m \frac{e(\theta m)}{W_m} \right\|_\infty > C\sqrt{k} \cdot 2^{k(\alpha-1)/2} \right) \le 2^{-k}, \tag{3.9}$$

for appropriate constant C. (We will recall a proof in Lemma 4.7 below.) We also need an estimate for the derivative in θ of the functions above, which is clearly of the form

$$\mathbb{P}\left(\left\| \sum_{m \,:\, 2^k \le |m| < 2^{k+1}} m Y_m \frac{e(\theta m)}{W_m} \right\|_\infty > C\sqrt{k} \cdot 2^{k(\alpha+1)/2} \right) \le 2^{-k}. \tag{3.10}$$

By the Borel–Cantelli lemma, we see that the union of these two events occur finitely often, almost surely.

Apply (2.5), with $a = \sqrt{k} \cdot 2^{k(\alpha-1)/2}$ and $A = \sqrt{k} \cdot 2^{k(\alpha+1)/2}$. We see that the conclusion of the Lemma holds provided

$$(\alpha - 1)(1 - d/2) + (\alpha + 1)d/2 < 0$$

where α is the constant associated to the selector random variables, and d is the Minkowski dimension of Λ. This inequality is true for $d < 1 - \alpha$. This completes the proof. $\qquad\square$

The previous lemma implies the ℓ^p-control of the term associated to (3.3), after a straightforward interpolation between (3.7) and (3.8). We turn to the sparse bound.

Corollary 3.11. *There is an $\eta > 0$, so that almost surely, there is a finite constant $C_\omega > 0$ so that we have for all integers k, intervals I of length 2^k, and functions f, g supported on I,*

$$|\langle P_k f, g \rangle| \leq C_\omega 2^{-\eta' k} \langle f \rangle_{3I,r} \langle g \rangle_{I,r} |I| \qquad 0 < 2 - r < c(\eta, \delta). \tag{3.12}$$

The constants $c(\eta, \delta)$ and $\eta' = \eta'(\eta, \delta, r)$ are positive and sufficiently small.

On the right above, we have a geometric decay in k, and a sum over intervals of fixed length which are disjoint. It is easy to see that with f and g fixed, we have

$$\sum_k 2^{-\eta' k} \sum_{I \in \mathcal{D} \,:\, |I| = 2^k} \langle f \rangle_{3I,r} \langle g \rangle_{I,r} |I| \lesssim \Pi_r(f, g),$$

for an appropriate sparse operator Π. This completes the sparse bound for the term associated to (3.3).

Proof. Observe that almost surely, there exists $C_\omega < \infty$, so that these inequalities hold.

$$|\langle P_k f, g \rangle| \leq C_\omega \begin{cases} 2^{-\eta k} \langle f \rangle_{3I,2} \langle g \rangle_{I,2} |I| \\ 2^{\alpha k} \langle f \rangle_{3I,1} \langle g \rangle_{I,1} |I| \end{cases} \tag{3.13}$$

The top line follows from (3.8). The last line is the inequality that is *below duality*. (And, geometric growth in k.) It follows from estimate

$$|\langle P_k f, g \rangle| \leq \sum_x \sum_{n \,:\, 2^k \leq |n| < 2^{k+1}} \frac{|Y_n|}{|n|^{1-\alpha}} |f(x - n)| \cdot |g(x)|.$$

Then, use the ℓ^1-norm on f, the same on g, and the ℓ^∞ norm on $|Y_n|$.

To conclude (3.12), interpolate between the top and bottom estimates in (3.13). The bottom line has a fixed positive geometric growth, while the ℓ^2 estimate has a fixed negative geometric growth. For a choice of $1 < r < 2$, with $2 - r$ sufficiently small, we will have the geometric decay claimed. $\qquad\square$

The control of the fourth term associated with (3.4), again requires no cancelation, as follows immediately from this next lemma.

Lemma 3.14. *Almost surely, we have*

$$\sum_{m \neq 0} \left| \frac{1}{S_{|m|}} - \frac{1}{W_{|m|}} \right| \cdot |f(x - m)| \in \mathrm{Sparse}_1. \tag{3.15}$$

Proof. We only discuss the case of $m > 0$. By the Law of the Iterated Logarithm, we have

$$S_m = W_m + O\big(m^{\frac{1-\alpha}{2}} \sqrt{\log \log m}\big).$$

And, recall that $W_m \sim m^{(1-\alpha)}$. It follows that

$$\left| \frac{1}{S_m} - \frac{1}{W_m} \right| \lesssim \frac{\sqrt{\log \log m}}{m^{\frac{3}{2} - \frac{\alpha}{2}}} \lesssim m^{-\beta}$$

where $\beta > 1$. The sparse bound is then immediate. $\square$

4. Proof of Theorem 1.11

The summands in (1.7) are rewritten as below, in which we assume that $m > 0$.

$$X_m e(\lambda m) \cdot \frac{f(x - S_m) - f(x + S_m)}{S_m} \tag{4.1}$$

$$= X_m e(\lambda m)(f(x - S_m) - f(x + S_m)) \Big\{ \frac{1}{S_m} - \frac{1}{W_m} \Big\} \tag{4.2}$$

$$+ \Big\{ \frac{\sigma_m}{W_m} - \frac{c_\alpha}{m} \Big\} e(\lambda m) \cdot (f(x - S_{m-1} - 1) - f(x + S_{m-1} + 1)) \tag{4.3}$$

$$+ Y_m e(\lambda m) \cdot \frac{f(x - S_{m-1} - 1) - f(x + S_{m-1} + 1)}{W_m} \tag{4.4}$$

$$+ c_\alpha e(\lambda m) \cdot \frac{f(x - S_{m-1} - 1) - f(x + S_{m-1} + 1)}{m}. \tag{4.5}$$

In the first stage we simply replace $\frac{1}{S_m}$ by $\frac{1}{W_m}$. But the remaining terms use the identity

$$X_m f(x + S_m) = X_m f(x + S_{m-1} + 1),$$

which step is motivated by a martingale argument below.

The term associated with (4.2) is controlled by the estimate (3.15), and that for (4.3) is entirely similar. The term in (4.4) is analogous to Corollary 3.11, for which we need this lemma. Define the maximal operator

$$Q_k f := \sup_{\lambda \in \Lambda} \left| \sum_{m \, : \, 2^k \leq |m| \leq 2^{k+1}} Y_m e(\lambda m) \cdot \frac{f(x - S_{m-1} - 1) - f(x + S_{m-1} + 1)}{W_m} \right|.$$

$$\tag{4.6}$$

Lemma 4.7. *Assume that $\Lambda \subset \mathbb{T}$ has Minkowski dimension at most $1 - \alpha$, then there is a $\eta > 0$ so that we have almost surely*

$$\sup_k \|Q_k \; : \; \ell^1 \to \ell^1\| + \|Q_k \; : \; \ell^\infty \to \ell^\infty\| < \infty, \tag{4.8}$$

$$\sup_k 2^{\eta k} \|Q_k \; : \; \ell^2 \to \ell^2\| < \infty. \tag{4.9}$$

Proof. The top line follows from the Strong Law of Large Numbers. We turn to the second line, where there is geometric decay. There are no cancellative effects between positive and negative translations, and so we only consider the positive ones. The ℓ^2 bound is a consequence of Lemma 2.2, and so we need to consider the multipliers

$$M(\lambda, \theta) := \sum_{m \; : \; 2^k \le m \le 2^{k+1}} Y_m \frac{e(\lambda m + \theta(S_{m-1} + 1))}{W_m}.$$

So to prove the lemma, it suffices to show that with probability at least $1 - 2^{-\epsilon k}$, we have the two inequalities

$$\|M(\lambda, \theta)\|_{L^\infty(\lambda, \theta)} \lesssim 2^{k(-(\alpha+1)/2+\epsilon)}, \tag{4.10}$$

$$\|\partial_\lambda M(\lambda, \theta)\|_{L^\infty(\lambda, \theta)} \lesssim 2^{k((1-\alpha)/2+\epsilon)}. \tag{4.11}$$

The summands in the definition of $M(\lambda, \beta)$, for fixed λ and β, form a bounded martingale difference sequence, with square function bounded in $L^\infty(\Omega)$ by

$$\left[\sum_{m \; : \; 2^k \le m \le 2^{k+1}} \frac{\sigma_m}{W_m^2} \right]^{1/2} \lesssim 2^{-k\frac{1-\alpha}{2}}.$$

It is well known that such martingale differences are sub-gaussian, hence, uniformly in λ and β, we have

$$\mathbb{P}(|M(\lambda, \beta)| > 2^{k(-\frac{1-\alpha}{2}+\epsilon)}) \lesssim \exp(-2^{2\epsilon k}). \tag{4.12}$$

But, $M(\lambda, \beta)$ clearly has gradient at most 2^k in norm. That means to test the $L^\infty(\lambda, \beta)$ norm, we apply the inequality (4.12) on a set of at most 2^{2k} choices of (λ, β). Therefore (4.10) clearly follows. The analysis for (4.11) is similar.   □

With this bound in hand, we can repeat the proof of Corollary 3.11, and conclude that almost surely we have

$$\sum_k \|Q_k \; : \; \ell^p \to \ell^p\| < \infty, \qquad 1 < p < \infty,$$

$$\sum_k |\langle Q_k f, g \rangle| \in \mathrm{Sparse}_{r,r}(f, g) \qquad 0 < r < 2 - c(d, \delta).$$

This completes the analysis of the term associated with (4.4).

The term associated with (4.5) is arithmetic in nature. Let a_j be the smallest positive integer such that $S_{a_j} = j$. It is a consequence of the Strong Law of Large

Numbers that we have

$$a_j = p_j + O(j^{\epsilon + \frac{1}{2(1-\alpha)}}) = \lfloor C_\alpha j^{\frac{1}{1-\alpha}} \rfloor + O(j^{\epsilon + \frac{1}{2(1-\alpha)}}), \tag{4.13}$$

where $0 < C_\alpha = (1-\alpha)^{\frac{1}{1-\alpha}} < \infty$. Now observe that for fixed λ, we have

$$\sum_{m>1} e(\lambda m) \cdot \frac{f(x - S_{m-1} - 1) - f(x + S_{m-1} + 1)}{m} \tag{4.14}$$

$$= \sum_{j=1}^{\infty} A_j(\lambda)(f(x-j) - f(x+j)), \tag{4.15}$$

where

$$A_j(\lambda) := \sum_{m=a_{j-1}}^{a_j - 1} \frac{e(\lambda m)}{m}, \tag{4.16}$$

and $a_0 := 1$ if $a_1 > 1$.

Attention turns to the coefficients $A_j(\lambda)$. The point below is that if j is small relative to λ, we have an excellent approximation to $A_j(\lambda)$, and otherwise, the coefficient is small for other reasons.

Lemma 4.17. *These two inequalities below hold uniformly over all compactly supported functions f, almost surely.*

$$\sup_{0<\lambda<1} \sum_{|j|<\lambda^{-\frac{\alpha}{1-\alpha}}} |\Delta_j f(x-j)| \lesssim Mf(x) \tag{4.18}$$

$$\text{where} \quad \Delta_j = A_j(\lambda) - \frac{e(p_j\lambda)}{j}, \tag{4.19}$$

$$\sup_{0<\lambda<1} \sum_{|j|\geq\lambda^{-\frac{\alpha}{1-\alpha}}} |A_j(\lambda)f(x-j)| \lesssim Mf(x). \tag{4.20}$$

Above, M is the maximal function, and it is in Sparse$_1$.

Proof. We begin with an elementary estimate. Set

$$r_j = p_j - p_{j-1} \simeq j^{\frac{1}{1-\alpha}-1} = j^{\frac{\alpha}{1-\alpha}}. \tag{4.21}$$

We use this notation to rewrite $A_j(\lambda)$ in terms of the Dirichlet kernel.

$$A_j(\lambda) = e(p_j\lambda) \sum_{m=0}^{r_j-1} \frac{e(-\lambda m)}{p_{j-1}+m} + O(j^{-\frac{1}{1-\alpha}}) \tag{4.22}$$

$$= \frac{e(p_j\lambda)}{p_{j-1}} \sum_{m=1}^{r_j} e(-\lambda m) + O\left(\frac{r_{j-1}}{p_{j-1}^2}\right) \tag{4.23}$$

$$= \frac{e(p_j\lambda)}{p_{j-1}} D_{r_j}(-\lambda) + O(j^{-1-\frac{1}{1-\alpha}}). \tag{4.24}$$

In the last line, D_n denotes the nth Dirichlet kernel. Clearly, convolution with the Big-Oh term is bounded by the maximal function, so that we continue with the term involving the Dirichlet kernel.

By the estimate $|D_n(\lambda) - n| \lesssim n\lambda$, for $0 < \lambda < 1$, we have

$$d_j = \left| A_j(\lambda) - \frac{e(p_j\lambda)r_j}{p_{j-1}} \right| \lesssim r_j\lambda \lesssim j^{\frac{\alpha}{1-\alpha}}\lambda.$$

It follows that for non-negative f,

$$\sup_{0<\lambda<1} \sum_{j\,:\,0\le|j|\le\lambda^{-\frac{\alpha}{1-\alpha}}} d_j f(x - j) \lesssim Mf(x).$$

This nearly completes the proof of (4.18). The last step is to observe that

$$\left| \frac{r_j}{p_{j-1}} - \frac{1}{j} \right| \lesssim j^{-2}.$$

And, convolution with respect to j^{-2} is bounded by the maximal function as well. This completes the proof of (4.18).

By the estimate $|D_n(\lambda)| \lesssim \lambda^{-1}$, for $0 < \lambda < 1$, we have

$$|A_j(\lambda)| \lesssim (p_j\lambda)^{-1} \simeq j^{-\frac{1}{1-\alpha}}\lambda. \tag{4.25}$$

Hence, we have, for non-negative f,

$$\sup_{0<\lambda<1} \sum_{|j|>\lambda^{-\frac{\alpha}{1-\alpha}}} |A_j(\lambda)|f(x - j) \lesssim Mf(x),$$

where M denotes the discrete Hardy–Littlewood maximal function. This proves (4.20). □

We see that the proof of our theorem is reduced to this deterministic, and trivial, result. Here, we simply use a crude bound, and the assumption that Λ has no points close to the origin. With the summation condition, this bound is trivial.

Lemma 4.26. *For $0 < \epsilon < 1/4$*

$$\sup_{\epsilon<|\lambda|<\frac{1}{2}} \left| \sum_{1<j<\lambda^{-\frac{\alpha}{1-\alpha}}} \frac{e(\lambda p_j)}{j}(f(x + j) - f(x - j)) \right| \lesssim (\log 1/\epsilon)Mf(x) \tag{4.27}$$

This last lemma is the one that uses the assumption in Theorem 1.11 that Λ is bounded away from the origin. If we remove this assumption, the lemma above shows that arithmetic issues become paramount. Indeed, we see that the main results of [7, 16] are relevant. But, here we note that

(a) the sparse variants of the main results in these papers are not known,
(b) these are very involved papers, and
(c) their main results would have to be extended.

In particular, [16] would have to be extended to the case of an arbitrary monomial in the oscillatory term, as well as incorporating maximal truncations into the main theorem. (We hope to address these in a future paper.) In discrete harmonic analysis, randomly formed operators typically do not inherit any difficult arithmetic structure. It is notable in these questions that they can.

References

[1] Mustafa Akcoglu, Alexandra Bellow, Roger L. Jones, Viktor Losert, Karin Reinhold-Larsson, and Máté Wierdl, *The strong sweeping out property for lacunary sequences, Riemann sums, convolution powers, and related matters*, Ergodic Theory Dynam. Systems **16** (1996), no. 2, 207–253. ↑320

[2] Frédéric Bernicot, Dorothee Frey, and Stefanie Petermichl, *Sharp weighted norm estimates beyond Calderón–Zygmund theory*, Anal. PDE **9** (2016), no. 5, 1079–1113. ↑323

[3] J. Bourgain, *On the pointwise ergodic theorem on L^p for arithmetic sets*, Israel J. Math. **61** (1988), no. 1, 73–84. ↑320

[4] Jean Bourgain, *Pointwise ergodic theorems for arithmetic sets*, Inst. Hautes Études Sci. Publ. Math. **69** (1989), 5–45. With an appendix by the author, Harry Furstenberg, Yitzhak Katznelson and Donald S. Ornstein. ↑320

[5] James Campbell and Karl Petersen, *The spectral measure and Hilbert transform of a measure-preserving transformation*, Trans. Amer. Math. Soc. **313** (1989), no. 1, 121–129. ↑318, 320

[6] Lennart Carleson, *On convergence and growth of partial sums of Fourier series*, Acta Math. **116** (1966), 135–157. ↑317

[7] L. Cladek, K. Henriot, B. Krause, I. Laba, and M. Pramanik, *A Discrete Carleson Theorem Along the Primes with a Restricted Supremum*, available at **1604.08695**. ↑318, 320, 329

[8] José M. Conde-Alonso and Guillermo Rey, *A pointwise estimate for positive dyadic shifts and some applications*, Math. Ann. (2015), 1–25. ↑321

[9] A. Culiuc, F. Di Plinio, and Y. Ou, *Domination of multilinear singular integrals by positive sparse forms*, available at http://arxiv.org/abs/1603.05317. ↑321

[10] Ciprian Demeter, Michael T. Lacey, Terence Tao, and Christoph Thiele, *Breaking the duality in the return times theorem*, Duke Math. J. **143** (2008), no. 2, 281–355. ↑320

[11] Nikos Frantzikinakis, *Uniformity in the polynomial Wiener–Wintner theorem*, Ergodic Theory Dynam. Systems **26** (2006), no. 4, 1061–1071. ↑320

[12] Richard A. Hunt, *On the convergence of Fourier series*, Orthogonal Expansions and their Continuous Analogues (Proc. Conf., Edwardsville, Ill., 1967), Southern Illinois Univ. Press, Carbondale, Ill., 1968, pp. 235–255. ↑318

[13] Alexandru D. Ionescu and Stephen Wainger, *L^p boundedness of discrete singular Radon transforms*, J. Amer. Math. Soc. **19** (2006), no. 2, 357–383 (electronic). ↑320

[14] Alexandru D. Ionescu, Elias M. Stein, Akos Magyar, and Stephen Wainger, *Discrete Radon transforms and applications to ergodic theory*, Acta Math. **198** (2007), no. 2, 231–298. ↑320

[15] Carlos E. Kenig and Peter A. Tomas, *Maximal operators defined by Fourier multipliers*, Studia Math. **68** (1980), no. 1, 79–83. ↑318

[16] Ben Krause and Michael T. Lacey, *A Discrete Quadratic Carleson Theorem on ℓ^2 with a Restricted Supremum*, IMRN, to appear (2015). ↑318, 320, 322, 329

[17] B. Krause and P. Zorin-Kranich, *A random pointwise ergodic theorem with Hardy field weights*, available at 1410.0806. ↑320

[18] Michael T. Lacey, *An elementary proof of the A_2 Bound*, Israel J. Math., to appear (2015). ↑321

[19] Michael T. Lacey and Scott Spencer, *Sparse Bounds for Oscillatory and Random Singular Integrals* (2016), available at http://arxiv.org/abs/1609.06364. ↑320, 321

[20] Michael Lacey, Karl Petersen, Máté Wierdl, and Dan Rudolph, *Random ergodic theorems with universally representative sequences*, Ann. Inst. H. Poincaré Probab. Statist. **30** (1994), no. 3, 353–395 (English, with English and French summaries). ↑320

[21] Andrei K. Lerner, *On pointwise estimates involving sparse operators*, New York J. Math. **22** (2016), 341–349. MR3484688 ↑323

[22] Victor Lie, *The (weak-L^2) boundedness of the quadratic Carleson operator*, Geom. Funct. Anal. **19** (2009), no. 2, 457–497. ↑318

[23] ———, *The Polynomial Carleson Operator*, available at http://arxiv.org/abs/ 1105.4504. ↑318

[24] Mariusz Mirek, *Weak type (1, 1) inequalities for discrete rough maximal functions*, J. Anal. Math. **127** (2015), 247–281. ↑320

[25] ———, *Square function estimates for discrete Radon transforms* (2015), available at arXiv:1512.07524. ↑320

[26] Mariusz Mirek, Bartosz Trojan, and E.M. Stein, *$L^p(\mathbb{Z}^d)$-estimates for discrete operators of Radon type: Variational estimates*, available at arXiv:1512.07523. ↑320

[27] ———, *$L^p(\mathbb{Z}^d)$-estimates for discrete operators of Radon type: Maximal functions and vector-valued estimates*, available at arXiv:1512.07518. ↑320

[28] Richard Oberlin, Andreas Seeger, Terence Tao, Christoph Thiele, and James Wright, *A variation norm Carleson theorem*, J. Eur. Math. Soc. (JEMS) **14** (2012), no. 2, 421–464. ↑320

[29] Lillian B. Pierce, *Personal Communication, American Institute of Mathematics* (May, 2015). ↑318

[30] ———, *Discrete fractional Radon transforms and quadratic forms*, Duke Math. J. **161** (2012), no. 1, 69–106. ↑320

[31] L.B. Pierce and Po-Lam Yung, *A polynomial Carleson operator along the paraboloid* (2015), available at 1505.03882. ↑318

[32] Joseph Rosenblatt, *Universally bad sequences in ergodic theory*, Almost everywhere convergence, II (Evanston, IL, 1989), Academic Press, Boston, MA, 1991, pp. 227–245. ↑320

[33] E.M. Stein and S. Wainger, *Discrete analogues of singular Radon transforms*, Bull. Amer. Math. Soc. (N.S.) **23** (1990), no. 2, 537–544. ↑318, 320

[34] Elias M. Stein and Stephen Wainger, *Discrete analogues in harmonic analysis. I. l^2 estimates for singular Radon transforms*, Amer. J. Math. **121** (1999), no. 6, 1291–1336. ↑320

[35] _____ , *Oscillatory integrals related to Carleson's theorem*, Math. Res. Lett. **8** (2001), no. 5-6, 789–800. ↑318

[36] Norbert Wiener and Aurel Wintner, *Harmonic analysis and ergodic theory*, Amer. J. Math. **63** (1941), 415–426. ↑320

Ben Krause
Department of Mathematics
The University of British Columbia
1984 Mathematics Road
Vancouver, B.C., Canada V6T 1Z2
e-mail: benkrause@math.ubc.ca

Michael T. Lacey
School of Mathematics
Georgia Institute of Technology
686 Cherry Street
Atlanta, GA 30332-0160, USA
e-mail: lacey@math.gatech.edu

Operator Theory:
Advances and Applications, Vol. 261, 333–344
© Springer International Publishing AG, part of Springer Nature 2018

Nodal Sets of Laplace Eigenfunctions: Estimates of the Hausdorff Measure in Dimensions Two and Three

Alexander Logunov and Eugenia Malinnikova

In memory of our teacher Victor Petrovich Havin

Abstract. Let Δ_M be the Laplace operator on a compact n-dimensional Riemannian manifold without boundary. We study the zero sets of its eigenfunctions $u : \Delta_M u + \lambda u = 0$. In dimension $n = 2$ we refine the Donnelly–Fefferman estimate by showing that $\mathcal{H}^1(\{u = 0\}) \leq C\lambda^{3/4-\beta}$ for some $\beta \in (0, 1/4)$. The proof employs the Donnelly–Fefferman estimate and a combinatorial argument, which also gives a lower (non-sharp) bound in dimension $n = 3$: $\mathcal{H}^2(\{u = 0\}) \geq c\lambda^\alpha$ for some $\alpha \in (0, 1/2)$. The positive constants c, C depend on the manifold, α and β are universal.

Mathematics Subject Classification (2010). Primary 31B05. Secondary 35R01, 58G25.

Keywords. Laplace eigenfunctions, nodal set, harmonic functions.

1. Introduction

Let Δ_M be the Laplace operator on a compact n-dimensional Riemannian manifold without boundary. It was conjectured by Yau, see [18], that the nodal sets $E_\lambda = \{u_\lambda = 0\}$ of Laplace eigenfunctions u_λ, $\Delta_M u_\lambda + \lambda u_\lambda = 0$ satisfy the following inequality

$$C_1\sqrt{\lambda} \leq \mathcal{H}^{n-1}(E_\lambda) \leq C_2\sqrt{\lambda}.$$

This conjecture was proved by Donnelly and Fefferman under the assumption that the Riemannian metric is real-analytic ([3]). The left-hand side estimate was also proved for smooth non-analytic surfaces by Brüning ([1]).

A. L. was supported in part by ERC Advanced Grant 692616 and ISF Grants 1380/13, 382/15.
Eu. M. was supported by Project 213638 of the Research Council of Norway.

The previous best known estimate from below for a non-analytic manifold in higher dimensions is

$$\mathcal{H}^{n-1}(E_\lambda) \geq C\lambda^{(3-n)/4},$$

which gives a constant for $n = 3$. The two known approaches are: (1) follow the ideas of Donnelly and Fefferman and find many balls on the wave-scale $\lambda^{-1/2}$ with bounded doubling index, as it is done in [2] or (2) use the Green formula $2\int_{E_\lambda} |\nabla_M u_\lambda| = \lambda \int_M |u_\lambda|$ and the estimate $\frac{\|u_\lambda\|_\infty}{\|u_\lambda\|_1} \leq C\lambda^{(n-1)/4}$, see [16]. The approach in [2] also exploits the Sogge–Zelditch estimates of L^p-norms of eigenfunctions. The following upper estimate in dimension two was established by Donnelly and Fefferman, see [4],

$$\mathcal{H}^1(E_\lambda) \leq C\lambda^{3/4}.$$

In this paper we obtain tiny improvements to the estimate from below in dimension three and to the estimate from above in dimension two. We show that in dimension 2

$$\mathcal{H}^1(E_\lambda) \leq C\lambda^{3/4-\beta}, \tag{1.1}$$

for some $\beta \in (0, 1/4)$. It gives a small refinement to the Donnelly–Fefferman estimate. The proof of (1.1) relies on the results and methods from [3, 4]. Roughly speaking, the Donnelly–Fefferman argument, which gives the estimate with $\frac{3}{4}$, is combined with a combinatorial argument presented below, which gives the β improvement. The same combinatorial argument shows that in dimension $n = 3$

$$\mathcal{H}^2(E_\lambda) \geq C\lambda^\alpha, \tag{1.2}$$

for some $\alpha > 0$. As far is we know it gives the first bound that grows to infinity as λ increases, but we note that the last result is not sharp and can be improved up to the bound $c\sqrt{\lambda} \leq \mathcal{H}^{n-1}(E_\lambda)$ conjectured by Yau.

This paper is the first part of the work, which consists of three parts. Polynomial upper estimates for the Hausdorff measure of the nodal sets in higher dimensions are proved in the second part [9] by a new technique of propagation of smallness. The lower bound in Yau's conjecture is proved in the third part [10] as well as its harmonic counterpart (Nadirashvili's conjecture). We remark that the results in [9, 10] do not give the estimate (1.1) and all three parts can be read independently.

2. Toolbox

2.1. Inequalities for solutions of elliptic equations

Let $(\mathcal{M}, g)$ be a smooth Riemannian manifold and $\Delta_\mathcal{M}$ the Laplace operator on $\mathcal{M}$, which is defined by the metric g. We always assume that the metric is fixed. In the sequel we consider $\mathcal{M} = M \times \mathbb{R}$, where M is a compact manifold with a given metric, on which we study the eigenfunctions, and $\mathcal{M}$ is endowed with the usual metric of the product. Although $\mathcal{M}$ is not compact itself, we will always work on the compact subset $P = M \times [-1, 1]$ of $\mathcal{M}$ where all our estimates are uniform.

A function h on $\mathcal{M}$ is called harmonic if it satisfies the elliptic equation

$$L(h) = \operatorname{div}(\sqrt{g}(g^{ij})\nabla h) = 0 \tag{2.1}$$

in local coordinates. More precisely, the Laplace operator on $\mathcal{M}$ is given by $\Delta_{\mathcal{M}}(f) = \frac{1}{\sqrt{g}}\operatorname{div}(\sqrt{g}(g^{ij})\nabla f)$. Harmonic functions satisfy the maximum and minimum principles and the standard elliptic gradient estimates, see for example [5, Chapter 3]. Further, there exists a constant C such that for any geodesic ball $B(x, r) \subset P$

$$|\nabla h(x)| \le \frac{C}{r} \sup_{B(x,r)} |h|. \tag{2.2}$$

The Harnack inequality holds: if h satisfies (2.1) and $h > 0$ in $B(x, r)$, then for any $y \in B(x, \frac{2}{3}r)$

$$\frac{1}{C}h(y) < h(x) < Ch(y). \tag{2.3}$$

The following consequence of the Harnack inequality will be also used later. If h satisfies (2.1) and $h(x) \ge 0$ then

$$\sup_{B(x,r)} h \ge c \sup_{B(x,\frac{2}{3}r)} |h| \tag{2.4}$$

for some $c = c(\mathcal{M}) > 0$ (it follows from the Harnack inequality applied to the function $\sup_{B(x,r)} h - h$).

2.2. Estimates on the wavelength scale

Now let (M, g_0) be a compact Riemannian manifold. We consider a Laplace eigenfunction u which satisfies $\Delta_M u = -\lambda u$. Adding a new variable, we consider the function $h(\xi, t) = u(\xi)e^{\sqrt{\lambda}t}$ on the product manifold $\mathcal{M} = M \times \mathbb{R}$. The function h turns out to be harmonic on $\mathcal{M}$. This observation can be used to claim that on the wave-scale $\lambda^{-1/2}$ the behavior of the Laplace eigenfunctions resembles that of harmonic functions. This well-known trick was successfully exploited for example in [8, 14, 11].

Let ξ be an arbitrary point on M. Denote by $B(\xi, r) = B_r(\xi)$ the geodesic ball with center at ξ of radius r, when the center of the ball is not important we will omit it in the notation and write B_r.

Lemma 2.1. *There exist a small number $\varepsilon = \varepsilon(M) > 0$ and constants $C_1 = C_1(M)$, $C_2 = C_2(M)$ such that for any eigenfunction u, $\Delta_M u = -\lambda u$, and any $r < \varepsilon\lambda^{-1/2}$ the following inequalities hold*

$$\text{(a)} \sup_{B_r} |u| \le 2 \max_{\partial B_r} |u|, \quad \text{(b)} \sup_{B_{\frac{1}{2}r}} |\nabla u| \le C_1 \max_{\partial B_r} |u|/r. \tag{2.5}$$

If, in addition, $u(\xi) \ge 0$ and (therefore) $A = \max_{\partial B_r(\xi)} u \ge 0$, then

$$\sup_{B_{\frac{2}{3}r}(\xi)} |u| \le C_2 A. \tag{2.6}$$

The inequalities (2.5) and (2.6) follow from the standard elliptic estimates, we provide the proofs for the convenience of the reader.

We work in local coordinates on $\mathcal{M} = M \times \mathbb{R}$ and consider the harmonic function $h(\eta, t) = u(\eta)e^{\sqrt{\lambda}t}$ on $\mathcal{M}$. The Laplace operator corresponds (locally) to an elliptic operator L defined on a bounded subdomain Ω of $\mathbb{R}^{n+1}$, see above. We choose local coordinates such that the distance on the manifold is equivalent to the Euclidean distance (for example by choosing normal coordinates). Denote by $G_{\Omega,L}$ the Green function for L on Ω. By $|x - y|$ we denote the ordinary Euclidean distance between points x and y, locally $|x - y|$ is comparable to the distance between the corresponding points in the Riemannian metric on $\mathcal{M}$. We use the following upper estimate (see [17], [15], [7]) of the Green function:

$$G_{\Omega,L}(x, y) \leq \frac{C}{|x - y|^{d-2}},$$

where $d = n+1$ is the dimension of $\mathcal{M}$. The constant C depends on the coordinate chart on $\mathcal{M}$, we consider a finite set of charts that covers $M \times [-1, 1]$.

Proof of Lemma 2.1. First we suppose that $\sup_{B_r(\xi)} u > 0$ and prove that

$$\sup_{B_r(\xi)} u \leq 2 \max_{\partial B_r(\xi)} u, \tag{2.7}$$

if $r < \varepsilon(M)\lambda^{-1/2}$, where $\varepsilon(M)$ is a sufficiently small positive number, which will be chosen later. Put $A = \max_{\partial B_r(\xi)} u$ and $K = \sup_{B_r(\xi)} u$. Let ξ_0 be a point in the closed ball $\overline{B_r}$, where K is attained.

We consider the cylinder $Q = B_r \times (\frac{-1}{2\sqrt{\lambda}}, \frac{1}{2\sqrt{\lambda}})$ on $\mathcal{M}$. We have $\sup_Q h \leq \sqrt{e}K$ on Q. Without loss of generality we assume that $Q \subset \Omega$. First, we prove $A > 0$.

Suppose that $A \leq 0$ and define the function $w(\xi, t) := A(1 - 4\lambda t^2)/\sqrt{e} + 4\sqrt{e}K\lambda t^2$ and note that $w \geq h$ on ∂Q and $|Lw| < C_3(K - A)\lambda$ on Q for some $C_3 = C_3(M)$. Now, consider the difference $v = h - w$. It is non-positive on ∂Q and satisfies $v(\xi_0, 0) = K - A/\sqrt{e}$ and $|Lv| \leq C_3(K - A)\lambda$ on Q.

We can decompose v into the sum $v = g_1 + g_2$, where g_1 is a non-positive harmonic function in Q with $g_1|_{\partial Q} = v|_{\partial Q}$ and $g_2(y) = \int_Q G_{Q,L}(x, y)Lv(x)dx$. Since $Q \subset \Omega$, the Green function satisfies

$$G_{Q,L}(x, y) \leq G_{\Omega,L}(x, y) \leq C_4/|x - y|^{d-2}.$$

Further, for any $y \in Q$ a simple estimate gives

$$\int_Q |x - y|^{2-d}dx \leq C_5\frac{r}{\sqrt{\lambda}} \leq C_5\frac{\varepsilon}{\lambda}.$$

Combining the estimates, we get

$$g_2(y) = \int_Q G_{Q,L}(x, y)Lv(x)dx \leq C_3C_4(K - A)\lambda \int_Q |x - y|^{2-d}dx \leq C_6(K - A)\varepsilon.$$

Hence $g_2(\xi_0, 0) \leq C_6(K - A)\varepsilon$. The function g_1 is non-positive in Q and therefore

$$K - A/\sqrt{e} = v(\xi_0, 0) = g_1(\xi_0, 0) + g_2(\xi_0, 0) \leq C_6(K - A)\varepsilon.$$

Thus $A(1/\sqrt{e} - C_6\varepsilon) \geq K(1 - C_6\varepsilon)$ and if ε is sufficiently small we obtain that $A > 0$.

Now, we know that $A \geq 0$ and we repeat the argument above with $\tilde{w}(\xi, t) = \sqrt{e}A + 4\sqrt{e}K\lambda t^2$ in place of w and obtain

$$K - \sqrt{e}A \leq C_6 K\varepsilon.$$

If ε is chosen sufficiently small, then (2.7) follows. Inequality (2.5 (a)) follows from (2.7) if one replaces u by $-u$. Finally, the inequalities (2.5 (b)) and (2.6) are obtained by combining (2.5 (a)) with (2.2) and (2.4) respectively, where the last two inequalities are applied to the harmonic function $h(\xi, t) = u(\xi)e^{\sqrt{\lambda}t}$. $\qquad\square$

2.3. Doubling index

Let h be a harmonic function on $\mathcal{M}$. Locally h can be regarded as a solution to the elliptic equation $Lh = 0$. We identify h with a function on the cube $\mathcal{K}_\rho^d = [-\rho, \rho]^d \subset \mathbb{R}^d, d = n + 1$. We choose local geodesic coordinates, then the metric is locally equivalent to the Euclidean one and L is a small perturbation of the Euclidean Laplace operator. Let l be a positive odd integer such that $l > 2\sqrt{d}$, $l = 2l_0 + 1$. For each cube q in $\mathcal{K}_\rho^d$ let lq denote the cube obtained from q by the homothety with the center at the center of q and coefficient l. Suppose that $lq \subset \mathcal{K}_\rho^d$, then we define the doubling index $N(h, q)$ by

$$\int_{lq} |h(x)|^2 dx = 2^{N(h,q)} \int_q |h(x)|^2 dx.$$

The notion of doubling index was used for estimates of the nodal sets in [3, 4, 8] and in many subsequent works. We will need the following properties of the doubling index.

Lemma 2.2.

(i) (L^∞-estimate) If a cube q is inscribed in a ball B (and therefore lq contains $2B$), then

$$\sup_{\frac{4}{3}B} |h| \leq C_7 2^{N(h,q)/2} \sup_B |h|,$$

for some positive $C_7 = C_7(M)$.

(ii) (Monotonicity property) There exists a positive integer $A = A(d)$, a constant $C_0 = C_0(d) > 1$ and a positive number $\rho = \rho(M)$ such that if q_1 and q are cubes that are contained in $\mathcal{K}_\rho^d$, and $Aq_1 \subset q$ then $N(h, q_1) \leq C_0 N(h, q)$.

Proof. (i) Indeed, we have

$$\int_{2B} |h|^2 \leq \int_{lq} |h|^2 = 2^{N(h,q)} \int_q |h|^2 \leq 2^{N(h,q)} \int_B |h|^2.$$

Clearly, $\int_B |h|^2 \leq (\sup_B |h|)^2 |B|$. Further, by an elliptic estimate for h, $\sup_{\frac{4}{3}B} |h| \leq C \left(\int_{2B} |h|^2 \right)^{1/2} |B|^{-1/2}$. The inequality follows.

(ii) The monotonicity property is left without proof. We refer to [6] and [11] for the proof of the monotonicity property of the doubling index defined through integrals over concentric geodesic spheres instead of cubes. Using this, it is not difficult to derive the monotonicity property of doubling index for cubes instead of spheres. □

3. Inscribed balls and a local estimate of the volume of the nodal set

The aim of this section is to estimate from below the volume of the nodal set of an eigenfunction u of the Laplace operator in a geodesic ball of radius comparable to the wavelength $\lambda^{-1/2}$, where $\Delta_M u + \lambda u = 0$. The estimates presented in this section are very far from being sharp.

Let us fix a point O on M and assume $u(O) = 0$. Denote by $|x|$ the distance from the point x to O. We will consider the geodesic ball B_r of radius $r \leq \varepsilon \lambda^{-1/2}$ and with center at O, where $\varepsilon = \varepsilon(M)$ is chosen so that the inequalities (2.5) and (2.6) hold.

Lemma 3.1. *Assume that* $\sup_{B_{\frac{r}{2}}} |u| \leq 2^N \sup_{B_{\frac{r}{4}}} |u|$, *where N is a positive integer,* $N \geq 4$. *Then*

$$\mathcal{H}^{n-1}\{|x| \leq r/2, u(x) = 0\} \geq c r^{n-1} N^{2-n}, \tag{3.1}$$

for some positive $c = c(M)$.

Proof. Applying (2.6), one can deduce

$$\frac{\max_{\partial B_{r/2}} u}{\max_{\partial B_{3r/8}} u} \leq C_2 \frac{\sup_{B_{r/2}} |u|}{\sup_{B_{r/4}} |u|} \leq C_2 2^N.$$

Let $S_j = \{x : |x| = r_j = r(\frac{3}{8} + \frac{j}{8N})\}$, $m_j^+ = \max_{S_j} u$ and $m_j^- = \min_{S_j} u$, $j = 0, 1, \ldots, N$. Recall that u is zero at O. It follows from the weak maximum principle (2.7) that

$$m_j^- < 0, \quad m_j^+ > 0 \quad \text{and} \quad m_j^+ \leq 2m_{j+1}^+, \quad |m_j^-| \leq 2|m_{j+1}^-|.$$

We consider the ratios $\tau_j = m_{j+1}^+/m_j^+$, $j = 0, \ldots, N-1$. Then each $\tau_j \geq 1/2$ and

$$\tau_0 \ldots \tau_{N-1} = \frac{\max_{\partial B_{r/2}} u}{\max_{\partial B_{3r/8}} u} \leq C_2 \frac{\sup_{B_{r/2}} |u|}{\sup_{B_{r/4}} |u|} \leq C_2 2^N.$$

Therefore at most $N/4$ of the ratios $\tau_0, \ldots, \tau_{N-1}$ are greater than a sufficiently large constant $C_3 = C_3(C_2)$. Similarly, at most $N/4$ of the ratios $|m_{j+1}^-|/|m_j^-|$ are greater than C_3. Hence there are at least $N/2$ numbers k, $0 \leq k \leq N-1$ such that $m_{k+1}^+ \leq C_3 m_k^+$ and $|m_{k+1}^-| \leq C_3 |m_k^-|$. We want to show that for each such k there is a ball of radius cr/N and centered on the sphere S_k where u is positive.

Indeed, let x_0 be such that $|x_0| = r_k$ and $u(x_0) = m_k^+ = \max_{\{|x|=r_k\}} u(x)$ and let b be the ball centered at x_0 with radius $\frac{r}{16N}$. Then

$$\sup_b u \leq \max_{\{|x| \leq r_{k+1}\}} u(x) \leq C_1 \max_{\{|x|=r_{k+1}\}} u(x) \leq C_4 m_k^+.$$

Applying (2.6) we see that $\max_{\frac{1}{2}b} |u| \leq C_5 m_k^+$. Taking into account (2.5) (b) and $u(x_0) = m_k^+$, we deduce that u is positive in a smaller ball of radius $c_1 r/N$ centered at x_0.

Similarly, we can find a ball of radius $c_1 r/N$ with center on S_k where u is negative. Thus the spherical layer $\{x : r_{k-1} < |x| < r_{k+1}\}$ contains two balls of radius $c_1 r/N$ where u has opposite signs. Then

$$\mathcal{H}^{n-1}\{x : r_{k-1} < |x| < r_{k+1} : u(x) = 0\} \geq c_2 \left(\frac{r}{N}\right)^{n-1}.$$

The last inequality holds for at least $N/2$ numbers k, so (3.1) follows. $\qquad \square$

4. Combinatorial argument

We need the following lemma about the doubling index defined in Section 2.3. This lemma holds for an arbitrary function $h \in L^2$, not necessarily harmonic.

Lemma 4.1. *Let a cube Q be partitioned into $(Kl)^d$ equal cubes q_i with side length $\frac{1}{Kl}$ (where l is the odd integer from the definition of the doubling index and K is an arbitrary positive integer). Put $N_{\min} = \min_i N(h, q_i)$, the minimum is taken over those cubes q_i of the partition for which $lq_i \subset Q$, and assume that $N_{\min} \geq 2d \ln l/ \ln 2$. Then $N(h, \frac{1}{l}Q) \geq \frac{1}{2}KN_{\min}$.*

Proof. Define $Q_j = \frac{K+j(l-1)}{Kl}Q$ for $j = 0, 1, \ldots, K$, in particular $Q_0 = \frac{1}{l}Q$ and $Q_K = Q$.

We know that $\int_{lq_i} |h|^2 \geq 2^{N_{\min}} \int_{q_i} |h|^2$ for each q_i and therefore

$$2^{N_{\min}} \int_{Q_j} |h|^2 \leq \sum_{q_i \subset Q_j} \int_{lq_i} |h|^2 \leq l^d \int_{Q_{j+1}} |h|^2,$$

since the union of the (open) cubes $lq_i, q_i \subset Q_j$, is contained in Q_{j+1} and covers each point of Q_{j+1} with multiplicity at most l^d.

Further, the inequality $N_{\min} \geq 2d \ln l/ \ln 2$ implies

$$2^{N_{\min}/2} \int_{Q_j} |h|^2 \leq \int_{Q_{j+1}} |h|^2.$$

Finally, multiplying the last inequalities for $j = 0, \ldots, K-1$, we obtain

$$\int_{Q_K} |h|^2 \geq 2^{KN_{\min}/2} \int_{Q_0} |h|^2 = 2^{KN_{\min}/2} \int_{\frac{1}{l}Q} |h|^2. \qquad \square$$

Suppose now that h is a harmonic function on $\mathcal{M} = M \times \mathbb{R}$. Given a cube c, define $\tilde{N}(h,c) = \sup_{c' \subset c} N(h,c')$, where the supremum is taken over all subcubes c' of the cube c. The monotonicity property implies

$$\tilde{N}\left(h, \frac{1}{A}c\right) \leq C_0 N(h,c),$$

when c is contained in $\mathcal{K}_\rho^{d+1}$ and ρ is sufficiently small. If a cube c contains a cube c', then $\tilde{N}(h,c) \geq \tilde{N}(h,c')$.

Our aim is to divide the cube $q = \mathcal{K}_{\rho/l}^{d+1}$ into small cubes and estimate the number of cubes with large doubling constants.

Lemma 4.2. *Let h be a solution to $Lh = 0$ in q. There exist constants $B_0 = B_0(d, L)$ and $\delta = \delta(d) > 0$ such that if the cube q is partitioned into $B > B_0$ equal subcubes, then at least half of these subcubes c satisfy*

$$\tilde{N}(h,c) \leq \max\left\{\frac{\tilde{N}(h,q)}{B^\delta}, \frac{2d\ln l}{\ln 2}\right\}.$$

Proof. Let $N_0 = \tilde{N}(h,q)$. We will do the partition step by step. At the beginning we have one cube q with $\tilde{N}(h,q) = N_0$. We fix A and C_0 from the monotonicity property of the doubling index and choose an integer K such that $K > 4C_0$.

On the first step we divide q into $Y = [lKA]^d$ subcubes. First, divide q into $[lK]^d$ subcubes. By Lemma 4.1 at least one subcube c satisfies $N(h,c) \leq 2N_0/K$ if N_0 is sufficiently large. Then $\tilde{N}(h, \frac{1}{A}c) \leq 2C_0 N_0/K \leq N_0/2$. Thus if we divide q into $[lKA]^d$ subcubes, then at least one subcube will have $\tilde{N} \leq N_0/2$ and all other subcubes will have $\tilde{N} \leq N_0$.

On the second step we will repeat the partition procedure in each subcube c from the first step. Then at least one subcube c' of c will have $\tilde{N}(h,c') \leq \tilde{N}(h,c)/2$. Also $\tilde{N}(h,c'') \leq \tilde{N}(h,c)$ for any other subcube c'' of c.

Going from the $(j-1)$st to the jth step, we take any cube c from the previous step and divide it into Y equal subcubes. In each cube with $\tilde{N}(h,c) \leq N_0/2^s$, $1 \leq s \leq j$, we get at least one cube c' with $\tilde{N}(h,c') \leq N_0/2^{s+1}$ and for other cubes in c we have $\tilde{N}(h,c') \leq N_0/2^s$.

Using the standard induction argument, one can see that on the jth step there is one cube with the doubling index less than or equal to $N_0/2^j$, $\binom{j}{1}(Y-1)$ other cubes with the indices less than or equal to $N_0/2^{j-1}$, and so on, with $\binom{j}{k}(Y-1)^{j-k}$ other cubes with the indices smaller than $N_0/2^k$, $k \geq 0$ (assuming that $N_0/2^j \geq 2d\ln l/\ln 2$). The sum $\sum_{k=0}^{j} \binom{j}{k}(Y-1)^{j-k} = (1 + (Y-1))^j$ is the number of all cubes on the jth step.

Let $\xi_1, \ldots, \xi_j$ be i.i.d. random variables such that $\mathcal{P}(\xi_1 = 1) = 1/Y$ and $\mathcal{P}(\xi_1 = 0) = (Y-1)/Y$. By the law of large numbers

$$\mathcal{P}\left(\frac{\sum_{i=1}^j \xi_i}{j} > \frac{1}{2Y}\right) \to 1 \quad \text{as } j \to \infty.$$

If j is sufficiently large, then

$$\frac{1}{2} \le P\left(\sum_{i=1}^{j} \xi_i \ge \frac{j}{2Y}\right) = \sum_{j \ge k \ge \frac{j}{2Y}} P\left(\sum_{i=1}^{j} \xi_i = k\right) = \sum_{k \ge \frac{j}{2Y}} \binom{j}{k} \frac{(Y-1)^{j-k}}{Y^j}.$$

We conclude that at least half of all cubes on the jth step have doubling indices bounded by $N_0/2^{\frac{j}{2Y}}$. Let $B = [lKA]^{jd} = Y^j$, then $N_0/2^{\frac{j}{2Y}} \le N_0/B^\delta$, where $\delta = \delta(Y)$ is a positive number such that $Y^\delta < 2^{\frac{1}{4Y}}$ and thus δ depends only on the dimension d. Here we have assumed that $j > j_0$ to apply the law of large numbers and we have also assumed that $N_0/B^\delta \ge 2d \ln l / \ln 2$ to apply Lemma 4.1.   □

5. Estimates of the nodal sets of eigenfunctions

5.1. Lower estimate in dimension three

Suppose now that u is a Laplace eigenfunction, $\Delta_M u + \lambda u = 0$ on M, where M is a smooth Riemannian three-dimensional manifold. Using the standard trick, we consider the manifold $\mathcal{M} = M \times \mathbb{R}$ and a new function $h(\xi, t) = u(\xi)e^{\sqrt{\lambda}t}$, which satisfies $\Delta_{\mathcal{M}} h = 0$. We therefore work on a four-dimensional manifold.

We fix a cube Q on M and consider the cube $\tilde{Q} = Q \times I$ on $\mathcal{M}$, where I is the interval centered at the origin with the length equal to the side length of Q, we choose Q sufficiently small such that a chart for Q in normal coordinates is contained in some $\mathcal{K}_\rho^4$.

The Donnelly–Fefferman estimate, see [3], implies that $\tilde{N}(u, Q) \le C\sqrt{\lambda}$ for some $C = C(M)$ if the diameter of Q is less than $c(M)$, and therefore $\tilde{N}(h, \tilde{Q}) \le C_1\sqrt{\lambda}$. See also [11] for the explanation of the Donnelly–Fefferman estimate via the three sphere theorem for harmonic functions.

We partition $\tilde{Q}$ into B smaller cubes $\tilde{q}$ with the side length of order $\lambda^{-1/2}$, such that for each small cube $\tilde{q}$ there is a zero of h within $\frac{1}{10}\tilde{q}$ (it is well known, see [3], that the nodal set of u is $c\lambda^{-1/2}$ dense on M). Then $B \sim [c\sqrt{\lambda}]^4 \sim c_1\lambda^2$ and B is large enough when $\lambda > \lambda_0$.

By Lemma 4.2, half of all small cubes have doubling indices bounded by $C\sqrt{\lambda}/B^\delta \le C_1\lambda^{1/2-2\delta}$. In each small cube of the wavelength size $C/\sqrt{\lambda}$ the doubling index for h is comparable to the doubling index for the function u on the projection of the cube to M, since $h(\xi, t) = u(\xi)e^{\sqrt{\lambda}t}$. Then at least one half of the small cubes of size $C/\sqrt{\lambda}$ in Q have doubling indices bounded by $C_2\lambda^{1/2-2\delta}$. In each such cube q we can find a smaller subcube q' with diameter $\frac{\varepsilon}{\sqrt{\lambda}}$ such that u is equal to 0 at the center of q'. Then combining Lemma 2.2 (i) and the estimate (3.1), we obtain

$$\mathcal{H}^2(\{u=0\} \cap q') \ge \frac{c_2}{\lambda N(u, q')} \ge c_3 \lambda^{-3/2} \lambda^{2\delta}.$$

The number of such cubes is comparable to $\lambda^{3/2}$. Thus $\mathcal{H}^2(\{u=0\}) \ge c_4 \lambda^{2\delta}$.

5.2. Upper estimate in dimension two

Following [4], using local isothermal coordinates in a geodesic disk of radius r, we transform the eigenfunction u, $\Delta_M u + \lambda u = 0$ to a function f in the unit ball of $\mathbb{R}^2$ that satisfies $\Delta_0 f + \lambda r^2 \psi f = 0$, where Δ_0 is the Euclidian Laplacian and ψ is a bounded function (the bound depends on the metric).

We will combine the combinatorial argument from Section 4 with the following estimate for the length of the nodal set by Donnelly and Fefferman, [4]. Let Q be the unit square.

Suppose that $g : Q \to \mathbb{R}$ satisfies $\tilde{N}(g, Q) \leq \Gamma$, $\Delta g = \Gamma \psi g$, where ψ is a function in Q with sufficiently small L^∞-norm. Then

$$\mathcal{H}^1(x \in \tfrac{1}{100}Q : g(x) = 0) \leq C\Gamma.$$

In [4] this estimate was applied on the scale $\lambda^{-1/4}$: for any square q on M with side $\sim \lambda^{-1/4}$ one can consider a function $u(\lambda^{-1/4}x)$ and apply the estimate with $\Gamma \sim \lambda^{1/2}$ (using that the doubling index for any cube is bounded by $C\lambda^{1/2}$) to see that $H^{n-1}(\{u = 0\} \cap q) \leq C\lambda^{1/4}$. Summing the estimates over such cubes covering M, one has $H^1(\{u = 0\}) \leq C\lambda^{3/4}$.

However a combinatorial argument will show that very few cubes with side $\sim \lambda^{-1/4}$ have doubling indices comparable to $\lambda^{1/2}$, in fact, most of the cubes have significantly smaller doubling indices. We are going to refine the global length estimate via combining the combinatorial argument and the Donnelly–Fefferman estimate on various scales.

Lemma 5.1. *Fix a geodesic ball B on the surface with isothermal coordinates and let q be a square in B with side-length $\sim \lambda^{-1/4}$. Then*

$$\mathcal{H}^1(\{u = 0\} \cap q) \leq C\tilde{N}(u, 100q)^{1/2}. \tag{5.1}$$

Proof. Denote $\tilde{N}(u, 100q)$ by N_0. Let us divide q into squares with side-length $\sim N_0^{1/2}\lambda^{-1/2}$. In each of those the doubling index is bounded by N_0 and rescaling such small squares to unit squares and applying the estimate of Donnelly and Fefferman with $\Gamma = N_0$, we bound the length of the nodal set in such a small square by $CN_0^{3/2}\lambda^{-1/2}$. The number of such squares is $\sim \frac{\lambda^{1/2}}{N_0}$. Then $\mathcal{H}^1(\{u = 0\} \cap q) \leq CN_0^{3/2}\lambda^{-1/2}\lambda^{1/2}N_0^{-1} = CN_0^{1/2}$ $\qquad\square$

Recall that $\mathcal{K}_\rho^d = [-\rho, \rho]^d$ and let $\mathcal{K} = \mathcal{K}_\rho^2$ be a square such that $100\mathcal{K}$ lies in (the chart for) B, the side-length of $\mathcal{K}$ depends only on the geometry of the surface M and does not depend on λ. We partition $\mathcal{K}$ into squares with side-length $\lambda^{-1/4}$, then for each such square q we have $\mathcal{H}^1(\{u = 0\} \cap q) \leq C\tilde{N}^{1/2}(u, 100q_0)$ and summing up over all squares q in the partition of $\mathcal{K}$, we obtain

$$\mathcal{H}^1(\{u = 0\} \cap \mathcal{K}) \leq C\sum_{q \subset \mathcal{K}} \tilde{N}^{1/2}(u, 100q).$$

Further, we consider the harmonic extension $h(\xi, t) = u(\xi)e^{\sqrt{\lambda}t}$ of u and let $\tilde{\mathcal{K}} = \mathcal{K}_\rho^3 = \mathcal{K} \times [-\rho, \rho]$. Note that $N(h, \tilde{q}) \geq N(u, q)$, whenever q is the projection of $\tilde{q}$ to M and then the same inequality holds for $\tilde{N}$.

Let Y be a sufficiently large integer defined in Section 4. Choose an integer j such that $Y^j \sim \lambda^{3/4}$. We partition $\tilde{\mathcal{K}}$ into $Y^j \sim \lambda^{3/4}$ subcubes with side-length $\sim \lambda^{-1/4}$. According to Section 4 these cubes can be divided into j groups $G_0, \ldots, G_j$ such that $\tilde{N}(h, \tilde{q}) \le N_0 2^{s-j}$ for each cube $\tilde{q} \in G_s$, where $N_0 := \tilde{N}(h, K_0) \le C\sqrt{\lambda}$ and the number of cubes in G_s is $\binom{j}{s}(Y-1)^s$. However we need to replace $\tilde{N}(h, \tilde{q})$ by $\tilde{N}(h, 100\tilde{q})$ in the estimate for a number a cubes in order to estimate the sum $\sum_{q \subset \mathcal{K}} \tilde{N}^{1/2}(u, 100q)$. It can be done by changing the parameter l in the definition of the doubling index in Section 2.3. The doubling index with a parameter l in a cube $100c$ can be estimated by the doubling index with a parameter $10000l$ in a cube c. We therefore have $\tilde{N}(h, 100\tilde{q}) \le CN_0 2^{s-j}$ for each cube $q \in G_s$, here we abuse the notation $\tilde{N}$ for a doubling index with the modified l and denote it by the same letter.

Finally, we apply the inequality $\tilde{N}(u, q) \le \tilde{N}(h, \tilde{q})$, where q is the projection of $\tilde{q}$, and estimate $\tilde{N}^{1/2}(u, 100q)$ by the average of the corresponding quantities over $Y^{j/3}$ cubes $\tilde{q}$ with the projection q. We obtain

$$\mathcal{H}^1(\{u=0\} \cap \mathcal{K}) \le C \sum_{q \subset \mathcal{K}} \tilde{N}^{1/2}(u, 100q) \le CY^{-j/3} \sum_{\tilde{q} \subset \tilde{\mathcal{K}}} \tilde{N}^{1/2}(h, 100\tilde{q}).$$

Further we partition all cubes $\tilde{q}$ into the groups G_s,

$$\sum_{\tilde{q} \subset \tilde{\mathcal{K}}} \tilde{N}^{1/2}(h, 100\tilde{q}) = \sum_{s=0}^{j} \sum_{\tilde{q} \in G_s} \tilde{N}^{1/2}(h, 100\tilde{q}) \le C\lambda^{1/4} \sum_{s=0}^{j} \binom{j}{s}(Y-1)^s 2^{(-1/2)(j-s)}$$

$$= C\lambda^{1/4}(Y - 1 + 2^{-1/2})^j.$$

We have $Y^j = c\lambda^{3/4}$, then $Y - 1 + 2^{-1/2} = Y^{1-\eta}$ for some $\eta = \eta(Y) > 0$ and $\mathcal{H}^1(\{u=0\}) \le C_M \lambda^{3/4(1-\eta)}$.

Acknowledgment

Both authors were encouraged to study the topic of nodal geometry by Mikhail Sodin, we are very grateful to Mikhail for his constant support, for the numerous ideas and advice that he was generous to share, and for his comments on the earlier versions of the text. We would also like to thank Lev Buhovsky, who read the draft of this paper and made very helpful suggestions and comments.

This work was started when the first author visited NTNU and the second author visited the Chebyshev Laboratory (SPBSU). The work was finished at TAU and Purdue University. We are grateful to these institutions for their hospitality and for great working conditions.

References

[1] J. Brüning, *Über Knoten von Eigenfunktionen des Laplace–Beltrami Operator*, Math Z. **158** (1978), 15–21.

[2] T.H. Colding, W.P. Minicozzi II, *Lower Bounds for Nodal Sets of Eigenfunctions*, Comm. Math. Phys. **306** (2011), 777–784.

[3] H. Donnelly, C. Fefferman, *Nodal sets of eigenfunctions on Riemannian manifolds*, Invent. Math. **93** (1988), 161–183.

[4] H. Donnelly, C. Fefferman, *Nodal sets for eigenfunctions of the Laplacian on surfaces*, J. Amer. Math. Soc. **3** (1990), 333–353.

[5] D. Gilbarg, N.S. Trudinger, *Elliptic Partial Differential Equations of Second Order*, Springer, 1998.

[6] N. Garofalo, F.-H. Lin, *Monotonicity properties of variational integrals, A_p-weights and unique continuation*, Indiana Univ. Math. J. **35** (1986), 245–268.

[7] H. Hueber, M. Sieveking, *Continuous bounds for quotients of Green functions*, Arch. Rational Mech. Anal. **89** (1985), no. 1, 57–82.

[8] F.-H. Lin, *Nodal sets of solutions of elliptic and parabolic equations*, Comm. Pure Appl. Math. **44** (1991), 287–308.

[9] A. Logunov, *Nodal sets of Laplace eigenfunctions: polynomial upper estimates of the Hausdorff measure*, arXiv:1605.02587, to appear in Annals of Math.

[10] A. Logunov, *Nodal sets of Laplace eigenfunctions: proof of Nadirashvili's conjecture and of the lower bound in Yau's conjecture*, arXiv:1605.02589, to appear in Annals of Math.

[11] D. Mangoubi, *The effect of curvature on convexity properties of harmonic functions and eigenfunctions*, J. Lond. Math. Soc. **87** (2013), 645–662.

[12] D. Mangoubi, *On the inner radius of a nodal domain*, Canad. Math. Bull. **51** (2008), no. 2, 249–260.

[13] D. Mangoubi, *Local asymmetry and the inner radius of nodal domains*, Comm. Partial Differential Equations **33** (2008), no. 9, 1611–1621.

[14] F. Nazarov, L. Polterovich, M. Sodin, *Sign and area in nodal geometry of Laplace eigenfunctions*, Amer. J. Math. **127** (2005), 879–910.

[15] J. Serrin, *On the Harnack inequality for linear elliptic equations*, J. Anal. Math. **4** (1954–1956) no. 1, 292–308.

[16] C.D. Sogge, S. Zelditch, *Lower bounds on the Hausdorff measure of nodal sets II*, Math. Res. Lett. **19** (2012), 1361–1364.

[17] K.-O. Widman, *Inequalities for the Green function and boundary continuity of the gradient of solutions of elliptic differential equations.* Math. Scand. **21** (1967), 17–37.

[18] S.-T. Yau, *Problem section, Seminar on Differential Geometry*, Annals of Mathematical Studies 102, Princeton, 1982, 669–706.

Alexander Logunov
School of Mathematical Sciences
Tel Aviv University
Tel Aviv 69978, Israel

and

Chebyshev Lab.
St. Petersburg State University
14th Line V.O., 29B
St. Petersburg 199178 Russia
e-mail: log239@yandex.ru

Eugenia Malinnikova
Dept. of Mathematical Sciences
Norwegian University
of Science and Technology
N-7491 Trondheim, Norway

and

Dept. of Mathematics
Purdue University
150 N. University st.
West Lafayette, IN 47907-2067, USA
e-mail:
eugenia.malinnikova@ntnu.no

Operator Theory:
Advances and Applications, Vol. 261, 345–387
© Springer International Publishing AG, part of Springer Nature 2018

Differentiability of Solutions to the Neumann Problem with Low-regularity Data via Dynamical Systems

Vladimir Maz'ya and Robert McOwen

In memory of Victor Havin

Abstract. We obtain conditions for the differentiability of weak solutions for a second-order uniformly elliptic equation in divergence form with a homogeneous co-normal boundary condition. The modulus of continuity for the coefficients is assumed to satisfy the square-Dini condition and the boundary is assumed to be differentiable with derivatives also having this modulus of continuity. Additional conditions for the solution to be Lipschitz continuous or differentiable at a point on the boundary depend upon the stability of a dynamical system that is derived from the coefficients of the elliptic equation.

Mathematics Subject Classification (2010). Primary 35A08.

Keywords. Differentiability, Lipschitz continuity, weak solution, elliptic equation, divergence form, co-normal boundary condition, modulus of continuity, square-Dini condition, dynamical system, asymptotically constant, uniformly stable.

0. Introduction

For $n \geq 2$, let U be a Lipschitz domain in $\mathbb{R}^n$ with exterior unit normal ν on ∂U. Given a point $p \in \partial U$, let B be an open ball centered at p. We want to consider solutions of the uniformly elliptic equation in divergence form in $U \cap B$ with homogeneous co-normal boundary condition on $\partial U \cap B$:

$$\partial_i(a_{ij}\,\partial_j u) = 0 \quad \text{in} \quad U \cap B,$$
$$\nu_i\, a_{ij}\,\partial_j u = 0 \quad \text{on} \quad \partial U \cap B. \tag{1}$$

(Here and throughout this paper we use the summation convention on repeated indices.)

Let $C^1_{\text{comp}}(\overline{U} \cap B) = \{u \in C^1(\overline{U} \cap B) : \operatorname{supp} u \text{ is compact in } \overline{U} \cap B\}$. Recall that a *weak solution* of (1) is a function $u \in H^{1,2}(U \cap B)$, i.e., ∇u is square-integrable on $U \cap B$, that satisfies

$$\int_U a_{ij}\, \partial_j u\, \partial_i \eta \, dx = 0 \quad \text{for all } \eta \in C^1_{\text{comp}}(\overline{U} \cap B). \tag{2}$$

However, for irregular coefficients a_{ij}, a weak solution of (1) need not have a well-defined normal derivative along ∂U, so the boundary condition in (1) is not meaningful, and we must only work with the variational formulation (2). When the coefficients a_{ij} are bounded and measurable, the classical results of Stampacchia [16] show that a solution of (2) is Hölder continuous on $\overline{U} \cap B$. We want to consider mild regularity conditions on a_{ij} and the boundary ∂U under which a solution of (2) must be Lipschitz continuous, or even differentiable, at a given point of $\overline{U} \cap B$.

We shall assume that the modulus of continuity ω for the coefficients satisfies the *square-Dini condition*

$$\int_0^1 \omega^2(r)\, \frac{dr}{r} < \infty. \tag{3}$$

Under this condition, the regularity of weak solutions at interior points of $U \cap B$ was investigated in [14], and found to also depend upon the stability of a first-order dynamical system derived from the coefficients. In this paper, we shall investigate the regularity of weak solutions at points on $\partial U \cap B$ and find somewhat analogous results. Without loss of generality, we may assume that the boundary point is the origin $x = 0$, and by a change of independent variables we may arrange $a_{ij}(0) = \delta_{ij}$. It turns out that the conditions for differentiability at a boundary point depend upon both the coefficients a_{ij} and the shape of the boundary ∂U in a rather complicated way, so for the purposes of describing our results in this introduction, let us consider two special cases: **I.** When the boundary is flat near 0. **II.** When the operator is just the Laplacian near 0.

I. Since our results are local in nature, we may assume the domain is the half-space

$$\mathbb{R}^n_+ = \{(\widetilde{x}, x_n) : x_n > 0\} = \{(x_1, \ldots, x_{n-1}, x_n) : x_n > 0\}.$$

We assume $u \in H^{1,2}_{\text{loc}}(\overline{\mathbb{R}^n_+})$, i.e., first-order derivatives are integrable over compact subsets of $\overline{\mathbb{R}^n_+}$. For $x \in \mathbb{R}^n_+$, let us write $x = r\,\theta$ where $r = |x|$ and $\theta \in S^{n-1}_+ = \{x \in \mathbb{R}^n_+ : |x| = 1\}$. We shall find that the relevant first-order dynamical system is

$$\frac{d\varphi}{dt} + R(e^{-t})\,\varphi = 0 \quad \text{for } T < t < \infty, \tag{4}$$

where $R(r)$ is the $(n-1) \times (n-1)$ matrix given by

$$[R(r)]_{\ell k} := \fint_{S^{n-1}_+} \left(a_{\ell k}(r\theta) - n \sum_{j=1}^n a_{\ell j}(r\theta)\, \theta_j\, \theta_k \right) ds_\theta \quad \text{for } \ell, k = 1, \ldots, n-1. \tag{5}$$

Here and throughout the paper, the slashed integral denotes mean value. Following [2], we say that (4) is *uniformly stable* as $t \to \infty$ if for every $\varepsilon > 0$ there exists a $\delta = \delta(\varepsilon) > 0$ such that any solution ϕ of (4) satisfying $|\phi(t_1)| < \delta$ for some $t_1 > 0$ satisfies $|\phi(t)| < \varepsilon$ for all $t \geq t_1$. (Since (4) is linear, an equivalent condition for uniform stability may be formulated in terms of the fundamental matrix; cf. Remark 2 in Appendix E.) Moreover, a solution of (4) is *asymptotically constant* as $t \to \infty$ if there is a constant vector ϕ_∞ such that $\phi(t) \to \phi_\infty$ as $t \to \infty$. As discussed in [14], these two stability conditions are independent of each other, but if $R(r)\,r^{-1} \in L^1(0,\varepsilon)$, then (4) is both uniformly stable and all solutions are asymptotically constant; in particular, if ω satisfies the Dini condition then both conditions are met. As we shall see in Theorem 1 in Section 2: *if* (4) *is uniformly stable as $t \to \infty$, then every solution $u \in H_{\mathrm{loc}}^{1,2}(\mathbb{R}_+^n)$ of* (2) *with $U = \mathbb{R}_+^n$ is Lipschitz continuous at $x = 0$; if, in addition, every solution of the dynamical system* (4) *is asymptotically constant as $t \to \infty$, then u is differentiable at $x = 0$.* Examples show (cf. Section 4) that solutions of (2) need not be Lipschitz continuous at $x = 0$ if the dynamical system (4) is not uniformly stable as $t \to \infty$.

II. We assume $a_{ij} = \delta_{ij}$ and U is a Lipschitz domain whose curved boundary ∂U contains $x = 0$, and let B denote a ball centered at 0. By a rotation of the independent coordinates, we may assume that ∂U is given near $x = 0$ as the graph of a Lipschitz function h, i.e., $x_n = h(\widetilde{x})$ where $h(\widetilde{0}) = 0$. Since Lipschitz functions are differentiable almost everywhere, its gradient $\widetilde{\nabla} h$ is well defined. We need $\widetilde{\nabla} h$ to satisfy the condition

$$\sup_{|\widetilde{x}|=r} |\widetilde{\nabla} h(\widetilde{x})| \leq \omega(r) \text{ as } r \to 0, \tag{6}$$

where $\omega(r)$ satisfies the square Dini condition (3). We again require stability properties of the dynamical system (4), but now the $(n-1) \times (n-1)$ matrix $R(r)$ is given by

$$[R(r)]_{\ell k} = n \int_{S_+^{n-1}} \frac{\partial h(r\theta)}{\partial x_\ell} \theta_n \theta_k \, ds_\theta. \tag{7}$$

As a special case of Theorem 2 in Section 3, we have: *if* (4), (7) *is uniformly stable as $t \to \infty$, then every solution $u \in H^{1,2}(U \cap B)$ of* (2) *is Lipschitz continuous at $x = 0$; if, in addition, every solution of the dynamical system* (4), (7) *is asymptotically constant as $t \to \infty$, then u is differentiable at $x = 0$.*

It is possible to obtain analytic conditions at p that imply the desired stability of (4); we can even obtain conditions under which a solution of (2) must have a critical point, i.e., $\nabla u(p) = 0$. For $n = 2$, of course, (4) is a scalar equation, so conditions for uniform stability and solutions being asymptotically constant are easily obtained; this is done in Section 4. For $n > 2$, conditions may be obtained in terms of the largest eigenvalue $\mu(r)$ of the symmetric matrix $S(r) = -\frac{1}{2}(R(r) + R^t(r))$, where R^t denotes the transpose of R. Let us mention two conditions on $\mu(r)$:

$$\int_{r_1}^{r_2} \mu(\rho) \frac{d\rho}{\rho} < K \quad \text{for all } 0 < r_1 < r_2 < \varepsilon \tag{8}$$

and

$$\int_r^\varepsilon \mu(\rho) \frac{d\rho}{\rho} \to -\infty \quad \text{as } r \to 0. \tag{9}$$

As an application to **I**, if R is defined by (5), then we show in Section 2 that: (8) *implies that every solution* $u \in H_{\mathrm{loc}}^{1,2}(\mathbb{R}_+^n)$ *of* (2) *with* $U = \mathbb{R}_+^n$ *is Lipschitz continuous at* $x = 0$ *and* (9) *implies that* u *is differentiable at* $x = 0$ *with* $\partial_j u(0) = 0$ *for* $j = 1, \ldots, n$. As an application to **II**, if $a_{ij} = \delta_{ij}$ and R is defined by (7), then the results in Section 3 show that: (8) *implies that every solution* $u \in H_{\mathrm{loc}}^{1,2}(U \cap B)$ *of* (2) *is Lipschitz continuous at* $x = 0$ *and* (9) *implies that* u *is differentiable at* $x = 0$ *with* $\partial_j u(0) = 0$ *for* $j = 1, \ldots, n$. Additional analytic conditions on the matrix R itself that imply the desired stability of (4) may be found in [14], but we shall not discuss them further since they apply in general to the dynamical system (4) and are not peculiar to the Neumann problem that we consider here.

Now let us say something about the methods used to prove these results. First we note that the modulus of continuity $\omega(r)$ is a continuous, nondecreasing function of r near $r = 0$, and we need to assume that ω does not vanish as fast as r when $r \to 0$, i.e., for some $\kappa > 0$

$$\omega(r) r^{-1+\kappa} \text{ is nonincreasing for } r \text{ near } 0. \tag{10}$$

Our analysis of regularity at $0 \in \partial \mathbb{R}_+^n$ is analogous to the analysis in [14] for an interior point, and we shall adopt similar notation to make the parallels clear. In particular, we use a decomposition

$$u(x) = u_0(r) + \tilde{v}(r) \cdot \tilde{x} + w(x), \tag{11a}$$

where the scalar function u_0 and $(n-1)$-vector function $\tilde{v} = (v_1, \ldots, v_{n-1})$ are given by

$$u_0(r) := \fint_{S_+^{n-1}} u(r\theta) \, ds_\theta, \quad v_k(r) := \frac{n}{r} \fint_{S_+^{n-1}} u(r\theta) \theta_k \, ds_\theta \text{ for } k = 1, \ldots, n-1. \tag{11b}$$

Note that the scalar function w has zero mean and first moments on the half-sphere:

$$\fint_{S_+^{n-1}} w(r\theta) \, ds_\theta = 0 = \fint_{S_+^{n-1}} w(r\theta) \theta_k \, ds_\theta \quad \text{for } k = 1, \ldots, n-1. \tag{11c}$$

As we shall see, the assumption that the dynamical system (4), (5) is uniformly stable as $t \to \infty$ not only implies that $\tilde{v}$ and $\tilde{x} \cdot \tilde{v}'$ are bounded as $r \to 0$, but that $|u_0(r) - u_0(0)|$ and $|w(x)|$ are both bounded by $r\,\omega(r)$ as $r \to 0$. Thus we have

$$u(x) = u(0) + \tilde{v}(r) \cdot \tilde{x} + O(r\,\omega(r)) \quad \text{as } r \to 0$$

with $\tilde{v}(r)$ bounded, which shows that u is Lipschitz at $x = 0$. If we also know that all solutions of (4), (5) are asymptotically constant as $t \to \infty$, then we shall show $\tilde{v}(r) = \tilde{v}(0) + o(1)$ as $r \to 0$, which proves that u is differentiable at $x = 0$.

It could be of interest to compare our results on the differentiability of so-lutions to the Neumann problem with asymptotic expansions that have been ob-tained for solutions of the Dirichlet problem (cf. [10] and [11] which more generally consider elliptic operators of order $2m$). It is important to observe that the anal-ysis at a boundary point for the Neumann problem is more complicated than it is for the Dirichlet problem. The reason for this can be clearly seen in the case of $\mathbb{R}^n_+$ for $n \geq 3$: the dynamical system that controls the behavior of $\tilde{v}$ in the decom-position (11a) is $(n-1)$-dimensional, while the corresponding decomposition for the Dirichlet problem involves only the coefficient of x_n, and so leads to a scalar ODE.

Let us mention that the square-Dini condition has been encountered in a va-riety of contexts: the differentiability of functions [18], Littlewood–Paley estimates for parabolic equations [5], and the absolute continuity of elliptic measure and L^2-boundary conditions for the Dirichlet problem [1], [4], [6], [9]. In addition, let us observe that the projection methods used here were not only used in [14] but also in [12] and [13].

1. A model problem for the Laplacian in a half-space

In this section, we consider (2) when the operator is the Laplacian and $U = \mathbb{R}^n_+$. However, in order for these results to be useful in our study of variable coefficients, we need to introduce some inhomogeneous terms to our variational problem. We assume that $\vec{f}, f_0 \in L^p_{\text{loc}}(\overline{\mathbb{R}^n_+})$ for some $p > n$, i.e., f is L^p-integrable over any compact set $K \subset \overline{\mathbb{R}^n_+}$. For $p' = p/(p-1)$ let

$$H^{1,p'}_{\text{comp}}(\overline{\mathbb{R}^n_+}) := \{\eta \in H^{1,p'}(\mathbb{R}^n_+) : \eta(x) = 0 \text{ for all sufficiently large } |x|\},$$

and define

$$F[\eta] = \int_{\mathbb{R}^n_+} (f_0\eta - \vec{f} \cdot \nabla\eta)\, dx \quad \text{for } \eta \in H^{1,p'}_{\text{comp}}(\overline{\mathbb{R}^n_+}). \tag{12a}$$

We now want to find a solution $u \in H^{1,p}_{\text{loc}}(\overline{\mathbb{R}^n_+})$ of the variational problem

$$\int_{\mathbb{R}^n_+} \nabla u \cdot \nabla\eta\, dx + F[\eta] = 0 \quad \text{for all } \eta \in H^{1,p'}_{\text{comp}}(\overline{\mathbb{R}^n_+}). \tag{12b}$$

We can obtain the solution using the Neumann function $N(x,y)$, which is a fun-damental solution for Δ satisfying $\partial N/\partial x_n = 0$ for $x \in \partial\mathbb{R}^n_+$ and $y \in \mathbb{R}^n_+$. Using the method of reflection, it can be written as

$$N(x,y) = \Gamma(x - y) + \Gamma(x - y^*), \tag{13}$$

where $\Gamma(x)$ is the standard fundamental solution for the Laplacian Δ and $y^* = (\tilde{y}, -y_n)$ is the reflection in the boundary of $y = (\tilde{y}, y_n) \in \mathbb{R}^n_+$. Using $N(x,y)$, we obtain the solution of (12) as follows. First, replace $\eta(y)$ in (12b) by $\chi_R(y)N(x,y)$ (for fixed x), where $\chi_R(y) = \chi(|y|/R)$ with a smooth cut-off function $\chi(t)$ satisfying $\chi(t) = 1$ for $t < 1$ and $\chi(t) = 0$ for $t > 2$. This can be done since $\nabla N(x,y) = O(|x-$

$y|^{1-n}$) as $|x - y| \to 0$ implies (for fixed x) we have $\chi_R(y)N(x,y) \in H^{1,q}_{\text{comp}}(\mathbb{R}^n_+)$ for all $q < n/(n-1)$. Since $p > n$ is equivalent to $p' < n/(n-1)$, we have $\chi_R(y)N(x,y) \in H^{1,p'}_{\text{comp}}(\mathbb{R}^n_+)$ and, provided the functions $\vec{f}$ and f_0 decay sufficiently as $|x| \to \infty$, we can let $R \to \infty$ to obtain the following solution formula for the problem (12):

$$u(x) = \int_{\mathbb{R}^n_+} \left(N(x,y)f_0(y) - \nabla_y N(x,y) \cdot \vec{f}(y) \right) dy. \tag{14}$$

For example, (14) is the solution for (12) if $f_0, \vec{f}$ have compact support in $\overline{\mathbb{R}^n_+}$.

In fact, we shall require a further refinement of (12), but first we need to discuss projections. For $g \in L^1_{\text{loc}}(\overline{\mathbb{R}^n_+} \setminus \{0\})$ and $r > 0$, let $Pg(r,\theta)$ denote the projection of $g(r\theta)$ to the functions on S^{n-1}_+ spanned by $1, \theta_1, \ldots, \theta_{n-1}$:

$$Pg(r,\theta) := \fint_{S^{n-1}_+} g(r\phi)\, ds_\phi + n \sum_{m=1}^{n-1} \theta_m \fint_{S^{n-1}_+} \phi_m\, g(r\phi)\, ds_\phi, \tag{15a}$$

where we have used

$$\fint_{S^{n-1}_+} \theta_m^2\, ds_\theta = \frac{1}{n} \quad \text{for } m = 1, \ldots, n.^{[1]} \tag{15b}$$

Note that $P1 = 1$ and $P\theta_m = \theta_m$ for $m = 1, \ldots, n-1$. For $k \geq 1$, if $g \in C^k(\overline{\mathbb{R}^n_+} \setminus \{0\})$ then we can easily check that $Pg \in C^k((0,\infty) \times S^{n-1}_+)$; moreover, we have $\partial(Pg)/\partial x_n = 0$ on $\mathbb{R}^{n-1} \setminus \{0\}$ since there is no θ_n-term in the definition of Pg. Let us summarize this last remark in the following:

Lemma 1. *For $k \geq 1$, if $u \in C^k(\overline{\mathbb{R}^n_+} \setminus \{0\})$ then $Pu \in C^k((0,\infty) \times S^{n-1}_+)$ and $\partial Pu/\partial x_n = 0$ on $\mathbb{R}^{n-1} \setminus \{0\}$.*

If $g \in L^1_{\text{loc}}(\overline{\mathbb{R}^n_+})$ and f is a bounded function with compact support in $\overline{\mathbb{R}^n_+}$, then Pf also has compact support and hence the product $g\, Pf$ is integrable on $\mathbb{R}^n_+$. In fact, it is easy to see by Fubini's theorem that

$$\int_{\mathbb{R}^n_+} g\, Pf\, dx = \int_{\mathbb{R}^n_+} f\, Pg\, dx. \tag{16}$$

Of course, (16) also holds if $g \in L^p_{\text{loc}}(\overline{\mathbb{R}^n_+})$ and $f \in L^{p'}_{\text{comp}}(\overline{\mathbb{R}^n_+})$. In particular, if $g, g_j \in L^p_{\text{loc}}(\overline{\mathbb{R}^n_+})$ satisfy $\int fg_j\, dx \to \int fg\, dx$ for every $f \in L^{p'}_{\text{comp}}(\overline{\mathbb{R}^n_+})$, then $\int fPg_j\, dx \to \int fPg\, dx$ for every $f \in L^{p'}_{\text{comp}}(\overline{\mathbb{R}^n_+})$. In fact, we claim more.

Lemma 2. *If $p > n$ and $g, g_j \in H^{1,p}_{\text{loc}}(\overline{\mathbb{R}^n_+})$ satisfy $\int_{\mathbb{R}^n_+} \vec{f} \cdot \nabla g_j\, dx \to \int_{\mathbb{R}^n_+} \vec{f} \cdot \nabla g\, dx$ for every $\vec{f} \in L^{p'}_{\text{comp}}(\overline{\mathbb{R}^n_+})$, then $\int_{\mathbb{R}^n_+} \vec{f} \cdot \nabla Pg_j\, dx \to \int_{\mathbb{R}^n_+} \vec{f} \cdot \nabla Pg\, dx$ for every $\vec{f} \in L^{p'}_{\text{comp}}(\overline{\mathbb{R}^n_+})$.*

---

[1] To verify (15b), note that $\theta_1^2 + \cdots + \theta_n^2 = 1$ implies $\int_{S^{n-1}} \theta_i^2 ds = |S^{n-1}|/n$ for $i = 1, \ldots, n$. In particular, $\int_{S^{n-1}_+} \theta_n^2 ds = |S^{n-1}|/2n = |S^{n-1}_+|/n$, and for $i = 1, \ldots, n-1$, $|S^{n-1}_+| = (n-1)\int_{S^{n-1}_+} \theta_i^2\, ds + |S^{n-1}_+|/n$, which yields (15b).

Proof. Since $p > n$ we have $H^{1,p}_{\mathrm{loc}}(\overline{\mathbb{R}^n_+}) \subset C(\overline{\mathbb{R}^n_+})$, and by density we may assume that $\vec{f} \in C^1_{\mathrm{comp}}(\overline{\mathbb{R}^n_+})$. We can integrate by parts, and apply the above argument with $f = \mathrm{div}\,\vec{f}$ in $\mathbb{R}^n_+$ and f_n in $\mathbb{R}^{n-1}$:

$$
\int_{\mathbb{R}^n_+} \vec{f} \cdot \nabla P(g_j)\, dx = -\int_{\mathbb{R}^n_+} \mathrm{div}\,\vec{f}\, P(g_j)\, dx + \int_{\mathbb{R}^{n-1}} (f_n\, P(g_j))|_{x_n=0}\, d\tilde{x}
$$

$$
= -\int_{\mathbb{R}^n_+} P(\mathrm{div}\,\vec{f})\, g_j\, dx + \int_{\mathbb{R}^{n-1}} (P(f_n)\, g_j)|_{x_n=0}\, d\tilde{x}
$$

$$
\to -\int_{\mathbb{R}^n_+} P(\mathrm{div}\,\vec{f})\, g\, dx + \int_{\mathbb{R}^{n-1}} (P(f_n)\, g)|_{x_n=0}\, d\tilde{x} \qquad (17)
$$

$$
= -\int_{\mathbb{R}^n_+} \mathrm{div}\,\vec{f}\, Pg\, dx + \int_{\mathbb{R}^{n-1}} (f_n\, Pg)|_{x_n=0}\, d\tilde{x}
$$

$$
= \int_{\mathbb{R}^n_+} \vec{f} \cdot \nabla Pg\, dx.
$$

$\qquad\qquad\qquad\qquad\qquad\qquad\qquad\qquad\qquad\qquad\qquad\qquad\qquad\qquad$ $\square$

Note that (16) also enables us to define P on distributions; for example, for F as in (12a) we have $PF[\eta] = F[P\eta]$ for any $\eta \in H^{1,p'}_{\mathrm{comp}}(\overline{\mathbb{R}^n_+})$. Now, for a function or distribution g, let us define

$$
g^\perp = (I - P)g. \qquad (18)
$$

In particular, for F as in (12a), we can define the functional $F^\perp$:

$$
F^\perp[\eta] := \int_{\mathbb{R}^n_+} (f_0\, \eta^\perp - \vec{f} \cdot \nabla(\eta^\perp))\, dx \quad \text{for} \ \ \eta \in H^{1,p'}_{\mathrm{comp}}(\overline{\mathbb{R}^n_+}). \qquad (19a)
$$

Now we can state the required refinement of (12): to find $w \in H^{1,p}_{\mathrm{loc}}(\overline{\mathbb{R}^n_+})$ satisfying $Pw = 0$ and

$$
\int_{\mathbb{R}^n_+} \nabla w \cdot \nabla \eta\, dx + F^\perp[\eta] = 0 \quad \text{for all } \eta \in H^{1,p'}_{\mathrm{comp}}(\overline{\mathbb{R}^n_+}). \qquad (19b)
$$

We will need the projection P of $N(x, y)$ with respect to y (i.e., for fixed x). To compute this, we first expand $N(x, y)$ in spherical harmonics $\{\tilde{\varphi}_{k,m} : m = 1, \ldots, \tilde{N}(k) \text{ and } k = 0, \ldots\}$ on S^{n-1}, where $\tilde{N}(k)$ is the dimension of the space of spherical harmonics that are even in x_n. In fact, as we show in Appendix A, assuming $n \geq 3$ this yields

$$
N(x, y) = \frac{a_0}{|y|^{n-2}} + \frac{a_0(n-2)}{c_n} \frac{|x|}{|y|^{n-1}} \sum_{m=1}^{n-1} \hat{x}_m \hat{y}_m
$$

$$
+ \sum_{k=2}^{\infty} \frac{|x|^k}{|y|^{n-2+k}} \sum_{m=1}^{\tilde{N}(k)} a_{k,m}\, \tilde{\varphi}_{k,m}(\hat{x})\, \tilde{\varphi}_{k,m}(\hat{y}) \quad \text{for } |x| < |y|, \qquad (20a)
$$

and

$$N(x,y) = \frac{a_0}{|x|^{n-2}} + \frac{a_0(n-2)}{c_n} \frac{|y|}{|x|^{n-1}} \sum_{m=1}^{n-1} \hat{x}_m \hat{y}_m$$

$$+ \sum_{k=2}^{\infty} \frac{|y|^k}{|x|^{n-2+k}} \sum_{m=1}^{\tilde{N}(k)} a_{k,m} \, \tilde{\varphi}_{k,m}(\hat{x}) \, \tilde{\varphi}_{k,m}(\hat{y}) \quad \text{for } |y| < |x|. \tag{20b}$$

The coefficients $a_0, a_{k,m}$ can be computed but their values are not important to us now; and we have here used the notation $\hat{x} = x/|x|$ and $\hat{y} = y/|y|$ (although elsewhere we have used $\theta = x/|x|$). If we denote the projection P of $N(x,y)$ with respect to y simply by $PN(x,y)$, then we have

$$PN(x,y) = \begin{cases} \frac{a_0}{|y|^{n-2}} + \frac{a_0(n-2)}{c_n} \frac{|x|}{|y|^{n-1}} \sum_{m=1}^{n-1} \hat{x}_m \hat{y}_m & \text{for } |x| < |y|, \\ \frac{a_0}{|x|^{n-2}} + \frac{a_0(n-2)}{c_n} \frac{|y|}{|x|^{n-1}} \sum_{m=1}^{n-1} \hat{x}_m \hat{y}_m & \text{for } |y| < |x|. \end{cases} \tag{21}$$

We can also define

$$N^{\perp}(x,y) = N(x,y) - PN(x,y)$$

$$= \begin{cases} \sum_{k=2}^{\infty} \frac{|x|^k}{|y|^{n-2+k}} \sum_{m=1}^{\tilde{N}(k)} a_{k,m} \, \tilde{\varphi}_{k,m}(\hat{x}) \, \tilde{\varphi}_{k,m}(\hat{y}) & \text{for } |x| < |y|, \\ \sum_{k=2}^{\infty} \frac{|y|^k}{|x|^{n-2+k}} \sum_{m=1}^{\tilde{N}(k)} a_{k,m} \, \tilde{\varphi}_{k,m}(\hat{x}) \, \tilde{\varphi}_{k,m}(\hat{y}) & \text{for } |y| < |x|. \end{cases} \tag{22}$$

Using the same argument as for (14), provided the functions $\vec{f}$ and f_0 decay sufficiently as $|x| \to 0$ and $|x| \to \infty$, we have the following solution formula for the problem (19):

$$w(x) = \int_{\mathbb{R}_+^n} \left(N^{\perp}(x,y) f_0(y) - \nabla_y N^{\perp}(x,y) \cdot \vec{f}(y) \right) dy. \tag{23}$$

For example, (23) holds if $f_0, \vec{f}$ have compact support in $\overline{\mathbb{R}_+^n}$.

Let us now obtain estimates on the solution of (19) given by (23) when we make certain assumptions about the decay of f_0 and $\vec{f}$ as $|x| \to 0$ and $|x| \to \infty$. We do so using the L^p-mean on annuli: for $r > 0$ define

$$M_p(w,r) = \left(\fint_{A_r^+} |w(x)|^p \, dx \right)^{1/p} \quad \text{where } A_r^+ = \{x \in \mathbb{R}_+^n : r < |x| < 2r\}. \tag{24a}$$

Using this, we can also define

$$M_{1,p}(w,r) = r M_p(\nabla w, r) + M_p(w,r). \tag{24b}$$

Proposition 1. *Suppose F is the distribution (12a) where $\vec{f}, f_0 \in L_{\text{loc}}^p(\overline{\mathbb{R}_+^n} \setminus \{0\})$ for $p > n$ satisfy*

$$\int_{\{x \in \mathbb{R}_+^n : |x| < 1\}} (|\vec{f}(x)| + |x f_0(x)|) |x| \, dx + \int_{\{x \in \mathbb{R}_+^n : |x| > 1\}} (|\vec{f}(x)| + |x f_0(x)|) \, |x|^{-1-n} \, dx < \infty.$$

Then (23) *defines a solution* $w \in H^{1,p}_{\text{loc}}(\mathbb{R}^n_+ \setminus \{0\})$ *of* (19) *that satisfies* $Pw = 0$ *and*

$$M_{1,p}(w,r) \le c\left(r^{-n} \int_0^r \left[M_p(\vec{f},\rho)\rho^n + M_p(f_0,\rho)\rho^{n+1}\right] d\rho\right.$$
$$\left. + r^2 \int_r^\infty \left[M_p(\vec{f},\rho)\rho^{-2} + M_p(f_0,\rho)\rho^{-1}\right] d\rho\right).$$

Proof. To obtain the desired estimates, let us assume that $n \ge 3$, $r < |x| < 2r$, and introduce the annulus $\widetilde{A}^+_r = \{x \in \mathbb{R}^n_+ : r/2 < |x| < 4r\}$. Then let us split the solution (23) into several parts:

$$w(x) = \int_{\widetilde{A}^+_r} \left(N(x,y)f_0(y) - \nabla_y N(x,y) \cdot \vec{f}(y)\right) dy$$
$$- \int_{r/2<|y|<|x|} \left(PN(x,y)f_0(y) - \nabla_y PN(x,y) \cdot \vec{f}(y)\right) dy$$
$$- \int_{|x|<|y|<4r} \left(PN(x,y)f_0(y) - \nabla_y PN(x,y) \cdot \vec{f}(y)\right) dy$$
$$+ \int_{|y|<r/2} \left(N^\perp(x,y)f_0(y) - \nabla_y N^\perp(x,y) \cdot \vec{f}(y)\right) dy$$
$$+ \int_{|y|>4r} \left(N^\perp(x,y)f_0(y) - \nabla_y N^\perp(x,y) \cdot \vec{f}(y)\right) dy$$
$$= w_1(x) + w_2(x) + w_3(x) + w_4(x) + w_5(x).$$

(Here, and subsequently, by an integral such as $\int_{|y|<r/2}$ we actually mean the integral over $\{y \in \mathbb{R}^n_+ : |y| < r/2\}$.) We estimate each of these terms separately.

The first term, w_1, can be estimated using classical results. For example, we can apply Theorem B* in [17] with $\lambda = n - 1$, $\alpha = 1$, $\beta = 0$, and $p = q > n$ (which implies $p' < n$) to obtain

$$\left\| \int_{\widetilde{A}^+_r} \nabla_y N(x,y) \cdot \vec{f}(y)\, dy \right\|_{L^p(A^+_r)} \le cr\, \|\vec{f}\|_{L^p(\widetilde{A}^+_r)}.$$

The same argument shows that

$$r \left\| \int_{\widetilde{A}^+_r} \frac{\partial}{\partial x_i} N(x,y) f_0(y)\, dy \right\|_{L^p(A^+_r)} \le cr^2\, \|f_0\|_{L^p(\widetilde{A}^+_r)}.$$

We can also apply Theorem B* in [17] with $\lambda = n-2$, $\alpha = 2$, $\beta = 0$, and $p = q > n$ to obtain

$$\left\| \int_{\widetilde{A}^+_r} N(x,y) f_0(y)\, dy \right\|_{L^p(A^+_r)} \le cr^2\, \|f_0\|_{L^p(\widetilde{A}^+_r)}.$$

Finally, we apply the L^p-boundedness of singular integral operators to obtain

$$\left\| \int_{\widetilde{A}^+_r} \frac{\partial}{\partial x_i} \nabla_y N(x,y) \cdot \vec{f}(y)\, dy \right\|_{L^p(A^+_r)} \le c\, \|\vec{f}\|_{L^p(\widetilde{A}^+_r)}.$$

We conclude that

$$M_{1,p}(w_1, r) \le c \left(r \widetilde{M}_p(\vec{f}, r) + r^2 \widetilde{M}_p(f_0, r) \right), \tag{25}$$

where the tilde in $\widetilde{M}_p$ denotes that the spherical mean is taken over $\widetilde{A}_r^+$ instead of A_r^+.

For the second term, we note that $r/2 < |y| < |x| < 2r$ implies $|PN(x, y)| \le c|x|^{2-n} \le cr^{-n}|y|^2$ and $|\nabla_y PN(x, y)| \le c|x|^{1-n} \le cr^{-n}|y|$, so

$$\left| \int_{r/2 < |y| < |x|} PN(x, y) f_0(y) \, dy \right| \le cr^{-n} \int_{r/2 < |y| < |x|} |y|^2 |f_0(y)| \, dy$$

$$\le cr^{-n} \int_{|y| < 2r} |y|^2 |f_0(y)| \, dy$$

and

$$\left| \int_{r/2 < |y| < |x|} \nabla_y PN(x, y) \cdot \vec{f}(y) \, dy \right| \le cr^{-n} \int_{r/2 < |y| < |x|} |y| |\vec{f}(y)| \, dy$$

$$\le cr^{-n} \int_{|y| < 2r} |y| |\vec{f}(y)| \, dy.$$

Similarly, we can estimate

$$\left| r \int_{r/2 < |y| < |x|} \frac{\partial}{\partial x_j} PN(x, y) f_0(y) \, dy \right| \le cr^{-n} \int_{|y| < 2r} |y|^2 |f_0(y)| \, dy$$

and

$$\left| r \int_{r/2 < |y| < |x|} \frac{\partial}{\partial x_j} \nabla_y PN(x, y) \cdot \vec{f}(y) \, dy \right| \le cr^{-n} \int_{|y| < 2r} |y| |\vec{f}(y)| \, dy.$$

From these estimates we easily obtain

$$M_{1,p}(w_2, r) \le cr^{-n} \int_{|y| < 2r} \left(|y| |\vec{f}(y)| + |y|^2 |f_0(y)| \right) dy. \tag{26}$$

For the third term we note that $r < |x| < |y| < 4r$ implies $|PN(x, y)| \le c|y|^{2-n} \le cr^2|y|^{-n}$ and $|\nabla_y PN(x, y)| \le c|y|^{1-n} \le cr^2|y|^{-n-1}$, so

$$\left| \int_{|x| < |y| < 4r} PN(x, y) f_0(y) \, dy \right| \le cr^2 \int_{|x| < |y| < 4r} |y|^{-n} |f_0(y)| \, dy$$

$$\le cr^2 \int_{r < |y|} |y|^{-n} |f_0(y)| \, dy$$

and

$$\left| \int_{|x|<|y|<4r} \nabla_y PN(x,y) \cdot \vec{f}(y) dy \right| \leq c r^2 \int_{|x|<|y|<4r} |y|^{-n-1} |\vec{f}(y)| dy$$

$$\leq r^2 \int_{r<|y|} |y|^{-n-1} |\vec{f}(y)| dy.$$

Similarly, we can estimate

$$\left| r \int_{|x|<|y|<4r} \frac{\partial}{\partial x_j} PN(x,y) f_0(y) \, dy \right| \leq c r^2 \int_{r<|y|} |y|^{-n} |f_0(y)| \, dy$$

and

$$\left| r \int_{|x|<|y|<4r} \frac{\partial}{\partial x_j} \nabla_y PN(x,y) \cdot \vec{f}(y) \, dy \right| \leq r^2 \int_{r<|y|} |y|^{-n-1} |\vec{f}(y)| \, dy.$$

From these we easily obtain

$$M_{1,p}(w_3, r) \leq c r^2 \int_{|y|>r} \left(|y|^{-n-1} |\vec{f}(y)| + |y|^{-n} |f_0(y)| \right) \, dy. \tag{27}$$

For the fourth term, we use $|y| < r/2 < |x|/2 < |x|$ to obtain $|N^{\perp}(x,y)| \leq c|y|^2/|x|^n \leq c r^{-n} |y|^2$ and $|\nabla_y N^{\perp}(x,y)| \leq c|y|/|x|^n \leq c r^{-n} |y|$. Consequently,

$$\left| \int_{|y|<r/2} N^{\perp}(x,y) f_0(y) dy \right| \leq c r^{-n} \int_{|y|<r/2} |y|^2 |f_0(y)| \, dy$$

$$\leq c r^{-n} \int_{|y|<r} |y|^2 |f_0(y)| \, dy$$

and

$$\left| \int_{|y|<r/2} \nabla_y N^{\perp}(x,y) \cdot \vec{f}(y) dy \right| \leq c r^{-n} \int_{|y|<r/2} |y| |\vec{f}(y)| \, dy$$

$$\leq c r^{-n} \int_{|y|<r} |y| |\vec{f}(y)| \, dy.$$

Similarly, we can estimate

$$\left| r \int_{|y|<r/2} \frac{\partial}{\partial x_i} N^{\perp}(x,y) f_0(y) dy \right| \leq c r^{-n} \int_{|y|<r} |y|^2 |f_0(y)| \, dy$$

and

$$\left| r \int_{|y|<r/2} \frac{\partial}{\partial x_i} \nabla_y N^{\perp}(x,y) \cdot \vec{f}(y) dy \right| \leq c r^{-n} \int_{|y|<r} |y| |\vec{f}(y)| \, dy.$$

From these we easily obtain

$$M_{1,p}(w_4, r) \leq c r^{-n} \int_{|y|<r} \left(|y|^2 |\vec{f}(y)| + |y| |f_0(y)| \right) \, dy. \tag{28}$$

For the fifth term, we use $|x| < 2r < 4r < |y|$ to show that $|N^\perp(x,y)| \le c\,|x|^2/|y|^n \le c\,r^2\,|y|^{-n}$ and $|\nabla_y N^\perp(x,y)| \le c\,|x|^2/|y|^{n+1} \le c\,r^2\,|y|^{-n-1}$. Hence

$$\left| \int_{|y|>4r} N^\perp(x,y) f_0(y)\,dy \right| \le c\,r^2 \int_{|y|>4r} |y|^{-n}|f_0(y)|\,dy$$

$$\le c\,r^2 \int_{|y|>r} |y|^{-n}|f_0(y)|\,dy$$

and

$$\left| \int_{|y|>4r} \nabla_y N^\perp(x,y) \cdot \vec{f}(y)\,dy \right| \le c\,r^2 \int_{|y|>4r} |y|^{-n-1}|\vec{f}(y)|\,dy$$

$$\le c\,r^2 \int_{|y|>r} |y|^{-n-1}|\vec{f}(y)|\,dy.$$

Similarly, we estimate the first-order derivatives, so we eventually obtain

$$M_{1,p}(w_5, r) \le c\,r^2 \int_{|y|>r} \left(|\vec{f}(y)| + |y||f_0(y)| \right) |y|^{-n-1}\,dy. \tag{29}$$

Putting these all together, we have

$$M_{1,p}(w, r) \le c\left(r\widetilde{M}_p(\vec{f}, r) + r^2 \widetilde{M}_p(f_0, r) + r^{-n} \int_{|y|<2r} (|\vec{f}(y)| + |y||f_0(y)|)|y|\,dy \right.$$

$$\left. + r^2 \int_{|y|>r} (|\vec{f}(y)| + |y||f_0(y)|)|y|^{-n-1}\,dy \right).$$

But

$$r^{-n} \int_{r<|y|<2r} (|\vec{f}(y)| + |y||f_0(y)|)|y|\,dy \approx r^2 \int_{r<|y|<2r} (|\vec{f}(y)| + |y||f_0(y)|)|y|^{-n-1}\,dy,$$

so we can write this as

$$M_{1,p}(w, r) \le c\left(r\widetilde{M}_p(\vec{f}, r) + r^2 \widetilde{M}_p(f_0, r) + r^{-n} \int_{|y|<r} (|\vec{f}(y)| + |y||f_0(y)|)|y|\,dy \right.$$

$$\left. + r^2 \int_{|y|>r} (|\vec{f}(y)| + |y||f_0(y)|)|y|^{-n-1}\,dy \right). \tag{30}$$

Finally, the integrals in (30) can be estimated in terms of M_p and combined with the $\widetilde{M}_p$ term. For example, we can replace $|y|^2$ by $c \int_{|y|/2}^{|y|} \rho\,d\rho$, let $\rho = |z|$, and then interchange the order of integration to obtain

$$\int_{|y|<r} |y|^2|f_0(y)|dy = c \int_{|y|<r} \int_{|y|/2<|z|<|y|} |z|^{2-n}|f_0(y)|\,dz\,dy$$

$$\le c \int_{|z|<r} |z|^{2-n} \int_{|z|<|y|<2|z|} |g(y)|\,dy\,dz.$$

But, by the Hölder inequality,

$$\int_{|z|<|y|<2|z|} |f_0(y)| dy \le \left(\int_{|z|<|y|<2|z|} |f_0(y)|^p dy \right)^{1/p} \left(\int_{|z|<|y|<2|z|} dy \right)^{1/p'}$$

$$= c \left(\int_{|z|<|y|<2|z|} |g(y)|^p dy \right)^{1/p} |z|^{n/p'} = c\, M_p(f_0, |z|)\, |z|^n.$$

Thus

$$\int_{|y|<r} |y|^2 |f_0(y)|\, dy \le c \int_{|z|<r} |z|^2\, M_p(f_0, |z|)\, dz = c \int_0^r \rho^{n+1}\, M_p(f_0, \rho)\, d\rho.$$

Similarly, we can show the inequality

$$\int_{|y|>r} |y|^{-n} |f_0(y)|\, dy \le c \int_r^\infty \rho^{-1} M_p(f_0, \rho)\, d\rho.$$

If we similarly estimate the analogous integrals involving $\vec{f}$, we will obtain the estimate in the proposition. $\qquad\qquad\qquad\qquad\qquad\qquad\qquad\qquad\qquad\qquad\quad\square$

2. Variable coefficients in the half-space problem

In this section we consider (2) when $U = \mathbb{R}^n_+$, i.e., we assume that $u \in H^{1,2}_{\mathrm{loc}}(\overline{\mathbb{R}^n_+})$ satisfies

$$\int_{\mathbb{R}^n_+} a_{ij}\, \partial_j u\, \partial_i \eta\, dx = 0 \quad \text{for all } \eta \in C^1_{\mathrm{comp}}(\overline{\mathbb{R}^n_+}). \tag{31}$$

We want to consider the regularity of u at a point on $\mathbb{R}^{n-1} = \partial\mathbb{R}^n_+$ which, for convenience, we take to be the origin. As shown in Appendix C, the continuity of the a_{ij} enables us to conclude that $u \in H^{1,p}_{\mathrm{loc}}(\overline{\mathbb{R}^n_+})$ for all $p > 2$. Let us fix $p \in (n, \infty)$. By a change of independent variables we may arrange $a_{ij}(0) = \delta_{ij}$, so we assume that the coefficients satisfy

$$\sup_{|x|=r} |a_{ij}(x) - \delta_{ij}| \le \omega(r) \quad \text{as } r \to 0, \tag{32}$$

where ω is a continuous, nondecreasing function satisfying (3) and (10). We shall also assume that we have scaled the independent variables so that for δ very small we have

$$\int_0^1 \frac{\omega^2(r)}{r}\, dr < \delta \quad \text{and} \quad \omega(1) = \delta \tag{33}$$

For convenience, we extend ω to satisfy $\omega(r) = \delta$ for $r > 1$.

Now let us introduce a smooth cut-off function $\chi(r)$ satisfying $\chi(r) = 1$ for $r < 1/4$ and $\chi(r) = 0$ for $r > 3/4$. Then $\chi(|x|)u(x)$ is a compactly supported function that agrees with $u(x)$ near $x = 0$. What equation does χu satisfy? If we replace η in (2) by $\chi\eta$ and rearrange, we obtain

$$\int_{\mathbb{R}^n_+} (a_{ij}\, \partial_j(\chi u)\, \partial_i\eta - f_i\partial_i\eta + f_0\eta)\, dx = 0 \quad \text{for all } \eta \in C^1_{\text{comp}}(\overline{\mathbb{R}^n_+}),$$

where $f_i := a_{ij}u\, \partial_j\chi$ and $f_0 := a_{ij}\partial_j u\, \partial_i\chi$ are known to be in $L^p_{\text{comp}}(\overline{\mathbb{R}^n_+})$. Since we are interested in the behavior of u near $x = 0$ where u and χu agree, after relabeling we can assume that $u \in H^{1,p}(\mathbb{R}^n_+)$ has support in $|x| < 1$ and satisfies

$$\int_{\mathbb{R}^n_+} (a_{ij}\, \partial_j u\, \partial_i\eta - f_i\partial_i\eta + f_0\eta)\, dx = 0 \quad \text{for all } \eta \in C^1_{\text{comp}}(\overline{\mathbb{R}^n_+}), \tag{34}$$

where $f_i, f_0 \in L^p$ have support in $1/4 \leq |x| \leq 3/4$. Since u vanishes outside $|x| < 1$, there is no harm in assuming that a_{ij} satisfies

$$a_{ij}(x) = \delta_{ij} \quad \text{for } |x| \geq 1. \tag{35}$$

Let us recall the decomposition $u(x) = u_0(r) + \tilde{v}(r) \cdot \tilde{x} + w(x)$ as defined in (11). Since we have assumed that u is supported in $|x| < 1$, we see that $u_0, \tilde{v}$, and w are all supported in $|x| < 1$. Moreover, as shown in Appendix B,

$$\nabla u \in L^2(B_+(1)) \Rightarrow$$
$$\int_0^1 \left[(u_0')^2 + |\tilde{v}|^2 + r^2|\tilde{v}'|^2\right] r^{n-1}\, dr < \infty \quad \text{and} \quad \nabla w \in L^2(B_+(1)). \tag{36}$$

To formulate the relationship between the decomposition and the dynamical system, let $r = e^{-t}$ and introduce

$$\varepsilon(t) := \omega(e^{-t}) \quad \text{for } -\infty < t < \infty. \tag{37a}$$

Notice that

$$\int_0^\infty \varepsilon^2(t)\, dt = \int_0^1 \frac{\omega^2(r)}{r}\, dr. \tag{37b}$$

To control the behavior of $\tilde{v}(r)$ and $r\tilde{v}'(r)$ as $r \to 0$, we need to control the behavior of $\tilde{v}(t)$ and $\tilde{v}_t(t)$ as $t \to \infty$. In Appendix D of this paper, we show that new dependent variables (φ, ψ) can be introduced that satisfy a $2(n-1)$-dimensional dynamical system

$$\frac{d}{dt}\begin{pmatrix} \varphi \\ \psi \end{pmatrix} + \begin{pmatrix} 0 & 0 \\ 0 & -nI \end{pmatrix}\begin{pmatrix} \varphi \\ \psi \end{pmatrix} + \mathcal{R}(t)\begin{pmatrix} \varphi \\ \psi \end{pmatrix} = g(t, \nabla w) + h(t) \quad \text{for } T < t < \infty, \tag{38a}$$

where the matrix $\mathcal{R}$ depends upon the coefficients a_{ij} and can be decomposed into blocks

$$\mathcal{R}(t) = \begin{pmatrix} R_1(t) & R_2(t) \\ R_3(t) & R_4(t) \end{pmatrix} \quad \text{with } |R_j(t)| \leq \varepsilon(t). \tag{38b}$$

The block R_1 satisfies

$$|R_1(t) - R(t)| \leq c\varepsilon^2(t) \quad \text{as } t \to \infty, \tag{38c}$$

where the $(n-1) \times (n-1)$ matrix $R(t)$ is given by (5). The term $g(t, \nabla w)$ in (38a) denotes a vector function of t that depends on ∇w (the gradient in the x-variables) in such a way that

$$|g(t, \nabla w)| \leq c\varepsilon(t) \fint_{S_+^{n-1}} |\nabla w| \, ds \quad \text{for } t > 0, \tag{38d}$$

and the term h in (38a) is a vector function in $L^1(0, \infty)$ with L^1-norm satisfying

$$\|h\|_1 \leq c \left(\|\vec{f}\|_p + \|f_0\|_p \right). \tag{38e}$$

Moreover, the difference between the new dependent variables (φ, ψ) and $(\widetilde{v}, \widetilde{v}_t)$ is estimated by

$$\left| \begin{pmatrix} \widetilde{v} \\ \widetilde{v}_t \end{pmatrix} - \begin{pmatrix} n(\varphi + \psi) \\ n^2 \psi \end{pmatrix} \right| \leq c\varepsilon(t) \left(|\varphi(t)| + |\psi(t)| + \fint |\nabla w| ds \right). \tag{38f}$$

We will use this and the stability of (φ, ψ) as $t \to \infty$ to control the behavior of $\widetilde{v}$ as $r \to 0$.

With these preliminaries, we are able to prove the following.

Theorem 1. *Suppose the a_{ij} satisfy (32) where ω satisfies (3) and (10), and the dynamical system (4) with matrix R given by (5) is uniformly stable as $t \to \infty$. Then every weak solution $u \in H_{\text{loc}}^{1,2}(\mathbb{R}_+^n)$ of (31) is Lipschitz continuous at $x = 0$. If, in addition, every solution of the dynamical system (4) is asymptotically constant as $t \to \infty$, then u is differentiable at $x = 0$, and*

$$\partial_j u(0) = \lim_{r \to 0} \frac{n}{r} \fint_{S_+^{n-1}} u(r\theta) \, \theta_j \, ds \quad \text{for } j = 1, \ldots, n-1, \qquad \partial_n u(0) = 0.$$

Proof. As indicated above, we may assume for some $p \in (n, \infty)$ that $u \in H^{1,p}(\mathbb{R}_+^n)$ is supported in $|x| < 1$ and satisfies (34). The strategy of the proof is to construct a solution u^* of (34) in the form (11). This is done by finding w as a fixed point for a certain map S on the Banach space Y, which is defined to be $w \in H_{\text{loc}}^{1,p}(\overline{\mathbb{R}_+^n} \setminus \{0\})$ with finite norm

$$\|w\|_Y = \sup_{0 < r < 1} \frac{M_{1,p}(w, r)}{\omega(r) \, r} + \sup_{r > 1} \frac{M_{1,p}(w, r)}{r^{-n}}. \tag{39}$$

Since $p > 2$ we see that $w \in Y$ implies $M_2(\nabla w, r) \leq C\omega(r)$ for $0 < r < 1$. As we shall see, finding w also yields $\widetilde{v}$ and $r\widetilde{v}_r$ from the solution of the dynamical system (38a). Moreover, u_0' can be found in terms of $\widetilde{v}$, $r\widetilde{v}_r$, and w, and we find that the stability properties of the dynamical system (38a) control the asymptotic behavior of $\widetilde{v}$ and $r\widetilde{v}_r$ as $r \to 0$, and hence also of u_0. Under the assumed stability of (38a), the constructed u^* has the required regularity, and it only remains to show that $u^* = u$; this is done using the uniqueness of solutions of (34) discussed in Appendix F. Let us now discuss the details of this argument.

For a given $w \in Y$, we want to solve (38a) with initial conditions $\phi(0) = 0 = \psi(0)$ to find (ϕ, ψ) and hence $\widetilde{v}$, $r\widetilde{v}_r$. To control the dependence of $\widetilde{v}$ on w, let us write $\widetilde{v} = \widetilde{v}^w + \widetilde{v}^0$ where $\widetilde{v}^w$ corresponds to solving (38a) with $h \equiv 0$ and

$\widetilde{v}^0$ corresponds to solving it with $g(t, \nabla w) \equiv 0$. In order to estimate $\widetilde{v}^w$ on $(0, \infty)$, we will use Proposition 2 in Appendix E. Consequently, we need $g = (g_1, g_2)$ to satisfy: i) $g_1 \in L^1(0, \infty)$ and ii) g_2 satisfies (97e). First, we use (38d) to conclude that

$$\int_0^\infty |g_1(t, \nabla w)| \, dt \leq c \left(\int_0^\infty \varepsilon^2(t) \, dt \right)^{1/2} \left(\int_0^\infty \int_{S_+^{n-1}} |\nabla w|^2 \, ds \, dt \right)^{1/2}$$

$$\leq c\sqrt{\delta} \left(\int_0^1 \fint_{S_+^{n-1}} |\nabla w|^2 ds \, \frac{d\rho}{\rho} \right)^{1/2}. \tag{40a}$$

We will deduce the finiteness of this bound below. Second, we use (38d) to conclude that

$$e^{\alpha t} \int_t^\infty |g_2(\tau, \nabla w)| \, e^{-\alpha \tau} \, d\tau \leq c \, \varepsilon(t) \int_t^\infty e^{\alpha(t-\tau)} \fint_{S_+^{n-1}} |\nabla w| \, ds \leq c_\alpha \, \varepsilon(t),$$

where

$$c_\alpha = \frac{c}{\sqrt{2\alpha}} \left(\int_0^1 \fint |\nabla w|^2 ds \frac{d\rho}{\rho} \right)^{1/2}. \tag{40b}$$

Now let us perform a calculation for $1 \leq p < \infty$: for $j = 0, 1, \ldots$, let $r_j = 2^{-j}$ so

$$\int_0^1 \int_{S_+^{n-1}} |\nabla w|^p \, ds \frac{d\rho}{\rho} = \sum_{j=1}^\infty \int_{r_j}^{2r_j} \int_{S_+^{n-1}} |\nabla w|^p \, ds \, \frac{d\rho}{\rho} = \sum_{j=1}^\infty \int_{A_{r_j}^+} |\nabla w|^p |x|^{-n} \, dx$$

$$\leq c \sum_{j=1}^\infty \fint_{A_{r_j}^+} |\nabla w|^p \, dx = c \sum_{j=1}^\infty M_p^p(\nabla w, r_j). \tag{40c}$$

As previously observed, $p > 2$ and $w \in Y$ imply $M_2(\nabla w, r) \leq C\,\omega(r)$ as $r \to 0$, so we may apply the above estimate with $p = 2$ and the following calculation

$$\sum_{j=1}^\infty \omega^2(r_j) = 2 \sum_{j=1}^\infty \omega^2(r_j) \frac{r_{j-1} - r_j}{r_{j-1}} \leq 2 \int_0^1 \omega^2(\rho) \frac{d\rho}{\rho} < 2\delta \tag{40d}$$

to deduce the finiteness of c_α and the bound (40a). Thus we have confirmed i) and ii).

Next, let us describe the variational PDE that w satisfies. As in [14] we introduce $\Omega_{ij} = a_{ij} - \delta_{ij}$, which satisfies $|\Omega_{ij}(x)| \leq \omega(r)$ for $0 < r = |x| < 1$ and $\Omega_{ij}(x) = 0$ for $|x| > 1$. Now the variational problem (34) can be written as

$$\int_{\mathbb{R}_+^n} (\nabla u \cdot \nabla \eta + \Omega_{ij} \partial_j u \, \partial_i \eta - f_i \partial_i \eta + f_0 \eta) \, dx = 0 \quad \text{for all } \eta \in C_{\text{comp}}^1(\overline{\mathbb{R}_+^n}).$$

Since this holds for all η, it holds for $\eta^\perp$:

$$\int_{\mathbb{R}_+^n} (\nabla u \cdot \nabla(\eta^\perp) + \Omega_{ij} \partial_j u \, \partial_i(\eta^\perp) - f_i \partial_i(\eta^\perp) + f_0 \, \eta^\perp) \, dx = 0.$$

Now we claim that

$$\int_{\mathbb{R}^n_+} \nabla(Pu) \cdot \nabla(\eta^\perp) \, dx = 0 = \int_{\mathbb{R}^n_+} \nabla w \cdot \nabla(P\eta) \, dx \quad \text{for all } \eta \in C^1_{\text{comp}}(\overline{\mathbb{R}^n_+}). \quad (41)$$

To prove this, let us first assume the inclusion $u \in C^2_{\text{comp}}(\overline{\mathbb{R}^n_+})$. Then, by Lemma 1, we have $Pu \in C^2_{\text{comp}}(\overline{\mathbb{R}^n_+})$, $P\eta \in C^1_{\text{comp}}(\overline{\mathbb{R}^n_+})$, and $\partial Pu/\partial x_n = 0 = \partial P\eta/\partial x_n$ on $\mathbb{R}^{n-1}$. Applying the divergence theorem, we conclude that

$$\int_{\mathbb{R}^n_+} \nabla(Pu) \cdot \nabla(\eta^\perp) \, dx = -\int_{\mathbb{R}^n_+} \Delta(Pu) \, \eta^\perp \, dx = -\int_{\mathbb{R}^n_+} (\Delta(Pu))^\perp \, \eta \, dx.$$

However, for fixed r, $Pu(r, \cdot) \in V = \text{span}(1, \theta_1, \ldots, \theta_{n-1})$ on S^{n-1}_+, and Δ preserves V, so $(\Delta(Pu))^\perp = 0$; this proves the first equality in (41) when $u \in C^2_{\text{comp}}(\overline{\mathbb{R}^n_+})$. In general, for $u \in H^{1,p}_{\text{comp}}(\overline{\mathbb{R}^n_+})$, we extend u by zero to $\mathbb{R}^n_-$, and mollify by $u_\varepsilon = \phi_\varepsilon \star u$, where $\phi_\varepsilon(x) = \varepsilon^{-n}\phi(|x|/\varepsilon)$ with $\phi \in C^\infty(\mathbb{R}^n)$ satisfying $\text{supp}(\phi) \subset B_1(0)$ and $\int \phi(x) \, dx = 1$. Then u_ε is smooth (on all of $\mathbb{R}^n$), and we can apply the above argument to obtain

$$\int_{\mathbb{R}^n_+} \nabla(P u_\varepsilon) \cdot \nabla(\eta^\perp) \, dx = 0.$$

Since we have assumed that $u \in H^{1,p}(\overline{\mathbb{R}^n_+})$, we can show in the standard way that $u_\varepsilon \to u$ in $H^{1,p}(\overline{\mathbb{R}^n_+})$ as $\varepsilon \to 0$. Then we can use Lemma 2 to conclude (even without the $\perp$ on η) that

$$\int_{\mathbb{R}^n_+} \nabla(P u_\varepsilon) \cdot \nabla\eta \, dx \to \int_{\mathbb{R}^n_+} \nabla(Pu) \cdot \nabla\eta \, dx \text{ as } \varepsilon \to 0 \text{ for all } \eta \in C^1_{\text{comp}}(\overline{\mathbb{R}^n_+} \setminus \{0\}).$$

This establishes the first equality in (41). The second equality in (41) follows by a similar argument.

But (41) means that

$$\int_{\mathbb{R}^n_+} \nabla u \cdot \nabla(\eta^\perp) \, dx = \int_{\mathbb{R}^n_+} \nabla w \cdot \nabla\eta \, dx \quad \text{for all } \eta \in C^1_{\text{comp}}(\overline{\mathbb{R}^n_+}). \quad (42)$$

Using this and the fact that u'_0 can be expressed in terms of $\tilde{v}^w$ and w (see (87a) in Appendix D), we see that the variational problem that w satisfies can be written as

$$\int_{\mathbb{R}^n_+} \nabla w \cdot \nabla\eta \, dx + F^\perp_{1,w}[\eta] + F^\perp_{1,0}[\eta] + F^\perp_0[\eta] = 0 \quad \text{for all } \eta \in C^1_{\text{comp}}(\overline{\mathbb{R}^n_+}), \quad (43)$$

where

$$F^\perp_{1,w}[\eta] = \int_{\mathbb{R}^n_+} \vec{f}^w \cdot \nabla(\eta^\perp) dx, \quad F^\perp_{1,0}[\eta] = \int_{\mathbb{R}^n_+} \vec{f}^0 \cdot \nabla(\eta^\perp) dx, \quad F^\perp_0[\eta] = \int_{\mathbb{R}^n_+} f_0 \eta^\perp dx, \quad (44)$$

with the vector functions $\vec{f}^w$ and $\vec{f}^0$ defined by

$$f_i^w = \Omega_{ij}\left(\partial_j w - \frac{r\tilde{\beta}\cdot(\tilde{v}^w)' + \tilde{\gamma}\cdot\tilde{v}^w + p[\nabla w]}{\alpha}\theta_j + \partial_j(\tilde{x}\cdot\tilde{v}^w)\right) \qquad (45)$$

$$f_i^0 = \Omega_{ij}\left(\frac{\vartheta(r) - r\tilde{\beta}\cdot(\tilde{v}^0)' - \tilde{\gamma}\cdot\tilde{v}^0}{\alpha}\theta_j + \partial_j(\tilde{x}\cdot\tilde{v}^0)\right) - f_i. \qquad (46)$$

Here, as in Appendix D, the functions α, $\tilde{\beta}$, $\tilde{\gamma}$, $p[\nabla w]$, and ϑ of r satisfy

$$|\alpha(r) - 1|, |\tilde{\beta}(r)|, |\tilde{\gamma}(r)| \leq \omega(r) \quad \text{for } 0 < r < 1,$$

$$|p[\nabla w](r)| \leq \omega(r)\fint_{S_+^{n-1}}|\nabla w(r\theta)|\,ds \quad \text{for } 0 < r < 1, \qquad (47)$$

$$|\vartheta(r)| \leq \fint_{S_+^{n-1}}(|\vec{f}(r\theta)| + |f_0(r\theta)|)\,ds \quad \text{for } 0 < r < 1,$$

and $\alpha(r) = 1$ and $\tilde{\beta}(r) = \tilde{\gamma}(r) = p[\nabla w](r) = \vartheta(r) = 0$ for $r > 1$.

For $w \in Y$, define $z = S(w)$ to be the solution of

$$\int_{\mathbb{R}_+^n}\nabla z\cdot\nabla\eta\,dx + F_{1,w}^{\perp}[\eta] + F_{1,0}^{\perp}[\eta] + F_0^{\perp}[\eta] = 0 \quad \text{for all } \eta \in C_{\text{comp}}^1(\overline{\mathbb{R}_+^n}) \qquad (48)$$

that is provided by Proposition 1, i.e.,

$$S(w) = z(x) = \int_{\mathbb{R}_+^n}\left(N^{\perp}(x,y)f_0(y) - \nabla_y N^{\perp}(x,y)\cdot\vec{f}^0(y) - \nabla_y N^{\perp}(x,y)\cdot\vec{f}^w(y)\right)dy. \qquad (49)$$

If we can show that $S : Y \to Y$ has a fixed point w, then this is the solution of (43) that we seek.

To show S has a fixed point, we write $Sw = \xi - Tw$ where

$$\xi(x) = \int_{\mathbb{R}_+^n}\left(N^{\perp}(x,y)f_0(y) - \nabla_y N^{\perp}(x,y)\cdot\vec{f}^0(y)\right)dy \qquad (50)$$

and

$$Tw(x) = \int_{\mathbb{R}_+^n}\nabla_y N^{\perp}(x,y)\cdot\vec{f}^w(y)\,dy. \qquad (51)$$

If we can show that $\xi \in Y$ and $T : Y \to Y$ has small norm, then we can solve the equation $w + Tw = \xi$ to find our fixed point $w = Sw$. To estimate $M_{1,p}(Tw,r)$ we will apply Proposition 1:

$$M_{1,p}(Tw,r) \leq c\left(r^{-n}\int_0^r M_p(\vec{f}^w,\rho)\rho^n\,d\rho + r^2\int_r^{\infty}M_p(\vec{f}^w,\rho)\rho^{-2}\,d\rho\right). \qquad (52)$$

So we only need to estimate $M_p(\vec{f}^w,r)$ and integrate.

Now let us assume that $\|w\|_Y \le 1$ and show that $\|Tw\|_Y$ is small. We split Tw into three terms:

$$T_1 w(x) = \int_{\mathbb{R}^n_+} \nabla_y N^{\perp}(x,y) \cdot \Omega \nabla w(y)\, dy,$$

$$T_2 w(x) = \int_{\mathbb{R}^n_+} \nabla_y N^{\perp}(x,y) \cdot \Omega \nabla(\tilde{y} \cdot \tilde{v}^w)\, dy,$$

$$T_3 w(x) = \int_{\mathbb{R}^n_+} \frac{1}{\alpha(r_y)} \left(r_y\, \tilde{\beta} \cdot (\tilde{v}^w)'(r_y) + \tilde{\gamma} \cdot \tilde{v}^w(r_y) + p[\nabla w](r_y) \right)$$
$$\cdot \nabla_y N^{\perp}(x,y) \cdot \Omega \theta(y)\, dy.$$

Here $r_y := |y|$ and we have written the vector $\Omega_{ij}\partial_j w$ simply as $\Omega \nabla w$; similarly for $\Omega \nabla(\tilde{y} \cdot \tilde{v}^w)$ and $\Omega \theta$. Let us first consider $T_1 w$. Recall that $|\Omega(r)| \le \omega(r)$ for $0 < r < 1$ and $\Omega(r) \equiv 0$ for $r > 1$, so

$$r^{-n} \int_0^r M_p(\Omega \nabla w, \rho)\rho^n d\rho + r^2 \int_r^\infty M_p(\Omega \nabla w, \rho)\rho^{-2} d\rho$$

$$\le \begin{cases} r^{-n}\int_0^r \omega(\rho)M_p(\nabla w, \rho)\rho^n d\rho + r^2\int_r^1 \omega(\rho)M_p(\nabla w, \rho)\rho^{-2} d\rho & \text{for } 0 < r < 1 \\ r^{-n}\int_0^1 \omega(\rho)M_p(\nabla w, \rho)\rho^n d\rho & \text{for } r > 1. \end{cases}$$

For $0 < r < 1$ we have assumed the inequality $\|w\|_Y \le 1$ and we have $M_{1,p}(w,r) \le \omega(r)\, r$, so $M_p(\nabla w, r) \le \omega(r)$ and we can estimate

$$r^{-n} \int_0^r \omega(\rho)M_p(\nabla w, \rho)\rho^n d\rho + r^2 \int_r^1 \omega(\rho)M_p(\nabla w, \rho)\rho^{-2} d\rho$$

$$\le r^{-n} \int_0^r \omega^2(\rho)\rho^n d\rho + r^2\delta \int_r^1 \omega(\rho)\rho^{-2} d\rho$$

$$\le c\left(\omega^2(r)r + \delta\omega(r)r(r^\kappa - 1)\right) \le c\,\delta\,\omega(r)\,r.$$

Consequently, for $0 < r < 1$ we see by (52) that

$$M_{1,p}(T_1 w, r) \le c\,\delta\,\omega(r)\,r.$$

Meanwhile, for $r > 1$, (52) implies

$$M_{1,p}(T_1 w, r) \le r^{-n} \int_0^1 \omega(\rho)M_p(\nabla w, \rho)\rho^n d\rho \le r^{-n}\int_0^1 \omega^2(\rho)\rho^n d\rho \le c\delta\,r^{-n}.$$

Thus $\|T_1 w\|_Y \le c\,\delta$ and, if we take δ sufficiently small, we can arrange that $T_1 : Y \to Y$ has norm less than $1/3$.

Next consider $T_2 w$. Again we use the relations $|\Omega(r)| \le \omega(r)$ for $0 < r < 1$ and $\Omega(r) \equiv 0$ for $r > 1$ to obtain

$$r^{-n} \int_0^r M_p(\Omega \nabla(\tilde{y} \cdot \tilde{v}^w), \rho)\rho^n d\rho + r^2 \int_r^\infty M_p(\Omega \nabla(\tilde{y} \cdot \tilde{v}^w), \rho)\rho^{-2} d\rho$$

$$\le \begin{cases} r^{-n}\int_0^r \omega(\rho)M_p(\nabla(\tilde{y} \cdot \tilde{v}^w), \rho)\rho^n d\rho \\ \quad + r^2\int_r^1 \omega(\rho)M_p(\nabla(\tilde{y} \cdot \tilde{v}^w), \rho)\rho^{-2} d\rho & \text{for } 0 < r < 1 \\ r^{-n}\int_0^1 \omega(\rho)M_p(\nabla(\tilde{y} \cdot \tilde{v}^w), \rho)\rho^n d\rho & \text{for } r > 1. \end{cases}$$

To estimate $\nabla(\tilde{y} \cdot \tilde{v}^w)$ we need to estimate $\tilde{v}^w$ and $r(\tilde{v}^w)'$. But, using (38f), these can be expressed in terms of the solution (ϕ, ψ) of the dynamical system (38a). Thus we find

$$\sup_{|y|<1} |\nabla(\tilde{y} \cdot \tilde{v}^w)| \leq c \sup_{r<1}(r|(\tilde{v}^w)'| + |\tilde{v}^w|)$$

$$\leq c \sup_{t>0}(|\phi(t)| + |\psi(t)|) \leq c(c_\alpha + \|g_1\|_1),$$

where we have used Proposition 2 in Appendix E for the last estimate. Now we can estimate c_α and $\|g_1\|_1$ as in (40) to find $c_\alpha \leq c\sqrt{\delta}$ and $\|g_1\|_1 \leq c\delta$. So we conclude that for $0 < \rho < 1$

$$M_p(\nabla(\tilde{y} \cdot \tilde{v}^w), \rho) \leq c \sup_{|y|<1} |\nabla(\tilde{y} \cdot \tilde{v}^w)| \leq c\sqrt{\delta}.$$

We can use this in (52) to estimate $M_{1,p}(T_2 w, r)$:

$$M_{1,p}(T_2 w, r)$$
$$\leq \begin{cases} c\left(r^{-n} \int_0^r \omega(\rho)\sqrt{\delta}\rho^n \, d\rho + r^2 \int_r^1 \omega(\rho)\sqrt{\delta}\rho^{-2} \, d\rho\right) \leq c\sqrt{\delta}\omega(r)r & \text{for } 0 < r < 1 \\ cr^{-n} \int_0^1 \omega(\rho)\sqrt{\delta}\rho^n \, d\rho \leq c\delta^{3/2} r^{-n} & \text{for } r > 1. \end{cases}$$

Thus, $\|T_2 w\|_Y \leq c(\sqrt{\delta} + \delta^{3/2})$ and, if we take δ sufficiently small, we can arrange that $T_2 : Y \to Y$ has norm less than $1/3$.

Finally we consider $T_3 w$. We first need to estimate $M_p(\alpha^{-1}(r\tilde{\beta} \cdot (\tilde{v}^w)' + \tilde{\gamma} \cdot \tilde{v}^w + p[\nabla w])\Omega\theta, r)$ for $0 < r < 1$. But, recalling the properties (47) and some of the estimates used for T_2, we have

$$M_p(\alpha^{-1}(r\tilde{\beta} \cdot (\tilde{v}^w)' + \tilde{\gamma} \cdot \tilde{v}^w + p[\nabla w])\Omega\theta, r)$$
$$\leq c\omega^2(r)\left[M_p(r(\tilde{v}^w)', r) + M_p(\tilde{v}^w, r) + M(\nabla w, r)\right]$$
$$\leq c\sqrt{\delta}\,\omega^2(r).$$

Applying Proposition 1, we obtain for $0 < r < 1$

$$M_{1,p}(T_3 w, r) \leq c \left(r^{-n} \int_0^r \sqrt{\delta}\omega^2(\rho)\rho^n \, d\rho + r^2 \int_r^1 \sqrt{\delta}\omega^2(\rho)\rho^{-2} \, d\rho\right) \leq c\delta^{3/2}\omega(r)r.$$

Meanwhile, for $r > 1$ we simply have

$$M_{1,p}(T_3 w, r) \leq c r^{-n} \int_0^1 \sqrt{\delta}\,\omega(\rho)\rho^n \, d\rho \leq c\,\delta^{3/2}\, r^{-n}.$$

Thus $\|T_3 w\|_Y \leq c\,\delta^{3/2}$, and if we take δ sufficiently small, we can arrange $T_3 : Y \to Y$ to have norm less than $1/3$. Consequently, $T = T_1 + T_2 - T_3 : Y \to Y$ has norm less than 1.

To show that ξ defined in (50) is in Y, let us split it up into several terms: $\xi = \xi_1 - \xi_2 + \xi_3 + \xi_4$, where

$$\xi_1(x) = \int_{\mathbb{R}^n_+} N^\perp(x,y)\, f_0(y)\, dy,$$

$$\xi_2(x) = \int_{\mathbb{R}^n_+} \nabla_y N^\perp(x,y) \cdot \vec{f}(y)\, dy,$$

$$\xi_3(x) = \int_{\mathbb{R}^n_+} \nabla_y N^\perp(x,y) \cdot \Omega \nabla(\tilde{x} \cdot \tilde{v}^0)\, dy,$$

$$\xi_4(x) = \int_{\mathbb{R}^n_+} \frac{1}{\alpha(r_y)} \left(\vartheta(r_y) - r_y\, \tilde{\beta} \cdot (\tilde{v}^0)'(r_y) - \tilde{\gamma} \cdot \tilde{v}^0(r_y) \right) \nabla_y N^\perp(x,y) \cdot \Omega \theta\, dy.$$

Since f_0 belongs to L^p and is supported on $|x| < 1$, we can apply Proposition 1:

$$M_{1,p}(\xi_1,r) \le c\left(r^{-n} \int_0^r M_p(f_0,\rho)\rho^{n+1}\, d\rho + r^2 \int_r^1 M_p(f_0,\rho)\, d\rho \right)$$

$$\le \begin{cases} c\, r^2\, \|f_0\|_p & \text{for } 0 < r < 1 \\ c\, r^{-n}\, \|f_0\|_p & \text{for } r > 1. \end{cases}$$

Since $r \le \omega(r)$ for $0 < r < 1$, we see that $\xi_1 \in Y$. To estimate ξ_2 recall that $\vec{f}$ is supported on $1/4 < |x| < 1$, so $|\vec{f}(x)| \le c\omega(|x|)\,|\vec{f}(x)|$, and from Proposition 1 we obtain the estimate

$$M_{1,p}(\xi_2,r) \le c\left(r^{-n} \int_0^r \omega(\rho)\, M_p(\vec{f},\rho)\rho^{n+1}\, d\rho + r^2 \int_r^1 \omega(\rho)\, M_p(\vec{f},\rho)\, d\rho \right)$$

$$\le \begin{cases} c\omega(r)\, r\, \|\vec{f}\|_p & \text{for } 0 < r < 1 \\ c\delta\, r^{-n}\, \|\vec{f}\|_p & \text{for } r > 1. \end{cases}$$

We see that $\xi_2 \in Y$. The proofs that ξ_3 and ξ_4 are in Y are quite similar to estimating T_2 and T_3 above, so we will not give the details. But we can conclude not only that $\xi \in Y$, but

$$\|\xi\|_Y \le c\,(\|\vec{f}\|_p + \|f_0\|_p). \tag{53}$$

Now we let $w \in Y$ be the fixed point of S, so w satisfies (43). We use w to find $\tilde{v}^w$ and then (87a) to find u_0'. Integrating (87a) to find u_0 (up to a constant) and letting $\tilde{v} = \tilde{v}^w$, we see that

$$u^*(x) := u_0(r) + \tilde{x} \cdot \tilde{v}(r) + w(x) \tag{54}$$

is a solution of (34). Now we want to show that u^* has the desired regularity properties. Since $w = (I + T)^{-1}\xi \in Y$, we know that $M_p(\nabla w, r) \le c\omega(r)$ as $r \to 0$. Moreover, $Pw = 0$ implies that $\int_{|x|<r} w\, dx = 0$ for every $r > 0$. Using this and $p > n$, Morrey's inequality (cf. [8]) implies

$$\sup_{|x|<r} |w(x)| \le c_n\, r \left(\fint_{|y|<r} |\nabla w|^p\, dy \right)^{1/p}.$$

(Recall that $|x| < r$ still refers to points $x \in \mathbb{R}^n_+$.) But for fixed $r \in (0,1)$ we can introduce $r_j = 2^{-j}\, r$ and compute

$$\fint_{|y|<r} |\nabla w|^p\, dy = \frac{n}{r^n |S^{n-1}_+|} \sum_{j=0}^{\infty} \int_{r_{j+1}<|y|<r_j} |\nabla w|^p\, dy \le c \sup_{0<\rho<r} M^p_p(\nabla w, \rho).$$

We conclude that

$$\sup_{|x|<r} |w(x)| \le c\, r\, \omega(r) \quad \text{as } r \to 0, \tag{55}$$

which implies that w is differentiable at $x = 0$ with $\partial_j w(0) = 0$ for $j = 1, \ldots, n$. Moreover, our assumption that (4) is uniformly stable as $t \to \infty$ implies by Proposition 2 in Appendix E that (ϕ, ψ) remains bounded as $t \to \infty$, and in fact $|\psi(t)| \le c\,\varepsilon(t)$ as $t \to \infty$. We now want to use (38f) to show that $\tilde{v}$ is bounded as $t \to \infty$. From the second component in (38f) we have

$$|\tilde{v}_t(t)| \le c_1\, \varepsilon(t) + c_2 \fint |\nabla w|\, ds.$$

Let us integrate this from T to $T + \ln 2$:

$$\int_T^{T+\ln 2} |\tilde{v}_t(t)|\, dt \le c_1\, \varepsilon(T) + c_2 \int_T^{T+\ln 2} \int_{S^{n-1}_+} |\nabla w|\, ds\, dt.$$

But letting $R = e^{-T-\ln 2}$ we find

$$\int_T^{T+\ln 2} \int_{S^{n-1}_+} |\nabla w|\, ds\, dt = \int_R^{2R} \int_{S^{n-1}_+} |\nabla w|\, ds\, \frac{dr}{r}$$

$$\le \frac{c}{R^n} \int_{A^+_R} |\nabla w|\, dx \le c\, M_p(\nabla w, R).$$

Since we have assumed that $M_p(\nabla w, r)$ is bounded by $\omega(r)$ as $r \to 0$, we have shown the estimate

$$\int_T^{T+\ln 2} |\tilde{v}_t(t)|\, dt \le c\,\varepsilon(T) \quad \text{as } T \to \infty. \tag{56a}$$

Using this in the first component in (38f), we have

$$\int_T^{T+\ln 2} |\tilde{v}(t)|\, dt \le C \quad \text{as } T \to \infty. \tag{56b}$$

But now we may use the elementary inequality

$$\sup_{a \le t \le b} |v(t)| \le c \int_a^b (|v(t)| + |v_t(t)|)\, dt \tag{57}$$

to conclude that $|\tilde{v}(T)|$ is bounded as $T \to \infty$. Of course, $\tilde{v}(t)$ is actually $\tilde{v}(e^{-t}) = \tilde{v}(r)$, so we see that $|\tilde{v}(r)|$ is bounded as $r \to 0$, and hence $\tilde{x} \cdot \tilde{v}$ is Lipschitz. Finally, using (87a), we can estimate

$$|u_0(r) - u_0(0)| \leq \int_0^r |u_0'(\rho)| \, d\rho$$

$$\leq c \int_0^r \left(|\vartheta(\rho)| + \omega(\rho)\rho|\tilde{v}'(\rho)| + \omega(\rho)|\tilde{v}(\rho)| + \omega(\rho) \fint |\nabla w| \, ds \right) d\rho.$$

From (87b) we have $\vartheta(r) = \overline{f}_1(r) + r^{1-n} \int_0^r \overline{f}_0(\rho)\rho^{n-1} \, d\rho$ where $\overline{f}_1, \overline{f}_0$ are given in (86). Since f_0 belongs to $L^p(\mathbb{R}^n)$ and vanishes for $r > 1$, we can estimate $\int_0^\rho |\overline{f}_0(\tau)|\tau^{n-1} \, d\tau = c \int_{B_\rho} |f_0(x)| \, dx \leq c\rho^n \|f_0\|_{L^p}$. Hence

$$\int_0^r \rho^{1-n} \int_0^\rho |\overline{f}_0(\tau)|\tau^{n-1} \, d\tau \leq c\|f_0\|_{L^p} \int_0^r \rho \, d\rho = c\,r^2 \, \|f_0\|_{L^p}.$$

Since $\overline{f}_1$ vanishes for $r > 1$ and also for $0 < r < 1/4$, we can even more easily verify that $\int_0^r |\overline{f}_1(\rho)| \, d\rho \leq c\,r^2 \, \|\overline{f}\|_{L^p}$. Since $\tilde{v}(r)$ is bounded as $r \to 0$, we see that $\int_0^r \omega(\rho)|\tilde{v}(\rho)| \, d\rho \leq c\,r\,\omega(r)$. To estimate the last term, we can use the Hölder inequality to obtain

$$\int_0^r \omega(\rho) \fint_{S_+^{n-1}} |\nabla w| \, ds \, d\rho \leq c \left(\int_0^r (\omega(\rho))^q d\rho \right)^{1/q} \left(\int_0^r \left(\fint_{S_+^{n-1}} |\nabla w| \, ds \right)^p d\rho \right)^{1/p}$$

$$\leq c\omega(r) r \left(\int_0^r \fint_{S_+^{n-1}} |\nabla w|^p \, ds \frac{d\rho}{\rho} \right)^{1/p},$$

and we can estimate the last line (much as in (40c) but with $r_j = 2^{-j} r$) by

$$c\omega(r) r \left(\sum_{j=0}^\infty M_p^p(\nabla w, r_j) \right)^{1/p} \leq c\omega(r) r \left(\int_0^r \omega^p(\rho) \, d\rho \right)^{1/p} \leq c\omega(r) r.$$

We have one more term to estimate (using $r_j = 2^{-j} r$):

$$\int_0^r \omega(\rho)\rho|\tilde{v}'(\rho)| \, d\rho = \sum_{j=0}^\infty \int_{r_{j+1}}^{r_j} \omega(\rho)\,\rho\,|\tilde{v}'(\rho)| \, d\rho \leq \sum_{j=0}^\infty \omega(r_j) \int_{t_j}^{t_j+\ln 2} |\tilde{v}_t|e^{-t} \, dt$$

$$\leq \sum_{j=0}^\infty r_j \, \omega^2(r_j) \leq c \int_0^r \omega^2(\rho) \, d\rho \leq c\omega^2(r)\,r,$$

where we have used (56a)). We conclude that

$$|u_0(r) - u_0(0)| \leq c\omega(r)\,r \quad \text{as } r \to 0, \tag{58}$$

which shows that u_0 is differentiable at $r = 0$ with $u_0'(0) = 0$. Since u_0 and w in (54) are differentiable and $\tilde{x} \cdot \tilde{v}$ is Lipschitz at $x = 0$, we conclude that u^* is Lipschitz at $x = 0$.

Next we need to confirm that $u = u^*$ in order to conclude that u is Lipschitz at $x = 0$. But u and u^* both satisfy (34) and the estimate $M_{1,p}(u,r) \leq c\, r^{-n}$ as $r \to \infty$. Then, by Corollary 6 in Appendix F, we see indeed that $u = u^*$.

Finally, let us also assume that all solutions of (4) are asymptotically constant. Then, by Proposition 2 in Appendix E, we know that $\phi(t) \to \phi_\infty$ as $t \to \infty$. Using (38f), we can apply the above arguments to $\widetilde{v} - n\phi_\infty$ to conclude that

$$\sup_{0 < \rho < r} |\widetilde{v}(\rho) - n\phi_\infty| \leq c\omega(r).$$

This shows that $\widetilde{x} \cdot \widetilde{v}(r)$ is differentiable at $x = 0$. Putting this together with the differentiability of u_0 and w at $x = 0$, we have completed the proof of Theorem 1.

□

We can use the results of [14] on the largest eigenvalue $\mu(r)$ of the symmetric matrix $S(r) = -\frac{1}{2}(R(r) + R^t(r))$ to obtain the following corollaries to Theorem 1; in both we assume the a_{ij} satisfy (32), where ω satisfies (3) and (10). To begin with, in [14] it was shown that (8) implies that (4) is uniformly stable; hence we obtain the following:

Corollary 1. *Suppose that $\mu(r)$ satisfies (8). Then every solution $u \in H^{1,2}_{\mathrm{loc}}(\mathbb{R}^n_+)$ of (31) is Lipschitz continuous at $x = 0$.*

Moreover, in [14] it was shown that (9) implies that the null solution of (4) is asymptotically stable, which in turn shows that $\widetilde{v}$ in (11) tends to zero as $r \to 0$. Consequently, we obtain the following:

Corollary 2. *Suppose that $\mu(r)$ satisfies (9). Then every solution $u \in H^{1,2}_{\mathrm{loc}}(\mathbb{R}^n_+)$ of (31) is differentiable at $x = 0$ and all derivatives are zero: $\partial_j u(0) = 0$ for $j = 1, \ldots, n$.*

3. Curved boundaries

In this section we consider the regularity of a weak solution of (1) near a point on ∂U. Since we are interested in the local behavior of solutions, we may assume that U is bounded, the point on ∂U is the origin in $\mathbb{R}^n$, and the boundary ∂U is given near the origin by $x_n = h(\widetilde{x})$ where $h(\widetilde{0}) = 0$. Recall our assumption (6), which implies that h is differentiable at $\widetilde{x} = 0$ and $\nabla h(\widetilde{0}) = 0$.

Let us introduce new independent variables

$$y_j = x_j \text{ for } j = 1, \ldots, n-1 \quad \text{and} \quad y_n = x_n - h(x_1, \ldots, x_{n-1}).$$

Notice that $\partial y_j / \partial x_k = \delta_{jk}$ for $j \neq n$ and $\partial y_n / \partial x_k = -\partial h / \partial x_k$ for $k = 1, \ldots, n-1$ and $\partial y_n / \partial x_n = 1$. Consequently, the Jacobian determinant for this change of variables is 1 and by the chain rule

$$\frac{\partial u}{\partial x_k} = \frac{\partial u}{\partial y_k} - \frac{\partial u}{\partial y_n}\frac{\partial h}{\partial x_k} \text{ for } k = 1, \ldots, n-1 \quad \text{and} \quad \frac{\partial u}{\partial x_n} = \frac{\partial u}{\partial y_n}.$$

We want to express (2) in terms of the y-coordinates. Let i' and j' be indices that range from 1 to $n-1$. Then

$$a_{ij}\frac{\partial u}{\partial x_j}\frac{\partial \eta}{\partial x_i} = a_{i'j'}\left(\frac{\partial u}{\partial y_{j'}} - \frac{\partial u}{\partial y_n}\frac{\partial h}{\partial x_{j'}}\right)\left(\frac{\partial \eta}{\partial y_{i'}} - \frac{\partial \eta}{\partial y_n}\frac{\partial h}{\partial x_{i'}}\right)$$

$$+ a_{i'n}\frac{\partial u}{\partial y_n}\left(\frac{\partial \eta}{\partial y_{i'}} - \frac{\partial \eta}{\partial y_n}\frac{\partial h}{\partial x_{i'}}\right) + a_{nj'}\left(\frac{\partial u}{\partial y_{j'}} - \frac{\partial u}{\partial y_n}\frac{\partial h}{\partial x_{j'}}\right)\frac{\partial \eta}{\partial y_n} + a_{nn}\frac{\partial u}{\partial y_n}\frac{\partial \eta}{\partial y_n}$$

$$= a_{i'j'}\frac{\partial u}{\partial y_{j'}}\frac{\partial \eta}{\partial y_{i'}} + \left(a_{i'n} - a_{i'j'}\frac{\partial h}{\partial x_{j'}}\right)\frac{\partial u}{\partial y_n}\frac{\partial \eta}{\partial y_{i'}} + \left(a_{nj'} - a_{i'j'}\frac{\partial h}{\partial x_{i'}}\right)\frac{\partial u}{\partial y_{j'}}\frac{\partial \eta}{\partial y_n}$$

$$+ \left(a_{nn} - a_{i'n}\frac{\partial h}{\partial x_{i'}} - a_{nj'}\frac{\partial h}{\partial x_{j'}} + a_{i'j'}\frac{\partial h}{\partial x_{j'}}\frac{\partial h}{\partial x_{i'}}\right)\frac{\partial u}{\partial y_n}\frac{\partial \eta}{\partial y_n}.$$

Now, if we let $U_0 = U \cap B_\varepsilon(0)$ for $\varepsilon > 0$ sufficiently small, then $x \in U_0$ satisfies $x_n > h(x_1, \ldots, x_{n-1})$, so if we let V_0 denote the corresponding domain in the y-variables, then $V_0 \subset \mathbb{R}^n_+$ and

$$\int_{U_0} a_{ij}\frac{\partial u}{\partial x_i}\frac{\partial \eta}{\partial x_j}\,dx = \int_{V_0} \widetilde{a}_{ij}\frac{\partial u}{\partial y_i}\frac{\partial \eta}{\partial y_j}\,dy,$$

where

$$\widetilde{a}_{ij} = \begin{cases} a_{ij} & \text{if } 1 \leq i,j \leq n-1, \\ a_{in} - a_{ij'}\frac{\partial h}{\partial x_{j'}} & \text{if } 1 \leq i \leq n-1,\ j = n, \\ a_{nj} - a_{i'j}\frac{\partial h}{\partial x_{i'}} & \text{if } 1 \leq j \leq n-1,\ i = n, \\ a_{nn} - a_{i'n}\frac{\partial h}{\partial x_{i'}} - a_{nj'}\frac{\partial h}{\partial x_{j'}} + a_{i'j'}\frac{\partial h}{\partial x_{j'}}\frac{\partial h}{\partial x_{i'}} & \text{if } i = j = n. \end{cases}$$

This enables us to view the original problem as one for the coefficients $\widetilde{a}_{ij}$ in the half-space $\{(y_1, \ldots, y_n) : y_n > 0\}$. In order to apply our results from the previous section, we need $\widetilde{a}_{ij}$ to be square-Dini continuous and satisfy (32); but these conditions follow from our assumption (6).

Now we can write down the 1st-order dynamical system (38) associated with the $\widetilde{a}_{ij}$ in $\mathbb{R}^{n-1}_+$ whose stability properties determine the differentiability of a weak solution. In particular, formula (5) for the $(n-1) \times (n-1)$ matrix R yields

$$[R(r)]_{\ell k} = \fint_{S^{n-1}_+}\left(a_{\ell k} - n\sum_{j=1}^{n} a_{\ell j}\theta_j\theta_k + n\sum_{j=1}^{n-1} a_{\ell j}\frac{\partial h}{\partial x_j}\theta_n\theta_k\right)ds_\theta. \tag{59}$$

In (59) we need to emphasize that the integrand is regarded as a function of $y \in \mathbb{R}^n_+$, even though the coefficients a_{ij} and h were originally defined in the x variables. Also, note that if $h \equiv 0$, then we are in the half-space case, and the formula for $R(r)$ in (59) agrees with (5).

Theorem 2. *Suppose that U is a bounded domain with Lipschitz boundary ∂U containing the point 0, near which the boundary can be represented as $x_n = h(x_1, \ldots, x_{n-1})$. Suppose the a_{ij} satisfy (32) and h satisfies (6), where ω satisfies (3) and (10). If the dynamical system (4) with matrix R given by (59) is* *uniformly stable as $t \to \infty$, then every solution $u \in H^{1,2}(U \cap B)$ of (2) is Lipschitz continuous at $x = 0$. If, in addition, every solution of the dynamical system (4), (59) is asymptotically constant as $t \to \infty$, then u is differentiable at $x = 0$.*

As in Section 2, the conditions (8) and (9) can be used to obtain corollaries to this theorem, but now μ is the largest eigenvalue of the matrix $S(r) = -\frac{1}{2}(R(r) + R^t(r))$, where R is given by (59). In the following results we assume the conditions on a_{ij}, h, and ω stated in the theorem; in the second one we note that $\partial u(0)/\partial x_j = \partial u(0)/\partial y_j$ since $\nabla h(\widetilde{0}) = 0$.

Corollary 3. *Suppose that $\mu(r)$ satisfies (8). Then every solution $u \in H^{1,2}_{\text{loc}}(U \cap B)$ of (2) is Lipschitz continuous at $x = 0$.*

Corollary 4. *Suppose that $\mu(r)$ satisfies (9). Then every solution $u \in H^{1,2}_{\text{loc}}(U \cap B)$ of (2) is differentiable at $x = 0$ and all derivatives are zero: $\partial_j u(0) = 0$ for $j = 1, \ldots, n$.*

4. Examples: $n = 2$

Let us first consider variable coefficients a_{ij} in $\mathbb{R}^2_+$. For $n = 2$ we have $\theta_1 = \cos \phi$ and $\theta_2 = \sin \phi$ for $0 < \phi < \pi$, so (5) yields a scalar function

$$R(r) = \frac{1}{\pi} \int_0^\pi \left(a_{11}(r\theta) - 2a_{11}(r\theta) \cos^2 \phi - 2a_{12}(r\theta) \cos \phi \sin \phi \right) d\phi. \tag{60a}$$

In this case, the dynamical system (4) is merely a single equation, and we easily find the general solution: $\phi(t) = C \exp[-\int_T^t R(e^{-\tau}) d\tau]$. Consequently (cf. Remark 2 in Appendix D), we know that (4) is uniformly stable if and only if $\int_s^t R(e^{-\tau}) d\tau$ is uniformly bounded below for $T < s < t < \infty$. Expressing this in terms of r rather than t, we see that uniform stability

$$\int_{r_1}^{r_2} \frac{R(\rho)}{\rho} d\rho > -K \quad \text{for all } 0 < r_1 < r_2 < \varepsilon \tag{60b}$$

implies that every weak solution of (31) is Lipschitz at the origin. Similarly, solutions of (4) are asymptotically constant when $\int_T^\infty R(e^{-\tau}) d\tau$ either converges to a finite number or diverges to ∞. In terms of $R(r)$, we find that (60b) together with the condition

$$\int_0^\varepsilon \frac{R(\rho)}{\rho} d\rho \text{ converges to an extended real number } > -\infty \tag{60c}$$

imply that every weak solution of (31) is differentiable at the origin.

To make all this more precise, let us turn to a class of operators considered in [7] and [14]:

$$a_{ij} = \delta_{ij} + g(r)\theta_i \theta_j, \tag{61a}$$

where $|g(r)| \leq c\omega(r)$. In this case we can calculate $R(r) = -\frac{1}{2}g(r)$ so that uniform stability

$$\int_{r_1}^{r_2} \frac{g(\rho)}{\rho}\,d\rho < K \quad \text{for all } 0 < r_1 < r_2 < \varepsilon \tag{61b}$$

implies that every weak solution of (31) is Lipschitz continuous at the origin; and if in addition

$$\int_0^{\varepsilon} \frac{g(\rho)}{\rho}\,d\rho \text{ converges to an extended real number} < \infty, \tag{61c}$$

then every weak solution of (31) is differentiable at the origin.

For (61a) we can construct explicit solutions of (31) by solving an ODE. For example, if we let

$$u(r,\phi) = U(r)\,\cos\phi, \tag{62}$$

then this is a solution provided U satisfies

$$\frac{1}{r}\left[(1+g(r))\,r\,U'\right]' - \frac{1}{r^2}U = 0. \tag{63}$$

Moreover, we can determine the behavior of $U(r)$ as $r \to 0$ from that of $g(r)$. To do this, it is simpler to again use the variable $t = -\log r$. Letting $\tilde{g}(t) = g(e^{-t})$, we want U to satisfy

$$\frac{d}{dt}\left[(1+\tilde{g}(t))\,\frac{dU}{dt}\right] - U = 0 \quad \text{as } t \to \infty. \tag{64}$$

We can apply standard results in the asymptotic theory of ODEs. For example, if $\tilde{g}(t)$ is C^1 and satisfies

$$\tilde{g}(t),\ \frac{d\tilde{g}}{dt} = o(1) \quad \text{as } t \to \infty, \tag{65}$$

then we can apply Theorem 2.2.1 in [3] to conclude that a solution $U(t)$ of (64) exists for which both $U(t)$ and $(1+\tilde{g}(t))dU/dt$ are asymptotic to

$$(1+\tilde{g}(t))^{-1/4} \exp\left(-\int_1^t \left(\frac{1}{1+\tilde{g}(s)} + \frac{(d\tilde{g}/ds)^2}{16(1+\tilde{g}(s))^2}\right)^{1/2} ds\right) \tag{66}$$

$$\sim e^{-t}\exp\left(\frac{1}{2}\int_1^t \tilde{g}(s)\,ds\right).$$

This solution satisfies the finite-energy condition $\int_1^{\infty}(U^2 + (U_t)^2)e^{-nt}\,dt < \infty$, so u is an $H^{1,2}$-solution of (31). However, if $g(r)$ does not satisfy the Dini condition at $r = 0$ then $\int_1^t \tilde{g}(s)\,ds \to \infty$ as $t \to \infty$ and u is not Lipschitz continuous at the origin. An example of such a function $g(r)$ is

$$g(r) = |\log r|^{-\alpha} \quad \text{where } 1/2 < \alpha \leq 1; \tag{67}$$

note that $\tilde{g}(t) = t^{-\alpha}$ satisfies (65) but (61b) is not satisfied. In particular, this example shows that a weak solution of (31) when the coefficients a_{ij} are square-Dini continuous need not be Lipschitz continuous if the associated dynamical system (4) is not uniformly stable.

Next let us suppose that the origin lies on the boundary ∂U, which locally has the form $x_2 = h(x_1)$, where $h(0) = 0$ and $|h'(r)| \leq c\omega(r)$ as $r \to 0$. Then we introduce new independent variables $y_1 = x_1$ and $y_2 = x_2 - h(x_1)$ and view a_{ij} as functions of $(y_1, y_2) \in \mathbb{R}_+^2$. We can calculate the scalar function $R(r)$ in (59):

$$\frac{1}{\pi} \int_0^\pi \left(a_{11}(r\theta) - 2\big(a_{11}(r\theta)\cos^2\phi + a_{12}(r\theta)\cos\phi\sin\phi\big) \right.$$
$$\left. + 2a_{11}(r\theta)h'(r\theta_1)\cos\phi\sin\phi \right) d\phi. \tag{68}$$

Again, we find that (60b) implies that every weak solution $u \in H^{1,2}(U)$ of (1) is Lipschitz at the origin, and if (60c) also holds then u is differentiable there.

Now let us consider the special case of (68) when the operator is the Laplacian, so that $a_{ij} = \delta_{ij}$. In this case, we have simply

$$R(r) = \frac{2}{\pi} \int_0^\pi h'(r\cos\phi)\cos\phi\sin\phi\, d\phi. \tag{69}$$

One way to make sure that (60b) and (60c) hold is to have $R(r) \geq 0$ for $0 < r < \varepsilon$. This will be the case, for example, if

$$h'(x) \leq 0 \quad \text{for } -\varepsilon < x < 0 \quad \text{and} \quad h'(x) \geq 0 \quad \text{for } 0 < x < \varepsilon. \tag{70}$$

Consequently, if the boundary function h satisfies (70), we can conclude that every weak solution $u \in H^{1,2}(U)$ of (1) is differentiable at the origin.

We should compare our results for the Laplacian with those of [19] concerning conformal maps. In [19], the hypotheses on the boundary are weaker than ours, and asymptotics are obtained, not just conclusions about differentiability. However, under the hypotheses on the boundary that we consider, Theorem XI(A) in [19] shows that the behavior of a conformal map as $z \to 0$ is dominated by

$$\exp\left[-\pi \int_{|z|}^a \frac{1}{r\,\Theta(r)}\, dr \right]. \tag{71}$$

Here $\Theta(r)$ measures the angle between the two arcs Γ_- and Γ_+ corresponding to $x_2 = h(x_1)$ for $x < 0$ and for $x > 0$ respectively. Consequently, $|\Theta(r) - \pi| \leq \omega(r)$ as $r \to 0$, and we can write

$$\frac{\pi}{r\Theta(r)} = \frac{1}{r}\left[1 - \left(1 - \frac{\Theta(\rho)}{\pi} \right) \right]^{-1} = \frac{1}{r}\left[1 + \left(1 - \frac{\Theta(r)}{\pi} \right) + O(\omega^2(r)) \right].$$

Thus, as $|z| \to 0$, (71) is asymptotic to

$$C\,|z| \exp\left[\frac{1}{\pi} \int_{|z|}^a \frac{\Theta(r) - \pi}{r}\, dr \right]. \tag{72}$$

This means, for example, that the convergence (or divergence to $-\infty$) of the intergal in (72) determines whether the conformal map is Lipschitz continuous at $z = 0$; this is an analogue to our condition (60b) for a harmonic function to be Lipschitz continuous at the origin.

Appendix A. Asymptotic expansion of the Neumann function

In this appendix we derive the asymptotic expansion of the Neumann function $N(x,y)$. We need to use an expansion of the fundamental solution Γ in spherical harmonics. Let $\mathcal{H}(k)$ denote the spherical harmonics of degree k and let $N(k) = \dim \mathcal{H}(k)$. For each k, choose a basis $\{\varphi_{k,m} : m = 1, \ldots, N(k)\}$ for $\mathcal{H}(k)$ that is orthonormal with respect to the spherical mean inner product:

$$\fint_{S^{n-1}} \varphi_{k,\ell}\, \varphi_{k,m}\, ds = \begin{cases} 1 & \ell = k \\ 0 & \ell \neq m \end{cases}.$$

For notational convenience, let $\hat{x} = x/|x|$ and $\hat{y} = y/|y|$. We also assume that $n \geq 3$, the case of $n = 2$ being analogous. For $|x| < |y|$ we can write $\Gamma(|x-y|)$ as a convergent series

$$\Gamma(|x-y|) = \sum_{k=0}^{\infty} \frac{|x|^k}{|y|^{n-2+k}} \sum_{m=1}^{N(k)} a_{k,m}\, \varphi_{k,m}(\hat{x})\, \varphi_{k,m}(\hat{y}), \tag{73a}$$

where $a_{k,m}$ are certain coefficients.[2] With $x = 0$ we know that $\Gamma(|y|) = a_0 |y|^{2-n}$ with $a_0 = (2-n)^{-1}\omega_n^{-1}$ where $\omega_n = |S^{n-1}|$. We can also use a Taylor series for $f_y(x) = |x-y|^{2-n}$, i.e.,

$$|x-y|^{2-n} = |y|^{2-n} + (n-2)|y|^{-n}\sum_j x_j y_j + \cdots$$

to compute the other coefficients. For example, we can write

$$\Gamma(|x-y|) = a_0 \left(\frac{1}{|y|^{n-2}} + (n-2)\frac{|x|}{|y|^{n-1}} \sum_{m=1}^{n} \hat{x}_m \hat{y}_m + \cdots \right). \tag{73b}$$

But to compute our Neumann function $N(x,y)$, we want the basis $\{\varphi_{km}\}$ for $k > 1$ to also possess certain symmetries with respect to the half-space.

Recall that the spherical harmonics of degree k are generated by the restriction to the unit sphere of the harmonic polynomials of degree k:

$$h(x) = \sum_{|\alpha|=k} c_\alpha x^\alpha \quad \text{is harmonic,} \tag{74}$$

where $\alpha = (\alpha_1, \ldots, \alpha_n)$ and $x^\alpha = x_1^{\alpha_1} \cdots x_n^{\alpha_n}$. For our half-space geometry, we want to distinguish those harmonic functions for which α_n is even or odd. Let $\mathcal{H}_e(k)$ be the spherical harmonics corresponding to even α_n and let $N_e(k)$ denote its dimension; choose an orthonormal basis $\{\varphi_{k,m}^e : m = 1, \ldots, N_e(k)\}$ for $\mathcal{H}_e(k)$. Similarly, let $\mathcal{H}_o(k)$ be the spherical harmonics corresponding to odd α_n and choose an orthonormal basis $\{\varphi_{k,m}^o : m = 1, \ldots, N_o(k)\}$ for $\mathcal{H}_o(k)$. Then $\{\varphi_{k,m}^e : m =$

---

[2]The expansion (73a) was used in [14], but the coefficients were unfortunately left out of the formula there.

$1, \ldots, N_e(k)\} \cup \{\varphi^o_{k,m} : m = 1, \ldots, N_o(k)\}$ is an orthonormal basis for $\mathcal{H}(k)$ which we may use to rewrite (73a) as:

$$
\Gamma(|x - y|) = \sum_{k=0}^{\infty} \frac{|x|^k}{|y|^{n-2+k}} \left(\sum_{m=1}^{N_e(k)} a_{k,m} \, \varphi^e_{k,m}(\hat{x}) \, \varphi^e_{k,m}(\hat{y}) \right.
$$
$$
\left. + \sum_{m=1}^{N_o(k)} b_{k,m} \, \varphi^o_{k,m}(\hat{x}) \, \varphi^o_{k,m}(\hat{y}) \right), \tag{75}
$$

But

$$
y^* = (\tilde{y}, -y_n) \quad \Rightarrow \quad \varphi^e_{k,m}(\hat{y}^*) = \varphi^e_{k,m}(\hat{y}) \quad \text{and} \quad \varphi^o_{k,m}(\hat{y}^*) = -\varphi^o_{k,m}(\hat{y})
$$

so

$$
\Gamma(|x - y^*|) = \sum_{k=0}^{\infty} \frac{|x|^k}{|y|^{n-2+k}} \left(\sum_{m=1}^{N_e(k)} a_{k,m} \, \varphi^e_{k,m}(\hat{x}) \, \varphi^e_{k,m}(\hat{y}) \right.
$$
$$
\left. - \sum_{m=1}^{N_o(k)} b_{k,m} \, \varphi^o_{k,m}(\hat{x}) \, \varphi^o_{k,m}(\hat{y}) \right).
$$

When we add $\Gamma(|x - y|)$ and $\Gamma(|x - y^*|)$ the terms involving $\varphi^o_{k,m}$ cancel, so we obtain

$$
N(x, y) = 2 \sum_{k=0}^{\infty} \frac{|x|^k}{|y|^{n-2+k}} \sum_{m=1}^{N_e(k)} a_{k,m} \, \varphi^e_{k,m}(\hat{x}) \, \varphi^e_{k,m}(\hat{y}) \quad \text{for } |x| < |y|.
$$

Restricting the $\varphi^e_{k,m}$ to S^{n-1}_+ yields spherical harmonics with zero normal derivative along the boundary ∂S^{n-1}_+, but we also want them to be orthonormal with respect to the spherical mean inner product on S^{n-1}_+. We easily calculate

$$
\fint_{S^{n-1}_+} \varphi^e_{k,m} \varphi^e_{k',m'} \, ds = \frac{1}{2} \fint_{S^{n-1}} \varphi^e_{k,m} \varphi^e_{k',m'} \, ds = \begin{cases} \frac{1}{2} & k = k' \text{ and } m = m', \\ 0 & k \neq k' \text{ or } m \neq m'. \end{cases}
$$

Consequently, we will have an orthonormal basis $\{\tilde{\varphi}_{k,m} : m = 1, \ldots, \tilde{N}(k)\}$ of spherical harmonics with zero normal derivative along the boundary ∂S^{n-1}_+ if we define:

$$
\tilde{\varphi}_{k,m} = \sqrt{2} \, \varphi^e_{k,m}|_{S^{n-1}_+} \quad \text{and} \quad \tilde{N}(k) = N_e(k). \tag{76}
$$

For $k = 1$, we want $\tilde{\varphi}_{1,m} = \bar{c} \theta_m$ for some constant $\bar{c}$ and all $m = 1, \ldots, n - 1$. Using (15b) and the fact that the $\varphi_{1,m}$ are orthonormal, we see that

$$
\tilde{\varphi}_{1,m} = \frac{1}{\sqrt{c_n}} \theta_m, \quad \text{for } m = 1, \ldots, n - 1. \tag{77}
$$

We therefore obtain (20a). By interchanging the roles of x and y we get the expansions of $\Gamma(|x - y|)$ and $\Gamma(|x - y^*|)$ for $|x| > |y|$, and add them together to obtain (20b).

Appendix B. Orthogonality properties

In this appendix we discuss orthogonality properties necessary to show (36). In fact, we first prove the following.

Lemma 3. *If* $f \in H^{1,1}_{\mathrm{loc}}(\overline{\mathbb{R}^n} \setminus \{0\})$ *and* $r > 0$, *then for* $i = 1, \ldots, n$ *we have*

$$\fint_{S^{n-1}_+} f(r\theta)\, ds = 0 \;\Rightarrow\; \fint_{S^{n-1}_+} \theta_i \partial_i f(r\theta)ds = 0, \tag{78}$$

and for $j = 1, \ldots, n-1$ *we have*

$$\fint_{S^{n-1}_+} \theta_j f(r\theta)\, ds = 0 \;\Rightarrow\; \fint_{S^{n-1}_+} \partial_j f(r\theta)ds = 0 = \fint_{S^{n-1}_+} \theta_j \theta_i \partial_i f(r\theta)ds. \tag{79}$$

Proof. To prove (78) we consider $\phi \in C^\infty_{\mathrm{comp}}(0, \infty)$ and write

$$\left\langle \fint \theta_i \partial_i f ds, \phi \right\rangle = \int_0^\infty \fint_{S^{n-1}_+} \theta_i\, \partial_i f(r\theta) ds\, \phi(r)\, dr$$

$$= \frac{1}{|S^{n-1}_+|} \int_{\mathbb{R}^n_+} x_i \partial_i f(x)\phi(|x|)|x|^{-n}\, dx.$$

Taking the divergence of $x_i f(x)\phi(|x|)|x|^{-n}$, we obtain

$$\partial_i(x_i f(x)\phi(|x|)|x|^{-n}) = \theta_i \partial_i f(x)\phi(|x|)|x|^{-n+1} + f(x)\phi'(|x|)|x|^{-n}.$$

By the divergence theorem,

$$\int_{\mathbb{R}^n_+} \partial_i(x_i f(x)\phi(|x|)|x|^{-n})dx = -\int_{\mathbb{R}^{n-1}} (x_n f(x)\phi(|x|)|x|^{-n})|_{x_n=0}\, d\tilde{x} = 0,$$

so

$$\int_{\mathbb{R}^n_+} x_i \partial_i f(x)\phi(|x|)|x|^{-n} dx = -\int_{\mathbb{R}^n_+} f(x)\phi'(|x|)|x|^{-n+1} dx$$

$$= -\int_0^\infty \fint_{S^{n-1}_+} f(r\theta)\, ds\, \phi'(|x|)\, dr.$$

By the hypothesis in (78), this last integral vanishes, which confirms the conclusion in (78).

To prove (79), we again consider $\phi \in C^\infty_{\mathrm{comp}}(0, \infty)$ and write

$$\left\langle \fint \partial_j f ds, \phi \right\rangle = \int_0^\infty \fint_{S^{n-1}_+} \partial_j f(r\theta)\, ds\, \phi(r)\, dr$$

$$= \frac{1}{|S^{n-1}_+|} \int_{\mathbb{R}^n_+} \partial_j f(x)|x|^{1-n}\phi(|x|)\, dx.$$

But we can integrate by parts in this last integral to obtain

$$\int_{\mathbb{R}^n_+} f(x)[r^{1-n}\phi(r)]'|_{r=|x|}\theta_j\, dx = \int_0^\infty \fint_{S^{n-1}_+} f(r\theta)\theta_j\, ds\, [r^{1-n}\phi(r)]' r^{n-1} dr.$$

This gives the first conclusion in (79). To obtain the second conclusion, we write

$$\left\langle \oint \theta_j \theta_i \partial_i f \, ds, \phi \right\rangle = \frac{1}{|S_+^{n-1}|} \int_{\mathbb{R}_+^n} x_j \, x_i \partial_i f(x) \phi(|x|) |x|^{-n-1} \, dx.$$

Take the divergence (for fixed j):

$$\partial_i(x_j \, x_i \, f(x) \phi(|x|)|x|^{-n-1}) = x_j f(x) \phi(|x|)|x|^{-n-1} + n x_j f(x) \phi(|x|)|x|^{-n-1}$$
$$+ x_j x_i \partial_i f(x) \phi(|x|)|x|^{-n-1} + x_j x_i f(x) \left(\phi(r) r^{-n-1}\right)' \theta_i.$$

So applying the divergence theorem yields

$$\int_{\mathbb{R}_+^n} x_j \, x_i \partial_i f(x) \phi(|x|)|x|^{-n-1} \, dx = - \int_{\mathbb{R}^{n-1}} \left. \left(x_j x_n f(x) \phi(|x|)|x|^{-n-1}\right)\right|_{x_n=0} d\widetilde{x}$$

$$- \int_{\mathbb{R}_+^n} \left((n+1)x_j f(x)\phi(|x|)|x|^{-n-1} + x_j x_i f(x)\left(\phi(r) r^{-n-1}\right)'\theta_i\right) dx.$$

The boundary integral clearly vanishes and the domain integral simplifies considerably to yield

$$\int_{\mathbb{R}_+^n} x_j \, x_i \partial_i f(x) \phi(|x|)|x|^{-n-1} \, dx = - \int_0^\infty \int_{S_+^{n-1}} \theta_j f(r\theta) \, ds \, \phi'(r) \, dr$$

By the hypothesis in (79), this last integral vanishes, which confirms the second conclusion in (79).　　　　　　　　　　　　　　　　　　　　　　　　　□

Now we are able to address (36).

Corollary 5. *If $u \in H^{1,2}(B_+(1))$ and we introduce the spectral decomposition (11), then there is a constant $c > 0$ such that*

$$\int_{B_+(1)} |\nabla u|^2 \, dx \geq c \int_0^1 \left[(u_0')^2 + |\widetilde{v}|^2 + r^2|\widetilde{v}'|^2\right] r^{n-1} \, dr + \int_{B_+(1)} |\nabla w|^2 \, dx.$$

Proof. We compute

$$\nabla_i u(x) = \begin{cases} u_0'(r)\theta_i + \widetilde{v}'(r) \cdot \widetilde{x} \, \theta_i + v_i(r) + \nabla_i w, & 1 \leq i \leq n-1 \\ u_0'(r)\theta_n + \widetilde{v}'(r) \cdot \widetilde{x} \, \theta_n + \nabla_n w, & i = n, \end{cases}$$

and

$$|\nabla u|^2 = (u_0')^2 + 2u_0'(\widetilde{v}' \cdot \widetilde{x}) + 2u_0'(\widetilde{\theta} \cdot \widetilde{v}) + 2u_0'(\theta \cdot \nabla w) + (\widetilde{v}' \cdot \widetilde{x})^2$$
$$+ 2(\widetilde{v}' \cdot \widetilde{x})(\widetilde{\theta} \cdot \widetilde{v}) + 2(\widetilde{v}' \cdot \widetilde{x})(\theta \cdot \nabla w) + |\widetilde{v}|^2 + 2\widetilde{v} \cdot \nabla w + |\nabla w|^2.$$

(In the formula for $|\nabla u|^2$, note that $\widetilde{\theta} = (\theta_1, \ldots, \theta_{n-1})$ and dot products involving $\widetilde{\theta}$ or $\widetilde{v}$ are summed only over $1, \ldots, n-1$.) The integral over S_+^{n-1} of some of these terms vanish due to $\int_{S_+^{n-1}} \theta_i \, ds = 0$ for $i = 1, \ldots, n-1$:

$$\int_{S_+^{n-1}} u_0' \left(\widetilde{v}' \cdot \widetilde{x}\right) ds = 0 = \int_{S_+^{n-1}} u_0' \left(\widetilde{\theta} \cdot \widetilde{v}\right) ds.$$

Other terms vanish by (78) and (79):

$$\int_{S_+^{n-1}} u_0'(\theta \cdot \nabla w)\,ds = 0 = \int_{S_+^{n-1}} \tilde{v} \cdot \nabla w\,ds = \int_{S_+^{n-1}} (\tilde{v}' \cdot \tilde{x})(\theta \cdot \nabla w)\,ds.$$

Still other terms simplify in view of (15b):

$$\int_{S_+^{n-1}} (\tilde{v}' \cdot \tilde{x})^2\,ds = r^2 \sum_{i,j=1}^{n-1} v_i' v_j' \int_{S_+^{n-1}} \theta_i \theta_j\,ds = \frac{r^2 |S_+^{n-1}|}{n} \sum_{i=1}^{n-1} (v_i')^2$$

$$\int_{S_+^{n-1}} (\tilde{v}' \cdot \tilde{x})(\tilde{\theta} \cdot \tilde{v})\,ds = r \sum_{i,j=1}^{n-1} v_i' v_j \int_{S_+^{n-1}} \theta_i \theta_j\,ds = \frac{r |S_+^{n-1}|}{n} \sum_{i=1}^{n-1} v_i' v_i.$$

So

$$\int_{B_+(1)} |\nabla u|^2\,dx = |S_+^{n-1}| \int_0^1 \left((u_0')^2 + \frac{r^2}{n} |\tilde{v}'|^2 + \frac{2r}{n} \tilde{v}' \cdot \tilde{v} \right.$$
$$\left. + |\tilde{v}|^2 \right) r^{n-1}\,dr + \int_{B_+(1)} |\nabla w|^2\,dx.$$

Using

$$\frac{2r}{n} \tilde{v} \cdot \tilde{v}' = \frac{2r}{n^{2/3} n^{1/3}} \tilde{v} \cdot \tilde{v}' \geq -\frac{r^2 |\tilde{v}'|^2}{n^{4/3}} - \frac{|\tilde{v}|^2}{n^{2/3}}$$

we find

$$\int_{|x|<1} |\nabla u|^2\,dx \geq |S^{n-1}| \int_0^1 \left((u_0')^2 + r^2 \left(\frac{1}{n} - \frac{1}{n^{4/3}} \right) |\tilde{v}'|^2 \right.$$
$$\left. + \left(1 - \frac{1}{n^{2/3}} \right) |\tilde{v}|^2 \right) r^{n-1}\,dr + \int_{|x|<1} |\nabla w|^2\,dx.$$

Since $n^{4/3} > n$ and $n^{2/3} > 1$, this completes the proof. □

Appendix C. Sobolev regularity of weak solutions

In this appendix we show that, if $u \in H^{1,2}_{\text{loc}}(\overline{\mathbb{R}^n_+})$ satisfies (31) where the a_{ij} are continuous functions then $u \in H^{1,p}_{\text{loc}}(\overline{\mathbb{R}^n_+})$ for any $p > 2$. Let us introduce the operator $\mathcal{L}$, which is defined on $v \in H^{1,q}_{\text{loc}}(\overline{\mathbb{R}^n_+})$ for any $q > 1$, and assigns a functional on $H^{1,q'}_{\text{comp}}(\overline{\mathbb{R}^n_+})$ defined by

$$\langle \mathcal{L}v, \eta \rangle = -\int_{\mathbb{R}^n_+} a_{ij} \partial_j v\, \partial_i \eta\,dx \quad \text{for all } \eta \in H^{1,q'}_{\text{comp}}(\overline{\mathbb{R}^n_+}). \tag{80}$$

In this context, we have assumed that $u \in H^{1,2}_{\text{loc}}(\overline{\mathbb{R}^n_+})$ is a solution of $\mathcal{L}u = 0$, and we want to conclude that $u \in H^{1,p}_{\text{loc}}(\overline{\mathbb{R}^n_+})$ for any $p > 2$.

The assertion $u \in H^{1,p}_{\text{loc}}(\overline{\mathbb{R}^n_+})$ is proved by localizing near a point in $\overline{\mathbb{R}^n_+}$. Since the issue is on the boundary, we assume the point is 0, so it suffices to show that $\phi_0 u \in H^{1,p}(B_+)$ for some $\phi_0 \in C_0^\infty(B)$ with $\phi_0 \equiv 1$ near 0; here

$B = \{x \in \mathbb{R}^n : |x| < 1\}$ and $B_+ = \{x \in \mathbb{R}^n_+ : |x| < 1\}$. By continuity, for any $\varepsilon > 0$ we can find a $\delta > 0$ so that $\sup_{|x| \le \delta} |a_{ij}(x) - \delta_{ij}| \le \varepsilon$. However, for notational convenience we simply assume a small oscillation condition in B_+:

$$\sup_{|x| \le 1} |a_{ij}(x) - \delta_{ij}| \le \varepsilon, \tag{81}$$

Let us denote by $\mathcal{L}_0$ the operator (80) with $a_{ij} = \delta_{ij}$. Let $N(x,y)$ be the Neumann function for the Laplacian on $\mathbb{R}^n_+$, and denote the associated integral operator by $\mathcal{N}$. Note that for $u \in C^1_{\text{comp}}(\overline{\mathbb{R}^n_+})$ we have by Green's identities

$$\mathcal{N}\mathcal{L}_0 u(x) = -\int_{\mathbb{R}^n_+} \nabla_y N(x,y) \cdot \nabla u(y)\, dy = \int_{\mathbb{R}^n_+} \Delta_y N(x,y) u(y)\, dy = u(x).$$

Since any $u \in H^{1,p}_{\text{comp}}(\overline{\mathbb{R}^n_+})$ can be approximated by $u_j \in C^1_{\text{comp}}(\overline{\mathbb{R}^n_+})$, we conclude that $\mathcal{N}\mathcal{L}_0$ is the identity on $H^{1,p}_{\text{comp}}(\overline{\mathbb{R}^n_+})$.

For $\phi_1 \in C^\infty_{\text{comp}}(B)$ satisfying $\phi_0 \phi_1 = \phi_0$ on B, let us write

$$\mathcal{L}_0(\phi_0 u) + (\mathcal{L} - \mathcal{L}_0)(\phi_0 u) = \phi_0 \mathcal{L} u + [\mathcal{L}, \phi_0](\phi_1 u) = [\mathcal{L}, \phi_0](\phi_1 u),$$

where we have used (31) to conclude that $\phi_0 \mathcal{L} u = 0$. Now we apply $\mathcal{N}$ to obtain

$$\phi_0 u + \mathcal{N}(\mathcal{L} - \mathcal{L}_0)(\phi_0 u) = \mathcal{N}[\mathcal{L}, \phi_0](\phi_1 u). \tag{82}$$

Taking $\varepsilon = \varepsilon(p)$ sufficiently small in (81), we can arrange that both

$$\mathcal{N}(\mathcal{L} - \mathcal{L}_0)\phi_1 : H^{1,p}(B_+) \to H^{1,p}(B_+) \quad \text{and}$$
$$\mathcal{N}(\mathcal{L} - \mathcal{L}_0)\phi_1 : H^{1,2}(B_+) \to H^{1,2}(B_+) \tag{83}$$

have operator norms less than $1/2$. If we can show that the right-hand side of (82) is in $H^{1,p}(B_+)$, then we can use a Neumann series to conclude that $\phi_0 u \in H^{1,p}(B_+)$. So we only need show

$$[\mathcal{L}, \phi_0](\phi_1 u) \in H^{-1,p}(B_+). \tag{84}$$

For $v \in H^{1,p'}(B_+)$, let us compute

$$\langle [\mathcal{L}, \phi_0](\phi_1 u), v \rangle = \int_{B_+} a_{ij}(x)\left(\partial_j(\phi_1 u)\partial_i(\phi_0 v) - \partial_j(\phi_0 u)\partial_i v\right) dx.$$

Using $\phi_0\, \partial_j u\, \partial_i v = \phi_1 \phi_0\, \partial_j u\, \partial_i v$, we find

$$|\langle [\mathcal{L}, \phi_0](\phi_1 u), v \rangle| \le C \int_{B_+} (|u|\,|\nabla v| + |\nabla u|\,|v|)\, dx.$$

Let us first assume $n > 2$. Then, by the Hölder and Sobolev inequalities,

$$\int_{B_+} |u|\,|\nabla v|\, dx \le \|u\|_{L^p(B_+)}\|v\|_{H^{1,p'}(B_+)} \le C\|u\|_{H^{1,2}(B_+)}\|v\|_{H^{1,p'}(B_+)}$$

provided $p \le 2n/(n-2)$. Similarly, we can use the Hölder and Sobolev inequalities to estimate

$$\int_{B_+} |\nabla u|\,|v|\, dx \le \|u\|_{H^{1,2}(B_+)}\|v\|_{L^2(B_+)} \le C\|u\|_{H^{1,2}(B_+)}\|v\|_{H^{1,p'}(B_+)}$$

provided $2 \le np'/(n-p')$. But we can easily see that $p \le 2n/(n-2)$ is equivalent to $2 \le np'/(n-p')$, so we have shown (84) for $p = 2(1+\alpha)$ where $\alpha = 2/(n-2)$. This is an improvement over $p = 2$, and we can iterate it a finite number of times to deduce (84) for any $p > 2$. If $n = 2$, then the above argument works for any $2 < p < \infty$.

Appendix D. Derivation of the dynamical system

In this appendix we provide the details behind the derivation of the dynamical system (38) for a given solution u of the variational problem (34). Starting with (11), we calculate

$$\partial_j u = u_0'(r)\,\theta_j + (\tilde{v}'(r) \cdot \tilde{x})\,\theta_j + \tilde{v}_j(r) + \partial_j w \quad \text{for } j = 1, \dots, n-1$$

and

$$\partial_n u = u_0'(r)\,\theta_n + (\tilde{v}'(r) \cdot \tilde{x})\,\theta_n + \partial_n w.$$

Now let us consider $\eta = \eta(r)$ in (34). Then $\partial_i \eta = \eta'(r)\theta_i$ for $i = 1, \dots, n$, and plugging this and (11) into (34), we find

$$\int_0^\infty \left[\left(\alpha\,u_0' + r\,\tilde{\beta} \cdot \tilde{v}' + \tilde{\gamma} \cdot \tilde{v} + p[\nabla w] - \overline{f}_1 \right) \eta' + \overline{f}_0\,\eta \right] r^{n-1}\,dr = 0, \qquad (85)$$

where

$$\alpha(r) = \fint_{S_+^{n-1}} \sum_{i,j=1}^n a_{ij}(r\theta)\theta_i\theta_j\,ds,$$

$$\tilde{\beta}_k(r) = \fint_{S_+^{n-1}} \sum_{i,j=1}^n a_{ij}(r\theta)\theta_i\theta_j\theta_k\,ds \quad (k = 1, \dots, n-1),$$

$$\tilde{\gamma}_j(r) = \fint_{S_+^{n-1}} \sum_{i=1}^n a_{ij}(r\theta)\,\theta_i\,ds \quad (j = 1, \dots, n-1),$$

$$p[\nabla w](r) = \fint_{S_+^{n-1}} \sum_{i,j=1}^n a_{ij}(r\theta)\,\partial_j w(r\theta)\,\theta_i\,ds,$$

$$\overline{f}_1(r) = \fint_{S_+^{n-1}} \sum_{i=1}^n f_i(r\theta)\theta_i\,ds, \quad \text{and} \quad \overline{f}_0(r) = \fint_{S_+^{n-1}} f_0(r\theta)\,ds. \qquad (86)$$

Note that α, $p[\nabla w]$, $\overline{f}_1$, and $\overline{f}_0$ are scalar-valued while $\tilde{\beta}$ and $\tilde{\gamma}$ are $(n-1)$-vector-valued. For $0 < r < 1/4$ we have $\overline{f}_1(r) = 0$, while using (32) and properties discussed in [14], we see that the others satisfy

$$|\alpha(r) - 1|, |\tilde{\beta}(r)|, |\tilde{\gamma}(r)| \le w(r) \quad \text{for } 0 < r < 1,$$

$$|p[\nabla w](r)| \le w(r) \fint_{S_+^{n-1}} |\nabla w(r\theta)|\,ds \quad \text{for } 0 < r < 1.$$

Using (35) and $u = 0$ for $|x| > 1$, we see that $\alpha(r) = 1$ and $\widetilde{\beta}(r) = \widetilde{\gamma}(r) = p[\nabla w](r) = \overline{f}_1(r) = \overline{f}_0(r) = 0$ for $r > 1$. Now if we integrate by parts in (85) we obtain

$$\int_0^\infty -[r^{n-1}(\alpha u_0' + r\widetilde{\beta}\cdot\widetilde{v}' + \widetilde{\gamma}\cdot\widetilde{v} + p[\nabla w] - \overline{f}_1)]'\eta + r^{n-1}\overline{f}_0\,\eta\,dr = 0,$$

which means

$$-[r^{n-1}(\alpha u_0' + r\widetilde{\beta}\cdot\widetilde{v}' + \widetilde{\gamma}\cdot\widetilde{v} + p[\nabla w] - \overline{f}_1)]' + r^{n-1}\overline{f}_0 = 0.$$

But we can integrate this to find

$$\alpha u_0' + r\widetilde{\beta}\cdot\widetilde{v}' + \widetilde{\gamma}\cdot\widetilde{v} + p[\nabla w] = \vartheta(r) \tag{87a}$$

where

$$\vartheta(r) = \overline{f}_1(r) + r^{1-n}\int_0^r \overline{f}_0(\rho)\rho^{n-1}\,d\rho. \tag{87b}$$

Since $\alpha(r) \geq \varepsilon > 0$, (87a) can be solved for u_0' in terms of $\widetilde{v}$ and w.

Similarly, we can let $\eta = \eta(r)x_\ell$ in (34) for $\ell = 1, \ldots, n-1$; this will give us a system of equations for the vector function $\widetilde{v}$. To begin with, we have $\partial_i\eta = r\eta'(r)\theta_i\theta_\ell + \eta(r)\delta_{i\ell}$. If we plug this and (11) into (34), we find

$$\int_0^\infty [(u_0'\widetilde{\beta} + rA\widetilde{v}' + B\widetilde{v} + \widetilde{\xi}[\nabla w] - \widetilde{f}^\#)r\eta' +$$

$$(u_0'\widetilde{\gamma} + rB\widetilde{v}' + C\widetilde{v} + \widetilde{\zeta}[\nabla w] + \widetilde{f}^\flat)\eta]r^{n-1}dr = 0,$$

where A, B, and C are $(n-1)\times(n-1)$ matrix-valued functions defined by

$$A_{\ell k}(r) = \fint_{S_+^{n-1}} a_{ij}(r\theta)\theta_i\theta_j\theta_\ell\theta_k ds \quad (\ell, k = 1, \ldots, n-1),$$

$$B_{\ell k}(r) = \fint_{S_+^{n-1}} a_{\ell j}(r\theta)\theta_j\theta_k ds \quad (\ell, k = 1, \ldots, n-1), \tag{88}$$

$$C_{\ell k}(r) = \fint_{S_+^{n-1}} a_{\ell k}(r\theta)\,ds \quad (\ell, k = 1, \ldots, n-1),$$

and $\widetilde{\xi}[\nabla w]$, $\widetilde{\zeta}[\nabla w]$, $f^\#$, and $f^\flat$ are $(n-1)$-vector-valued functions defined by

$$\widetilde{\xi}_\ell[\nabla w](r) = \fint_{S_+^{n-1}} \sum_{i,j=1}^n a_{ij}\,\theta_i\,\theta_\ell\,\partial_j w\,ds_\theta, \quad \widetilde{\zeta}_\ell[\nabla w](r) = \fint_{S_+^{n-1}} \sum_{j=1}^n a_{\ell j}\,\partial_j w\,ds,$$

$$\widetilde{f}_\ell^\#(r) = \fint_{S_+^{n-1}} \sum_{i=1}^n f_i(r\theta)\theta_i\theta_\ell\,ds \quad \text{and} \quad \widetilde{f}_\ell^\flat(r) = \fint_{S_+^{n-1}} f_0(r\theta)\theta_\ell\,ds. \tag{89}$$

Using (32) and properties discussed in [14], we see that these functions satisfy

$$|A - n^{-1}I_{n-1}|, \ |B - n^{-1}I_{n-1}|, \ |C - I_{n-1}| \leq \omega(r) \quad \text{for } 0 < r < 1$$

$$|\widetilde{\xi}[\nabla w](r)|, \ |\widetilde{\zeta}[\nabla w](r)| \leq \omega(r)\fint_{S_+^{n-1}}|\nabla w|\,ds \quad \text{for } 0 < r < 1. \tag{90}$$

(Here I_{n-1} denotes the $(n-1) \times (n-1)$ identity matrix.) For $r > 1$ we use (35) and $u = 0$ to deduce the relations $A(r) = n^{-1}I = B(r)$, $C(r) = I$, and $\widetilde{\xi}[\nabla w](r) = 0 = \widetilde{\zeta}[\nabla w](r) = \widetilde{f}^{\#}(r) = \widetilde{f}^{\flat}(r)$. Now, using integration by parts, we obtain the 2nd-order system of ODEs

$$
- \left[r^n (u_0' \widetilde{\beta} + rA\widetilde{v}' + B\widetilde{v} + \widetilde{\xi}[\nabla w] - \widetilde{f}^{\#}) \right]'
$$
$$
+ r^{n-1}(u_0'\widetilde{\gamma} + rB\widetilde{v}' + C\widetilde{v} + \widetilde{\zeta}[\nabla w] + \widetilde{f}^{\flat}) = 0. \tag{91}
$$

At this point we can use (87a) to eliminate u_0' from (91), and then use the change of variables $r = e^{-t}$. After the change of variables we have

$$
\left[e^{-nt} \left(-A\widetilde{v}_t + B\widetilde{v} + \widetilde{\xi}[\nabla w] - \widetilde{f}^{\#} - \frac{1}{\alpha}(-\widetilde{\beta} \cdot \widetilde{v}_t + \widetilde{\gamma} \cdot \widetilde{v} + p[\nabla w] - \vartheta)\widetilde{\beta}) \right) \right]_t
$$
$$
+ e^{-nt} \left(-B\widetilde{v}_t + C\widetilde{v} + \widetilde{\zeta}[\nabla w] + \widetilde{f}^{\flat} - \frac{1}{\alpha}(-\widetilde{\beta} \cdot \widetilde{v}_t + \widetilde{\gamma} \cdot \widetilde{v} + p[\nabla w] - \vartheta)\widetilde{\gamma}) \right) = 0,
$$

which after some rearrangement can be written as

$$
\left[-A\widetilde{v}_t + B\widetilde{v} + \widetilde{\xi}[\nabla w] - \widetilde{f}^{\#} + \frac{\widetilde{\beta} \cdot \widetilde{v}_t - \widetilde{\gamma} \cdot \widetilde{v} - p[\nabla w] + \vartheta}{\alpha} \widetilde{\beta} \right]_t
$$
$$
- (B - nA)\widetilde{v}_t + \frac{\widetilde{\beta} \cdot \widetilde{v}_t}{\alpha}(\widetilde{\gamma} - n\widetilde{\beta}) + (C - nB)\widetilde{v} - \frac{\widetilde{\gamma} \cdot \widetilde{v}}{\alpha}(\widetilde{\gamma} - n\widetilde{\beta})
$$
$$
= n \left[\widetilde{\xi}[\nabla w] - \frac{p[\nabla w] - \vartheta}{\alpha}\widetilde{\beta} \right] + \frac{p[\nabla w] - \vartheta}{\alpha}\widetilde{\gamma} - \widetilde{\zeta}[\nabla w] - \widetilde{f}^{\flat}.
$$

To avoid differentiating the coefficient matrices, let us convert this to a first-order system for the $2(n-1)$-vector function $V = (V_1, V_2)$ where $V_1 = \widetilde{v}$ and

$$
V_2 = -A\widetilde{v}_t + B\widetilde{v} + \widetilde{\xi}[\nabla w] - \widetilde{f}^{\#} + \frac{\widetilde{\beta} \cdot \widetilde{v}_t - \widetilde{\gamma} \cdot \widetilde{v} - p[\nabla w] + \vartheta}{\alpha}\widetilde{\beta}.
$$

Notice that the matrix A is invertible near $x = 0$ and for $|x| > 1$, so we may assume that the variables were rescaled to make A invertible for all x. Thus we may solve for the t-derivatives of V_1 and V_2 (which we now denote by the dot notation) to find:

$$
\dot{V}_1 - A^{-1}BV_1 + A^{-1}V_2 - \frac{\widetilde{\beta} \cdot \dot{V}_1 - \widetilde{\gamma} \cdot V_1}{\alpha}A^{-1}\widetilde{\beta}
$$
$$
= A^{-1}\left[\widetilde{\xi}[\nabla w] - \widetilde{f}^{\#} - \frac{p[\nabla w] - \vartheta}{\alpha}\widetilde{\beta} \right]
$$
$$
\dot{V}_2 + (C - BA^{-1}B)V_1 + (BA^{-1} - n)V_2 \tag{92}
$$
$$
+ \frac{\widetilde{\beta} \cdot \dot{V}_1}{\alpha}(\widetilde{\gamma} - (n + A^{-1})\widetilde{\beta}) + \frac{\widetilde{\gamma} \cdot V_1}{\alpha}((n + A^{-1})\widetilde{\beta} - \widetilde{\gamma})
$$
$$
= n\left[\widetilde{\xi}[\nabla w] - \frac{p[\nabla w] - \vartheta}{\alpha}\widetilde{\beta} \right] + \frac{p[\nabla w] - \vartheta}{\alpha}\widetilde{\gamma} - \widetilde{\zeta}[\nabla w] - \widetilde{f}^{\flat}.
$$

Now (92) is still pretty complicated, but notice that the terms involving $\dot{V}_1$ and $\dot{V}_2$ in (92) are of the form $(I + D(t))\dot{V}$ where I is the $2(n-1) \times 2(n-1)$-identity matrix and the matrix $D(t)$ has matrix norm satisfying $|D(t)| \leq c\varepsilon^2(t)$. Consequently, we can multiply (92) by $(I + D(t))^{-1}$ and, after some calculations, see that V satisfies a 1st-order system in the form

$$\frac{dV}{dt} + M(t)V = F(t, \nabla w) + F_0(t), \tag{93a}$$

where $M(t)$ is a $2(n-1) \times 2(n-1)$ matrix of the form

$$M(t) = M_\infty + S_1(t) + S_2(t), \quad \text{where}$$

$$M_\infty = \begin{pmatrix} -I & n\,I \\ \frac{n-1}{n}I & (1-n)\,I \end{pmatrix} \quad \text{and} \tag{93b}$$

$$S_1(t) = \begin{pmatrix} I - A^{-1}B & A^{-1} - nI \\ C - BA^{-1}B + \frac{1-n}{n}I & BA^{-1} - I \end{pmatrix}.$$

The S_i satisfy

$$|S_1(t)| \leq \varepsilon(t) \text{ and } |S_2(t)| \leq c\varepsilon^2(t) \text{ as } t \to \infty,$$
$$S_1(t) = 0 = S_2(t) \text{ for } t < 0, \tag{93c}$$

while the vector $F(t, \nabla w)$ satisfies

$$|F(t, \nabla w)| \leq c\varepsilon(t) \oint_{S_+^{n-1}} |\nabla w|\, ds \text{ as } t \to \infty \text{ and } F(t, \nabla w) \equiv 0 \text{ for } t < 0, \tag{93d}$$

and the vector $F_0(t)$ has support in $t > 0$ with L^1-norm satisfying

$$\|F_0\|_{L^1(\mathbb{R})} \leq c\,(\|\vec{f}\|_p + \|f_0\|_p). \tag{93e}$$

Note that $M(t)$ and $F_0(t)$ depend on a_{ij}, $\vec{f}$, and f_0, but not on w.

We can further simplify our dynamical systems by another change of dependent variables. We can calculate the eigenvalues of M_∞ to be $\lambda = 0$ and $\lambda = -n$ (each occurring $n-1$ times). The matrix

$$J = \begin{pmatrix} nI & nI \\ I & (1-n)I \end{pmatrix}$$

diagonalizes M_∞, i.e., $J^{-1}M_\infty J = \text{diag}(0, \ldots, 0, -n, \ldots, n)$, so let us introduce new dependent variables $V \to (\phi, \psi)$ by

$$V = J \begin{pmatrix} \phi \\ \psi \end{pmatrix}. \tag{94}$$

We find that the dynamical system (93a) now takes the form (38a), where the conditions (38c) and (38d) follow from (93d) and (93e) respectively, and $\mathcal{R}$ is of the form (38b) with

$$R_1 = \frac{n-1}{n^2}A^{-1} - \frac{n-1}{n}A^{-1}B + C - BA^{-1}B + \frac{1}{n}BA^{-1} - I. \tag{95}$$

To simplify this expression for R_1, let us write

$$A = n^{-1}(1 + \widetilde{A}), \qquad B = n^{-1}(1 + \widetilde{B}), \qquad \text{and} \quad C = n^{-1}(1 + \widetilde{C}),$$

where $|\widetilde{A}|, |\widetilde{B}|, |\widetilde{C}| \leq c\varepsilon(t)$ as $t \to \infty$. Then $A^{-1} \approx n(I - \widetilde{A})$, and a calculation shows that

$$R_1 \approx \widetilde{C} - \widetilde{B} = C - nB \quad \text{as } t \to \infty,$$

which gives formula (38c). Finally, if we follow our changes of dependent variables from $(\widetilde{v}, \widetilde{v}_r)$ to (φ, ψ), we easily see that (38f) holds.

Now our original assumption that $u \in H^{1,2}_{\text{loc}}(\mathbb{R}^n_+)$ has implications for $V(t)$ as $t \to \infty$. In fact, by orthogonality properties in the decomposition (11), we find that $\nabla u \in L^2(B^n_+)$, where $B^n_+ = B_1(0) \cap \mathbb{R}^n_+$, implies

$$\int_0^1 \left((u_0')^2 + |\widetilde{v}|^2 + r^2|\widetilde{v}'|^2 \right) r^{n-1} \, dr < \infty \quad \text{and} \quad \nabla w \in L^2(B^n_+).$$

In particular, by $V_1 = \widetilde{v}$ and the second equation in (92) for $\dot{V}_2$, this implies

$$\int_0^\infty \left(|V|^2 + |\dot{V}|^2 + |\overline{\nabla w}|^2 \right) e^{-nt} \, dt < \infty, \tag{96}$$

where

$$\overline{\nabla w} = \fint_{S^{n-1}_+} \nabla w \, ds.$$

Thus $V(t)$ and its first-order derivative cannot grow too rapidly as $t \to \infty$.

Appendix E. Stability properties of dynamical systems

Here we recall a result on stability properties of dynamical systems that was obtained in [14]. Let $\varepsilon(t)$ be a positive, nonincreasing continuous function satisfying

$$\int_0^\infty \varepsilon^2(\tau) \, d\tau < \infty.$$

Consider a $2k \times 2k$-dimensional dynamical system in the form

$$\frac{d}{dt} \begin{pmatrix} \varphi \\ \psi \end{pmatrix} + \begin{pmatrix} 0 & 0 \\ 0 & -nI \end{pmatrix} \begin{pmatrix} \varphi \\ \psi \end{pmatrix} + \mathcal{R}(t) \begin{pmatrix} \varphi \\ \psi \end{pmatrix} = g(t) \quad \text{for } t > 0, \tag{97a}$$

where $n > 0$ and $\mathcal{R}$ can be written as a matrix of $k \times k$-blocks

$$\mathcal{R}(t) = \begin{pmatrix} R_1(t) & R_2(t) \\ R_3(t) & R_4(t) \end{pmatrix} \quad \text{with } |R_j(t)| \leq \varepsilon(t) \text{ on } 0 < t < \infty, \tag{97b}$$

with the block R_1 satisfying

$$|R_1(t) - R(t)| \leq c\varepsilon^2(t) \quad \text{as } t \to \infty, \tag{97c}$$

for a certain $k \times k$ matrix $R(t)$. We also assume that the vector function $g(t) = (g_1(t), g_2(t))$ satisfies the following conditions:

$$g_1 \in L^1(0, \infty) \tag{97d}$$

and there exists $\delta > 0$ such that for any choice of $\alpha \in [n - \delta, n)$ there is a constant c_α so that

$$e^{\alpha t} \int_t^\infty |g_2(s)| \, e^{-\alpha s} \, ds \le c_\alpha \varepsilon(t) \quad \text{for } 0 < t < \infty. \tag{97e}$$

We want to relate the stability for (97) to that for

$$\frac{d\varphi}{dt} + R\varphi = 0 \quad \text{for } t > 0, \tag{98a}$$

and the "finite-energy" condition on ψ

$$\int_0^\infty \left(|\psi|^2 + |\psi_t|^2 \right) e^{-nt} \, dt < \infty. \tag{98b}$$

Proposition 2. *Suppose that $\mathcal{R}$ and $g = (g_1, g_2)$ satisfy (97d) and (97e). Assume also that (98a) is uniformly stable. Then all solutions (ϕ, ψ) of (97a) that satisfy (98b) will remain bounded as $t \to \infty$, and $\psi(t) \to 0$. In fact, for $\alpha = n - \delta$ with $\delta > 0$ sufficiently small, we will have the estimates*

$$\sup_{0 < t < \infty} |\varphi(t)| \le c \left(c_\alpha + |\varphi(0)| + \|g_1\|_1 \right), \tag{99a}$$

$$|\psi(t)| \le c \, \varepsilon(t) (c_\alpha + \sup_{t < \tau < \infty} |\varphi(\tau)|). \tag{99b}$$

In addition, if all solutions of (98a) are asymptotically constant as $t \to \infty$, then the solution (φ, ψ) of (97) also has a limit:

$$(\varphi(t), \psi(t)) \to (\varphi_\infty, 0) \quad \text{as } t \to \infty. \tag{100}$$

Remark 1. In [14], Proposition 2 was stated and proved for the special case $k = n$. However, the proof in [14] does not use the condition $k = n$, so it proves the above proposition. This is important for the application in this paper since we need to take $k = n - 1$.

Remark 2. Since our dynamical system (98a) is linear, the condition that it is uniformly stable is equivalent to the condition $|\Phi(t)\Phi^{-1}(s)| \le K$ for $t > s > 0$, where Φ denotes the fundamental matrix for (98a). (Cf. [2].)

Appendix F. Uniqueness of solutions

In this appendix we discuss uniqueness for solutions of our variational equation. Suppose $u \in H^{1,p}_{\text{loc}}(\mathbb{R}^n_+)$ for $p \ge 2$ satisfies

$$\int_{\mathbb{R}^n_+} a_{ij} \partial_j u \, \partial_i \eta \, dx = 0 \quad \text{for } \eta \in C^1_{\text{comp}}(\overline{\mathbb{R}^n_+}), \tag{101}$$

and

$$M_{1,p}(u, r) \le C \, r^{-\alpha} \quad \text{for } r > 1. \tag{102}$$

The following proposition describes that values of $\alpha > 0$ for which we can conclude that $u \equiv 0$.

Proposition 3. *Suppose $u \in H^{1,p}_{\mathrm{loc}}(\overline{\mathbb{R}^n_+})$ for $p \geq 2$ satisfies (101) and (102) where α satisfies*

$$\alpha > n(p-2)/2p. \tag{103}$$

Then $u \equiv 0$.

As a special case we obtain the uniqueness result that is useful in our proof of Theorem 1.

Corollary 6. *If $u \in H^{1,p}_{\mathrm{loc}}(\overline{\mathbb{R}^n_+})$ for $p \geq 2$ is a solution of (34) that satisfies*

$$M_{1,p}(u,r) \leq C r^{-n} \quad for\ r > 1,$$

then u is unique.

Proof of Proposition 3. The strategy is to show that (101) holds with $\eta = u$, i.e.,

$$\int_{\mathbb{R}^n_+} a_{ij}\partial_j u\,\partial_i u\,dx = 0. \tag{104}$$

The ellipticity of a_{ij} then implies $\nabla u \equiv 0$, i.e., that u is constant. Finally, $\alpha > 0$ in (102) implies that $u \equiv 0$.

First let us determine the values of $\alpha > 0$ that imply that $\nabla u \in L^2(\mathbb{R}^n_+)$. Since $p \geq 2$, we know that $\nabla u \in L^2_{\mathrm{loc}}(\overline{\mathbb{R}^n_+})$, so the question is whether

$$\int_{x \in \mathbb{R}^n_+,\,|x|>1} |\nabla u|^2\,dx < \infty. \tag{105}$$

But

$$\int_{\substack{x \in \mathbb{R}^n_+,\\ |x|>1}} |\nabla u|^2\,dx = \int_1^\infty \int_{S^{n-1}_+} |\nabla u(r\theta)|^2\,r^{n-1}\,ds\,dr = \sum_{j=0}^\infty \int_{2^j}^{2^{j+1}} \int_{S^{n-1}_+} |\nabla u|^2\,r^{n-1}\,ds\,dr,$$

and by Hölder's inequality

$$\int_{2^j}^{2^{j+1}} \int_{S^{n-1}_+} |\nabla u|^2\,r^{n-1}\,ds\,dr = \int_{2^j<|x|<2^{j+1}} |\nabla u|^2\,dx$$

$$\leq \left(\int_{2^j<|x|<2^{j+1}} |\nabla u|^p\,dx \right)^{2/p} \left(\int_{2^j<|x|<2^{j+1}} dx \right)^{(p-2)/p}$$

$$\leq |S^{n-1}_+|\,M_{1,p}(u,2^j)^2\,2^{nj(p-2)/p} \leq C\,2^{j(-2\alpha+\frac{n(p-2)}{p})}.$$

Thus, we see that (103) implies (105).

Now let us verify that (105) is sufficient to enable us to take $u = \eta$ in (101), i.e., that (104) holds. It suffices to show that there exist $u_m \in C^1_{\mathrm{comp}}(\overline{\mathbb{R}^n_+})$ with $\nabla u_m \to \nabla u$ in $L^2(\mathbb{R}^n_+)$ as $m \to \infty$. But, using mollifiers, it suffices to show that this can be achieved with $u_m \in H^{1,2}_{\mathrm{comp}}(\overline{\mathbb{R}^n_+})$. So let $\chi(t)$ be a smooth function for $t > 0$ with $|\chi'(t)| \leq 2$ and

$$\chi(t) = \begin{cases} 0 & \text{if } t > 2 \\ 1 & \text{if } 0 < t < 1. \end{cases}$$

Then, for $m = 1, 2, \ldots$, define $u_m \in H^{1,2}_{\text{comp}}(\overline{\mathbb{R}^n_+})$ by

$$u_m(x) = u(x) \cdot \chi_m(|x|) \quad \text{where } \chi_m(t) = \chi(t/m).$$

For $i = 1, \ldots, n$ we compute

$$\nabla_i u(x) - \nabla_i u_m(x) = (1 - \chi_m(|x|))\nabla_i u(x) + u(x) \cdot \chi'_m(|x|) \cdot \frac{x_i}{|x|}.$$

We want to show both terms on the right tend to zero in $L^2(\mathbb{R}^n_+)$ as $m \to \infty$. If we assume (103), then we know that $\nabla u \in L^2(\mathbb{R}^n_+)$, and hence

$$\int_{\mathbb{R}^n_+} (1 - \chi_m)^2 |\nabla u|^2 \, dx \leq \int_{\mathbb{R}^n_+, |x|>m} |\nabla u|^2 \, dx \to 0 \quad \text{as } m \to \infty.$$

To estimate the second term we use

$$\int_{\mathbb{R}^n_+} (\chi'_m)^2 |u|^2 \, dx \leq \frac{4}{m^2} \int_{m<|x|<2m} |u(x)|^2 \, dx$$

$$\leq \frac{C}{m^2} \left(\int_{m<|x|<2m} |u|^p \, dx \right)^{2/p} (m^n)^{(p-2)/p} \leq C \, m^{n-2-\frac{2n}{p}} M_{1,p}(u,m)^2,$$

which tends to zero as $m \to \infty$ provided $n - 2 - (2n/p) - 2\alpha < 0$, i.e.,

$$\alpha > \frac{n(p-2)}{2p} - 1.$$

But this condition on α is certainly implied by (103), so we are done. □

Acknowledgment

This research was partly supported by the Research Project of the Italian Ministry of University and Research (MIUR) Prin 2015 n.2015HY8JCC "Partial differential equations an related analytic-geometric inequalities", by GNAMPA of the Italian INdAM (National Institute of High Mathematics), and by the Ministry of Education and Science of the Russian Federation, agreement n. 02.a03.21.0008.

References

[1] J. Chabrowski, G.M. Lieberman, *On the Dirichlet problem with L^2-boundary values in a half-space*, Indiana Univ. Math. J. **35** (1986), 623–642.

[2] W.A. Coppel, *Stability and Asymptotic Behavior of Differential Equations*, Heath & Co, 1965.

[3] M.S.P. Eastham, *The Asymptotic Solution of Linear Differential Systems*, Clarendon Press, 1989.

[4] E.B. Fabes, D. Jerison, C.E. Kenig, *Necessary and sufficient conditions for absolute continuity of elliptic harmonic measure*, Ann. Math. **119** (1984), 121–141.

[5] E. Fabes, S. Sroka, K.O. Widman, *Littlewood–Paley a priori estimates for parabolic equations with sub-Dini continuous coefficients*, Ann. Scuola Norm. Sup. Pisa Cl. Sci. (4) **6** (1979), no. 2, 305–334.

[6] R. Fefferman, *A criterion for the absolute continuity for the harmonic measure associated with an elliptic operator*, J. American Mathematical Society **2** (1989), 127–135.

[7] D. Gilbarg, J. Serrin, *On isolated singularities of solutions of second-order elliptic equations*, J. Analyse Math. **4** (1955/56), 309–340.

[8] D. Gilbarg, N. Trudinger, *Elliptic Partial Differential Equations of Second Order*, 2nd edition, Springer-Verlag, 1983.

[9] C.E. Kenig, J. Pipher, *The absolute continuity of elliptic measure revisited*, J. Fourier analysis and applications **4** (1998), 463–468.

[10] V. Kozlov, V. Maz'ya, *Asymptotic formula for solutions to the Dirichlet problem for elliptic equations with discontinuous coefficients near the boundary*, Ann. Scuola Norm. Sup. Pisa Cl. Sci. (5) Vol. II (2003), 551–600.

[11] V. Kozlov, V. Maz'ya, *Asymptotic formula for solutions to elliptic equations near the Lipschitz boundary*, Ann. Mat. Pura Appl. (4) **184** (2005), no. 2, 185–213.

[12] V. Maz'ya, R. McOwen, *Asymptotics for solutions of elliptic equations in double divergent form*, Comm. Part. Diff. Equat. **32** (2007), 1–17.

[13] V. Maz'ya, R. McOwen, *On the fundamental solution of an elliptic equation in non-divergence form*, AMS Translations: special volume dedicated to Nina Uraltseva **229** (2010), 145–172.

[14] V. Maz'ya, R. McOwen, *Differentiability of solutions to second-order elliptic equations via dynamical systems*. J. Differential Equations **250** (2011), 1137–1168.

[15] V. Maz'ya, M. Mitrea, T. Shaposhnikova, *The Dirichlet problem in Lipschitz domains for higher order elliptic systems with rough coefficients*, Journal d'Analyse Mathématique, **110** (2010), 167–239.

[16] G. Stampacchia, *Problemi al contorno ellitici, con dati discontinui, dotati di soluzionie hölderiane* (*in Italian*), Ann. Mat. Pura Appl. (4) **51** (1960), 1–37.

[17] E. Stein, G. Weiss *Fractional integrals on n-dimensional Euclidean space*, J. Math. and Mech. **7** (1958), 503–514.

[18] E. Stein, A. Zygmund, *On the differentiability of functions*, Studia Math **23** (1964), 247–283.

[19] S.E. Warschawski, *On conformal mapping of infinite strips*, Transactions AMS **51** no. 2 (1942), 280–335.

Vladimir Maz'ya
Department of Mathematics
Linköping University
SE-581 83 Linköping, Sweden

and

RUDN University
6 Miklukho-Maklay St
Moscow, 117198, Russia
email: vladimir.mazya@liu.se

Robert McOwen
Department of Mathematics
Northeastern University
Boston, MA 02115, USA
email: r.mcowen@northeastern.edu

Operator Theory:
Advances and Applications, Vol. 261, 389–393
© Springer International Publishing AG, part of Springer Nature 2018

A Function with Support of Finite Measure and "Small" Spectrum

Fedor Nazarov and Alexander Olevskii

Dedicated to the memory of V.P. Havin

Abstract. We give a simple construction of a function on $\mathbb{R}$ supported on a set of finite measure whose spectrum has density zero.

Mathematics Subject Classification (2010). Primary 42B10.

Keywords. Uncertainty principle, Schwartz function, uniqueness theorem, spectral gap.

1. The result

Let F be a function in $L^2(\mathbb{R})$. We say that it is supported on S if

$$F = 0 \text{ almost everywhere on } \mathbb{R} \setminus S.$$

Suppose the set $S \subset \mathbb{R}$ is of finite Lebesgue measure. Then the Fourier transform $\widehat{F}$ of F is a continuous function, so the spectrum of F is naturally defined as the closure of the set where $\widehat{F}$ takes non-zero values.

According to the uncertainty principle, the support and the spectrum of a (non-trivial) function F cannot be both "small sets". This principle has various versions (see, e.g., [4]).

In particular, the classical uniqueness theorem for analytic functions implies that if F is supported on an interval and it has a "spectral gap" (that is, $\widehat{F} = 0$ on an interval) then $F = 0$.

Another important result says that if the support S and the spectrum Q of F are both of finite measure then $F = 0$ [3], [2].

On the other hand, F may have a support of finite measure and a spectral gap; see [5], where such an example was constructed with $F = 1_S$.

Answering a question posed by Benedicks, Kargaev and Volberg [6] constructed an example of an L^1 function F such that

$$|S| < +\infty, |\mathbb{R} \setminus Q| = +\infty$$

(here and below by $|A|$ we denote the Lebesgue measure of the set A).

Another line of related research goes back to Men'shov's classical "correction" theorem, saying that for every (measurable) function F on the unit circle $\mathbb{T}$, there is a function F_ε such that

$$|\{t \in \mathbb{T} : F_\varepsilon(t) \neq F(t)\}| < \varepsilon \,,$$

and the Fourier series of F_ε converges uniformly on $\mathbb{T}$.

The approach of Men'shov also allows one to make the spectrum of F_ε "small" in $\mathbb{Z}$.

More precisely, Arutunyan [1] proved that in fact the spectrum can be localized on any symmetric set in $\mathbb{Z}$ that contains arbitrarily long intervals. In the non-periodic case, such a result (in the general setting of non-compact abelian groups) was obtained by Kislyakov [7].

The proofs of these results are quite involved. Our goal in this note is to present a simple construction of an indicator function with small spectrum.

Theorem. *There is a function $F \in L^2(\mathbb{R})$ supported on a set S of finite measure and such that*

$$|Q \cap (-R, R)| = o(R) \ as \ R \to \infty \,.$$

In addition, F can be chosen as the indicator function of S.

2. Proof

2.1. Take a Schwartz function F_0 such that

$$0 \leq F_0(t) \leq 1 \qquad (t \in \mathbb{R})$$

and its Fourier transform $\widehat{F_0}$ is positive on $(-1, 1)$ and vanishes outside that interval. Define a sequence of functions F_n recursively by

$$F_n := F_{n-1} + G_n \qquad (n = 1, 2, \dots) \,,$$

where

$$G_n(t) := F_{n-1}(t)[1 - F_{n-1}(t)] \cos k_n t \tag{1}$$

We are going to prove that if the numbers k_n grow sufficiently fast, then the sequence F_n converges to a function F satisfying the requirements of the theorem.

2.2. Clearly, F_n and G_n are Schwartz functions.

A simple induction shows that for every $t \in \mathbb{R}$, we have

$$|G_n(t)| \le \max\{F_{n-1}(t), 1 - F_{n-1}(t)\}$$

and

$$0 \le F_n(t) \le 1.$$

The Fourier transforms of $F_{n-1}[1-F_{n-1}]$, $F_{n-1}^2[1-F_{n-1}]$, and $F_{n-1}^2[1-F_{n-1}]^2$ vanish outside a compact interval, so for each $n \ge 1$, we have:

$$\int_{\mathbb{R}} G_n = \int_{\mathbb{R}} F_{n-1} G_n = 0$$

and

$$\int_{\mathbb{R}} G_n^2 = \frac{1}{2} \int_{\mathbb{R}} F_{n-1}^2 [1 - F_{n-1}]^2,$$

provided that k_n is chosen sufficiently large. It follows that

$$\int_{\mathbb{R}} F_n = \int_{\mathbb{R}} F_0 =: C \tag{2}$$

and, thereby,

$$I_n := \int_{\mathbb{R}} F_n(1 - F_n) \le C$$

(here, as usual, by C we denote a positive constant that may vary from line to line).

Observe also that

$$I_n = \int_{\mathbb{R}} [F_{n-1} + G_n][1 - F_{n-1} - G_n] = I_{n-1} - \int_{\mathbb{R}} G_n^2,$$

which implies that

$$\sum_{n \in [1,N]} \int_{\mathbb{R}} G_n^2 \le I_0 - I_N \le C,$$

and so

$$\sum_n \int_{\mathbb{R}} G_n^2 \le C \tag{3}$$

2.3. Define the sequence Q_n of intervals on (another copy of) $\mathbb{R}$ recursively as follows:

$$Q_0 := [-1, 1],$$
$$Q_n := \mathrm{conv}(Q_{n-1} \cup [k_n + 2Q_{n-1}] \cup [-k_n + 2Q_{n-1}])$$

(here conv E denotes the convex hull of a set $E \subset \mathbb{R}$). Clearly, for every n,

$$\mathrm{spec}\, F_{n-1} \subset Q_{n-1};$$
$$\mathrm{spec}\, G_n \subset [k_n + 2Q_{n-1}] \cup [-k_n + 2Q_{n-1}].$$

Set $Q := Q_0 \cup \bigcup_n ([k_n + 2Q_{n-1}] \cup [-k_n + 2Q_{n-1}])$.

Choosing k_n growing sufficiently fast we can ensure that the spectra of G_n are pairwise disjoint and

$$|Q \cap (-R, R)| = o(R) \text{ as } R \to \infty. \tag{4}$$

2.4. Consider the series $F_0 + G_1 + G_2 + \cdots$ Since the spectra of the terms are pairwise disjoint, this series is orthogonal in $L^2(\mathbb{R})$. Then (3) implies that it converges in $L^2(\mathbb{R})$ to some non-trivial function F. The partial sums of this series are F_n. Take a subsequence F_{n_ℓ} such that

$$F_{n_\ell} \to F \text{ almost everywhere on } \mathbb{R} \text{ as } \ell \to \infty.$$

Recall that all F_n are non-negative functions, so (2) implies that

$$F \geq 0 \text{ almost everywhere and } \int_{\mathbb{R}} F < \infty. \tag{5}$$

It follows from (1) and (3) that

$$\sum_n \int_{\mathbb{R}} [F_n(1 - F_n)]^2 = 2\sum_n \int_{\mathbb{R}} G_n^2 < +\infty,$$

so we must have

$$F(1 - F) = \lim_{\ell \to \infty} F_{n_\ell}(1 - F_{n_\ell}) = 0 \text{ almost everywhere,}$$

which implies that F is the indicator function of a set S. According to (5), this set has finite measure. Clearly the spectrum of F is a subset of Q. Due to (4) it has density zero. This finishes the proof.

Remark. Consider the function

$$h(R) := |Q \cap (-R, R)|.$$

Under the conditions of the theorem, it cannot be bounded. However the proof above shows that it may increase arbitrarily slowly. It remains an open question, however, if Q can have *uniform* density 0, i.e., if it is possible that

$$\lim_{R \to \infty} \sup_{x \in \mathbb{R}} \frac{1}{2R} |Q \cap (x - R, x + R)| = 0.$$

References

[1] F.G. Arutyunyan *Representation of functions by multiple series.* (Russian) Doklady AN ArmSSR, **64**, No. 2 (1977), 72–76.

[2] W.O. Amrein, A.M. Bertier *On support properties of L^p-functions and their Fourier transforms.* J. Funct. Anal. **24** (1977), 258–267.

[3] M. Benedicks *On Fourier transforms of functions supported on sets of finite Lebesgue measure.* – Royal Institute of Technology, Stockholm (1974), preprint; – J. Math. Anal. Appl. **106** (1985), 180–183.

[4] V.P. Havin, B. Jöricke *The Uncertainty Principle in Harmonic Analysis.* Springer-Verlag, Berlin, Heidelberg, 1994.

[5] P.P. Kargaev, *The Fourier transform of the characteristic function of a set vanishing on an interval.* (Russian), Mat. Sb. (N.S.) **117** (1982), 397–411. English translation in Math. USSR-Sb **45** (1983), 397–411.

[6] P.P. Kargaev, A.L. Volberg *Three results concerning the support of functions and their Fourier transforms.* Indiana Univ. Math. J. **41** (1992), 1143–1164.

[7] S.V. Kisliakov *A new correction theorem.* (Russian) Izvestiya AN SSSR, Ser. Math., **48**, No. 2 (1984), 305–330.

Fedor Nazarov
Department of Mathematics
Kent State University
Kent, OH, USA
e-mail: nazarov@math.kent.edu

Alexander Olevskii
Tel Aviv University
School of Mathematical Sciences
Tel Aviv, Israel
e-mail: olevskii@post.tau.ac.il

Operator Theory:
Advances and Applications, Vol. 261, 395–416
© Springer International Publishing AG, part of Springer Nature 2018

An Elementary Approach to Operator Lipschitz Type Estimates

V.V. Peller

To the memory of Viktor Petrovich Havin,
a wonderful mathematician and a wonderful man,
who greatly influenced my mathematical taste

Abstract. The paper brings an elementary approach to Lipschitz type estimates for functions of operators. The approach is based on the reduction to the case of operators on finite-dimensional inner product spaces. This allows us to avoid double and triple operator integrals and consider instead double and triple operator sums. Unlike in the case of double and triple operator integrals, double and triple operator sums can be defined for arbitrary functions which completely eliminates the problem of definitions of double operator integrals and various types of triple operator integrals.

Mathematics Subject Classification (2010). Primary 47A55. Secondary 47A60, 47A63, 47B10, 46E39.

Keywords. Functions of commuting operators; functions of noncommuting operators; Lipschitz type estimates; functions of perturbed operators; double operator sums; triple operator sums; Besov spaces.

1. Introduction

The purpose of this paper is to give an elementary approach to Lipschitz type estimates that does not involve double or triple operator integrals. The approach is based on the observation that to obtain Lipschitz type estimates for functions of operators on Hilbert space, it suffices to obtain such estimates for operators on finite-dimensional inner product spaces that do not depend on the dimension of the space.

An important problem of perturbation theory is to study the behavior of functions of operators under perturbation. In the framework of such problems a

The author is partially supported by NSF grant DMS 1300924.

significant role is played by the problem of obtaining Lipschitz type estimates for functions of operators. A function f on the real line $\mathbb{R}$ is called *operator Lipschitz* if

$$\|f(A) - f(B)\| \leq \text{const} \, \|A - B\| \tag{1.1}$$

for arbitrary self-adjoint operators A and B. It turns out (see Theorem 3.2.1 of [4]) that if inequality (1.1) holds for all bounded self-adjoint operators A and B, then it also holds for unbounded self-adjoint operators A and B with bounded difference $A - B$.

Obviously, if f is an operator Lipschitz function, then it must be a *Lipschitz function*, i.e., $|f(x) - f(y)| \leq \text{const} \, |x - y|$ for arbitrary x and y in $\mathbb{R}$. However, the converse is false. This was proved by Farforovskaya in [9].

Similarly, a function f on $\mathbb{R}^2 = \mathbb{C}$ is called *operator Lipschitz* if

$$\|f(N_1) - f(N_2)\| \leq \text{const} \, \|N_1 - N_2\| \tag{1.2}$$

for arbitrary normal operators N_1 and N_2. Again, if inequality (1.2) holds for all bounded normal operators, then it also holds for unbounded normal operators N_1 and N_2 with bounded $N_1 - N_2$ (see Theorem 3.2.1 of [4]).

It follows from the results of [11] that an operator Lipschitz function f on $\mathbb{R}$ must be differentiable everywhere on $\mathbb{R}$ and differentiable at infinity, i.e., the limit $\lim_{|x|\to\infty} x^{-1} f(x)$ exists. On the other hand, an operator Lipschitz function does not have to be continuously differentiable which was shown in [12]. In particular, the function $x \mapsto x^2 \sin(1/x)$ is operator Lipschitz on $\mathbb{R}$ (see [12] and [4]).

Necessary conditions for operator Lipschitzness were obtained in [17] and [18]. Those necessary conditions are based on the trace class criterion for Hankel operators obtained in [16] (see also [19]). In particular, it follows from the results of [17] that if f is an operator Lipschitz function on $\mathbb{R}$, then f belongs locally to the Besov class $B_{1,1}^1(\mathbb{R})$, see [15] for an introduction to Besov classes.

There are various sufficient conditions for operator Lipschitzness. It follows from the results of [17] and [18] that if f belongs to the (homogeneous) Besov class $B_{\infty,1}^1(\mathbb{R})$, then f is operator Lipschitz. In [1] operator Lipschitzness was established under a weaker condition on f in terms of Cauchy integrals (see also the survey [4]). Note that operator Lipschitzness for functions on $\mathbb{R}$ is equivalent to trace class Lischitzness:

$$\|f(A) - f(B)\|_{\boldsymbol{S}_1} \leq \text{const} \, \|A - B\|_{\boldsymbol{S}_1}$$

(see [4]). Here $\boldsymbol{S}_1$ stands for trace class. If we consider Lipschitzness in the Schatten–von Neumann norm $\boldsymbol{S}_p$ with $1 < p < \infty$, the corresponding inequality

$$\|f(A) - f(B)\|_{\boldsymbol{S}_p} \leq \text{const} \, \|A - B\|_{\boldsymbol{S}_p}$$

holds even for arbitrary Lipschitz function. This was established in [22].

It turns out that for functions on $\mathbb{C}$ the condition that f belongs to the Besov class $B_{\infty,1}^1(\mathbb{R}^2)$ is also sufficient for operator Lipschitzness and for trace class Lipschitzness. This was proved in [7].

In the paper [6] Lipschitz type estimates were studied for functions of *not necessarily commuting* self-adjoint operators. It was shown in [6] that if $f \in B^1_{\infty,1}(\mathbb{R}^2)$ and $1 \le p \le 2$, then the Lipschitz type estimate

$$\|f(A_1, B_1) - f(A_2, B_2)\|_{\boldsymbol{S}_p} \le \text{const} \max \left\{ \|A_1 - A_2\|_{\boldsymbol{S}_p}, \|B_1 - B_2\|_{\boldsymbol{S}_p} \right\}$$

holds for arbitrary pairs (A_1, A_2) and (B_1, B_2) of not necessarily commuting bounded self-adjoint operators. On the other hand, it was established in [6] that there is no such a Lipschitz type estimate in the norm of $\boldsymbol{S}_p$ with $p > 2$ as well as in the operator norm.

To obtain the above sufficient conditions double operator integrals and triple operator integrals were used. We refer the reader to the survey [20] on applications of multiple operator integrals in perturbation theory. We also refer the reader to the survey [4] on operator Lipschitz functions.

In this paper we reduce the problem of Lipschitz type estimates to the case of operators on finite-dimensional spaces. In this case double operator integrals and multiple operator integrals become finite sums. This completely eliminates the problem how to define such multiple operator integrals.

In § 3 we define double and triple operator sums and estimate their norm. In § 4 we obtain representations of the differences of functions of the initial and perturbed operators in terms of double and triple operator sums. This allows us to obtain in § 5 Lipschitz type estimates for functions of operators on finite-dimensional spaces. Finally, in § 6 we deduce from finite-dimensional estimates the Lipschitz type estimates for functions of operators on Hilbert space. In § 2 we give a brief introduction to Besov classes and functions of noncommuting self-adjoint operators.

Note that the same technique of the reduction to the finite-dimensional case can be used to prove the Hölder type estimates in the operator norm and Schatten–von Neumann norms that were obtained in [2], [3] and [7]. This method can also be used for commutator Lipschitz estimates, see [4].

2. Preliminaries

2.1. An introduction to Besov classes

In this paper we deal only with the Besov spaces $B^1_{\infty,1}(\mathbb{R}^d)$, $d = 1, 2$. We give here a brief introduction to such spaces and we refer the reader to [15] for detailed information about Besov classes.

Let w be an infinitely differentiable function on $\mathbb{R}$ such that

$$w \ge 0, \quad \text{supp}\, w \subset \left[\frac{1}{2}, 2\right], \quad \text{and} \quad w(s) = 1 - w\left(\frac{s}{2}\right) \quad \text{for} \quad s \in [1, 2]. \quad (2.1)$$

We define the functions W_n, $n \in \mathbb{Z}$, on $\mathbb{R}^d$ by

$$(\mathscr{F}W_n)(x) = w\left(\frac{\|x\|_2}{2^n}\right), \quad n \in \mathbb{Z}, \quad x = (x_1, \ldots, x_d), \quad \|x\|_2 \stackrel{\text{def}}{=} \left(\sum_{j=1}^{d} x_j^2\right)^{1/2},$$

where $\mathscr{F}$ is the *Fourier transform* defined on $L^1(\mathbb{R}^d)$ by

$$(\mathscr{F}f)(t) = \int_{\mathbb{R}^d} f(x)e^{-\mathrm{i}(x,t)}\,dx, \quad x = (x_1,\ldots,x_d), \quad t = (t_1,\ldots,t_d),$$

$$(x,t) \overset{\text{def}}{=} \sum_{j=1}^{d} x_j t_j.$$

Clearly,

$$\sum_{n\in\mathbb{Z}}(\mathscr{F}W_n)(t) = 1, \quad t \in \mathbb{R}^d \setminus \{0\}.$$

With each tempered distribution $f \in \mathscr{S}'(\mathbb{R}^d)$, we associate the sequence $\{f_n\}_{n\in\mathbb{Z}}$,

$$f_n \overset{\text{def}}{=} f * W_n. \tag{2.2}$$

The formal series $\sum_{n\in\mathbb{Z}} f_n$ is a Littlewood–Paley type expansion of f. This series does not necessarily converge to f.

Initially we define the (homogeneous) Besov class $\dot{B}^1_{\infty,1}(\mathbb{R}^d)$ as the space of all $f \in \mathscr{S}'(\mathbb{R}^n)$ such that

$$\{2^n\|f_n\|_{L^\infty}\}_{n\in\mathbb{Z}} \in \ell^1(\mathbb{Z}) \tag{2.3}$$

and put

$$\|f\|_{B^1_{\infty,1}} \overset{\text{def}}{=} \big\|\{2^n\|f_n\|_{L^\infty}\}_{n\in\mathbb{Z}}\big\|_{\ell^1(\mathbb{Z})}.$$

According to this definition, the space $\dot{B}^1_{\infty,1}(\mathbb{R}^n)$ contains all polynomials and all polynomials f satisfy the equality $\|f\|_{B^s_{p,q}} = 0$. Moreover, the distribution f is determined by the sequence $\{f_n\}_{n\in\mathbb{Z}}$ uniquely up to a polynomial. It is easy to see that the series $\sum_{n\geq 0} f_n$ converges in $\mathscr{S}'(\mathbb{R}^d)$. However, the series $\sum_{n<0} f_n$ can diverge in general. It can easily be proved that the series

$$\sum_{n<0} \frac{\partial f_n}{\partial x_j}, \quad \text{where} \quad 1 \leq j \leq d, \tag{2.4}$$

converges uniformly on $\mathbb{R}^d$.

Now we can define the modified (homogeneous) Besov class $B^1_{\infty,1}(\mathbb{R}^d)$. We say that a distribution f belongs to $B^1_{\infty,1}(\mathbb{R}^d)$ if (2.3) holds and

$$\frac{\partial f}{\partial x_j} = \sum_{n\in\mathbb{Z}} \frac{\partial f_n}{\partial x_j}, \quad 1 \leq j \leq d,$$

in the space $\mathscr{S}'(\mathbb{R}^d)$ (equipped with the weak-$*$ topology). Now the function f is determined uniquely by the sequence $\{f_n\}_{n\in\mathbb{Z}}$ up to a constant polynomial and a polynomial g belongs to $B^1_{\infty,1}(\mathbb{R}^d)$ if and only if g is constant.

Note that the functions f_n have the following properties: $f_n \in L^\infty(\mathbb{R}^d)$ and $\operatorname{supp}\mathscr{F}f \subset \{\xi \in \mathbb{R}^d : \|\xi\| \leq 2^{n+1}\}$. Bounded functions whose Fourier transforms are supported in $\{\xi \in \mathbb{R}^d : \|\xi\| \leq \sigma\}$ can be characterized by the following

Paley–Wiener–Schwartz type theorem (see [23], Theorem 7.23 and Exercise 15 of Chapter 7):

Let f be a continuous function on $\mathbb{R}^d$ and let $M, \sigma > 0$. The following statements are equivalent:

(i) $|f| \leq M$ *and* $\operatorname{supp} \mathscr{F} f \subset \{\xi \in \mathbb{R}^d : \|\xi\| \leq \sigma\}$;

(ii) f *is a restriction to $\mathbb{R}^d$ of an entire function on $\mathbb{C}^d$ such that*

$$|f(z)| \leq M e^{\sigma \|\operatorname{Im} z\|}$$

for all $z \in \mathbb{C}^d$.

2.2. Besov classes of periodic functions

Studying periodic functions on $\mathbb{R}^d$ is equivalent to studying functions on the d-dimensional torus $\mathbb{T}^d$. To define Besov spaces on $\mathbb{T}^d$, we consider a function w satisfying (2.1) and define the trigonometric polynomials W_n, $n \geq 0$, by

$$W_n(\zeta) \stackrel{\text{def}}{=} \sum_{j \in \mathbb{Z}^d} w\left(\frac{|j|}{2^n}\right) \zeta^j, \quad n \geq 1, \quad W_0(\zeta) \stackrel{\text{def}}{=} \sum_{\{j : |j| \leq 1\}} \zeta^j,$$

where

$$\zeta = (\zeta_1, \ldots, \zeta_d) \in \mathbb{T}^d, \quad j = (j_1, \ldots, j_d), \quad \text{and} \quad |j| = \left(|j_1|^2 + \cdots + |j_d|^2\right)^{1/2}.$$

For a distribution f on $\mathbb{T}^d$ we put

$$f_n = f * W_n, \quad n \geq 0, \tag{2.5}$$

and we say that f belongs to the Besov class $B^s_{\infty,1}(\mathbb{T}^d)$ if

$$\sum_{n \geq 0} 2^n \|f_n\|_{L^\infty} < \infty. \tag{2.6}$$

Note that locally the Besov space $B^1_{\infty,1}(\mathbb{R}^d)$ coincides with the Besov space $B^1_{\infty,1}$ of periodic functions on $\mathbb{R}^d$.

2.3. Functions of noncommuting self-adjoint operators

Let A and B be bounded not necessarily commuting self-adjoint operators on Hilbert space. The standard way to define functions $f(A, B)$ of A and B is to consider double operator integrals (see [6]). In this paper we are going to use the method of Fourier expansion.

Suppose that f is a continuous function of two variables that is defined on a square whose interior contains $\sigma(A) \times \sigma(B)$, where $\sigma(A)$ and $\sigma(B)$ stand for the spectra of A and B. Without changing the values of f on $\sigma(A) \times \sigma(B)$ we may replace f with a periodic function. Clearly, we can rescale the problem and assume that our functions are 2π-periodic in each variable.

Consider the Fourier expansion of f:

$$f(x, y) \sim \sum_{j,k \in \mathbb{Z}} \hat{f}(j, k) e^{\mathrm{i}jx} e^{\mathrm{i}ky}.$$

If f is a trigonometric polynomial of degree N, we can represent f in the form

$$f(x,y) = \sum_{j=-N}^{N} e^{ijx} \left(\sum_{k=-N}^{N} \widehat{f}(j,k)e^{iky} \right)$$

and put

$$f(A,B) \overset{\text{def}}{=} \sum_{j=-N}^{N} e^{ijA} \left(\sum_{k=-N}^{N} \widehat{f}(j,k)e^{ikB} \right).$$

Clearly,

$$\|f(A,B)\| \le \sum_{k=-N}^{N} \left\| \sum_{k=-N}^{N} \widehat{f}(j,k)e^{ikB} \right\| \le (1+2N)\|f\|_{L^\infty}. \tag{2.7}$$

Suppose now that f is a 2π-periodic function in $B_{\infty,1}^1(\mathbb{T}^2)$. Then we can define $f(A,B)$ by

$$f(A,B) = \sum_{n\ge 0} f_n(A,B),$$

where f_n is the trigonometric polynomial defined by (2.5). By (2.6) and (2.7), the series on the right-hand side converges absolutely in the norm and

$$\|f(A,B)\| \le \text{const} \sum_{n\ge 0} 2^n \|f_n\|_{L^\infty} \le \text{const}\, \|f\|_{B_{\infty,1}^1(\mathbb{T}^2)}.$$

If A and B are self-adjoint operators with finite spectra (in particular, if A and B are self-adjoint operators on a finite-dimensional inner product space), then the functions $f(A,B)$ can be defined for arbitrary functions f on $\mathbb{R}^2$ by the formula:

$$f(A,B) = \sum_{\lambda\in\sigma(A)} \sum_{\mu\in\sigma(B)} f(\lambda,\mu)E_A\{\lambda\}E_B\{\mu\},$$

where $E_A\{\lambda\}$ is the orthogonal projection onto the eigenspace of A that corresponds to the eigenvalue λ and $E_B\{\mu\}$ is the orthogonal projection onto the eigenspace of B that corresponds to the eigenvalue μ.

3. Estimates of double and triple operator sums

Let $\mathscr{H}$ be a finite-dimensional inner product space and let Λ be a finite set. By a *spectral measure on* Λ we mean a function E defined on the one point subsets of Λ such that $E\{\lambda\}$ is an orthogonal projection on $\mathscr{H}$ for every $\lambda \in \Lambda$,

$$E\{\lambda\}E\{\mu\} = \mathbf{0} \quad \text{if} \quad \lambda \ne \mu \quad \text{and} \quad \sum_{\lambda\in\Lambda} E\{\lambda\} = I.$$

Suppose that Λ and M are finite sets, and E_1 and E_2 are spectral measures on Λ and M. By *double operator sums* we mean expressions of the form

$$\sum_{\lambda \in \Lambda} \sum_{\mu \in M} \Phi(\lambda, \mu) E_1\{\lambda\} T E_2\{\mu\},$$

where Φ is a complex function on $\Lambda \times M$ and T is a linear operator on $\mathscr{H}$.

The following assertion is a special case of the well-known result [8] for double operator integrals (see also [17] and [4]). We prove it here for completeness.

Theorem 3.1. *Let E_1 and E_2 be as above. Suppose that σ is a σ-finite measure on a set Ω, φ_λ, $\lambda \in \Lambda$, and ψ_μ, $\mu \in M$, are functions in $L^2(\sigma)$, and Φ is the function on $\Lambda \times M$ defined by*

$$\Phi(\lambda, \mu) = \int_\Omega \varphi_\lambda(\omega) \psi_\mu(\omega) \, d\sigma(\omega).$$

Then

$$\left\| \sum_{\lambda \in \Lambda} \sum_{\mu \in M} \Phi(\lambda, \mu) E_1\{\lambda\} T E_2\{\mu\} \right\| \leq \max_{\lambda \in \Lambda} \|\varphi_\lambda\|_{L^2(\sigma)} \max_{\mu \in M} \|\psi_\mu\|_{L^2(\sigma)} \|T\|$$

for every linear operator T on $\mathscr{H}$.

Note that by duality we can obtain the same estimate in the trace norm

$$\left\| \sum_{\lambda \in \Lambda} \sum_{\mu \in M} \Phi(\lambda, \mu) E_1\{\lambda\} T E_2\{\mu\} \right\|_{\boldsymbol{S}_1} \leq \max_{\lambda \in \Lambda} \|\varphi_\lambda\|_{L^2(\sigma)} \max_{\mu \in M} \|\psi_\mu\|_{L^2(\sigma)} \|T\|_{\boldsymbol{S}_1}. \quad (3.1)$$

In particular, in the case when σ is the counting measure, we obtain the following estimate.

Theorem 3.2. *Let*

$$\Phi(\lambda, \mu) = \sum_n \varphi_\lambda(n) \psi_\mu(n),$$

where $\varphi_\lambda = \{\varphi_\lambda(n)\}_n$ and $\psi_\mu = \{\psi_\mu(n)\}_n$ are complex sequences in ℓ^2. Then

$$\left\| \sum_{\lambda \in \Lambda} \sum_{\mu \in M} \Phi(\lambda, \mu) E_1\{\lambda\} T E_2\{\mu\} \right\| \leq \max_{\lambda \in \Lambda} \|\varphi_\lambda\|_{\ell^2} \max_{\mu \in M} \|\psi_\mu\|_{\ell^2} \|T\|.$$

Proof of Theorem 3.1. Let x and y be vectors in $\mathscr{H}$. Put

$$x_\mu \stackrel{\text{def}}{=} E_2\{\mu\} x, \quad \mu \in M, \quad \text{and} \quad y_\lambda \stackrel{\text{def}}{=} E_2\{\lambda\} y, \quad \lambda \in \Lambda.$$

We have

$$\left| \left(\sum_{\lambda, \mu} \Phi(\lambda, \mu) E_1\{\lambda\} T E_2\{\mu\} x, y \right) \right| = \left| \left(\sum_{\lambda, \mu} \Phi(\lambda, \mu) (T x_\mu, y_\lambda) \right) \right|$$

$$= \left| \left(\sum_{\lambda,\mu} \int_{\Omega} \varphi_{\lambda}(\omega) \psi_{\mu}(\omega) (Tx_{\mu}, y_{\lambda}) \, d\sigma(\omega) \right) \right|$$

$$= \left| \int_{\Omega} \left(T \sum_{\mu} \psi_{\mu}(\omega) x_{\mu}, \sum_{\lambda} \varphi_{\lambda}(\omega) y_{\lambda} \right) d\sigma(\omega) \right|$$

$$\le \|T\| \int_{\Omega} \left\| \sum_{\mu} \psi_{\mu}(\omega) x_{\mu} \right\| \cdot \left\| \sum_{\lambda} \varphi_{\lambda}(\omega) y_{\lambda} \right\| d\sigma(\omega)$$

$$\le \|T\| \left(\int_{\Omega} \left\| \sum_{\mu} \psi_{\mu}(\omega) x_{\mu} \right\|^2 \right)^{1/2} \left(\int_{\Omega} \left\| \sum_{\lambda} \varphi_{\lambda}(\omega) y_{\lambda} \right\|^2 \right)^{1/2}$$

$$= \|T\| \left(\sum_{\mu} \|x_{\mu}\|^2 \int_{\Omega} |\psi_{\mu}(\omega)|^2 \, d\sigma(\omega) \right)^{1/2} \left(\sum_{\lambda} \|y_{\lambda}\|^2 \int_{\Omega} |\varphi_{\lambda}(\omega)|^2 \, d\sigma(\omega) \right)^{1/2}$$

$$= \|T\| \max_{\mu \in M} \|\psi_{\mu}\|_{L^2(\sigma)} \left(\sum_{\mu} \|x_{\mu}\|^2 \right)^{1/2} \max_{\lambda \in \Lambda} \|\varphi_{\lambda}\|_{L^2(\sigma)} \left(\sum_{\lambda} \|y_{\lambda}\|^2 \right)^{1/2}$$

$$= \|T\| \cdot \|x\|_{\mathscr{H}} \|y\|_{\mathscr{H}} \max_{\lambda \in \Lambda} \|\varphi_{\lambda}\|_{\ell^2} \max_{\mu \in M} \|\psi_{\mu}\|_{\ell^2}. \qquad \square$$

We proceed now to estimates of triple operator sums. Let $\mathscr{H}$ be a finite-dimensional inner product space and let Λ, M and N be finite sets. Suppose that E_1, E_2 and E_3 are spectral measures on Λ, M and N. By *triple operator sums* we mean expressions of the form

$$\sum_{\lambda \in \Lambda} \sum_{\mu \in M} \sum_{\nu \in N} \Phi(\lambda, \mu, \nu) E_1\{\lambda\} T E_2\{\mu\} R E_3\{\nu\},$$

where Φ is a complex function on $\Lambda \times M \times N$, and T and R are linear operators on $\mathscr{H}$. The following estimate for triple operator integrals in terms of the Haagerup tensor product norm was proved in [10] in the case of the operator norm and in [5] (see also [6]) in the case of Schatten–von Neumann norms $\boldsymbol{S}_p$. It is convenient to use the notation $\|T\|_{\boldsymbol{S}_{\infty}}$ for the operator norm of a linear operator T.

We denote by $\mathcal{B}$ the space of infinite matrices $\{\beta_{j,k}\}_{j,k}$ equipped with the operator norm.

Theorem 3.3. *Let E_1, E_2 and E_3 be as above and let Φ be the function on $\Lambda \times M \times N$ defined by*

$$\Phi(\lambda, \mu, \nu) = \sum_{j,k} \alpha_{\lambda}(j) \beta_{\mu}(j, k) \gamma_{\nu}(k),$$

where $\alpha_{\lambda} \in \ell^2$, $\lambda \in \Lambda$, $\beta_{\mu} \in \mathcal{B}$, $\mu \in M$, and $\gamma_{\nu} \in \ell^2$, $\nu \in N$. Suppose that $2 \le p \le \infty$, $2 \le q \le \infty$ and $1/r = 1/p + 1/q$.

Then

$$\left\| \sum_{\lambda \in \Lambda} \sum_{\mu \in M} \sum_{\nu \in N} \Phi(\lambda, \mu, \nu) E_1\{\lambda\} T E_2\{\mu\} R E_3\{\nu\} \right\|_{S_r}$$
$$\leq \max_{\lambda \in \Lambda} \|\alpha_\lambda\|_{\ell^2} \max_{\mu \in M} \|\beta_\mu\|_{\mathcal{B}} \max_{\nu \in N} \|\gamma_\nu\|_{\ell^2} \|T\|_{S_p} \|R\|_{S_q}.$$

We need the following lemma whose proof can be found in [5].

Lemma 3.4. *Let* $2 \leq p \leq \infty$. *Suppose that* T *is an operator on Hilbert space of class* S_p *and* $\{A_j\}_{j\geq 0}$ *is a sequence of bounded linear operators such that*

$$\left\| \sum_{j \geq 0} A_j^* A_j \right\| \leq 1 \quad \text{and} \quad \left\| \sum_{j \geq 0} A_j A_j^* \right\| \leq 1.$$

Then the row operator matrix

$$A(T) \overset{\text{def}}{=} \left(A_0 T \ A_1 T \ A_2 T \ \cdots \right),$$

belongs to S_p *and*

$$\|A(T)\|_{S_p} \leq \|T\|_{S_p}.$$

Proof of Theorem 3.3. Put

$$A_j \overset{\text{def}}{=} \sum_{\lambda \in \Lambda} \alpha_\lambda(j) E_1\{\lambda\}, \qquad B_{jk} \overset{\text{def}}{=} \sum_{\mu \in M} \beta_\mu(j, k) E_2\{\mu\} \quad \text{and}$$

$$\Gamma_k \overset{\text{def}}{=} \sum_{\nu \in N} \gamma_\nu(k) E_3(\nu).$$

Let $A(T)$ be the infinite row defined by

$$A(T) \overset{\text{def}}{=} \left(A_0 T \quad A_1 T \quad A_2 T \quad \cdots \right),$$

let B be the infinite matrix $B \overset{\text{def}}{=} \{B_{jk}\}_{j,k\geq 0}$ and let $\Gamma(R)$ be the infinite column defined by

$$\Gamma(R) \overset{\text{def}}{=} \begin{pmatrix} R\Gamma_0 \\ R\Gamma_1 \\ R\Gamma_2 \\ \vdots \end{pmatrix}.$$

Clearly,

$$\left\| \sum_{j \geq 0} |A_j|^2 \right\| = \max_{\lambda \in \Lambda} \|\alpha_\lambda\|_{\ell^2}^2, \quad \left\| \sum_{j \geq 0} |\Gamma_k|^2 \right\| = \max_{\nu \in N} \|\gamma_\nu\|_{\ell^2}^2$$

and

$$\|B\| = \max_{\mu \in M} \left\| \{\beta_\mu(j, k)\}_{j,k\geq 0} \right\|_{\mathcal{B}}.$$

It is easy to see that

$$\sum_{\lambda \in \Lambda} \sum_{\mu \in M} \sum_{\nu \in N} \Phi(\lambda, \mu, \nu) E_1\{\lambda\} T E_2\{\mu\} R E_3\{\nu\} = A(T) B \Gamma(R). \qquad (3.2)$$

By Lemma 3.4,

$$\|A(T)\|_{\boldsymbol{S}_p} \le \max_{\lambda \in \Lambda} \|\alpha_\lambda\|_{\ell^2} \|T\|_{\boldsymbol{S}_p}.$$

Passing to the adjoint operator, we see that it follows from Lemma (3.4) that

$$\|\Gamma(R)\|_{\boldsymbol{S}_q} \le \max_{\nu \in N} \|\gamma_\nu\|_{\ell^2} \|R\|_{\boldsymbol{S}_q}.$$

The result follows now from (3.2).   □

We proceed now to estimates of triple operator sums in terms of Haagerup-like tensor product norms that were introduced in [6]. The following result for triple operator integrals was obtained in [4].

Theorem 3.5. *Let E_1, E_2 and E_3 be as above and let Φ be the function on $\Lambda \times M \times N$ defined by*

$$\Phi(\lambda, \mu, \nu) = \sum_{j,k} \alpha_\lambda(j) \beta_\mu(k) \gamma_\nu(j,k),$$

where $\alpha_\lambda \in \ell^2$, $\lambda \in \Lambda$, $\beta_\mu \in \ell^2$, $\mu \in M$, and $\gamma_\nu \in \mathcal{B}$, $\nu \in N$. Suppose that $2 \le q \le \infty$, $1/2 \le 1/p + 1/q \le 1$ and $1/r = 1/p + 1/q$. Then

$$\left\| \sum_{\lambda \in \Lambda} \sum_{\mu \in M} \sum_{\nu \in N} \Phi(\lambda, \mu, \nu) E_1\{\lambda\} T E_2\{\mu\} R E_3\{\nu\} \right\|_{\boldsymbol{S}_r} \tag{3.3}$$
$$\le \max_{\lambda \in \Lambda} \|\alpha_\lambda\|_{\ell^2} \max_{\mu \in M} \|\beta_\mu\|_{\ell^2} \max_{\nu \in N} \|\gamma_\nu\|_{\mathcal{B}} \|T\|_{\boldsymbol{S}_p} \|R\|_{\boldsymbol{S}_q}.$$

Proof. To prove the theorem, we use the fact that the $\boldsymbol{S}_r$ norm is dual to the norm in $\boldsymbol{S}_{r'}$, $1/r + 1/r' = 1$. Let Q be a linear operator on $\mathscr{H}$ and let Ψ be the function defined by $\Psi(\mu, \nu, \lambda) = \Phi(\lambda, \mu, \nu)$.

We have

$$\left| \mathrm{trace}\left(\sum_{\lambda,\mu,\nu} \Phi(\lambda, \mu, \nu) E_1\{\lambda\} T E_2\{\mu\} R E_3\{\nu\} Q \right) \right|$$

$$= \left| \mathrm{trace}\left(\sum_{\lambda,\mu,\nu} \Phi(\lambda, \mu, \nu) E_2\{\mu\} R E_3\{\nu\} Q E_1\{\lambda\} T \right) \right|$$

$$\le \|T\|_{\boldsymbol{S}_p} \left\| \sum_{\lambda,\mu,\nu} \Phi(\lambda, \mu, \nu) E_2\{\mu\} R E_3\{\nu\} Q E_1\{\lambda\} \right\|_{\boldsymbol{S}_{p'}}$$

$$= \|T\|_{\boldsymbol{S}_p} \left\| \sum_{\lambda,\mu,\nu} \Psi(\mu, \nu, \lambda) E_2\{\mu\} R E_3\{\nu\} Q E_1\{\lambda\} \right\|_{\boldsymbol{S}_{p'}}.$$

Applying Theorem 3.3 to the function Ψ, we obtain

$$\left\| \sum_{\lambda,\mu,\nu} \Psi(\mu, \nu, \lambda) E_2\{\mu\} R E_3\{\nu\} Q E_1\{\lambda\} \right\|_{\boldsymbol{S}_{p'}}$$

$$\le \max_{\mu \in M} \|\beta_\mu\|_{\ell^2} \max_{\nu \in N} \|\gamma_\nu\|_{\mathcal{B}} \max_{\lambda \in \Lambda} \|\alpha_\lambda\|_{\ell^2} \|R\|_{\boldsymbol{S}_q} \|Q\|_{\boldsymbol{S}_{r'}}.$$

By duality, this implies inequality (3.3).   □

Remark. In particular, p and q satisfy the assumptions of Theorem 3.5 if $1 \le p \le 2$ and $1/p + 1/q \le 1$.

In a similar way one can prove the following result, which was established in [5] for triple operator integrals.

Theorem 3.6. *Let E_1, E_2 and E_3 be as above and let Φ be the function on $\Lambda \times M \times N$ defined by*

$$\Phi(\lambda, \mu, \nu) = \sum_{j,k} \alpha_\lambda(j, k) \beta_\mu(j) \gamma_\nu(k),$$

where $\alpha_\lambda \in \mathcal{B}$, $\lambda \in \Lambda$, $\beta_\mu \in \ell^2$, $\mu \in M$, and $\gamma_\nu \in \ell^2$, $\nu \in N$. Suppose that $2 \le p \le \infty$, $1/2 \le 1/p + 1/q \le 1$ and $1/r = 1/p + 1/q$. Then

$$\left\| \sum_{\lambda \in \Lambda} \sum_{\mu \in M} \sum_{\nu \in N} \Phi(\lambda, \mu, \nu) E_1\{\lambda\} T E_2\{\mu\} R E_3\{\nu\} \right\|_{S_r}$$

$$\le \max_{\lambda \in \Lambda} \|\alpha_\lambda\|_{\mathcal{B}} \max_{\mu \in M} \|\beta_\mu\|_{\ell^2} \max_{\nu \in N} \|\gamma_\nu\|_{\ell^2} \|T\|_{S_p} \|R\|_{S_q}.$$

Remark. In particular, p and q satisfy the assumptions of Theorem 3.6 if $1 \le q \le 2$ and $1/p + 1/q \le 1$.

4. A representation of differences of functions of operators in terms of operator sums

We start with functions of normal operators. Let N_1 and N_2 be normal operators on a finite-dimensional inner product space. Then

$$N_1 = \sum_{\lambda \in \sigma_1} \lambda E_1\{\lambda\} \quad \text{and} \quad N_2 = \sum_{\mu \in \sigma_2} \mu E_2\{\mu\},$$

where $\sigma_1 \overset{\text{def}}{=} \sigma(N_1)$ and $\sigma_2 \overset{\text{def}}{=} \sigma(N_2)$ are the spectra of N_1 and N_2, and $E_1\{\lambda\}$ and $E_2\{\mu\}$ are the spectral projections of N_1 and N_2 onto the corresponding eigenspaces of N_1 and N_2. Consider their real and imaginary parts:

$$A_1 = \operatorname{Re} N_1 = \sum_{\lambda \in \sigma_1} (\operatorname{Re} \lambda) E_1\{\lambda\}, \quad B_1 = \operatorname{Im} N_1 = \sum_{\lambda \in \sigma_1} (\operatorname{Im} \lambda) E_1\{\lambda\},$$

$$A_2 = \operatorname{Re} N_2 = \sum_{\mu \in \sigma_2} (\operatorname{Re} \mu) E_2\{\mu\} \quad \text{and} \quad B_2 = \operatorname{Im} N_2 = \sum_{\mu \in \sigma_2} (\operatorname{Im} \mu) E_2\{\mu\}.$$

The following theorem is an analog of formula (5.4) of [7] for normal operators on Hilbert space. However, in the case of finite-dimensional spaces we do not need any assumptions on the function f on $\mathbb{C}$.

Theorem 4.1. *Let N_1 and N_2 be as above and let f be a complex function on $\mathbb{C}$. Suppose that Φ_1 and Φ_2 are functions on $\sigma_1 \times \sigma_2$ such that*

$$\Phi_1(\lambda, \mu) = \frac{f(\operatorname{Re} \lambda, \operatorname{Im} \mu) - f(\operatorname{Re} \mu, \operatorname{Im} \mu)}{\operatorname{Re} \lambda - \operatorname{Re} \mu} \quad \text{if} \quad \operatorname{Re} \lambda \ne \operatorname{Re} \mu, \qquad (4.1)$$

and

$$\Phi_2(\lambda, \mu) = \frac{f(\operatorname{Re}\lambda, \operatorname{Im}\lambda) - f(\operatorname{Re}\lambda, \operatorname{Im}\mu)}{\operatorname{Im}\lambda - \operatorname{Im}\mu} \quad if \quad \operatorname{Im}\lambda \neq \operatorname{Im}\mu. \qquad (4.2)$$

Then the following formula holds:

$$f(N_1) - f(N_2) = \sum_{\lambda \in \sigma_1} \sum_{\mu \in \sigma_2} \Phi_1(\lambda, \mu) E_1\{\lambda\}(A_1 - A_2) E_2\{\mu\}$$

$$+ \sum_{\lambda \in \sigma_1} \sum_{\mu \in \sigma_2} \Phi_2(\lambda, \mu) E_1\{\lambda\}(B_1 - B_2) E_2\{\mu\}.$$

Remark. Note that Φ_1 can take arbitrary values at those points (λ, μ) for which $\operatorname{Re}\lambda = \operatorname{Re}\mu$, while Φ_2 can take arbitrary values at those points (λ, μ) for which $\operatorname{Im}\lambda = \operatorname{Im}\mu$.

Proof. We have

$$\sum_{\lambda \in \sigma_1} \sum_{\mu \in \sigma_2} \Phi_1(\lambda, \mu) E_1\{\lambda\}(A_1 - A_2) E_2\{\mu\}$$

$$= \sum_{\lambda \in \sigma_1} \sum_{\mu \in \sigma_2} \Phi_1(\lambda, \mu) E_1\{\lambda\} A_1 E_2\{\mu\} - \sum_{\lambda \in \sigma_1} \sum_{\mu \in \sigma_2} \Phi_1(\lambda, \mu) E_1\{\lambda\} A_2 E_2\{\mu\}.$$

Clearly,

$$\sum_{\lambda \in \sigma_1} \sum_{\mu \in \sigma_2} \Phi_1(\lambda, \mu) E_1\{\lambda\} A_1 E_2\{\mu\} = \sum_{\lambda \in \sigma_1} \sum_{\mu \in \sigma_2} (\operatorname{Re}\lambda) \Phi_1(\lambda, \mu) E_1\{\lambda\} E_2\{\mu\}$$

and

$$\sum_{\lambda \in \sigma_1} \sum_{\mu \in \sigma_2} \Phi_1(\lambda, \mu) E_1\{\lambda\} A_2 E_2\{\mu\} = \sum_{\lambda \in \sigma_1} \sum_{\mu \in \sigma_2} (\operatorname{Re}\mu) \Phi_1(\lambda, \mu) E_1\{\lambda\} E_2\{\mu\},$$

and so

$$\sum_{\lambda \in \sigma_1} \sum_{\mu \in \sigma_2} \Phi_1(\lambda, \mu) E_1\{\lambda\}(A_1 - A_2) E_2\{\mu\}$$

$$= \sum_{\lambda \in \sigma_1} \sum_{\mu \in \sigma_2} (\operatorname{Re}\lambda - \operatorname{Re}\mu) \Phi_1(\lambda, \mu) E_1\{\lambda\} E_2\{\mu\}$$

$$= \sum_{\lambda \in \sigma_1} \sum_{\mu \in \sigma_2} \big(f(\operatorname{Re}\lambda, \operatorname{Im}\mu) - f(\operatorname{Re}\mu, \operatorname{Im}\mu) \big) E_1\{\lambda\} E_2\{\mu\}.$$

Similarly,

$$\sum_{\lambda \in \sigma_1} \sum_{\mu \in \sigma_2} \Phi_2(\lambda, \mu) E_1\{\lambda\}(B_1 - B_2) E_2\{\mu\}$$

$$= \sum_{\lambda \in \sigma_1} \sum_{\mu \in \sigma_2} \big(f(\operatorname{Re}\lambda, \operatorname{Im}\lambda) - f(\operatorname{Re}\lambda, \operatorname{Im}\mu) \big) E_1\{\lambda\} E_2\{\mu\}.$$

Thus,

$$\sum_{\lambda \in \sigma_1} \sum_{\mu \in \sigma_2} \Phi_1(\lambda, \mu) E_1\{\lambda\}(A_1 - A_2) E_2\{\mu\}$$

$$+ \sum_{\lambda \in \sigma_1} \sum_{\mu \in \sigma_2} \Phi_2(\lambda, \mu) E_1\{\lambda\}(B_1 - B_2) E_2\{\mu\}$$

$$= \sum_{\lambda \in \sigma_1} \sum_{\mu \in \sigma_2} \big(f(\operatorname{Re}\lambda \operatorname{Im}\lambda) - f(\operatorname{Re}\mu, \operatorname{Im}\mu)\big) E_1\{\lambda\} E_2\{\mu\} = f(N_1) - f(N_2). \quad \square$$

Consider now the case of functions of self-adjoint operators. Let A and B be self-adjoint operators on a finite-dimensional inner product space. Then

$$A = \sum_{\lambda \in \sigma(A)} \lambda E_A\{\lambda\} \quad \text{and} \quad B = \sum_{\mu \in \sigma(B)} \mu E_B\{\mu\},$$

where $\sigma_A \overset{\text{def}}{=} \sigma(A)$, $\sigma_B \overset{\text{def}}{=} \sigma(B)$, and $E_A\{\lambda\}$ and $E_B\{\mu\}$ are the spectral projections of A and B onto the corresponding eigenspaces of A and B.

Theorem 4.2. *Let A and B be as above and let f be a complex function on $\mathbb{C}$. Suppose that Φ is a function on $\sigma_A \times \sigma_B$ such that*

$$\Phi(\lambda, \mu) = \frac{f(\lambda) - f(\mu)}{\lambda - \mu} \quad \text{if} \quad \lambda \neq \mu.$$

Then

$$f(A) - f(B) = \sum_{\lambda \in \sigma_A} \sum_{\mu \in \sigma_B} \Phi(\lambda, \mu) E_A\{\lambda\}(A - B) E_B\{\mu\}.$$

Again, Φ can take arbitrary values at those points (λ, μ), for which $\lambda \neq \mu$. It is easy to see that Theorem 4.2 is a special case of Theorem 4.1.

We proceed now to functions of pairs of noncommuting self-adjoint operators.

Let (A_1, B_1) and (A_2, B_2) be pairs of not necessarily commuting self-adjoint operators on a finite-dimensional inner product space. We denote by

$$E_{A_1}\{\lambda\}, \quad \lambda \in \sigma(A_1), \qquad E_{B_1}\{\mu\}, \quad \mu \in \sigma(B_1),$$

$$E_{A_2}\{\lambda\}, \quad \lambda \in \sigma(A_2), \qquad E_{B_2}\{\mu\}, \quad \mu \in \sigma(B_2),$$

the spectral projections of A_1, B_1, A_2, B_2 onto the corresponding eigenspaces.

Suppose that Ψ_1 is a function on $\sigma(A_1) \times \sigma(A_2) \times \sigma(B_1)$ and Ψ_2 is a function on $\sigma(A_2) \times \sigma(B_1) \times \sigma(B_2)$ such that

$$\Psi_1(\lambda_1, \lambda_2, \mu) = \frac{f(\lambda_1, \mu) - f(\lambda_2, \mu)}{\lambda_1 - \lambda_2} \quad \text{if} \quad \lambda_1 \neq \lambda_2 \qquad (4.3)$$

and

$$\Psi_2(\lambda, \mu_1, \mu_2) = \frac{f(\lambda, \mu_1) - f(\lambda, \mu_2)}{\mu_1 - \mu_2} \quad \text{if} \quad \mu_1 \neq \mu_2. \qquad (4.4)$$

Again, Ψ_1 is allowed to take arbitrary values at those points $(\lambda_1, \lambda_2, \mu)$, for which $\lambda_1 = \lambda_2$, while Ψ_2 is allowed to take arbitrary values at those points (λ, μ_1, μ_2), for which $\mu_1 = \mu_2$.

Theorem 4.3. *Let A_1, B_1, A_2, B_2, Ψ_1 and Ψ_2 be as above and let f be a complex function on $\mathbb{R}^2$. Then*

$$f(A_1, B_1) - f(A_2, B_2)$$
$$= \sum_{\lambda_1, \lambda_2, \mu} \Psi_1(\lambda_1, \lambda_2, \mu) E_{A_1}\{\lambda_1\}(A_1 - A_2) E_{A_2}\{\lambda_2\} E_{B_1}\{\mu\}$$
$$+ \sum_{\lambda, \mu_1, \mu_2} \Psi_2(\lambda, \mu_1, \mu_2) E_{A_2}\{\lambda\}(B_1 - B_2) E_{B_1}\{\mu_1\} E_{B_2}\{\mu_2\}.$$

Proof. We have

$$\sum_{\lambda_1, \lambda_2, \mu} \Psi_1(\lambda_1, \lambda_2, \mu) E_{A_1}\{\lambda_1\}(A_1 - A_2) E_{A_2}\{\lambda_2\} E_{B_1}\{\mu\}$$
$$= \sum_{\lambda_1, \lambda_2, \mu} \Psi_1(\lambda_1, \lambda_2, \mu) E_{A_1}\{\lambda_1\} A_1 E_{A_2}\{\lambda_2\} E_{B_1}\{\mu\}$$
$$- \sum_{\lambda_1, \lambda_2, \mu} \Psi_1(\lambda_1, \lambda_2, \mu) E_{A_1}\{\lambda_1\} A_2 E_{A_2}\{\lambda_2\} E_{B_1}\{\mu\}$$
$$= \sum_{\lambda_1, \lambda_2, \mu} \lambda_1 \Psi_1(\lambda_1, \lambda_2, \mu) E_{A_1}\{\lambda_1\} E_{A_2}\{\lambda_2\} E_{B_1}\{\mu\}$$
$$- \sum_{\lambda_1, \lambda_2, \mu} \lambda_2 \Psi_1(\lambda_1, \lambda_2, \mu) E_{A_1}\{\lambda_1\} E_{A_2}\{\lambda_2\} E_{B_1}\{\mu\}$$
$$= \sum_{\lambda_1, \lambda_2, \mu} \big(f(\lambda_1, \mu) - f(\lambda_2, \mu)\big) E_{A_1}\{\lambda_1\} E_{A_2}\{\lambda_2\} E_{B_1}\{\mu\}$$
$$= f(A_1, B_1) - f(A_2, B_1).$$

Similarly,

$$\sum_{\lambda, \mu_1, \mu_2} \Psi_2(\lambda, \mu_1, \mu_2) E_{A_2}\{\lambda\}(B_1 - B_2) E_{B_1}\{\mu_1\} E_{B_2} = f(A_2, B_1) - f(A_2, B_2)$$

which completes the proof. $\square$

5. Lipschitz type estimates for operators on finite-dimensional spaces

In this section we obtain Lipschitz type estimates for operators on finite-dimensional spaces. The following result was established in [7] for normal operators on Hilbert space.

Theorem 5.1. *There exists a positive number C such that for an arbitrary function f in the Besov class $B^1_{\infty,1}(\mathbb{R}^2)$ and for arbitrary normal operators N_1 and N_2 on a finite-dimensional inner product space, the following inequality holds:*

$$\|f(N_1) - f(N_2)\| \leq C\|f\|_{B^1_{\infty,1}}\|N_1 - N_2\|. \tag{5.1}$$

Proof. It follows form the definition of the Besov class $B^1_{\infty,1}(\mathbb{R}^2)$ given in § 2 that it suffices to prove inequality (5.1) for the functions $f_n = f * W_n$ (see (2.2)). Hence, to prove inequality (5.1), it suffices to show that if f is a bounded continuous function on $\mathbb{R}^2$ and the Fourier transform $\mathscr{F}f$ of f is supported in $\{\zeta \in \mathbb{C} : |\zeta| < \sigma\}$, then

$$\|f(N_1) - f(N_2)\| \leq \mathrm{const}\,\sigma \|f\|_{L^\infty(\mathbb{R}^2)} \|N_1 - N_2\|.$$

We prove this inequality for $\sigma = 1$. To establish this inequality for arbitrary $\sigma > 0$, it suffices to use rescaling. We are going to use the following formulae:

$$\frac{f(x_1, y_2) - f(x_2, y_2)}{x_1 - x_2} = \sum_{n \in \mathbb{Z}} \frac{f(n\pi, y_2) - f(x_2, y_2)}{n\pi - x_2} \cdot \frac{\sin(x_1 - n\pi)}{x_1 - n\pi} \tag{5.2}$$

and

$$\frac{f(x_1, y_1) - f(x_1, y_2)}{y_1 - y_2} = \sum_{n \in \mathbb{Z}} \frac{f(x_1, y_1) - f(x_1, n\pi)}{y_1 - n\pi} \cdot \frac{\sin(y_2 - n\pi)}{y_2 - n\pi}. \tag{5.3}$$

Formulae (5.2) and (5.3) were deduced in the proof of Theorem 6.2 of [7] from the Kotel'nikov–Shannon formula, see [13], Lect. 20.2. Moreover, (see Theorem 6.1 of [7])

$$\sum_{n \in \mathbb{Z}} \frac{|f(n\pi, y) - f(x, y)|^2}{(n\pi - x)^2} \leq 3\|f\|^2_{L^\infty(\mathbb{R}^2)}, \quad x, y \in \mathbb{R}, \tag{5.4}$$

and

$$\sum_{n \in \mathbb{Z}} \frac{\sin^2(x - n\pi)}{(x - n\pi)^2} = 1, \quad x \in \mathbb{R}, \tag{5.5}$$

(the last formula is well known, see, e.g., [24], 3.3.2, Example IV).

By Theorem 4.1, we should estimate the norms of

$$\sum_{\lambda \in \sigma_1} \sum_{\mu \in \sigma_2} \Phi_1(\lambda, \mu) E_1\{\lambda\}(A_1 - A_2) E_2\{\mu\}$$

and

$$\sum_{\lambda \in \sigma_1} \sum_{\mu \in \sigma_2} \Phi_2(\lambda, \mu) E_1\{\lambda\}(B_1 - B_2) E_2\{\mu\} \tag{5.6}$$

where Φ_1 and Φ_2 satisfy (4.1) and (4.2).

By (5.2),

$$\sum_{\lambda \in \sigma_1} \sum_{\mu \in \sigma_2} \Phi_1(\lambda, \mu) E_1\{\lambda\}(A_1 - A_2) E_2\{\mu\}$$

$$= \sum_{n \in \mathbb{Z}} \frac{f(n\pi, \mathrm{Im}\,\mu) - f(\mathrm{Re}\,\mu, \mathrm{Im}\,\mu)}{n\pi - \mathrm{Re}\,\mu} \cdot \frac{\sin(x_1 - \pi n)}{x_1 - \pi n} E_1\{\lambda\}(A_1 - A_2) E_2\{\mu\}.$$

By Theorem 3.2, inequality (5.4) and equality (5.5), we obtain

$$\left\| \sum_{\lambda \in \sigma_1} \sum_{\mu \in \sigma_2} \Phi_1(\lambda, \mu) E_1\{\lambda\}(A_1 - A_2) E_2\{\mu\} \right\|$$

$$\leq \sqrt{3} \|f\|_{L^\infty(\mathbb{R}^2)} \|A_1 - A_2\| \leq \sqrt{3} \|f\|_{L^\infty(\mathbb{R}^2)} \|N_1 - N_2\|.$$

In a similar way one can estimate the operator norm of (5.6). □

Theorem 5.1 implies that if f is a function on $\mathbb{R}$ in the Besov class $B^1_{\infty,1}(\mathbb{R})$, then

$$\|f(A) - f(B)\| \leq \text{const} \|f\|_{B^1_{\infty,1}} \|A - B\|$$

for arbitrary self-adjoint operators on a finite-dimensional inner product space. This inequality was established in [17] and [18] for self-adjoint operators on Hilbert space.

In [1] a Lipschitz type estimate for functions of self-adjoint operators was obtained under weaker assumptions on the function f. We obtain here the result of [1] for operators on finite-dimensional spaces.

We denote by $\widehat{\mathscr{M}}$ the set of functions on $\mathbb{R}$ of the form

$$f(x) = \int_{\mathbb{C}\backslash\mathbb{R}} \left(\frac{1}{\zeta - x} - \frac{1}{\zeta} \right) d\varkappa(\zeta) + c, \qquad (5.7)$$

where $c \in \mathbb{C}$ and $\varkappa$ is a complex Radon measure on $\mathbb{C} \setminus \mathbb{R}$ satisfying

$$\|\mu\|_{\mathscr{M}(\mathbb{C}\backslash\mathbb{R})} \overset{\text{def}}{=} \sup_{x \in \mathbb{R}} \int_{\mathbb{C}\backslash\mathfrak{F}} \frac{d|\varkappa|(\zeta)}{|\zeta - x|^2} < \infty. \qquad (5.8)$$

Theorem 5.2. *Let f be a function on $\mathbb{R}$ given by (5.7) with $\varkappa$ satisfying (5.8). Then*

$$\|f(A) - f(B)\| \leq \|\varkappa\|_{\mathscr{M}(\mathbb{C}\backslash\mathbb{R})} \|A - B\| \qquad (5.9)$$

for arbitrary self-adjoint operators A and B on a finite-dimensional inner product space.

Proof. Let f be given by (5.7). Then

$$\frac{f(x) - f(y)}{x - y} = \int_{\mathbb{C}\backslash\mathbb{R}} \frac{d\varkappa(\zeta)}{(\zeta - x)(\zeta - y)} \qquad \text{if} \quad x \neq y.$$

By Theorem 4.2,

$$f(A) - f(B) = \sum_{\lambda \in \sigma(A)} \sum_{\mu \in \sigma(B)} \int_{\mathbb{C}\backslash\mathbb{R}} \frac{d\varkappa(\zeta)}{(\zeta - \lambda)(\zeta - \mu)}.$$

By Theorem 3.1,

$$\|f(A) - f(B)\| \leq \max_{\lambda \in \Lambda} \left(\int_{\mathbb{C}\backslash\mathbb{R}} \frac{d|\varkappa|(\zeta)}{|\zeta - \lambda|^2} \right)^{1/2} \max_{\mu \in M} \left(\int_{\mathbb{C}\backslash\mathbb{R}} \frac{d|\varkappa|(\zeta)}{|\zeta - \mu|^2} \right)^{1/2} \|A - B\|$$

$$\leq \|\varkappa\|_{\mathscr{M}(\mathbb{C}\backslash\mathbb{R})} \|A - B\|. \qquad \square$$

If we use inequality (3.1), we can obtain the following Lipschitz type estimate in trace norm.

Theorem 5.3. *Suppose that f, A and B are as in the statement of Theorem 5.2. Then*

$$\|f(A) - f(B)\|_{\boldsymbol{S}_1} \le \|\varkappa\|_{\mathscr{M}(\mathbb{C}\backslash\mathbb{R})}\|A - B\|_{\boldsymbol{S}_1}.$$

We proceed now to Lipschitz type estimates of functions of pairs of not necessarily commuting self-adjoint operators on finite-dimensional inner product spaces.

Theorem 5.4. *There exists a positive number C such that*

$$\|f(A_1, B_1) - f(A_2, B_2)\|_{\boldsymbol{S}_p} \le C\|f\|_{B^1_{\infty,1}} \max\big\{\|A_1 - A_2\|_{\boldsymbol{S}_p}, \|B_1 - B_2\|_{\boldsymbol{S}_p}\big\} \quad (5.10)$$

for every $p \in [1,2]$, for an arbitrary function f in $B^1_{\infty,1}(\mathbb{R}^2)$ and for arbitrary pairs (A_1, B_1) and (A_2, B_2) of not necessarily commuting self-adjoint operators on a finite-dimensional inner product space.

Proof. As in the proof of Theorem 5.1, it suffices to prove inequality (5.10) for the functions $f_n = f * W_n$ (see (2.2)). Thus, it is sufficient to prove that for bounded continuous functions f on $\mathbb{R}^2$ whose Fourier transform $\mathscr{F}f$ is supported in the ball $\{\zeta \in \mathbb{C} : |\zeta| < \sigma\}$, the following inequality holds

$$\|f(A_1, B_1) - f(A_2, B_2)\|_{\boldsymbol{S}_p}$$
$$\le \text{const}\,\sigma\|f\|_{L^\infty(\mathbb{R})} \max\big\{\|A_1 - A_2\|_{\boldsymbol{S}_p}, \|B_1 - B_2\|_{\boldsymbol{S}_p}\big\}.$$

It suffices to prove this inequality for $\sigma = 1$. The general case reduces to the case $\sigma = 1$ by rescaling.

By Theorem 4.3, we have to estimate the $\boldsymbol{S}_p$ norms of

$$\sum_{\lambda_1 \in \sigma(A_1)} \sum_{\lambda_2 \in \sigma(A_2)} \sum_{\mu \in \sigma(B_1)} \Psi_1(\lambda_1, \lambda_2, \mu) E_{A_1}\{\lambda_1\}(A_1 - A_2)E_{A_2}\{\lambda_2\}E_{B_1}\{\mu\}$$

and

$$\sum_{\lambda \in \sigma(A_2)} \sum_{\mu_1 \in \sigma(B_1)} \sum_{\mu_2 \in \sigma(B_2)} \Psi_2(\lambda, \mu_1, \mu_2)E_{A_2}\{\lambda\}(B_1 - B_2)E_{B_1}\{\mu_1\}E_{B_2}\{\mu_2\}, \quad (5.11)$$

where Ψ_1 and Ψ_2 are the functions satisfying (4.3) and (4.4).

By Theorem 6.1 of [6],

$$\frac{f(x_1, y) - f(x_2, y)}{x_1 - x_2} = \sum_{j,k \in \mathbb{Z}} \frac{\sin(x_1 - j\pi)}{x_1 - j\pi} \cdot \frac{\sin(x_2 - k\pi)}{x_2 - k\pi} \gamma_{jk}(y),$$

where

$$\gamma_{jk}(y) = \begin{cases} \frac{f(j\pi, y) - f(k\pi, y)}{j\pi - k\pi}, & j \ne k \\ \frac{\partial f(s, y)}{\partial s}\big|_{(j\pi, y)}, & j = k. \end{cases}$$

Moreover,

$$\big\|\{\gamma_{jk}(y)\}_{j,k \in \mathbb{Z}}\big\|_{\boldsymbol{B}} \le \text{const}\,\|f\|_{L^\infty(\mathbb{R})}$$

(see Theorem 6.1 of [6]).

Keeping (5.5) in mind, we see that it follows now from Theorem 3.5 that

$$\left\| \sum_{\lambda_1,\lambda_2,\mu} \Psi_1(\lambda_1,\lambda_2,\mu) E_{A_1}\{\lambda_1\}(A_1 - A_2) E_{A_2}\{\lambda_2\} E_{B_1}\{\mu\} \right\|_{\boldsymbol{S}_p}$$

$$\leq \text{const}\, \|f\|_{L^\infty(\mathbb{R})} \|A - B\|_{\boldsymbol{S}_p}.$$

Similarly, to estimate the $\boldsymbol{S}_p$ norm of (5.11), we can use the expansion

$$\frac{f(x,y_1) - f(x,y_2)}{y_1 - y_2} = \sum_{j,k\in\mathbb{Z}} \alpha_{jk}(x) \cdot \frac{\sin(y_1 - j\pi)}{y_1 - j\pi} \cdot \frac{\sin(y_2 - k\pi)}{y_2 - k\pi},$$

where

$$\alpha_{jk}(x) = \begin{cases} \frac{f(j\pi,x)-f(k\pi,x)}{j\pi-k\pi}, & j \neq k \\ \left.\frac{\partial f(s,x)}{\partial s}\right|_{(j\pi,s)}, & j = k \end{cases}$$

(see Theorem 6.2 of [6]) and apply Theorem 3.6. $\qquad\qquad\qquad\square$

6. Reduction to the finite-dimensional case

In this section we reduce Lipschitz type estimates for operators on Hilbert space to the corresponding estimates for operators on finite-dimensional inner product spaces.

It follows from Theorem 3.2.1 and Lemma 3.1.12 of [4] that for a continuous function f on $\mathbb{R}^2$, the following statements are equivalent:

(1) $\|f(N_1) - f(N_2)\| \leq C\|N_1 - N_1\|$ for arbitrary bounded normal operators N_1 and N_2 on Hilbert space;

(2) $\|f(N_1) - f(N_2)\| \leq C\|N_1 - N_1\|$ for arbitrary not necessarily bounded normal operators N_1 and N_2 on Hilbert space with bounded difference;

(3) $\|f(N_1) - f(N_2)\| \leq C\|N_1 - N_1\|$ for arbitrary normal operators N_1 and N_2 on finite-dimensional inner product spaces.

Note that the constant C is the same.

Recall that such functions are called *operator Lipschitz*.

This allows us to extend Theorem 5.1 to operators on infinite-dimensional Hilbert spaces.

Theorem 6.1. *Let* $f \in B^1_{\infty,1}(\mathbb{R}^2)$. *Then*

$$\|f(N_2) - f(N_2)\| \leq \text{const}\, \|f\|_{B^1_{\infty,1}} \|N_1 - N_2\|$$

for arbitrary not necessarily bounded normal operators on Hilbert space with bounded difference.

Recall that the original proof of this result was obtained in [7].

We can deduce now from Theorem 6.1 the following result of [17] and [18].

Corollary 6.2. *Let f be a function on $\mathbb{R}$ in the Besov class $B^1_{\infty,1}(\mathbb{R})$. Then*

$$\|f(A) - f(B)\| \leq \text{const}\, \|f\|_{B^1_{\infty,1}} \|A - B\|$$

for arbitrary not necessarily bounded self-adjoint operators A and B.

Let us proceed to the proof of inequality (5.9) for operators on Hilbert space. The following result was established in [1], see also [4].

Theorem 6.3. *Let f be a function on $\mathbb{R}$ given by (5.7) with $\varkappa$ satisfying (5.8). Then*

$$\|f(A) - f(B)\| \leq \|\varkappa\|_{\mathscr{M}(\mathbb{C}\backslash\mathbb{R})} \|A - B\|$$

for arbitrary not necessarily bounded self-adjoint operators A and B on Hilbert space with bounded difference.

Proof. The result follows immediately from Theorem 5.2 and from the equivalence of statements (1)–(3) in the beginning of this section.   □

Let us establish a Lipschitz type inequality in trace norm.

Theorem 6.4. *Let f be a function on $\mathbb{R}$ given by (5.7) with $\varkappa$ satisfying (5.8). Then*

$$\|f(A) - f(B)\|_{\boldsymbol{S}_1} \leq \|\varkappa\|_{\mathscr{M}(\mathbb{C}\backslash\mathbb{R})} \|A - B\|_{\boldsymbol{S}_1} \tag{6.1}$$

for arbitrary not necessarily bounded self-adjoint operators A and B on Hilbert space with trace class difference.

Proof. First of all, it follows from Theorem 3.2.1 of [4] that it suffices to prove the desired inequality for bounded self-adjoint operators A and B.

Suppose that A and B are bounded self-adjoint operators on a Hilbert space $\mathscr{H}$. Consider a sequence $\{P_n\}_{n\geq 1}$ of finite rank orthogonal projections such that $\lim_{n\to\infty} P_n = I$ in the strong operator topology. Put $A_n \overset{\text{def}}{=} P_n A P_n$ and $B_n \overset{\text{def}}{=} P_n B P_n$.

It follows from Theorem 5.3 that

$$\|f(A_n) - f(B_n)\|_{\boldsymbol{S}_1} \leq \|\varkappa\|_{\mathscr{M}(\mathbb{C}\backslash\mathbb{R})} \|A_n - B_n\|_{\boldsymbol{S}_1} \leq \|\varkappa\|_{\mathscr{M}(\mathbb{C}\backslash\mathbb{R})} \|A - B\|_{\boldsymbol{S}_1}.$$

On the other hand, it is easy to see that

$$\lim_{n\to\infty} f(A_n) = f(A) \quad \text{and} \quad \lim_{n\to\infty} f(B_n) = f(B)$$

in the strong operator topology (it suffices to approximate f by polynomials). This proves inequality (6.1).   □

We proceed now to Lipschitz type estimate for functions of pairs of not necessarily commuting self-adjoint operators. The following result was obtained in [6].

Theorem 6.5. *There exists a positive number C such that*

$$\|f(A_1, B_1) - f(A_2, B_2)\|_{\boldsymbol{S}_p} \leq C \|f\|_{B^1_{\infty,1}} \max\left\{\|A_1 - A_2\|_{\boldsymbol{S}_p}, \|B_1 - B_2\|_{\boldsymbol{S}_p}\right\} \tag{6.2}$$

for every $p \in [1, 2]$, *for an arbitrary function* f *in* $B^1_{\infty,1}(\mathbb{R}^2)$, *and for arbitrary pairs* (A_1, B_1) *and* (A_2, B_2) *of bounded not necessarily commuting self-adjoint operators on Hilbert space.*

Proof. As before, it suffices to prove that

$$\|f(A_1, B_1) - f(A_2, B_2)\|_{\boldsymbol{S}_p}$$
$$\leq \operatorname{const} \sigma \|f\|_{L^\infty(\mathbb{R}^2)} \max \big\{ \|A_1 - A_2\|_{\boldsymbol{S}_p}, \|B_1 - B_2\|_{\boldsymbol{S}_p} \big\}$$

for an arbitrary bounded continuous function f on $\mathbb{R}^2$ whose Fourier transform is supported in $\{\xi : \|\xi\| \leq \sigma\}$. In this case f is the restriction to $\mathbb{R}^2$ of an entire function (which is denoted by the same symbol f) on $\mathbb{C}^2$. Let

$$f(z_1, z_2) = \sum_{j,k \geq 0} \widehat{f}(j,k) z_1^j z_2^k.$$

Let $R > 2 \max\{\|A\|, \|B\|\}$. Then

$$\sum_{j,k \geq 0} |\widehat{f}(j,k)| R^{j+k} < \infty. \tag{6.3}$$

As in the proof of the previous theorem, consider a sequence $\{P_n\}_{n \geq 1}$ of finite rank orthogonal projections such that $\lim_{n \to \infty} P_n = I$ in the strong operator topology. Put $A_{1,n} \overset{\text{def}}{=} P_n A_1 P_n$, $B_{1,n} \overset{\text{def}}{=} P_n B_1 P_n$ $A_{2,n} \overset{\text{def}}{=} P_n A_2 P_n$ and $B_{2,n} \overset{\text{def}}{=} P_n B_2 P_n$.

It follows from Theorem 5.4 that

$$\|f(A_{1,n}, B_{1,n}) - f(A_{2,n}, B_{2,n})\|_{\boldsymbol{S}_p}$$
$$\leq \operatorname{const} \sigma \|f\|_{L^\infty(\mathbb{R}^2)} \max \big\{ \|A_{1,n} - A_{2,n}\|_{\boldsymbol{S}_p}, \|B_{1,n} - B_{2,n}\|_{\boldsymbol{S}_p} \big\}$$
$$\leq \operatorname{const} \sigma \|f\|_{L^\infty(\mathbb{R}^2)} \max \big\{ \|A_1 - A_2\|_{\boldsymbol{S}_p}, \|B_1 - B_2\|_{\boldsymbol{S}_p} \big\}.$$

It follows from (6.3) that

$$f(A_{1,n}, B_{1,n}) = \sum_{j,k \geq 0} \widehat{f}(j,k) A_{1,n}^j B_{1,n}^k \longrightarrow \sum_{j,k \geq 0} \widehat{f}(j,k) A_1^j B_1^k = f(A_1, B_1)$$

and

$$f(A_{2,n}, B_{2,n}) = \sum_{j,k \geq 0} \widehat{f}(j,k) A_{2,n}^j B_{2,n}^k \longrightarrow \sum_{j,k \geq 0} \widehat{f}(j,k) A_2^j B_2^k = f(A_2, B_2)$$

in the strong operator topology. Thus,

$$\|f(A_1, B_1) - f(A_2, B_2)\|_{\boldsymbol{S}_p}$$
$$\leq \limsup_{n \to \infty} \|f(A_{1,n}, B_{1,n}) - f(A_{2,n}, B_{2,n})\|_{\boldsymbol{S}_p}$$
$$\leq \operatorname{const} \sigma \|f\|_{L^\infty(\mathbb{R}^2)} \max \big\{ \|A_1 - A_2\|_{\boldsymbol{S}_p}, \|B_1 - B_2\|_{\boldsymbol{S}_p} \big\}. \qquad \square$$

Theorem 6.5 allows us to obtain Lipschitz type estimates in trace norm for functions of normal operators. The following result was obtained in [7].

Theorem 6.6. *Let* $f \in B^1_{\infty,1}(\mathbb{R}^2)$. *Then*

$$\|f(N_2) - f(N_2)\|_{\boldsymbol{S}_1} \leq \text{const}\,\|f\|_{B^1_{\infty,1}}\|N_1 - N_2\|_{\boldsymbol{S}_1}$$

for arbitrary not necessarily bounded normal operators on Hilbert space with trace class difference.

Proof. It follows from Theorem 3.2.1 of [6] that it suffices to consider the case of bounded operators N_1 and N_2.

Consider now the self-adjoint operators

$$A_1 \overset{\text{def}}{=} \operatorname{Re} N_1, \quad B_1 \overset{\text{def}}{=} \operatorname{Im} N_1, \quad A_2 \overset{\text{def}}{=} \operatorname{Re} N_2 \quad \text{and} \quad B_2 \overset{\text{def}}{=} \operatorname{Im} N_2.$$

By Theorem 6.5,

$$
\begin{aligned}
\|f(N_2) - f(N_2)\|_{\boldsymbol{S}_1} &= \|f(A_1, B_1) - f(A_2, B_2)\|_{\boldsymbol{S}_1} \\
&\leq \text{const}\,\|f\|_{B^1_{\infty,1}} \max\left\{\|A_1 - A_2\|_{\boldsymbol{S}_1}, \|B_1 - B_2\|_{\boldsymbol{S}_1}\right\} \\
&\leq \text{const}\,\|f\|_{B^1_{\infty,1}}\|N_1 - N_2\|_{\boldsymbol{S}_1}. \qquad\qquad \square
\end{aligned}
$$

References

[1] A.B. Aleksandrov, *Operator Lipschitz functions and model spaces.* Zap. Nauchn. Sem. POMI, **416**, 2013, 5–58 (Russian); English transl., J. Math. Sci. (N.Y.), **202**:4 (2014), 485–518.

[2] A.B. Aleksandrov and V.V. Peller, *Operator Hölder–Zygmund functions.* Advances in Math. **224** (2010), 910–966.

[3] A.B. Aleksandrov and V.V. Peller, *Functions of operators under perturbations of class* $\boldsymbol{S}_p$. J. Funct. Anal. **258** (2010), 3675–3724.

[4] A.B. Aleksandrov and V.V. Peller, *Operator Lipschitz functions.* Russian Mathematical Surveys, 71:4 (2016), 605–702.

[5] A.B. Aleksandrov and V.V. Peller, *Multiple operator integrals, Haagerup and Haagerup-like tensor products, and operator ideals,* Bulletin London Math. Soc. **49** (2017), 463–479.

[6] A.B. Aleksandrov, F.L. Nazarov and V.V. Peller, *Functions of noncommuting self-adjoint operators under perturbation and estimates of triple operator integrals.* Adv. Math. **295** (2016), 1–52.

[7] A.B. Aleksandrov, V.V. Peller, D. Potapov, and F. Sukochev, *Functions of normal operators under perturbations.* Advances in Math. **226** (2011), 5216–5251.

[8] M.S. Birman and M.Z. Solomyak, *Double Stieltjes operator integrals. III.* Problems of Math. Phys., Leningrad. Univ. **6** (1973), 27–53 (Russian).

[9] Yu.B. Farforovskaya, *The connection of the Kantorovich–Rubinshtein metric for spectral resolutions of selfadjoint operators with functions of operators.* Vestnik Leningrad. Univ. **19** (1968), 94–97 (Russian).

[10] K. Juschenko, I.G. Todorov and L. Turowska, *Multidimensional operator multipliers.* Trans. Amer. Math. Soc. **361** (2009), 4683–4720.

[11] B.E. Johnson and J.P. Williams, *The range of a normal derivation.* Pacific J. Math. **58** (1975), 105–122.

[12] E. Kissin and V.S. Shulman, *On a problem of J.P. Williams.* Proc. Amer. Math. Soc. **130** (2002), 3605–3608.

[13] B.Ya. Levin, *Lectures on entire functions.* Translation of Math. Monogr., vol. 150, 1996.

[14] A. McIntosh, *Counterexample to a question on commutators.* Proc. Amer. Math. Soc. **29** (1971) 337–340.

[15] J. Peetre, *New thoughts on Besov spaces.* Duke Univ. Press., Durham, NC, 1976.

[16] V.V. Peller, *Hankel operators of class $\mathbf{S}_p$ and their applications (rational approximation, Gaussian processes, the problem of majorizing operators).* Mat. Sbornik, **113** (1980), 538–581. English Transl. in Math. USSR Sbornik, **41** (1982), 443–479.

[17] V.V. Peller, *Hankel operators in the theory of perturbations of unitary and self-adjoint operators.* Funktsional. Anal. i Prilozhen. **19:2** (1985), 37–51 (Russian). English transl.: Funct. Anal. Appl. **19** (1985), 111–123.

[18] V.V. Peller, *Hankel operators in the perturbation theory of unbounded self-adjoint operators.* Analysis and partial differential equations, 529–544, Lecture Notes in Pure and Appl. Math., **122**, Dekker, New York, 1990.

[19] V.V. Peller, *Hankel operators and their applications.* Springer-Verlag, New York, 2003.

[20] V.V. Peller, *Multiple operator integrals in perturbation theory.* Bull. Math. Sci. **6** (2016), 15–88.

[21] G. Pisier, *Similarity problems and completely bounded maps.* Second, expanded edition. Includes the solution to "The Halmos problem". Lecture Notes in Mathematics, 1618. Springer-Verlag, Berlin, 2001.

[22] D. Potapov and F. Sukochev, *Operator-Lipschitz functions in Schatten–von Neumann classes.* Acta Math. **207** (2011), 375–389.

[23] W. Rudin, *Functional analysis.* McGraw Hill, 1991.

[24] E.C. Titchmarsh, *The theory of functions.* Oxford University Press, Oxford, 1958.

V.V. Peller
Department of Mathematics
Michigan State University
East Lansing
Michigan 48824, USA
e-mail: peller@math.msu.edu

Operator Theory:
Advances and Applications, Vol. 261, 417–465
© Springer International Publishing AG, part of Springer Nature 2018

Spectral Gap Properties of the Unitary Groups: Around Rider's Results on Non-commutative Sidon Sets

Gilles Pisier

Dedicated to the memory of V.P. Havin

Abstract. We present a proof of Rider's unpublished result that the union of two Sidon sets in the dual of a non-commutative compact group is Sidon, and that randomly Sidon sets are Sidon. Most likely this proof is essentially the one announced by Rider and communicated in a letter to the author around 1979 (lost by him since then). The key fact is a spectral gap property with respect to certain representations of the unitary groups $U(n)$ that holds uniformly over n. The proof crucially uses Weyl's character formulae. We survey the results that we obtained 30 years ago using Rider's unpublished results. Using a recent different approach valid for certain orthonormal systems of matrix-valued functions, we give a new proof of the spectral gap property that is required to show that the union of two Sidon sets is Sidon. The latter proof yields a rather good quantitative estimate. Several related results are discussed with possible applications to random matrix theory.

Mathematics Subject Classification (2010). Primary 43A46. Secondary 47A56, 22D10.

Keywords. Sidon set, spectral gap, random Fourier series, random matrix theory, irreducible representation.

A subset Λ of a discrete Abelian group $\widehat{G}$ is called Sidon if every continuous function on G with Fourier transform supported in Λ has an absolutely convergent Fourier series.

The study of Sidon sets in discrete Abelian groups was actively developed in the 1970s and 1980s, after Drury's remarkable proof of the stability of Sidon sets under finite unions (see [32]). Rider [47] connected Sidon sets to random Fourier series. This led the author to a new characterization of Sidon sets as $\Lambda(p)$-sets (in Rudin's sense) with constants $O(\sqrt{p})$ and eventually to an arithmetic characterization of Sidon sets (see [37, 34, 39]). Bourgain [1] gave a different proof of this. The 2013 book [18] by Graham and Hare gives an account of this subject, updating the 1975 one [32] by Lopez and Ross. See also [30] for connections with Banach space theory.

Throughout this, the main example always remains the integers $\widehat{G} = \mathbb{Z}$ (with $G = \mathbb{T} = \mathbb{R}/\mathbb{Z}$), and Sidon sets are defined by the properties of Fourier series on $\mathbb{T}$ with coefficients supported in the set. The classical example of a Sidon set is a set formed of a sequence $\{n(k)\}$ such that $\inf n(k+1)/n(k) > 1$ (such sets are called "Hadamard lacunary"). While the theory was initially inspired by this first example, much of it rests on another one, where $\mathbb{T}$ is replaced by $G = \mathbb{T}^{\mathbb{N}}$ (or by $\{-1,1\}^{\mathbb{N}}$), and the fundamental Sidon set in its dual $\widehat{G}$ is the one formed by the coordinate functions on G. In particular, the connections with random Fourier series are closely related to this second example.

Sidon sets are the analogue for discrete groups of the so-called "Helson sets" in continuous groups. The latter subject was actively studied in the late 1960s and 1970s notably by Kahane and Varopoulos in Orsay, Körner in Cambridge and many more (see [25, 26, 19]). Indeed, Sidon sets were then quite popular in harmonic analysis: in the Polish school following an old tradition (Banach, Kaczmarz, Steinhaus, Hartman, ...), in the US after Hewitt and Ross, but also in the Italian (around Figà-Talamanca) and Australian schools (around Edwards and Gaudry).

The harmonic analysis of thin sets was extended already in the late 1960s to subsets of the dual "object" $\widehat{G}$ of any non-commutative compact group G, with Fourier series replaced by the Peter-Weyl orthogonal development of functions on G. In this setting pioneering work was done by Figà-Talamanca and Rider ([15, 16, 12]) on generalized random Fourier series. There was initially a lot of excitement around the opening that non-commutative compact groups offered as a substitute for $\mathbb{T}$. However, the subject was given a cold shower when it was discovered (see [48, 49, 8, 24]) that even for the simplest example $G = SU(2)$ infinite Sidon sets do not exist. Since finite Sidon sets were considered trivial, this brought this whole direction to a full stop and probably gave a bad reputation to Sidon sets in the duals of non-commutative compact groups. After that, many in the next generation of researchers, in particular in the Polish school (Bożejko, Pytlik, Szwarc, ...) and the Italian one (Figà-Talamanca, Picardello, ...), turned to harmonic analysis on free groups (see, e.g., [13, 14]). In this setting free sets, or "almost free" sets, such as the so-called Leinert sets (see, e.g., [29]) or L-sets in the sense of [42], can be viewed as analogous in some sense to Sidon sets in discrete non-commutative groups.

This context probably explains why Rider, when he published in [47] his theorem connecting Sidon sets and random Fourier series decided not to include the details on the proof of the same result for subsets of the duals of non-commutative compact groups. In the commutative case, full details could be included without any special technical difficulty because the key ingredient was a variant of Drury's interpolation trick (by then well known), invented to prove that the union of two

Sidon sets is Sidon, and actually Rider's theorem could be viewed as a generalization of Drury's union theorem. However, the extension of the latter to the non-commutative case was far from obvious (see Remark 1.11), and in fact it was still open until Rider's [47]. Nevertheless, Rider chose to only announce there that he had settled it and promised to include the details, which involved a delicate estimate based on Weyl's character formula for the unitary groups (see Theorem 2.1), in a later publication, but he never did.

In the late 1970s the author proved a series of results on Sidon sets all based initially on Rider's breakthrough from [47]. It turned out that essentially all these results could be extended for subsets of $\widehat{G}$ when G is a non-commutative compact group [34, 38]. However, the latter extension required the non-commutative unpublished version of Rider's [47]. At the author's request at the time, Rider kindly communicated to him a detailed handwritten proof of his key result in the non-commutative case. Unfortunately, although a copy of this letter was kept for a long time, it seems now to have been lost. Perhaps the successive moves of the Jussieu Math. Inst. are an excuse, but the guilt is on the author. The more so since Daniel Rider passed away in 2008.

The main goal of this paper is to present the details of a proof of Rider's Theorem for subsets of $\widehat{G}$ when G is a general (a priori non-commutative) compact group. Toward the end we give another proof, quite different, that we recently obtained in a more general framework not requiring any group structure.

The main point of Rider's proof is a spectral gap property of the family $\{U(n) \mid n \geq 1\}$ formed of *all* the unitary groups. The property involves the embedding $U(n) \to U(2n)$ obtained by adding 1's on the main diagonal, but the relevant estimate has to be uniform over n. We feel that this property is of independent interest, likely to find applications in random matrix theory, now that the latter field has become part of the main stream (much more so now than 40 years ago!).

This motivated us to include the full details of (what most likely was) Rider's proof. We then describe in §3 how Rider derived from his spectral gap result the stability of Sidon sets under finite unions and the fact the Sidon property is equivalent to a weaker one involving random Fourier series that we name "randomly Sidon".

In §4 we survey the non-commutative results that we obtained in the 1980s using Rider's unpublished work. Actually we take special care and give detailed proofs because we detected some exaggerated claims there (in [38]) that we no longer believe are true. See Remark 4.14.

In §5, we single out several natural inequalities for random unitaries, related to the classical ones of Khintchine for random signs. We review what is known and discuss the problem of finding the best constants for these.

We seize this occasion to try to revive a bit the whole subject of Sidon sets in duals of non-Abelian compact groups in the light of the recent surge of interest in random matrix theory and Voiculescu's free probability (see [60]). Indeed, although finite sets $\Lambda \subset \widehat{G}$ are a trivial example of Sidon set, in the non-commutative setting one is led to consider sequences of compact groups (G_n) and sequences of subsets $\Lambda_n \subset \widehat{G}_n$ with uniformly bounded Sidon constants. Then even if the cardinality of the subsets Λ_n is uniformly bounded (and in fact even if it is equal to 1!) the notion is interesting. The simplest (and prototypical) example of this situation with $|\Lambda_n| = 1$ is the case when $G_n = U(n)$ the group of unitary $n \times n$-matrices, and Λ_n is the singleton formed of the irreducible representation (in short irrep) defining $U(n)$ as acting on $\mathbb{C}^n$. Sets of this kind and various generalizations were tackled early on by Rider under the name "local lacunary sets" (see [50]), but we suspect that this setting of sequences of groups, with uniform estimates, which is nowadays commonly accepted, was viewed as not so natural at the time.

We illustrate this in Theorem 4.15. There we consider a sequence of compact groups G_n and a sequence of unitary irreps $\pi_n \in \widehat{G}_n$ with unbounded dimensions, and we focus on the situation when the singletons $\{\pi_n\}$ have uniformly bounded Sidon constants. We give several equivalent characterizations of this situation, in terms of the character $t \mapsto \operatorname{tr}(\pi_n(t))$ of π_n. Surprisingly, this becomes void if one uses a sequence of finite groups, or of groups that are amenable as discrete groups. In that case the dimensions must remain bounded. E. Breuillard opened our eyes to this phenomenon. We refer the reader to the forthcoming paper [5] for more on this.

1. Notation. Background. Spectral gaps

Throughout this section, let G be a compact group. We denote by $\widehat{G}$ the dual object formed as usual of all the (equivalence classes of) irreducible representations (irreps in short) on G. We identify two irreps when they are unitarily equivalent. We denote by $M(G)$ the space of Radon measures on G equipped as usual with the total variation norm $\mu \mapsto \|\mu\|_{M(G)} = |\mu|(G)$.

We denote by M_d the space of all complex matrices of size $d \times d$ with the usual operator norm as acting on ℓ_2^d.

We denote by $U(d) \subset M_d$ the compact group formed of all unitary matrices of size $d \times d$.

For any measure μ on G and any irrep $\pi : G \to U(d_\pi)$ we define the Fourier transform by

$$\widehat{\mu}(\pi) = \int \overline{\pi(t)} \mu(dt) \in M_{d_\pi}. \tag{1.1}$$

Note that $\forall \mu_1, \mu_2 \in M(G)$

$$\widehat{\mu_1 * \mu_2}(\rho) = \widehat{\mu_1}(\rho)\widehat{\mu_2}(\rho). \tag{1.2}$$

We denote by m_G the normalized Haar measure and by $t_G \in \widehat{G}$ the trivial representation on G.

We denote $L_p(G) = L_p(G, m_G)$. We view $L_1(G)$ as isometrically embedded in $M(G)$ via $f \mapsto f m_G$. In particular, the Fourier transform of any $f \in L_1(G)$ is defined as

$$\widehat{f}(\pi) = \int \overline{\pi(t)} f(t) m_G(dt). \tag{1.3}$$

For any $f \in L_2(G)$ we have (Parseval)

$$\|f\|_2 = \left(\sum_{\rho \in \widehat{G}} d_\rho \operatorname{tr} |\widehat{f}(\rho)|^2 \right)^{1/2},$$

and the Fourier expansion of f takes the form

$$f = \sum_{\rho \in \widehat{G}} d_\rho \operatorname{tr}(^t \widehat{f}(\rho) \rho).$$

Remark. Note that our definitions of $\widehat{\mu}$ and $\widehat{f}$ in (1.1) and (1.3) differ from that of [23], where $\widehat{\mu}(\pi)$ is defined as $\int \pi(t)^* \mu(dt)$ and similarly for $\widehat{f}$. Thus the Fourier coefficient in the sense of [23] is the transpose of what it is in our sense. The advantage is that we have (1.2) while the convention of [23] requires to reverse the order of the factors on the right-hand side of (1.2).

We denote by χ_π the character of π, i.e., we have $\chi_\pi(x) = \operatorname{tr}(\pi(x))$ for any $x \in G$. A measure $\mu \in M(G)$ (resp. a function $f \in L_1(G)$) is called *central* if

$$\forall g \in G \quad \mu = \delta_g * \mu * \delta_{g^{-1}}$$

(resp. $f = \delta_g * f * \delta_{g^{-1}}$). Then the Fourier transform $\widehat{\mu}$ (resp. $\widehat{f}$) is scalar-valued, i.e., $\widehat{\mu}(\pi)$ or $\widehat{f}(\pi)$ belong to the space of scalar multiples of the identity matrix of size d_π. Thus the subspace of central functions in L_p ($1 \le p < \infty$) coincides with the closed linear span of the characters $\{\chi_\pi \mid \pi \in \widehat{G}\}$.

There is a bounded linear projection P from $M(G)$ onto the subspace of all central measures, defined simply by

$$P(\mu) = \int \delta_g * f * \delta_{g^{-1}} m_G(dg). \tag{1.4}$$

Clearly $\|P(\mu)\| \le \|\mu\|$. We denote by $A(G)$ the Banach space formed of those $f: G \to \mathbb{C}$ such that $\sum_{\pi \in \widehat{G}} d_\pi \operatorname{tr} |\widehat{f}(\pi)| < \infty$, and we equip it with the norm

$$\|f\|_{A(G)} = \sum_{\pi \in \widehat{G}} d_\pi \operatorname{tr} |\widehat{f}(\pi)|.$$

Definition 1.1 (Sidon sets). A subset $\Lambda \subset \widehat{G}$ is called Sidon if there is a constant C such that

$$\|f\|_{A(G)} \le C \|f\|_{C(G)}$$

for any $f \in C(G)$ with Fourier transform supported in Λ. More explicitly, this means that for any finitely supported family (a_π) with $a_\pi \in M_{d_\pi}$ ($\pi \in \Lambda$) we have

$$\sum_{\pi \in \Lambda} d_\pi \operatorname{tr} |a_\pi| \le C \| \sum_{\pi \in \Lambda} d_\pi \operatorname{tr}(\pi a_\pi) \|_\infty.$$

For any pair $f, h \in L_2(G)$, the convolution $f * h$ belongs to $A(G)$ and

$$\|f * h\|_{A(G)} \le \|f\|_{L_2(G)} \|h\|_{L_2(G)}. \tag{1.5}$$

Moreover, for any $f \in A(G)$ and any $\nu \in M(G)$ we have

$$\int f d\nu = \sum_{\pi \in \widehat{G}} d_\pi \, \mathrm{tr} \left({}^t \widehat{f}(\pi) \widehat{\nu}(\overline{\pi})\right) = \sum_{\pi \in \widehat{G}} d_\pi \sum_{i,j \le d_\pi} \widehat{f}(\pi)_{ij} \widehat{\nu}(\overline{\pi})_{ij}. \tag{1.6}$$

and hence

$$\left| \int f(g) \nu(dg) \right| \le \|f\|_{A(G)} \| \sup_{\pi \in \widehat{G}} \|\widehat{\nu}(\pi)\|. \tag{1.7}$$

More generally, let $f, h \in L_\infty(G; M_d)$ $(d \ge 1)$. We define the convolution $F = f * h$ using the matrix product in M_d, so that $F_{ij} = \sum_k f_{ik} * h_{kj}$. Let x, y be in the unit ball of ℓ_2^d. We have then

$$\|\langle Fx, y\rangle\|_{A(G)} \le \|f\|_{L_\infty(G;M_d)} \|h\|_{L_\infty(G;M_d)}. \tag{1.8}$$

Indeed, this follows easily from (here we use (1.5))

$$\|\langle Fx, y\rangle\|_{A(G)} \le \sum_k \left\| \sum_i \bar{x}_i f_{ik} \right\|_2 \left\| \sum_j y_j h_{kj} \right\|_2$$

$$\le \left(\sum_k \left\| \sum_i \bar{x}_i f_{ik} \right\|_2^2 \right)^{1/2} \left(\sum_k \left\| \sum_j y_j h_{kj} \right\|_2^2 \right)^{1/2}$$

$$= \left(\int \sum_k \left| \sum_i \bar{x}_i f_{ik} \right|_2^2 dm_G \right)^{1/2} \left(\int \sum_k \left| \sum_j y_j h_{kj} \right|_2^2 dm_G \right)^{1/2}$$

$$\le \|f\|_{L_2(G;M_d)} \|h\|_{L_2(G;M_d)}.$$

A fortiori, we obtain by (1.7)

$$\left| \int \langle F(g)x, y\rangle \nu(dg) \right| \le \|f\|_{L_\infty(G;M_d)} \|h\|_{L_\infty(G;M_d)} \sup_{\pi \in \widehat{G}} \|\widehat{\nu}(\pi)\|. \tag{1.9}$$

Taking the sup over x, y, we find

$$\left\| \int F(g) \nu(dg) \right\|_{M_d} \le \|f\|_{L_\infty(G;M_d)} \|h\|_{L_\infty(G;M_d)} \sup_{\pi \in \widehat{G}} \|\widehat{\nu}(\pi)\|. \tag{1.10}$$

Notation: Let $\mathcal{G} = \prod_{\pi \in \widehat{G}} U(d_\pi)$. Let $u \mapsto u_\pi \in U(d_\pi)$ denote the coordinates on $\mathcal{G}$.

Definition 1.2 (Randomly Sidon). A subset $\Lambda \subset \widehat{G}$ is called randomly Sidon if there is a constant C such that for any finitely supported family (a_π) with $a_\pi \in M_{d_\pi}$ $(\pi \in \Lambda)$ we have

$$\sum_{\pi \in \Lambda} d_\pi \, \mathrm{tr} \, |a_\pi| \le C \int \left\| \sum_{\pi \in \Lambda} d_\pi \, \mathrm{tr}(u_\pi \pi a_\pi) \right\|_\infty m_{\mathcal{G}}(du).$$

Note that in Lemma 4.6 we give a simple general argument showing that replacing the random unitaries (u_π) by standard complex Gaussian random matrices (with the usual normalization) leads to the same notion of "randomly Sidon".

Clearly Sidon implies randomly Sidon (with the same constant).

We denote by $\mathcal{P}(G) \subset M(G)$ the set of probability measures on G. We say that $\Lambda \subset \widehat{G}$ is symmetric if $\bar{\pi} \in \Lambda$ for any $\pi \in \Lambda$.

Definition 1.3 (Spectral gap). Let $0 \leq \gamma < \delta \leq 1$. We will say that a probability measure $\mu \in \mathcal{P}(G)$ has a (δ, γ)-spectral gap with respect to a symmetric subset $\Lambda \subset \widehat{G}$ if $\widehat{\mu}(\pi) = \delta I$ for any $\pi \in \Lambda$ and $\|\widehat{\mu}(\rho)\| \leq \gamma$ for any nontrivial $\rho \notin \Lambda$.

Remark 1.4 (Spectral gap as an inequality). Let $E \subset L_2(G)$ be the subspace formed of those $f \in L_2(G)$ such that $\widehat{f}(\pi) = 0$ for any non-trivial $\pi \notin \Lambda$. Let $P : L_2(G) \to E$ denote the orthogonal projection. Note $Pf = \int f \, dm_G + \sum_{\pi \in \Lambda} d_\pi \operatorname{tr}({}^t \widehat{f}(\pi)\pi)$ for any $f \in L_2(G)$. Let $P_\delta f = \int f \, dm_G + \delta \sum_{\pi \in \Lambda} d_\pi \operatorname{tr}({}^t\widehat{f}(\pi)\pi)$. Then μ has a (δ, γ)-spectral gap with respect to Λ if and only if

$$\forall f \in L_2(G) \quad \|\mu * f - P_\delta f\|_2 \leq \gamma \|f - Pf\|_2.$$

Definition 1.5 $((\delta, \gamma)$-isolated). We will say that $\Lambda \subset \widehat{G}$ is (δ, γ)-isolated if there is $\mu \in \mathcal{P}(G)$ that has a (δ, γ)-spectral gap with respect to Λ.

Remark 1.6. Using the central projection (1.4) we may always assume in the preceding that μ is a central measure.

The basic example is the set $\Lambda = \{-1, 1\} \subset \mathbb{Z}$. The measure $\mu = (1 + \cos(t))m_\mathbb{T}(dt)$ has a $(1/2, 0)$-spectral gap with respect to Λ. On $G = \{-1, 1\}$ the measure $\mu = (1 + \xi)m_G$ does the same with respect to the set formed of the character $\xi \in \widehat{G}$ associated with the identity map. More generally, Riesz products give more sophisticated examples. Let G be a compact Abelian group. Let $\{\gamma_n \mid n \in \mathbb{N}\} \subset \widehat{G}$ be "quasi-independent", i.e., such that there is no nontrivial choice of $(\xi_n) \in \{-1, 0, 1\}^\mathbb{N}$ finitely supported such that $\prod \gamma_n^{\xi_n} = 1$. Assume $-1 \leq \delta_n \leq 1$. Then the probability measures $\nu_k = \prod_{n \leq k}(1 + \delta_n \operatorname{Re}(\gamma_n))m_G$ converge weakly when $k \to \infty$ to a probability ν on G. We refer to ν as the Riesz product associated with $\prod(1 + \delta_n \operatorname{Re}(\gamma_n))$.

If we assume that $\delta_n = \delta$ for all n and $0 < \delta < 1$, then the Riesz product ν has a (δ, δ^2)-spectral gap with respect to $\Lambda = \{\gamma_n\} \cup \{\bar{\gamma}_n\}$. For instance, this holds for $G = \mathbb{R}/2\pi\mathbb{Z}$ when $\Lambda = \{\gamma_n\}$ is identified with the subset $\{2^n\} \subset \mathbb{Z}$ by $\gamma_n(t) = \exp(i2^n t)$. This also holds for $G = \{-1, 1\}^\mathbb{N}$ (resp. $G = \mathbb{T}^\mathbb{N}$) when $\Lambda \subset \widehat{G}$ is the set $\{\xi_n\}$ (resp. $\{\xi_n\} \cup \{\bar{\xi}_n\}$) with (ξ_n) denoting the coordinates on G.

Let $\sigma_n : U(n) \to M_n$ be the "defining" irrep, i.e., the identity map on $U(n)$.

Lemma 1.7. *Let $n \geq 1$. For any $0 < \delta \leq 1/(2n)$, let*

$$\varphi_n^\delta = 1 + \delta(\chi_{\sigma_n} + \overline{\chi_{\sigma_n}}) = 1 + \delta(\operatorname{tr}(\sigma_n) + \overline{\operatorname{tr}(\sigma_n)}).$$

Let $\nu_n^\delta \in M(U(n))$ be the probability measure defined by $\nu_n^\delta = \varphi_n^\delta \, m_{U(n)}$. Then ν_n^δ has a $(\delta/n, 0)$-spectral gap with respect to $\{\sigma_n, \bar{\sigma}_n\}$.

Proof. Obviously $\widehat{\varphi_n^\delta}(\sigma_n) = \widehat{\varphi_n^\delta}(\overline{\sigma_n}) = \delta/n$ and $\widehat{\varphi_n^\delta}(\pi) = 0$ for any other nontrivial irrep π. $\square$

Definition 1.8 (Peak sets). Let $0 < \varepsilon < 1$. We say that $\Lambda \subset \widehat{G}$ is an ε-peak set with constant w if there is $\nu \in M(G)$ with $\|\nu\|_{M(G)} \leq w$ such that $\widehat{\nu}(\pi) = I$ for any $\pi \in \Lambda$ and $\sup_{\rho \notin \Lambda} \|\widehat{\nu}(\rho)\| \leq \varepsilon$.

Remark 1.9. If ν is as in Definition 1.8 for some $0 < \varepsilon < 1$ then ν^{*k} satisfies the same with ε^k, w^k in place of ε, w. Therefore, if Λ is an ε-peak set for some $0 < \varepsilon < 1$, then it is so for all $0 < \varepsilon < 1$.

Definition 1.10 (Peaking Sidon sets). We say that a Sidon set $\Lambda \subset \widehat{G}$ is peaking if for any $0 < \varepsilon < 1$ and any $u \in \mathcal{G}$ (or merely for any $u \in \prod_{\pi \in \Lambda} U(d_\pi)$) there is a measure $\mu_\varepsilon^u \in M(G)$ such that

$$\widehat{\mu_\varepsilon^u}(\pi) = u_\pi \ \forall \pi \in \Lambda, \quad \sup_{\pi \notin \Lambda} \|\widehat{\mu_\varepsilon^u}(\pi)\| \leq \varepsilon \text{ and } \|\widehat{\mu_\varepsilon^u}\| \leq w(\varepsilon)$$

where $w(\varepsilon)$ depends only on ε.

Remark 1.11 (The main difficulty of the non-Abelian case). Note that one of our main goals will be to prove that actually any Sidon set is peaking. This will be reached in Theorem 3.5 and Remark 3.9. Once this goal is attained, it follows as an easy corollary that the union of two Sidon sets is also one (see Corollary 3.6). In the Abelian case, Drury's (or Rider's) proof made crucial use of the Riesz product $\prod(1 + \delta(z_n + \bar{z}_n)/2)$ on $\mathbb{T}^\mathbb{N}$ ($0 \leq \delta < 1$). With the notation in Lemma 1.7 this is the same as the infinite product of the probability ν_1^δ on $\mathbb{T}$. The latter has a (δ, δ^2)-spectral gap with respect to the Sidon set formed of the coordinates on $\mathbb{T}^\mathbb{N}$, which is the fundamental example in the Abelian case. The proof that Sidon sets are peaking uses a certain transplantation trick due to Drury to pass from the fundamental example to the general case. It is not really difficult to adapt that trick to the non-Abelian case (see the proof of Theorem 3.5). However, in the non-Abelian case the fundamental example is the product $\prod_{n \geq 1} U(n)$ but the product of the probabilities ν_n^δ *fails* to have the required spectral gap, whence the need for a substitute for the Riesz product. This is precisely the role of Theorem 2.1 in the next section.

The preceding definitions are connected by the following simple result.

Proposition 1.12. *Let $0 < \gamma < \delta < 1$. Any (δ, γ)-isolated symmetric set $\Lambda \subset \widehat{G}$ is an ε-peak set with constant w for some $0 < \varepsilon < 1$ and $w \geq 0$ depending only on γ, δ. Any Sidon set $\Lambda \subset \widehat{G}$ that is also an ε-peak set with constant w for some $0 < \varepsilon < 1$ and $w \geq 0$ is peaking.*

Proof. Let μ be as in Definition 1.3. Let $\nu = \delta^{-1}(\mu - m_G)$ with $\varepsilon = \gamma/\delta$ and $w = d^{-1}(\|\mu\| + 1)$. Then ν satisfies the property in Definition 1.8. If Λ is Sidon with constant C, by (i) in Lemma 3.3 (Hahn–Banach), for any $u \in \mathcal{G}$ there is a measure $\mu^u \in M(G)$ such that

$$\widehat{\mu^u}(\pi) = u_\pi \ \forall \pi \in \Lambda \text{ and } \|\widehat{\mu^u}\| \leq C.$$

Let ν be as in Definition 1.8. Then $\mu_\varepsilon^u = \mu^u * \nu$ is as in Definition 1.10 with $w(\varepsilon) = Cw$. This gives the announced result for some $0 < \varepsilon < 1$, but replacing ν by its convolution powers we obtain a similar result for any $0 < \varepsilon < 1$. $\qquad\square$

Proposition 1.13. *Let $G = \prod_{n \in \mathbb{N}} G_n$ be the product of a sequence of compact groups, let (μ_n) be a sequence with $\mu_n \in \mathcal{P}(G_n)$ and let (Λ_n) be a sequence of symmetric subsets with $\Lambda_n \subset \widehat{G_n}$ for each n. Let $0 < \gamma < \delta < 1$. Let $\gamma' = \max\{\gamma, \delta^2\} < \delta$. If μ_n has a (δ, γ)-spectral gap with respect to Λ_n for each n, then the product $\mu = \otimes_{n \in \mathbb{N}} \mu_n$ has a (δ, γ')-spectral gap with respect to the subset $\Lambda \subset \widehat{G}$, denoted by $\dot{\Sigma}\Lambda_n$, consisting of all the irreps π on G of the following form: for some n there is $\pi_n \in \Lambda_n$ such that*

$$\forall x = (x_n) \in G \quad \pi(x) = \pi_n(x_n).$$

Proof. Let $\pi \in \dot{\Sigma}\Lambda_n$. Then $\widehat{\mu}(\pi) = \widehat{\mu_n}(\pi_n)$. Any nontrivial $\pi \in \widehat{G}$ is of the form $\pi(x) = \otimes_{n \in \mathbb{N}} \pi_n(x_n)$ for some sequence (π_n) with $\pi_n \in \widehat{G_n}$ containing some but only finitely many nontrivial terms. If at least one of these nontrivial terms π_n is not in Λ_n, then $\|\widehat{\mu}(\pi)\| \leq \gamma$. If they are all in Λ_n and $\pi \notin \Lambda$, there must be at least two of them and then $\|\widehat{\mu}(\pi)\| \leq \delta^2$. The result is then immediate. $\qquad\square$

Remark 1.14. Let $G_k = U(d_k)$ and $G = \prod G_k$. Assume that $N = \sup_k d_k < \infty$. Let $0 < \delta \leq 1/(2N)$. Let $\varphi_n \in L_1(G)$ be defined for $x = (x_k) \in G$ by $\varphi_n(x) = \prod_{k \leq n}(1 + \delta(\operatorname{tr}(x_k) + \operatorname{tr}(\overline{x_k})))$, and let $\nu_n = \varphi_n m_G$. As for Riesz products, $\nu_n \in \mathcal{P}(G)$, ν_n converges weakly to some $\nu \in \mathcal{P}(G)$, and it is easy to check, similarly, that ν has a $(\delta/N, \delta^2/N^2)$-spectral gap. This can also be seen as a particular case of the preceding proposition with $\gamma = 0$ and δ replaced by δ/N.

2. The unitary groups

The main difficulty Rider had to overcome to establish his main result is the following spectral gap (and interpolation) property of the sequence of the unitary groups $\{U(n) \mid n \geq 1\}$, which in our opinion, is quite deep. Note however that, for the applications to Sidon sets, any probability with the same gap property as the one denoted below by ν_n would do (see §6).

Let $1 \leq k \leq n$. Let $\Gamma(k) \subset U(n)$ be the copy of $U(k)$ embedded in $U(n)$ via $a \mapsto a \oplus I$. Let $\mu_{k,n}$ be the central symmetric probability measure defined by

$$\mu_{k,n} = \int \delta_s * m_{\Gamma(k)} * \delta_{s^{-1}} \, m_{U(n)}(ds). \tag{2.1}$$

We denote by $\sigma_n \in \widehat{U(n)}$ the defining representation of $U(n)$.

We denote by $S_n \subset \widehat{U(n)}$ the set

$$S_n = \{\sigma_n, \overline{\sigma_n}\}.$$

For emphasis: it is crucial in the next statement that $\gamma < 1/2$ be *independent of n*.

Theorem 2.1 (Rider, circa 1975, unpublished). *For any even $n \geq 2$, let $k = n/2$ and let $\nu_n = \mu_{k,n}$. For any odd n, let $k_+ = n/2 + 1/2$, $k_- = n/2 - 1/2$ and $\nu_n = 1/2(\mu_{k_-,n} + \mu_{k_+,n})$. There is a positive constant $\gamma < 1/2$ such that for any $n \geq 4$, the symmetric central probability measure ν_n has a $(1/2, \gamma)$-spectral gap with respect to S_n. More precisely, for any $1/4 < \gamma < 1/2$ this holds for all sufficiently large n.*

Remark 2.2. The case $n = 1$, $G = \mathbb{T} = \mathbb{R}/2\pi\mathbb{Z}$ is classical. Then the probability measure

$$\mu(dt) = (1 + \cos t)m_{\mathbb{T}}(dt)$$

(which is the building block for Riesz products) satisfies the analogous interpolation property, with $\gamma = 0$.

Corollary 2.3. *Let $(d_k)_{k \in I}$ be an arbitrary collection of integers. Let $G = \prod_{k \in I} U(d_k)$. Let $S \subset \widehat{G}$ be the subset formed of all representations π that, for some $k \in I$, are of the form $\pi(g) = \sigma_{d_k}(g_k)$ $(g \in G)$. For any $0 < \varepsilon < 1$ there is a measure $\mu_\varepsilon \in M(G)$ such that*

$$\widehat{\mu_\varepsilon}(\pi) = I \ \forall \pi \in S, \quad \sup_{\pi \notin S} \|\widehat{\mu_\varepsilon}(\pi)\| \leq \varepsilon \ and \ \|\mu\| \leq w(\varepsilon)$$

where $w(\varepsilon)$ depends only on ε.

Proof. By Theorem 2.1, there is N (e.g., $N = 4$) and $0 < \gamma < 1/2$ such that S_n has a $(1/2, \gamma)$-spectral gap for any $n \geq N$. Let $G = G_1 \times G_2$ with $G_1 = \prod_{d_k < N} U(d_k)$ and $G_2 = \prod_{d_k \geq N} U(d_k)$. Let $S_1 \subset \widehat{G_1}$ and $S_2 \subset \widehat{G_2}$ be the corresponding subsets and let $\Lambda_j = S_j \cup \overline{S_j}$ $(j = 1, 2)$. By Remark 1.14, Λ_1 is (λ, λ^2)-isolated for any $0 < \lambda \leq 1/2N$. We may clearly assume $\gamma \geq 1/4$. Then by Proposition 1.13, Λ_2 is $(1/2, \gamma)$-isolated. Taking convolution powers, we see that it is also $(1/2^m, \gamma^m)$-isolated for any integer $m \geq 1$. Choose m minimal but large enough so that $1/2^m \leq 1/(2N)$. Let $\delta = 1/2^m$ and $\gamma' = \max\{\gamma^m, \delta^2\}$. Then both Λ_1 and Λ_2 are (δ, γ')-isolated. Therefore, by Proposition 1.13 $S \cup \overline{S}$ is also (δ, γ')-isolated. By Proposition 1.12, $S \cup \overline{S}$ is an ε-peak set for some $0 < \varepsilon < 1$. Let $\nu_1 \in M(G)$ be such that $\widehat{\nu_1} = I$ on $S \cup \overline{S}$ but $\|\widehat{\nu_1}\| \leq \varepsilon$ outside $S \cup \overline{S}$. It remains to show the same but with S in place of $S \cup \overline{S}$. For any $z \in \mathbb{T}$, let $Z(z) \in G$ be the element such that $Z(z)_k = zI_k$. Note $\widehat{\delta_{Z(z)}}(\sigma_k) = \bar{z}I_k$. Then let

$$\nu_2 = \int z(\delta_{Z(z)} * \nu_1)m_{\mathbb{T}}(dz).$$

Now $\widehat{\nu_2} = I$ on S, and $\widehat{\nu_2} = 0$ on $\bar{S}$. Also $\|\widehat{\nu_2}\| \leq \|\widehat{\nu_1}\|$ on all of $\widehat{G}$. Thus $\|\widehat{\nu_1}\| \leq \varepsilon$ outside S and $\|\nu_2\| \leq \|\nu_1\|$. By Remark 1.9 this completes the proof. $\qquad \square$

We will need some background on irreps of the unitary groups. The ultra-classical reference is Hermann Weyl's [62]. See, e.g., [45, 52, 55] for more recent accounts on the combinatorics of this rich subject. We greatly benefited from the expositions in [11] and [17].

Recall that for any compact group G, the set $\widehat{G}$ consists of irreps on G with exactly one representative, up to unitary equivalence, of each irrep. Let $G = U(n)$. Then $\widehat{G}$ is in 1-1 correspondence with the set of n-tuples $m = (m_1, m_2, \ldots, m_n)$ in $\mathbb{Z}^n$ such that $m_1 \geq \cdots \geq m_n$. Let $t = (t_1, \ldots, t_n) \in \mathbb{C}^n$. Let $A_m(t)$ denote the determinant of the $n \times n$-matrix $a_m(t)$ defined by

$$a_m(t)_{ij} = t_i^{m_j}.$$

Let $\delta = (n-1, n-2, \ldots, 1, 0)$. Let π_m be the irrep corresponding to m, and let χ_m denote its character. Then for any unitary $g \in U(n)$ with eigenvalues $t = (t_1, \ldots, t_n) \in \mathbb{T}^n$, g is unitarily equivalent to the diagonal matrix $D(t)$ with coefficients t. This implies that $\chi_m(g) = \operatorname{tr}(\pi_m(g)) = \operatorname{tr}(\pi_m(D(t))) = \chi_m(D(t))$. For simplicity, we will identify t with $D(t)$ and we set $\chi_m(t) = \chi_m(D(t))$. We can now state Weyl's fundamental character formula, which goes back to [62]:

$$\chi_m(t) = \frac{A_{m+\delta}(t)}{A_\delta(t)}. \tag{2.2}$$

Note that $A_\delta(t)$ is but the classical Vandermonde determinant

$$A_\delta(t) = \prod_{i<j}(t_i - t_j).$$

We observe that for any $d \in \mathbb{Z}$ we have

$$A_{m+(d,\ldots,d)}(t) = (t_1 t_2 \ldots t_n)^d A_m(t)$$

and hence for any $g \in G$

$$\chi_{m+(d,\ldots,d)}(g) = \det(g)^d \chi_m(g).$$

Thus if we choose $d = -m_n$, and set $\lambda_j = m_j + d$, we have $\lambda_1 \geq \cdots \geq \lambda_{n-1} \geq \lambda_n = 0$, and

$$\chi_m(g) = \det(g)^{m_n} \chi_\lambda(g). \tag{2.3}$$

Remark 2.4 (Distinguished representations of $U(n)$). The trivial representation of $U(n)$ corresponds to $m_1 = \cdots = m_n = 0$, so that $d = 0$ and $\lambda_1 = \cdots = \lambda_n = 0$, and then $\chi_m(t) = 1$ for all $t \in U(n)$. The representation $\sigma_n(t) = t$ corresponds to $m = \lambda = (1, 0, \ldots, 0)$ and $d = 0$. Then

$$\chi_m(t) = t_1 + \cdots + t_n.$$

The representation $\sigma_n(t) = \bar{t}$ corresponds to $m = (0, \ldots, 0, -1)$ or equivalently to $\lambda = (1, \ldots, 1, 0)$ and $d = 1$. Then

$$\chi_m(t) = \bar{t}_1 + \cdots + \bar{t}_n = \left(\prod_{j\neq 1} t_j + \cdots + \prod_{j\neq n} t_j\right) \det(t)^{-1}.$$

In the sequel, we denote

$$\lambda_+ = (1, 0, \ldots, 0) \quad \text{and} \quad \lambda_- = (1, \ldots, 1, 0).$$

The point of (2.3) is that now λ can be identified with a Young diagram with a first row of λ_1 boxes, sitting as usual above a second row of λ_2 boxes, and so on. This will allow us to take advantage of the so-called Jacobi–Trudi formula (see [17, p. 75]):

$$\chi_\lambda(t) = s_\lambda(t), \tag{2.4}$$

where s_λ is the famous Schur symmetric polynomial in $t = (t_1, \ldots, t_n)$, which can be defined for $\lambda \neq 0$ as the sum

$$s_\lambda(t) = \sum t^T \tag{2.5}$$

running over all the admissible fillings (or "tableaux") T of the diagram λ with the numbers $1, 2, \ldots, n$. Here an admissible filling assigns to any box a number in $1, 2, \ldots, n$ so that the numbers are strictly increasing when running down a column and weakly increasing along each row, and

$$t^T = \prod_{1 \leq i \leq n} t_i^{r_i}$$

where $r_i \geq 0$ is the number of times i is used in the filling T. By convention, for the case $\lambda_1 = \cdots = \lambda_n = 0$, we set $s_0(t) = 1$.

Let $1^n = (1, \ldots, 1)$ where 1 is repeated n times. Then (2.5) implies

$$s_\lambda(1^n) = |\{T\}|, \tag{2.6}$$

i.e., $s_\lambda(1^n)$ is the number of admissible fillings of λ with the numbers $1, 2, \ldots, n$. Then for any $\lambda = (\lambda_1, \ldots, \lambda_n)$ with $\lambda_1 \geq \cdots \geq \lambda_n \geq 0$ we have

$$\chi_\lambda(1^n) = s_\lambda(1^n) = \prod_{i<j} \frac{\lambda_i - \lambda_j + j - i}{j - i}. \tag{2.7}$$

Note that $\frac{\lambda_i - \lambda_j + j - i}{j - i} \geq 1$ for all $i < j$.

This classical formula can be deduced from (2.2): by setting $t = (1, x, x^2, \ldots, x^{n-1})$, and observing that $A_{\lambda+\delta}(1, x, x^2, \ldots, x^{n-1})$ is a Vandermonde determinant, we have

$$\chi_\lambda(1, x, x^2, \ldots, x^{n-1}) = x^{\sum (i-1)\lambda_i} \prod_{i<j} \frac{x^{\lambda_i - \lambda_j + j - i} - 1}{x^{j-i} - 1}.$$

Then letting x tend to 1, and making the obvious common division in the numerator and denominator, (2.7) follows.

The preceding definition of the Schur symmetric polynomial s_λ is classically given as a function of k variables with k not necessarily equal to the number of rows n of λ: one sets

$$s_\lambda(t_1, \ldots, t_k) = \sum t^T$$

where the sum runs over all the admissible fillings of the Young diagram λ by the numbers $1, 2, \ldots, k$, with t^T as before.

If $\lambda_n > 0$ and $k < n$, then the first column has length $> k$, so there are no admissible fillings by $(1, \ldots, k)$ and $s_\lambda(t_1, \ldots, t_k) = 0$ in that case.

We now fix $1 \leq k < n$. We wish to compute the restriction of χ_λ to the subgroup $U(k)$ viewed as embedded in $U(n)$ via $a \mapsto a \oplus I$ or equivalently $a \mapsto \left(\begin{smallmatrix} a & 0 \\ 0 & I_{n-k} \end{smallmatrix}\right)$. In other words we are after a formula for $\chi_\lambda(t_1, \ldots, t_k, 1^{n-k})$. We find it convenient to use (2.4) and (2.5). Note that any admissible filling of λ by $(1, \ldots, n)$ induces by restricting it to $(1, \ldots, k)$ a filling of a diagram $\mu \leq \lambda$, in the sense that $\mu_i \leq \lambda_i$ for all $1 \leq i \leq n$. The remaining set of boxes, denoted by $\lambda \setminus \mu$ is (in general) no longer a diagram, it is only what is called a skew diagram, but the rule for filling it is respected by the induced numbering on its rows and columns, so that we can extend to $\lambda \setminus \mu$ the notation (2.5). Thus to any admissible filling of λ by $(1, \ldots, n)$ we associate $\mu \leq \lambda$ with a filling by $(1, \ldots, k)$ and $\lambda \setminus \mu$ with a filling by $(k+1, \ldots, n)$. Conversely, a moment of thought shows that separate admissible fillings of μ by $(1, \ldots, k)$ and $\lambda \setminus \mu$ by $(k+1, \ldots, n)$ can be joined to form a filling of λ by $(1, \ldots, n)$. This leads to the identity (see [52, p. 175])

$$s_\lambda(t) = \sum_{\mu \leq \lambda} s_\mu(t_1, \ldots, t_k) s_{\lambda \setminus \mu}(t_{k+1}, \ldots, t_n), \tag{2.8}$$

where again we set by convention $s_{\lambda \setminus \mu}(t_{k+1}, \ldots, t_n) = 1$ if $\mu = \lambda$. Moreover, we write $\mu \subset \lambda$ when $\mu_i \leq \lambda_i$ for all $1 \leq i \leq n$.

Lemma 2.5. *Recall that $\mu_{k,n}$ is the central symmetric probability measure defined by (2.1). Let $m = (m_1, \ldots, m_n) \in \mathbb{Z}^n$, and let $\lambda_j = m_j - m_n$ $(1 \leq j \leq n)$. The Fourier transform of $\mu_{k,n}$ is as follows.*
If $m_n > 0$ we have $\widehat{\mu_{k,n}}(\pi_m) = 0$.
If $m_n \leq 0$, let $d = -m_n$ and let $[d]^k = (d, \ldots, d, 0, \ldots, 0)$ with d repeated k-times. Then $\widehat{\mu_{k,n}}(\pi_m) = 0$ unless $[d]^k \subset \lambda$ in which case we have

$$\widehat{\mu_{k,n}}(\pi_m) = \frac{s_{\lambda \setminus [d]^k}(1^{n-k})}{s_\lambda(1^n)}. \tag{2.9}$$

Proof. We denote by $(t_1, \ldots, t_k, 1^{n-k})$ the eigenvalues of $g \in \Gamma(k)$, with $t = (t_1, \ldots, t_k) \in \mathbb{T}^k$. Then

$$\widehat{\mu_{k,n}}(\pi_m) = \frac{1}{\dim(\pi_m)} F_{k,n}(\pi_m) I$$

where by (2.3)

$$F_{k,n}(\pi_m) = \int \overline{\det(g)^{-d} \chi_\lambda(g)} m_{\Gamma(k)}(dg)$$

$$= \int (t_1 \ldots t_k)^d \overline{\chi_\lambda(t_1 \ldots t_k, 1^{n-k})} m_{\Gamma(k)}(dg).$$

By (2.8) we have

$$\chi_\lambda(t_1 \ldots t_k, 1^{n-k}) = \sum_{\mu \leq \lambda} s_\mu(t_1, \ldots, t_k) s_{\lambda \setminus \mu}(1^{n-k}).$$

Since the characters of $\Gamma(k)$ are orthonormal in $L_2(m_{\Gamma(k)})$ the integral

$$\int (t_1 \ldots t_k)^d \overline{s_\mu(t_1, \ldots, t_k)} m_{\Gamma(k)}(dg)$$

is equal to 1 if π_μ is equivalent to the irrep $g \mapsto \det(g)^d$ on $\Gamma(k)$, and is equal to 0 otherwise. Since $g \mapsto \det(g)^d$ on $\Gamma(k)$ corresponds to $(d, \ldots, d)$ (k times) on $U(k)$, we have

$$F_{k,n}(\pi) = \sum_{\mu \leq \lambda} \int (t_1 \ldots t_k)^d \overline{s_\mu(t_1, \ldots, t_k)} m_{\Gamma(k)}(dg)\, s_{\lambda \backslash \mu}(1^{n-k})$$

$$= s_{\lambda \backslash [d]^k}(1^{n-k}).$$

More precisely, $\widehat{\mu_{k,n}}(\pi) = 0$ for all $d < 0$, and also $\widehat{\mu_{k,n}}(\pi) = 0$ whenever $[d]^k \not\leq \lambda$. Thus, if $[d]^k \leq \lambda$ and $0 \leq d \leq \lambda_k$, we have

$$F_{k,n}(\pi) = s_{\lambda \backslash [d]^k}(1^{n-k}).$$

Moreover

$$\dim(\pi_m) = \dim(\pi_\lambda) = \chi_\lambda(1_G) = s_\lambda(1^n). \tag{2.10}$$

This proves (2.9). $\qquad\square$

Lemma 2.6. *Let $1 \leq k < n$. Let $\lambda = (\lambda_1, \ldots, \lambda_n)$ with $\lambda_1 \geq \cdots \geq \lambda_n = 0$. Assume $[d]^k \subset \lambda$ or equivalently $0 \leq d \leq \lambda_k$. Let $\lambda' = (\lambda_1, \ldots, \lambda_k) \backslash [d]^k$ and $\lambda'' = (\lambda_{k+1}, \ldots, \lambda_n)$. Then*

$$s_{\lambda \backslash [d]^k}(1^{n-k}) \leq s_{\lambda'}(1^{n-k}) s_{\lambda''}(1^{n-k}).$$

We have equality if $\lambda_{k+1} \leq d$. Moreover, $s_{\lambda \backslash [d]^k}(1^{n-k}) = 0$ if $d < \lambda_{n-k+1}$ (and a fortiori if $k + 1 > n - k$ and $d < \lambda_{k+1}$).

Proof. To any admissible filling of $\lambda \backslash [d]^k$ we may associate, by restriction, an admissible filling of λ' and one of λ''. Since this correspondence is clearly injective, the inequality follows from (2.6). Equality holds if it is surjective. Consider a pair of separate fillings of λ' and λ''. If $\lambda_{k+1} \leq d$ there is no problem to join them into a filling of $\lambda \backslash [d]^k$, so we have surjectivity. If $\lambda_{k+1} > d$ there may be an obstruction, however $s_{\lambda \backslash [d]^k}(1^{n-k}) = 0$ if $d < \lambda_{n-k+1}$, because one cannot fill the $(d+1)$th column strictly increasingly by $1, \ldots, n - k$ (that column being of length $\geq n - k + 1$ is too long for that). $\qquad\square$

Lemma 2.7. *With the same notation as in Lemma 2.6:*

(i) *if $d = 0$ then $\widehat{\mu_{k,n}}(\pi_m) = 0$ if $\lambda_{n-k+1} > 0$, and*

$$\widehat{\mu_{k,n}}(\pi_m) = \left(\prod_{i < j, j > n-k} \frac{\lambda_i + j - i}{j - i} \right)^{-1} \quad \text{if } \lambda_{n-k+1} = 0;$$

(ii) *if $d \geq 1$, $[d]^k \subset \lambda$ and $n - k \leq k$ then*

$$\widehat{\mu_{k,n}}(\pi_m) \leq \left(\prod_{i \leq k < j} \frac{\lambda_i - \lambda_j + j - i}{j - i} \right)^{-1};$$

(iii) *moreover, $\widehat{\mu_{k,n}}(\pi_m) = 0$ if $\lambda_k < d$ or if $d < \lambda_{n-k+1}$.*

Proof. We will use (2.9). Recall $\lambda_n = 0$.

(i) Assume $d = 0$. Clearly, $s_\lambda(1^{n-k}) = 0$ if $\lambda_{n-k+1} > 0$, because then we cannot fill the first column. Now assume $\lambda_{n-k+1} = 0$ ($= \lambda_n$). By (2.7) we have then $s_\lambda(1^{n-k}) = \prod_{i < j \leq n-k} \frac{\lambda_i - \lambda_j + j - i}{j - i}$, and hence by (2.9) and (2.7)

$$\widehat{\mu_{k,n}}(\pi_m) = \left(\prod_{i < j, \; j > n-k} \frac{\lambda_i + j - i}{j - i} \right)^{-1}.$$

(ii) Let $\mu = [d]^k$. Note that $[d]^k \subset \lambda$ implies $\lambda_k \geq d$. With the notation of Lemma 2.6, since by (2.6) $n - k \leq k$ clearly implies $s_{\lambda'}(1^{n-k}) \leq s_{\lambda'}(1^k)$, we have

$$s_{\lambda \backslash [d]^k}(1^{n-k}) \leq s_{\lambda'}(1^k) s_{\lambda''}(1^{n-k}).$$

We note that $\lambda'_i = \lambda_i - d$ for $i \leq k$ and $\lambda''_i = \lambda_{k+i}$ for $i \leq n - k$. Therefore, by (2.7) on one hand

$$s_{\lambda'}(1^k) = \prod_{i < j \leq k} \frac{\lambda'_i - \lambda'_j + j - i}{j - i} = \prod_{i < j \leq k} \frac{\lambda_i - \lambda_j + j - i}{j - i}, \qquad (2.11)$$

and on the other hand

$$s_{\lambda''}(1^{n-k}) = \prod_{i < j \leq n-k} \frac{\lambda''_i - \lambda''_j + j - i}{j - i} = \prod_{i < j \leq n-k} \frac{\lambda_{k+i} - \lambda_{k+j} + k + j - k + i}{k + j - k + i}$$

or equivalently

$$s_{\lambda''}(1^{n-k}) = \prod_{k < i < j \leq n} \frac{\lambda_i - \lambda_j + j - i}{j - i}. \qquad (2.12)$$

Dividing the product of (2.11) and (2.12) by $s_\lambda(1^n)$ as given by (2.7), we obtain our claim (ii).

(iii) If $\lambda_k < d$, then $[d]^k \subset \lambda$ is impossible, and if $d < \lambda_{n-k+1}$ the $(d+1)$th column of λ has length $\geq n - k + 1$ and hence cannot be filled strictly increasingly by $(1, \ldots, n - k)$, so that $s_{\lambda \backslash [d]^k}(1^{n-k}) = 0$. Thus $\widehat{\mu_{k,n}}(\pi_m) = 0$ by (2.9). □

Lemma 2.8. *With the same notation as in Lemma 2.6:*

(i) *if $d = 0$ then*

$$\widehat{\mu_{k,n}}(\pi_m) \leq \frac{(n-k)(n-k+1)}{n(n+1)} \quad \text{if } \lambda_1 \geq 2, \qquad (2.13)$$

$$\widehat{\mu_{k,n}}(\pi_m) \leq \frac{(n-k)(n-k-1)}{n(n-1)} \quad \text{if } \lambda_1 = 1 \text{ and } \lambda \neq \lambda_+; \qquad (2.14)$$

(ii) *if $d \geq 1$ and $n - k \leq k$ then*

$$\widehat{\mu_{k,n}}(\pi_m) \leq \frac{(n-k)(n-k+1)}{n(n+1)} \text{ if } \lambda_k \geq 2, \tag{2.15}$$

$$\widehat{\mu_{k,n}}(\pi_m) \leq \frac{(n-k)(n-k-1)}{n(n-1)} \text{ if } \lambda_k = \lambda_1 = 1 \text{ but } \lambda \neq \lambda_-, \tag{2.16}$$

$$\widehat{\mu_{k,n}}(\pi_m) \leq \frac{(n-k)(n-k-1)}{n(n-1)} \text{ if } n-1 > k, \ \lambda_k = 1, \lambda_1 \geq 2 \text{ and } \lambda_{n-1} = 0, \tag{2.17}$$

$$\widehat{\mu_{k,n}}(\pi_m) \leq \frac{k(n-k)}{(n+1)(n-1)} \text{ if } \lambda_k = 1, \lambda_1 \geq 2 \text{ and } \lambda_{n-1} \geq 1. \tag{2.18}$$

Proof. (i) We will use Lemma 2.7 (i). Note that if $\lambda_i \geq \mu_i \geq 0$ for all $i \leq n$ we must have

$$\left(\prod_{i<j,j>n-k} \frac{\lambda_i + j - i}{j - i} \right)^{-1} \leq \left(\prod_{i<j,j>n-k} \frac{\mu_i + j - i}{j - i} \right)^{-1}.$$

Assume first that $\lambda_1 \geq 2$. We compare λ with $\mu = (2, 0, \ldots, 0)$. Then

$$\prod_{i<j,j>n-k} \frac{\mu_i + j - i}{j - i} \geq \prod_{j>n-k} \frac{\mu_1 + j - 1}{j - 1} = \frac{n(n+1)}{(n-k)(n-k+1)}.$$

Now assume $\lambda_1 = 1$. Then $\lambda = (1, \ldots, 1, 0, 0, \ldots)$ where 1 appears r times. If $\lambda \neq \lambda_+$ (see Remark 2.4) we must have $r \geq 2$. Then comparing λ with $\mu = (1, 1, 0, \ldots, 0)$, we obtain

$$\prod_{i<j,j>n-k} \frac{\lambda_i + j - i}{j - i} \geq \prod_{j>n-k} \frac{\mu_1 + j - 1}{j - 1} \prod_{j>n-k} \frac{\mu_2 + j - 2}{j - 2} = \frac{n}{n-k} \frac{n-1}{n-k-1}.$$

This proves (i).

We now turn to (ii). Assume $d \geq 1$. We use Lemma 2.7 (ii) but we distinguish several subcases.

† Assume first that $\lambda_k \geq 2$. Then, since $\lambda_n = 0$

$$\prod_{i \leq k} \frac{\lambda_i - \lambda_n + n - i}{n - i} \geq \prod_{i \leq k} \frac{2 + n - i}{n - i} = \frac{n(n+1)}{(n-k)(n-k+1)}.$$

This proves (2.15).

†† Now assume $\lambda_k = 1$, so that $d = 1$. Then the case $\lambda_1 = 1$ is easy. Indeed, let $k \leq s < n$ be such that $\lambda_j = 1$ for $j \leq s$ and $\lambda_j = 0$ for $j > s$. Since we exclude λ_-, we know that $s < n - 1$ (see Remark 2.4), and hence $\lambda_{n-1} = 0$. When $n = 2$ this is impossible. When $n = 3$, the only possibility is $k = 1$ and then $\lambda \setminus [1]^k = 0$, and hence $\widehat{\mu_{k,n}}(\pi_m) = 0$. Therefore, we may restrict to $n \geq 4$. Note that $s < n - 1$

guarantees $k < n - 1$. Then using both $j = n$ and $j = n - 1$ we find

$$\prod_{i \leq k < j} \frac{\lambda_i - \lambda_j + j - i}{j - i} \geq \prod_{i \leq k} \frac{\lambda_i + n - i}{n - i} \prod_{i \leq k} \frac{\lambda_i + n - 1 - i}{n - 1 - i}$$

$$\geq \prod_{i \leq k} \frac{1 + n - i}{n - i} \prod_{i \leq k} \frac{n - i}{n - 1 - i} = \frac{n(n-1)}{(n-k)(n-k-1)}.$$

This proves (2.16).

$\dagger\dagger\dagger$ Now assume $\lambda_k = 1$ (and hence $d = 1$) and $\lambda_1 \geq 2$.

Case 1. Assume first that $\lambda_{n-1} = 0$. Then, assuming $n - 1 > k$

$$\prod_{i \leq k < j} \frac{\lambda_i - \lambda_j + j - i}{j - i} \geq \prod_{i \leq k, j \in \{n, n-1\}} \frac{\lambda_i - \lambda_j + j - i}{j - i}$$

$$\geq \prod_{i \leq k} \frac{\lambda_i + n - i}{n - i} \prod_{i \leq k} \frac{\lambda_i + n - 1 - i}{n - 2}$$

$$\geq \prod_{i \leq k} \frac{1 + n - i}{n - i} \prod_{i \leq k} \frac{1 + n - 1 - i}{n - 1 - i} = \frac{n}{n - k} \frac{n - 1}{n - k - 1}.$$

Case 2. Now assume $\lambda_{n-1} \geq 1$. Since we still assume $\lambda_k = 1$ and $\lambda_1 \geq 2$, we can compare λ with μ defined by $\mu_1 = 2$, $\mu_i = 1$ for all $i < n$ and $\mu_n = 0$. Since $\lambda \geq \mu$ and $\mu_j = \lambda_j$ for all $j > k$ we have

$$\prod_{i \leq k < j} \frac{\lambda_i - \lambda_j + j - i}{j - i} \geq \prod_{i \leq k < j} \frac{\mu_i - \mu_j + j - i}{j - i}$$

$$= \prod_{k < j < n} \frac{\mu_1 - \mu_j + j - 1}{j - 1} \prod_{i \leq k} \frac{\mu_i - \mu_n + n - i}{n - i}$$

but

$$\prod_{k < j < n} \frac{\mu_1 - \mu_j + j - 1}{j - 1} = \frac{n - 1}{k}$$

and

$$\prod_{i \leq k} \frac{\mu_i - \mu_n + n - i}{n - i} = \frac{n + 1}{n - 1} \prod_{1 < i \leq k} \frac{n - i + 1}{n - i} = \frac{n + 1}{n - k}$$

and hence

$$\prod_{i \leq k < j} \frac{\mu_i - \mu_j + j - i}{j - i} \geq \frac{(n - 1)(n + 1)}{k(n - k)}.$$

This proves (2.17). $\qquad\square$

Remark 2.9. In the proof of part (ii) in the preceding Lemma 2.8 the majorizations of $\widehat{\mu_{k,n}}(\pi_m)$ appearing there are all proved actually for

$$\left(\prod_{i \leq k < j} \frac{\lambda_i - \lambda_j + j - i}{j - i} \right)^{-1}.$$

Lemma 2.10. *With the same notation as in Lemma 2.6, let $n > 3$ be an odd integer and let $k = (n-1)/2 > 1$ so that $n = 2k + 1$. If $d \geq 1$ then*

$$\widehat{\mu_{k,n}}(\pi_m) \leq \frac{(n-k)(n-k-1)}{n(n+1)} \qquad \text{if } \lambda_k \geq 2, \tag{2.19}$$

$$\widehat{\mu_{k,n}}(\pi_m) \leq \frac{(n-k-1)(n-k-2)}{n(n-1)} \qquad \text{if } \lambda_k = 1 \text{ and } \lambda_1 = 1 \text{ but } \lambda \neq \lambda_-, \tag{2.20}$$

$$\widehat{\mu_{k,n}}(\pi_m) \leq \frac{(n-k-1)(n-k-2)}{n(n-1)} \qquad \text{if } \lambda_k = 1, \lambda_1 \geq 2 \text{ and } \lambda_{n-1} = 0, \tag{2.21}$$

$$\widehat{\mu_{k,n}}(\pi_m) \leq \frac{(k+1)(n-k-1)}{(n+1)(n-1)} \qquad \text{if } \lambda_k = 1, \lambda_1 \geq 2 \text{ and } \lambda_{n-1} \geq 1. \tag{2.22}$$

Proof. We again decompose λ into λ' and λ'', but we will modify the definition of λ'. Now λ' will have $k+1$ rows. Its first k rows are as before the same as those of $\lambda \setminus [d]^k$, and the $(k+1)$th row is like this: if $\lambda_{k+1} < d$ we set $\lambda'_{k+1} = 0$, while if $\lambda_{k+1} \geq d$ we set $\lambda'_{k+1} = \lambda_{k+1} - d$. As for λ'' it is formed as before of the last $n-k$ rows of λ. Then arguing as in Lemma 2.6 we find

$$s_{\lambda \setminus [d]^k}(1^{n-k}) \leq s_{\lambda'}(1^{k+1}) s_{\lambda''}(1^{n-k}).$$

By (2.9) we have

$$\widehat{\mu_{k,n}}(\pi_m) \leq \frac{s_{\lambda'}(1^{k+1}) s_{\lambda''}(1^{n-k})}{s_\lambda(1^n)}.$$

We now use (2.7) for λ', λ'' and λ. This gives us

$$\widehat{\mu_{k,n}}(\pi_m) \leq \left(\prod_{i \leq k+1 < j} \frac{\lambda_i - \lambda_j + j - i}{j - i} \right)^{-1} \prod_{i \leq k} \frac{\lambda'_i - \lambda'_{k+1} + k + 1 - i}{\lambda_i - \lambda_{k+1} + k + 1 - i}.$$

Now if $\lambda_{k+1} \geq d$ the second factor is equal to 1 and if $\lambda_{k+1} < d$ we have $\lambda'_i - \lambda'_{k+1} = \lambda_i - d < \lambda_i - \lambda_{k+1}$. Thus we may remove that second factor. Therefore

$$\widehat{\mu_{k,n}}(\pi_m) \leq \left(\prod_{i \leq k+1 < j} \frac{\lambda_i - \lambda_j + j - i}{j - i} \right)^{-1}.$$

Thus it suffices to majorize $\left(\prod_{i \leq k+1 < j} \frac{\lambda_i - \lambda_j + j - i}{j - i} \right)^{-1}$ by the bounds appearing in Lemma 2.10. We now invoke Remark 2.9. Observing that $n - (k+1) \leq k + 1$ we may apply part (ii) of Lemma 2.8 with $k+1$ taking the place of k. Then replacing k by $k+1$ in the upper bounds appearing in part (ii) in Lemma 2.8 and using Remark 2.9 we obtain the desired bounds for $\left(\prod_{i \leq k+1 < j} \frac{\lambda_i - \lambda_j + j - i}{j - i} \right)^{-1}$. □

Proof of Theorem 2.1. We apply first part (i) in Lemma 2.8 to settle the case $d = 0$. Thus we may assume $d \geq 1$. We apply then part (ii) from that same Lemma 2.8 to settle the cases either $n = 2k$ or $n = 2k - 1$, with the restriction $n - 1 > k$ which requires $n > 3$. Then Lemma 2.10 settles the remaining case $n = 2k + 1$. Note that $k/n \to 1/2$ when $n \to \infty$ if either $k = n/2$, $k = k_+$

or $k = k_-$, and all the bounds appearing in Lemmas 2.8 and 2.10 tend to $1/4$. Therefore, for any $1/4 < \gamma < 1/2$ there is $n(\gamma)$ such that for any $n \geq n(\gamma)$

$$\sup_{\pi \notin S_n} \|\widehat{\nu_n}(\pi)\| \leq \gamma.$$

Since $\widehat{\mu_{k,n}}(\pi) = k/n$ when $\pi = \sigma_n$ or $\pi = \overline{\sigma_n}$ (and since $(k_+ + k_-)/2n = 1/2$) we have $\widehat{\nu_n}(\pi) = 1/2$. Thus ν_n has a $(1/2, \gamma)$-spectral gap for any $n \geq n(\gamma)$, which settles the last assertion in Theorem 2.1. Checking the bounds for small values of n, actually we can find a $\gamma < 1/2$ whenever $n \geq 4$.   □

Remark 2.11 (A natural question). Assume that $k = [\theta n]$ where $0 < \theta < 1$ is fixed. Then $\widehat{\mu_{k,n}}(\sigma_n) = \widehat{\mu_{k,n}}(\overline{\sigma_n}) = (n - k)/n \approx 1 - \theta$. By Lemmas 2.8 and 2.10, if we assume $\theta \leq 1/2$ (to ensure that $k \leq n - k$) then $\mu_{k,n}$ has a (δ_n, γ_n)-spectral gap with $\delta_n \approx 1 - \theta$ and $\gamma_n \approx (1 - \theta)^2$ when $n \to \infty$. We do not know whether this (or any similar spectral gap) holds when $1/2 < \theta < 1$.

3. Rider's results on Sidon sets

We now turn to the applications of the spectral gap obtained in Corollary 2.3 to Sidon sets. We start with two simple lemmas. Their proof is not too different from their commutative version.

Lemma 3.1. *Let G be any compact group. Let $\Lambda \subset \widehat{G}$ be randomly Sidon with constant C. Then for any finitely supported family (b_π) with $b_\pi \in C(G; M_{d_\pi})$ $(\pi \in \Lambda)$ we have*

$$\left| \sum_{\pi \in \Lambda} d_\pi \operatorname{tr}\left(\int \pi(g) b_\pi(g) m_G(dg) \right) \right| \leq C \int \left\| \sum_{\pi \in \Lambda} d_\pi \operatorname{tr}(u_\pi b_\pi) \right\|_\infty m_G(du). \quad (3.1)$$

Proof. Let $f_\pi(t) = \int \pi(g) b_\pi(t^{-1}g) m_G(dg)$. Then by the translation invariance of m_G, $t \mapsto \pi(t^{-1}) f_\pi(t)$ is constant. Let

$$a_\pi = f_\pi(1) = \int \pi(g) b_\pi(g) m_G(dg).$$

Thus $f_\pi(t) = \pi(t) f_\pi(1) = \pi(t) a_\pi$. Let us write for short $\mathbb{E}$ for the integral with respect to m_G. For any fixed $g \in G$, by translation invariance of the norm in $C(G)$ and since (u_π) and $(u_\pi \pi(g))$ have the same distribution, we have

$$\mathbb{E} \sup_{t \in G} \left| \sum_{\pi \in \Lambda} d_\pi \operatorname{tr}(u_\pi b_\pi(t)) \right| = \mathbb{E} \sup_{t \in G} \left| \sum_{\pi \in \Lambda} d_\pi \operatorname{tr}(u_\pi \pi(g) b_\pi(t^{-1}g)) \right|,$$

and hence

$$= \int \mathbb{E} \sup_{t \in G} \left| \sum_{\pi \in \Lambda} d_\pi \operatorname{tr}(u_\pi \pi(g) b_\pi(t^{-1}g)) \right| m_G(dg)$$

and by Jensen this is

$$\geq \mathbb{E} \sup_{t\in G} \left| \sum_{\pi\in\Lambda} d_\pi \operatorname{tr}(u_\pi f_\pi(t)) \right| = \mathbb{E} \sup_{t\in G} \left| \sum_{\pi\in\Lambda} d_\pi \operatorname{tr}(u_\pi \pi(t)a_\pi) \right|.$$

Since Λ is assumed randomly Sidon, this last term is

$$\geq C^{-1} \sum_{\pi\in\Lambda} d_\pi \operatorname{tr}|a_\pi| \geq C^{-1}|\sum_{\pi\in\Lambda} d_\pi \operatorname{tr}(a_\pi)|.$$

This completes the proof. $\qquad\square$

Remark 3.2. Let $b_\pi(g) = \pi(g^{-1})a_\pi$. In that case (3.1) implies

$$\left| \sum_{\pi\in\Lambda} d_\pi \operatorname{tr}(a_\pi) \right| \leq C \int \left\| \sum_{\pi\in\Lambda} d_\pi \operatorname{tr}(u_\pi \pi a_\pi) \right\|_\infty m_G(du).$$

This shows that (3.1) generalizes the randomly Sidon property.

Lemma 3.3.

(i) *Let $\Lambda \subset \widehat{G}$ be a Sidon set with constant C. For any $u \in \mathcal{G}$ (or merely for any $u \in \prod_{\pi\in\Lambda} U(d_\pi)$) there is $\mu^u \in M(G)$ with $\|\mu^u\| \leq C$ such that $\widehat{\mu^u}(\pi) = u_\pi$ for any $\pi \in \Lambda$.*

(ii) *Let $\Lambda \subset \widehat{G}$ be a randomly Sidon set with constant C. Then, there is a functional $\varphi \in L_1(\mathcal{G}; C(G))^*$ with norm $\leq C$ such that for any $\pi \in \Lambda$ and any $b_\pi \in C(G; M_{d_\pi})$*

$$\varphi(\operatorname{tr}(u_\pi b_\pi)) = \int \operatorname{tr}(\pi(g)b_\pi(g))m_G(dg).$$

(iii) *Assuming (ii) and assuming $C(G)$ separable, there is a weak* measurable (in the sense of the following remark) bounded function $u \mapsto \mu^u \in M(G)$ with $\sup \|\mu^u\|_{M(G)} \leq C$ such that for any $\pi \in \Lambda$*

$$\mathbb{E}(u_\pi \mu^u) = \pi m_G.$$

The latter is an equality between matrix-valued measures (or matrices with entries in $M(G)$) by which we mean that for any $f \in C(G)$ we have

$$\mathbb{E}\left(u_\pi \int f(g)\mu^u(dg) \right) = \int f(g)\pi(g)m_G(dg) = \widehat{f}(\bar{\pi}).$$

Proof. Both (i) and (ii) are immediate consequences of Hahn–Banach: for (i) we use the definition of Sidon sets and for (ii) we use Lemma 3.1. To check (iii), as explained in the next remark, we note that $\varphi \in L_1(\mathcal{G}; C(G))^*$ defines a μ^u such that $\varphi(f(u)h(g)) = \mathbb{E}(f(u) \int h(g)\mu^u(dg))$ (here $f \in L_1(\mathcal{G})$ $h \in C(G)$), with ess sup $\|\mu^u\|_{M(G)} = \|\varphi\|$. Then (ii) can be rephrased as saying that the action of the $d_\pi \times d_\pi$-matrix (with entries in $M(G)$) $\mathbb{E}(u_\pi \mu^u)$ on an arbitrary $b_\pi \in C(G; M_{d_\pi})$ coincides with that of $\pi(g)m_G$. Then (iii) becomes clear. $\qquad\square$

Remark 3.4 (On the dual of $L_1(\mathcal{G}; C(G))$). In the present paragraph $(\mathcal{G}, m_\mathcal{G})$ can be any probability space. It is a well-known fact that $L_1(\mathcal{G}; C(G))$ is the projective tensor product of $L_1(\mathcal{G})$ and $C(G)$, so that its dual can be identified isometrically to the space $B(C(G), L_\infty(\mathcal{G}))$ of bounded linear maps from $C(G)$ to $L_\infty(\mathcal{G})$. Explicitly, to any linear form $\varphi \in L_1(\mathcal{G}; C(G))^*$ we naturally associate a bounded linear map $T_\varphi : C(G) \to L_\infty(\mathcal{G})$ with $\|T_\varphi\| = \|\varphi\|$ such that $\varphi(f \otimes x) = \int (T_\varphi(f))(\omega) x(\omega) m_\mathcal{G}(d\omega)$ for any $f \in C(G), x \in L_1(\mathcal{G})$.

Assume $C(G)$ separable. Then $\widehat{G}$ is countable and $L_1(\mathcal{G})$ is also separable. Let D be a dense countable subset of $C(G)$, and let V be its linear span. Then any $\xi \in C(G)^*$ is determined by its values on D, and also (by linearity) by its values on V. Clearly we can find a measurable subset $\Omega_0 \subset \mathcal{G}$ with full measure on which all the maps $\omega \mapsto |(T_\varphi(f))(\omega)|$ are bounded by $\|T_\varphi\| \|f\|$ for any $f \in D$, and such that $f \mapsto T_\varphi(f)(\omega)$ extends to a linear form of norm $\leq \|T_\varphi\|$ on $C(G)$ (for this one way is to consider linearity over the rationals). This allows us to define on Ω_0 a function $\omega \mapsto \mu^\omega \in M(G)$ bounded by $\|T_\varphi\|$ such that $\omega \mapsto \mu^\omega(f) = \int f(g)\mu^\omega(dg)$ is measurable for any $f \in D$ and hence for any $f \in C(G)$ (this is what we mean by "weak* measurability") with $\sup_{\Omega_0} \|\mu^\omega\| \leq \|T_\varphi\|$, that represents φ in the sense that for a.e. ω

$$\int f(g)\mu^\omega(dg) = (T_\varphi(f))(\omega). \tag{3.2}$$

We denote by $\mathcal{L}_\infty(\mathcal{G}; M(G))$ the space of all equivalence classes (modulo equality a.e.) of bounded weak* measurable functions $\omega \mapsto \mu^\omega \in M(G)$ equipped with the norm ess $\sup_\omega \|\mu^\omega\|$. Conversely, for any such $\omega \mapsto \mu^\omega \in M(G)$ we can associate a bounded linear map $T : C(G) \to L_\infty(\mathcal{G})$ with $\|T\| \leq$ ess $\sup_\omega \|\mu^\omega\|$ that takes $f \in C(G)$ to the function $\omega \mapsto \int f(g)\mu^\omega(dg)$. Thus we obtain an isometric isomorphism between $B(C(G), L_\infty(\mathcal{G}))$ and $\mathcal{L}_\infty(\mathcal{G}; M(G))$.

The preceding discussion shows that $\mathcal{L}_\infty(\mathcal{G}; M(G))$ can be identified isometrically with the space $L_1(\mathcal{G}; C(G))^*$.

We now deduce Rider's version of Drury's Theorem:

Theorem 3.5. *Let G be any compact group. Let $\Lambda \subset \widehat{G}$ be a randomly Sidon set with constant C. For any $0 < \varepsilon < 1$ there is a measure $\mu_\varepsilon \in M(G)$ such that*

$$\sup_{\pi \in \Lambda} \|\widehat{\mu}_\varepsilon(\pi) - I\| \leq \varepsilon \ \forall \pi \in \Lambda, \quad \sup_{\pi \notin \Lambda} \|\widehat{\mu}_\varepsilon(\pi)\| \leq \varepsilon \text{ and } \|\mu_\varepsilon\| \leq w(\varepsilon) \tag{3.3}$$

where $w(\varepsilon)$ depends only on ε and C.

More generally, for any $z \in \mathcal{G}$ (or merely for any $z \in \prod_{\pi \in \Lambda} U(d_\pi)$), there is $\mu_\varepsilon^z \in M(G)$ such that

$$\sup_{\pi \in \Lambda} \|\widehat{\mu_\varepsilon^z}(\pi) - z_\pi\| \leq \varepsilon \ \forall \pi \in \Lambda, \quad \sup_{\pi \notin \Lambda} \|\widehat{\mu_\varepsilon^z}(\pi)\| \leq \varepsilon \text{ and } \|\mu_\varepsilon^z\| \leq w(\varepsilon).$$

Proof. We have all the ingredients to reproduce the Drury–Rider trick. To avoid all irrelevant convergence and/or measurability issues, we assume that Λ is finite and that $C(G)$ is separable. It is easy to pass from the finite case to the general one by a simple compactness argument (in the unit ball of $M(G)$ equipped with the weak* topology). Let μ^u be as in Lemma 3.3 (iii). Let $\Lambda' \subset \widehat{\mathcal{G}}$ be the set formed by the coordinates $\{u_\pi \mid \pi \in \Lambda\}$. Note that here we abuse the notation: we still denote simply by u_π the irreducible representation $u \mapsto u_\pi$ on $\mathcal{G}$.

By Corollary 2.3 there is $\nu \in M(\mathcal{G})$ with $\|\nu\| \leq w(\varepsilon)$ such that $\widehat{\nu}(u_\pi) = \int \overline{u_\pi}\nu(du) = I$ for $\pi \in \Lambda$ and $\|\widehat{\nu}(r)\| = \|\int \bar{r}(u)\nu(du)\| \leq \varepsilon$ for any representation $r \notin \Lambda'$. Let

$$\Phi^u = \int \mu^{uu'} * \mu^{u'^{-1}} m_{\mathcal{G}}(du') \in \mathcal{L}_\infty(\mathcal{G}; M(G)).$$

Denoting $\bar{z} = (\overline{z_\pi}) \in \mathcal{G}$, we then define

$$\mu_\varepsilon^z = \int \Phi^{\bar{z}u}\nu(du).$$

Note that $\|\mu_\varepsilon^z\| \leq C^2 w(\varepsilon)$. A simple verification (using $(\pi m_G) * (\pi m_G) = \pi m_G$) shows that (iii) in Lemma 3.3 is preserved, i.e., we have

$$\mathbb{E}(u_\pi \Phi^u) = \pi m_G,$$

and hence for each fixed $z \in \mathcal{G}$

$$\overline{z_\pi}\mathbb{E}(u_\pi \Phi^{\bar{z}u}) = \mathbb{E}((\bar{z}u)_\pi \Phi^{\bar{z}u}) = \pi m_G.$$

Therefore

$$\mathbb{E}(u_\pi \Phi^{\bar{z}u}) = {}^t z_\pi \pi m_G.$$

More explicitly, for any fixed $f \in C(G)$ if we denote $\varphi_f(u) = \int f(g)\Phi^u(dg)$ we have

$$\forall \pi \in \Lambda \quad \mathbb{E}(u_\pi \varphi_f(\bar{z}u)) = {}^t z_\pi \widehat{f}(\pi), \tag{3.4}$$

and hence taking the trace of both sides

$$\forall \pi \in \Lambda \quad \mathbb{E}(\operatorname{tr}(u_\pi)\varphi_f(\bar{z}u)) = \operatorname{tr}({}^t z_\pi \widehat{f}(\pi)). \tag{3.5}$$

By the definition of μ_ε^z

$$\int f d\mu_\varepsilon^z = \int \varphi_f(\bar{z}u)\nu(du). \tag{3.6}$$

More generally, we can extend the definition of φ_f to any matrix-valued $f \in C(G; M_d)$: we simply set again

$$\varphi_f(u) = \int f(g)\Phi^u(dg).$$

Let $\rho \in \widehat{G}$. Note that $\varphi_{\bar{\rho}} = \widehat{\Phi^u}(\rho)$. Since $\widehat{\Phi^u}(\rho) = \int \widehat{\mu^{uu'}}(\rho)\widehat{\mu^{u'^{-1}}}(\rho)m_{\mathcal{G}}(du')$ and $\operatorname{ess\,sup}_u \|\mu^u\| \leq C$, the matrix-valued function $u \mapsto \varphi_\rho(u) = \widehat{\Phi^u}(\rho)$ (being the convolution on $\mathcal{G}$ of two M_{d_ρ}-valued functions bounded by C) has its coefficients in the space of absolutely convergent Fourier series $A(\mathcal{G})$, so that we can apply (1.10) (with $\mathcal{G}$ in place of G) to it.

Consider the "pseudo-measure" ν' on $\mathcal{G}$ defined a priori by its formal Fourier expansion

$$\nu' = \sum_{r \notin \Lambda'} d_r \operatorname{tr}({}^t \widehat{\nu}(r)r).$$

Since we assume that Λ is finite $\nu' \in M(G)$, and since $\widehat{\nu}(\pi) = I$ when $\pi \in \Lambda$ we have

$$\nu = \left(\sum_{\pi \in \Lambda} d_\pi \, \mathrm{tr}(u_\pi) \right) m_G + \nu'. \tag{3.7}$$

Recall that by our choice of ν we have $\sup_{r \in \widehat{G}} \|\widehat{\nu'}(r)\| \le \varepsilon$. By (1.10) we have for any $\rho \in \widehat{G}$

$$\left\| \int \varphi_\rho(\bar{z}u)\nu'(du) \right\| \le C^2 \sup_{r \in \widehat{G}} \|\widehat{\nu'}(r)\| \le C^2 \varepsilon. \tag{3.8}$$

We claim that

$$\forall \pi \in \Lambda \quad \widehat{\mu_\varepsilon^z}(\pi) - z_\pi = \int \varphi_\pi(\bar{z}u) d\nu'(u)$$

and

$$\forall \rho \notin \Lambda \quad \widehat{\mu_\varepsilon^z}(\rho) = \int \varphi_\rho(\bar{z}u) d\nu'(u).$$

From this claim and (3.8) we deduce the conclusion, except that we obtain it $(C^2\varepsilon, C^2 w(\varepsilon))$ in place of $(\varepsilon, w(\varepsilon))$.

Thus it only remains to justify the claim. By (3.6), (3.7) and (3.5) we have for any $f \in C(G)$

$$\int f d\mu_\varepsilon^z - \int \varphi_f(\bar{z}u) d\nu'(u) = \int \varphi_f(\bar{z}u) \left(\sum_{\pi \in \Lambda} d_\pi \, \mathrm{tr}(u_\pi) \right) dm_G(u)$$
$$= \sum_{\pi \in \Lambda} d_\pi \, \mathrm{tr}({}^t z_\pi \widehat{f}(\bar{\pi})). \tag{3.9}$$

Consider now the case $f = \overline{\rho_{ij}}, 1 \le i, j \le d_\rho$. We have $\widehat{f}(\bar{\pi}) = 0$ if $\rho \ne \pi$ and $\widehat{f}(\bar{\pi}) = d_\pi^{-1} e_{ij}$ if $\rho = \pi$. Therefore we find

$$\sum_{\pi \in \Lambda} d_\pi \, \mathrm{tr}({}^t z_\pi \widehat{f}(\bar{\pi})) = 1_{\rho \in \Lambda}(z_\pi)_{ij},$$

which by (3.9) implies our claim.   □

Corollary 3.6 (Rider, circa 1975, unpublished). *The union of two Sidon sets is a Sidon set.*

Proof. Let $\Lambda \subset \widehat{G}$ be a Sidon set. In the situation of Theorem 3.5, for any $f \in C(G)$ we have by the triangle inequality

$$\|f * \mu_\varepsilon\|_\infty \ge \left\| \sum_{\pi \in \Lambda} d_\pi \, \mathrm{tr}({}^t(\widehat{f * \mu_\varepsilon})(\pi)\pi) \right\|_\infty - \left\| \sum_{\pi \notin \Lambda} d_\pi \, \mathrm{tr}({}^t(\widehat{f * \mu_\varepsilon})(\pi)\pi) \right\|_\infty$$

and hence

$$w(\varepsilon)\|f\|_\infty \ge \|f * \mu_\varepsilon\|_\infty \ge ((1-\varepsilon)/C) \sum_{\pi \in \Lambda} d_\pi \, \mathrm{tr} \, |\widehat{f}(\pi)| - \varepsilon\|f\|_{A(G)}.$$

Let $\Lambda_j \subset \widehat{G}$ be two disjoint Sidon sets with Sidon constants C_j $(j = 1, 2)$. Let $f = f_1 + f_2 \in C(G)$ be a function with $\widehat{f}_j$ supported in Λ_j. By the preceding inequality

$$w_1(\varepsilon)\|f\|_\infty \geq (1 - \varepsilon)C_1^{-1} \sum_{\pi \in \Lambda_1} d_\pi \operatorname{tr}|\widehat{f}(\pi)| - \varepsilon\|f_2\|_{A(G)},$$

$$w_2(\varepsilon)\|f\|_\infty \geq (1 - \varepsilon)C_2^{-1} \sum_{\pi \in \Lambda_2} d_\pi \operatorname{tr}|\widehat{f}(\pi)| - \varepsilon\|f_1\|_{A(G)},$$

and hence adding both

$$(w_1(\varepsilon) + w_2(\varepsilon))\|f\|_\infty \geq ((1 - \varepsilon)\min\{C_1^{-1}, C_2^{-1}\} - \varepsilon)(\|f_1\|_{A(G)} + \|f_2\|_{A(G)}).$$

Then if we choose ε small enough so that $C_\varepsilon = ((1 - \varepsilon)\min\{C_1^{-1}, C_2^{-1}\} - \varepsilon) > 0$ we deduce that $\Lambda_1 \cup \Lambda_2$ is Sidon with constant at most $(w_1(\varepsilon) + w_2(\varepsilon))C_\varepsilon^{-1}$. $\square$

Corollary 3.7 (Rider, circa 1975, unpublished). *Any randomly Sidon set is a Sidon set.*

Proof. In the situation of Theorem 3.5, for any $z = (z_\pi) \in \mathcal{G}$ we have for any $f \in C(G)$ with $\widehat{f}$ supported in Λ

$$w(\varepsilon)\|f\|_\infty \geq \|f * \mu_\varepsilon^z\|_\infty \geq \left|\sum_{\pi \in \Lambda} d_\pi \operatorname{tr}(\widehat{f}(\pi)z_\pi)\right| - \varepsilon\|f\|_{A(G)}$$

and hence taking the sup over z

$$w(\varepsilon)\|f\|_\infty \geq (1 - \varepsilon)\|f\|_{A(G)}.$$

Thus, for any $\varepsilon < 1$, Λ is Sidon with constant at most $(1 - \varepsilon)^{-1}w(\varepsilon)$. $\square$

Remark 3.8. Actually, Corollary 3.7 implies Corollary 3.6, because it is easy to see that randomly Sidon sets are stable under finite unions.

Remark 3.9. Let $\Lambda \subset \widehat{G}$ be Sidon with constant C. Assume that for all $0 < \varepsilon < 1$ there is $\mu_\varepsilon \in M(G)$ such that (3.3) holds. Then Λ is peaking. Indeed, by Hahn–Banach, for any $z \in \prod_{\pi \in \Lambda} M_{d_\pi}$ with $\sup_{\pi \in \Lambda}\|z_\pi\| < \infty$ there is $\nu \in M(G)$ with $\|\nu\|_{M(G)} \leq C \sup_{\pi \in \Lambda}\|z_\pi\|$ such that $\widehat{\nu}(\pi) = z_\pi$ for any $\pi \in \Lambda$. Since $\|\widehat{\mu}_\varepsilon(\pi) - I\| \leq \varepsilon < 1$, $\widehat{\mu}_\varepsilon(\pi)$ is invertible and $\|(\widehat{\mu}_\varepsilon(\pi))^{-1}\| \leq (1 - \varepsilon)^{-1}$ for any $\pi \in \Lambda$. Let $z_\pi = (\widehat{\mu}_\varepsilon(\pi))^{-1}$. Let ν be the measure (given by Hahn–Banach) such that $\|\nu\|_{M(G)} \leq C(1 - \varepsilon)^{-1}$ and $\widehat{\nu}(\pi) = (\widehat{\mu}_\varepsilon(\pi))^{-1}$ for any $\pi \in \Lambda$. Let $\nu_\varepsilon = \nu * \mu_\varepsilon$. Then by (1.2) $\widehat{\nu}_\varepsilon(\pi) = 1$ for $\pi \in \Lambda$ and $\|\widehat{\nu}_\varepsilon(\pi)\| \leq \|\nu\|_{M(G)}\|\widehat{\mu}_\varepsilon(\pi)\| \leq C\varepsilon(1 - \varepsilon)^{-1}$ for $\pi \notin \Lambda$. Also $\|\nu_\varepsilon\|_{M(G)} \leq \|\nu\|_{M(G)}\|\mu_\varepsilon\|_{M(G)} \leq C(1 - \varepsilon)^{-1}w(\varepsilon)$. This shows that Λ is an ε'-peak set for $\varepsilon' = C\varepsilon(1 - \varepsilon)^{-1}$. By Proposition 1.12 this shows that Λ is peaking.

Remark 3.10. In [63], Wilson managed to prove the union theorem in $\widehat{G}$ when G is a *connected* compact group. His proof uses the structure theory of continuous compact groups and Lie groups. Apparently, it does not extend to general compact groups, and does not give any quantitative estimate.

4. Gaussian and sub-Gaussian random Fourier series

In this section we survey (with sketches of proofs) the main results of [37, 38]. We will take special care of Theorem 4.13 because unfortunately we detected a gap and probably an erroneous claim made by us in [38] concerning that statement (see Remark 4.14).

All the Gaussian variables we consider are always assumed (implicitly) to have mean 0. A Gaussian random variable g will be called normalized if $\mathbb{E}|g|^2 = 1$. We use this for either the real-valued case or the complex-valued one. We deliberately avoid the term "normal", which usually implies that $\mathbb{E}|g|^2 = 2$ in the complex case. By a complex-valued Gaussian variable, we mean a variable of the form $g = g_1 + ig_2$ such that g_1, g_2 are independent (real-valued) Gaussian variables with the same L_2-norm (and hence the same distribution).

Let (g_n) be an i.i.d. sequence of real- (resp. complex-) valued normalized Gaussian variables. Then for any nonzero real (resp. complex) sequence $x = (x_n) \in \ell_2$, the variable $g = (\sum |x_n|^2)^{-1/2} \sum x_n g_n$ is a normalized Gaussian variable. Therefore

$$\|\sum x_n g_n\|_p = \|g_1\|_p \left(\sum |x_n|^2\right)^{1/2}. \tag{4.1}$$

and also in the real (resp. complex) case

$$\mathbb{E}\exp\left(\sum x_n g_n\right) = \exp\left(\sum |x_n|^2/2\right) \quad \text{(resp.}$$
$$\mathbb{E}\exp\left(\mathrm{Re}\left(\sum x_n g_n\right)\right) = \exp\left(\sum |x_n|^2/2\right)\text{).} \tag{4.2}$$

We now turn to the behaviour of Sidon sets in L_p for $p < \infty$. In many cases the growth of the L_p-norms of a function when $p \to \infty$ is equivalent to its exponential integrability, as in the following elementary and well-known lemma.

We start by recalling the definition of certain Orlicz spaces. Let $(\Omega, \mathbb{P})$ be a probability space. Let $0 < a < \infty$. Let

$$\forall x \geq 0 \quad \psi_a(x) = \exp x^a - 1.$$

We denote by $L_{\psi_a}(\mathbb{P})$, or simply by L_{ψ_a} the space of those $f \in L_0(\Omega, \mathbb{P})$ for which there is $t > 0$ such that $\mathbb{E}\exp|f/t|^a < \infty$ and we set

$$\|f\|_{\psi_a} = \inf\{t > 0 \mid \mathbb{E}\exp|f/t|^a \leq e\}.$$

In the next two lemmas (and Remark 4.2) we recall several well-known properties of these spaces.

Lemma 4.1. *Fix a number $a > 0$. The following properties of a (real or complex) random variable f are equivalent:*

(i) $f \in L_p$ *for all* $p < \infty$ *and* $\sup_{p \geq 1} p^{-1/a}\|f\|_p < \infty$;
(ii) $f \in L_{\psi_a}$;
(iii) *there is* $t > 0$ *such that* $\sup_{c > 0} \exp(tc^a)\mathbb{P}\{|f| > c\} < \infty$.

(iv) *Let (f_n) be an i.i.d. sequence of copies of f. Then*

$$\sup_n (\log(n+1))^{-1/a}|f_n| < \infty \ a.s.$$

Moreover, there is a positive constant C_a such that for any $f \geq 0$ we have

$$C_a^{-1} \sup_{p \geq 1} p^{-1/a}\|f\|_p \leq \|f\|_{\psi_a} \leq C_a \sup_{p \geq 1} p^{-1/a}\|f\|_p, \tag{4.3}$$

and this still holds if we restrict the sup over $p \geq 1$ to be over all even integers.

Proof. First observe that the conditions

$$\sup_{p \geq 1} p^{-1/a}\|f\|_p < \infty \quad \text{and} \quad \sup_{p \geq a} p^{-1/a}\|f\|_p < \infty$$

are obviously equivalent. Assume that $\sup_{p \geq a} p^{-1/a}\|f\|_p \leq 1$. Then

$$\mathbb{E}\exp|f/t|^a = 1 + \sum_1^\infty \mathbb{E}|f/t|^{an}(n!)^{-1} \leq 1 + \sum_1^\infty (an)^n t^{-an}(n!)^{-1}$$

hence by Stirling's formula for some constant C

$$\leq 1 + C\sum_1^\infty (an)^n t^{-an} n^{-n} e^n = 1 + C\sum_1^\infty (at^{-a}e)^n$$

from which it becomes clear (since $1 < e$) that (i) implies (ii). Conversely, if (ii) holds we have a fortiori for all $n \geq 1$

$$(n!)^{-1}\|f/t\|_{an}^{an} \leq \mathbb{E}\exp|f/t|^a \leq e$$

and hence

$$\|f\|_{an} \leq e^{\frac{1}{an}}(n!)^{\frac{1}{an}}t \leq e^{\frac{1}{a}}n^{\frac{1}{a}}t = (an)^{\frac{1}{a}}t(e/a)^{1/a},$$

which gives $\|f\|_p \leq p^{1/a}t(e/a)^{1/a}$ for the values $p = an$, $n = 1, 2, \ldots$. One can then easily interpolate (using Hölder's inequality) to obtain (i). The equivalences of (ii) with (iii) and (iv) are elementary exercises. The last assertion is a simple recapitulation left to the reader. □

Remark 4.2. Let

$$\|f\|_{\psi_a,\infty} = \inf\{t \mid \sup_{c>0}(\psi_a(c)\mathbb{P}(\{|f/t| > c\})) \leq \psi_a(1)\}.$$

In addition to (ii) ⇔ (iii), it is easy to check that $\|\ \|_{\psi_a,\infty}$ and $\|\ \|_{\psi_a}$ are equivalent norms on L_{ψ_a}. This is in sharp contrast with the case of L_p-spaces (when we replace ψ_a by $c \mapsto c^p$) for which weak-L_p is a strictly larger space than L_p.

When $\mathbb{E}f = 0$ (in the case $a = 2$) the following variant explains why the variables such that $\|f\|_{L_{\psi_2}} < \infty$ are usually called sub-Gaussian. Indeed, by (4.2) if f is a normalized real-valued Gaussian random variable, then the number $sg(f)$ defined below is equal to 1 and equality holds in (4.4) when s=1. Although our terminology is slightly different, it is more customary to call sub-Gaussian any variable satisfying (4.4) below.

Lemma 4.3. *If f is real-valued, the following are equivalent:*

(i) $f \in L_{\psi_2}$ *and* $\mathbb{E}f = 0$;

(ii) *there is constant $s \geq 0$ such that for any $t \in \mathbb{R}$*

$$\mathbb{E} \exp tf \leq \exp s^2 t^2/2. \tag{4.4}$$

Moreover, assuming $\mathbb{E}f = 0$, $\|f\|_{\psi_2}$ is equivalent to the number $sg(f)$ defined as the smallest $s \geq 0$ for which this holds.

Proof. Assume that $f \in L_{\psi_2}$ with $\|f\|_{\psi_2} \leq 1$. Let f' be an independent copy of f. Let $F = f - f'$. Note that since the distribution of F is symmetric all its odd moments vanish, and hence

$$\mathbb{E} \exp xF = 1 + \sum_{n \geq 1} \frac{x^{2n}}{2n!} \mathbb{E}F^{2n}.$$

We have $\|F\|_{\psi_2} \leq \|f\|_{\psi_2} + \|f'\|_{\psi_2} \leq 2$. Therefore $\mathbb{E}\left(\frac{F}{2}\right)^{2n} \leq n! \mathbb{E} \exp\left(\frac{F}{2}\right)^2 \leq en!$ and hence

$$\mathbb{E} \exp xF \leq 1 + \sum_{n \geq 1} \frac{(2x)^{2n}}{2n!} en! \leq 1 + \sum_{n \geq 1} \frac{(2\sqrt{e}x)^{2n}}{n!} \leq \exp\left(4ex^2\right).$$

But since $t \mapsto \exp -xt$ is convex for any $x \in \mathbb{R}$, and $\mathbb{E}f' = 0$ we have $1 = e^0 \leq \mathbb{E} \exp -xf'$ and hence $\mathbb{E} \exp xF = \mathbb{E} \exp xf \mathbb{E} \exp -xf' \geq \mathbb{E} \exp xf$. Thus we conclude $sg(f) \leq (8e)^{1/2}$. By homogeneity this shows $sg(f) \leq (8e)^{1/2}\|f\|_{\psi_2}$.

Conversely, assume $sg(f) \leq 1$. Clearly (4.4) implies $\mathbb{E}f = 0$. Then for any $x, t > 0$

$$\mathbb{P}(\{f > x\})e^{tx} \leq \mathbb{E}e^{tf} \leq e^{x^2/2}.$$

taking $x = t$ we find $\mathbb{P}(\{f > t\}) \leq e^{-t^2/2}$, and since $sg(-f) = sg(f) \leq 1$ we also have $\mathbb{P}(\{-f > t\}) \leq e^{-t^2/2}$, and hence

$$\mathbb{P}(\{|f| > t\}) \leq 2e^{-t^2/2}.$$

Fix $c > \sqrt{2}$. Let $\theta = 1/2 - 1/c^2$. Note $\theta > 0$.

$$\mathbb{E} \exp (f/c)^2 - 1 = \int_0^\infty (2t/c^2) \exp (t/c)^2 \mathbb{P}(\{|f| > t\}) \, dt$$

$$\leq \int_0^\infty (4t/c^2)e^{-\theta t^2} \, dt = 2/\theta c^2.$$

Elementary calculation shows that if $c_0 = (2(e+1)(e-1)^{-1})^{1/2}$ we have $1 + 2/\theta c_0^2 = e$. Thus we conclude $\|f\|_{\psi_2} \leq c_0$. By homogeneity, this shows $\|f\|_{\psi_2} \leq c_0 sg(f)$. $\square$

The next result was repeatedly used in [34]. It shows that independent random unitary matrices are dominated in a strong sense by their Gaussian analogues.

Lemma 4.4. *Let $(d_k)_{k \in I}$ be an arbitrary collection of integers. Let $\mathbf{G} = \prod_{k \in I} U(d_k)$. Let $u \mapsto u_k$ denote the coordinates on $\mathbf{G}$, and $u_k(i, j)$ $(1 \leq i, j \leq d_k)$ the entries of u_k. Let $\{g_k(i, j)\}$ $(1 \leq i, j \leq d_k)$ be a collection of independent complex-valued Gaussian random variables such that $\mathbb{E}(g_k(i, j)) = 0$ and $\mathbb{E}|g_k(i, j)|^2 = 1/d_k$,* *on a probability space* $(\Omega, \mathbb{P})$. *For some* $C_0 > 0$ *there is a positive operator* $T :$ $L_1(\Omega, \mathbb{P}) \to L_1(\mathbf{G}, m_{\mathbf{G}})$ *with* $\|T : L_p(\Omega, \mathbb{P}) \to L_p(\mathbf{G}, m_{\mathbf{G}})\| \le C_0$ *for all* $1 \le p \le \infty$ *such that*

$$\forall k \forall i, j \le d_k \quad T(g_k(i,j)) = u_k(i,j).$$

Sketch. Let $g_k = v_k|g_k|$ be the polar decomposition of g_k. Let $\mathcal{E}$ be the conditional expectation with respect to (v_k). Since (v_k) and $(|g_k|)$ are independent random variables, we have $\mathcal{E}(g_k) = v_k \mathbb{E}|g_k|$. By known results $\mathbb{E}|g_k| = \delta_k I$ for some $\delta_k > 0$ such that $\delta = \inf_k \delta_k > 0$. Thus $\mathcal{E}(g_k) = v_k \delta_k$. Since $0 < \delta/\delta_k < 1$ for all k, it is easy to see that there is a (positive) operator $W : L_p(\mathbf{G}, m_{\mathbf{G}}) \to L_p(\mathbf{G}, m_{\mathbf{G}})$ with $\|W\| \le 1$ for any $1 \le p \le \infty$ such that $W(v_k) = (\delta/\delta_k)v_k$ and hence $\delta^{-1} W \mathcal{E}(g_k) = v_k$. Thus, since (u_k) and (v_k) have the same distribution, $T = \delta^{-1} W \mathcal{E}$ gives us the desired operator. $\qquad\square$

Remark 4.5 (Matricial contraction principle). Let (u_k) and (g_k) be as in Lemma 4.4. Let $\{x_k(i,j) \mid k \ge 1, 1 \le i, j \le d_k\}$ be a finitely supported family in an arbitrary Banach space B. For any matrix $a \in M_{d_k}$ with complex entries, we denote by ax and xa the matrix products (with entries in B). By convention, we write $\operatorname{tr}(u_k x_k) = \sum_{ij} u_k(i,j) x_k(j,i)$. With this notation, the following "contraction principle" holds

$$\int \left\| \sum d_k \operatorname{tr}(a_k u_k b_k x_k) \right\| dm_{\mathbf{G}}$$

$$\le \sup_k \|a_k\|_{M_{d_k}} \sup_k \|b_k\|_{M_{d_k}} \int \left\| \sum d_k \operatorname{tr}(u_k x_k) \right\| dm_{\mathbf{G}}.$$

Indeed, this is obvious by the translation invariance of $m_{\mathbf{G}}$ if a_k, b_k are all unitary. Then the result follows since the unit ball of M_{d_k} is the closed convex hull of its extreme points, namely its unitary elements.

The same inequality holds if we replace (u_k) by any sequence of variables (z_k) such that for any unitary matrices $a_k, b_k \in U(d_k)$ the sequences (z_k) and $(a_k z_k b_k)$ have the same distribution. In particular this holds for the Gaussian sequence (g_k).

Notation. Let G be any compact group. We denote by (g_π) an independent family indexed by $\widehat{G}$, defined like this: g_π is a random $d_\pi \times d_\pi$-matrix the entries of which are independent complex Gaussian random variables with L_2-norm $= (1/d_\pi)^{1/2}$. All our random variables are assumed defined on a suitable probability space $(\Omega, \mathbb{P})$.

In the sequel, we similarly think of (u_π) as an independent family of unitary $d_\pi \times d_\pi$-matrices indexed by $\widehat{G}$, on the probability space $(\mathcal{G}, m_{\mathcal{G}})$. For simplicity we denote the integral on $\mathcal{G}$ by $\mathbb{E}$.

The following basic fact compares the notions of randomly Sidon for (g_π) and (u_π). It is proved by same truncation trick that was used in [37]. See [34, Chap. V and VI] for further details and more general facts.

Lemma 4.6. *For a subset* $\Lambda \subset \widehat{G}$, *the following are equivalent.*

(i) *There is a constant α_1 such that for any finitely supported family $(a_\pi) \in \prod_{\pi \in \Lambda} M_{d_\pi}$*

$$\sum_{\pi \in \Lambda} d_\pi \operatorname{tr} |a_\pi| \le \alpha_1 \mathbb{E} \left\| \sum_{\pi \in \Lambda} d_\pi \operatorname{tr}(g_\pi \pi a_\pi) \right\|_\infty .$$

(ii) *There is a constant α_2 such that for any finitely supported family $(a_\pi) \in \prod_{\pi \in \Lambda} M_{d_\pi}$*

$$\sum_{\pi \in \Lambda} d_\pi \operatorname{tr} |a_\pi| \le \alpha_2 \mathbb{E} \left\| \sum_{\pi \in \Lambda} d_\pi \operatorname{tr}(u_\pi \pi a_\pi) \right\|_\infty .$$

Sketch. From Lemma 4.4 it is easy to deduce that

$$\mathbb{E} \left\| \sum_\Lambda d_\pi \operatorname{tr}(u_\pi \pi a_\pi) \right\|_\infty m_{\mathbf{G}}(du) \le C_0 \mathbb{E} \left\| \sum_\Lambda d_\pi \operatorname{tr}(g_\pi \pi a_\pi) \right\|_\infty ,$$

and hence (ii) $\Rightarrow$ (i). To check the converse, recall the well-known fact that $c_4 = \sup \mathbb{E} \|g_\pi\|^2 < \infty$, from which it is easy to deduce by Chebyshev's inequality that there exists $c_5 > 0$ such that

$$\sup \mathbb{E}(\|g_\pi\| 1_{\{\|g_\pi\| > c_5\}} \le (2\alpha_1)^{-1}.$$

We may assume that the sequences (u_π) and (g_π) are mutually independent, so that the sequences (g_π) and $(u_\pi g_\pi)$ have the same distribution. Then by the triangle inequality and by Remark 4.5

$$\mathbb{E} \left\| \sum_\Lambda d_\pi \operatorname{tr}(g_\pi \pi a_\pi) \right\|_\infty = \mathbb{E} \left\| \sum_\Lambda d_\pi \operatorname{tr}(u_\pi g_\pi \pi a_\pi) \right\|_\infty$$

$$\le \mathbb{E} \left\| \sum_\Lambda d_\pi \operatorname{tr}(u_\pi g_\pi 1_{\{\|g_\pi\| \le c_5\}} \pi a_\pi) \right\|_\infty + \mathbb{E} \left\| \sum_\Lambda d_\pi \operatorname{tr}(u_\pi g_\pi 1_{\{\|g_\pi\| > c_5\}} \pi a_\pi) \right\|_\infty$$

$$\le c_5 \mathbb{E} \left\| \sum_\Lambda d_\pi \operatorname{tr}(u_\pi \pi a_\pi) \right\|_\infty + \mathbb{E} \sum_\Lambda d_\pi \|g_\pi\| 1_{\{\|g_\pi\| > c_5\}} \operatorname{tr} |a_\pi|$$

$$\le c_5 \mathbb{E} \left\| \sum_\Lambda d_\pi \operatorname{tr}(u_\pi \pi a_\pi) \right\|_\infty + (2\alpha_1)^{-1} \sum_\Lambda d_\pi \operatorname{tr} |a_\pi|.$$

Using this we see that (i) implies

$$\sum_\Lambda d_\pi \operatorname{tr} |a_\pi| \le \alpha_1 c_5 \mathbb{E} \left\| \sum_\Lambda d_\pi \operatorname{tr}(u_\pi \pi a_\pi) \right\|_\infty + \frac{1}{2} \sum_\Lambda d_\pi \operatorname{tr} |a_\pi|,$$

and hence (i) $\Rightarrow$ (ii) with $\alpha_2 \le 2\alpha_1 c_5$. $\qquad\square$

Remark 4.7 (Comparison of randomizations). Actually, Lemma 4.6 follows from a much more general fact proved in [34]. Let (a_π) be a finitely supported family indexed by $\widehat{G}$ with $a_\pi \in M_{d_\pi}$ $(\pi \in \widehat{G})$. In [34], the random Fourier series

$$R(x) = \sum_{\pi \in \widehat{G}} d_\pi \operatorname{tr}(u_\pi \pi(x) a_\pi) \quad (x \in G)$$

randomized by $u = (u_\pi)$ on $(\mathcal{G}, m_{\mathcal{G}})$ is compared to

$$\widetilde{R}(x) = \sum_{\pi \in \widehat{G}} d_\pi \operatorname{tr}(g_\pi \pi(x) a_\pi) \quad (x \in G)$$

randomized by g_π on $(\Omega, \mathbb{P})$. By [34, p. 97] there is a universal constant $c > 0$ such that

$$c^{-1} \mathbb{E} \sup_{x \in G} |\widetilde{R}(x)| \le \mathbb{E} \sup_{x \in G} |R(x)| \le c \mathbb{E} \sup_{x \in G} |\widetilde{R}(x)|. \tag{4.5}$$

In particular, a set is randomly Sidon if and only if it so when we replace the random unitaries (u_π) by the Gaussian variables g_π, so we recover Lemma 4.6.

Remark 4.8. A similar comparison holds for the random Fourier series

$$L(x) = \sum d_\pi \operatorname{tr}(u_\pi a_\pi \pi(x)) \ (x \in G) \text{ and } \widetilde{L}(x) = \sum d_\pi \operatorname{tr}(g_\pi a_\pi \pi(x)) \ (x \in G),$$

where the randomization is on the other side of π, but this can be easily derived from the case of R and $\widetilde{R}$ by observing that

$$|L(x)| = |\overline{L(x)}| = \left| \sum d_\pi \operatorname{tr}((u_\pi a_\pi \pi(x))^*) \right| = \left| \sum d_\pi \operatorname{tr}(u_\pi^* \pi(x^{-1}) a_\pi^*) \right|,$$

and the last series can be treated as $R(x^{-1})$ for a suitable R.

Remark 4.9. By passing to the series $\widetilde{R}$, we allow ourselves the use of the rich theory of Gaussian processes. We will use these ideas to prove the next statement. Let us briefly outline this. Let $f(x) = \sum_{\pi \in \widehat{G}} d_\pi \operatorname{tr}(\pi(x) a_\pi)$ so that ${}^t\widehat{f}(\pi) = a_\pi$. Let $f_t(g) = f(gt)$. Let

$$d_f(s,t) = \|\widetilde{R}(s) - \widetilde{R}(t)\|_2 = \|f_s - f_t\|_2 = \left(\sum d_\pi \operatorname{tr} |(\pi(s) - \pi(t))^t \widehat{f}(\pi)|^2 \right)^{1/2}.$$

The metric entropy integral associated to f is usually defined as

$$\int_0^\infty (\log N_f(\varepsilon))^{1/2} \, d\varepsilon$$

where $N_f(\varepsilon)$ is the smallest number of open balls of d_f-radius ε that suffice to cover G.

Since the measure and the distance are both (left) translation invariant, one checks easily that

$$m_G(\{t \mid d_f(t,1) < \varepsilon\})^{-1} \le N_f(\varepsilon) \le m_G(\{t \mid d_f(t,1) < \varepsilon/2\})^{-1}. \tag{4.6}$$

Thus we may work with the following quantity equivalent to the metric entropy integral:

$$\mathcal{I}_2(f) = \int_0^\infty \left(\log \frac{1}{m_G(\{t \mid d_f(t,1) < \varepsilon\})} \right)^{1/2} d\varepsilon.$$

The metric entropy integral was originally introduced in the subject in a 1967 paper of Dudley to give new upper bounds for general Gaussian processes. In the stationary case, Fernique showed that the same integral is also a lower bound. The latter bound implies that there is an absolute constant c such that

$$\mathcal{I}_2(f) \le c\mathbb{E} \left\| \sum d_\pi \operatorname{tr}(g_\pi \pi^t \widehat{f}(\pi)) \right\|_\infty. \tag{4.7}$$

A fortiori, this implies Sudakov's minoration (see, e.g., [41, p. 69] or [58]): there is a numerical constant c' such that

$$\sup_{\varepsilon>0} \varepsilon(\log N_f(\varepsilon))^{1/2} \le c'\mathbb{E} \sup_{x \in G} |\widetilde{R}(x)|,$$

and hence

$$\sup_{\varepsilon>0} \varepsilon \left(\log \frac{1}{m_G(\{x \mid d(x,1) < \varepsilon\})} \right)^{1/2} \le cc'\mathbb{E} \sup_{x \in G} |R(x)|. \tag{4.8}$$

The next two theorems essentially come from [37, 38]. They show that a set is Sidon if and only if it is a $\Lambda(p)$-set (in Rudin's sense [51]) for all $p > 2$ with a constant growing at most like $\sqrt{p}$.

Theorem 4.10 (Sidon versus $\Lambda(p)$-sets). *Let $\Lambda \subset \widehat{G}$. The following three assertions are equivalent.*

(i) *Λ is a Sidon set.*

(ii) *There is a constant C such that for any $f \in L_2(G)$ with $\widehat{f}$ supported in Λ we have*

$$\|f\|_{\psi_2} \le C\|f\|_2.$$

(ii)' *There is a constant C such that for any any finitely supported family (a_π) $(a_\pi \in M_{d_\pi})$ we have for any $p \ge 2$*

$$\left\| \sum_{\pi \in \Lambda} d_\pi \operatorname{tr}(\pi a_\pi) \right\|_p \le Cp^{1/2} \left(\sum_{\pi \in \Lambda} d_\pi \operatorname{tr} |a_\pi|^2 \right)^{1/2}.$$

Sketch. The equivalence between (ii) and (ii)' is immediate by (4.3). The proof that (i) $\Rightarrow$ (ii) follows a classical argument due to Rudin that Figà-Talamanca and Rider adapted to the non-Abelian case. The quicker argument in [34] avoids their moment computations by using instead Lemma 4.4, but first we use (i) in Lemma 3.3. With the notation in that lemma, assuming Λ Sidon, the operator of convolution by μ^u has norm of at most C on $L_p(G)$ for any $1 \le p \le \infty$. Therefore, for any $f = \sum_{\pi \in \Lambda} d_\pi \operatorname{tr}(\pi^t \widehat{f}(\pi))$ (finite sum) we have

$$\left\| \sum_{\pi \in \Lambda} d_\pi \operatorname{tr}(\pi^t(u_\pi \widehat{f}(\pi))) \right\|_p \le C\|f\|_p.$$

As before let $\mathcal{G} = \prod_{\pi \in \widehat{G}} U(d_\pi)$ (actually we could work simply with $\prod_{\pi \in \Lambda} U(d_\pi)$). Let $u = (u_\pi) \in \mathcal{G}$. Let $F_u = \sum_{\pi \in \Lambda} d_\pi \operatorname{tr}(\pi^t(u_\pi^* \widehat{f}(\pi)))$. Applying this with F_u in place of f we find

$$\|f\|_p \le C\|F_u\|_p$$

and hence

$$\|f\|_p \le C(\int \|F_u\|_p^p m_{\mathcal{G}}(du))^{1/p}.$$

Note $F_u = \sum_{\pi \in \Lambda} d_\pi \operatorname{tr}(u_\pi^* \widehat{f}(\pi)^t \pi)$. By Lemma 4.4

$$\left(\int \|F_u\|_p^p m_{\mathcal{G}}(du) \right)^{1/p} \le C_0 \left(\mathbb{E} \left\| \sum_{\pi \in \Lambda} d_\pi \operatorname{tr}(g_\pi \widehat{f}(\pi)^t \pi) \right\|_p^p \right)^{1/p}$$

and since $({}^t\pi(x)g_\pi(\omega))$ (on $G \times \Omega$) and (g_π) (on Ω) have the same distribution, we have using (4.1)

$$\left(\mathbb{E} \left\| \sum_{\pi \in \Lambda} d_\pi \operatorname{tr}(g_\pi \widehat{f}(\pi)^t \pi) \right\|_p^p \right)^{1/p} = \left(\mathbb{E} \left| \sum_{\pi \in \Lambda} d_\pi \operatorname{tr}(g_\pi \widehat{f}(\pi)) \right|^p \right)^{1/p} = \gamma(p)\|f\|_2$$

where $\gamma(p)$ is the L_p-norm of a normalized complex Gaussian variable. This gives us

$$\|f\|_p \le CC_0\gamma(p)\|f\|_2,$$

and since $\gamma(p) = O(\sqrt{p})$, we obtain (ii) by (4.3).

The proof that (ii) $\Rightarrow$ (i) in [37, 34] uses the metric entropy characterization of the Gaussian random Fourier series that are continuous a.s.. We merely outline the original argument. Fix $f = \sum_{\pi \in \Lambda} d_\pi \operatorname{tr}(\pi^t \widehat{f}(\pi))$ (finite sum). We will use Gaussian process theory through the *minorization* (4.7). But, by another result from that theory (a variant of Dudley's upper bound), the integral $\mathcal{I}_2(f)$ *majorizes* the sub-Gaussian processes that are suitably dominated in the metric sense by d_f. More specifically, since (ii) implies $\|f_s - f_t\|_{\psi_2} \le Cd_f(s,t)$ the said majorization implies (assuming $\int f dm_G = 0$) that (c' is here an absolute constant)

$$\|f\|_\infty \le c'C\mathcal{I}_2(f). \tag{4.9}$$

Therefore, we obtain

$$\|f\|_\infty \le c'C\mathcal{I}_2(f) \le cc'C\mathbb{E} \left\| \sum d_\pi \operatorname{tr}(g_\pi \pi^t \widehat{f}(\pi)) \right\|_\infty, \tag{4.10}$$

and hence

$$\left| \sum d_\pi \operatorname{tr}({}^t\widehat{f}(\pi)) \right| = |f(1)| \le cc'C\mathbb{E} \left\| \sum d_\pi \operatorname{tr}(g_\pi \pi^t \widehat{f}(\pi)) \right\|_\infty. \tag{4.11}$$

But by the distributional invariance property of (g_π) we have for any $z_\pi \in U(d_\pi)$

$$\mathbb{E}\left\|\sum d_\pi \operatorname{tr}(g_\pi \pi^t \widehat{f}(\pi))\right\|_\infty = \mathbb{E}\left\|\sum d_\pi \operatorname{tr}(z_\pi g_\pi \pi^t \widehat{f}(\pi))\right\|_\infty$$

$$= \mathbb{E}\left\|\sum d_\pi \operatorname{tr}(g_\pi \pi(^t \widehat{f}(\pi) z_\pi))\right\|_\infty,$$

and hence (4.11) applied to $\sum d_\pi \operatorname{tr}(g_\pi \pi(^t \widehat{f}(\pi) z_\pi))$ implies after taking the sup over z_π

$$\sum_{\pi \in \Lambda} d_\pi \operatorname{tr}(|\widehat{f}(\pi)|) = \sum_{\pi \in \Lambda} d_\pi \operatorname{tr}(|^t \widehat{f}(\pi)|) \leq cc'C\mathbb{E}\left\|\sum d_\pi \operatorname{tr}(g_\pi \pi^t \widehat{f}(\pi))\right\|_\infty.$$

In other words, provided we can replace (g_π) by (u_π), we conclude that Λ is randomly Sidon and hence Sidon by Corollary 3.7. The replacement of (g_π) by (u_π) is justified by Lemma 4.6 (see also the discussion around (4.5)).   □

Remark 4.11. Let $f \in C(G)$. Note $\|f\|_\infty = \int \sup_{x\in G} |f(tx)| m_G(dt)$. For proper perspective, we use this observation to rewrite (4.10) as

$$\|f\|_\infty = \int \left\|\sum d_\pi \operatorname{tr}(\pi(t)\pi^t \widehat{f}(\pi))\right\|_\infty m_G(dt) \leq cc'C\mathbb{E}\left\|\sum d_\pi \operatorname{tr}(g_\pi \pi^t \widehat{f}(\pi))\right\|_\infty.$$
(4.12)

Let $Y_x(t) = \sum d_\pi \operatorname{tr}(\pi(t)\pi(x)^t \widehat{f}(\pi))$ and $X_x(\omega) = \sum d_\pi \operatorname{tr}(g_\pi(\omega)\pi(x)^t \widehat{f}(\pi))$. Then (4.12) means

$$\int \sup_{x\in G} |Y_x| \, dm_G \leq cc'C\mathbb{E}\sup_{x\in G} |X_x|.$$
(4.13)

In the preceding proof the Dudley–Fernique metric entropy bounds were used only to prove (4.12) or equivalently (4.13). These require a certain group invariance (namely the process (X_x) must be a stationary Gaussian process). Inspired by the latter bounds, Talagrand [57] managed to prove a general version of (4.13) that does not require any group invariance. More precisely, he proved that there is an absolute constant τ_0 such that:

If (φ_n) are variables such that for any finitely supported scalar sequence (a_n) we have

$$\left\|\sum a_n \varphi_n\right\|_{\psi_2} \leq \left(\sum |x_n|^2\right)^{1/2}$$

and if (f_n) are arbitrary functions on a set S then we have

$$\mathbb{E}\sup_{x\in S}\left|\sum \varphi_n f_n(x)\right| \leq \tau_0 \mathbb{E}\sup_{x\in S}\left|\sum g_n f_n(x)\right|.$$

We used this in our recent paper [44] to prove a version of the implication sub-Gaussian $\Rightarrow$ Sidon for general uniformly bounded orthonormal systems, that improves an earlier breakthrough due to Bourgain and Lewko [3]. We also give in [44] an analogue of (ii) $\Rightarrow$ (i) in Theorem 4.10 to the case when the system $\{d_\pi^{1/2}\pi_{ij} \mid \pi \in \Lambda, 1 \leq i, j \leq d_\pi\}$ is replaced by an orthonormal system on a probability space (T, m) indexed by a set Λ such that the norms of the $d_\pi \times d_\pi$ matrices $[\pi_{ij}(t)]$ are uniformly bounded over $t \in T$ and $\pi \in \Lambda$. In the same framework, we also give an analogue of the equivalence between Sidon and randomly Sidon. See §6 for a related application of these ideas.

The following refinement of Theorem 4.10 proved in [38] will be useful.

Lemma 4.12. *Assume that G is Abelian (so that $d_\pi = 1$ for all π). Let $1 < p < 2 < p' < \infty$ such that $1/p + 1/p' = 1$. Assume that there is a constant C such that for any $f \in L_2(G)$ with $\widehat{f}$ supported in Λ we have*

$$\|f\|_{\psi_{p'}} \le C \left(\sum_{\pi \in \Lambda} |\widehat{f}(\pi)|^p \right)^{1/p}. \tag{4.14}$$

Then Λ is Sidon.

Proof. Let

$$d_{p,f}(t,s) = \left(\sum_{\pi \in \widehat{G}} |\widehat{f}(\pi)(\pi(t) - \pi(s))|^p \right)^{1/p}.$$

We will use a variant of the metric entropy integral $\mathcal{I}_2(f)$, namely

$$\mathcal{I}_p(f) = \int_0^\infty \left(\log \frac{1}{m_G(\{t \mid d_{p,f}(t,1) < \varepsilon\})} \right)^{1/p'} d\varepsilon.$$

Schematically, the proof can be described like this. By a generalization of the Dudley majorization (4.9) we see (assuming still $\int f \, dm_G = 0$) that if we assume

$$\forall t, s \in G \quad \|f_t - f_s\|_{\psi_{p'}} \le C d_{p,f}(t,s)$$

then we have

$$\|f\|_\infty \le C c'_p \mathcal{I}_p(f),$$

and replacing f by $\sum_{\pi \in \widehat{G}} |\widehat{f}(\pi)|\pi$ (which leaves $d_{p,f}$ and hence also $\mathcal{I}_p(f)$ invariant) we find

$$\sum_{\pi \in \widehat{G}} |\widehat{f}(\pi)| \le C c'_p \mathcal{I}_p(f). \tag{4.15}$$

This shows that if Λ satisfies the assumption (4.14) then any $f \in L_2(G)$ with $\widehat{f}$ supported in $\Lambda \setminus \{0\}$ satisfies

$$\sum_{\pi \in \Lambda} |\widehat{f}(\pi)| \le C c'_p \mathcal{I}_p(f). \tag{4.16}$$

We may assume $0 \notin \Lambda$ for simplicity. The conclusion will follow from the inequality

$$\mathcal{I}_p(f) \le c'' \left(\sum_{\pi \in \widehat{G}} |\widehat{f}(\pi)| \right)^{1-\theta} \mathcal{I}_2(f)^\theta, \tag{4.17}$$

where c'' depends only on p and where $0 < \theta < 1$.

Indeed, (4.17) combined with (4.16) implies

$$\sum_{\pi \in \Lambda} |\widehat{f}(\pi)| \leq C c_p' c'' \left(\sum_{\pi \in \widehat{G}} |\widehat{f}(\pi)| \right)^{1-\theta} \mathcal{I}_2(f)^{\theta},$$

and after a suitable division we find

$$\sum_{\pi \in \Lambda} |\widehat{f}(\pi)| \leq (C c_p' c'')^{1/\theta} \mathcal{I}_2(f).$$

But now using Fernique's lower bound (4.7), we conclude as in the preceding proof that Λ is Sidon.

It remains to justify (4.17). Let $N_p(\varepsilon)$ denote the smallest number of sets of $d_{p,f}$-diameter $\leq \varepsilon$ that suffice to cover G. Let $e_n(d_{p,f})$ be the smallest number ε such that G can be covered by 2^n sets of $d_{p,f}$-diameter $\leq \varepsilon$ (i.e., such that $N_p(\varepsilon) \leq 2^n$). We first note that $\mathcal{I}_p(f)$ is equivalent to

$$\int_0^{\infty} (\log N_p(\varepsilon))^{1/p'} d\varepsilon.$$

Then, since $\int_0^{\infty} (\log N_p(\varepsilon))^{1/p'} d\varepsilon = \sum_n \int_{e_n}^{e_{n-1}} (\log N_p(\varepsilon))^{1/p'} d\varepsilon$ one checks easily that the latter quantity is equivalent to the following one:

$$\Sigma_p(f) = \sum_0^{\infty} e_n(d_{p,f}) n^{-1/p'}.$$

Let $1 < q < p < 2$. Let $0 < \theta < 1$ be such that $(1-\theta)/q + \theta/2 = 1/p$. By Hölder's inequality,

$$d_{p,f} \leq d_{q,f}^{1-\theta} d_{2,f}^{\theta}.$$

Thus if A_0 has $d_{q,f}$-diameter $\leq r_0$ and if A_1 has $d_{2,f}$-diameter $\leq r_1$, then $A_0 \cap A_1$ has $d_{p,f}$-diameter $\leq r_0^{1-\theta} r_1^{\theta}$. From this it is clear (taking intersections) that G can be covered by $2^n \times 2^n$ sets with $d_{p,f}$-diameter $\leq (e_n(d_{q,f}))^{1-\theta}(e_n(d_{2,f}))^{\theta}$. In other words

$$e_{2n}(d_{p,f}) \leq (e_n(d_{q,f}))^{1-\theta}(e_n(d_{2,f}))^{\theta}.$$

Therefore by Hölder

$$\sum_0^{\infty} e_{2n}(d_{p,f}) n^{-1/p'} \leq \sum_0^{\infty} (e_n(d_{q,f}) n^{-1/q'})^{1-\theta} (e_n(d_{2,f}) n^{-1/2})^{\theta}$$

$$\leq (\Sigma_q(f))^{1-\theta} (\Sigma_2(f))^{\theta}.$$

But since the numbers $e_n(d_{p,f})$ (and also $e_n(d_{p,f}) n^{-1/p'}$) are obviously non-increasing we have $\sum_0^{\infty} e_n(d_{p,f}) n^{-1/p'} \leq 2 \sum_0^{\infty} e_{2n}(d_{p,f})(2n)^{-1/p'}$ and hence we obtain

$$\Sigma_p(f) \leq 2^{1/p} (\Sigma_q(f))^{1-\theta} (\Sigma_2(f))^{\theta}.$$

Lastly, we invoke a result from approximation theory, that tells us that for any $1 < q < \infty$ there is a constant β_q such that

$$\Sigma_q(f) \leq \beta_q \sum_{\pi \in \widehat{G}} |\widehat{f}(\pi)|.$$

See [33] or [7, Prop. 2, p. 142]. Since $\Sigma_p(f)$ is equivalent to $\mathcal{I}_p(f)$, this gives us (4.17). □

Theorem 4.13 (Sidon versus <u>central</u> $\Lambda(p)$-sets). *Let $\Lambda \subset \widehat{G}$. Recall $\mathcal{G} = \prod_{\pi \in \Lambda} U(d_\pi)$. Consider the following assertions in addition to* (i) *and* (ii) *in Theorem 4.10.*

(iii) *Same as* (ii) *for all (central) functions f of the form $f = \sum_{\pi \in A} d_\pi \chi_\pi$ where $A \subset \Lambda$ is an arbitrary finite subset.*

(iii)′ *There is a constant C such that for any even integer $2 \leq p < \infty$ and any finite subset $A \subset \Lambda$ we have*

$$\left\| \sum_{\pi \in A} d_\pi \chi_\pi \right\|_p \leq C \sqrt{p} \left\| \sum_{\pi \in A} d_\pi \chi_\pi \right\|_2 = C \sqrt{p} \left(\sum_{\pi \in A} d_\pi^2 \right)^{1/2}.$$

(iv) *For any $0 < \delta < 1$ there is $0 < \beta < \infty$ such that for any finite subset $A \subset \Lambda$ we have*

$$m_G \left(\left\{ t \in G \mid \sum_{\pi \in A} d_\pi \operatorname{Re}(\chi_\pi) > \delta \sum_{\pi \in A} d_\pi^2 \right\} \right) \leq e \exp - \left(\beta \sum_{\pi \in A} d_\pi^2 \right)$$

(v) *There are $0 < \delta < 1$ and $0 < \beta < \infty$ such that for any finite subset $A \subset \Lambda$ we have*

$$m_G \left(\left\{ t \in G \mid \sum_{\pi \in A} d_\pi \operatorname{Re}(\chi_\pi) > \delta \sum_{\pi \in A} d_\pi^2 \right\} \right) \leq e \exp - \left(\beta \sum_{\pi \in A} d_\pi^2 \right)$$

(vi) *There is a constant C such that for any finite $A \subset \Lambda$*

$$\sum_{\pi \in A} d_\pi^2 \leq C \int_{\mathcal{G}} \sup_{g \in G} \left| \sum_{\pi \in A} d_\pi \operatorname{tr}(u_\pi \pi(g)) \right| m_{\mathcal{G}}(du).$$

(vii) *There is $0 < \delta < 1$ such that any finite subset $A \subset \Lambda$ contains a further subset $B \subset A$ with Sidon constant at most $1/\delta$ and such that $\sum_{\pi \in B} d_\pi^2 \geq \delta \sum_{\pi \in A} d_\pi^2$.*

(viii) *There is a constant C such that for any finite subset $A \subset \Lambda$, and any $f \in L_2(G)$ with $\widehat{f}$ supported in A we have*

$$\|f\|_{\psi_2} \leq C (\sum_{\pi \in A} d_\pi^2)^{1/2} \sup_{\pi \in A} \|\widehat{f}(\pi)\|. \tag{4.18}$$

Then (i) $\Rightarrow$ (ii) $\Rightarrow$ (iii) $\Leftrightarrow$ (iii)′ $\Rightarrow$ (iv) $\Rightarrow$ (v) $\Rightarrow$ (vi) $\Rightarrow$ (vii) $\Rightarrow$ (viii). *Moreover,* (viii) $\Rightarrow$ (i) *if G is Abelian, or more generally if the dimensions $\{d_\pi \mid \pi \in \Lambda\}$ are uniformly bounded.*

Proof. Recall (i) ⇔ (ii) by Theorem 4.10, (ii) ⇒ (iii) is trivial and (iii) ⇔ (iii)′ follows from (4.3).

Assume (iii). In the rest of the proof, we follow [38] except for the correction indicated in Remark 4.14. Let $A \subset \Lambda$ be a finite subset. Let $N(A) = \sum_{\pi \in A} d_\pi^2$. (Incidentally, $N(A)$ is the Plancherel measure of A.) By (4.3) $\| \sum_{\pi \in A} d_\pi \chi_\pi \|_{L_{\psi_2}} \le CC_2(N(A))^{1/2}$. Therefore for any $\delta > 0$ we have

$$m_G \left(\left\{ t \in G \mid \left| \sum_{\pi \in A} d_\pi \chi_\pi \right| > \delta N(A) \right\} \right) \le e \exp -(\delta^2 N(A)/(CC_2)^2).$$

A fortiori, (iv) holds and (iv) ⇒ (v) is trivial.

Assume (v). Let $A \subset \Lambda$ be a finite subset. Consider the random Fourier series

$$S_A(g) = \sum_{\pi \in A} d_\pi \operatorname{tr}(u_\pi \pi(g))$$

defined for $u = (u_\pi) \in \mathcal{G}$ as in Remark 4.7. The associated metric d_A is given by

$$d_A(g, g')^2 = \sum_{\pi \in A} d_\pi \operatorname{tr} |\pi(g) - \pi(g')|^2 = 2 \sum_{\pi \in A} d_\pi^2 - 2 \sum_{\pi \in A} d_\pi \operatorname{Re}(\chi_\pi(g'g^{-1})).$$

Therefore

$$\{ g \in G \mid d_A(g, 1) < \varepsilon N(A)^{1/2} \}$$

$$= \left\{ g \in G \mid \sum_{\pi \in A} d_\pi \operatorname{Re}(\chi_\pi(g)) > (1 - \varepsilon^2/2) N(A) \right\}.$$

Thus (v) implies that for some $\varepsilon > 0$ (chosen so that $1 - \varepsilon^2/2 = \delta$) we have

$$m_G \left(\left\{ g \in G \mid d_A(g, 1) < \varepsilon N(A)^{1/2} \right\} \right) \le \exp (1 - \beta N(A)).$$

Then by (4.8) we find

$$\varepsilon N(A)^{1/2} (\beta N(A) - 1)^{1/2} \le cc' \mathbb{E} \sup_{g \in G} |S_A(g)|,$$

from which (vi) is immediate.

Assume (vi). Let V_A denote the linear space formed of all random functions of the form $F(u)(g) = \sum_{\pi \in A} d_\pi \operatorname{tr}(u_\pi a_\pi \pi(g))$ with a_π arbitrary in M_{d_π}. Let $\| \cdot \|_A$ be the norm induced on it by $L_1(\mathcal{G}; C(G))$, i.e.,

$$\|F\|_A = \int_{\mathcal{G}} \sup_{g \in G} |F(u)(g)| m_{\mathcal{G}}(du).$$

With S_A, $N(A)$ as before, (v) tells us that

$$\|S_A\|_A \ge N(A)/C.$$

By Hahn–Banach, there is $y \in A^*$ with $\|y\|_A^* \le 1$ such that $\langle y, S_A \rangle = \|S_A\|_A \ge N(A)/C$. Identifying $y \in A^*$ with a family (y_π) with $y_\pi \in M_{d_\pi}$ ($\pi \in A$), we may assume that $\langle y, F \rangle = \sum_{\pi \in A} d_\pi \operatorname{tr}(y_\pi a_\pi)$. Then $\langle y, S_A \rangle = \sum_{\pi \in A} d_\pi \operatorname{tr}(y_\pi)$. Moreover, by the translation invariance of the norm $\| \cdot \|_A$ (on $\mathcal{G}$ and on G), for any fixed u', g' we have $\|F\|_A = \|F(\cdot u')(g'\cdot)\|_A$. By duality this implies $\|y\|_A^* = \|(\pi(g')y_\pi u'_\pi)\|_A^*$, and hence

$$\left\|\left(\int \pi(g')y_\pi\pi(g'^{-1})m_G(dg')\right)\right\|_A^* \leq 1.$$

But since the π's are irreducible, $\int \pi(g')y_\pi\pi(g'^{-1})m_G(dg') = I_{d_\pi}\,\mathrm{tr}(y_\pi)/d_\pi$, and hence

$$\|(I_{d_\pi}\,\mathrm{tr}(y_\pi)/d_\pi)\|_A^* \leq 1,$$

which means that for any F

$$\left|\sum_{\pi\in A}\mathrm{tr}(y_\pi)\,\mathrm{tr}(a_\pi)\right| \leq \|F\|_A.$$

Since $\|F\|_A$ is invariant if we replace a_π by $|a_\pi|\,|\mathrm{tr}(y_\pi)|(\mathrm{tr}(y_\pi))^{-1}$ we also have

$$\left|\sum_{\pi\in A}|\mathrm{tr}(y_\pi)|\,\mathrm{tr}\,|a_\pi|\right| \leq \|F\|_A. \tag{4.19}$$

In particular, in the case $F(u)(g) = d_\pi\,\mathrm{tr}(u_\pi\pi(g))$ for some $\pi \in A$, this implies

$$|d_\pi\,\mathrm{tr}(y_\pi)| \leq d_\pi^2. \tag{4.20}$$

Now recalling that $\langle y, S_A\rangle = \|S_A\|_A \geq N(A)/C$ we have $\sum_{\pi\in A}d_\pi\,\mathrm{tr}(y_\pi) \geq N(A)/C$, and hence there is a subset $B \subset A$ (namely $B = \{\pi \mid |\mathrm{tr}(y_\pi)| > d_\pi/2C\}$) such that $|\sum_{\pi\in B}d_\pi\,\mathrm{tr}(y_\pi)| \geq N(A)/2C$ and $|\mathrm{tr}(y_\pi)| > d_\pi/2C$ for any $\pi \in B$. By (4.20)

$$\sum_{\pi\in B}d_\pi^2 \geq \left|\sum_{\pi\in B}d_\pi\,\mathrm{tr}(y_\pi)\right| \geq N(A)/2C,$$

and by (4.19) the randomly Sidon constant of B is at most $2C$, so that (vii) holds by Corollary 3.7.

Assume (vii). Let $N(A) = \sum_{\pi\in A}d_\pi^2$. We will show that there is C_δ depending only on the δ appearing in (vii) such that for any finite subset $A \subset \Lambda$, and any $f \in L_2(G)$ with $\hat{f}$ supported in A we have

$$\|f\|_{\psi_2} \leq C_\delta N(A)^{1/2}\sup_{\pi\in A}\|\hat{f}(\pi)\|. \tag{4.21}$$

To prove this we may assume that Λ is finite. Let C_Λ be the smallest constant for which (4.21) holds for all $A \subset \Lambda$.

Fix $A \subset \Lambda$ and let $B \subset A$ as in (vii). Let $f \in L_2(G)$ with $\hat{f}$ supported in A. We will show that (vii) implies that

$$\|f\|_{\psi_2} \leq C'_\delta N(A)^{1/2}\sup_{\pi\in A}\|\hat{f}(\pi)\| + C_\Lambda(1-\delta)^{1/2}N(A)^{1/2}\sup_{\pi\in A}\|\hat{f}(\pi)\|, \tag{4.22}$$

where C'_δ is a constant depending only on δ. Indeed, by the triangle inequality we have

$$\|f\|_{\psi_2} \le \left\| \sum_{\pi \in B} d_\pi \operatorname{tr}({}^t\widehat{f}(\pi)\pi) \right\|_{\psi_2} + \left\| \sum_{\pi \in A \setminus B} d_\pi \operatorname{tr}({}^t\widehat{f}(\pi)\pi) \right\|_{\psi_2}.$$

Note

$$\left\| \sum_{\pi \in B} d_\pi \operatorname{tr}({}^t\widehat{f}(\pi)\pi) \right\|_2 = \left(\sum_{\pi \in B} d_\pi \operatorname{tr} |\widehat{f}(\pi)|^2 \right)^{1/2}$$

$$\le N(B)^{1/2} \sup_{\pi \in B} \|\widehat{f}(\pi)\| \le N(A)^{1/2} \sup_{\pi \in A} \|\widehat{f}(\pi)\|.$$

Thus, by Theorem 4.10 applied to the set B there is C'_δ such that

$$\left\| \sum_{\pi \in B} d_\pi \operatorname{tr}({}^t\widehat{f}(\pi)\pi) \right\|_{\psi_2} \le C'_\delta \left\| \sum_{\pi \in B} d_\pi \operatorname{tr}({}^t\widehat{f}(\pi)\pi) \right\|_2 \le C'_\delta N(A)^{1/2} \sup_{\pi \in A} \|\widehat{f}(\pi)\|,$$

and by the definition of C_Λ we have

$$\left\| \sum_{\pi \in A \setminus B} d_\pi \operatorname{tr}({}^t\widehat{f}(\pi)\pi) \right\|_{\psi_2} \le C_\Lambda N(A \setminus B)^{1/2} \sup_{A \setminus B} \|\widehat{f}(\pi)\|$$

$$\le C_\Lambda (1 - \delta)^{1/2} N(A)^{1/2} \sup_A \|\widehat{f}(\pi)\|,$$

from which (4.22) is immediate.

Equivalently, (4.22) means $C_\Lambda \le C'_\delta + C_\Lambda(1 - \delta)^{1/2}$ and hence

$$C_\Lambda \le (1 - (1 - \delta)^{1/2}) C'_\delta,$$

which proves (4.21). Thus we have proved (vii) $\Rightarrow$ (viii).

Now assume (viii) but *we also assume that* $d_\pi = 1$ *for all* $\pi \in \Lambda$. Let us denote by $\ell_{2,1}(\Lambda)$ the classical Lorentz space of scalar sequences indexed by Λ. Explicitly, given a scalar family $a = (a_\pi)$ (say, tending to 0 at ∞), we denote by (a_n^*) the non-increasing rearrangement of the numbers $\{|\widehat{f}(\pi)| \mid \pi \in \Lambda\}$. Let

$$\|a\|_{2,1} = \sum_1^\infty a_n^*/n^{1/2}.$$

The space $\ell_{2,1}(\Lambda)$ is defined as formed of those a for which this sum is finite. It is well known that $\| \ \|_{2,1}$ is equivalent to a norm on $\ell_{2,1}(\Lambda)$ (we will not use this). Note that (4.18) simply means $\|f\|_{\psi_2} \le C|A|^{1/2} \sup_A |\widehat{f}(\pi)|$. This implies

$$\|f\|_{\psi_2} \le 3C\|(\widehat{f}(\pi))\|_{2,1}.$$

Indeed, using the disjoint decomposition of Λ associated to $\{a_n^*\} = \cup_{k \geq 0}\{a_n^* \mid 2^k \leq n < 2^{k+1}\}$, we find

$$\|f\|_{\psi_2} \leq C \sum_{k \geq 0} 2^{k/2} a_{2^k}^* \leq 3C \sum_1^\infty a_n^*/n^{1/2} = 3C\|(\widehat{f}(\pi))\|_{2,1}.$$

Let $1 < p < 2$. Let $2 < p' < \infty$ be the conjugate, so that $1/p + 1/p' = 1$. Let

$$\|(\widehat{f}(\pi))\|_p = \left(\sum_\Lambda |\widehat{f}(\pi)|^p\right)^{1/p}.$$

We claim that there is a constant χ depending only on p and C such that for any f with $\widehat{f}$ supported in Λ

$$\|f\|_{\psi_{p'}} \leq \chi\|(\widehat{f}(\pi))\|_p. \tag{4.23}$$

This follows from a rather simple interpolation argument.

Indeed, we have $\|(\widehat{f}(\pi))\|_p = (\sum a_n^{*\,p})^{1/p}$. Fix a number $N \geq 1$. Let $f = f_0 + f_1$ be the decomposition of f associated to $\{a_n^*\} = \{a_n^* \mid 1 \leq n \leq N\} \cup \{a_n^* \mid n > N\}$, so that

$$\|f_0\|_\infty \leq \sum_1^N a_n^* \quad \text{and} \quad \|f_1\|_{\psi_2} \leq 3C \sum_{n > N} a_n^*/n^{1/2}.$$

By homogeneity we may assume $\|(\widehat{f}(\pi))\|_p = 1$. Then $a_n^* \leq n^{-1/p}$ for all $n \geq 1$. Therefore

$$\sum_1^N a_n \leq \sum_1^N n^{-1/p} \leq p' N^{1/p'}$$

and

$$\sum_{n>N} a_n^*/n^{1/2} \leq \sum_{n>N} n^{-1/p-1/2} \leq \tfrac{2p'}{p'-2} N^{1/p'-1/2}.$$

Let $c = p' N^{1/p'}$ so that $\|f_0\|_\infty \leq c$. We have

$$\mathbb{P}(\{|f| > 2c\}) \leq \mathbb{P}(\{|f_0| > c\}) + \mathbb{P}(\{|f_1| > c\}) = \mathbb{P}(\{|f_1| > c\}).$$

But we have

$$\mathbb{P}(\{|f_1| > c\}) \leq e \exp -c^2/\|f_1\|_{\psi_2}^2$$

and since $\|f_1\|_{\psi_2} \leq 3C\tfrac{2p'}{p'-2} N^{1/p'-1/2} = 3C\tfrac{2p'}{p'-2}(c/p')^{1-p'/2}$ we find after substituting

$$\mathbb{P}(\{|f| > 2c\}) \leq \mathbb{P}(\{|f_1| > c\}) \leq e \exp -(\chi' c^{p'}),$$

where χ' is a constant depending only on p and C. This has been established for c's of the form $c = p' N^{1/p'}$, but it is easy to obtain all values by interpolating between two such values. From this, our claim (4.23) is now immediate (recall that (iii) $\Rightarrow$ (ii) in Lemma 4.1 and Remark 4.2). From this claim, we obtain that Λ is Sidon by Lemma 4.12. The case when the dimensions d_π are uniformly bounded by a fixed number D follows by a straightforward modification of the same argument (but all the resulting bounds will depend on D). In any case, this shows that (viii) $\Rightarrow$ (i) in the latter case. $\square$

Remark 4.14. In [38] it is erroneously claimed that (viii) $\Rightarrow$ (i) in full generality in the non-Abelian case. However we recently noticed that the proof has a serious gap, and we now believe that the result does not hold. Indeed, if $\Lambda = \{\pi_n \mid n \in \mathbb{N}\}$ and if the dimensions of the representations in Λ form a sequence such that $d_n^2 \geq d_1^2 + \cdots + d_{n-1}^2$, then the mere knowledge that the individual singletons $\{\pi_n\}$ are Sidon with a fixed constant (Rider [48] called this "local Sidon property") is sufficient to guarantee that (vii) holds, but it seems unlikely that this is enough to force Λ to be Sidon.

Although we state it in full generality, the next result is significant only if the dimensions of the irreps π_n are unbounded.

Theorem 4.15 (Characterizing sub-Gaussian characters). *Let G_n be a sequence of compact groups, let $\pi_n \in \widehat{G_n}$ be nontrivial irreps and let $\chi_n = \chi_{\pi_n}$ as well as $d_n = d_{\pi_n}$. The following are equivalent.*

(i) *There is a constant C such that the singletons $\{\pi_n\} \subset \widehat{G_n}$ are Sidon with constant C, i.e., we have*

$$\forall n \forall a \in M_{d_n} \quad \operatorname{tr}|a| \leq C \sup_{g \in G} |\operatorname{tr}(a\pi_n(g))|.$$

(ii) *There is a constant C such that*

$$\forall n \quad \|\chi_n\|_{\psi_2} \leq C.$$

(ii)′ *There is $\beta > 0$ such that*

$$\forall n \quad \int \exp\left(\beta|\chi_n|^2\right) dm_{G_n} \leq e.$$

(ii)″ *There is a constant C such that for any $t \in \mathbb{R}$*

$$\forall n \quad \int \exp\left(t\chi_n - Ct^2\right) dm_{G_n} \leq 1.$$

(iii) *For each $0 < \delta < 1$ there is $0 < \theta < 1$ such that*

$$\forall n \quad m_{G_n}\{\operatorname{Re}(\chi_n) > \delta d_n\} \leq e\theta^{d_n^2}.$$

(iii)′ *For each $0 < \delta < 1$ there is $0 < \theta < 1$ and $D > 0$ such that for any n with $d_n > D$ we have*

$$m_{G_n}\{\operatorname{Re}(\chi_n) > \delta d_n\} \leq \theta^{d_n^2}.$$

(iii)″ *There are $0 < \delta < 1$ and $0 < \theta < 1$ such that*

$$\forall n \quad m_{G_n}\{\operatorname{Re}(\chi_n) > \delta d_n\} \leq e\theta^{d_n^2}.$$

(iv) *There is a constant C such that*

$$\forall n \quad d_n \leq C \int_{U(d_n)} \sup_{g \in G_n} |\operatorname{tr}(u\pi_n(g))| m_{U(d_n)}(du).$$

458   G. Pisier

Proof. Note that the properties (ii) (ii)' and (ii)'' are simply reformulations of each other by Lemmas 4.1 and 4.3. Note that the content of (iii) and (iii)'' is void when $e\theta^{d_n^2} \geq 1$. Thus (iii) $\Rightarrow$ (iii)' $\Rightarrow$ (iii)'' are trivial. The implication (i) $\Rightarrow$ (ii) $\Rightarrow$ (iii) is a special case of (i) $\Rightarrow$ (ii) $\Rightarrow$ (iv) in Theorem 4.13 and (iii)'' $\Rightarrow$ (iv) is a special case of (v) $\Rightarrow$ (vi) in Theorem 4.13. Moreover, we may invoke the implication (vi) $\Rightarrow$ (vii) in Theorem 4.13 for our special case of singletons. Then the corollary boils down to the observation that if Λ is a singleton the implication (vii) $\Rightarrow$ (i) in Theorem 4.13 trivially holds (take $A = \Lambda$, then necessarily $B = \Lambda$).   □

Although I never had concrete examples, I believed naively for many years that Theorem 4.15 could be applied to finite groups. To my surprise, Emmanuel Breuillard showed me that it is not so (and he pointed out Turing's paper [59] that already emphasized that general phenomenon, back in 1938). It turns out that, when the groups G_n are finite (or amenable as discrete groups), Theorem 4.15 can hold only if the dimensions d_n remain bounded. The reason lies in the presence of large Abelian subgroups with index of order $\exp o(d_n^2)$. The latter follows from the quantitative refinements in [61, 10] of a classical theorem of Camille Jordan on finite linear groups. See the forthcoming paper [5] for details.

5. Some questions about best constants

We denote by A_p, B_p the best possible constants in the classical Khintchine inequalities. These inequalities say that for any scalar sequence $x \in \ell_2$ we have

$$A_p \left(\sum |x_j|^2\right)^{1/2} \leq \left(\int \left|\sum \varepsilon_j x_j\right|^p d\mathbb{P}\right)^{1/p} \leq B_p \left(\sum |x_j|^2\right)^{1/2}.$$

After much effort by many authors, the exact values of A_p, B_p were obtained by Szarek and Haagerup (see [21, 56]). Let $p_0 = 1.87\ldots$ be the only solution in the interval $]1, 2[$ of the equation $2^{1/2-1/p} = \gamma_p$ (or explicitly $\Gamma((p+1)/2) = \sqrt{\pi}/2$), then Haagerup (see [21]) proved:

$$A_p = 2^{1/2-1/p} \qquad 0 < p \leq p_0, \tag{5.1}$$

$$A_p = \gamma_p \qquad p_0 \leq p \leq 2, \tag{5.2}$$

$$B_p = \gamma_p \qquad 2 \leq p < \infty. \tag{5.3}$$

The bounds $A_p \leq \gamma_p$ for $p \leq 2$ and $B_p \geq \gamma_p$ for $p \geq 2$ are easy consequences of the Central Limit Theorem, applied to $\lim_{n\to\infty}(\varepsilon_1 + \cdots + \varepsilon_n)/\sqrt{n}$. The bound $A_p \leq 2^{1/2-1/p}$ is immediate by considering the function $(\varepsilon_1 + \varepsilon_2)/\sqrt{2}$.

For the complex analogue of these inequalities, the best constants are also known: if we replace the sequence (ε_n) (independent choices of signs) by an i.i.d. sequence (z_n) uniformly distributed over $\{z \in \mathbb{C} \mid |z| = 1\}$, then the same inequalities hold but now the best constants, that we denote $A_p[\mathbb{T}], B_p[\mathbb{T}]$, are $A_p[\mathbb{T}] = \gamma_p^{\mathbb{C}}$ if $1 \leq p \leq 2$ and $B_p[\mathbb{T}] = \gamma_p^{\mathbb{C}}$ if $p \geq 2$, where $\gamma_p^{\mathbb{C}}$ is the L_p-norm of a standard complex-valued Gaussian variable normalized in L_2. Indeed, in analogy with

Haagerup's result, Sawa [53, 54] proved that there is a phase transition at a number $p_0^{\mathbb{C}}$, but now $0 < p_0^{\mathbb{C}} = 0.475 \cdots < 1$!

Let G be a matrix group, such as $U(d)$, $SU(d)$, $O(d)$, $SO(d)$. Let $\pi_n : G^{\mathbb{N}} \to G$ denote the nth coordinate. Let $E[G]$ be the linear span of the matrix coefficients of $\Lambda = \{\pi_n \mid n \in \mathbb{N}\}$. Thus a typical element of $E[G]$ can be written as a finite sum $f = \sum \mathrm{tr}(\pi_n x_n)$, where (x_n) is a finitely supported family in M_d. Then $\|f\|_2 = (d^{-1} \sum \mathrm{tr} |x_n|^2)^{1/2}$.

We denote by $A_p[G], B_p[G]$ the best (positive) constants A, B in the following inequality

$$\forall f \in E[G] \qquad A\|f\|_2 \le \|f\|_p \le B\|f\|_2. \tag{5.4}$$

Let $\Gamma^u = \prod_{d \ge 1} U(d)$, $\Gamma^o = \prod_{d \ge 1} O(d)$. We set $\mathcal{G}^u = (\Gamma^u)^{\mathbb{N}}$ and $\mathcal{G}^o = (\Gamma^o)^{\mathbb{N}}$. We define similarly $\mathcal{G}^{su}$ and $\mathcal{G}^{so}$.

Problem: *Let $1 \le p \ne 2 < \infty$. What are the values of $A_p[G], B_p[G]$ for $G = U(d)$ for $d > 1$?*

Same question for $SU(d), O(d), SO(d)$.

It is natural to consider also the best constants $A_p^c[G], B_p^c[G]$ for which (5.4) holds for all central functions f, i.e., all f of the form $f = \sum \mathrm{tr}(\pi_n) x_n$ where (x_n) is a finitely supported family in $\mathbb{C}$.

Another natural question is to find the best $A_p[G], B_p[G]$ for $G = \mathcal{G}^u$ and similarly when G is either $\mathcal{G}^o$, $\mathcal{G}^{su}$ or $\mathcal{G}^{so}$.

The constants $A_p[\mathcal{G}^u], B_p[\mathcal{G}^u]$ can equivalently be viewed as the best constants in (5.4) when f is any finite sum of the form

$$f(\omega) = \sum \mathrm{tr}(\rho_n(\omega) x_n) \quad (x_n \in M_{d_n})$$

where $\Omega = \prod U(d_n)$ is equipped with its uniform (Haar) probability, $\rho_n : \Omega \to U(d_n)$ is the nth coordinate and (d_n) is an arbitrary sequence of integers (and similarly for o, su, so). Then $\|f\|_2 = (\sum d_n^{-1} \mathrm{tr} |x_n|^2)^{1/2}$.

Consider a (real or complex) Banach space B. Recall that a B-valued random variable X is called Gaussian if for any real linear form $\xi : B \to \mathbb{R}$, the real-valued variable $\xi(X)$ is Gaussian. By definition, the covariance of a B-valued random variable X is the bilinear form $(\xi, \xi') \mapsto \mathbb{E}(\xi(X)\xi'(X))$. Let $g^{(d)}$ be a Gaussian random matrix with the same covariance as $x \mapsto \pi_n(x)$ (the latter does not depend on n), so that, by the central limit theorem (CLT in short), $n^{-1/2}(\pi_1 + \cdots + \pi_n)$ tends in distribution to $g^{(d)}$. In particular, when $G = SO(d)$ or $O(d)$ (resp. $G = SU(d)$ or $U(d)$) $n^{-1/2}(\mathrm{tr}(\pi_1) + \cdots + \mathrm{tr}(\pi_n))$ tends in distribution to a standard real (resp. complex) Gaussian random variable normalized in L_2. It follows that $B_p \ge B_p^c \ge \gamma_p^{\mathbb{R}}$ (resp. $B_p \ge B_p^c \ge \gamma_p^{\mathbb{C}}$) for all $p \ge 2$ and $A_p \le A_p^c \le \gamma_p^{\mathbb{R}}$ (resp. $A_p \le A_p^c \le \gamma_p^{\mathbb{C}}$) for all $p \le 2$.

In [23, §36, p. 390] it was proved that

$$\forall p \in 2\mathbb{N} \quad B_p[\mathcal{G}^u] \le 2((p/2)!)^{1/p}$$

with an improved bound for $p = 4$ namely $B_4[\mathcal{G}^u] \leq 2$. A fortiori, $B_4[U(d)] \leq 2$ for all $d \geq 1$. Since $\gamma_4^{\mathbb{C}} = 2$ this implies

$$B_4[U(d)] = B_4^c[U(d)] = B_4[\mathcal{G}^u] = B_4^c[\mathcal{G}^u] = 2.$$

Hewitt and Ross quote [16] but they also credit Rider and quote another paper of his entitled "Continuity of random Fourier series" that apparently never appeared. Moreover, by a result due to Helgason [22]

$$A_1[\mathcal{G}^u] \geq 1/\sqrt{2}.$$

Let $G = U(d)$ (resp. $G = O(d)$). Let $(g_n^{(d)})$ be an i.i.d. sequence of copies of $g^{(d)}$. Following [34], we describe in Lemma 4.4 a very general comparison principle showing that for some absolute constant C_0 the family of coefficients $\{\pi_n(i,j)\}$ is the image of $\{g_n^{(d)}(i,j)\}$ under a positive operator of norm at most C_0 on L_p. In the proof of Lemma 4.4 we show this with

$$C_0 \leq \chi = \sup_d (d^{-1}\mathbb{E}\operatorname{tr}|g^{(d)}|)^{-1} < \infty,$$

but we do not know the best value of C_0. In any case, this reasoning implies

$$\forall p \geq 2 \quad B_p[U(d)] \leq (d^{-1}\mathbb{E}\operatorname{tr}|g^{(d)}|)^{-1}\gamma_p^{\mathbb{C}} \text{ and } B_p[\mathcal{G}^u] \leq \chi\gamma_p^{\mathbb{C}},$$

and similarly for $O(d)$ with the analogue of $g^{(d)}$ that has real-valued Gaussian entries.

Remark 5.1. Let G be a compact group, let $\Lambda \subset \widehat{G}$, and let E_Λ be the linear span of the matrix coefficients of the representations in Λ. Let A_p^Λ and B_p^Λ be the best constants for which (5.4) holds for any $f \in E_\Lambda$. Then, if $p > 2$, B_p^Λ can be interpreted as the constant of Λ as a $\Lambda(p)$-set in Rudin's sense [51]. See [1] for a rather recent survey on $\Lambda(p)$-sets. A similar interpretation is valid for A_p^Λ and $\Lambda(p)$-sets when $1 < p < 2$, but "true" examples of such sets are lacking for $1 < p < 2$.

Remark 5.2. One can also ask what are the best constants in (5.4) with respect to the usual non-commutative L_p-spaces when f is in the linear span of free Haar unitaries in the sense of [60]. Now semicircular (or circular) variables replace the Gaussian ones, when invoking the CLT, so that $B_p \geq \|x\|_p$ where x is a semicircular (or circular) variable in the sense of [60] normalized in L_2 (note that $\|x\|_\infty = 2$). For these free Haar unitaries, Bożejko's inequality in [4] implies that for any even integer $p = 2n$ we have $B_{2n} = (\frac{1}{n+1}\binom{2n}{n})^{1/2n}$. The latter number is again the L_p-norm of a "free Gaussian", i.e., a semicircular variable normalized in L_2. In particular (for this see also Haagerup's [20]) we have $B_p \leq 2$ for all $p \geq 2$. Related results appear in [46, Lemma 7]. See [6] for interesting results on this theme.

6. A new approach to Rider's spectral gap

We now show how the new method presented in [44] yields another proof of Rider's spectral gap estimate. We do not obtain the nice precise description of the measure $\mu_{k,n}$ that possesses the desired spectral gap property, as in Theorem 2.1, but we do get a more refined quantitative bound.

We need to recall the definitions of the projective and injective tensor product norms $\| \ \|_\wedge$ and $\| \ \|_\vee$ on the algebraic tensor product $L_1(m_1) \otimes L_1(m_2)$ of two arbitrary L_1-spaces. Let $T = \sum x_j \otimes y_j \in L_1(m_1) \otimes L_1(m_2)$. Then

$$\|T\|_\wedge = \int \left| \sum x_j(t_1)y_j(t_2) \right| dm_1(t_1)\, dm_2(t_2)$$

$$\|T\|_\vee = \sup \left\{ \left| \sum \langle x_j, \psi_1 \rangle \langle y_j, \psi_2 \rangle \right| \ \Big| \ \|\psi_1\|_\infty \leq 1, \|\psi_2\|_\infty \leq 1 \right\}.$$

Let $(d_k)_{k \in I}$ be an arbitrary collection of integers. Let $G = \prod_{k \in I} U(d_k)$. Let $u \mapsto u_k \in U(d_k)$ denote the coordinates on G. We know that the family $\{d_k^{1/2} u_k(i,j)\}$ is sub-Gaussian (see Lemma 4.4 or (i) $\Rightarrow$ (ii) in Theorem 4.10). Let

$$S = \sum_{k,i,j} (d_k^{1/2} u_k(i,j)) \otimes (d_k^{1/2} u_k(j,i)).$$

Actually, by Lemma 4.4, in the terminology of [44], the family $\{d_k^{1/2} u_k(i,j)\}$ is C_0-dominated by $\{d_k^{1/2} g_k(i,j)\}$. Therefore, by [44, Theorem 1.10], for any $0 < \varepsilon < 1$ there is a decomposition

$$S = t + r$$

for some $t, r \in L_1(G) \otimes L_1(G)$ satisfying

$$\|t\|_\wedge \leq w(\varepsilon) \quad \text{and} \quad \|r\|_\vee \leq \varepsilon,$$

where $w(\varepsilon)$ depends only on ε and $w(\varepsilon) = O(\log(1/\varepsilon))$ when $\varepsilon \to 0$.

Consider the mapping $P : L_1(G) \otimes L_1(G) \to L_1(G)$ defined by $P(x \otimes y) = x * y$.

A simple verification shows that since $u_k = u_k * u_k$ or equivalently $u_k(i,j) = \sum_\ell u_k(i,\ell) * u_k(\ell,j)$

$$P(S) = \sum_k d_k \sum_{i,j} u_k(i,j) * u_k(j,i) = \sum_k d_k \sum_i u_k(i,i) = \sum_k d_k \operatorname{tr}(u_k).$$

Moreover, for any $t, r \in L_1(G) \otimes L_1(G)$ we have

$$\|P(t)\|_1 \leq \|t\|_\wedge$$

and

$$\|P(r)\|_* \leq \|r\|_\vee$$

where

$$\forall f \in L_1(G) \quad \|f\|_* = \sup_{\pi \in \widehat{G}} \|\widehat{f}(\pi)\|.$$

Indeed, note that $\|P(f)\|_* = \|T_{P(f)}\|_{B(L_2(G))}$ where $T_{P(f)}$ is the convolutor $x \mapsto x * P(f)$. Then by a well-known consequence of Grothendieck's theorem (obtained using translation invariance), there is a constant K such that

$$(K)^{-1}\|T_{P(f)}\|_{B(L_2(G))} \le \|T_{P(f)}\|_{B(L_\infty(G),L_1(G))} = \|f\|_\vee.$$

Here K is the complex Grothendieck constant. Actually (see [43]) K is not really needed here in view of the bound $\|r\|_{\gamma_2^*} \le \varepsilon$ directly obtained in [44]). Indeed, if $f = \sum x_k \otimes y_k$ we have for any φ, ψ in $L_\infty(G)$ | $\sum \langle \varphi, x_k \rangle \langle \psi, y_k \rangle| \le \|f\|_\vee \|\varphi\|_\infty \|\psi\|_\infty$ and hence by a suitable averaging (replacing φ, ψ by suitable translates)

$$\sup_{s,t \in G} \left| \sum (\varphi * x_k)(s)\, (y_k * \psi)(t) \right| \le K\|f\|_\vee \|\varphi\|_2 \|\psi\|_2$$

and hence

$$\sup_{t \in G} \left| \sum \varphi * \left(\sum x_k * y_k \right) * \psi(t) \right| \le K\|f\|_\vee \|\varphi\|_2 \|\psi\|_2$$

from which $\|P(f)\|_* = \|\sum x_k * y_k\|_* \le K\|f\|_\vee$ follows immediately.

Thus we obtain

Theorem 6.1. *If the index set I is finite there is a decomposition $\sum_{k \in I} d_k \operatorname{tr}(u_k) = T + R$ with $T, R \in L_1(G)$ such that $\|T\|_1 \le w(\varepsilon)$ and $\|R\|_* \le K\varepsilon$. If I is infinite there is a similar decomposition within formal Fourier series with $T \in M(G)$ such that $\|T\|_{M(G)} \le w(\varepsilon)$ and $\|R\|_* \le K\varepsilon$.*

Let $\pi_k(u) = u_k$ for any $u \in G$. Note $\widehat{T}(\pi_k) = I - \widehat{R}(\pi_k)$. Thus, if (say) $K\varepsilon < 1/2$ then $\|\widehat{T}(\pi_k) - I\| \le 1/2$ and hence $\|(\widehat{T}(\pi_k))^{-1}\| \le 2$. Since the set $\Lambda = \{\pi_k\}$ is Sidon with constant $= 1$, the argument in Remark 3.9 shows:

Corollary 6.2. *For any $\varepsilon < (2K)^{-1}$ there is a measure $\mu \in M(G)$ with $\|\mu\|_{M(G)} \le 2w(\varepsilon)$ such that $\widehat{\mu}(\pi_k) = I$ for all k and $\sup_{\pi \notin \{\pi_k\}} \|\widehat{\mu}(\pi)\| \le 2\varepsilon$.*

The preceding proof yields the estimate $w(\varepsilon) = O(\log(1/\varepsilon))$ that does not seem accessible by Rider's original approach. This logarithmic bound follows from [35, Lemma 3]. See [44, Remark 1.16] for a detailed deduction.

References

[1] J. Bourgain, *Sidon sets and Riesz products*. Ann. Inst. Fourier (Grenoble) **35** (1985), 137–148.

[2] J. Bourgain, *Λ_p-sets in analysis: results, problems and related aspects*. Handbook of the geometry of Banach spaces, Vol. I, 195–232, North-Holland, Amsterdam, 2001.

[3] J. Bourgain and M. Lewko, *Sidonicity and variants of Kaczmarz's problem*. Annales Inst. Fourier **67** (2017), 1321–1352.

[4] M. Bożejko, *On $\Lambda(p)$ sets with minimal constant in discrete noncommutative groups*. Proc. of the Amer. Math. Soc. **51** (1975), 407–412.

[5] E. Breuillard and G. Pisier, *Random unitaries and amenable linear groups*, in preparation.

[6] A. Buchholz, *Optimal constants in Khintchine type inequalities for Fermions, Rademachers and q-Gaussian operators*, Bull. Polish Acad. Sci. Math. **53** (2005), 315–321.

[7] B. Carl, *Entropy numbers of diagonal operators with an application to eigenvalue problems*. J. Approx. Theory **32** (1981) 135–150.

[8] C. Cecchini, *Lacunary Fourier series on compact Lie groups*, J. Funct. Anal. 11 (1972) 191–203.

[9] S. Chevet, *Séries de variables aléatoires gaussiennes à valeurs dans $E \otimes_\varepsilon F$, applications aux espaces de Wiener abstraits*. Séminaire sur la géométrie des espaces de Banach 1977–1978, École Polytechnique, Exp. XIX, 1978. (available on www.numdam.org).

[10] M. Collins, *On Jordan's theorem for complex linear groups*. J. Group Theory **10** (2007), 411–423.

[11] J. Faraut, *Analyse sur les groupes de Lie*. Calvage & Mounet, 2006.

[12] A. Figà-Talamanca, *Random Fourier series on compact groups. Theory of Group Representations and Fourier Analysis* (C.I.M.E., II Ciclo, Montecatini Terme, 1970) pp. 1–63 Edizioni Cremonese, Rome, 1971.

[13] A. Figà-Talamanca and C. Nebbia, *Harmonic analysis and representation theory for groups acting on homogeneous trees*. Cambridge University Press, Cambridge, 1991.

[14] A. Figà-Talamanca and M. Picardello, *Harmonic analysis on free groups*. Marcel Dekker, New York, 1983.

[15] A. Figà-Talamanca and D. Rider, *A theorem of Littlewood and lacunary series for compact groups*. Pacific J. Math. **16** (1966) 505–514.

[16] A. Figà-Talamanca and D. Rider, *A theorem on random Fourier series on noncommutative groups*. Pacific J. Math. **21** (1967) 487–492.

[17] W. Fulton, *Young tableaux*. Cambridge University Press, 1997.

[18] C. Graham and K. Hare, *Interpolation and Sidon sets for compact groups*. Springer, New York, 2013. xviii+249 pp.

[19] C. Graham and O.C. McGehee, *Essays in commutative harmonic analysis*. Springer-Verlag, New York-Berlin, 1979.

[20] U. Haagerup, *An example of a non-nuclear C^*-algebra which has the metric approximation property*. Inventiones Mat. **50** (1979), 279–293.

[21] U. Haagerup, *The best constants in the Khintchine inequality*. Studia Math. **70** (1981), 231–283 (1982).

[22] S. Helgason, *Topologies of Group Algebras and a Theorem of Littlewood*. Trans. Amer. Math. Soc. **86** (1957), 269–283.

[23] E. Hewitt and K. Ross, *Abstract harmonic analysis, Volume II, Structure and Analysis for Compact Groups, Analysis on Locally Compact Abelian Groups*. Springer, Heidelberg, 1970.

[24] M. Hutchinson, *Local Λ sets for profinite groups*, Pacific J. Math. 80 (1980) 81–88.

[25] J.-P. Kahane, *Séries de Fourier absolument convergentes*. Springer, 1970.

[26] J.-P. Kahane, *Some random series of functions*. Second edition. Cambridge University Press, 1985.

[27] M. Ledoux, *The concentration of measure phenomenon.* Mathematical Surveys and Monographs, 89. American Mathematical Society, Providence, RI, 2001.

[28] M. Ledoux and M. Talagrand, *Probability in Banach Spaces. Isoperimetry and Processes.* Springer-Verlag, Berlin, 1991.

[29] F. Lehner, *A characterization of the Leinert property.* Proc. Amer. Math. Soc. **125** (1997), 3423–3431.

[30] D. Li and H. Queffélec, *Introduction l'étude des espaces de Banach.* Société Mathématique de France, Paris, 2004.

[31] J. Lindenstrauss and H.P. Rosenthal, *The $\mathcal{L}_p$ spaces.* Israel J. Math. **7** (1969), 325–349.

[32] J. López and K.A. Ross, *Sidon sets.* Lecture Notes in Pure and Applied Mathematics, Vol. 13. Marcel Dekker, Inc., New York, 1975.

[33] M.B. Marcus, *The ε-entropy of some compact subsets of ℓ_p.* J. Approx. Theory **10** (1974) 304–312.

[34] M.B. Marcus and G. Pisier, *Random Fourier series with Applications to Harmonic Analysis.* Annals of Math. Studies, no 101, Princeton Univ. Press, 1981.

[35] J.-F. Méla, *Mesures ε-idempotentes de norme bornée.* Studia Math. **72** (1982), 131–149.

[36] W.A. Parker, *Central Sidon and central Λ_p sets.* J. Austral. Math. Soc. **14**, 62–74 (1972).

[37] G. Pisier, *Ensembles de Sidon et processus gaussiens.* C. R. Acad. Sc. Paris, t. A **286** (1978) 671–674.

[38] G. Pisier, *De nouvelles caractérisations des ensembles de Sidon.* Advances in Maths. Supplementary studies, vol 7B (1981) 685–726.

[39] G. Pisier, *Arithmetic characterizations of Sidon sets.* Bull. A.M.S. **8** (1983), 87–90.

[40] G. Pisier, *Probabilistic methods in the geometry of Banach spaces.* Probability and analysis (Varenna, 1985), 167–241, Lecture Notes in Math. 1206, Springer-Verlag, Berlin, 1986.

[41] G. Pisier, *The volume of Convex Bodies and Banach Space Geometry.* Cambridge University Press, 1989.

[42] G. Pisier, *Multipliers and lacunary sets in non amenable groups.* Amer. J. Math. **117** (1995) 337–376.

[43] G. Pisier, *Grothendieck's Theorem, past and present.* Bull. Amer. Math. Soc. **49** (2012), 237–323.

[44] G. Pisier, *On uniformly bounded orthonormal Sidon systems.* Math. Res. Letters **24** (2017), 893–932.

[45] A. Prasad, *Representation theory. A combinatorial viewpoint.* Cambridge University Press, Delhi, 2015.

[46] E. Ricard and Q. Xu, *A noncommutative martingale convexity inequality.* Annals of Probability **44** (2016), 867–882.

[47] D. Rider, *Randomly continuous functions and Sidon sets.* Duke Math. J. **42** (1975) 752–764.

[48] D. Rider, *$SU(n)$ has no infinite local Λ_p sets.* Boll. Un. Mat. Ital. (4) **12** (1975), 155–160.

[49] D. Rider, *Norms of characters and central Λ_p sets for $U(n)$*. Conference on Harmonic Analysis (Univ. Maryland, College Park, Md., 1971), pp. 287–294. Lecture Notes in Math., Vol. 266, Springer, Berlin, 1972.

[50] D. Rider, *Central lacunary sets*. Monatsh. Math. **76** (1972), 328–338.

[51] W. Rudin, *Trigonometric series with gaps*. J. Math. and Mech. **9** (1960) 203–227.

[52] B. Sagan, *The symmetric group*. Springer, second edition, New York, 2001.

[53] J. Sawa, *The best constant in the Khintchine inequality for complex Steinhaus variables, the case $p = 1$*. Studia Math. **81** (1985) 105–126.

[54] J. Sawa, *Some remarks on the Khintchine inequality for complex Steinhaus variables*.

[55] R. Stanley, *Enumerative combinatorics, vol. 2*. Cambridge Univ. Press

[56] S. Szarek, *On the best constants in the Khinchine inequality*. Studia Math. **58** (1976), 197–208.

[57] M. Talagrand, *Regularity of Gaussian processes*. Acta Math., **159** (1987), 99–149.

[58] M. Talagrand, *Upper and Lower Bounds for Stochastic Processes*. Springer, Berlin, 2014.

[59] A. Turing, *Finite approximations to Lie groups*. Annals of Math. **39** (1938), 105–111.

[60] D. Voiculescu, K. Dykema and A. Nica, *Free random variables*. Amer. Math. Soc., Providence, RI, 1992.

[61] B. Weisfeiler, *Post-classification version of Jordan's theorem on finite linear groups*. Proc. Natl. Acad. Sci. USA **81** (1984), 5278–5279.

[62] H. Weyl, *The classical groups*. Princeton Univ. Press, 1939. Reprinted by Dover.

[63] D.C. Wilson, *On the structure of Sidon sets*. Monatsh. Math. **101** (1986), 67–74.

Gilles Pisier
Mathematics Department
Texas A&M University
College Station, TX 77843-3368, USA
e-mail: pisier@math.tamu.edu

Operator Theory:
Advances and Applications, Vol. 261, 467–484
© Springer International Publishing AG, part of Springer Nature 2018

Sublinear Equations and Schur's Test for Integral Operators

Igor E. Verbitsky

Dedicated to the memory of V.P. Havin

Abstract. We study weighted norm inequalities of (p, r)-type,

$$\|\mathbf{G}(f\,d\sigma)\|_{L^r(\Omega, d\sigma)} \leq C\|f\|_{L^p(\Omega, \sigma)}, \quad \text{for all } f \in L^p(\sigma),$$

for $0 < r < p$ and $p > 1$, where $\mathbf{G}(fd\sigma)(x) = \int_\Omega G(x, y)f(y)d\sigma(y)$ is an integral operator associated with a nonnegative kernel $G(x, y)$ on $\Omega \times \Omega$, and σ is a locally finite positive measure in Ω.

We show that this embedding holds if and only if

$$\int_\Omega (\mathbf{G}\sigma)^{\frac{pr}{p-r}}\,d\sigma < +\infty,$$

provided G is a quasi-symmetric kernel which satisfies the weak maximum principle.

In the case $p = \frac{r}{q}$, where $0 < q < 1$, we prove that this condition characterizes the existence of a non-trivial solution (or supersolution) $u \in L^r(\Omega, \sigma)$, for $r > q$, to the sublinear integral equation

$$u - \mathbf{G}(u^q\,d\sigma) = 0, \quad u \geq 0.$$

We also give some counterexamples in the end-point case $p = 1$, which corresponds to solutions $u \in L^q(\Omega, \sigma)$ of this integral equation studied recently in [19, 20]. These problems appear in the investigation of weak solutions to the sublinear equation involving the (fractional) Laplacian,

$$(-\Delta)^\alpha u - \sigma\,u^q = 0, \quad u \geq 0,$$

for $0 < q < 1$ and $0 < \alpha < \frac{n}{2}$ in domains $\Omega \subseteq \mathbb{R}^n$ with a positive Green function.

Mathematics Subject Classification (2010). Primary 35J61. Secondary 31B15, 42B37, 42B25.

Keywords. Weighted norm inequalities, sublinear elliptic equations, Green's function, weak maximum principle, fractional Laplacian.

1. Introduction

Let Ω be a locally compact, Hausdorff space. For a positive, lower semicontinuous kernel $G\colon \Omega \times \Omega \to (0, +\infty]$, we denote by

$$\mathbf{G}(f\,d\sigma)(x) = \int_\Omega G(x,y)\,f(y)\,d\sigma(y), \quad x \in \Omega,$$

the corresponding integral operator, where $\sigma \in \mathcal{M}^+(\Omega)$, the class of positive locally finite Radon measures in Ω.

We study the (p,r)-weighted norm inequalities

$$\|\mathbf{G}(f d\sigma)\|_{L^r(\Omega,\sigma)} \le C \, \|f\|_{L^p(\Omega,\sigma)}, \tag{1.1}$$

in the case $0 < r < p$ and $p \ge 1$, where C is a positive constant which does not depend on $f \in L^p(\Omega,\sigma)$.

The main goal of this paper is to find explicit characterizations of (1.1) in terms of $\mathbf{G}\sigma$ under certain assumptions on G. We also study the relationship of inequality (1.1) with $p = \frac{r}{q}$, where $0 < q < 1$, to the existence of a positive function $u \in L^r(\Omega,\sigma)$ such that

$$u \ge \mathbf{G}(u^q\sigma) \quad d\sigma\text{-a.e. in } \Omega, \tag{1.2}$$

in the case $r > q$. In other words, u is a supersolution for the sublinear integral equation

$$u - \mathbf{G}(u^q\sigma) = 0, \quad 0 < u < +\infty \quad d\sigma\text{-a.e. in } \Omega, \tag{1.3}$$

where $0 < q < 1$.

In this paper, we assume that the kernel G of the integral operator is quasi-symmetric, and satisfies a weak maximum principle (see Sec. 2). Such restrictions are satisfied by the Green kernel associated with many elliptic operators, including the fractional Laplacian $(-\Delta)^\alpha$, as well as quasi-metric kernels, and radially symmetric, decreasing convolution kernels $G(x,y) = k(|x-y|)$ on $\mathbb{R}^n$ (see, e.g., [1, 2, 18, 19, 20] and the literature cited there).

If G is Green's kernel associated with the Laplacian in an open domain $\Omega \subseteq \mathbb{R}^n$, (1.3) is equivalent to the sublinear elliptic boundary value problem

$$\begin{cases} -\Delta u - \sigma u^q = 0, & u > 0 \text{ in } \Omega, \\ u = 0 & \text{on } \partial\Omega, \end{cases} \tag{1.4}$$

where $0 < q < 1$.

We observe that solutions $u \in L^r(\Omega,\sigma)$ to (1.4) in the case $r = 1 + q$ correspond to finite energy solutions $u \in L_0^{1,2}(\Omega)$ in the Dirichlet space, i.e.,

$$\int_\Omega |\nabla u|^2 dx < +\infty,$$

where u has zero boundary values (see [5, 21]).

The more difficult end-point case $p = 1$ of (1.1), along with solutions $u \in L^q(\Omega,\sigma)$ in the case $r = q$, was studied recently in [19, 20]. After a certain modification, it leads to solutions $u \in L_{\mathrm{loc}}^q(\Omega,\sigma)$, i.e., all solutions to (1.3), or (1.4) understood in a weak sense (see [16]). For Riesz kernels on $\Omega = \mathbb{R}^n$ such $(1, q)$-weighted norm inequalities, along with weak solutions to the sublinear problem

$$\begin{cases} (-\Delta)^\alpha u - \sigma u^q = 0, & u > 0 \text{ in } \mathbb{R}^n, \\ \liminf_{x \to \infty} u = 0, & u \in L^q_{loc}(\sigma), \end{cases} \tag{1.5}$$

for $0 < \alpha < \frac{n}{2}$, were treated earlier in [5, 6, 7].

Our main result is the following theorem.

Theorem 1.1. *Let $\sigma \in \mathcal{M}^+(\Omega)$. Suppose G is a positive quasi-symmetric, lower semicontinuous kernel which satisfies the weak maximum principle.*

(i) *If $1 < p < +\infty$ and $0 < r < p$, then the (p, r)-weighted norm inequality (1.1) holds if and only if*

$$\int_\Omega (\mathbf{G}\sigma)^{\frac{pr}{p-r}} d\sigma < +\infty. \tag{1.6}$$

(ii) *If $0 < q < 1$ and $q < r < \infty$, then there exists a positive (super)solution $u \in L^r(\Omega, d\sigma)$ to (1.3) if and only if (1.1) holds with $p = \frac{r}{q}$, or equivalently,*

$$\int_\Omega (\mathbf{G}\sigma)^{\frac{r}{1-q}} d\sigma < +\infty. \tag{1.7}$$

Remark 1.2. We observe that the "if" parts of statements (i) and (ii) of Theorem 1.1 fail if $p = 1$, and $r = q$, respectively. The "only if" parts hold for all $0 < r < p$ in statement (i), and $r > 0$ in statement (ii).

Remark 1.3. It is known that inequality (1.1) with $p = \frac{r}{q} \geq 1$ in the case $0 < q < 1$ yields the existence of a positive supersolution $u \in L^r(\Omega, \sigma)$ for (1.2). This statement follows from a lemma due to Gagliardo [12], and does not require G to be quasi-symmetric or to satisfy the WMP (see Sec. 3 below). However, the converse statement does not hold without the WMP (see [20] in the case $r = q$).

Remark 1.4. Without the assumption that G satisfies the WMP, the "only if" parts of statement (i) (with $p = \frac{r}{q} \geq 1$) and statement (ii) (with $r \geq q$) hold only for $0 < r \leq 1 - q^2$ (see Lemma 3.1 below).

In particular, if there exists a positive (super)solution $u \in L^q(\Omega, \sigma)$, then (1.7) holds with $r = q$ for $0 < q \leq q_0$, where $q_0 = \frac{\sqrt{5}-1}{2} = 0.61\ldots$ is the conjugate golden ratio. However, (1.7) with $r = q$ generally fails (even for symmetric kernels) in the case $q_0 < q < 1$; the cut-off $q = q_0$ here is sharp [20].

In Section 4 below, we discuss related results, and provide some counterexamples in the case $p = 1$.

2. Kernels and potential theory

Let $G \colon \Omega \times \Omega \to (0, +\infty]$ be a positive kernel. We will assume that Ω is a locally compact Hausdorff space, and G is lower semicontinuous, so that we can apply elements of the classical potential theory developed for such kernels (see [3, 11]).

Most of our results hold for *non-negative* kernels $G(x,y) \geq 0$. In that case, some statements concerning the existence of positive solutions (rather than supersolutions) require the additional assumption that G is non-degenerate; see [20].

By $\mathcal{M}^+(\Omega)$ we denote the class of all nonnegative, locally finite, Borel measures on Ω. We use the notation $\mathrm{supp}(\nu)$ for the support of $\nu \in \mathcal{M}^+(\Omega)$ and $\|\nu\| = \nu(\Omega)$ if ν is a finite measure.

For $\nu \in \mathcal{M}^+(\Omega)$, the potential of ν is defined by

$$\mathbf{G}\nu(x) := \int_\Omega G(x,y)d\nu(y), \quad \forall x \in \Omega,$$

and the potential with the adjoint kernel is given by

$$\mathbf{G}^*\nu(y) := \int_\Omega G(x,y)\,d\nu(x), \quad \forall y \in \Omega.$$

A positive kernel G on $\Omega \times \Omega$ is said to satisfy the *weak maximum principle* (WMP) with constant $h \geq 1$ if, for any $\nu \in \mathcal{M}^+(\Omega)$,

$$\sup\left\{\mathbf{G}\nu(x) : x \in \mathrm{supp}(\nu)\right\} \leq M \implies \sup\left\{\mathbf{G}\nu(x) : x \in \Omega\right\} \leq h\,M, \quad (2.1)$$

for any constant $M > 0$. When $h = 1$, G is said to satisfy the *strong maximum principle*. It holds for Green's kernels associated with the classical Laplacian, or the fractional Laplacian $(-\Delta)^\alpha$ in the case $0 < \alpha \leq 1$, for all domains Ω with positive Green's function. The WMP holds for Riesz kernels on $\mathbb{R}^n$ associated with $(-\Delta)^\alpha$ in the full range $0 < \alpha < \frac{n}{2}$, and more generally for all radially non-increasing kernels on $\mathbb{R}^n$ (see [1]).

The WMP also holds for the so-called quasi-metric kernels (see [8, 9, 15, 20]). We say that $d(x,y)\colon \Omega \times \Omega \to [0,+\infty)$ satisfies the quasimetric triangle inequality with quasimetric constant κ if

$$d(x,y) \leq \kappa[d(x,z) + d(z,y)], \quad (2.2)$$

for any $x,y,z \in \Omega$. We say that G is a *quasimetric* kernel (with quasimetric constant $\kappa > 0$) if G is symmetric and $d(x,y) = \frac{1}{G(x,y)}$ satisfies (2.2).

A kernel $G\colon \Omega \times \Omega \to (0,+\infty]$ is said to be *quasi-symmetric* if there exists a constant a such that

$$a^{-1}G(y,x) \leq G(x,y) \leq a\,G(y,x), \quad \forall x,y \in \Omega.$$

Many kernels associated with elliptic operators are quasi-symmetric and satisfy the WMP (see [2]).

For $0 < q < 1$ and $\sigma \in \mathcal{M}^+(\Omega)$, we are interested in *positive solutions* $u \in L^r(\sigma)$ $(r > 0)$ to the integral equation

$$u = \mathbf{G}(u^q\sigma), \quad u > 0 \quad d\sigma\text{-a.e.} \quad (2.3)$$

and *positive supersolutions* $u \in L^r(\sigma)$ to the integral inequality

$$u \geq \mathbf{G}(u^q\sigma), \quad u > 0 \quad d\sigma\text{-a.e.} \quad (2.4)$$

In [20], we characterized the existence of positive solutions $u \in L^q(\Omega, \sigma)$ and $u \in L^q_{\text{loc}}(\sigma)$. The latter correspond to the so-called "very weak" solutions to the sublinear boundary value problem (1.4) (see [9, 16]). It is easy to see that the condition $u \in L^q_{\text{loc}}(\sigma)$ is necessary for the existence of any positive (super)solution, since otherwise $u \equiv +\infty$ $d\sigma$-a.e. (see [20]).

For a measure $\lambda \in \mathcal{M}^+(\Omega)$, the *energy of* λ is given by

$$\mathcal{E}(\lambda) := \int_\Omega \mathbf{G}\lambda \, d\lambda.$$

The notion of energy is closely related to another major tool of potential theory, the capacity of a set, and the associated equilibrium measure.

For a kernel $G \colon \Omega \times \Omega \to (0, +\infty]$, we consider the *Wiener capacity*

$$\text{cap}(K) := \sup \left\{ \mu(K) \; : \; \mathbf{G}^*\mu(y) \leq 1 \text{ on } \text{supp}(\mu), \; \mu \in \mathcal{M}^+(K) \right\}, \qquad (2.5)$$

defined for compact sets $K \subset \Omega$.

The extremal measure μ for which the supremum in (2.5) is attained is called the *equilibrium measure*. Alternatively, capacity can be defined as a solution to the following extremal problem involving energy:

$$\text{cap}(K) := \left[\inf \left\{ \mathcal{E}(\mu) \; : \; \mu \in \mathcal{M}^+(K), \quad \mu(K) = 1 \right\} \right]^{-1}. \qquad (2.6)$$

We say a property holds *nearly everywhere* (or n.e.) on K when the exceptional set $Z \subset K$ has zero capacity, $\text{cap}(Z) = 0$.

We will use the following theorem [3, 11] which is fundamental to potential theory.

Theorem 2.1. *Let G be a positive symmetric kernel, and $K \subset \Omega$ a compact set. The two extremal problems*

$$\max \left\{ \lambda(K) \; : \; \mathbf{G}\lambda \leq 1 \text{ on } \text{supp}(\lambda), \; \lambda \in \mathcal{M}^+(K) \right\},$$

$$\max \left\{ 2\lambda(K) - \mathcal{E}(\lambda) \; : \; \lambda \in \mathcal{M}^+(K) \right\},$$

always have solutions, which are precisely the same, and each maximum coincides with the Wiener capacity $\text{cap}\, K$. The class of all solutions consists of measures $\lambda \in \mathcal{M}^+(K)$ for which

$$\mathcal{E}(\lambda) = \lambda(\Omega) = \text{cap}(K).$$

The potential of any solution has the following properties.

1. $\mathbf{G}\lambda(x) \geq 1$ *n.e. in* K,
2. $\mathbf{G}\lambda(x) \leq 1$ *on* $\text{supp}(\lambda)$,
3. $\mathbf{G}\lambda(x) = 1$ *$d\lambda$-a.e. in* Ω.

The extremal measure λ in Theorem 2.1 is the equilibrium measure for the set K. We observe that since G is a positive kernel, the capacity of all compact sets K is finite. (This is true even for non-negative kernels if $G(x, x) > 0$ for all $x \in \Omega$; see [11].)

3. Weighted norm inequalities, supersolutions, and energy estimates

We begin this section with a proof of Theorem 1.1.

Proof of Theorem 1.1. We first prove statement (i). If the (p,r)-inequality (1.1) holds for $0 < r < p$, then assuming that $f = (\mathbf{G}\sigma)^{\frac{r}{p-r}} \in L^p(\Omega, \sigma)$ and using it as a test function, we deduce that

$$\int_\Omega \left[\mathbf{G}\big((\mathbf{G}\sigma)^{\frac{r}{p-r}} d\sigma \big) \right]^r d\sigma \le C^r \left[\int_\Omega (\mathbf{G}\sigma)^{\frac{pr}{p-r}} d\sigma \right]^{\frac{r}{p}},$$

where C is the embedding constant in (1.1). We now use the pointwise inequality

$$\big[\mathbf{G}\sigma(x) \big]^s \le s\, h^{s-1} \, \mathbf{G}\big((\mathbf{G}\sigma)^{s-1} d\sigma \big)(x), \quad x \in \Omega, \tag{3.1}$$

for all $s \ge 1$, established in [14] for kernels satisfying the WMP with constant $h \ge 1$. Applying this inequality with $s = \frac{p}{p-r}$, we obtain

$$\int_\Omega (\mathbf{G}\sigma)^{\frac{pr}{p-r}} d\sigma \le c\Big(h, \frac{p}{p-r}\Big) C^r \left[\int_\Omega (\mathbf{G}\sigma)^{\frac{pr}{p-r}} d\sigma \right]^{\frac{r}{p}}.$$

Since $0 < r < p$, this estimate yields

$$\int_\Omega (\mathbf{G}\sigma)^{\frac{pr}{p-r}} d\sigma \le c\Big(h, \frac{p}{p-r}\Big) C^{\frac{pr}{p-r}}.$$

The extra assumption that $f = (\mathbf{G}\sigma)^{\frac{r}{p-r}} \in L^p(\Omega, \sigma)$ is easy to remove by using $\chi_K f$ in place of f, where K is a compact subset of Ω on which $\mathbf{G}\sigma(x) \le n$, and then letting $n \to +\infty$ (see details in [20]).

In the opposite direction, suppose that (1.6) holds for $0 < r < p$ and $p > 1$. Without loss of generality we may assume that $f \ge 0$. By Hölder's inequality,

$$\int_\Omega [\mathbf{G}(f d\sigma)]^r d\sigma = \int_\Omega \left[\frac{\mathbf{G}(f d\sigma)}{\mathbf{G}\sigma} \right]^r (\mathbf{G}\sigma)^r d\sigma$$

$$\le \left[\int_\Omega \left(\frac{\mathbf{G}(f d\sigma)}{\mathbf{G}\sigma} \right)^p d\sigma \right]^{\frac{r}{p}} \left[\int_\Omega (\mathbf{G}\sigma)^{\frac{pr}{p-r}} d\sigma \right]^{1 - \frac{r}{p}}.$$

We next sketch a proof of a $(1,1)$-weak type estimate obtained in a more general context in [20], Lemma 5.11:

$$\left\| \frac{\mathbf{G}(f d\sigma)}{\mathbf{G}\sigma} \right\|_{L^{1,\infty}(\Omega, d\sigma)} \le c \|f\|_{L^1(\Omega, d\sigma)}, \tag{3.2}$$

where $c = c(h, a)$ depends only on the constants $h \ge 1$ in the weak maximum principle, and $a > 0$ in the quasi-symmetry condition.

Since G is quasi-symmetric, we can assume without loss of generality that it is symmetric by replacing G with $\frac{1}{2}(G + G^*)$. Let $E_t = \big\{ x \in \Omega : \frac{\mathbf{G}(f d\sigma)}{\mathbf{G}\sigma}(x) > t \big\}$, where $t > 0$.

For an arbitrary compact set $K \subset E_t$, we denote by $\mu \in \mathcal{M}^+(K)$ an equilibrium measure on K (see Sec. 2 above) such that $\mathbf{G}\mu \geq 1$ n.e. on K and $\mathbf{G}\mu \leq 1$ on $\mathrm{supp}(\mu)$.

It is easy to see that in fact

$$\mathbf{G}\mu \geq 1 \quad d\sigma\text{-a.e. on } K. \tag{3.3}$$

Indeed, from (1.6) it follows that $\mathbf{G}\sigma < +\infty$ $d\sigma$-a.e. Since $\mathbf{G}\mu \geq 1$ n.e. on K, the set $Z = \{x \in K : \mathbf{G}\mu(x) < 1\}$ has zero capacity, and consequently,

$$
\begin{aligned}
\sigma(Z) &= \sigma(\{x \in Z : \mathbf{G}\sigma(x) < +\infty\}) \\
&\leq \sum_{n=1}^{+\infty} \sigma(\{x \in Z : \mathbf{G}\sigma(x) \leq n\}) \\
&\leq \sum_{n=1}^{+\infty} n \, \mathrm{cap}(\{x \in Z : \mathbf{G}\sigma(x) \leq n\}) = 0.
\end{aligned}
$$

Thus, $\sigma(Z) = 0$, which proves (3.3).

Since $\mathbf{G}\mu \leq 1$ on $\mathrm{supp}(\mu)$, it follows that $\mathbf{G}\mu \leq h$ on Ω by the WMP. From this and (3.3), using Fubini's theorem, we deduce the inequalities

$$
\begin{aligned}
\sigma(K) &\leq \int_K \mathbf{G}\mu \, d\sigma = \int_K \mathbf{G}\sigma_K \, d\mu \leq \int_K \frac{\mathbf{G}(f \, d\sigma)}{t} \, d\mu = \frac{1}{t} \int_K \mathbf{G}\mu \, f \, d\sigma \\
&\leq \frac{1}{t} \int_\Omega h \, f \, d\sigma = \frac{h}{t} \, \|f\|_{L^1(\Omega, \sigma)}.
\end{aligned}
$$

Taking the supremum over all $K \subset E_t$, we obtain

$$\sigma(E_t) \leq \frac{h}{t} \, \|f\|_{L^1(\Omega, \sigma)},$$

which proves (3.2).

The corresponding L^∞ estimate is obvious:

$$\left\| \frac{\mathbf{G}(f d\sigma)}{\mathbf{G}\sigma} \right\|_{L^\infty(\Omega, d\sigma)} \leq C \, \|f\|_{L^\infty(\Omega, d\sigma)}.$$

Thus, for $1 < p < +\infty$, by the Marcinkiewicz interpolation theorem we obtain

$$\left\| \frac{\mathbf{G}(f d\sigma)}{\mathbf{G}\sigma} \right\|_{L^p(\Omega, d\sigma)} \leq C \, \|f\|_{L^p(\Omega, d\sigma)},$$

for all $f \in L^p(\Omega, d\sigma)$. Hence,

$$\int_\Omega [\mathbf{G}(f d\sigma)]^r \, d\sigma \leq C \, \|f\|_{L^p(\Omega, \sigma)}^r \left[\int_\Omega (\mathbf{G}\sigma)^{\frac{pr}{p-r}} \, d\sigma \right]^{1 - \frac{r}{p}}.$$

This proves statement (i).

We now prove statement (ii). Let $0 < q < 1$. Suppose there exists a positive supersolution $u \in L^r(\Omega, \sigma)$ with $r > q$. As shown in [14], if G satisfies the WMP, then any nontrivial supersolution u satisfies the global pointwise bound

$$u(x) \geq c\,[\mathbf{G}\sigma(x)]^{\frac{1}{1-q}} \quad d\sigma - \text{a.e.,} \tag{3.4}$$

where c depends only on h and q. Thus, (1.7) holds.

Conversely, by statement (i), (1.7) with $r > q$ implies the (p,r)-inequality (1.1) with $p = \frac{r}{q}$. Letting $u_0 = c\,[\mathbf{G}\sigma(x)]^{\frac{1}{1-q}}$ where $c > 0$ is a positive constant, we get a sequence of iterations

$$u_{j+1} = \mathbf{G}(u_j^q\, d\sigma), \quad j = 0,1,\dots,$$

where by induction we see that $u_{j+1} \geq u_j$, provided the constant c is sufficiently small. Here the initial step $u_1 \geq u_0$ follows from (3.1) with $s = \frac{1}{1-q}$, since

$$u_1 = \mathbf{G}(u_0^q d\sigma) = c^q\, \mathbf{G}\big[(\mathbf{G}\sigma)^{\frac{q}{1-q}}\, d\sigma\big] \geq c\,[\mathbf{G}\sigma(x)]^{\frac{1}{1-q}} = u_0,$$

for an appropriate choice of $c = c(q,h)$. By (1.1) with $p = \frac{r}{q}$ and $f = u_j$, we have by induction,

$$\|u_{j+1}\|_{L^r(\Omega,\sigma)} = \big\|\mathbf{G}(u_j^q\, d\sigma)\big\|_{L^r(\Omega,\sigma)} \leq C\,\|u_j\|_{L^r(\Omega,\sigma)}^q < +\infty.$$

Since $0 < q < 1$ and $u_j \leq u_{j+1}$, it follows that

$$\|u_{j+1}\|_{L^r(\Omega,\sigma)} \leq C, \quad j = 0,1,\dots.$$

Using the monotone convergence theorem, we obtain a positive solution

$$u = \lim_{j\to\infty} u_j, \quad u \in L^r(\Omega,\sigma). \qquad \square$$

Theorem 1.1 makes use of energy conditions of the type

$$\int_\Omega (\mathbf{G}\sigma)^s d\sigma < \infty, \tag{3.5}$$

for some $s > 0$. Note that when $s = 1$, this gives the energy $\mathcal{E}(\sigma)$ introduced above.

In the next lemma, we deduce (3.5) for $s = \frac{r}{1-q}$ provided there exists a positive supersolution $u \in L^r(\Omega,\sigma)$ to (1.2), for general non-negative quasi-symmetric kernels G, without assuming that (1.1) holds, or that G satisfies the WMP. In the special case $r = q$ this was proved in [20], Lemma 5.1.

Lemma 3.1. *Let $\sigma \in \mathcal{M}^+(\Omega)$, and let $0 < q < 1$. Suppose G is a non-negative quasi-symmetric kernel on $\Omega \times \Omega$. Suppose there is a positive supersolution $u \in L^r(\Omega,\sigma)$ $(r > 0)$, i.e., $\mathbf{G}(u^q\, d\sigma) \leq u$ $d\sigma$-a.e. Let $0 < q \leq 1 - r^2$. Then*

$$\int_\Omega (\mathbf{G}\sigma)^{\frac{r}{1-q}} d\sigma \leq a^{\frac{rq}{(1-q)(1-r+q)}} \int_\Omega u^r\, d\sigma < +\infty. \tag{3.6}$$

Proof. Suppose $u \in L^r(\Omega,\sigma)$, where $0 < r < 1$ is a positive supersolution. Let $\gamma \geq 1$. By Hölder's inequality with exponents γ and $\gamma' = \frac{\gamma}{\gamma-1}$, we estimate

$$\mathbf{G}\sigma(x) = \int_\Omega u^{\frac{q}{\gamma}} u^{-\frac{q}{\gamma}} G(x,y)\, d\sigma(y) \leq \big[\mathbf{G}(u^q\, d\sigma(x)\big]^{\frac{1}{\gamma}} \big[\mathbf{G}(u^{-\frac{q}{\gamma-1}}\, d\sigma)(x)\big]^{\frac{1}{\gamma'}}$$

$$\leq [u(x)]^{\frac{1}{\gamma}} \big[\mathbf{G}(u^{-\frac{q}{\gamma-1}}\, d\sigma)(x)\big]^{\frac{1}{\gamma'}}.$$

Let $\gamma = 1 + \frac{q}{1-r}$, where $0 < r \le 1 - q^2$. Then $\frac{(1-q)\gamma'}{r} \ge 1$. Using the preceding inequality, along with Hölder's inequality with the conjugate exponents

$$\frac{(1-q)(1-r+q)}{1-r-q^2} > 1 \quad \text{and} \quad \frac{(1-q)(1-r+q)}{rq} \ge 1,$$

and Fubini's theorem, we estimate

$$\int_\Omega (\mathbf{G}\sigma)^{\frac{r}{1-q}} d\sigma \le \int_\Omega u^{\frac{r}{(1-q)\gamma}} \left[\mathbf{G}(u^{r-1}d\sigma)\right]^{\frac{r}{(1-q)\gamma'}} d\sigma$$

$$= \int_\Omega u^{\frac{r(1-r-q^2)}{(1-q)(1-r+q)}} \left[u^q\, \mathbf{G}(u^{r-1}\,d\sigma)\right]^{\frac{rq}{(1-q)(1-r+q)}} d\sigma$$

$$\le \left[\int_\Omega u^r\, d\sigma\right]^{\frac{1-r-q^2}{(1-q)(1-r+q)}} \left[\int_\Omega \mathbf{G}(u^{r-1}\,d\sigma)\,u^q\, d\sigma\right]^{\frac{rq}{(1-q)(1-r+q)}}$$

$$= \left[\int_\Omega u^r\, d\sigma\right]^{\frac{1-r-q^2}{(1-q)(1-r+q)}} \left[\int_\Omega \mathbf{G}^*(u^q d\sigma)u^{r-1}d\sigma\right]^{\frac{rq}{(1-q)(1-r+q)}}$$

$$\le \left[\int_\Omega u^r\, d\sigma\right]^{\frac{1-r-q^2}{(1-q)(1-r+q)}} a\left[\int_\Omega u^r\, d\sigma\right]^{\frac{rq}{(1-q)(1-r+q)}}$$

$$= a^{\frac{rq}{(1-q)(1-r+q)}} \left[\int_\Omega u^r\, d\sigma\right]^{\frac{1-r-q^2+rq}{(1-q)(1-r+q)}}.$$

In the last estimate we have used the inequality $\mathbf{G}(u^q d\sigma) \le u$. Since $1-r-q^2+rq = (1-q)(1-r+q)$, this completes the proof of (3.6). $\qquad\square$

We next show that, for general non-negative kernels G, the (p,r)-weighted norm inequality (1.1) with $p = \frac{r}{q} \ge 1$ yields the existence of a supersolution $u \in L^r(\Omega, \sigma)$ to (1.2). This is deduced from Gagliardo's lemma [12] (see also [23]), as in the special case $r = q$ in [20].

It will be convenient for us to construct a measurable function ϕ such that

$$0 < [\mathbf{G}(\phi\, d\sigma)]^q \le \phi < +\infty \quad d\sigma - \text{a.e.,} \tag{3.7}$$

for $0 < q < 1$. Clearly, if ϕ satisfies the above estimate, then $u = \phi^{\frac{1}{q}}$ satisfies (1.2). Moreover, $u \in L^r(\Omega, \sigma)$ if $\phi \in L^p(\Omega, \sigma)$, where $p = \frac{r}{q} \ge 1$.

We recall that a convex cone $P \subset B$ is *strictly convex at the origin* if, for any $\phi, \psi \in P$, $\alpha\phi + \beta\psi = 0$ implies $\phi = \psi = 0$, for any $\alpha, \beta > 0$ such that $\alpha + \beta = 1$.

Lemma 3.2 (Gagliardo [12]). *Let B be a Banach space, and let $P \subset B$ be a convex cone which is strictly convex at the origin and such that if $(\phi_n) \subset P$, $\phi_{n+1} - \phi_n \in P$, and $\|\phi_n\| \le M$ for all $n = 1, 2, \ldots$, then there exists $\phi \in P$ so that $\|\phi_n - \phi\| \to 0$.*

Let $S: P \to P$ be a continuous mapping with the following properties.

1. *For $\phi, \psi \in P$ such that $\phi - \psi \in P$, we have $S\phi - S\psi \in P$.*
2. *If $\|\phi\| \le 1$ and $\phi \in P$, then $\|Su\| \le 1$.*

Then for every $\lambda > 0$ there exists $\phi \in P$ so that $(1+\lambda)\phi - S\phi \in P$ and $0 < \|\phi\| \leq 1$. Moreover, for every $\psi \in P$ such that $0 < \|\psi\|_B \leq \frac{\lambda}{1+\lambda}$, ϕ can be chosen so that $\phi = \psi + \frac{1}{1+\lambda} S\phi$.

We will apply this lemma to $B = L^p(\sigma)$, $p \geq 1$, and the cone of non-negative functions P in B. In this case obviously one can ensure that $\phi > 0$ $d\sigma$-a.e.

Lemma 3.3. *Let (Ω, σ) be a σ-finite measure space, and let G be a non-negative kernel on $\Omega \times \Omega$. Let $0 < r < +\infty$ and $0 < q < 1$. Suppose (1.1) holds for $p = \frac{r}{q} \geq 1$ with an embedding constant $C = \varkappa > 0$. Then, for every $\lambda > 0$, there is a positive $\phi \in L^p(\sigma)$ satisfying (3.7) so that*

$$\|\phi\|_{L^p(\sigma)} \leq (1+\lambda)^{\frac{1}{1-q}} \varkappa^{\frac{q}{1-q}}.$$

Proof. The supersolution ϕ is constructed using Lemma 3.2. Define

$$S\colon L^p(\sigma) \to L^p(\sigma) \quad \text{by} \quad S\phi := \left[\frac{1}{\varkappa^q} \mathbf{G}(\phi \, d\sigma)\right]^q,$$

for all $\phi \in L^p(\sigma)$, $\phi \geq 0$. Inequality (1.1) gives that S is a bounded continuous operator. In fact, by (1.1) we see that if $\|\phi\|_{L^p(\sigma)} \leq 1$, then

$$\|S(\phi)\|_{L^p(\sigma)}^p = \frac{1}{\varkappa^r} \int_\Omega [\mathbf{G}(\phi\sigma)]^r \, d\sigma = \frac{1}{\varkappa^r} \varkappa^r \left(\int_\Omega \phi^p \, d\sigma\right)^q \leq 1.$$

Therefore, by Lemma 3.2, there exists $\phi \in L^p(\sigma)$ such that

$$(1+\lambda)\phi \geq \frac{1}{\varkappa^q}[\mathbf{G}(\phi\sigma)]^q,$$

$\|\phi\|_{L^p(\sigma)} \leq 1$, and $\phi > 0$ $d\sigma$-a.e. Setting $\phi_0 = c\,\phi$, where

$$c = \left[\frac{1}{(1+\lambda)\varkappa^q}\right]^{\frac{1}{1-q}},$$

we deduce that $\phi > 0$ $d\sigma$-a.e., and

$$\phi_0 \geq \mathbf{G}(\phi_0\sigma)^q, \quad \|\phi_0\|_{L^p(\sigma)} \leq (1+\lambda)^{\frac{1}{1-q}} \varkappa^{\frac{q}{1-q}}. \qquad \square$$

Remark 1.3 follows immediately from Lemma 3.3.

Remark 3.4. For $p = \frac{r}{q}$, a counterexample in [20] demonstrates that, without the WMP, the existence of a supersolution $u \in L^r(\Omega, \sigma)$ to (1.2) in the case $r = q$ does not imply the (p, r)-weighted norm inequality (1.1). A slight modification of that counterexample shows that the same is true in the case $r > q$ as well.

4. A counterexample in the end-point case $p = 1$

In the case $p = 1$, $0 < q < 1$, the $(1,q)$-weighted norm inequality (1.1) with $r = q$ follows from a similar inequality for the space of measures $\mathcal{M}^+(\Omega)$ in place of $L^1(\Omega, \sigma)$,

$$\|\mathbf{G}\nu\|_{L^q(\Omega,\sigma)} \leq C\,\|\nu\|, \quad \forall \nu \in \mathcal{M}^+(\Omega), \tag{4.1}$$

where $\|\nu\| = \nu(\Omega)$. This inequality was shown in [20] to be equivalent to the existence of a positive supersolution $u \in L^q(\Omega, \sigma)$ to (1.2) for quasi-symmetric kernels G satisfying the WMP. In this case, (4.1) is equivalent to (1.1) with $r = q$ and $p = 1$ in view of Lemma 3.3.

However, a characterization of (4.1) or (1.1), with $r = q$ and $p = 1$, in terms of the energy estimate (1.7) with $r = q$ is not available, contrary to the case $r > q$: the condition

$$\int_\Omega (\mathbf{G}\sigma)^{\frac{q}{1-q}}\,d\sigma < +\infty \tag{4.2}$$

is not sufficient for (4.1).

On the other hand, it is not difficult to see that (4.1) holds for all $\nu \in \mathcal{M}^+(\Omega)$ if and only if it holds for all finite linear combinations of point masses, $\nu = \sum_{j=1}^n a_j\,\delta_{x_j}$, $a_j > 0$. It had been conjectured that, for $0 < q < 1$, condition (4.2) combined with (4.1) for single point masses $\nu = \delta_x$, i.e.,

$$\int_\Omega G(x,y)^q\,d\sigma(y) \leq C < +\infty, \quad \forall x \in \Omega, \tag{4.3}$$

was not only necessary, but also sufficient for (4.1). (Notice that in the case $q \geq 1$ (4.3) is obviously necessary and sufficient for (4.1); see [20].)

In this section, we give a counterexample to this conjecture for Riesz potentials on $\mathbb{R}^n$,

$$\mathbf{I}_{2\alpha}\nu(x) = \int_{\mathbb{R}^n} \frac{d\nu(y)}{|x-y|^{n-2\alpha}}, \quad x \in \mathbb{R}^n,$$

where $\nu \in \mathcal{M}^+(\mathbb{R}^n)$, and $0 < 2\alpha < n$. Clearly, the Riesz kernels $|x-y|^{2\alpha-n}$ are symmetric, and satisfy the WMP.

Suppose $0 < q < 1$, $n \geq 1$, and $0 < 2\alpha < n$. We construct $\sigma \in \mathcal{M}^+(\mathbb{R}^n)$ such that

$$\mathcal{E}(\sigma) = \mathcal{E}_{\alpha,q}(\sigma) := \int_{\mathbb{R}^n} \left(\mathbf{I}_{2\alpha}\sigma\right)^{\frac{q}{1-q}}\,d\sigma < +\infty, \tag{4.4}$$

and

$$\mathcal{K}(\sigma) = \mathcal{K}_{\alpha,q}(\sigma) := \sup_{x \in \mathbb{R}^n} \int_{\mathbb{R}^n} \frac{d\sigma(y)}{|x-y|^{(n-2\alpha)q}} < +\infty, \tag{4.5}$$

but

$$\kappa(\sigma) = \kappa(\sigma)_{\alpha,q} := \sup\left\{ \frac{\|\mathbf{I}_{2\alpha}\nu\|_{L^q(\sigma)}}{\|\nu\|_{\mathcal{M}^+(\mathbb{R}^n)}} : \nu \in \mathcal{M}^+(\mathbb{R}^n),\ \nu \neq 0 \right\} = +\infty. \tag{4.6}$$

In other words, we need to construct a measure σ such that $\mathcal{E}(\sigma) < +\infty$ (in the special case $q = \frac{1}{2}$ this means that σ has finite energy), and (4.6) holds for all

δ-functions $\nu = \delta_x$ ($x \in \mathbb{R}^n$), but (4.6) fails for a linear combination of δ-functions

$$\nu = \sum_{j=1}^{\infty} a_j \, \delta_{x_j}, \quad \text{where} \quad \sum_{j=1}^{\infty} a_j < +\infty, \quad a_j > 0. \tag{4.7}$$

We will use a modification of the example considered in [6] for other purposes.

We will need the following lemma and its corollary in the radially symmetric case (see [6]).

Lemma 4.1. *Let $0 < q < 1$ and $0 < 2\alpha < n$. If $d\sigma = \sigma(|x|)\, dx$ is radially symmetric, then $\kappa(\sigma) < +\infty$ if and only if $\mathcal{K}(\sigma) < +\infty$. Moreover, there exists a constant $c = c(q, \alpha, n) > 0$ such that $\kappa(\sigma)$ satisfies*

$$\mathcal{K}(\sigma) \leq \kappa(\sigma)^q \leq c\,\mathcal{K}(\sigma), \tag{4.8}$$

where in the this case

$$\mathcal{K}(\sigma) = \int_{\mathbb{R}^n} \frac{d\sigma(y)}{|y|^{(n-2\alpha)q}}. \tag{4.9}$$

Remark 4.2. For radially symmetric σ, condition $\mathcal{K}(\sigma) < +\infty$ is equivalent to $\sigma \in L^{\frac{1}{1-q},1}(\mathbb{R}^n, \sigma)$, which is necessary and sufficient for (4.1) in this case; see [19, 20]. Here $L^{s,1}(\mathbb{R}^n, \sigma)$ denotes the corresponding Lorentz space with respect to the measure σ.

Corollary 4.3. *Let $\sigma_{R,\gamma} = \chi_{B(0,R)} |x|^{-\gamma}$, where $0 \leq \gamma < n - q(n - 2\alpha)$ and $R > 0$. Then*

$$\mathcal{K}(\sigma) = \frac{\omega_n \, R^{n-\gamma-q(n-2\alpha)}}{n - \gamma - q(n - 2\alpha)}, \tag{4.10}$$

and

$$\frac{\omega_n}{n - \gamma - q(n - 2\alpha)} \leq \frac{\kappa(\sigma_{R,\gamma})^q}{R^{n-\gamma-q(n-2\alpha)}} \leq \frac{c}{n - \gamma - q(n - 2\alpha)}, \tag{4.11}$$

where $c = c(q, \alpha, n)$, and $\omega_n = |S^{n-1}|$ is the surface area of the unit sphere.

Let

$$\sigma = \sum_{k=1}^{\infty} c_k \sigma_k, \tag{4.12}$$

where

$$\sigma_k = \sigma_{R_k, \gamma_k}(x + x_k), \quad R_k = |x_k| = k, \quad \gamma_k = n - q(n - 2\alpha) - \epsilon_k, \tag{4.13}$$

and the positive scalars c_k, ϵ_k are picked so that $\sum_{k=1}^{\infty} c_k < \infty$, $\epsilon_k \to 0$, and $0 < \gamma_k < n$. Notice that $\gamma_k \to n - q(n - 2\alpha)$ as $k \to \infty$, which is a critical exponent for the inequality (4.17) (with σ_k in place of σ) discussed below.

More precisely, for $0 < q < 1$ and $0 < \delta < +\infty$, we set

$$a_k = \frac{1}{k(\log(k+1))^{\frac{1}{q}}}, \quad c_k = \frac{1}{k^{2-q+\delta}}, \quad \epsilon_k = \frac{1}{k^{1+\delta}}, \quad k = 1, 2, \ldots, \tag{4.14}$$

so that

$$\sum_{k=1}^{+\infty} a_k < +\infty, \quad \sup_{k \geq 1} \frac{c_k}{\epsilon_k} < +\infty, \quad \sum_{k=1}^{+\infty} \frac{c_k}{\epsilon_k^{1-q}} < +\infty, \quad \text{but} \quad \sum_{k=1}^{+\infty} \frac{c_k a_k^q}{\epsilon_k} = +\infty.$$
(4.15)

We first verify condition (4.4). Notice that

$$c_1 A \leq [\mathcal{E}_{\alpha,q}(\sigma)]^{1-q} \leq c_2 A,$$
(4.16)

where A is the least constant in the inequality (see [4, 5], Lemma 3.3)

$$\int_{\mathbb{R}^n} |\mathbf{I}_\alpha f|^{1+q} \, d\sigma \leq A \, \|f\|_{L^2(dx)}^{1+q}, \quad \text{for all } f \in L^2(\mathbb{R}^n, dx),$$
(4.17)

or, equivalently,

$$\int_{\mathbb{R}^n} |\mathbf{I}_{2\alpha}(g \, d\sigma)|^{1+q} \, d\sigma \leq A^2 \, \|g\|_{L^{\frac{1+q}{q}}(d\sigma)}^{1+q}, \quad \text{for all } g \in L^2(\mathbb{R}^n, \sigma),$$
(4.18)

where the constants of equivalence c_1, c_2 in (4.16) depend only on α, q, and n.

Consequently, $[\mathcal{E}_{\alpha,q}(\sigma)]^{1-q}$ is equivalent to a *norm* on a subset of $\mathcal{M}^+(\mathbb{R}^n)$, so that

$$\left[\mathcal{E}_{\alpha,q}\left(\sum_k \sigma_k\right)\right]^{1-q} \leq c \sum_k [\mathcal{E}_{\alpha,q}(\sigma_k)]^{1-q},$$
(4.19)

where $c = c(\alpha, q, n)$ is a positive constant which depends only on α, q, and n.

We claim that

$$\mathcal{E}_{\alpha,q}(\sigma_k) \leq \frac{C \, R_k^{\frac{\epsilon_k}{1-q}}}{\epsilon_k}, \quad k = 1, 2, \ldots,$$
(4.20)

where $C = C(\alpha, q, n)$.

Indeed, by the semigroup property of Riesz kernels,

$$\mathbf{I}_{2\alpha}\sigma_k(x) = c(\alpha, n) \int_{B(0,R_k)} \frac{dt}{|x - t|^{n-2\alpha}|t + x_k|^{\gamma_k}}$$

$$\leq c(\alpha, n) \int_{\mathbb{R}^n} \frac{dt}{|x - t|^{n-2\alpha}|t + x_k|^{\gamma_k}} = c \, |x + x_k|^{2\alpha - \gamma_k},$$

where $c = c(n, 2\alpha + n - \gamma_k)$ remains bounded by a constant $C(\alpha, q, n)$ as $k \to +\infty$, since $\lim_{k \to +\infty}(2\alpha + n - \gamma_k) = 2\alpha + q(n - 2\alpha) < n$.

Notice that $(\gamma_k - 2\alpha)\frac{q}{1-q} + \gamma_k = n - \frac{\epsilon_k}{1-q}$. Hence, by the preceding estimate,

$$\mathcal{E}_{\alpha,q}(\sigma_k) = \int_{\mathbb{R}^n} \left(\mathbf{I}_{2\alpha}\sigma_k\right)^{\frac{q}{1-q}} d\sigma_k \leq c^{\frac{q}{1-q}} \int_{|x+x_k|<R_k} \frac{dx}{|x + x_k|^{n-\frac{\epsilon_k}{1-q}}}$$

$$= c^{\frac{q}{1-q}} \omega_n \int_0^{R_k} r^{\frac{\epsilon_k}{1-q}-1} \, dr \leq \frac{C(\alpha, q, n) \, R_k^{\frac{\epsilon_k}{1-q}}}{\epsilon_k},$$

which proves (4.20).

It follows from (4.19) and the preceding estimate that, for σ defined by (4.12),

$$\left[\mathcal{E}_{\alpha,q}(\sigma)\right]^{1-q} \le c(\alpha,q,n) \sum_k c_k \left[\mathcal{E}_{\alpha,q}(\sigma_k)\right]^{1-q}$$

$$\le c(\alpha,q,n)\, C(\alpha,q,n)^{1-q} \sum_k \frac{c_k\, R_k^{\epsilon_k}}{\epsilon_k^{1-q}} < +\infty, \tag{4.21}$$

by (4.15), since obviously $\sup_{k\ge 1} R_k^{\epsilon_k} < +\infty$ by (4.14). This proves (4.4).

To prove (4.5), we will need the following lemma.

Lemma 4.4. *Let $R > 0$, $0 < \beta < n$, and $0 < \epsilon < n - \beta$. For $\gamma = n - \beta - \epsilon > 0$, we have*

$$\phi_{R,\gamma}(x) := \int_{|t|<R} \frac{dt}{|x-t|^\beta |t|^\gamma} \approx \begin{cases} \dfrac{R^\epsilon - |x|^\epsilon}{\epsilon} & \text{if } |x| \le \frac{R}{2}, \\[2mm] R^\epsilon \left(\dfrac{R}{|x|}\right)^\beta & \text{if } |x| > \frac{R}{2}, \end{cases} \tag{4.22}$$

where the constants of equivalence depend only on β and n.

Proof. Suppose first that $|x| > \frac{R}{2}$. Then

$$\phi_{R,\gamma}(x) = \int_{|t|<\frac{R}{4}} \frac{dt}{|x-t|^\beta |t|^\gamma} + \int_{\frac{R}{4}<|t|<R} \frac{dt}{|x-t|^\beta |t|^\gamma}$$

$$:= I + II.$$

Clearly, in the first integral $\frac{|x|}{2} \le |x-t| \le \frac{3|x|}{2}$, and so I is bounded above and below by

$$\frac{\omega_n\, c(\beta)}{|x|^\beta} \int_0^R r^{n-1-\gamma}\, dr = \frac{c(\beta,n) R^{n-\gamma}}{|x|^\beta}.$$

To estimate the second term, notice that, for $|x| > 2R$ and $|t| < R$, we have $|x-t| > \frac{|x|}{2}$, so that

$$II \le \frac{c(\beta,n)}{R^\gamma |x|^\beta} \int_{\frac{R}{4}<|t|<R} dt = \frac{c(\beta,n) R^{n-\gamma}}{|x|^\beta}.$$

For $\frac{R}{2} < |x| < 2R$ and $|t| < R$, we have $|x-t| < 3R$, and consequently

$$II \le \frac{c(\beta,n)}{R^\gamma} \int_{|x-t|<3R} \frac{dt}{|x-t|^\beta} = \frac{\omega_n c(\beta,n)}{R^\gamma} \int_0^{3R} r^{n-1-\beta}\, dr$$

$$= C(\beta,n) R^{n-\beta-\gamma} \le \frac{C(\beta,n) R^{n-\gamma}}{|x|^\beta}.$$

Thus, $II \le c(n,\beta)\, I$, which proves (4.22) in the case $|x| \ge \frac{R}{2}$.

Suppose now that $|x| \le \frac{R}{2}$. Then

$$\phi_{R,\gamma}(x) = \int_{|t|<\frac{|x|}{2}} \frac{dt}{|x-t|^\beta |t|^\gamma} + \int_{\frac{|x|}{2}<|t|<2|x|} \frac{dt}{|x-t|^\beta |t|^\gamma} + \int_{2|x|<|t|<R} \frac{dt}{|x-t|^\beta |t|^\gamma}$$

$$:= III + IV + V.$$

Clearly, in the first integral $\frac{|x|}{2} < |x-t| < \frac{3|x|}{2}$, and so III is bounded above and below by

$$\frac{c(\beta)}{|x|^\beta} \int_{|t|<\frac{|x|}{2}} \frac{dt}{|t|^\gamma} = \frac{\omega_n c(\beta)}{|x|^\beta} \int_0^{\frac{|x|}{2}} r^{n-1-\gamma}\,dr = \frac{c(\beta,n)}{(n-\gamma)2^{n-\gamma}}|x|^\epsilon.$$

The second integral IV is bounded above and below by

$$\frac{c(\gamma)}{|x|^\gamma} \int_{\frac{|x|}{2}<|t|<2|x|} \frac{dt}{|x-t|^\beta}.$$

Clearly,

$$IV \le \frac{c(\gamma)}{|x|^\gamma} \int_{|x-t|<3|x|} \frac{dt}{|x-t|^\beta} = \frac{\omega_n c(\gamma)}{|x|^\gamma} \int_0^{3|x|} r^{n-1-\beta}\,dr$$

$$= \frac{\omega_n c(\gamma)}{|x|^{\beta+\gamma-n}} = c(\beta,\gamma,n)|x|^\epsilon,$$

so that $IV \le c(\beta,n)\,III$.

Finally, the integral V is bounded above and below by

$$c(\beta) \int_{2|x|<|t|<R} \frac{dt}{|t|^{\gamma+\beta}} = c(\beta)\int_{2|x|}^R r^{n-1-\gamma-\beta}\,dr = c(\beta)\frac{R^\epsilon - (2|x|)^\epsilon}{\epsilon}.$$

Combining these estimates we complete the proof of (4.22). $\qquad\square$

By Lemma 4.4 with $\beta = (n-2\alpha)q$, $R = R_k$, $\epsilon = \epsilon_k$, and $\gamma = \gamma_k = n-\beta-\epsilon_k$, we obtain, for $k = 2,3,\ldots$,

$$\phi_{R_k,\gamma_k}(x-x_k) = \int_{|t+x_k|<R_k} \frac{dt}{|x-t|^{(n-2\alpha)q}|t+x_k|^\gamma}$$

$$\le C(\alpha,q,n)\begin{cases} \dfrac{R_k^{\epsilon_k}}{\epsilon_k} & \text{if } |x-x_k| < 1, \\[2mm] \dfrac{R_k^{\epsilon_k}-1}{\epsilon_k} & \text{if } 1 \le |x-x_k| \le \frac{R_k}{2}, \\[2mm] R_k^{\epsilon_k} & \text{if } |x-x_k| > \frac{R_k}{2}. \end{cases} \qquad (4.23)$$

In the case $k = 1$, we use the estimate $\phi_{R_1,\gamma_1}(x-x_1) \le C(\alpha,q,n)\frac{R_1^{\epsilon_1}}{\epsilon_1}$ for all $x \in \mathbb{R}^n$.

We next estimate

$$\mathcal{K}(\sigma) = \sup_{x\in\mathbb{R}^n} \sum_{k=1}^{+\infty} c_k\,\phi_{R_k,\gamma_k}(x-x_k)$$

$$\le \sup_{x\in\mathbb{R}^n} \sum_{|x-x_k|\le 1} c_k\,\phi_{R_k,\gamma_k}(x-x_k) + \sup_{x\in\mathbb{R}^n} \sum_{1<|x-x_k|<\frac{R_k}{2}} c_k\,\phi_{R_k,\gamma_k}(x-x_k)$$

$$+ \sup_{x\in\mathbb{R}^n} \sum_{|x-x_k|\ge\frac{R_k}{2}} c_k\,\phi_{R_k,\gamma_k}(x-x_k)$$

$$:= I + II + III. \qquad (4.24)$$

Suppose that $j \leq |x| \leq j+1$ for some $j = 0, 1, \ldots$. We first estimate I. Since $|x - x_k| \leq 1$, and $|x_k| = k$, it follows that

$$k = |x_k| \leq 1 + |x| \leq 1 + |x - x_k| + |x_k| = k + 2.$$

Consequently, $j - 1 \leq k \leq j + 2$ if $j \geq 2$, and $1 \leq k \leq 3$ if $j = 0, 1, 2$. Hence, the corresponding sum contains no more than four terms, and therefore

$$I := \sup_{x \in \mathbb{R}^n} \sum_{|x - x_k| \leq 1} c_k \, \phi_{R_k, \gamma_k}(x - x_k)$$

$$\leq C(\alpha, q, n) \sup_{j \geq 0} \sum_{\max(j-1,1) \leq k \leq \max(j+2,3)} \frac{c_k \, R_k^{\epsilon_k}}{\epsilon_k}$$

$$\leq C(\alpha, q, n),$$

since by (4.13) and (4.15),

$$\sup_{k \geq 1} R_k^{\epsilon_k} < +\infty, \quad \text{and} \quad \sup_{k \geq 1} \frac{c_k}{\epsilon_k} < +\infty.$$

To estimate II, notice that $0 < \epsilon_k \log R_k \leq C$, and consequently

$$\frac{R_k^{\epsilon_k} - 1}{\epsilon_k} \leq C \log R_k.$$

Hence, by (4.23) and (4.14),

$$II := \sup_{x \in \mathbb{R}^n} \sum_{1 < |x - x_k| < \frac{R_k}{2}} c_k \, \phi_{R_k, \gamma_k}(x - x_k)$$

$$\leq C(\alpha, q, n) \sup_{x \in \mathbb{R}^n} \sum_{1 < |x - x_k| < \frac{R_k}{2}} \frac{c_k \, (R_k^{\epsilon_k} - 1)}{\epsilon_k}$$

$$\leq C(\alpha, q, n) \sum_{k=1}^{+\infty} c_k \log R_k < +\infty.$$

Finally, we estimate III using (4.23) and (4.14). Since $\sup_k R_k^{\epsilon_k} < +\infty$, we deduce that

$$III := \sup_{x \in \mathbb{R}^n} \sum_{|x - x_k| \geq \frac{R_k}{2}} c_k \, \phi_{R_k, \gamma_k}(x - x_k)$$

$$\leq C(\alpha, q, n) \sup_{x \in \mathbb{R}^n} \sum_{|x - x_k| \geq \frac{R_k}{2}} c_k \, R_k^{\epsilon_k}$$

$$\leq C(\alpha, q, n) \sum_{k=1}^{+\infty} c_k \leq C(\alpha, q, n).$$

This proves (4.5).

It remains to verify (4.6) for $\sigma = \sum_{k=1}^{+\infty} c_k \sigma_k$ and $\nu = \sum_{j=1}^{+\infty} a_j \delta_{x_j}$ defined above. We estimate

$$\|\mathbf{I}_{2\alpha}\nu\|_{L^q(\sigma)}^q = \sum_{k=1}^{+\infty} c_k \int_{\mathbb{R}^n} \left(\sum_{j=1}^{+\infty} \frac{a_j}{|x + x_j|^{n-2\alpha}} \right)^q d\sigma_k$$

$$\geq \sum_{k=1}^{+\infty} c_k \int_{\mathbb{R}^n} \frac{a_k^q}{|x + x_k|^{q(n-2\alpha)}} \, d\sigma_k$$

$$= \sum_{k=1}^{+\infty} c_k \, a_k^q \int_{|x+x_k|<R_k} \frac{dx}{|x + x_k|^{(n-2\alpha)q + \gamma_k}}.$$

Since

$$\int_{|x+x_k|<R_k} \frac{dx}{|x + x_k|^{(n-2\alpha)q + \gamma_k}} = \int_{|x|<R_k} \frac{dx}{|x|^{(n-2\alpha)q + \gamma_k}}$$

$$= \omega_n \int_0^{R_k} r^{-1+\epsilon_k} \, dr = \omega_n \frac{R_k^{\epsilon_k}}{\epsilon_k},$$

and $R_k^{\epsilon_k} \geq 1$, it follows by (4.14) that

$$\|\mathbf{I}_{2\alpha}\nu\|_{L^q(\sigma)}^q \geq \omega_n \sum_{k=1}^{+\infty} \frac{c_k \, a_k^q}{\epsilon_k} = \omega_n \sum_{k=1}^{+\infty} \frac{1}{k \log(k+1)} = +\infty.$$

References

[1] D.R. Adams and L.I. Hedberg, *Function Spaces and Potential Theory*, Grundlehren der math. Wissenschaften **314**, Berlin–Heidelberg–New York, Springer, 1996.

[2] A. Ancona, *Some results and examples about the behaviour of harmonic functions and Green's functions with respect to second order elliptic operators*, Nagoya Math. J. **165** (2002), 123–158.

[3] M. Brelot, *Lectures on Potential Theory*, Lectures on Math. **19**, Tata Institute, Bombay, 1960.

[4] C. Cascante, J.M. Ortega, and I.E. Verbitsky, *On L^p-L^q trace inequalities*, J. London Math. Soc. **74** (2006), 497–511.

[5] C.T. Dat and I.E. Verbitsky, *Finite energy solutions of quasilinear elliptic equations with sub-natural growth terms*, Calc. Var. PDE **52** (2015), 529–546.

[6] C.T. Dat and I.E. Verbitsky, *Nonlinear elliptic equations and intrinsic potentials of Wolff type*, J. Funct. Analysis **272** (2017), 112–165.

[7] C.T. Dat and I.E. Verbitsky, *Pointwise estimates of Brezis–Kamin type for solutions of sublinear elliptic equations*, Nonlin. Analysis Ser. A: Theory, Methods & Appl. **146** (2016), 1–19.

[8] M. Frazier, F. Nazarov, and I. Verbitsky, *Global estimates for kernels of Neumann series and Green's functions*, J. London Math. Soc. **90** (2014), 903–918.

[9] M. Frazier and I. Verbitsky, *Positive solutions to Schrödinger's equation and the exponential integrability of the balayage*, Ann. Inst. Fourier (Grenoble) **67** (2017), 1393–1425.

[10] O. Frostman, *Potentiel de masses à somme algébrique nulle*, Kungl. Fysiogr. Sällskapets i Lund Förhandlingar [Proc. Roy. Physiog. Soc. Lund] **20** (1950), 1–21.

[11] B. Fuglede, *On the theory of potentials in locally compact spaces*, Acta Math. **103** (1960), 139–215.

[12] E. Gagliardo, *On integral transformations with positive kernel*, Proc. Amer. Math. Soc. **16** (1965), 429–434.

[13] A. Grigor'yan and I.E. Verbitsky, *Pointwise estimates of solutions to semilinear elliptic equations and inequalities*, J. d'Analyse Math. (to appear) arXiv:1511.03188.

[14] A. Grigor'yan and I.E. Verbitsky, *Pointwise estimates of solutions to nonlinear equations for nonlocal operators*, arXiv:1707.09596.

[15] W. Hansen and I. Netuka, *On the Picard principle for* $\Delta + \mu$, Math. Z. **270** (2012) 783–807.

[16] M. Marcus and L. Véron, *Nonlinear Second Order Elliptic Equations Involving Measures*, Walter de Gruyter, Berlin–Boston, 2014.

[17] B. Maurey, *Théorèmes de factorisation pour les opérateurs linéaires à valeurs dans les espaces* L^p, Astérisque **11** (1974) Soc. Math. France, Paris.

[18] V. Maz'ya, *Sobolev Spaces, with Applications to Elliptic Partial Differential Equations*, 2nd, Augmented Edition. Grundlehren der math. Wissenschaften **342**, Springer, Berlin, 2011.

[19] S. Quinn and I.E. Verbitsky, *Weighted norm inequalities of* $(1,q)$*-type for integral and fractional maximal operators*, Harmonic Analysis, Partial Differential Equations and Applications, in Honor of Richard L. Wheeden, eds. S. Chanillo et al., Ser. Applied and Numerical Harmonic Analysis, Birkhäuser, 2016.

[20] S. Quinn and I.E. Verbitsky, *A sublinear version of Schur's lemma and elliptic PDE*, Analysis & PDE (to appear), arXiv:1702.02682.

[21] A. Seesanea and I.E. Verbitsky, *Finite energy solutions to inhomogeneous nonlinear elliptic equations with sub-natural growth terms*, Adv. Calc. Var. (to appear), arXiv:1709.02048.

[22] E. Stein and G. Weiss, *Introduction to Fourier Analysis on Euclidean Spaces*, Princeton Univ. Press, Princeton, N.J., 1971.

[23] P. Szeptycki, *Notes on Integral Transformations*, Dissert. Math. (Rozprawy Mat.), 231 (1984), p. 48.

Igor E. Verbitsky
Department of Mathematics
University of Missouri
Columbia, MO 65211, USA
e-mail: verbitskyi@missouri.edu

Printed in the United States
By Bookmasters